Mathematics for Machine Technology

Robert D. Smith

John C. Peterson

Australia • Brazil • Japan • Korea • Mexico • Singapore • Spain • United Kingdom • United States

Mathematics for Machine Technology, Sixth Edition
Robert D. Smith and John C. Peterson

Vice President, Career and Professional Editorial: Dave Garza

Director of Learning Solutions: Sandy Clark

Acquisitions Editor: James Devoe

Managing Editor: Larry Main

Product Manager: Mary Clyne

Editorial Assistant: Tom Best

Vice President, Career and Professional Marketing: Jennifer McAvey

Marketing Director: Deborah S. Yarnell

Marketing Manager: Mark Pierro

Marketing Coordinator: Jonathan Sheehan

Production Director: Wendy Troeger

Production Manager: Mark Bernard

Content Project Manager: Michael Tubbert

Library of Congress Control Number: 2008928933

ISBN-13: 978-1-4283-3656-8

ISBN-10: 1-4283-3656-7

Delmar
5 Maxwell Drive
Clifton Park, NY 12065-2919
USA

Cengage Learning products are represented in Canada by Nelson Education, Ltd.

For your lifelong learning solutions, visit **delmar.cengage.com**

Visit our corporate website at **cengage.com**

Notice to the Reader

Publisher does not warrant or guarantee any of the products described herein or perform any independent analysis in connection with any of the product information contained herein. Publisher does not assume, and expressly disclaims, any obligation to obtain and include information other than that provided to it by the manufacturer. The reader is expressly warned to consider and adopt all safety precautions that might be indicated by the activities described herein and to avoid all potential hazards. By following the instructions contained herein, the reader willingly assumes all risks in connection with such instructions. The publisher makes no representations or warranties of any kind, including but not limited to, the warranties of fitness for particular purpose or merchantability, nor are any such representations implied with respect to the material set forth herein, and the publisher takes no responsibility with respect to such material. The publisher shall not be liable for any special, consequential, or exemplary damages resulting, in whole or part, from the readers' use of, or reliance upon, this material.

Printed in the United States of America
3 4 5 6 7 12 11 10

Contents

Preface

Mathematics for Machine Technology is written to overcome the often mechanical "plug in" approach found in many trade-related mathematics textbooks. An understanding of mathematical concepts is stressed in all topics ranging from general arithmetic processes to oblique trigonometry, compound angles, and numerical control.

Both content and method are those that have been used by the authors in teaching applied machine technology mathematics classes for apprentices in machine, tool-and-die, and tool design occupations. Each unit is developed as a learning experience based on preceding units—making prerequisites unnecessary.

Presentation of basic concepts is accompanied by realistic industry-related examples and actual industrial applications. The applications progress from the simple to those with solutions that are relatively complex. Many problems require the student to work with illustrations such as are found in machine technology handbooks and engineering drawings.

Great care has been taken in presenting explanations clearly and in providing easy-to-follow procedural steps in solving exercise and problem examples. The book contains a sufficient number of exercises and problems to permit the instructor to selectively plan assignments.

An analytical approach to problem solving is emphasized in the geometry, trigonometry, compound angle, and numerical control sections. This approach is necessary in actual practice in translating engineering drawing dimensions to machine working dimensions. Integration of algebraic and geometric principles with trigonometry by careful sequence and treatment of material also helps the student in solving industrial application. The Instructor's Guide provides answers and solutions for all problems.

A majority of instructors state that their students are required to perform basic arithmetic operations on fractions and decimals prior to calculator usage. Thereafter, the students use the calculator almost exclusively in problem-solving computations. The structuring of calculator instructions and examples in this text reflect the instructors' preferences. Calculator instructions and examples have been updated and greatly expanded in this edition. The scientific calculator is introduced in the Preface. Extensive calculator instruction and examples are given directly following the units on fractions and mixed numbers and the units on decimals. Further calculator instruction and examples are given throughout the text wherever calculator applications are appropriate to the material presented. A Calculator Applications Index is provided at the end of the Preface. It provides a convenient reference for all the material in the text for which calculator usage is presented. Often there are differences in the methods of computation among various makes and models of calculators. Where there are two basic ways of performing calculations, both ways are shown.

Changes from the previous edition have been made to improve the presentation of topics and to update material.

A survey of instructors using the fifth edition was conducted. Based on their comments and suggestions, changes were made. The result is an updated and improved sixth edition that includes the following revisions:

- The content has been reviewed and revised to clarify and update wherever relevant.
- Ratio and proportion have been moved to the first section and placed before the material on percents.
- The material on percents has been rewritten to follow a unified proportion approach.
- The metric and the customary systems of measure have been placed in separate units.
- New material on conversion between the metric and the customary systems of measure has been added to the chapter on the metric system and to Appendix A.

- Units on hexadecimal and BCD (Binary Coded Decimal) numeration systems have been added.
- A unit on polar coordinates in the CNC section has been added.
- The solutions to *most* trigonometry examples and exercises that used the cot, sec, or csc functions have been rewritten.

A note of explanation may be needed for the last bullet. Many trigonometry examples and exercises use the cot, sec, or csc functions. Because scientific calculators do not have keys for these functions, the solutions for most of these types of problems were rewritten in terms of the sin, cos, or tan functions. This should help students get better numerical results.

About the Authors

Robert D. Smith, was Associate Professor Emeritus of Industrial Technology at Central Connecticut State University, New Britain, Connecticut. Mr. Smith had experience in the manufacturing industry as tool designer, quality control engineer, and chief manufacturing engineer. He also taught applied mathematics, physics, and industrial materials and processes on the secondary technical school level and machine technology applied mathematics for apprentices in machine, tool-and-die, and tool design occupations. He was the author of *Technical Mathematics 4e,* also published by Cengage/Delmar Learning.

John C. Peterson is a retired professor of mathematics at Chattanooga State Technical Community College, Chattanooga, Tennessee. Before he began teaching, he worked on several assembly lines in industry. He has taught at the middle school, high school, two-year college, and university levels. Dr. Peterson is the author or coauthor of four other Delmar/Cengage Learning books: *Introductory Technical Mathematics* (with Robert D. Smith), *Technical Mathematics, Technical Mathematics with Calculus,* and *Math for the Automotive Trade* (with William J. deKryger). In addition, he has had over 80 papers published in various journals, has given over 200 presentations, and has served as a vice president of the American Mathematical Association of Two-Year Colleges.

Acknowledgments

The development of this edition included a survey of teachers of mathematics for machine technology. The publisher wishes to acknowledge the contributions of hundreds of professionals who responded to that survey with their ideas on how to make this text a better classroom resource. In addition, the following instructors completed detailed reviews of this manuscript:

Matthew Brady, Greenville Technical College

Carolyn Chapel, Western Technical College

Chris Edmonds, Tennessee Technical College

Sandy Fines, Morristown Technical College

Fred Fulkerson, Conestoga College

Jeff Joseph, R. G. Drage Career Technical Center

Michael Williams, Fanshawe College

Christine Zimmerman, Fanshawe College

Introduction to the Scientific Calculator

A scientific calculator is to be used in conjunction with the material presented in this textbook. Complex mathematical calculations can be made quickly, accurately, and easily with a scientific calculator.

Although most functions are performed in the same way, there are some differences among different makes and models of scientific calculators. In this book, generally, where there are two basic ways of performing a function, or sequencing, both ways are shown. However, not all of the differences among the various makes and models of calculators can be shown. It is very important that you become familiar with the operation of your scientific calculator. An owner's manual or user's guide is included with the purchase of a scientific calculator; it explains the essential features and keys of the specific calculator, as well as providing information on the proper use. *It is essential that the owner's manual or user's guide be studied and referred to whenever there is a question regarding calculator usage.* Also, information can be obtained from the manufacturer's Internet Web site, which is often listed in the user's guide.

For use in this textbook, examples are shown and problems are solved with calculators having EOS™ (Equation Operating System), V.P.A.M. (Visually Perfect Algebraic Method), or D.A.L. (Direct Algebraic Logic). Key operations are performed following the mathematical expressions exactly as they are written.

Most scientific calculator keys can perform more than one function. Depending on the calculator, generally the 2nd key or SHIFT key enable you to use alternate functions. The alternate functions are marked above the key. Alternate functions are shown and explained in the book where their applications are appropriate to specific content.

Decisions Regarding Calculator Use

The exercises and problems presented throughout the text are well suited for solutions using the calculator. However, it is felt that decisions regarding calculator usage should be left to the discretion of the course classroom or shop instructor. The instructor best knows the unique learning environment and objectives to be achieved by the students in a course. Judgments should be made by the instructor as to the degree of emphasis to be placed on

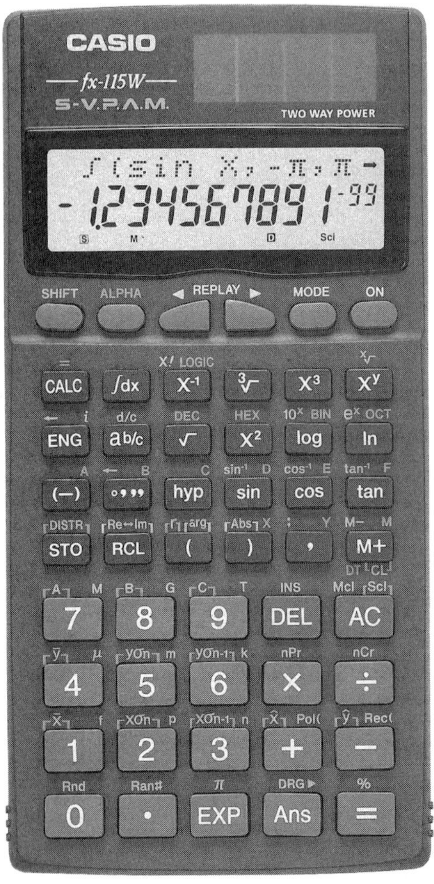

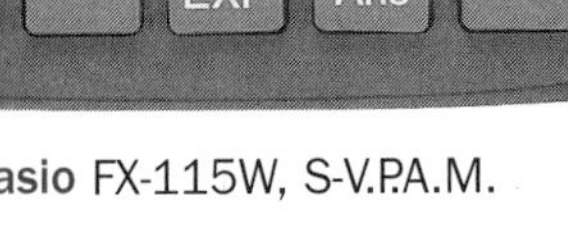
Casio FX-115W, S-V.P.A.M.

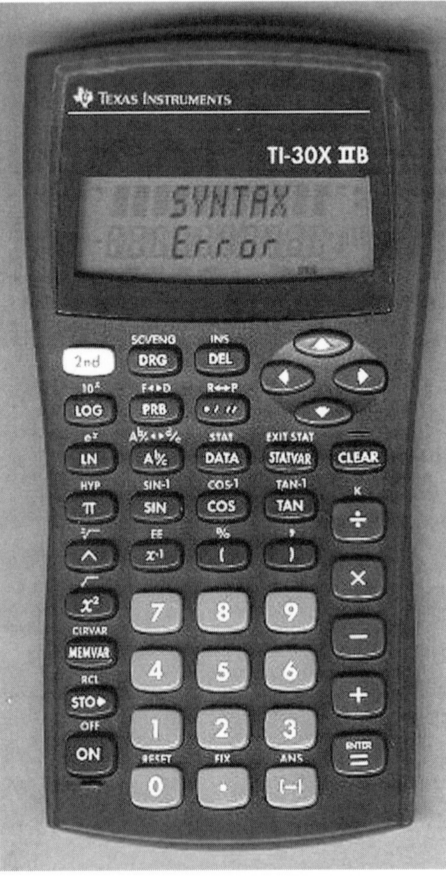

Texas Instruments TI-30XIIB

Sharp EL-506W, Advanced D.A.L.

calculator applications, when and where a calculator is to be used, and the selection of specific problems for solution by calculator. Therefore, exercises and problems in this text are *not* specifically identified as calculator applications.

Calculator instruction and examples of the basic operations of addition, subtraction, multiplication, and division of fractions are presented in Unit 7. They are presented for decimals in Unit 16. Further calculator instruction and examples of mathematics operations and functions are given throughout the text wherever calculator applications are appropriate to the material presented.

The index that follows lists the mathematics operations or functions and the pages on which the calculator instruction is first given for the operations or functions. It provides a convenient reference for all material in the text for which calculator usage is presented. The operations and functions are listed in the order in which material is presented in the text.

CALCULATOR APPLICATION INDEX

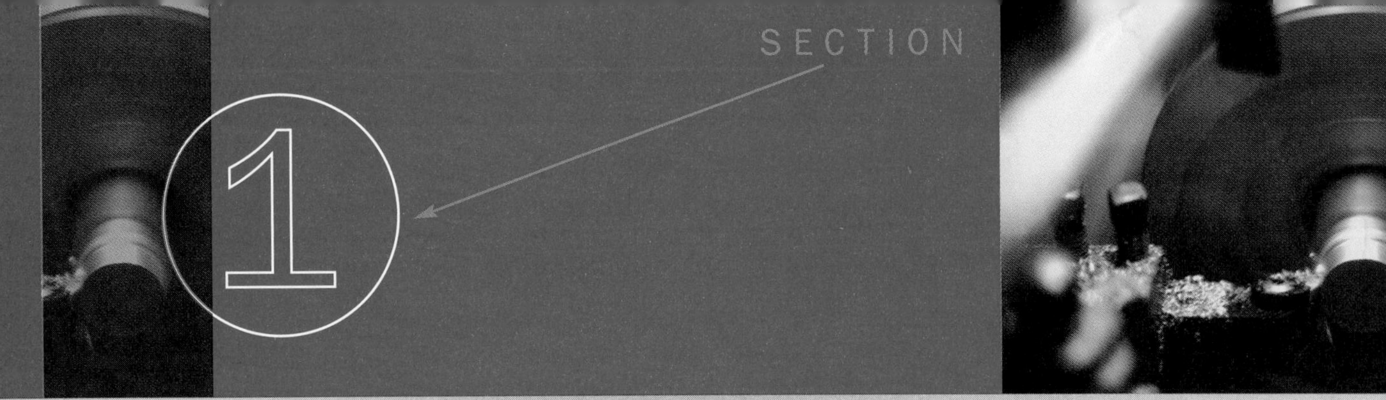

Common Fractions and Decimal Fractions

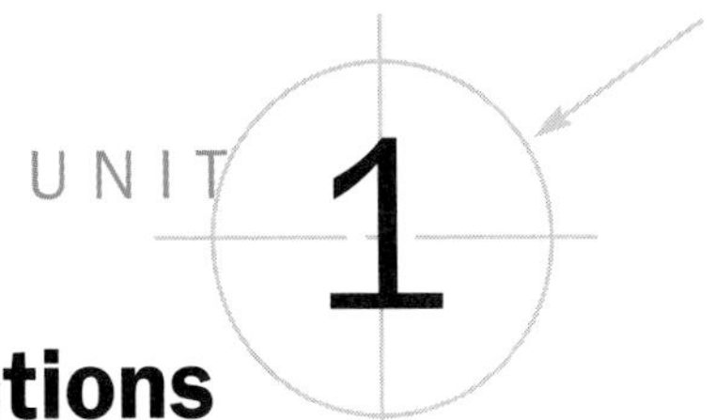

Introduction to Common Fractions and Mixed Numbers

Objectives *After studying this unit you should be able to*

- Express fractions in lowest terms.
- Express fractions as equivalent fractions.
- Express mixed numbers as improper fractions.
- Express improper fractions as mixed numbers.

Most measurements and calculations made by a machinist are not limited to whole numbers. Dimensions are sometimes given as fractions and certain measuring tools are graduated in fractional units. The machinist must be able to make calculations using fractions and to measure fractional values.

Fractional Parts

A *fraction* is a value that shows the number of equal parts taken of a whole quantity or unit. The symbols used to indicate a fraction are the bar (—) and the slash (/).

Line segment AB as shown in Figure 1-1 is divided into 4 equal parts.

$$1 \text{ part} = \frac{1 \text{ part}}{\text{total parts}} = \frac{1 \text{ part}}{4 \text{ parts}} = \frac{1}{4} \text{ of the length of the line segment.}$$

$$2 \text{ parts} = \frac{2 \text{ parts}}{\text{total parts}} = \frac{2 \text{ parts}}{4 \text{ parts}} = \frac{2}{4} \text{ of the length of the line segment.}$$

$$3 \text{ parts} = \frac{3 \text{ parts}}{\text{total parts}} = \frac{3 \text{ parts}}{4 \text{ parts}} = \frac{3}{4} \text{ of the length of the line segment.}$$

$$4 \text{ parts} = \frac{4 \text{ parts}}{\text{total parts}} = \frac{4 \text{ parts}}{4 \text{ parts}} = \frac{4}{4} = 1\text{, or unity (4 parts make up the whole).}$$

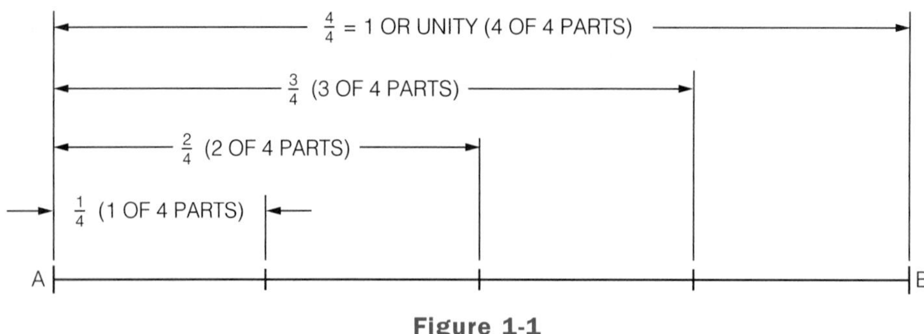

Figure 1-1

Each of the 4 equal parts of the line segment AB in Figure 1-2 is divided into 8 equal parts. There is a total of 4 × 8 or 32 parts.

1 part = $\frac{1}{32}$ of the total length.

7 parts = $\frac{7}{32}$ of the total length.

12 parts = $\frac{12}{32}$ of the total length.

23 parts = $\frac{23}{32}$ of the total length.

32 parts = $\frac{32}{32}$ of the total length.

$\frac{1}{2}$ of 1 part = $\frac{1}{2} \times \frac{1}{32} = \frac{1}{64}$ of the total length.

$\frac{32}{32}$ = 1 OR UNITY (32 OF 32 PARTS)

$\frac{23}{32}$ (23 OF 32 PARTS)

$\frac{12}{32}$ (12 OF 32 PARTS)

$\frac{7}{32}$ (7 OF 32 PARTS)

$\frac{1}{32}$ (1 OF 32 PARTS)

A B

$\frac{1}{2}$ OF $\frac{1}{32} = \frac{1}{64}$

$\frac{8}{32}$ OR $\frac{1}{4}$

Figure 1-2

➤ Note: 8 parts $= \dfrac{8}{32}$ of the total length and also $\dfrac{1}{4}$ of the total length. Therefore, $\dfrac{8}{32} = \dfrac{1}{4}$.

Definitions of Fractions

A *fraction* is a value that shows the number of equal parts taken of a whole quantity or unit. Some examples of fractions are $\frac{3}{4}, \frac{5}{8}, \frac{99}{100}$, and $\frac{17}{12}$. These same fractions written with a slash are $^{3}/_{4}$, $^{5}/_{8}$, $^{99}/_{100}$, and $^{17}/_{12}$.

The *denominator* of a fraction is the number that shows how many equal parts are in the whole quantity. The denominator is written below the bar.

The *numerator* of a fraction is the number that shows how many equal parts of the whole are taken. The numerator is written above the bar.

The numerator and denominator are called the *terms* of the fraction.

$$\frac{3}{4}\begin{array}{l}\leftarrow \text{numerator}\\ \leftarrow \text{denominator}\end{array}$$

A *common fraction* consists of two whole numbers. $\frac{5}{5}, \frac{3}{7}$, and $\frac{13}{4}$ are all examples of common fractions.

A *proper fraction* has a numerator that is smaller than its denominator. Examples of proper fractions are $\frac{3}{4}, \frac{5}{11}, \frac{91}{92}$, and $\frac{247}{961}$.

An *improper fraction* is a fraction in which the numerator is larger than or equal to the denominator, as in $\frac{3}{2}, \frac{5}{4}, \frac{11}{8}, \frac{6}{6}, \frac{17}{17}$.

A *mixed number* is a number composed of a whole number and a fraction, as in $3\frac{7}{8}, 7\frac{1}{2}$.

➤ **Note:** $3\frac{7}{8}$ means $3 + \frac{7}{8}$. It is read as three and seven-eighths. $7\frac{1}{2}$ means $7 + \frac{1}{2}$. It is read as seven and one-half.

Writing fractions with a slash can cause people to misread a number. For example, some people might think that $1^1/_4$ means $^{11}/_4 = \frac{11}{4}$ rather than $1\frac{1}{4}$. For this reason, the slash notation for fractions will not be used in this book.

A *complex fraction* is a fraction in which one or both of the terms are fractions or mixed numbers, as in $\frac{\frac{3}{4}}{6}, \frac{32}{\frac{15}{4}}, \frac{8\frac{3}{4}}{3}, \frac{\frac{7}{16}}{2\frac{2}{5}}, \frac{4\frac{1}{4}}{7\frac{5}{8}}$.

Expressing Fractions as Equivalent Fractions

The numerator and denominator of a fraction can be multiplied or divided by the same number without changing the value. For example, $\frac{1}{2} = \frac{1 \times 4}{2 \times 4} = \frac{4}{8}$. Both the numerator and denominator are multiplied by 4. Because $\frac{1}{2}$ and $\frac{4}{8}$ have the same value, they are *equivalent.* Also, $\frac{8}{12} = \frac{8 \div 4}{12 \div 4} = \frac{2}{3}$. Both numerator and denominator are divided by 4. Since $\frac{8}{12}$ and $\frac{2}{3}$ have the same value, they are equivalent.

A fraction is in its *lowest terms* when the numerator and denominator do not contain a common factor, as $\frac{5}{9}, \frac{7}{8}, \frac{3}{4}, \frac{11}{12}, \frac{15}{32}, \frac{9}{11}$. *Factors* are the numbers used in multiplying. For example, 2 and 5 are each factors of 10; $2 \times 5 = 10$. Expressing a fraction in lowest terms is often called *reducing* a fraction to lowest terms.

Procedure To reduce a fraction to lowest terms

- Divide both numerator and denominator by the greatest common factor (GCF).

Example Reduce $\frac{12}{42}$ to lowest terms.

Both terms can be divided by 2. $\frac{12 \div 2}{42 \div 2} = \frac{6}{21}$

➤ **Note:** The fraction is reduced, but not to lowest terms.

Further reduce $\frac{6}{21}$.

Both terms can be divided by 3. $\frac{6 \div 3}{21 \div 3} = \frac{2}{7}$ Ans

➤ **Note:** The value $\frac{2}{7}$ may be obtained in one step if each term of $\frac{12}{42}$ is divided by 2×3 or 6. Six is the greatest common factor (GCF).

$$\frac{12 \div 6}{42 \div 6} = \frac{2}{7} \quad \text{Ans}$$

Procedure To express a fraction as an equivalent fraction with an indicated denominator that is larger than the denominator of the fraction

- Divide the indicated denominator by the denominator of the fraction.
- Multiply both the numerator and denominator of the fraction by the value obtained.

Example Express $\frac{3}{4}$ as an equivalent fraction with 12 as the denominator.

Divide 12 by 4. $12 \div 4 = 3$

Multiply both 3 and 4 by 3. $\frac{3 \times 3}{4 \times 3} = \frac{9}{12}$ Ans

Expressing Mixed Numbers as Improper Fractions

Procedure To express a mixed number as an improper fraction

- Multiply the whole number by the denominator.
- Add the numerator to obtain the numerator of the improper fraction.
- The denominator is the same as that of the original fraction.

Example 1 Express $4\frac{1}{2}$ as an improper fraction.

Multiply the whole number by the denominator.

Add the numerator to obtain the numerator for the improper fraction.

The denominator is the same as that of the original fraction.

$$\frac{4 \times 2 + 1}{2} = \frac{9}{2} \quad \text{Ans}$$

Example 2 Express $12\frac{3}{16}$ as an improper fraction.

$$\frac{12 \times 16 + 3}{16} = \frac{195}{16} \quad \text{Ans}$$

Expressing Improper Fractions as Mixed Numbers

Procedure To express an improper fraction as a mixed number

- Divide the numerator by the denominator.
- Express the remainder as a fraction.

Examples Express the following improper fractions as mixed numbers.

$$\frac{11}{4} = 11 \div 4 = 2\frac{3}{4} \quad \text{Ans}$$

$$\frac{43}{3} = 43 \div 3 = 14\frac{1}{3} \quad \text{Ans}$$

$$\frac{931}{8} = 931 \div 8 = 116\frac{3}{8} \quad \text{Ans}$$

Application

Fractional Parts

1. Write the fractional part that each length, A through F, represents of the total shown on the scale in Figure 1-3.

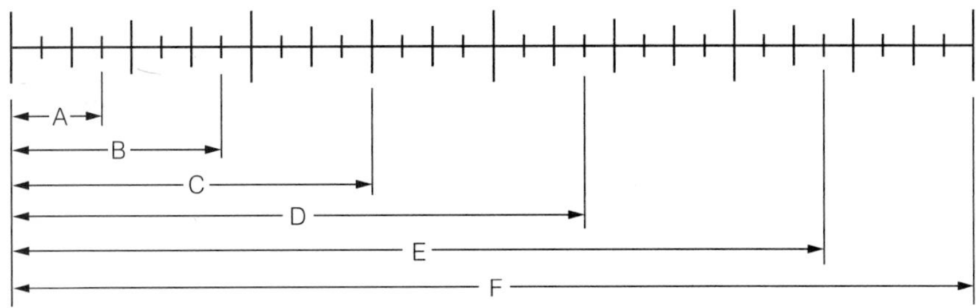

Figure 1-3

A = ______

B = ______

C = ______

D = ______

E = ______

F = ______

2. A welded support base is cut into four pieces as shown in Figure 1-4. What fractional part of the total length does each of the four pieces represent? All dimensions are in inches.

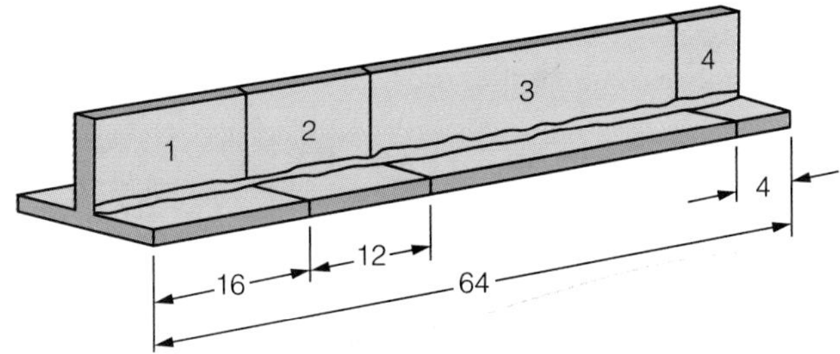

Figure 1-4

Piece 1: ______

Piece 2: ______

Piece 3: ______

Piece 4: ______

3. The circle in Figure 1-5 is divided into equal parts. Write the fractional part represented by each of the following:

a. 1 part ______

b. 3 parts ______

c. 7 parts ______

d. 5 parts ______

e. 16 parts ______

f. $\frac{1}{2}$ of 1 part ______

g. $\frac{1}{3}$ of 1 part ______

h. $\frac{3}{4}$ of 1 part ______

i. $\frac{1}{10}$ of 1 part ______

j. $\frac{1}{16}$ of 1 part ______

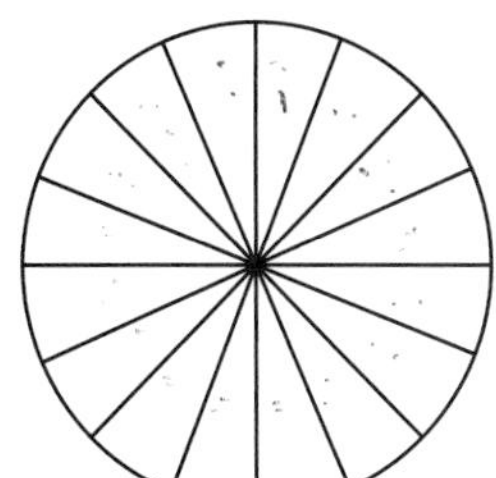

Figure 1-5

Expressing Fractions as Equivalent Fractions

4. Reduce to halves.

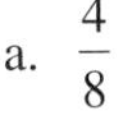
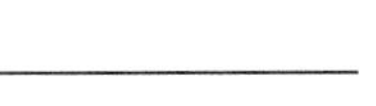

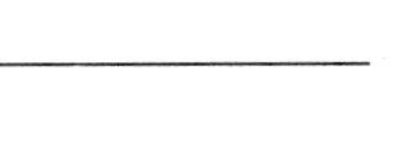
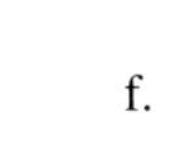

a. $\frac{4}{8}$ ______

b. $\frac{9}{18}$ ______

c. $\frac{100}{200}$ ______

d. $\frac{121}{242}$ ______

e. $\frac{25}{10}$ ______

f. $\frac{18}{12}$ ______

g. $\frac{126}{36}$ ______

h. $\frac{225}{50}$ ______

5. Reduce to lowest terms.

a. $\frac{6}{8}$ ____________
b. $\frac{12}{4}$ ____________
c. $\frac{6}{10}$ ____________
d. $\frac{30}{5}$ ____________
e. $\frac{11}{44}$ ____________
f. $\frac{14}{6}$ ____________
g. $\frac{24}{8}$ ____________
h. $\frac{65}{15}$ ____________
i. $\frac{25}{150}$ ____________
j. $\frac{14}{105}$ ____________

6. Express as thirty-seconds.

a. $\frac{1}{4}$ ____________
b. $\frac{3}{4}$ ____________
c. $\frac{11}{8}$ ____________
d. $\frac{7}{16}$ ____________
e. $\frac{21}{16}$ ____________
f. $\frac{19}{2}$ ____________
g. $\frac{197}{16}$ ____________
h. $\frac{21}{8}$ ____________

7. Express as equivalent fractions as indicated.

a. $\frac{3}{4} = \frac{?}{8}$ ____________
b. $\frac{7}{12} = \frac{?}{36}$ ____________
c. $\frac{6}{15} = \frac{?}{60}$ ____________
d. $\frac{17}{14} = \frac{?}{42}$ ____________
e. $\frac{20}{9} = \frac{?}{45}$ ____________
f. $\frac{14}{3} = \frac{?}{18}$ ____________
g. $\frac{7}{16} = \frac{?}{128}$ ____________
h. $\frac{13}{8} = \frac{?}{48}$ ____________
i. $\frac{21}{16} = \frac{?}{160}$ ____________

Mixed Numbers and Improper Fractions

8. Express the following mixed numbers as improper fractions.

a. $2\frac{2}{3}$ ____________
b. $1\frac{7}{8}$ ____________
c. $5\frac{2}{5}$ ____________
d. $3\frac{3}{8}$ ____________
e. $5\frac{9}{32}$ ____________
f. $8\frac{3}{7}$ ____________
g. $10\frac{1}{3}$ ____________
h. $9\frac{4}{5}$ ____________
i. $100\frac{1}{2}$ ____________
j. $4\frac{63}{64}$ ____________
k. $49\frac{3}{8}$ ____________
l. $408\frac{13}{16}$ ____________

9. Express the following improper fractions as mixed numbers.

a. $\frac{5}{3}$ ________ g. $\frac{127}{32}$ ________

b. $\frac{21}{2}$ ________ h. $\frac{57}{15}$ ________

c. $\frac{9}{8}$ ________ i. $\frac{150}{9}$ ________

d. $\frac{87}{4}$ ________ j. $\frac{235}{16}$ ________

e. $\frac{72}{9}$ ________ k. $\frac{514}{4}$ ________

f. $\frac{127}{124}$ ________ l. $\frac{401}{64}$ ________

10. Express the following mixed numbers as improper fractions. Then express the improper fractions as the equivalent fractions indicated.

a. $2\frac{1}{2} = \frac{?}{8}$ ________ d. $12\frac{2}{3} = \frac{?}{18}$ ________

b. $3\frac{3}{8} = \frac{?}{16}$ ________ e. $9\frac{7}{8} = \frac{?}{64}$ ________

c. $7\frac{4}{5} = \frac{?}{15}$ ________ f. $15\frac{1}{2} = \frac{?}{128}$ ________

11. Sketch and redimension the plate shown in Figure 1-6. Reduce all proper fractions to lowest terms. Reduce all improper fractions to lowest terms and express as mixed numbers. All dimensions are in inches.

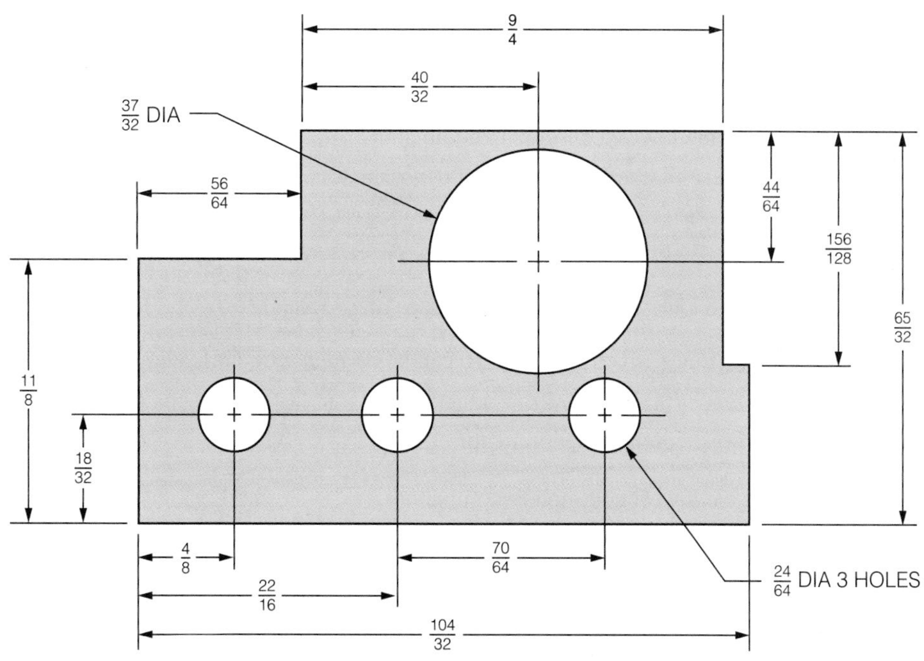

Figure 1-6

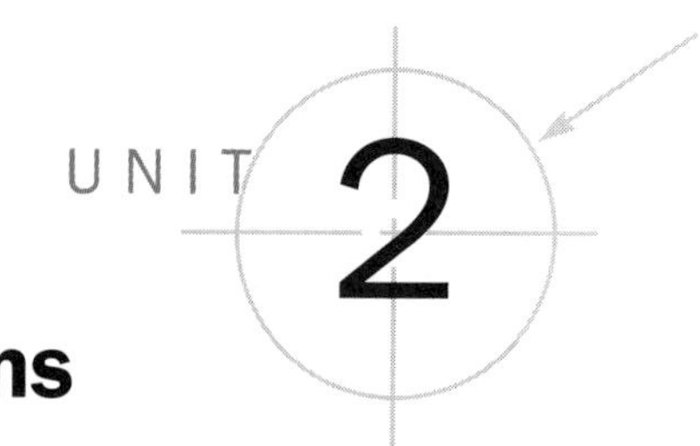

Addition of Common Fractions and Mixed Numbers

Objectives ***After studying this unit you should be able to***

- Determine lowest common denominators.
- Express fractions as equivalent fractions having lowest common denominators.
- Add fractions and mixed numbers.

A machinist must be able to add fractions and mixed numbers in order to determine the length of stock required for a job, the distances between various parts of a machined piece, and the depth of holes and cutouts in a workpiece.

Lowest Common Denominators

Fractions cannot be added unless they have a common denominator. *Common denominator* means that the denominators of each of the fractions are the same, as in $\frac{5}{8}, \frac{7}{8}, \frac{15}{8}$.

In order to add fractions that do not have common denominators, such as $\frac{3}{8} + \frac{1}{4} + \frac{7}{16}$, it is necessary to change to equivalent fractions with common denominators.

The *lowest common denominator* is the smallest denominator that is evenly divisible by each of the denominators of the fractions being added. Or, stated in another way, the *lowest common denominator* is the smallest denominator into which each denominator can be divided without leaving a remainder.

Procedure To find the lowest common denominator

- Determine the smallest number into which all denominators can be divided without leaving a remainder.
- Use this number as a common denominator.

Example 1 Find the lowest common denominator of $\frac{3}{8}, \frac{1}{4}$, and $\frac{7}{16}$.

The smallest number into which 8, 4, and 16 can be divided without leaving a remainder is 16.

Write 16 as the lowest common denominator.

Example 2 Find the lowest common denominator of $\frac{3}{4}, \frac{1}{3}, \frac{7}{8}$, and $\frac{5}{12}$.

The smallest number into which 4, 3, 8, and 12 can be divided is 24.

Write 24 as the lowest common denominator.

➤ **Note:** In this example, denominators such as 48, 72, and 96 are common denominators because 4, 3, 8, and 12 divide evenly into these numbers, but they are not the lowest common denominators.

Although any common denominator can be used when adding fractions, it is generally easier and faster to use the lowest common denominator.

Expressing Fractions as Equivalent Fractions with the Lowest Common Denominator

Procedure To change fractions into equivalent fractions having the lowest common denominator

- Divide the lowest common denominator by each denominator.
- Multiply both the numerator and denominator of each fraction by the value obtained.

Example 1 Express $\frac{2}{3}, \frac{7}{15},$ and $\frac{1}{2}$ as equivalent fractions having a lowest common denominator.

The lowest common denominator is 30. $30 \div 3 = 10; \frac{2 \times 10}{3 \times 10} = \frac{20}{30}$ Ans

Divide 30 by each denominator. $30 \div 15 = 2; \frac{7 \times 2}{15 \times 2} = \frac{14}{30}$ Ans

Multiply each term of the fraction by the value obtained. $30 \div 2 = 15; \frac{1 \times 15}{2 \times 15} = \frac{15}{30}$ Ans

Example 2 Change $\frac{5}{8}, \frac{15}{32}, \frac{3}{4},$ and $\frac{9}{16}$ to equivalent fractions having a lowest common denominator.

The lowest common denominator is 32.

$32 \div 8 = 4; \frac{5 \times 4}{8 \times 4} = \frac{20}{32}$ Ans

$32 \div 4 = 8; \frac{3 \times 8}{4 \times 8} = \frac{24}{32}$ Ans

$32 \div 32 = 1; \frac{15 \times 1}{32 \times 1} = \frac{15}{32}$ Ans

$32 \div 16 = 2; \frac{9 \times 2}{16 \times 2} = \frac{18}{32}$ Ans

Adding Fractions

Procedure To add fractions

- Express the fractions as equivalent fractions having the lowest common denominator.
- Add the numerators and write their sum over the lowest common denominator.
- Express an improper fraction as a mixed number when necessary and reduce the fractional part to lowest terms.

Example 1 Add $\frac{1}{2} + \frac{3}{5} + \frac{7}{10} + \frac{5}{6}$.

Express the fractions as equivalent fractions with 30 as the denominator.

$$\frac{1}{2} = \frac{15}{30}$$
$$\frac{3}{5} = \frac{18}{30}$$
$$\frac{7}{10} = \frac{21}{30}$$
$$+\frac{5}{6} = \frac{25}{30}$$

Add the numerators and write their sum over the lowest common denominator, 30.

$$= \frac{15 + 18 + 21 + 25}{30}$$

Express the fraction as a mixed number.

$$= \frac{79}{30} = 2\frac{19}{30} \text{ Ans}$$

Example 2 Determine the total length of the shaft shown in Figure 2-1. All dimensions are in inches.

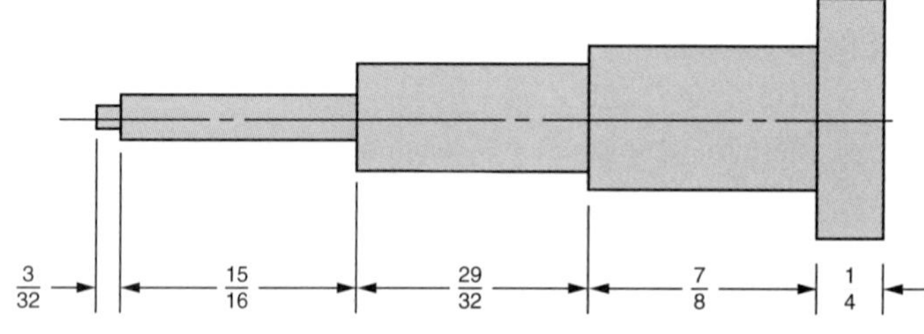

Figure 2-1

Express the fractions as equivalent fractions with 32 as the denominator.

$$\frac{3}{32} = \frac{3}{32}$$
$$\frac{15}{16} = \frac{30}{32}$$
$$\frac{29}{32} = \frac{29}{32}$$
$$\frac{7}{8} = \frac{28}{32}$$
$$+\frac{1}{4} = \frac{8}{32}$$

Add the numerators and write their sum over the lowest common denominator, 32.

$$= \frac{3 + 30 + 29 + 28 + 8}{32}$$

Express $\frac{98}{32}$ as a mixed number and reduce to lowest terms.

$$= \frac{98}{32} = 3\frac{2}{32} = 3\frac{1}{16}$$

Total length $= 3\frac{1}{16}''$ Ans

Adding Fractions, Mixed Numbers, and Whole Numbers

Procedure To add fractions, mixed numbers, and whole numbers

- Add the whole numbers.
- Add the fractions.
- Combine whole number and fraction.

Example 1 Add $\frac{1}{3} + 7 + 3\frac{1}{2} + \frac{5}{12} + 2\frac{19}{24}$.

Express the fractional parts as equivalent fractions with 24 as the denominator.

$$\frac{1}{3} = \frac{8}{24}$$
$$7 = 7$$
$$3\frac{1}{2} = 3\frac{12}{24}$$
$$\frac{5}{12} = \frac{10}{24}$$
$$+2\frac{19}{24} = +2\frac{19}{24}$$

Add the whole numbers.

$$= 7 + 3 + 2 = 12$$

Add the fractions. $= \frac{8 + 12 + 10 + 19}{24} = \frac{49}{24}$

Combine the whole number and the fractions. $= 12\frac{49}{24}$

Express the answer in lowest terms. $= 12\frac{49}{24} = 14\frac{1}{24}$ Ans

Example 2 Find the distance between the two $\frac{1}{2}$-inch diameter holes in the plate shown in Figure 2-2. All dimensions are in inches.

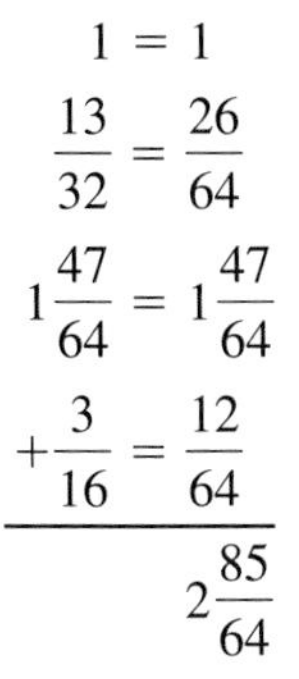

$$\begin{aligned} 1 &= 1 \\ \frac{13}{32} &= \frac{26}{64} \\ 1\frac{47}{64} &= 1\frac{47}{64} \\ +\frac{3}{16} &= \frac{12}{64} \\ \hline & \quad 2\frac{85}{64} \end{aligned}$$

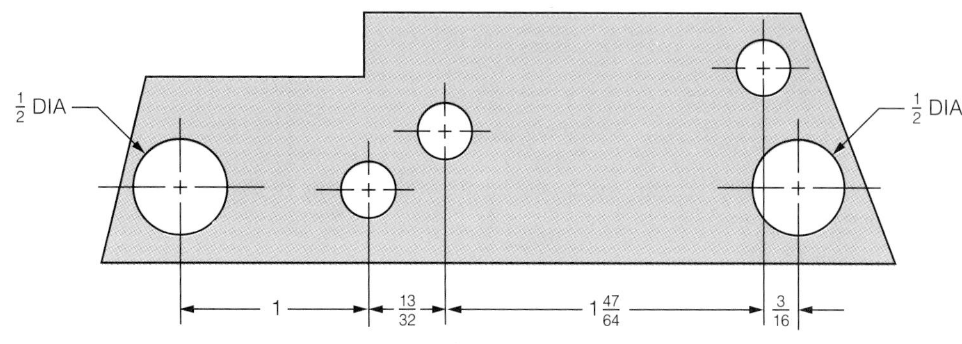

Figure 2-2

Distance $= 3\frac{21}{64}$" Ans

Application

Lowest Common Denominators

Determine the lowest common denominators of the following sets of fractions.

1. $\frac{2}{3}, \frac{1}{6}, \frac{5}{12}$ ________

2. $\frac{3}{5}, \frac{9}{10}, \frac{5}{6}$ ________

3. $\frac{5}{6}, \frac{7}{12}, \frac{3}{16}, \frac{19}{24}$ ________

4. $\frac{4}{5}, \frac{3}{4}, \frac{7}{10}, \frac{1}{2}$ ________

Equivalent Fractions with Lowest Common Denominators

Express these fractions as equivalent fractions having the lowest common denominator.

5. $\frac{1}{2}, \frac{3}{4}, \frac{5}{12}$ ________

6. $\frac{7}{16}, \frac{3}{8}, \frac{1}{2}$ ________

7. $\frac{9}{10}, \frac{1}{4}, \frac{3}{5}, \frac{1}{5}$ ________

8. $\frac{3}{16}, \frac{7}{32}, \frac{17}{64}, \frac{3}{4}$ ________

Adding Fractions

9. Determine the dimensions A, B, C, D, E, and F of the profile gage in Figure 2-3. All dimensions are in inches.

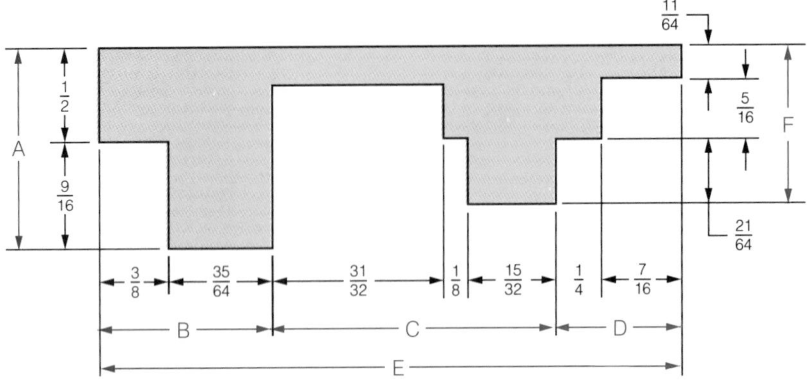

Figure 2-3

A = ________

B = ________

C = ________

D = ________

E = ________

F = ________

10. Determine the overall length, width, and height of the casting in Figure 2-4. All dimensions are in inches.

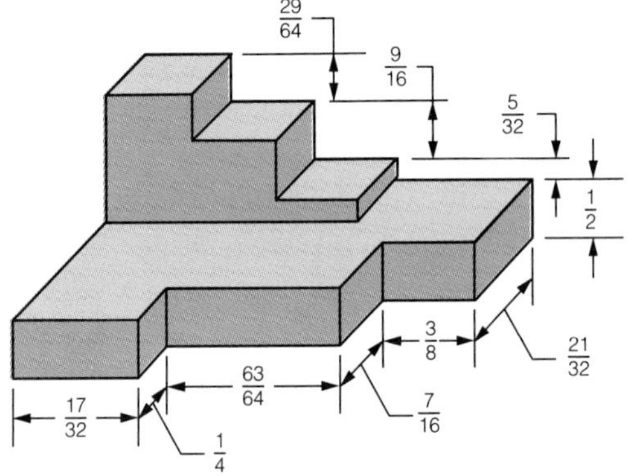

Figure 2-4

length = ________

width = ________

height = ________

Adding Fractions, Mixed Numbers, and Whole Numbers

11. Determine dimensions A, B, C, D, E, F, and G of the plate in Figure 2-5. Reduce to lowest terms where necessary. All dimensions are in inches.

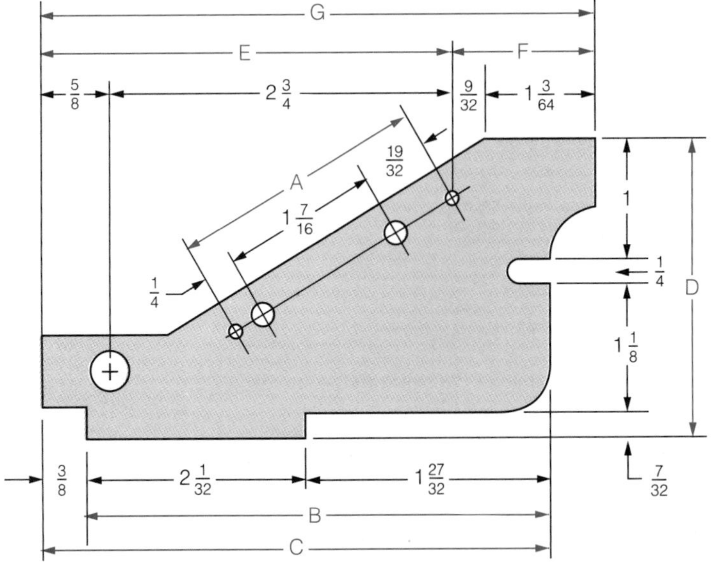

Figure 2-5

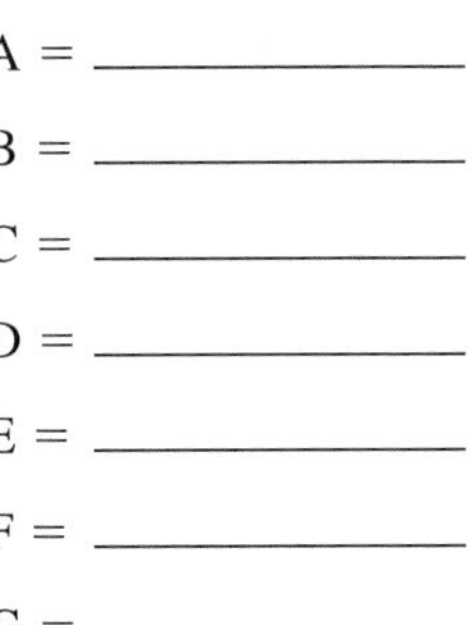

12. Determine dimensions A, B, C, and D of the pin in Figure 2-6. All dimensions are in inches.

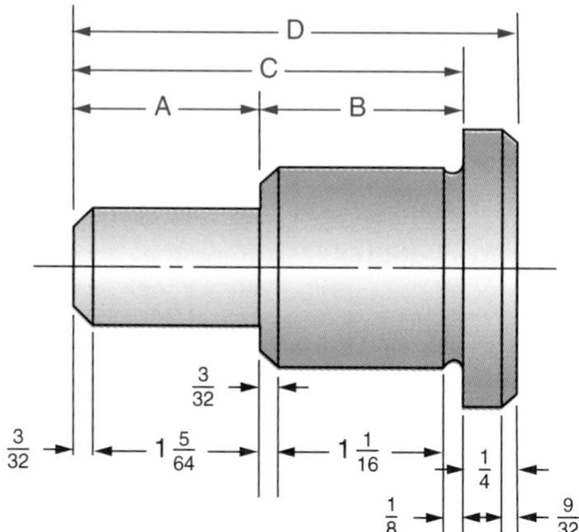

Figure 2-6

A = ____________

B = ____________

C = ____________

D = ____________

13. The operation sheet for machining an aluminum housing specifies 1 hour for facing, $2\frac{3}{4}$ hours for milling, $\frac{5}{6}$ hour for drilling, $\frac{3}{10}$ hour for tapping, and $\frac{2}{5}$ hour for setting up. What is the total time allotted for this job? ____________

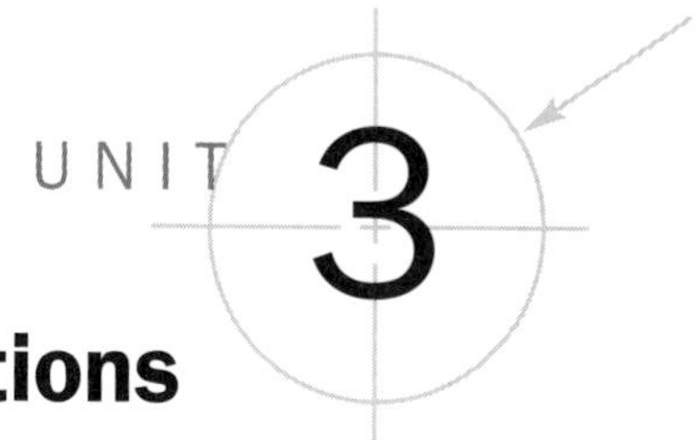

Subtraction of Common Fractions and Mixed Numbers

Objectives ***After studying this unit you should be able to***

- Subtract fractions.
- Subtract mixed numbers.

While making a part from an engineering drawing, a machinist often finds it necessary to express drawing dimensions as working dimensions. Subtraction of fractions and mixed numbers is sometimes required in order to properly position a part on a machine, to establish hole locations, and to determine depths of cut.

Subtracting Fractions

Procedure To subtract fractions

- Express the fractions as equivalent fractions having the lowest common denominator.
- Subtract the numerators.
- Write their difference over the lowest common denominator.
- Reduce the fraction to lowest terms.

Example 1 Subtract $\frac{3}{8}$ from $\frac{9}{16}$.

The lowest common denominator is 16.

Express $\frac{3}{8}$ as 16ths. $\frac{9}{16} = \frac{9}{16}$

Subtract the numerators. $-\frac{3}{8} = \frac{6}{16}$

Write their difference over the lowest common denominator. $\frac{9-6}{16} = \frac{3}{16}$ Ans

Example 2 Subtract $\frac{2}{5}$ from $\frac{3}{4}$.

Express the fractions as equivalent fractions with 20 as the denominator.

$$\frac{3}{4} = \frac{15}{20}$$

$$-\frac{2}{5} = \frac{8}{20}$$

Subtract the numerators and write the difference over the common denominator, 20.

$$= \frac{15 - 8''}{20} = \frac{7''}{20} \text{ Ans}$$

Example 3 Find the distances x and y between the centers of the pairs of holes in the strap shown in Figure 3-1. All dimensions are in inches.

To find distance x:

$$\frac{7}{8} = \frac{28}{32}$$

$$-\frac{11}{32} = \frac{11}{32}$$

$$\frac{28 - 11}{32} = \frac{17}{32}$$

$$x = \frac{17''}{32} \text{ Ans}$$

To find distance y:

$$\frac{63}{64} = \frac{63}{64}$$

$$-\frac{1}{4} = \frac{16}{64}$$

$$\frac{63 - 16}{64} = \frac{47}{64}$$

$$y = \frac{47''}{64} \text{ Ans}$$

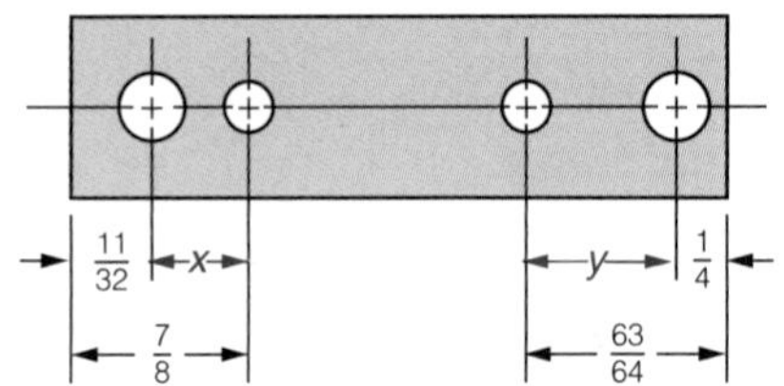

Figure 3-1

Subtracting Mixed Numbers

Procedure To subtract mixed numbers

- Subtract the whole numbers.
- Subtract the fractions.
- Combine whole number and fraction.

Example 1 Subtract $2\frac{1}{4}$ from $9\frac{3}{8}$.

Subtract the whole numbers. $9\frac{3}{8} = 9\frac{3}{8}$

Subtract the fractions. $-2\frac{1}{4} = 2\frac{2}{8}$

$$= \frac{3 - 2}{8} = \frac{1}{8}$$

Subtract the whole numbers. $= 9 - 2 = 7$

Combine the whole number and the fractions. $= 7\frac{1}{8}$ Ans

Example 2 Find the length of thread x of the bolt shown in Figure 3-2. All dimensions are in inches.

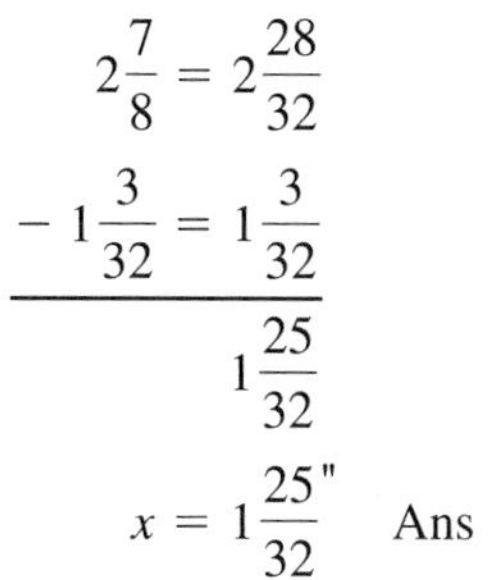

$$\begin{array}{r} 2\frac{7}{8} = 2\frac{28}{32} \\ -\ 1\frac{3}{32} = 1\frac{3}{32} \\ \hline 1\frac{25}{32} \end{array}$$

$$x = 1\frac{25}{32}'' \quad \text{Ans}$$

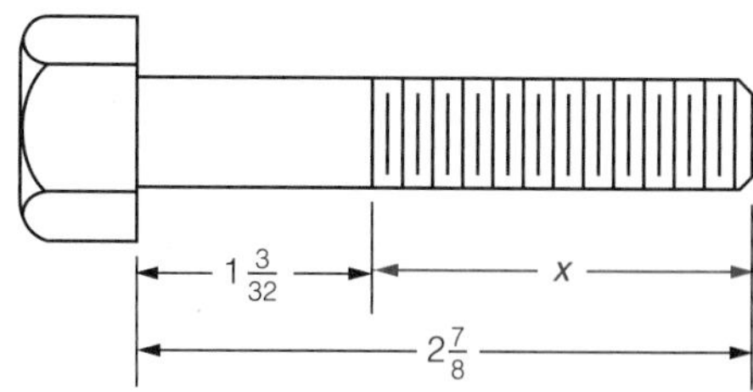

Figure 3-2

Example 3 Subtract $7\frac{15}{16}$ from $12\frac{5}{8}$.

$$\begin{array}{r} 12\frac{5}{8} = 12\frac{10}{16} = 11\frac{26}{16} \\ -\ 7\frac{15}{16} = 7\frac{15}{16} = 7\frac{15}{16} \\ \hline 4\frac{11}{16} \end{array} \quad \text{Ans}$$

➤ **Note:** Since $\frac{15}{16}$ cannot be subtracted from $\frac{10}{16}$, one unit of the whole number 12 is expressed as a fraction with the common denominator 16 and added to the fractional part of the mixed number.

Example 4 Subtract $52\frac{31}{64}$ from 75.

$$\begin{array}{r} 75 = 74\frac{64}{64} \\ -\ 52\frac{31}{64} = 52\frac{31}{64} \\ \hline 22\frac{33}{64} \end{array} \quad \text{Ans}$$

Example 5 Find dimension y of the counterbored block shown in Figure 3-3. All dimensions are in inches.

$$\begin{array}{r} 2\frac{3}{8} = 2\frac{12}{32} = 1\frac{44}{32} \\ -\ \frac{29}{32} = \frac{29}{32} = \frac{29}{32} \\ \hline 1\frac{15}{32} \end{array}$$

$$y = 1\frac{15}{32}'' \quad \text{Ans}$$

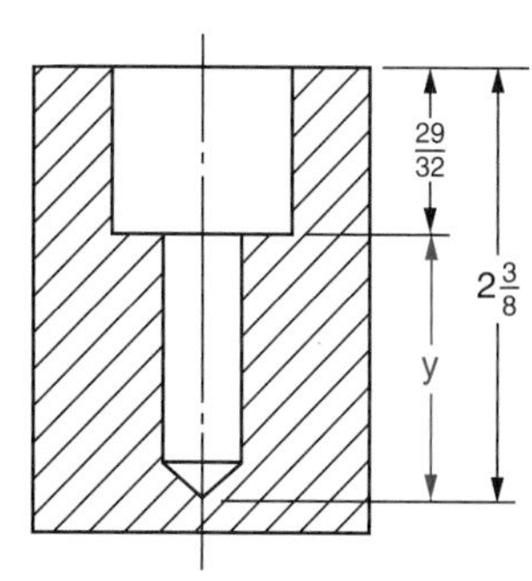

Figure 3-3

Application

Subtracting Fractions

1. Subtract each of the following fractions. Reduce to lowest terms where necessary.

a. $\frac{5}{8} - \frac{9}{32}$ ______

b. $\frac{7}{8} - \frac{5}{8}$ ______

c. $\frac{9}{10} - \frac{19}{50}$ ______

d. $\frac{5}{8} - \frac{9}{64}$ ______

e. $\frac{9}{16} - \frac{13}{64}$ ______

f. $\frac{19}{24} - \frac{3}{16}$ ______

2. Determine dimensions A, B, C, and D of the casting in Figure 3-4. All dimensions are in inches.

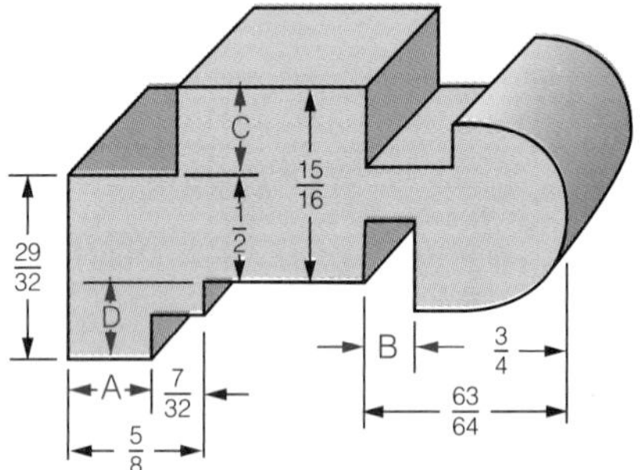

Figure 3-4

A = ______
B = ______
C = ______
D = ______

3. Determine dimensions A, B, C, D, E, and F of the drill jig in Figure 3-5. All dimensions are in inches.

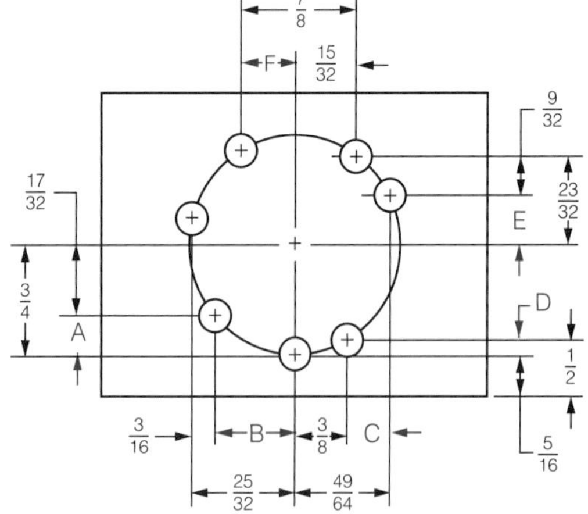

Figure 3-5

A = ______
B = ______
C = ______
D = ______
E = ______
F = ______

Subtracting Mixed Numbers

4. Determine dimensions A, B, C, D, E, F, and G of the tapered pin in Figure 3-6. All dimensions are in inches.

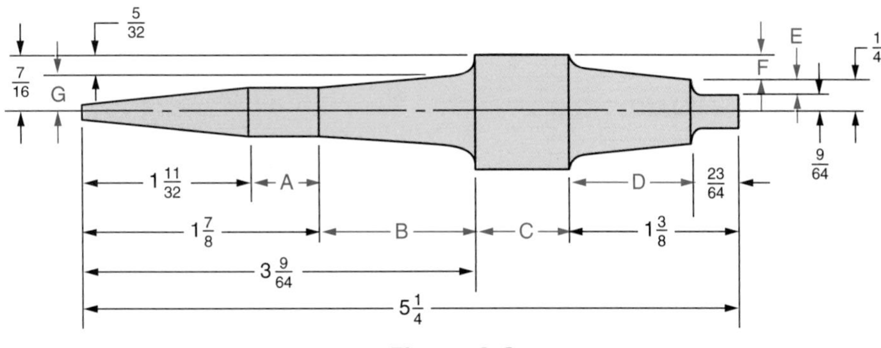

Figure 3-6

A = ______________

B = ______________

C = ______________

D = ______________

E = ______________

F = ______________

G = ______________

5. Determine dimensions A, B, C, D, E, F, G, H, and I of the plate in Figure 3-7. All dimensions are in inches.

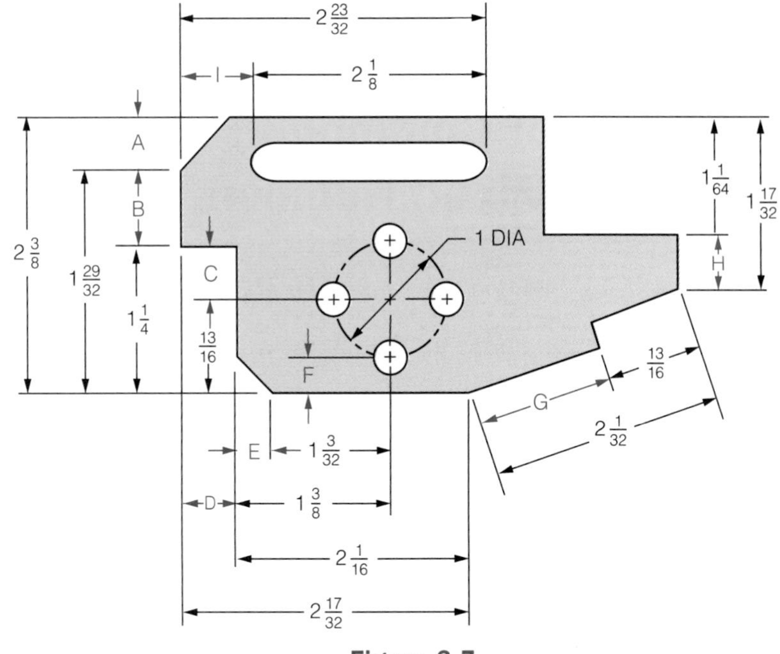

Figure 3-7

A = ______________

B = ______________

C = ______________

D = ______________

E = ______________

F = ______________

G = ______________

H = ______________

I = ______________

6. Three holes are bored in a checking gage. The lower left edge of the gage is the reference point for the hole locations. Sketch the hole locations and determine the missing distances. From the reference point:

Hole #1 is $1\frac{3}{32}''$ to the right, and $1\frac{5}{8}''$ up.

Hole #2 is $2\frac{1}{64}''$ to the right, and $2\frac{3}{16}''$ up.

Hole #3 is $3\frac{1}{4}''$ to the right, and $3\frac{1}{2}''$ up.

Determine:

a. The horizontal distance between hole #1 and hole #2. ______

b. The horizontal distance between hole #2 and hole #3. ______

c. The horizontal distance between hole #1 and hole #3. ______

d. The vertical distance between hole #1 and hole #2. ______

e. The vertical distance between hole #2 and hole #3. ______

f. The vertical distance between hole #1 and hole #3. ______

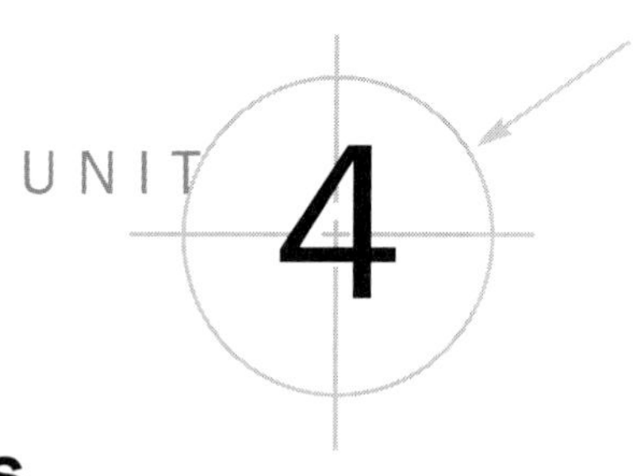

Multiplication of Common Fractions and Mixed Numbers

Objectives *After studying this unit you should be able to*

- Multiply fractions.
- Multiply mixed numbers.
- Divide by common factors (cancellation).

In machine technology, multiplying fractions and mixed numbers can be used to determine the area and volume of a piece of material or the amount of material that will be needed to produce a certain number of parts.

Multiplying Fractions

Procedure To multiply two or more fractions

- Multiply the numerators of the fractions to get the numerator of the product.
- Multiply the denominators of the fractions to get the denominator of the product.
- Write the product of the numerators over the product of the denominators.
- Reduce the resulting fraction to lowest terms.

Example 1 Multiply $\frac{3}{4}$ by $\frac{8}{9}$.

Multiply the numerators. $3 \times 8 = 24$

Multiply the denominators. $4 \times 9 = 36$

Write the product of the numerators over the product of the denominators. $\frac{24}{36}$

Reduce the resulting fraction to lowest terms. $\frac{24}{36} = \frac{24 \div 12}{36 \div 12} = \frac{2}{3}$ Ans

Example 2 Multiply $\frac{2}{3} \times \frac{5}{6} \times \frac{3}{10}$.

$$\frac{2 \times 5 \times 3}{3 \times 6 \times 10} = \frac{30}{180} = \frac{30 \div 30}{180 \div 30} = \frac{1}{6} \quad \text{Ans}$$

To multiply a whole number by a fraction, write the whole number as a fraction with a denominator of 1.

Example 3 Find the distance between the centers of the first and last holes shown in Figure 4-1. All dimensions are in inches.

Multiply $6 \times \frac{7}{16} = \frac{6}{1} \times \frac{7}{16} = \frac{6 \times 7}{1 \times 16} = \frac{42}{16}$

Reduce $\frac{42}{16} = 2\frac{10}{16} = 2\frac{5}{8}$

Distance $= 2\frac{5}{8}"$ Ans

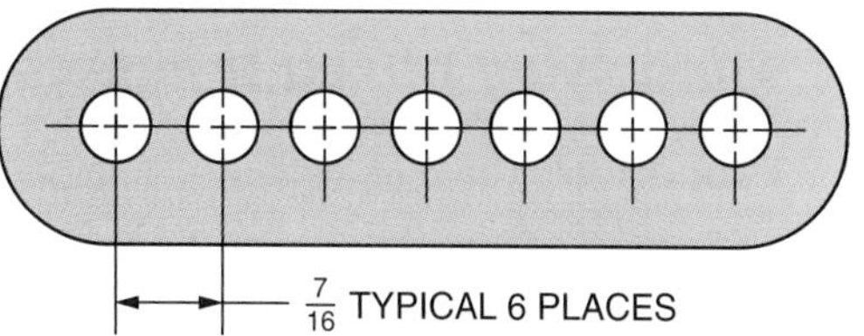

Figure 4-1

➤ **Note:** The value of a number remains unchanged when the number is placed over a denominator of 1. For Example, $6 = \frac{6}{1}$.

Dividing by Common Factors (Cancellation)

Problems involving multiplication of fractions are generally solved more quickly and easily if a numerator and denominator are divided by any common factors before the fractions are multiplied. This process of first dividing by common factors is commonly called *cancellation.* Cancellation allows you to avoid using large numbers in the numerator or denominator and reduces, or eliminates, the need to reduce the fraction after multiplying.

Example 1 Multiply by cancellation method. $\frac{3}{4} \times \frac{8}{9}$

Divide by 3, which is the factor common to both the numerator 3 and the denominator 9.

$3 \div 3 = 1$
$9 \div 3 = 3$

$$\frac{3}{4} \times \frac{8}{9} = \frac{\overset{1}{\cancel{3}}}{4} \times \frac{8}{\underset{3}{\cancel{9}}}$$

Divide by 4, which is the factor common to both the denominator 4 and the numerator 8.

$4 \div 4 = 1$
$8 \div 4 = 2$

$$\frac{\overset{1}{\cancel{3}}}{\underset{1}{\cancel{4}}} \times \frac{\overset{2}{\cancel{8}}}{\underset{3}{\cancel{9}}} = \frac{1 \times 2}{1 \times 3} = \frac{2}{3} \quad \text{Ans}$$

Multiply reduced fractions.

Example 2 Multiply $\frac{4}{7} \times \frac{5}{18} \times \frac{14}{15}$.

Divide 4 and 18 by 2.
Divide 7 and 14 by 7.
Divide 5 and 15 by 5.
Multiply.

$$\frac{\overset{2}{\cancel{4}}}{\underset{1}{\cancel{7}}} \times \frac{\overset{1}{\cancel{5}}}{\underset{9}{\cancel{18}}} \times \frac{\overset{2}{\cancel{14}}}{\underset{3}{\cancel{15}}} = \frac{2 \times 1 \times 2}{1 \times 9 \times 3} = \frac{4}{27} \quad \text{Ans}$$

Example 3 Multiply $\frac{5}{14} \times \frac{8}{9} \times \frac{7}{10}$.

Divide 5 and 10 by 5.
Divide 14 and 8 by 2.

$$\frac{\overset{1}{\cancel{5}}}{\underset{\underset{1}{\cancel{7}}}{\cancel{14}}} \times \frac{\overset{\overset{2}{\cancel{4}}}{\cancel{8}}}{9} \times \frac{\overset{1}{\cancel{7}}}{\underset{\underset{1}{\cancel{2}}}{\cancel{10}}} = \frac{1 \times 2 \times 1}{1 \times 9 \times 1} = \frac{2}{9} \quad \text{Ans}$$

The process is continued by dividing 7 and 7 by 7 and by dividing 2 and 4 by 2.
Multiply.

Multiplying Mixed Numbers

Procedure To multiply mixed numbers

- Express the mixed numbers as improper fractions.
- Follow the procedure for multiplying proper fractions.

Example 1 Multiply $2\frac{2}{5} \times 6\frac{7}{8}$.

Write the mixed number $2\frac{2}{5}$ as the fraction $\frac{12}{5}$.

Write the mixed number $6\frac{7}{8}$ as the fraction $\frac{55}{8}$.

Divide 5 and 55 by 5.
Divide 12 and 8 by 4.

$$\frac{\overset{3}{\cancel{12}}}{\underset{1}{\cancel{5}}} \times \frac{\overset{11}{\cancel{55}}}{\underset{2}{\cancel{8}}} = \frac{3}{1} \times \frac{11}{2}$$

Multiply numerators.
Multiply denominators.

$$\frac{3}{1} \times \frac{11}{2} = \frac{3 \times 11}{1 \times 2} = \frac{33}{2}$$

Express as a mixed number in lowest terms.

$$\frac{33}{2} = 16\frac{1}{2} \quad \text{Ans}$$

Example 2 The block of steel shown in Figure 4-2 is to be machined. The block measures $8\frac{3}{4}$ inches long, $4\frac{9}{16}$ inches wide, and $\frac{7}{8}$ inch thick. Find the volume of the block. All dimensions are in inches. (Volume = length × width × thickness.)

$$8\frac{3}{4} \times 4\frac{9}{16} \times \frac{7}{8} = \frac{35}{4} \times \frac{73}{16} \times \frac{7}{8} = \frac{35 \times 73 \times 7}{4 \times 16 \times 8}$$

$$= \frac{17885}{512} = 34\frac{477}{512}$$

$$\text{Volume} = 34\frac{477}{512} \text{ cubic inches} \quad \text{Ans}$$

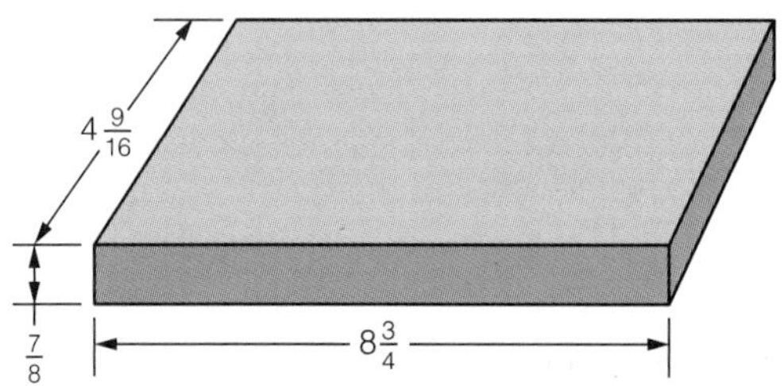

Figure 4-2

Application

Multiplying Fractions

1. Multiply these fractions. Reduce to lowest terms where necessary.

a. $\frac{2}{3} \times \frac{1}{6}$ ________

b. $\frac{1}{2} \times \frac{1}{4}$ ________

c. $\frac{5}{8} \times \frac{13}{64}$ ________

d. $\frac{3}{4} \times \frac{3}{5} \times \frac{2}{3}$ ________

e. $7 \times \frac{9}{14} \times 3$ ________

f. $\frac{7}{15} \times \frac{3}{8} \times \frac{5}{7}$ ________

2. Determine dimensions A, B, C, D, and E of the template shown in Figure 4-3. All dimensions are in inches.

A = ________

B = ________

C = ________

D = ________

E = ________

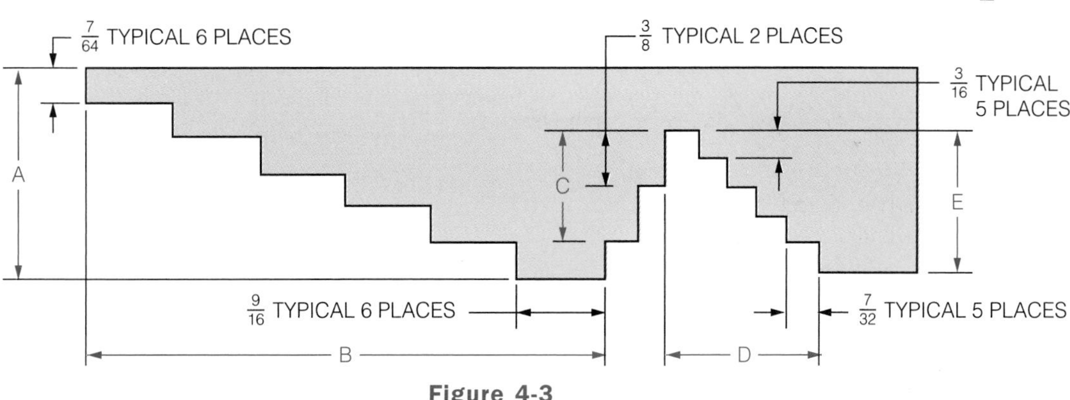

Figure 4-3

3. A special washer-faced nut is shown in Figure 4-4. All dimensions are in inches.

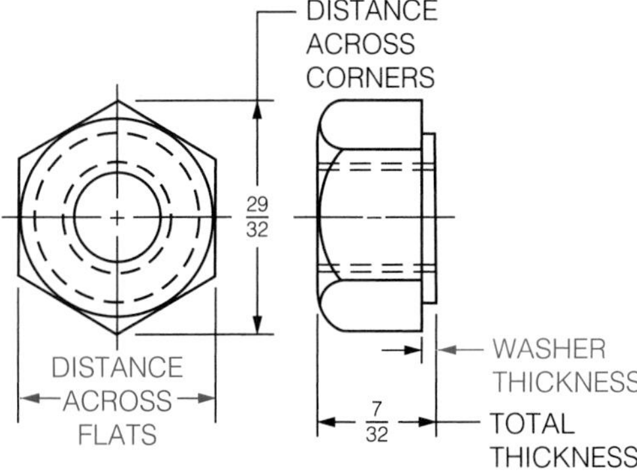

Figure 4-4

a. Determine the distance across flats.

Distance across flats $= \frac{55}{64} \times$ Distance across corners ________

b. Determine the washer thickness.

Washer thickness $= \frac{1}{8} \times$ Total thickness ________

4. The Unified Thread may have either a flat or rounded crest or root. If the sides of the Unified Thread are extended, a sharp V-thread is formed. H is the height of a sharp V-thread. The pitch, P, is the distance between two adjacent threads.

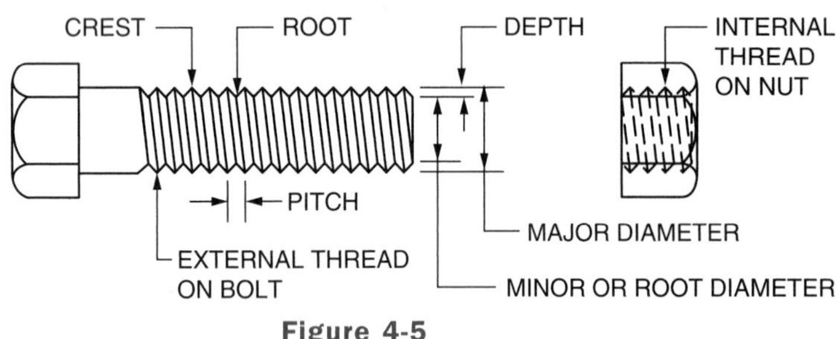

Figure 4-5

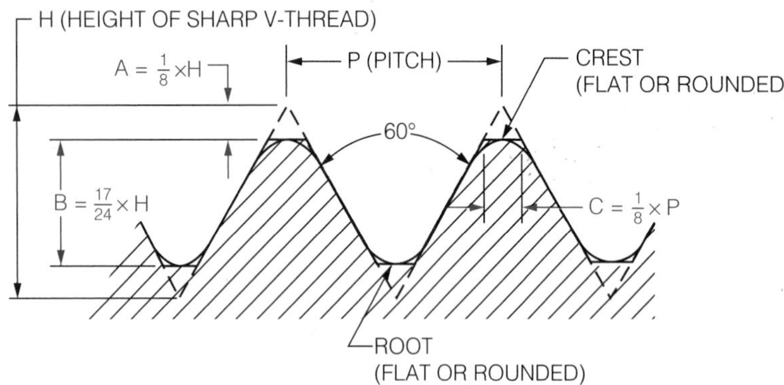

Figure 4-6

Find dimensions A, B, and C as indicated.

a. $H = \frac{7}{16}''$, A = ________, B = ________ f. $P = \frac{1}{4}''$, C = ________

b. $H = \frac{3}{8}''$, A = ________, B = ________ g. $P = \frac{3}{32}''$, C = ________

c. $H = \frac{15}{16}''$, A = ________, B = ________ h. $P = \frac{1}{20}''$, C = ________

d. $H = \frac{21}{32}''$, A = ________, B = ________ i. $P = \frac{1}{28}''$, C = ________

e. $H = \frac{3}{4}''$, A = ________, B = ________ j. $P = \frac{3}{16}''$, C = ________

Multiplying Mixed Numbers

5. Multiply these mixed numbers. Reduce to lowest terms where necessary.

a. $1\frac{2}{3} \times 6\frac{3}{10}$ ________ d. $1\frac{2}{3} \times 10\frac{1}{4} \times \frac{3}{8}$ ________

b. $3\frac{5}{16} \times 7\frac{3}{4}$ ________ e. $2\frac{3}{32} \times 3 \times \frac{1}{8}$ ________

c. $4\frac{5}{8} \times 2\frac{1}{2}$ ________ f. $2\frac{2}{3} \times 2\frac{2}{3} \times 5\frac{1}{4}$ ________

6. How many inches of drill rod are required in order to make 20 drills each $3\frac{3}{16}''$ long? Allow $\frac{3}{32}''$ waste for each drill.

7. A hole is cut in a rectangular metal plate as shown in Figure 4-7. To find the area of a rectangle, multiply the length by the width. Determine the area of the plate after the hole has been removed. All dimensions are in inches. The area will be in square inches.

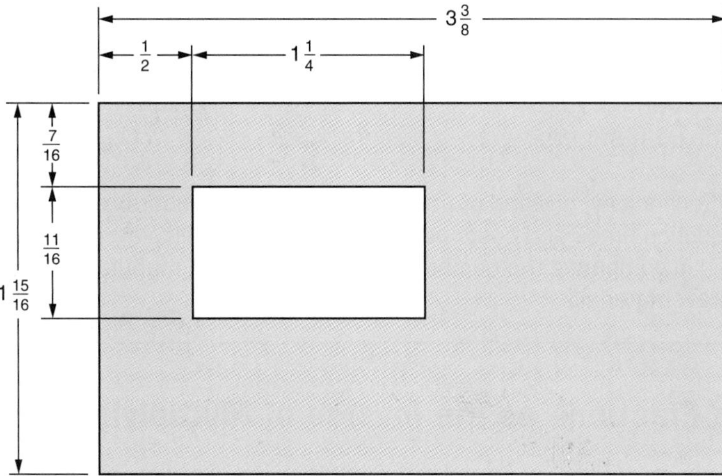

Figure 4-7

8. Six identical square holes are cut in a rectangular metal plate as shown in Figure 4-8. To find the area of a rectangle, multiply the length by the width. Determine the area of the plate after the holes have been removed. All dimensions are in inches. The area will be in square inches.

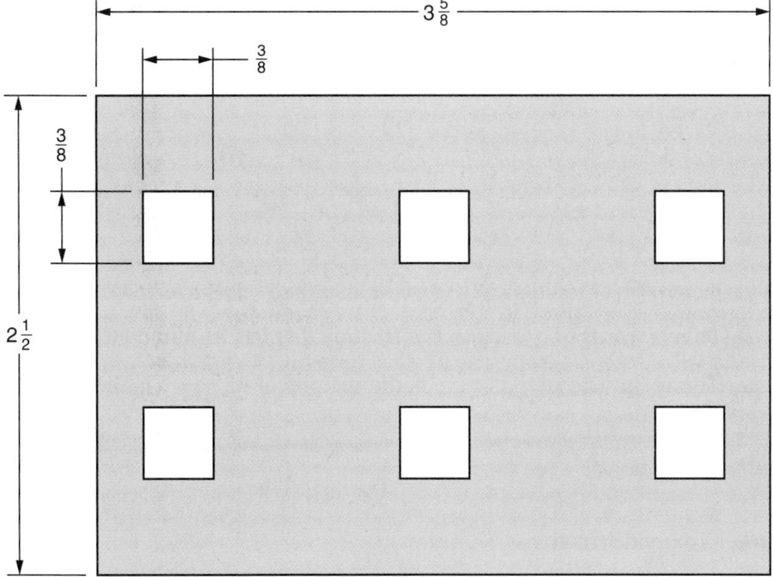

Figure 4-8

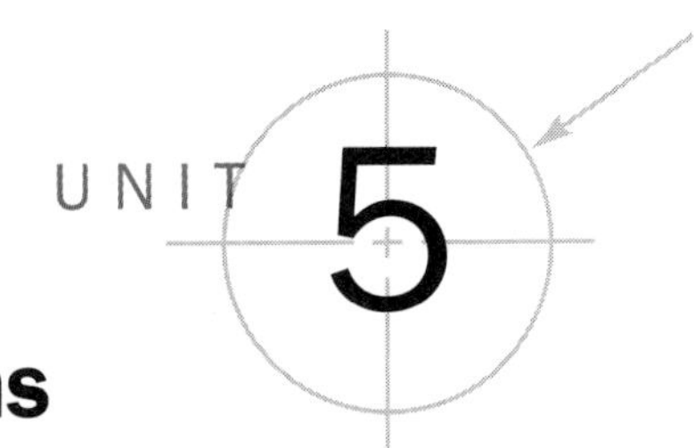

Division of Common Fractions and Mixed Numbers

Objectives ***After studying this unit you should be able to***

- Divide fractions.
- Divide mixed numbers.

In machine technology, division of fractions and mixed numbers can be used in determining production times and costs per machined unit, in calculating the pitch of screw threads, and in computing the number of parts that can be manufactured from a given amount of raw material.

Dividing Fractions as the Inverse of Multiplying Fractions

Division can be represented in several different ways. For example, 21 divided by 3 can be shown as $21 \div 3$, $3\overline{)21}$, $\frac{21}{3}$, and 21/3. Here the number being divided, 21, is called the *dividend,* the number used to divide, 3, is called the *divisor,* and the answer, 7, is called the *quotient.* In the problem $\frac{5}{8} \div \frac{3}{4}$, the divisor is $\frac{3}{4}$ and the dividend is $\frac{5}{8}$. As you will see in Example 1, the quotient is $\frac{5}{6}$.

Division is the inverse of multiplication. Dividing by 2 is the same as multiplying by $\frac{1}{2}$.

$$5 \div 2 = 2\frac{1}{2}$$

$$5 \times \frac{1}{2} = 2\frac{1}{2}$$

$$5 \div 2 = 5 \times \frac{1}{2}$$

Two is the reciprocal, or multiplicative inverse, of $\frac{1}{2}$, and $\frac{1}{2}$ is the reciprocal, or multiplicative inverse, of 2. The *reciprocal* of a fraction is a fraction that has its numerator and denominator interchanged. The reciprocal of $\frac{1}{3}$ is $\frac{3}{1}$, $\frac{8}{7}$ is the reciprocal of $\frac{7}{8}$, $\frac{63}{64}$ is the reciprocal of $\frac{64}{63}$, and the reciprocal of $\frac{9}{16}$ is $\frac{16}{9}$.

Procedure To divide fractions

- Determine the reciprocal of the divisor.
- Multiply the dividend by the reciprocal of the divisor.

Example 1 Divide $\frac{5}{8}$ by $\frac{3}{4}$.

The divisor is $\frac{3}{4}$ and its reciprocal is $\frac{4}{3}$.

Multiply the dividend, $\frac{5}{8}$, by $\frac{4}{3}$.

Follow the procedure for multiplication.

$$\frac{5}{8} \div \frac{3}{4} = \frac{5}{\cancel{8}_2} \times \frac{\cancel{4}^1}{3} = \frac{5}{6} \text{ Ans}$$

Example 2 The machine bolt shown in Figure 5-1 has a pitch of $\frac{1}{16}$". The pitch is the distance between two adjacent threads. Find the number of threads in $\frac{7}{8}$". All dimensions are in inches.

Divide $\frac{7}{8}$ by $\frac{1}{16}$.

$$\frac{7}{8} \div \frac{1}{16} = \frac{7}{\cancel{8}_1} \times \frac{\cancel{16}^2}{1} = 14 \text{ Ans}$$

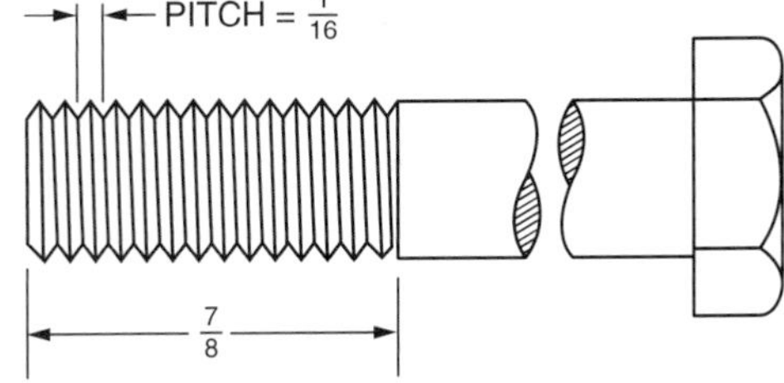

Figure 5-1

Dividing Mixed Numbers

Procedure To divide mixed numbers

- Express the mixed numbers as improper fractions.
- Follow the procedure for dividing fractions.

Example 1 Divide $7\frac{1}{2}$ by $2\frac{3}{8}$.

Express $7\frac{1}{2}$ and $2\frac{3}{8}$ as improper fractions. $\qquad 7\frac{1}{2} \div 2\frac{3}{8} = \frac{15}{2} \div \frac{19}{8}$

The reciprocal of $\frac{19}{8}$ is $\frac{8}{19}$.

Multiply by the reciprocal.

Multiply.

$$\frac{15}{\cancel{2}_1} \times \frac{\cancel{8}^4}{19} = \frac{60}{19} = 3\frac{3}{19} \text{ Ans}$$

➤ **Note:** The reciprocal of a mixed number such as $2\frac{3}{8}$ is NOT $2\frac{8}{3}$. To find the reciprocal of a mixed number, first change the number to an improper fraction. Thus, to find the reciprocal of $2\frac{3}{8}$, first rewrite it as $\frac{19}{8}$.

Example 2 A section of strip stock is shown in Figure 5-2 with five equally spaced holes. Determine the distance between two consecutive holes. All dimensions are in inches.

➤ **Note:** The number of spaces between the holes is one less than the number of holes.

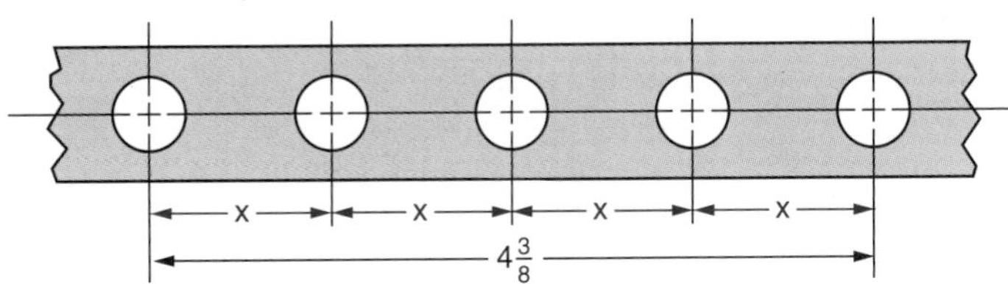

Figure 5-2

Express as improper fractions. $4\frac{3}{8} \div 4 = \frac{35}{8} \div \frac{4}{1}$

Multiply by the reciprocal. $\frac{35}{8} \times \frac{1}{4} = \frac{35}{32} = 1\frac{3}{32}$

$x = 1\frac{3}{32}''$ Ans

Application

Inverting Fractions

Invert each of the following.

1. $\frac{7}{8}$ __________
2. $\frac{1}{4}$ __________
3. $\frac{25}{8}$ __________
4. 6 __________

Dividing Fractions

5. This casting in Figure 5-3 shows seven tapped holes, A–G. The number of threads is determined by dividing the depth of the thread by the thread pitch. Find the number of threads in each of the tapped holes. All dimensions are in inches.

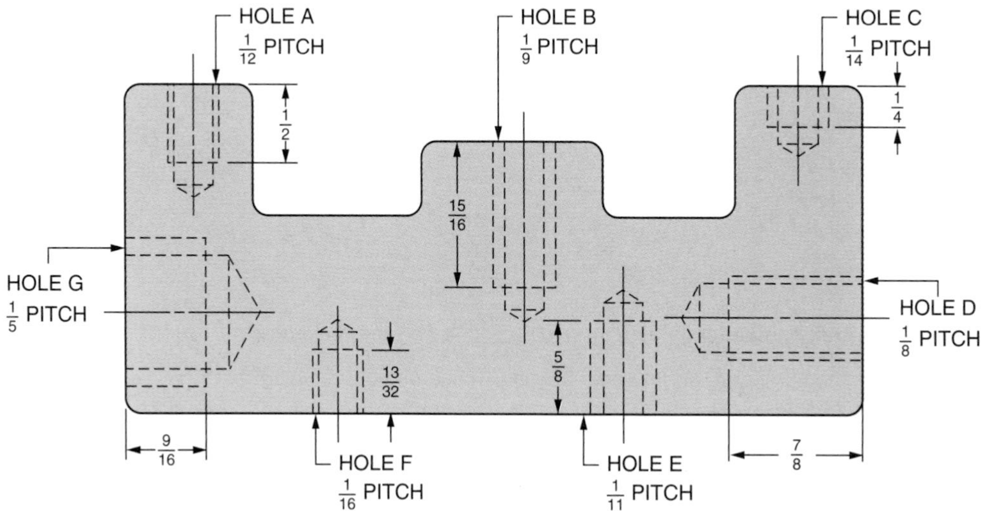

Figure 5-3

A = __________

B = __________

C = __________

D = __________

E = __________

F = __________

G = __________

6. Bar stock is being cut on a lathe. The tool feeds (advances) $\frac{3}{64}$ inch each time the stock turns once (1 revolution). How many revolutions will the stock make when the tool advances $\frac{3}{4}$ inch? ______

7. A groove $\frac{15}{16}$ inch deep is to be miled in a steel plate. How many cuts are required if each cut is $\frac{3}{16}$ inch deep? ______

Dividing Mixed Numbers

8. This sheet metal section shown in Figure 5-4 has 5 sets of drilled holes: A, B, C, D, and E. The holes within a set are equally spaced in the horizontal direction. Compute the horizontal distance between 2 consecutive holes for each set. All dimensions are in inches.

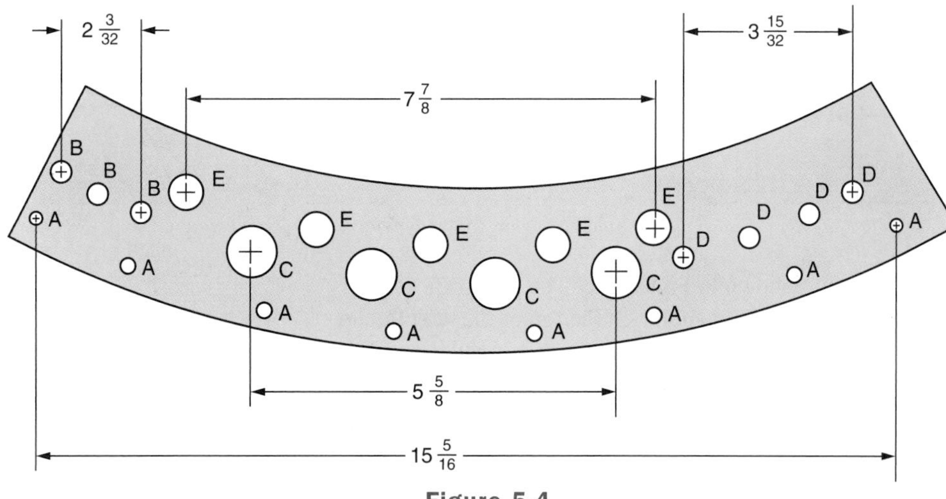

A = ______

B = ______

C = ______

D = ______

E = ______

Figure 5-4

9. The feed on a lathe is set for $\frac{1}{64}$ inch. How many revolutions does the work make when the tool advances $3\frac{3}{4}$ inches? ______

10. How many complete pieces can be blanked from a strip of steel $27\frac{1}{4}$ feet long if each stamping requires $2\frac{3}{16}$ inches of material plus an allowance of $\frac{5}{16}$ inch at one end of the strip? (12 inches = 1 foot) ______

11. A slot is milled the full length of a steel plate that is $3\frac{1}{4}$ feet long. This operation takes a total of $4\frac{1}{16}$ minutes. How many feet of steel are cut in one minute? ______

12. How many binding posts can be cut from a brass rod $42\frac{1}{2}$ inches long if each post is $1\frac{7}{8}$ inches long? Allow $\frac{3}{32}$ inch waste for each cut. ______

13. A bar of steel $23\frac{1}{4}$ feet long weighs $110\frac{1}{2}$ pounds. How much does a one-foot length of bar weigh?

14. A single-threaded (or single-start) square-thread screw is shown in Figure 5-5. The lead of a screw is the distance that the screw advances in one turn (revolution). The lead is equal to the pitch in a single-threaded screw. Given the number of turns and the amount of screw advance, determine the leads.

	Screw Advance	Number of Turns	Lead
a.	$2\frac{1}{4}''$	10	
b.	$7\frac{37}{64}''$	$24\frac{1}{4}$	
c.	$2\frac{7}{16}''$	$6\frac{1}{2}$	
d.	$1\frac{1}{2}''$	15	
e.	$6\frac{3}{10}''$	$12\frac{3}{5}$	

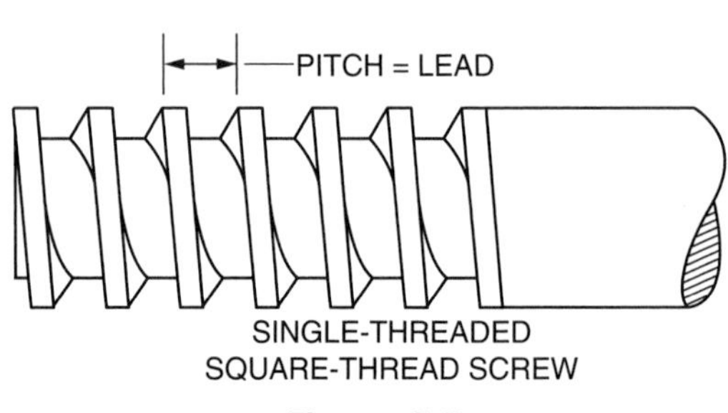

Figure 5-5

15. A double-threaded square-thread screw is shown in Figure 5-6. The pitch of a screw is the distance from the top of one thread to the same point on the top of the next thread. The lead is the distance the screw advances for each complete turn or revolution of the screw. In a double-threaded screw, the lead is twice the pitch. Given the number of turns and the amount of screw advance, determine the lead and pitch.

	Screw Advance	Number of Turns	Lead	Pitch
a.	$2\frac{5}{8}''$	10		
b.	$6\frac{61}{64}''$	$22\frac{1}{4}$		
c.	$10\frac{5}{16}''$	$16\frac{1}{2}$		
d.	$3\frac{9}{16}''$	12		

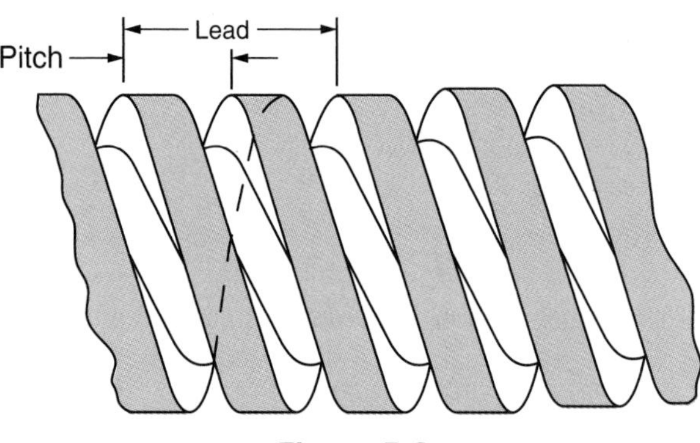

Figure 5-6

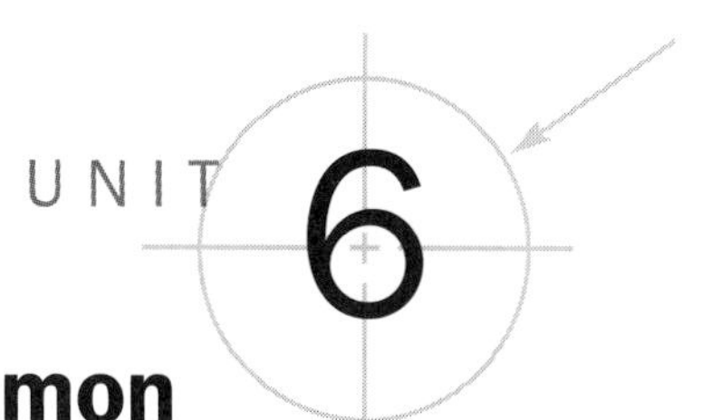

Combined Operations of Common Fractions and Mixed Numbers

Objectives *After studying this unit you should be able to*

- Solve problems that involve combined operations of fractions and mixed numbers.
- Solve complex fractions.

Before a part is machined, the sequence of machining operations, the machine setup, and the working dimensions needed to produce the part must be determined. In actual practice, calculations of machine setup and working dimensions require not only the individual operations of addition, subtraction, multiplication, and division, but a combination of two or more of these operations.

Order of Operations for Combined Operations

Procedure

- **Do all the work in the parentheses first.** Parentheses are used to group numbers. In a problem expressed in fractional form, the numerator and the denominator are each considered as being enclosed in parentheses. Brackets, [ ], and braces, { }, are used for "nesting" one group within another. They are treated the same as parentheses. On your calculator, use the parentheses, ( ), symbols.

Example 1 Compute $\{5 + [7 - 3(6 - 4) + 2] - 6\} + 1$.

Begin with the innermost grouping symbols, the parentheses.

$$\{5 + [7 - 3(6 - 4) + 2] - 6\} + 1$$
$$= \{5 + [7 - 3(2) + 2] - 6\} + 1$$
$$= \{5 + [7 - 6 + 2] - 6\} + 1$$

Next, work within the brackets.

$$= \{5 + [3] - 6\} + 1$$
$$= \{5 + 3 - 6\} + 1$$

Finally, do the operations inside the braces.

$$= \{2\} + 1$$
$$= 3 \quad \text{Ans}$$

$$\frac{4\frac{3}{4} - \frac{1}{2}}{10 + 6\frac{5}{8}} = \left(4\frac{3}{4} - \frac{1}{2}\right) \div \left(10 + 6\frac{5}{8}\right)$$

If an expression contains nested parentheses, do the work within the innermost parentheses first.

- **Do multiplication and division next.** Perform multiplication and division in order from left to right.
- **Do addition and subtraction last.** Perform addition and subtraction in order from left to right.

Example 2 Find the value of $\left(1\frac{2}{5} + \left(\frac{7}{3} - \frac{9}{5}\right)\right) + \frac{1}{4}$.

There are two sets of parentheses with one set nested inside the other.

Begin with the innermost parentheses: $\frac{7}{3} - \frac{9}{5} = \frac{8}{15}$

The result is $\left(1\frac{2}{5} + \frac{8}{15}\right) + \frac{1}{4}$.

Next, perform the operation in the remaining parentheses: $1\frac{2}{5} + \frac{8}{15} = 1\frac{14}{15}$

The result is $1\frac{14}{15} + \frac{1}{4}$.

Add these two fractions: $1\frac{14}{15} + \frac{1}{4} = 2\frac{11}{60}$ Ans

Combining Addition and Subtraction

Example 1 Find the value of $3\frac{1}{2} - \frac{3}{8} + \frac{5}{16}$.

There are no parentheses and there is no multiplication or division. So, perform addition and subtraction in order from left to right.

Subtract $\frac{3}{8}$ from $3\frac{1}{2}$. $3\frac{1}{2} - \frac{3}{8} = 3\frac{1}{8}$

Add $3\frac{1}{8}$ to $\frac{5}{16}$. $3\frac{1}{8} + \frac{5}{16} = 3\frac{7}{16}$ Ans

Example 2 Find x, the distance from the base of the plate in Figure 6-1 to the center of hole #2. All dimensions are in inches.

$$x = \frac{9}{16}'' + 2\frac{1}{8}'' - \frac{13}{32}''$$

Add. $\frac{9}{16}'' + 2\frac{1}{8}'' = 2\frac{11}{16}''$

Subtract. $2\frac{11}{16}'' - \frac{13}{32}'' = 2\frac{9}{32}''$ Ans

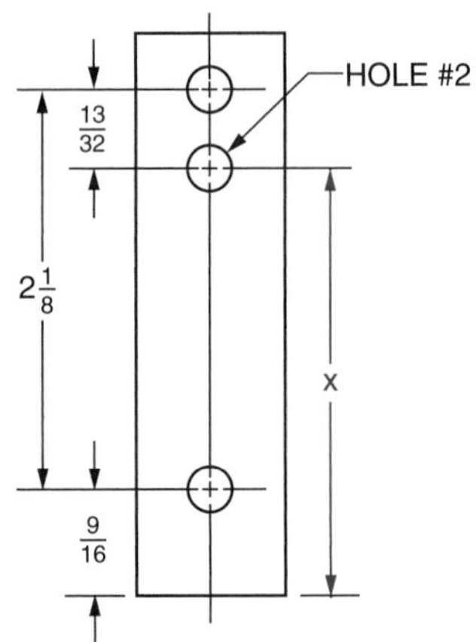

Figure 6-1

Combining Multiplication and Division

Example 1 Find the value of $\frac{2}{3} \times 8 \div 2\frac{1}{2}$.

There are no parentheses. Perform multiplication and division in order from left to right.

Multiply. $\frac{2}{3} \times 8 = \frac{2 \times 8}{3 \times 1} = \frac{16}{3}$

Divide. $\frac{16}{3} \div 2\frac{1}{2} = \frac{16}{3} \div \frac{5}{2} = \frac{16}{3} \times \frac{2}{5} = \frac{32}{15} = 2\frac{2}{15}$ Ans

Example 2 The stainless-steel plate shown in Figure 6-2 has slots that are of uniform length and equally spaced within a distance of $33\frac{1}{2}$ inches. The time required to rough and finish mill a one-inch length of slot is $\frac{7}{10}$ minute. How many minutes are required for the tool to cut all the slots? Disregard the time required to reposition the part. All dimensions are in inches.

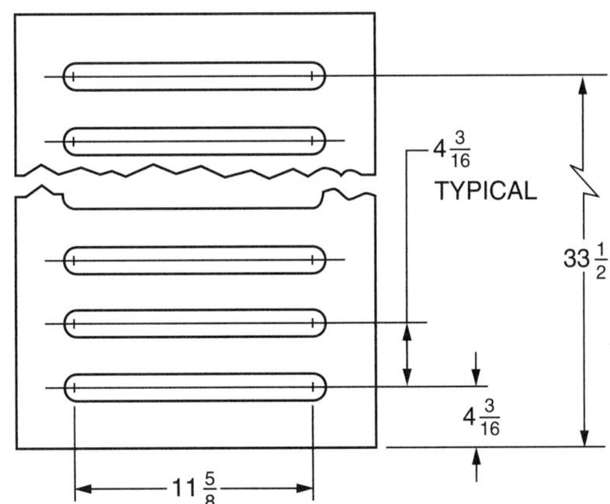

Figure 6-2

The number of grooves in $33\frac{1}{2}'' = 33\frac{1}{2} \div 4\frac{3}{16}$.

The time required to cut 1 groove $= \frac{7}{10} \times 11\frac{5}{8}$.

Total time equals the number of grooves multiplied by the time for each groove.

$$33\frac{1}{2} \div 4\frac{3}{16} \times \frac{7}{10} \times 11\frac{5}{8}$$

Divide. $33\frac{1}{2} \div 4\frac{3}{16} = \frac{67}{2} \times \frac{16}{67} = 8$

Multiply. $8 \times \frac{7}{10} \times 11\frac{5}{8} = \frac{8}{1} \times \frac{7}{10} \times \frac{93}{8} = 65\frac{1}{10}$

Total time $= 65\frac{1}{10}$ minutes Ans

Combining Addition, Subtraction, Multiplication, and Division

Example 1 Find the value of $7\frac{5}{6} + 5\frac{1}{2} \div \frac{3}{4} - 10 \times \frac{7}{16}$.

First divide and multiply.

$$7\frac{5}{6} + 5\frac{1}{2} \div \frac{3}{4} - 10 \times \frac{7}{16}$$

$$7\frac{5}{6} + 5\frac{1}{2} \times \frac{4}{3} - \frac{10}{1} \times \frac{7}{16}$$

Next add and subtract. $7\frac{5}{6} + 7\frac{1}{3} - 4\frac{3}{8} = 10\frac{19}{24}$ Ans

Example 2 Find the value of $\left(7\frac{5}{6} + 5\frac{1}{2}\right) \div \frac{3}{4} - 10 \times \frac{7}{16}$.

First do the work in parentheses. $\left(7\frac{5}{6} + 5\frac{1}{2}\right) = 7\frac{5}{6} + 5\frac{3}{6} = 13\frac{1}{3}$

Next divide and multiply. $13\frac{1}{3} \div \frac{3}{4} = \frac{40}{3} \times \frac{4}{3} = \frac{160}{9} = 17\frac{7}{9}$

$$10 \times \frac{7}{16} = 4\frac{3}{8}$$

Then add and subtract. $17\frac{7}{9} - 4\frac{3}{8} = 13\frac{29}{72}$ Ans

➤ **Note:** This example is the same as the preceding example except for the parentheses.

Complex Fractions

A *complex fraction* is an expression in which either the numerator or denominator or both are fractions or mixed numbers. A fraction indicates a division operation. Therefore, complex fractions can be solved by dividing the numerator by the denominator.

$$\frac{\frac{5}{9}}{\frac{1}{3}} = \frac{5}{9} \div \frac{1}{3}$$

Example Find the value of $\dfrac{5\frac{7}{8} + 2\frac{3}{4}}{3\frac{15}{16} - 1\frac{1}{8}}$.

➤ **Note:** The complete numerator is divided by the complete denominator. Therefore, parentheses are used to indicate that addition in the numerator and subtraction in the denominator must be performed before division.

$$\frac{5\frac{7}{8} + 2\frac{3}{4}}{3\frac{15}{16} - 1\frac{1}{8}} = \left(5\frac{7}{8} + 2\frac{3}{4}\right) \div \left(3\frac{15}{16} - 1\frac{1}{8}\right)$$

$$= \frac{69}{8} \div \frac{45}{16}$$

$$= \frac{\overset{23}{\cancel{69}}}{\underset{1}{\cancel{8}}} \times \frac{\overset{2}{\cancel{16}}}{\underset{15}{\cancel{45}}} = 3\frac{1}{15} \quad \text{Ans}$$

Application

Order of Operations for Combined Operations

1. Solve the following examples of combined operations.

a. $\frac{1}{2} + \frac{3}{16} - \frac{1}{4}$ __________

b. $3\frac{7}{8} - 2\frac{3}{16} + \frac{3}{8}$ __________

c. $\frac{3}{10} + 8\frac{2}{5} - 3\frac{1}{25}$ __________

d. $27 - 2\frac{2}{3} + 4\frac{1}{6}$ __________

e. $32\frac{1}{8} + 2\frac{3}{16} \times \frac{3}{4}$ __________

f. $\frac{7}{9} \times \left(\frac{2}{3} + 3\frac{5}{6}\right)$ __________

g. $12 - 4\frac{1}{2} \div \frac{1}{2} + 2\frac{3}{4}$ __________

h. $\left(16 - 4\frac{1}{2}\right) \div \frac{1}{2} + 5\frac{3}{8}$ __________

i. $\left(16 - 4\frac{1}{2}\right) \div \left(\frac{1}{2} + 2\frac{1}{8}\right)$ __________

j. $15\frac{1}{4} \times 1\frac{1}{3} + 2\frac{2}{3} \div 4\frac{5}{6}$ __________

Complex Fractions

2. Find the value of the following complex fractions.

a. $\dfrac{\frac{3}{4}}{\frac{1}{2}}$ ______

b. $\dfrac{3\frac{7}{8}}{5}$ ______

c. $\dfrac{\frac{15}{16}}{2\frac{1}{8}}$ ______

d. $\dfrac{\frac{1}{3}+\frac{5}{6}}{3\frac{3}{4}}$ ______

e. $\dfrac{6\frac{3}{4}-2\frac{7}{8}}{3\frac{1}{2}+1\frac{1}{16}}$ ______

f. $\dfrac{10\frac{1}{2}\times\frac{1}{2}}{4\div 2\frac{1}{4}}$ ______

Related Problems

3. Refer to the shaft shown in Figure 6-3. Determine the missing dimensions in the table using the dimensions given. All dimensions are in inches.

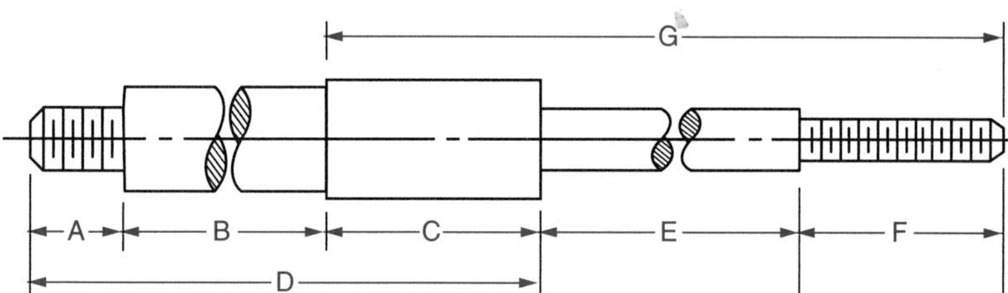

Figure 6-3

	A	B	C	D	E	F	G
a.	$\frac{1}{2}$		$1\frac{3}{8}$	$6\frac{3}{4}$		$\frac{15}{16}$	$7\frac{3}{8}$
b.		$3\frac{13}{32}$	$1\frac{5}{8}$	$5\frac{37}{64}$	$4\frac{3}{8}$	$\frac{3}{4}$	
c.	$\frac{7}{16}$	$4\frac{3}{32}$			$5\frac{1}{8}$	$\frac{27}{32}$	$7\frac{1}{32}$
d.	$\frac{5}{8}$		$1\frac{7}{16}$	$5\frac{31}{32}$		$\frac{7}{8}$	$7\frac{15}{16}$
e.		$3\frac{3}{4}$	$1\frac{11}{16}$	$6\frac{1}{32}$	$4\frac{61}{64}$	$\frac{25}{32}$	
f.	$\frac{11}{16}$	$4\frac{3}{16}$			$5\frac{3}{16}$	$\frac{7}{8}$	$7\frac{3}{64}$

4. The outside diameter of an aluminum tube is $3\frac{1}{16}$ inches. The wall thickness is $\frac{5}{32}$ inch. What is the inside diameter? ____________

5. Four studs of the following lengths in inches are to be machined from bar stock: $1\frac{3}{4}$", $1\frac{7}{8}$", $2\frac{5}{16}$", and $1\frac{11}{32}$". Allow $\frac{1}{8}$ inch waste for each cut and $\frac{1}{32}$ inch on each end of each stud for facing. What is the shortest length of bar stock required so that only three cuts are needed? ____________

6. Find dimensions A, B, C, and D of the idler bracket in Figure 6-4. All dimensions are in inches.

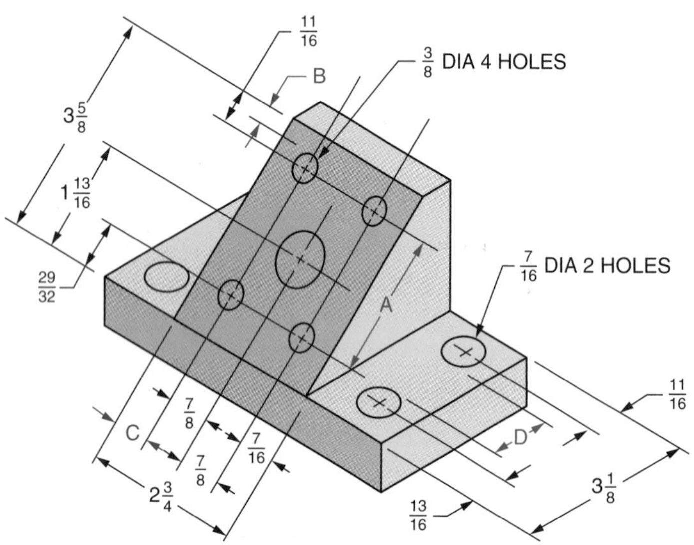

Figure 6-4

A = ____________

B = ____________

C = ____________

D = ____________

7. How long does it take to cut a distance of $1\frac{1}{4}$ feet along a shaft that turns 150 revolutions per minute with a tool feed of $\frac{1}{32}$ inch per revolution? ____________

8. An angle iron $47\frac{1}{2}$ inches long has two drilled holes that are equally spaced from the center of the piece. The center distance between the two holes is $19\frac{7}{8}$ inches. What is the distance from each end of the piece to the center of the closest hole? ____________

9. A tube has an inside diameter of $\frac{3}{4}$ inch and a wall thickness of $\frac{1}{16}$ inch. The tube is to be fitted in a drilled hole in a block. What diameter hole should be drilled in the block to give $\frac{1}{64}$ inch total clearance? ____________

10. Two views of a mounting block are shown in Figure 6-5. Determine dimensions A–G. All dimensions are in inches.

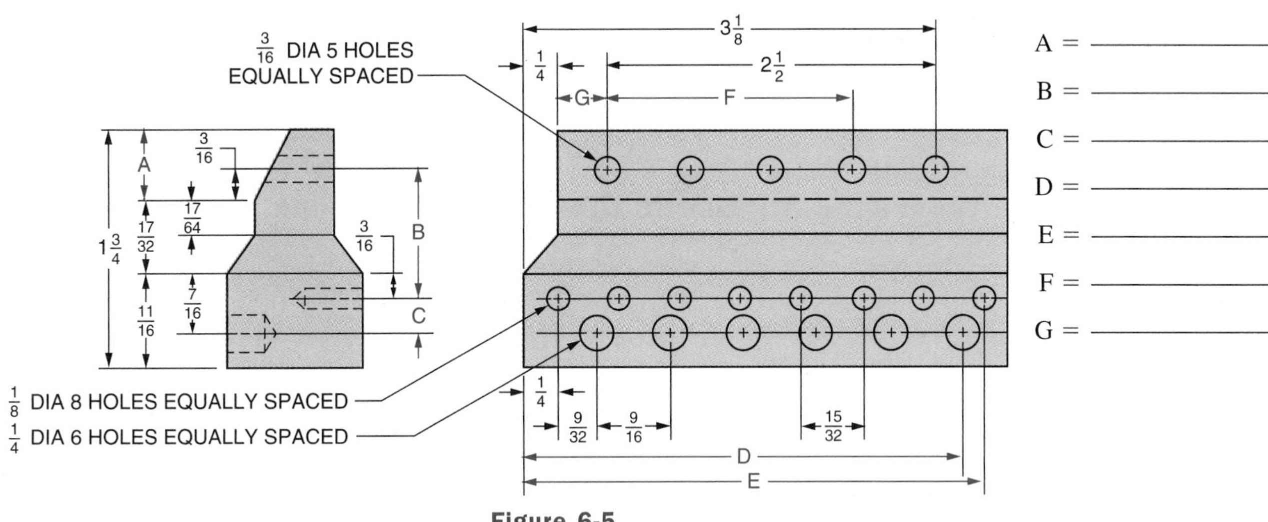

Figure 6-5

11. The composition of an aluminum alloy by weight is $\frac{19}{20}$ aluminum and $\frac{1}{50}$ copper. The only other element in the alloy is magnesium. How many pounds of magnesium are required for casting 125 pounds of alloy? ________

12. Pieces of the following lengths are cut from a 15-inch steel bar: $2\frac{1}{2}"$, $1\frac{3}{4}"$, $1\frac{7}{8}"$, and $\frac{5}{16}"$. Allowing $\frac{1}{8}$ inch waste for each cut, what is the length of bar left after the pieces are cut? ________

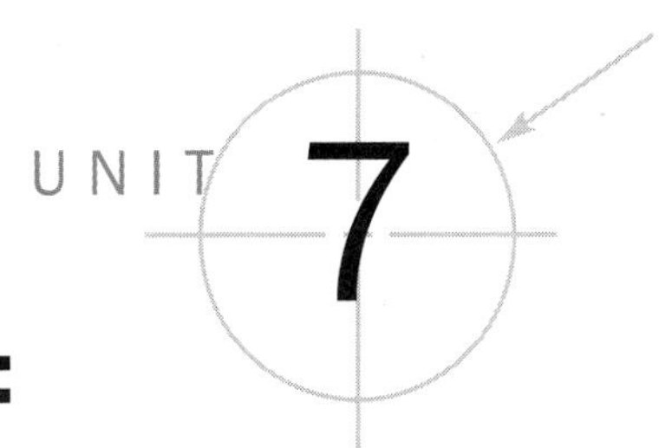

Computing with a Calculator: Fractions and Mixed Numbers

Objectives ***After studying this unit you should be able to***

- Perform individual operations of addition, subtraction, multiplication, and division with fractions using a calculator.
- Perform combinations of operations with fractions using a calculator.

Fractions

Depending on the calculator, the fraction key ($a\frac{b}{c}$) (or $A^b/_c$) is used when entering fractions and mixed numbers in a calculator. If your calculator has $A^b/_c$, substitute it for all of the examples shown. The answers to expressions entered as fractions will be given as fractions or mixed numbers with the fraction in lowest terms.

Enter the numerator, press $a\frac{b}{c}$, and enter the denominator. The fraction is displayed with the symbol ⌟ or ⌜ between the numerator and denominator.

Example Enter $\frac{3}{4}$.

3 [a b/c] 4, 3 ⌟ 4 is displayed.

Individual Arithmetic Operations: Fractions

The operations of addition, subtraction, multiplication, and division are performed with the four arithmetic keys and the equals key. The equals key completes all operations entered and readies the calculator for additional calculations. Certain makes and models of calculators have the execute key, [EXE], or enter keys, [ENTER] and [ENTER], instead of the equals key [=]. If your calculator has one of those keys, substitute it for [=] for all examples shown throughout this book.

Examples of each of the four arithmetic operations of addition, subtraction, multiplication, and division are presented. Following the individual operation problems, combined operations expressions are given with calculator solutions. An answer to a problem should be checked by doing the problem a second time to ensure that improper data was not entered in its solution. Remember to clear or erase previously recorded data and calculations before doing a problem. Depending on the make and model of the calculator, generally, answers are displayed with either the symbol ⌟, /, or ⌜ between the numerator and denominator. The following examples are shown with ⌟. Substitute / or ⌜ for ⌟ if necessary.

Example 1 Add. $\frac{3}{16} + \frac{19}{32}$

3 [a b/c] 16 [+] 19 [a b/c] 32 [=] 25 ⌟ 32, $\frac{25}{32}$ Ans

Example 2 Subtract. $\frac{7}{8} - \frac{5}{64}$

7 [a b/c] 8 [−] 5 [a b/c] 64 [=] 51 ⌟ 64, $\frac{51}{64}$ Ans

Example 3 Multiply. $\frac{3}{32} \times \frac{11}{16}$

3 [a b/c] 32 [×] 11 [a b/c] 16 [=] 33 ⌟ 512, $\frac{33}{152}$ Ans

Example 4 Divide. $\frac{5}{8} \div \frac{13}{15}$

5 [a b/c] 8 [÷] 13 [a b/c] 15 [=] 75 ⌟ 104, $\frac{75}{104}$ Ans

Mixed Numbers

Enter the whole number, press [a b/c], enter the fraction numerator, press [a b/c], and enter the denominator. The symbol ⌟ or ⌜ is displayed between the whole number and fraction.

Example Enter $15\frac{7}{16}$.

15 [a b/c] 7 [a b/c] 16

15 ⌟ 7 ⌟ 16 is displayed.

Individual Arithmetic Operations: Mixed Numbers

The following examples are of mixed numbers with individual arithmetic operations. Depending on the particular calculator, generally, answers are displayed with either the symbol ⌟, ⌜, or ⊔ between the whole number and fraction. The following examples are shown with ⌟. Substitute ⊔ or ⌜ for ⌟ if necessary.

Example 1 Add. $7\frac{3}{64} + 23\frac{5}{8}$

7 $\boxed{a\,b/c}$ 3 $\boxed{a\,b/c}$ 64 $\boxed{+}$ 23 $\boxed{a\,b/c}$ 5 $\boxed{a\,b/c}$ 8 $\boxed{=}$ 30 ⌟ 43 ⌟ 64, $30\frac{43}{64}$ Ans

Example 2 Subtract. $43\frac{7}{8} - 36\frac{29}{32}$

43 $\boxed{a\,b/c}$ 7 $\boxed{a\,b/c}$ 8 $\boxed{-}$ 36 $\boxed{a\,b/c}$ 29 $\boxed{a\,b/c}$ 32 $\boxed{=}$ 6 ⌟ 31 ⌟ 32, $6\frac{31}{32}$ Ans

Example 3 Multiply. $38\frac{5}{6} \times 14\frac{13}{16}$

38 $\boxed{a\,b/c}$ 5 $\boxed{a\,b/c}$ 6 $\boxed{\times}$ 14 $\boxed{a\,b/c}$ 13 $\boxed{a\,b/c}$ 16 $\boxed{=}$ 575 ⌟ 7 ⌟ 32, $575\frac{7}{32}$ Ans

Example 4 Divide. $159\frac{17}{64} \div 3\frac{7}{8}$

159 $\boxed{a\,b/c}$ 17 $\boxed{a\,b/c}$ 64 $\boxed{\div}$ 3 $\boxed{a\,b/c}$ 7 $\boxed{a\,b/c}$ 8 $\boxed{=}$ 41 ⌟ 25 ⌟ 248, $41\frac{25}{248}$ Ans

Practice Exercises, Individual Basic Operations with Fractions and Mixed Numbers

Evaluate the following expressions. The expressions are basic arithmetic operations. Remember to check your answers by doing each problem twice. The solutions to the problems directly follow the practice exercises. Compare your answers to the given solutions.

1. $\frac{5}{8} + \frac{11}{16}$
2. $\frac{31}{32} - \frac{7}{8}$
3. $\frac{9}{16} \times \frac{5}{8}$
4. $\frac{23}{25} \div \frac{4}{5}$
5. $85\frac{7}{64} + 107\frac{3}{4}$
6. $125\frac{7}{8} - 67\frac{63}{64}$
7. $62\frac{13}{16} \times 47\frac{1}{6}$
8. $785\frac{27}{32} \div 2\frac{3}{4}$
9. $\frac{59}{64} + 46\frac{27}{32}$
10. $37\frac{3}{8} - \frac{45}{64}$

Solutions to Practice Exercises, Individual Basic Operations with Fractions and Mixed Numbers

1. 5 $\boxed{a\,b/c}$ 8 $\boxed{+}$ 11 $\boxed{a\,b/c}$ 16 $\boxed{=}$ 1 ⌟ 5 ⌟ 16, $1\frac{5}{16}$ Ans
2. 31 $\boxed{a\,b/c}$ 32 $\boxed{-}$ 7 $\boxed{a\,b/c}$ 8 $\boxed{=}$ 3 ⌟ 32, $\frac{3}{32}$ Ans
3. 9 $\boxed{a\,b/c}$ 16 $\boxed{\times}$ 5 $\boxed{a\,b/c}$ 8 $\boxed{=}$ 45 ⌟ 128, $\frac{45}{128}$ Ans

4. 23 [a b/c] 25 [÷] 4 [a b/c] 5 [=] 1 ⌋ 3 ⌋ 20, $1\frac{3}{20}$ Ans

5. 85 [a b/c] 7 [a b/c] 64 [+] 107 [a b/c] 3 [a b/c] 4 [=] 192 ⌋ 55 ⌋ 64, $192\frac{55}{64}$ Ans

6. 125 [a b/c] 7 [a b/c] 8 [−] 67 [a b/c] 63 [a b/c] 64 [=] 57 ⌋ 57 ⌋ 64, $57\frac{57}{64}$ Ans

7. 62 [a b/c] 13 [a b/c] 16 [×] 47 [a b/c] 1 [a b/c] 6 [=] 2962 ⌋ 21 ⌋ 32, $2962\frac{21}{32}$ Ans

If your calculator showed the answer as 2692.65625, write the entire number on your paper. The whole number part of the answer is to the left of the decimal point, 2962. Now, for the fraction part of the answer, enter 0.65625 in your calculator and press the 2nd key and the [F◂▸D] key. (On a TI-30 calculator, the [F◂▸D] is over the PRB key.) Press [ENTER]. The calculator should display 21 ⌋ 32 or $\frac{21}{32}$. Combining the whole number and the fraction parts gives the mixed number $2962\frac{21}{32}$. On a Sharp™ calculator, pressing the [a b/c] key toggles the display between fraction and decimal expressions.

8. 785 [a b/c] 27 [a b/c] 32 [÷] 2 [a b/c] 3 [a b/c] 4 [=] 285 ⌋ 67 ⌋ 88, $285\frac{67}{88}$ Ans

9. 59 [a b/c] 64 [+] 46 [a b/c] 27 [a b/c] 32 [=] 47 ⌋ 49 ⌋ 64, $47\frac{49}{64}$ Ans

10. 37 [a b/c] 3 [a b/c] 8 [−] 45 [a b/c] 64 [=] 36 ⌋ 43 ⌋ 64, $36\frac{43}{64}$ Ans

Combined Operations

The expressions are solved by entering numbers and operations into the calculator in the same order as the expressions are written. Remember to check your answers by doing each problem twice.

Example 1 Evaluate. $275\frac{17}{32} + \frac{7}{8} \times 26\frac{3}{4}$

275 [a b/c] 17 [a b/c] 32 [+] 7 [a b/c] 8 [×] 26 [a b/c] 3 [a b/c] 4 [=] 298 ⌋ 15 ⌋ 16, $298\frac{15}{16}$ Ans

Because the calculator has algebraic logic, the multiplication operation $\left(\frac{7}{8} \times 26\frac{3}{4}\right)$ was performed before the addition operation $\left(\text{adding } 275\frac{17}{32}\right)$ was performed.

Example 2 Evaluate. $\frac{35}{64} - \frac{5}{8} + 18 \div 10\frac{2}{3}$

35 [a b/c] 64 [−] 5 [a b/c] 8 [+] 18 [÷] 10 [a b/c] 2 [a b/c] 3 [=] 1 ⌋ 39 ⌋ 64, $1\frac{39}{64}$ Ans

Example 3 Evaluate. $380\frac{29}{32} - \left(\frac{3}{16} + 9\frac{15}{64}\right) \times 12$

As previously discussed in Unit 6, operations enclosed within parentheses are done first. A calculator with algebraic logic performs the operations within parentheses before performing other operations in a combined operations expression. If an expression contains parentheses, enter the expression in the calculator in the order in which it is written. The parentheses keys must be used.

380 [a b/c] 29 [a b/c] 32 [−] [(] 3 [a b/c] 16 [+] 9 [a b/c] 15 [a b/c] 64 [)] [×] 12 [=]

267 ⌟ 27 ⌟ 32, $267\frac{27}{32}$ Ans

Example 4 Evaluate. $\dfrac{25\frac{47}{64} + 7 \times \frac{5}{8}}{\frac{3}{16} \times 2 + \frac{1}{8}}$

Recall that for a problem expressed in fractional form, the fraction bar is also used as a grouping symbol. The numerator and denominator are each considered as being enclosed in parentheses and these must be used when entering the problem in the calculator.

[(] 25 [a b/c] 47 [a b/c] 64 [+] 7 [×] 5 [a b/c] 8 [)] [÷] [(] 3 [a b/c] 16 [×] 2 [+] 1 [a b/c] 8 [)] [=] 60 ⌟ 7 ⌟ 32, $60\frac{7}{32}$ Ans

The expression may also be evaluated by using the [=] key to simplify the numerator without having to enclose the entire numerator in parentheses. However, parentheses must be used to enclose the denominator.

25 [a b/c] 47 [a b/c] 64 [+] 7 [×] 5 [a b/c] 8 [=] [÷] [(] 3 [a b/c] 16 [×] 2 [+] 1 [a b/c] 8 [)] [=] 60 ⌟ 7 ⌟ 32, $60\frac{7}{32}$ Ans

Practice Exercises, Combined Operations with Fractions and Mixed Numbers

Evaluate the following combined operations expressions. Remember to check your answers by doing each problem twice. The solutions to the problems directly follow the practice exercises. Compare your answers to the given solutions.

1. $\left(\frac{11}{16} + 12\frac{31}{32}\right) \div \frac{1}{8}$
2. $\dfrac{108}{\frac{3}{8}} - 3\frac{5}{64}$
3. $\dfrac{43\frac{9}{10} - 17\frac{3}{5} + \frac{7}{20}}{5}$
4. $120\frac{13}{16} + 98\frac{5}{8} \times \left(6 - \frac{3}{4}\right)$
5. $\dfrac{56\frac{3}{4} + 20 \times \frac{7}{8}}{4 \times \frac{2}{3}}$
6. $\left(\frac{25}{32} - \frac{3}{4}\right) \div \frac{1}{2} \times \frac{3}{4}$
7. $50 \times \left(28\frac{4}{5} - 17\frac{9}{10} + 27\right) \times \frac{3}{5}$
8. $\dfrac{40\frac{1}{2}}{1\frac{1}{8}} - \left(15\frac{5}{64} + 8\frac{29}{32}\right)$
9. $\dfrac{270 - 175\frac{1}{2} \times \frac{7}{8}}{\frac{1}{64} \times 128}$

Solutions to Practice Exercises, Combined Operations with Fractions and Mixed Numbers

1. [(] 11 [ab/c] 16 [+] 12 [ab/c] 31 [ab/c] 32 [)] [÷] 1 [ab/c] 8 [=] 109 ⌋ 1 ⌋ 4, $109\frac{1}{4}$ Ans

   or 11 [ab/c] 16 [+] 12 [ab/c] 31 [ab/c] 32 [=] [÷] 1 [ab/c] 8 [=] 109 ⌋ 1 ⌋ 4, $109\frac{1}{4}$ Ans

2. 108 [÷] 3 [ab/c] 8 [−] 3 [ab/c] 5 [ab/c] 64 [=] 284 ⌋ 59 ⌋ 64, $284\frac{59}{64}$ Ans

3. [(] 43 [ab/c] 9 [ab/c] 10 [−] 17 [ab/c] 3 [ab/c] 5 [+] 7 [ab/c] 20 [)] [÷] 5 [=] 5 ⌋ 33 ⌋ 100, $5\frac{33}{100}$ Ans

   or 43 [ab/c] 9 [ab/c] 10 [−] 17 [ab/c] 3 [ab/c] 5 [+] 7 [ab/c] 20 [=] [÷] 5 [=] 5 ⌋ 33 ⌋ 100, $5\frac{33}{100}$ Ans

4. 120 [ab/c] 13 [ab/c] 16 [+] 98 [ab/c] 5 [ab/c] 8 [×] [(] 6 [−] 3 [ab/c] 4 [)] [=] 638 ⌋ 19 ⌋ 32, $638\frac{19}{32}$ Ans

5. [(] 56 [ab/c] 3 [ab/c] 4 [+] 20 [×] 7 [ab/c] 8 [)] [÷] [(] 4 [×] 2 [ab/c] 3 [)] [=] 27 ⌋ 27 ⌋ 32, $27\frac{27}{32}$ Ans

   or 56 [ab/c] 3 [ab/c] 4 [+] 20 [×] 7 [ab/c] 8 [=] [÷] [(] 4 [×] 2 [ab/c] 3 [)] [=] 27 ⌋ 27 ⌋ 32, $27\frac{27}{32}$ Ans

6. [(] 25 [ab/c] 32 [−] 3 [ab/c] 4 [)] [÷] 1 [ab/c] 2 [×] 3 [ab/c] 4 [=] 3 ⌋ 64, $\frac{3}{64}$ Ans

   or 25 [ab/c] 32 [−] 3 [ab/c] 4 [=] [÷] 1 [ab/c] 2 [×] 3 [ab/c] 4 [=] 3 ⌋ 64, $\frac{3}{64}$ Ans

7. 50 [×] [(] 28 [ab/c] 4 [ab/c] 5 [−] 17 [ab/c] 9 [ab/c] 10 [+] 27 [)] [×] 3 [ab/c] 5 [=] 1137, 1137 Ans

8. 40 [ab/c] 1 [ab/c] 2 [÷] 1 [ab/c] 1 [ab/c] 8 [−] [(] 15 [ab/c] 5 [ab/c] 64 [+] 8 [ab/c] 29 [ab/c] 32 [)] [=] 12 ⌋ 1 ⌋ 64, $12\frac{1}{64}$ Ans

9. [(] 270 [−] 175 [ab/c] 1 [ab/c] 2 [×] 7 [ab/c] 8 [)] [÷] [(] 1 [ab/c] 64 [×] 128 [)] [=] 58 ⌋ 7 ⌋ 32, $58\frac{7}{32}$ Ans

   or 270 [−] 175 [ab/c] 1 [ab/c] 2 [×] 7 [ab/c] 8 [=] [÷] [(] 1 [ab/c] 64 [×] 128 [)] [=] 58 ⌋ 7 ⌋ 32, $58\frac{7}{32}$ Ans

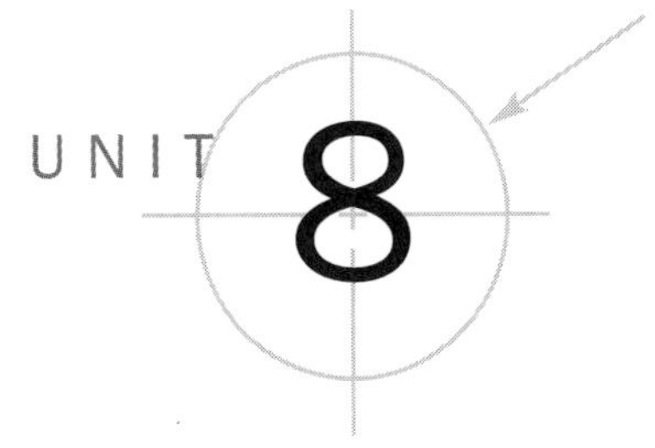

Introduction to Decimal Fractions

Objectives ***After studying this unit you should be able to***

- Locate decimal fractions on a number line.
- Express common fractions having denominators of powers of ten as equivalent decimal fractions.
- Write decimal numbers in word form.
- Write numbers expressed in word form as decimal fractions.

Most engineering drawings are dimensioned with decimal fractions rather than common fractions. The dials that are used in establishing machine settings and movement are graduated in decimal units. Tool speeds and travel are determined in decimal units and machined parts are usually measured in decimal units.

Explanation of Decimal Fractions

A decimal fraction is not written as a common fraction with a numerator and denominator. The denominator is omitted and replaced by a decimal point placed to the left of the numerator. *Decimal fractions* are equivalent to common fractions having denominators that are powers of 10, such as 10; 100; 1000; 10,000; 100,000; and 1,000,000. *Powers of 10* are numbers that are obtained by multiplying 10 by itself a certain number of times.

Meaning of Fractional Parts

The line segment shown in Figure 8-1 is one unit long. It is divided into ten equal smaller parts. The locations of common fractions and their decimal fraction equivalents are shown on the line.

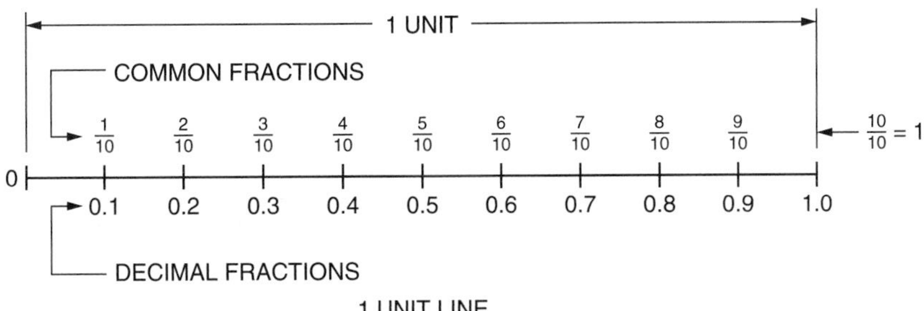

1 UNIT LINE

Figure 8-1

One of the ten equal small parts, $\frac{1}{10}$ or 0.1 of the 1 unit line, is shown enlarged in Figure 8-2.

The $\frac{1}{10}$ or 0.1 unit is divided into ten equal smaller units. The locations of common fractions and their decimal fraction equivalents are shown on the line in Figure 8-2.

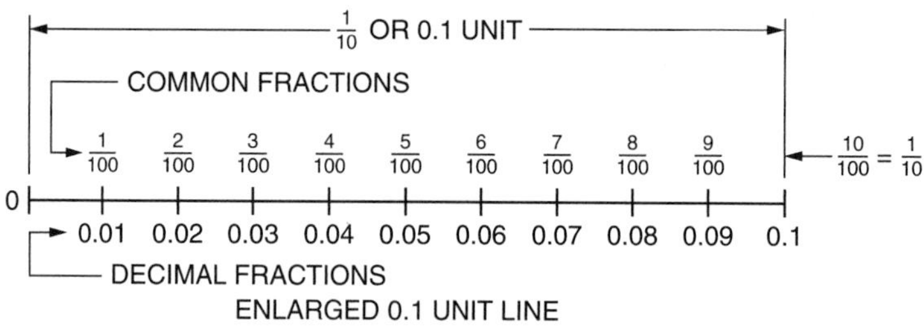

Figure 8-2

If the $\frac{1}{100}$ or 0.01 division is divided into ten equal smaller parts, the resulting parts are $\frac{1}{1000}$ or 0.001; $\frac{2}{1000}$ or 0.002; $\frac{3}{1000}$ or 0.003; . . . $\frac{9}{1000}$ or 0.009; $\frac{10}{1000} = \frac{1}{100}$ or 0.01.

- Each time the decimal point is moved one place to the left, a value $\frac{1}{10}$ or 0.1 times the previous value is obtained.
- Each time a decimal point is moved one place to the right, a value 10 times greater than the previous value is obtained.

Each time a decimal fraction is multiplied by 10 the decimal point is moved one place to the right. Each step in the following table shows both the decimal fraction and its equivalent common fraction.

Decimal Fraction	Common Fraction
0.000003 × 10 = 0.00003	3/1,000,000 × 10 = 3/100,000
0.00003 × 10 = 0.0003	3/100,000 × 10 = 3/10,000
0.0003 × 10 = 0.003	3/10,000 × 10 = 3/1000
0.003 × 10 = 0.03	3/1000 × 10 = 3/100
0.03 × 10 = 0.3	3/100 × 10 = 3/10
0.3 × 10 = 3.	3/10 × 10 = 3

➤ **Note:** As we will see later, the metric (or SI) system uses a small space rather than a comma to separate groups of three digits. In the metric system 10,000 is written 10 000. This applies to decimal fractions as well with 0.00003 written as 0.000 03.

Reading and Writing Decimal Fractions

The following chart gives the names of the parts of a number with respect to the positions from the decimal point.

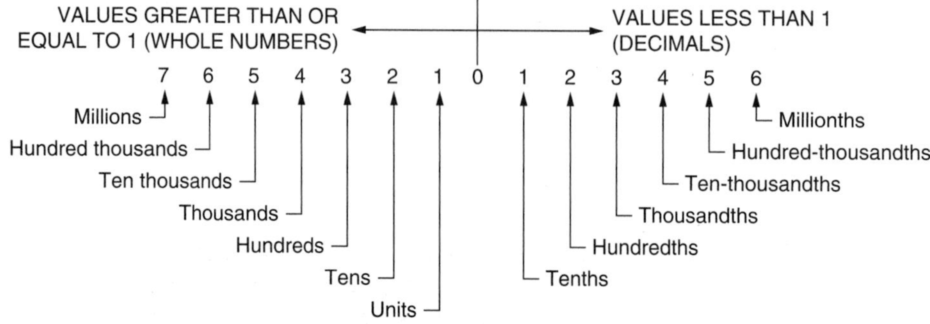

Figure 8-3

To read a decimal, read the number as a whole number. Then say the name of the decimal place of the last digit to the right.

Examples
1. 0.5 is read "five tenths."
2. 0.07 is read "seven hundredths."
3. 0.011 is read "eleven thousandths."

To write a decimal fraction from a word statement, write the number using a decimal point and zeros before the number as necessary for the given place value.

Examples
1. Two hundred nineteen ten-thousandths is written as 0.0219.
2. Forty-three hundred-thousandths is written as 0.00043.
3. Eight hundred seventeen millionths is written as 0.000817.

A number that consists of a whole number and a decimal fraction is called a *mixed decimal*. To read a mixed decimal, read the whole number, read the word *and* at the decimal point, and read the decimal.

Examples
1. 3.4 is read "three and four tenths."
2. 1.002 is read "one and two thousandths."
3. 16.0793 is read "sixteen and seven hundred ninety-three ten-thousandths."
4. 8.00032 is read "eight and thirty-two hundred-thousandths."

Simplified Method of Reading Decimal Fractions

Usually a simplified method of reading decimal fractions is used in the machine trades. This method is generally quicker, easier, and less likely to be misinterpreted. A tool-and-die maker reads 0.0265 inches as point zero, two, six, five inches. A machinist reads 4.172 millimeters as four, point one, seven, two millimeters.

Writing Decimal Fractions from Common Fractions Having Denominators That Are Powers of Ten

A common fraction with a denominator that is a power of ten can be written as a decimal fraction. For a common fraction with a numerator smaller than the denominator, replace the denominator with a decimal point. The decimal point is placed to the left of the first digit of the numerator. There are as many decimal places as there are zeros in the denominator. When writing a decimal fraction, place a zero to the left of the decimal point.

Examples

1. $\frac{9}{10} = 0.9$ Ans — There is 1 zero in 10 and 1 decimal place in 0.9.

2. $\frac{381}{1000} = 0.381$ Ans — There are 3 zeros in 1000 and 3 decimal places in 0.381.

3. $\frac{7}{10{,}000} = 0.0007$ Ans — There are 4 zeros in 10,000 and 4 decimal places in 0.0007. In order to maintain proper place value, 3 zeros are written between the decimal point and the 7.

Application

Meaning of Fractional Parts

1. Find the decimal value of each of the distances A, B, C, D, and E in Figure 8-4. Note the total unit value of the line.

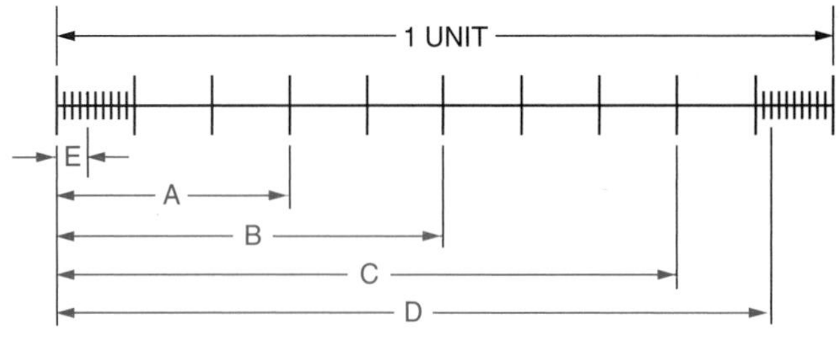

Figure 8-4

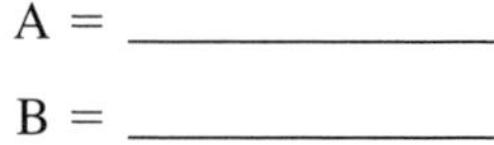

A = ____________

B = ____________

C = ____________

D = ____________

E = ____________

2. Find the decimal value of each of the distances A, B, C, D, and E in Figure 8-5. Note the total unit value of the lines.

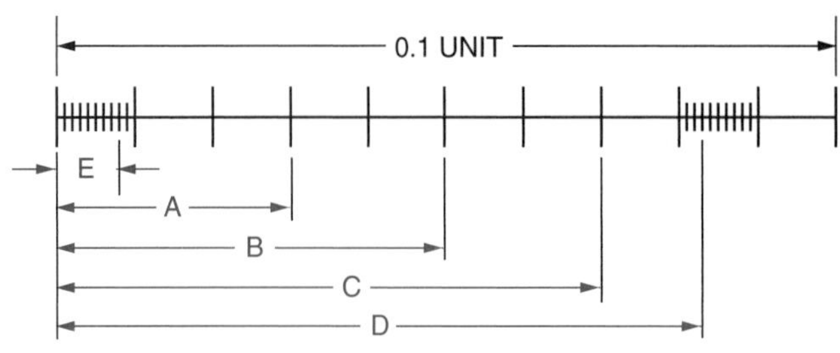

Figure 8-5

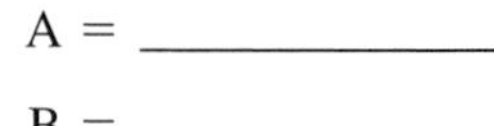

A = ____________

B = ____________

C = ____________

D = ____________

E = ____________

3. Find the decimal value of each of the distances A, B, C, D, and E in Figure 8-6. Note the total unit value of the line.

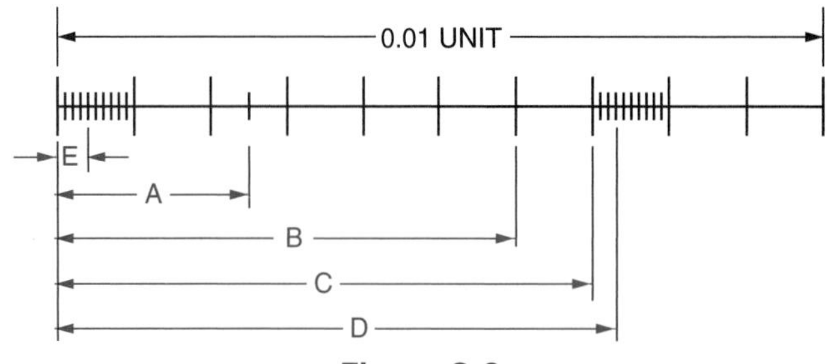

Figure 8-6

A = ____________

B = ____________

C = ____________

D = ____________

E = ____________

In each of the following problems, the value on the left must be multiplied by one of the following numbers: 0.0001; 0.001; 0.01; 0.1; 10; 100; 1000; or 10,000 in order to obtain the value on the right of the equal sign. Determine the proper number.

4. 0.9 × ____________ = 0.0009
5. 0.7 × ____________ = 0.007
6. 0.03 × ____________ = 0.3
7. 0.0003 × ____________ = 0.003
8. 0.135 × ____________ = 0.00135
9. 4 × ____________ = 0.4
10. 0.0643 × ____________ = 0.000643
11. 0.0643 × ____________ = 6.43
12. 0.00643 × ____________ = 64.3
13. 643 × ____________ = 0.643

Reading and Writing Decimal Fractions

Write these numbers as words.

14. 0.064 ________	19. 1.5 ________
15. 0.007 ________	20. 10.37 ________
16. 0.132 ________	21. 16.0007 ________
17. 0.0035 ________	22. 4.0012 ________
18. 0.108 ________	23. 13.103 ________

Write these words as numbers.

24. eighty-four ten-thousandths ________	28. thirty-five ten-thousandths ________
25. three tenths ________	29. ten and two tenths ________
26. forty-three and eight hundredths ________	30. five and one ten-thousandth ________
27. four and five hundred-thousandths ________	31. twenty and seventy-one hundredths ________

Each of the following common fractions has a denominator that is a power of 10. Write the equivalent decimal fraction for each.

32. $\frac{9}{10}$ ________	35. $\frac{43}{100}$ ________	38. $\frac{73}{1000}$ ________
33. $\frac{7}{10{,}000}$ ________	36. $\frac{61}{1000}$ ________	39. $\frac{1973}{100{,}000}$ ________
34. $\frac{17}{100}$ ________	37. $\frac{999}{10{,}000}$ ________	40. $\frac{47{,}375}{100{,}000}$ ________

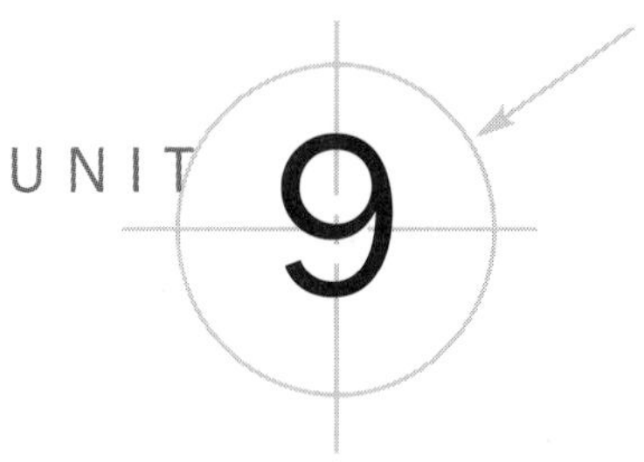

Rounding Decimal Fractions and Equivalent Decimal and Common Fractions

Objectives ***After studying this unit you should be able to***

- Round decimal fractions to any required number of places.
- Express common fractions as decimal fractions.
- Express decimal fractions as common fractions.

When engineering drawing dimensions of a part are given in fractional units, a machinist is usually required to express these fractional values as decimal working dimensions. In computing material requirements and in determining stock waste and scrap allowances, it is sometimes more convenient to express decimal values as approximate fractional equivalents.

Rounding Decimal Fractions

When working with decimals, the computations and answers may contain more decimal places than are required. The number of decimal places needed depends on the degree of precision desired. The degree of precision depends on how the decimal value is going to be used. The tools, machines, equipment, and materials determine the degree of precision obtainable. For example, a length of 0.875376 inch cannot be cut on a milling machine. In cutting to the nearer thousandths of an inch, the machinist would consider 0.875376 inch as 0.875 inch. *Rounding a decimal* means expressing the decimal with a fewer number of decimal places.

Procedure To round a decimal fraction

- Determine the number of decimal places required in an answer.
- If the digit directly following the last decimal place required is less than 5, drop all digits that follow the required number of decimal places.
- If the digit directly following the last decimal place required is 5 or larger, add 1 to the last required digit and drop all digits that follow the required number of decimal places.

Example 1 Round 0.873429 to three decimal places.

The digit following the third decimal place is 4. 0.873④29

Because 4 is less than 5, drop all digits after the third decimal place. 0.873 Ans

Example 2 Round 0.36845 to two decimal places.

The digit following the second decimal place is 8. 0.36⑧45

Because 8 is greater than 5, add 1 to the 6. 0.37 Ans

Example 3 Round 18.738257 to four decimal places.

The digit following the fourth decimal place is 5. 18.7382⑤7

Add 1 to the 2. 18.7383 Ans

Expressing Common Fractions as Decimal Fractions

A common fraction is an indicated division. For example, $\frac{3}{4}$ is the same as $3 \div 4$; $\frac{5}{16}$ is the same as $5 \div 16$; $\frac{99}{171}$ is the same as $99 \div 171$.

Because both the numerator and the denominator of a common fraction are whole numbers, expressing a common fraction as a decimal fraction requires division with whole numbers.

Procedure To express a common fraction as a decimal fraction

- Divide the numerator by the denominator.

  A common fraction that divides without a remainder is called a *terminating decimal.* A common fraction that does not terminate is expressed as a *repeating* or *nonterminating decimal.*

The division should be carried out to one more place than the number of places required in the answer, then rounded one place.

Example 1 Express $\frac{2}{3}$ as a 4-place decimal.

Divide the numerator by the denominator.
After the 2, add one more zero than the required number of decimal places. (Add 5 zeros.)

$$3\overline{)2.00000}\quad 0.66666$$

Round 0.66666 to 4 places. 0.6667 Ans

Example 2 Express $\frac{5}{7}$ as a 2-place decimal.

Add 3 zeros after the 5. $7\overline{)5.000}\quad 0.714$

Round to 2 places. 0.71 Ans

Expressing Decimal Fractions as Common Fractions

Procedure To express a decimal fraction as a common fraction

- Write the number after the decimal point as the numerator of a common fraction.
- Write the denominator as 1 followed by as many zeros as there are digits to the right of the decimal point.
- Express the common fraction in lowest terms.

Example 1 Express 0.375 as a common fraction.

Write 375 as the numerator.

There are three digits to the right of the decimal point. Write the denominator as 1 followed by 3 zeros. The denominator is 1000.

Reduce $\frac{375}{1000}$ to lowest terms. $0.375 = \frac{375}{1000} = \frac{3}{8}$ Ans

Example 2 Express 0.27 as a common fraction.

The numerator is 27. $0.27 = \frac{27}{100}$ Ans

The denominator is 100.

Example 3 Express 0.03125 as a common fraction.

The numerator is 3125.

The denominator is 100,000. $0.03125 = \frac{3125}{100{,}000} = \frac{1}{32}$ Ans

Reduce $\frac{3125}{100{,}000}$ to lowest terms.

Application

Rounding Decimal Fractions

Round the following decimals to the indicated number of decimal places.

1. 0.63165 (3 places) ______
2. 0.1247 (2 places) ______
3. 0.23975 (3 places) ______
4. 0.01723 (3 places) ______
5. 0.03894 (2 places) ______
6. 0.90039 (2 places) ______
7. 0.72008 (4 places) ______
8. 0.0006 (3 places) ______
9. 0.0003 (3 places) ______
10. 0.099 (3 places) ______

Express Common Fractions as Decimal Fractions

Express the common fractions as decimal fractions. Express the answer to 4 decimal places.

11. $\frac{11}{16}$ ______
12. $\frac{7}{8}$ ______
13. $\frac{5}{8}$ ______
14. $\frac{3}{4}$ ______
15. $\frac{2}{3}$ ______
16. $\frac{10}{11}$ ______
17. $\frac{2}{25}$ ______
18. $\frac{47}{64}$ ______
19. $\frac{7}{32}$ ______
20. $\frac{1}{2}$ ______
21. $\frac{4}{7}$ ______
22. $\frac{3}{8}$ ______

Solve the following.

23. In Figure 9-1, what decimal fraction of distance B is distance A? Express the answer to 4 decimal places. All dimensions are in inches. ______

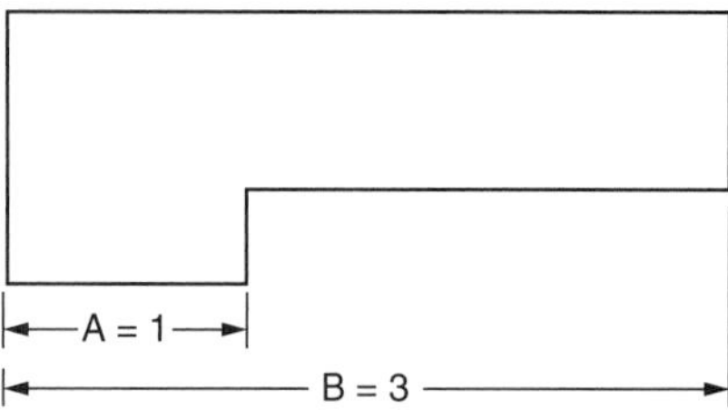

Figure 9-1

24. Five pieces are cut from the length of round stock shown in Figure 9-2. After the pieces are cut, the remaining length is thrown away. What decimal fraction of the original length of round stock (17") is the length that is thrown away? All dimensions are in inches. ______

Figure 9-2

25. Dimensions in Figure 9-3 are in feet and inches.

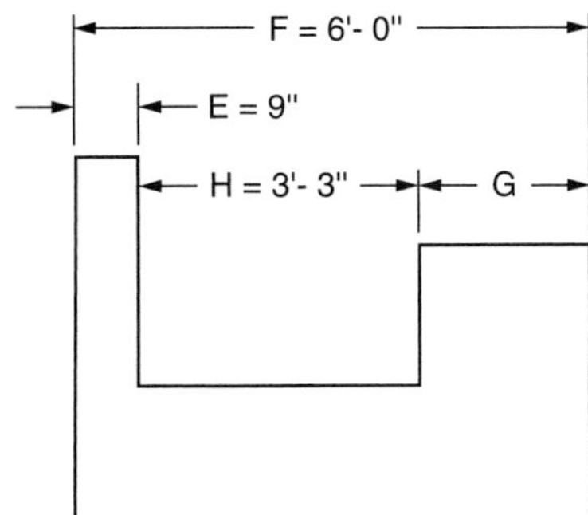

Figure 9-3

a. What decimal fraction of distance F is distance E? *Note:* Both the numerator and denominator of a common fraction must be in the same units before the value is expressed as a decimal fraction. Use 1 foot = 12 inches. __________

b. What decimal fraction of distance H is distance G? Express the answer to 4 decimal places. __________

Expressing Decimal Fractions as Common Fractions

Express the following decimal fractions as common fractions. Reduce to lowest terms.

26. 0.875 __________
27. 0.125 __________
28. 0.4 __________
29. 0.75 __________
30. 0.6 __________
31. 0.6875 __________
32. 0.67 __________
33. 0.003 __________
34. 0.008 __________
35. 0.502 __________
36. 0.99 __________
37. 0.4375 __________
38. 0.2113 __________
39. 0.8717 __________
40. 0.0005 __________
41. 0.03 __________
42. 0.09375 __________
43. 0.237 __________
44. 0.45 __________
45. 0.045 __________
46. 0.0045 __________

Solve the following.

47. In Figure 9-4, what common fractional part of distance B is distance A? All dimensions are in inches. ____________

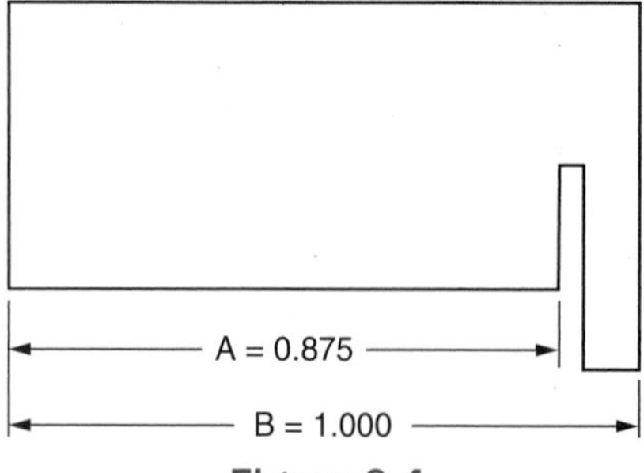

Figure 9-4

48. In Figure 9-5, what common fractional part of diameter C is diameter D? All dimensions are in feet. ____________

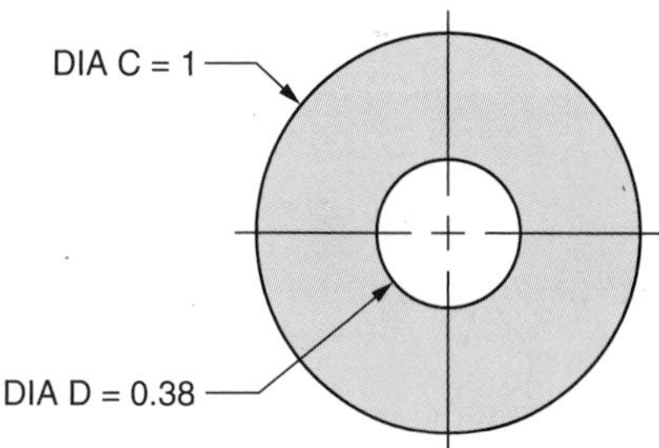

Figure 9-5

49. What common fractional part of distance A is each distance listed in Figure 9-6? All dimensions are in inches.

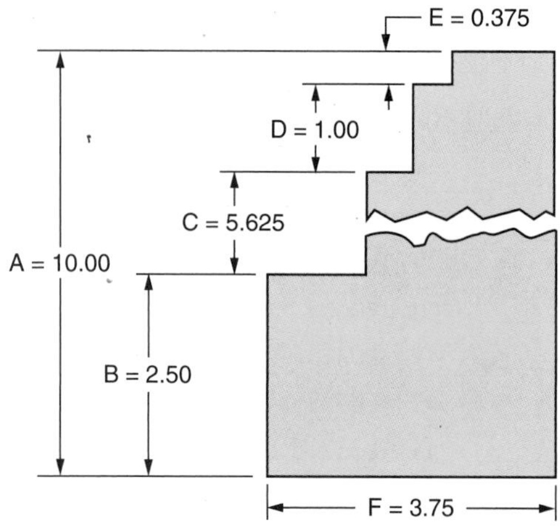

Figure 9-6

a. Distance B ____________

b. Distance C ____________

c. Distance D ____________

d. Distance E ____________

e. Distance F ____________

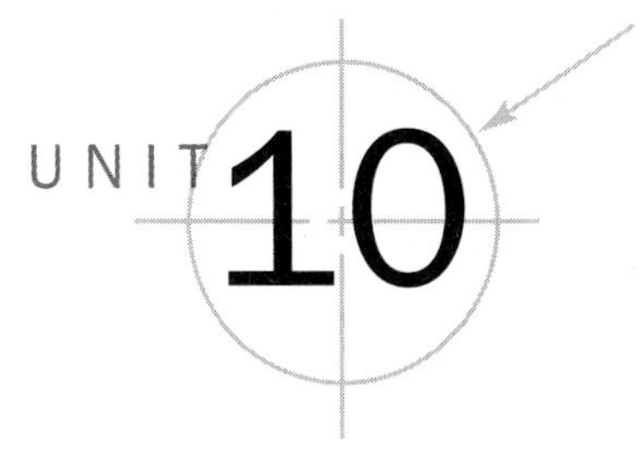

Addition and Subtraction of Decimal Fractions

Objectives ***After studying this unit you should be able to***

- Add decimal fractions.
- Add combinations of decimals, mixed decimals, and whole numbers.
- Subtract decimal fractions.
- Subtract combinations of decimals, mixed decimals, and whole numbers.

Adding and subtracting decimal fractions are required at various stages in the production of most products and parts. It is necessary to add and subtract decimals in order to estimate machining costs and production times, to compute stock allowances and tolerances, to determine locations and lengths of cuts, and to inspect finished parts.

Adding Decimal Fractions

Procedure To add decimal fractions

- Arrange the numbers so that the decimal points are directly under each other.
- Add each column as with whole numbers.
- Place the decimal point in the sum directly under the other decimal points.

Example 1 Add. 7.35 + 114.075 + 0.3422 + 0.003 + 218.7

➤ **Note:** To reduce the possibility of error, add zeros to decimals so that all the values have the same number of places to the right of the decimal point. Zeros added in this manner do not affect the value of the number.

Arrange the numbers so that the decimal points are directly under each other.

Add each column as with whole numbers.

Place the decimal point in the sum directly under the other decimal points.

$$\begin{array}{r} 7.3500 \\ 114.0750 \\ 0.3422 \\ 0.0030 \\ +\ 218.7000 \\ \hline 340.4702 \end{array} \quad \text{Ans}$$

The decimal point location of a whole number is directly to the right of the last digit.

Example 2 Find the length x of the swivel bracket shown in Figure 10-1. All dimensions are in millimeters.

Add.

$$\begin{array}{r} 8.78 \\ 25.40 \\ 12.80 \\ 30.00 \\ 3.90 \\ +\ 9.25 \\ \hline 90.13 \end{array}$$

$x = 90.13$ Ans

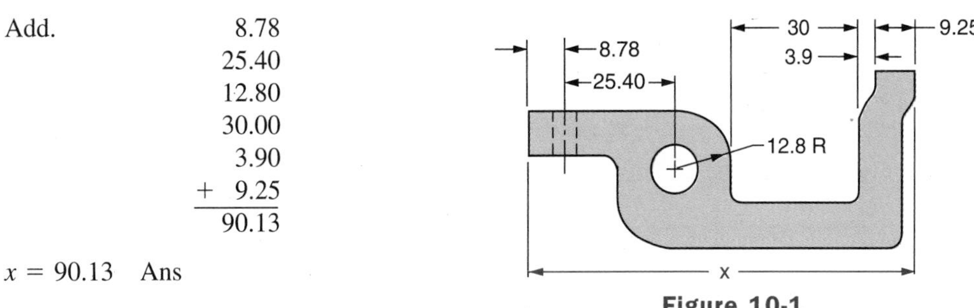

Figure 10-1

Subtracting Decimal Fractions

Procedure To subtract decimal fractions

- Arrange the numbers so that the decimal points are directly under each other.
- Subtract each column as with whole numbers.
- Place the decimal point in the difference directly under the other decimal points.

Example 1 Subtract 13.261 from 25.6.

Arrange the numbers so that the decimal points are directly under each other.

Add two zeros to 25.6 so that it has the same number of decimal places as 13.261.

Subtract.

$$\begin{array}{r} 25.600 \\ -\,13.261 \\ \hline 12.339 \end{array} \quad \text{Ans}$$

Subtract each column as with whole numbers.
Place the decimal point in the difference directly under the other decimal points.

Example 2 Determine dimensions A, B, C, and D of the support bracket shown in Figure 10-2. All dimensions are given in inches.

Solve for A:
A = 0.505 − 0.18
A = 0.325" Ans

$$\begin{array}{r} 0.505 \\ -\,0.180 \\ \hline 0.325 \end{array}$$

Solve for B:
B = 1.4 − 0.301
B = 1.099" Ans

$$\begin{array}{r} 1.400 \\ -\,0.301 \\ \hline 1.099 \end{array}$$

Solve for C:
C = 1.74 − 0.365
C = 1.375" Ans

$$\begin{array}{r} 1.740 \\ -\,0.365 \\ \hline 1.375 \end{array}$$

Solve for D:
D = 0.746 − 0.46
D = 0.286" Ans

$$\begin{array}{r} 0.746 \\ -\,0.460 \\ \hline 0.286 \end{array}$$

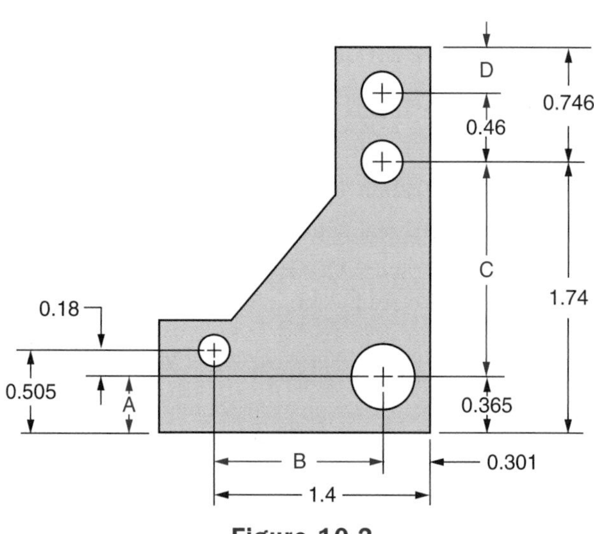

Figure 10-2

Application

Adding Decimal Fractions

1. Add the following numbers.

a. 0.375 + 10.4 + 5 __________

b. 0.003 + 0.13795 __________

c. 0.375 + 0.8 + 0.12 __________

d. 4.187 + 0.932 + 0.01 __________

e. 363.13 + 18.2 + 0.027 __________

f. 4 + 0.4 + 0.04 + 0.004 __________

g. 87 + 0.0239 + 7.23 __________

h. 0.0001 + 0.1 + 0.01 __________

i. 4.705 + 0.0937 + 0.98 __________

j. 0.063 + 4.9 + 324 __________

2. Determine dimensions A, B, C, D, E, and F of the profile gauge shown in Figure 10-3. All dimensions are in inches.

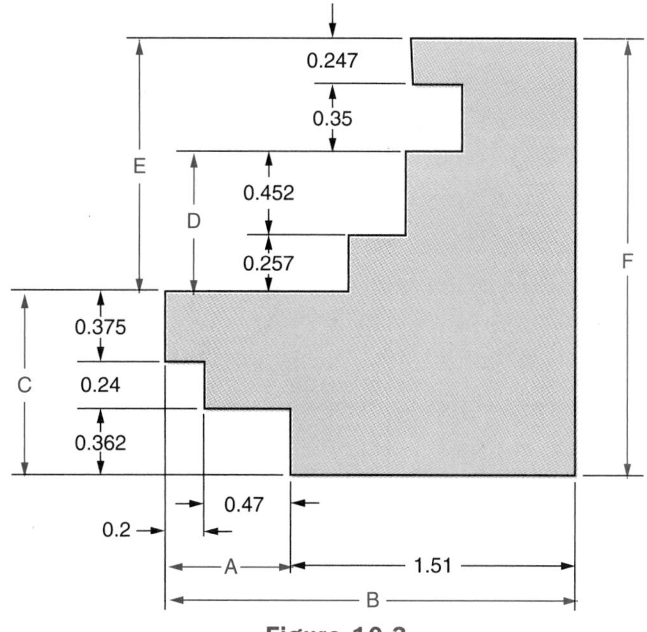

Figure 10-3

A = ____________
B = ____________
C = ____________
D = ____________
E = ____________
F = ____________

3. A sine plate is to be set to a desired angle by using size (gauge) blocks of the following thicknesses: 3.000 inches, 0.500 inch, 0.250 inch, 0.125 inch, 0.100 inch, 0.1007 inch, and 0.1001 inch. Determine the total height that the sine plate is raised. ____________
4. Three cuts are required to turn a steel shaft. The depths of the cuts, in millimeters, are 6.25, 3.18, and 0.137. How much stock has been removed per side? Round answer to 2 decimal places. ____________
5. A thickness or feeler gauge is shown in Figure 10-4. Thickness gauges are widely used in manufacturing and machine service and repair occupations. Find the smallest combination of gauge leaves that total each of the following thicknesses: (more than one combination may total certain thicknesses). ____________

a. 0.014"
b. 0.033"
c. 0.021"
d. 0.038"
e. 0.011"
f. 0.042"
g. 0.029"
h. 0.049"

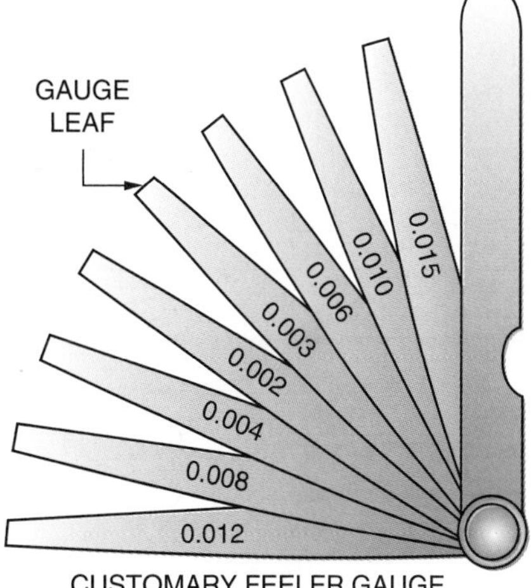

Figure 10-4

Subtracting Decimal Fractions

6. Subtract the following numbers. Where necessary, round answers to 3 decimal places.

a. 0.527 − 0.4136 __________

b. 0.319 − 0.0127 __________

c. 2.308 − 0.7859 __________

d. 0.3 − 0.299 __________

e. 0.4327 − 0.412 __________

f. 23.062 − 0.973 __________

g. 0.313 − 0.2323 __________

h. 4.697 − 0.0002 __________

i. 5.923 − 3.923 __________

7. The front and right side views of a sliding shoe are shown in Figure 10-5. Determine dimensions A, B, C, D, E, and F. All dimensions are in millimeters.

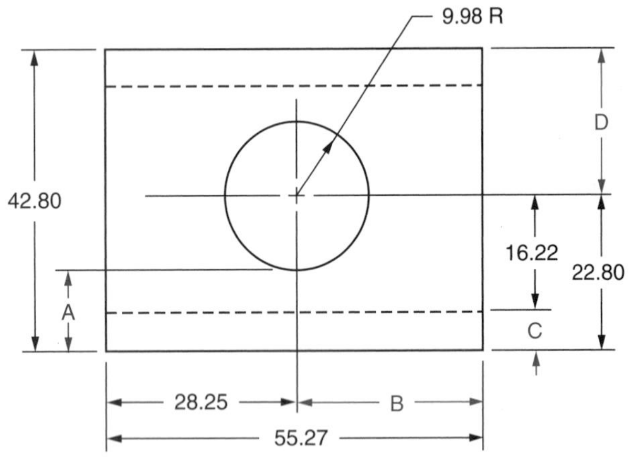

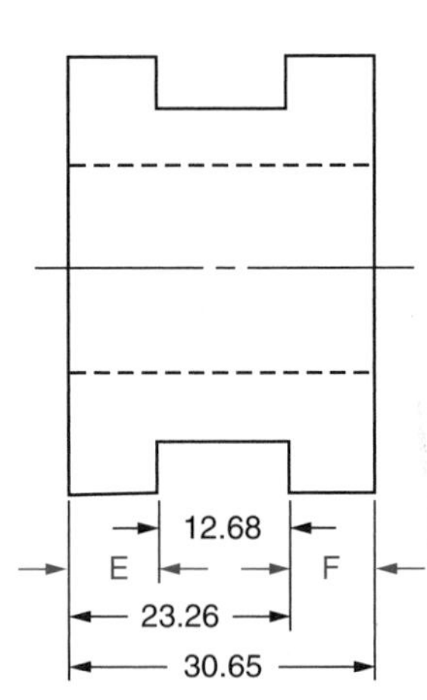

A = __________

B = __________

C = __________

D = __________

E = __________

F = __________

Figure 10-5

8. Refer to the plate shown in Figure 10-6 and determine the following distances. All dimensions are in inches.

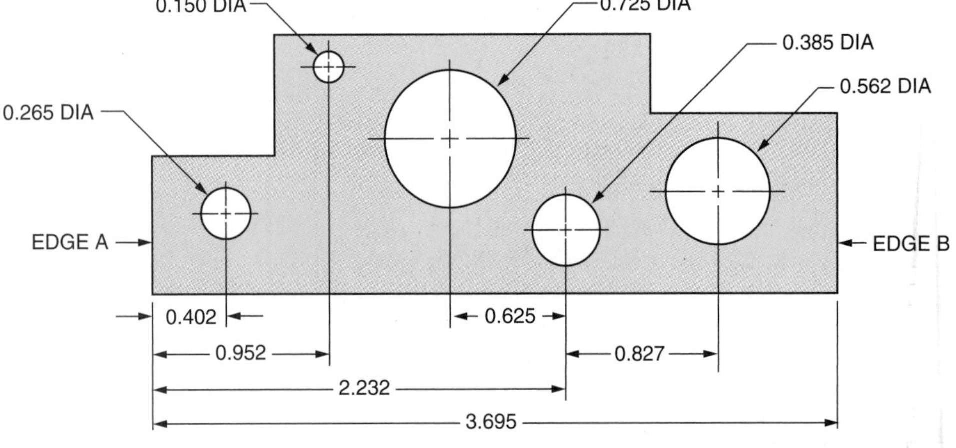

Figure 10-6

a. The horizontal center distance between the 0.265" diameter hole and the 0.150" diameter hole. __________

b. The horizontal center distance between the 0.385" diameter hole and the 0.150" diameter hole. __________

c. The distance between edge A and the center of the 0.725" diameter hole. ________

d. The distance between edge B and the center of the 0.385" diameter hole. ________

e. The distance between edge B and the center of the 0.562" diameter hole. ________

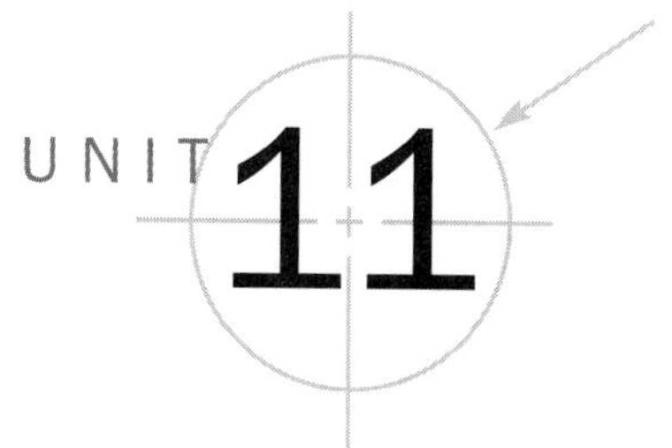

Multiplication of Decimal Fractions

Objectives ***After studying this unit you should be able to***

- Multiply decimal fractions.
- Multiply combinations of decimals, mixed decimals, and whole numbers.

A machinist must readily be able to multiply decimal fractions for computing machine feeds and speeds, for determining tapers, and for determining lengths and stock sizes. Multiplication of decimal fractions is also required in order to solve problems that involve geometry and trigonometry.

Multiplying Decimal Fractions

Procedure To multiply decimal fractions

- Multiply using the same procedure as with whole numbers.
- Beginning at the right of the product, point off the same number of decimal places as there are in the multiplicand and the multiplier combined.

Example 1 Multiply 50.123 by 0.87.

Multiply the same as with whole numbers.

Beginning at the right of the product, point off as many decimal places as there are in both the multiplicand and the multiplier.

Multiplicand	→	50.123 (3 places)
Multiplier	→	× 0.87 (2 places)
		3 50861
		40 0984
Product	→	43.60701 (5 places) Ans

Example 2 Compute the lengths of thread on each end of the shaft shown in Figure 11-1. All dimensions are in inches.

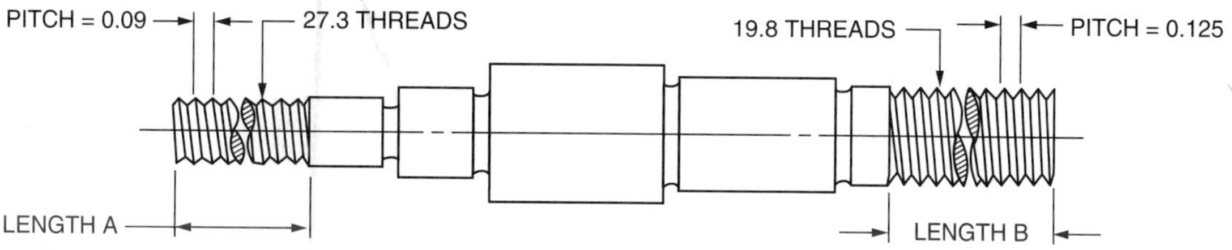

Figure 11-1

Compute Length A:
A = 2.457" Ans

```
   27.3 (1 place)
×  0.09 (2 places)
  2.457 (3 places)
```

Compute Length B:
B = 2.4750" Ans

```
    19.8 (1 place)
×  0.125 (3 places)
     990
    396
   1 98
  2.4750 (4 places)
```

When multiplying certain decimal fractions, the product has a smaller number of digits than the number of decimal places required. For these products add as many zeros to the left of the product as are necessary to give the required number of decimal places.

Example Multiply 0.0237 by 0.04. Round the answer to 5 decimal places.

The multiplicand, 0.0237, has 4 decimal places, and the multiplier, 0.04, has 2 decimal places. Therefore, the product must have 6 decimal places.

Multiply.

```
   0.0237 (4 places)
×    0.04 (2 places)
 0.000948 (6 places)
```

Add 3 zeros to the left of the product.
Round 0.000948 to 5 places.

0.00095 Ans

Application

Multiplying Decimal Fractions

1. Multiply these numbers. Where necessary, round the answers to 4 decimal places.

   a. 4.693 × 0.012 __________

   b. 2.2 × 1.5 __________

   c. 40 × 0.15 __________

   d. 6.43 × 0.26 __________

2. A section of a spur gear is shown in Figure 11-2. Given the circular pitches for various gear sizes, determine the working depths, clearances, and tooth thicknesses. Round the answers to 4 decimal places.

Working depth = 0.6366 × Circular pitch
Clearance = 0.05 × Circular pitch
Tooth thickness = 0.5 × Circular pitch

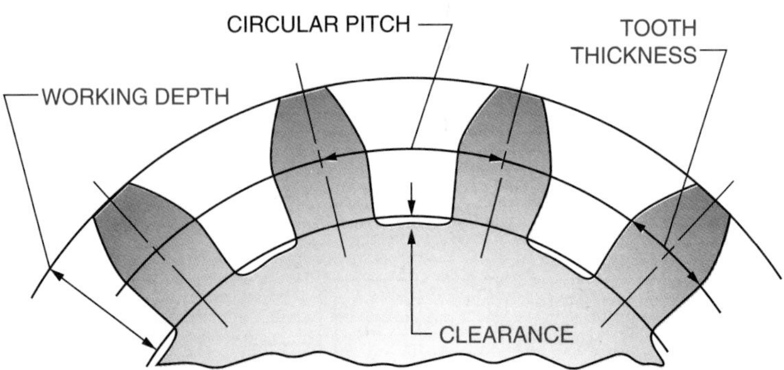

Figure 11-2

	Circular Pitch (inches)	Working Depth (inches)	Clearance (inches)	Tooth Thickness (inches)
a.	0.3925			
b.	0.1582			
c.	0.8069			
d.	1.2378			
e.	1.5931			

3. Determine diameters A, B, C, D, and E of the shaft in Figure 11-3. All dimensions are in millimeters.

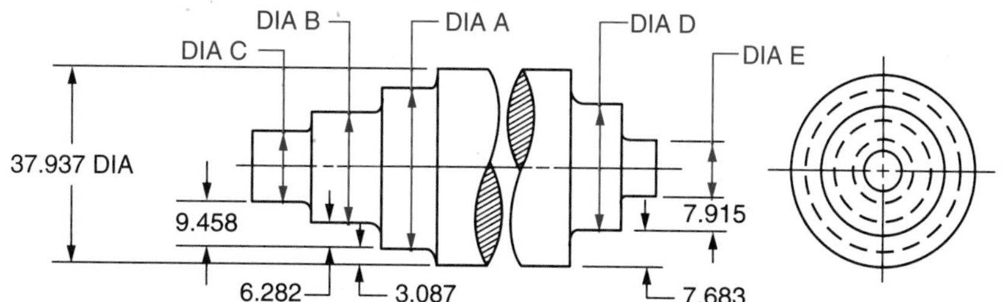

Figure 11-3

A = ______

B = ______

C = ______

D = ______

E = ______

4. Determine dimension x for each of these figure.

a. All dimensions are in inches. ______

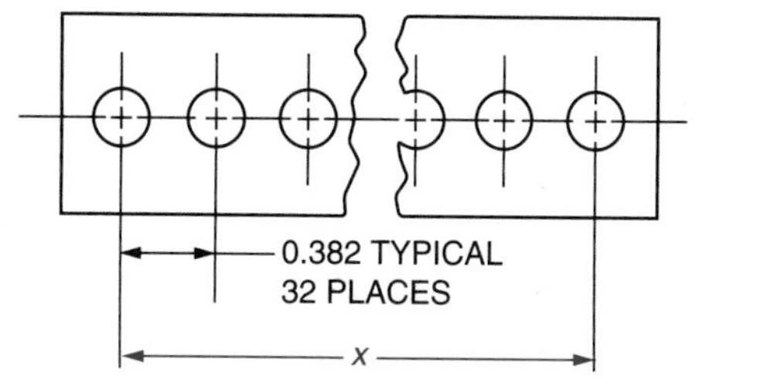

b. All dimensions are in millimeters. ______

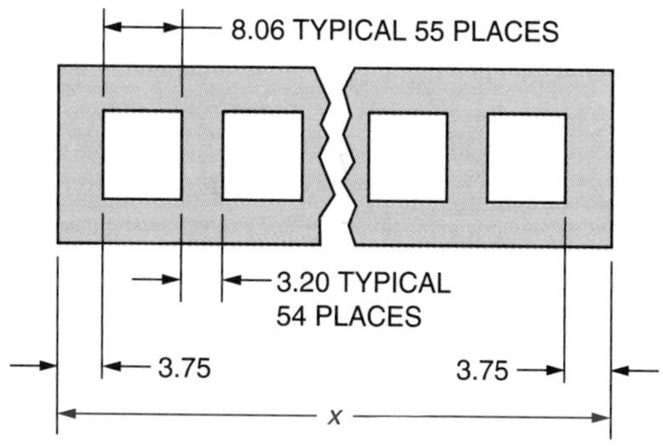

c. Round the answer to 3 decimal places. All dimensions are in inches. ______

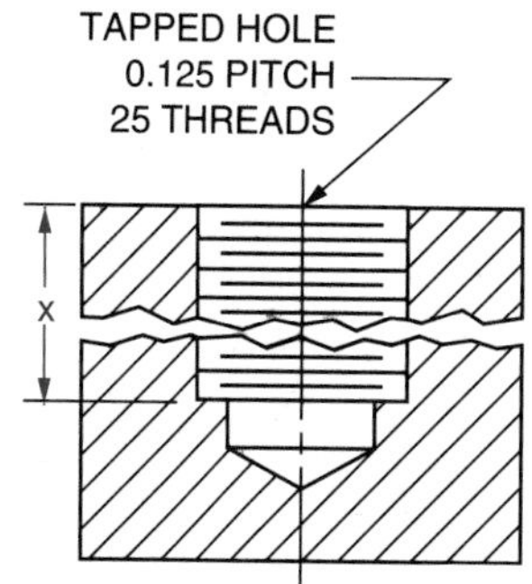

d. Round the answer to 3 decimal places. All dimensions are in inches. ______

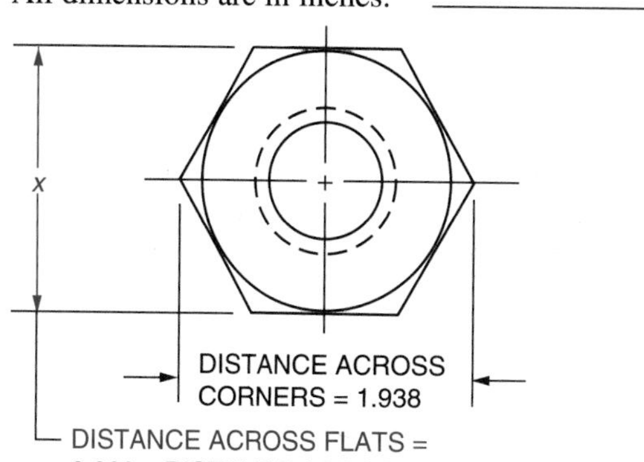

5. The length, L, of the point on any standard 118° included-angle drill, as shown in Figure 11-4, can be calculated using the formula $L = 0.3Ø$, where Ø represents the diameter of the drill. Determine the lengths of the following drill points with the given diameters. Round to 3 decimal places for inches and 1 decimal place for millimeters.

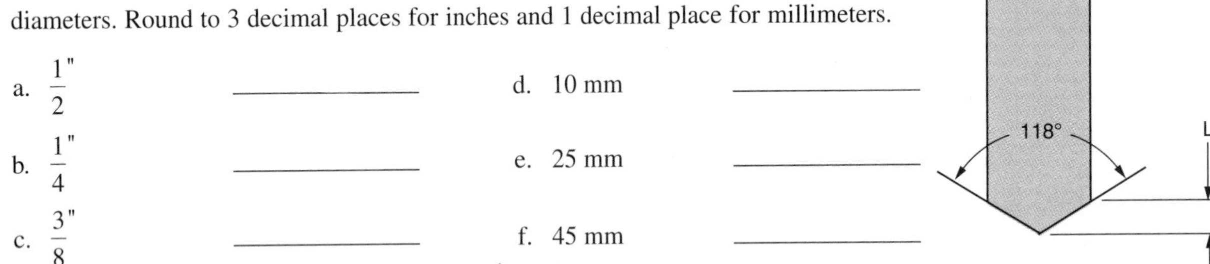

a. $\frac{1"}{2}$ ________ d. 10 mm ________

b. $\frac{1"}{4}$ ________ e. 25 mm ________

c. $\frac{3"}{8}$ ________ f. 45 mm ________

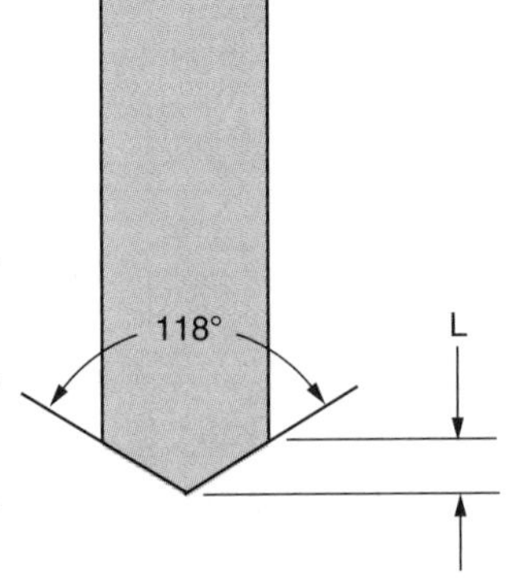

Figure 11-4

6. The length, L, of the point on any standard 82° included angle drill can be calculated using the formula $L = 0.575Ø$, where Ø represents the diameter of the drill. Determine the lengths of the following drill points with the given diameters. Round to 3 decimal places for inches and 1 decimal place for millimeters.

a. $\frac{1"}{2}$ ________ d. 10 mm ________

b. $\frac{1"}{4}$ ________ e. 25 mm ________

c. $\frac{3"}{8}$ ________ f. 45 mm ________

UNIT 12

Division of Decimal Fractions

Objectives ***After studying this unit you should be able to***

- Divide decimal fractions.
- Divide decimal fractions with whole numbers.
- Divide decimal fractions with mixed decimals.

Division with decimal fractions is used for computing the manufacturing cost and time per piece after total production costs and times have been determined. Division with decimal fractions is also required in order to compute thread pitches, gear tooth thicknesses and depths, cutting speeds, and depths of cut.

Dividing Decimal Fractions

Moving a decimal point to the right is equivalent to multiplying the decimal by a power of 10.

$0.237 \times 10 = 2.37$ $\qquad$ $0.237 \times 1000 = 237.$

$0.237 \times 100 = 23.7$ $\qquad$ $0.237 \times 10{,}000 = 2370.$

When dividing decimal fractions, the value of the answer (quotient) is not changed if the decimal points of both the divisor and the dividend are moved the same number of places to the right. It is the same as multiplying both divisor and dividend by the same number.

$$0.9375 \div 0.612 = (0.9375 \times 1000) \div (0.612 \times 1000) = 937.5 \div 612.$$

$$14.203 \div 6.87 = (14.203 \times 100) \div (6.87 \times 100) = 1420.3 \div 687.$$

Procedure To divide decimal fractions

- Move the decimal point of the divisor as many places to the right as are necessary to make the divisor a whole number.
- Move the decimal point of the dividend the same number of places to the right as were moved in the divisor.
- Place the decimal point in the quotient directly above the decimal point in the dividend.
- Add zeros to the dividend if necessary.
- Divide as with whole numbers.

Example 1 Divide 0.643 by 0.28. Round the answer to 3 decimal places.

To make the divisor a whole number move the decimal point 2 places to the right, 28.

The decimal point in the dividend is also moved 2 places to the right, 64.3.

Add 3 zeros to the dividend. One extra place is necessary in order to round the answer to 3 decimal places.

Place the decimal point of the quotient directly above the decimal point of the dividend.

Divide as with whole numbers.

```
     2.2964 ≈ 2.296  Ans
28)64.3000
   56
    83
    56
    270
    252
     180
     168
      120
      112
        8
```

Example 2 3.19 ÷ 0.072. Round the answer to 2 decimal places.

Move the decimal point 3 places to the right in the divisor, and 3 places to the right in the dividend.

Add 3 zeros to the dividend.

Place the decimal point of the quotient directly above the decimal point of the dividend.

Divide.

```
      44.305 ≈ 44.31  Ans
72)3190.000
   288
    310
    288
     220
     216
      400
      360
       40
```

When dividing a decimal fraction or a mixed decimal by a whole number, it is not necessary to move the decimal point of either the divisor or the dividend. Add zeros to the right of the dividend, if necessary, to obtain the desired number of decimal places in the answer.

Examples

1. Divide 0.63 by 12 to 4 decimal places.

```
   0.0525
12)0.6300  Ans
```

2. Divide 33.97 by 5 to 3 decimal places.

```
  6.794
5)33.970  Ans
```

Application

Dividing Decimal Fractions

1. Divide the following numbers. Express the answers to the indicated number of decimal places.

a. 0.69 ÷ 0.432 (3 places) ______

b. 0.92 ÷ 0.36 (2 places) ______

c. 0.001 ÷ 0.1 (4 places) ______

d. 10 ÷ 0.001 (3 places) ______

e. 1.023 ÷ 0.09 (3 places) ______

f. $\frac{16.3}{3.8}$ (2 places) ______

g. $\frac{37}{0.273}$ (2 places) ______

h. $\frac{0.005}{0.81}$ (4 places) ______

2. Rack sizes are given according to diametral pitch. Given 4 different diametral pitches, find the linear pitch and the whole depth of each rack to 4 decimal places. All dimensions are in inches.

$$\text{Linear Pitch} = \frac{3.1416}{\text{Diametral Pitch}} \qquad \text{Whole Depth} = \frac{2.157}{\text{Diametral Pitch}}$$

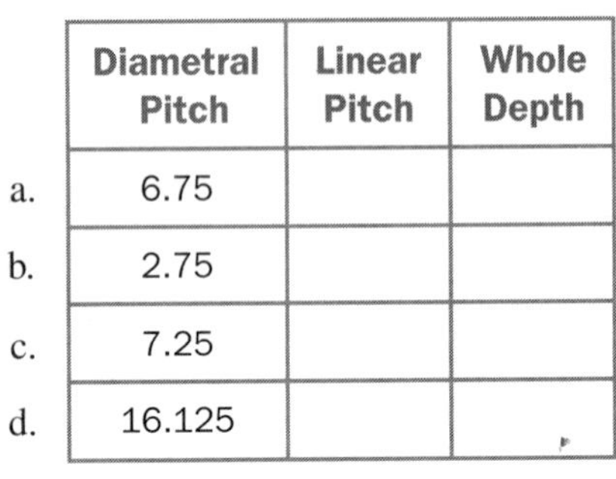

	Diametral Pitch	Linear Pitch	Whole Depth
a.	6.75		
b.	2.75		
c.	7.25		
d.	16.125		

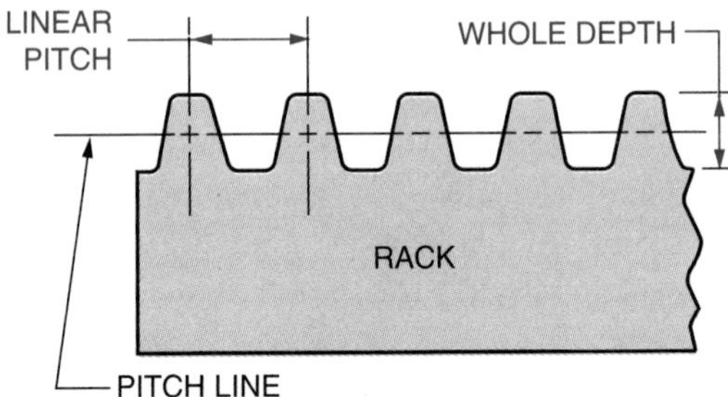

Figure 12-1

3. Four sets of equally spaced holes are shown in the machined plate in Figure 12-2. Determine dimensions A, B, C, and D to 2 decimal places. All dimensions are in millimeters.

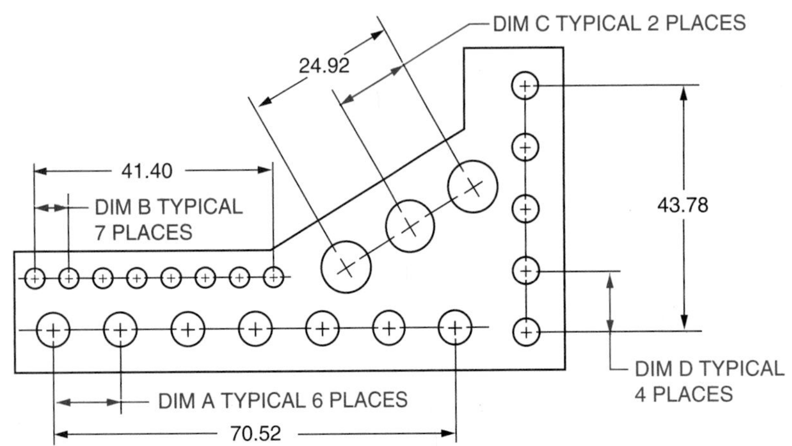

Figure 12-2

A = ______

B = ______

C = ______

D = ______

4. A cross-sectional view of a bevel gear is shown in Figure 12-3. Given the diametral pitch and the number of gear teeth, determine the pitch diameter, the addendum, and the dedendum. Round the answers to 4 decimal places.

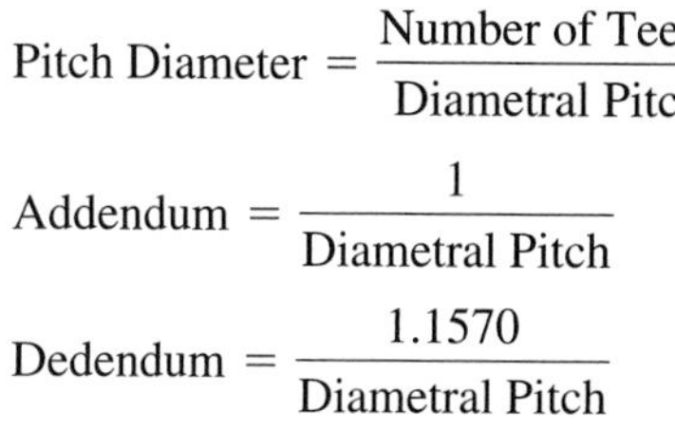

$$\text{Pitch Diameter} = \frac{\text{Number of Teeth}}{\text{Diametral Pitch}}$$

$$\text{Addendum} = \frac{1}{\text{Diametral Pitch}}$$

$$\text{Dedendum} = \frac{1.1570}{\text{Diametral Pitch}}$$

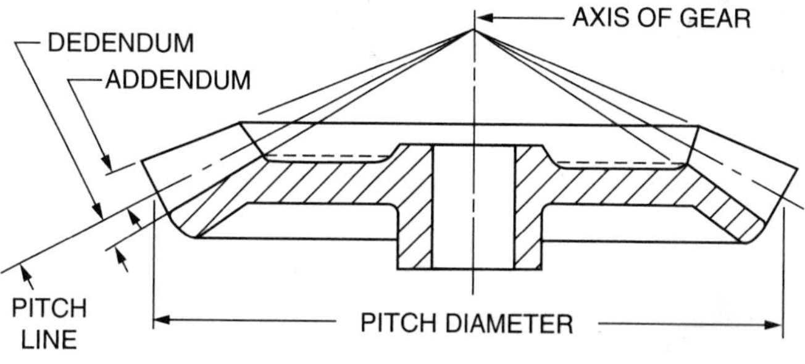

Figure 12-3

	Diametral Pitch	Number of Teeth	Pitch Diameter (inches)	Addendum (inches)	Dedendum (inches)
a.	4	45			
b.	6	75			
c.	8	44			
d.	3	54			

5. How many complete bushings each 14.60 millimeters long can be cut from a bar of bronze that is 473.75 millimeters long? Allow 3.12 millimeters waste for each piece. __________
6. A shaft is being cut in a lathe. The tool feeds (advances) 0.015 inch each time the shaft turns once (1 revolution). How many revolutions will the shaft turn when the tool advances 3.120 inches? Round the answer to 2 decimal places. __________
7. How much stock per stroke is removed by the wheel of a surface grinder if a depth of 4.725 millimeters is reached after 75 strokes? Round the answer to 3 decimal places. __________
8. An automatic screw machine is capable of producing one piece in 0.02 minute. How many pieces can be produced in 1.25 hours? __________
9. The bolt in Figure 12-4 has 7.7 threads. Determine the pitch to 3 decimal places. All dimensions are in inches. __________

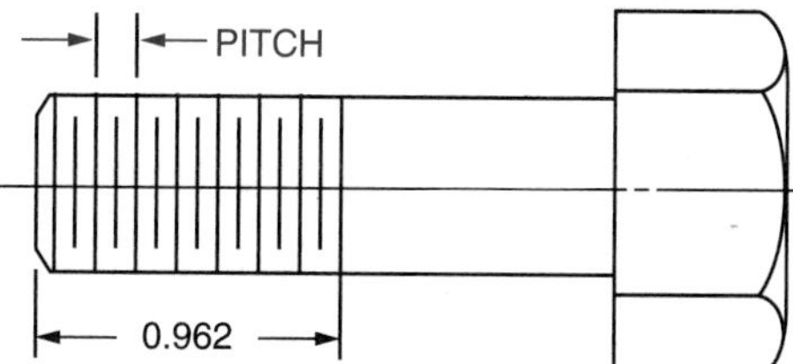

Figure 12-4

10. The block in Figure 12-5 has a threaded hole with a 0.0625-inch pitch. Determine the number of threads for the given depth to 1 decimal place. All dimensions are in inches. __________

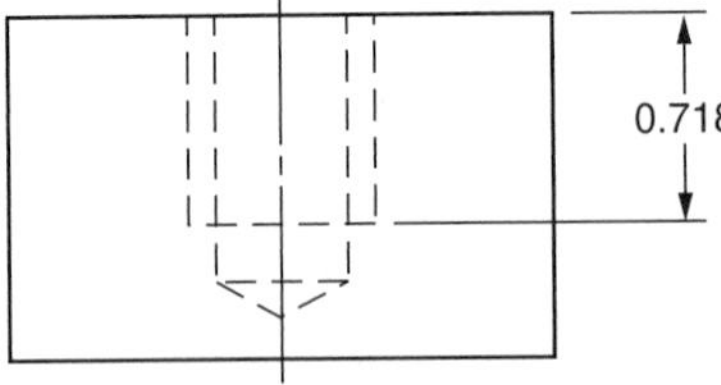

Figure 12-5

11. The length of a side of a square equals the distance from point A to point B divided by 1.4142. Determine the length of a side of the square plate in Figure 12-6 to 2 decimal places. All dimensions are in millimeters. __________

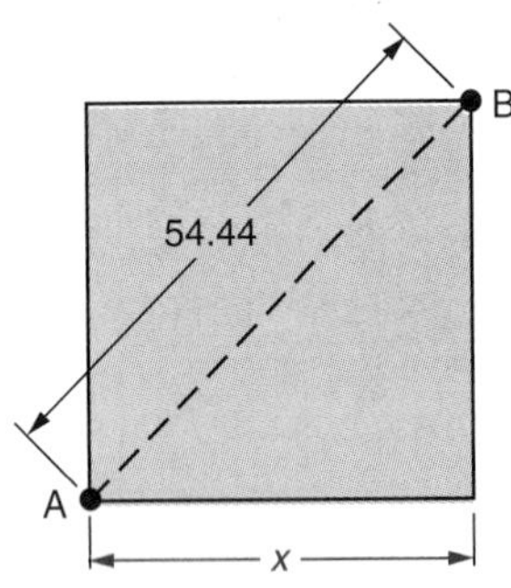

Figure 12-6

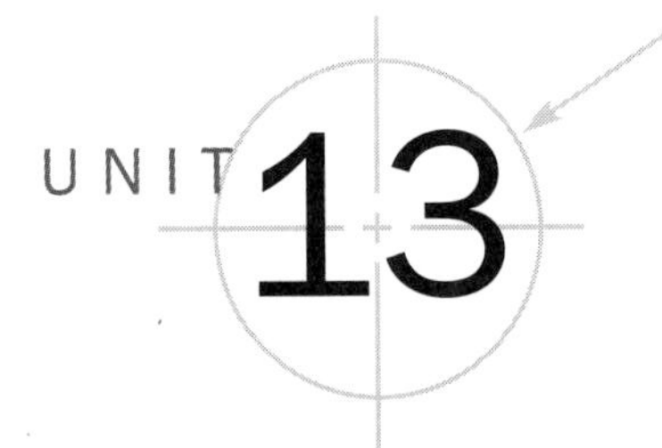

Powers

Objectives ***After studying this unit you should be able to***

- Raise numbers to indicated powers.
- Solve problems that involve combinations of powers with other basic operations.

Powers of numbers are used to compute areas of square plates and circular sections and to compute volumes of cubes, cylinders, and cones. Use of powers is particularly helpful in determining distances in problems that require applications of geometry and trigonometry.

Description of Powers

Two or more numbers multiplied to produce a given number are *factors* of the given number. Two factors of 8 are 2 and 4. The factors of 15 are 3 and 5. A *power* is the product of two or more equal factors. The third power of 5 is $5 \times 5 \times 5$ or 125. An *exponent* shows how many times a number is taken as a factor. It is written smaller than the number, above the number, and to the right of the number. The expression 3^2 means 3×3. The exponent 2 shows that 3 is taken as a factor twice. It is read as 3 to the second power or 3 squared.

Examples Find the indicated powers.

1. 2^5 Two to the fifth power means $2 \times 2 \times 2 \times 2 \times 2$ or 32. Ans
2. 3^3 Three to the third power or cubed means $3 \times 3 \times 3$ or 27. Ans
3. 0.72^2 0.72 to the second power or squared means 0.72×0.72 or 0.5184. Ans

$A = s^2$ is called a *formula*. A formula is a short method of expressing an arithmetic relationship by the use of symbols. Known values may be substituted for the symbols and other values can be found.

Example 1 Determine the area of the square shown in Figure 13-1. The area of a square equals the length of a side squared. The answer is given in square units. All dimensions are in inches.

$A = s^2$

$A = \left(\frac{7}{8}\text{ in}\right)^2$

$A = \frac{7}{8}\text{ in} \times \frac{7}{8}\text{ in}$

$A = \frac{49}{64}\text{ sq in}$ Ans

SIDE = $\frac{7}{8}$

SIDE = $\frac{7}{8}$

Figure 13-1

The symbols in^2, in.^{2}, sq in, and sq in. are all used for square inches. Some people write the symbol for inch as "in." so that the period will keep it from being confused with the word "in."

➤ **Note:** In linear measurement we often use the symbol ' for feet and " in inches. For example, 3' = 3 ft = 3 feet. These symbols are *not* to be used for area or volume. To indicate an area of 3 square feet, you may write 3 sq ft or 3 ft^2 but NEVER write $3'^2$.

Example 2 Find the volume of the cube shown in Figure 13-2. The volume of a cube equals the length of a side cubed. The answer is given in cubic units. All dimensions are in millimeters. Round answer to 1 decimal place.

$V = s^3$

$V = (1.6\text{ mm})^3$

$V = 1.6\text{ mm} \times 1.6\text{ mm} \times 1.6\text{ mm}$

$V = 4.096\text{ mm}^3$ or 4.1 mm^3 Ans

The symbol mm^3 is used for cubic millimeters.

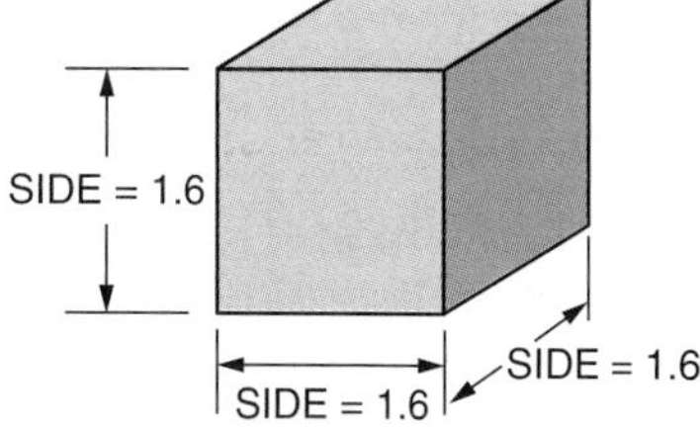

Figure 13-2

Use of Parentheses

In this example, only the numerator is squared.

$$\frac{2^2}{3} = \frac{2 \times 2}{3} = \frac{4}{3} = 1\frac{1}{3} \quad \text{Ans}$$

In this example, only the denominator is squared.

$$\frac{2}{3^2} = \frac{2}{3 \times 3} = \frac{2}{9} \quad \text{Ans}$$

Parentheses are used as grouping symbols. Parentheses indicate that both the numerator and the denominator of a fraction are raised to the given power.

$$\left(\frac{2}{3}\right)^2 = \frac{2^2}{3^2} = \frac{2 \times 2}{3 \times 3} = \frac{4}{9} \quad \text{Ans}$$

Procedure To solve problems that involve operations within parentheses

- Perform the operations within the parentheses.
- Raise to the indicated power.

Examples

1. $(1.2 \times 0.6)^2 = 0.72^2 = 0.72 \times 0.72 = 0.5184$ Ans
2. $(0.5 + 2.4)^2 = 2.9^2 = 2.9 \times 2.9 = 8.41$ Ans
3. $(0.75 - 0.32)^2 = 0.43^2 = 0.43 \times 0.43 = 0.1849$ Ans
4. $\left(\frac{14.4}{3.2}\right)^2 = 4.5^2 = 4.5 \times 4.5 = 20.25$ Ans

When solving power problems that also require addition, subtraction, multiplication, or division, perform the power operation first.

Examples

1. $5 \times 3^2 - 12 = 5 \times 9 - 12 = 45 - 12 = 33$ Ans
2. $33.5 - 5.5^2 + 8.7 = 33.5 - 30.25 + 8.7 = 11.95$ Ans
3. $\frac{2.2^3 - 5.608}{1.4} = \frac{10.648 - 5.608}{1.4} = \frac{5.040}{1.4} = 3.6$ Ans

The symbol π (pi) represents a constant value used in mathematical relationships involving circles. Depending upon the specific problem to be solved, generally, the value of pi used is $3\frac{1}{7}$, 3.14, or 3.1416. Most calculators have a π key.

Example Compute the volume of the cylinder shown to 2 decimal places. The answer is given in cubic units. All dimensions are in inches.

$V = \pi \times r^2 \times h$

$V = 3.14 \times (0.85 \text{ in})^2 \times 1.25 \text{ in}$

$V = 3.14 \times 0.7225 \text{ sq in} \times 1.25 \text{ in}$

$V = 2.8358 \text{ cu in}, \quad 2.84 \text{ cu in} \quad$ Ans

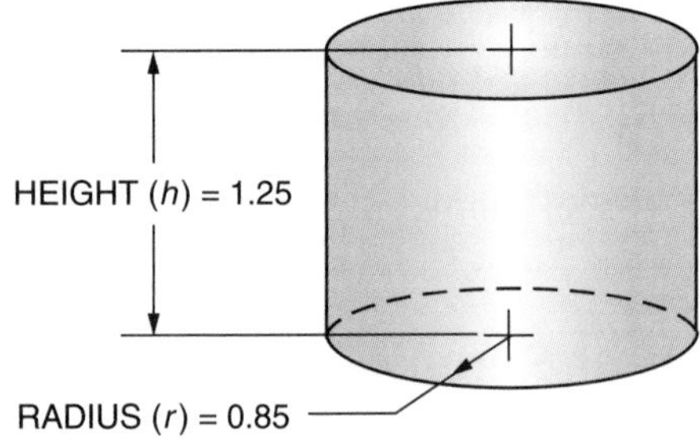

Figure 13-3

The symbols in^3, in.^3, cu in, and cu in. are all used for cubic inches.

Many problems require the application of the same formula more than once or the application of two different formulas in the solutions.

Example Find the metal area of this square plate. Round the answer to 2 decimal places. All dimensions are in inches. $A = s^2$

The metal area equals the area of the large square, A_1, minus the area of the removed square, A_2.

$A_1 = (5.250 \text{ in})^2$

$A_1 = 27.5625 \text{ sq in}$

$A_2 = (2.500 \text{ in})^2$

$A_2 = 6.2500 \text{ sq in}$

$A_3 = 27.5625 \text{ sq in} - 6.2500 \text{ sq in} \approx 21.31 \text{ sq in}$ Ans (rounded)

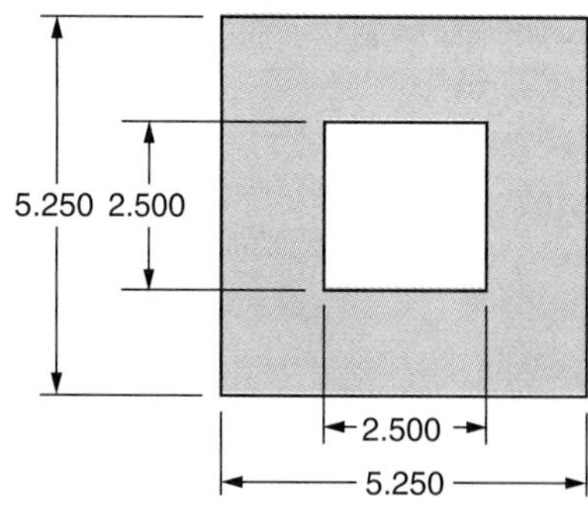

Figure 13-4

Application

Raising a Number to a Power

Raise the following numbers to the indicated power.

1. 3.4^3 ______
2. 1^8 ______
3. 100^4 ______
4. $\left(\frac{2}{3}\right)^3$ ______
5. $\frac{2^3}{3}$ ______
6. $\frac{3}{4^3}$ ______
7. $(0.3 \times 7)^2$ ______
8. $(20.7 + 7.2)^2$ ______
9. $\left(\frac{28.8}{7.2}\right)^3$ ______

Related Problems

In the following table the lengths of the sides of squares are given. Determine the areas of the squares. Round the answers to 2 decimal places where necessary.

	Side	Area
10.	1.25 in	
11.	23.070 mm	
12.	0.17 in	
13.	10.70 mm^2	
14.	0.02 in	

	Side	Area
15.	$\frac{3}{4}$ in	
16.	$\frac{7}{8}$ in	
17.	$3\frac{3}{4}$ in	
18.	$\frac{13}{16}$ in	
19.	$13\frac{3}{4}$ in	

$A = s^2$ where A = area

s = side

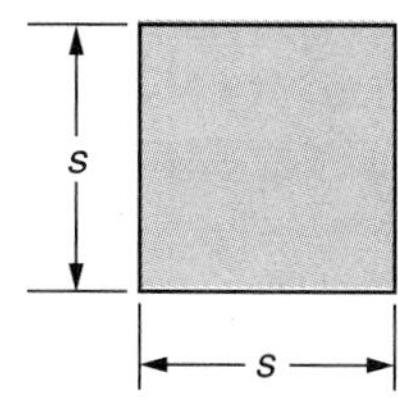

In the following table the lengths of the sides of cubes are given. Determine the volumes of the cubes. Round answers to 2 decimal places where necessary.

	Side	Volume
20.	0.29 in	
21.	20.60 mm	
22.	3.930 in	
23.	14.00 mm	
24.	0.075 in	

	Side	Volume
25.	$\frac{1}{3}$ in	
26.	$\frac{7}{8}$ in	
27.	$1\frac{1}{2}$ in	
28.	$9\frac{1}{8}$ in	
29.	$\frac{3}{4}$ in	

$V = s^3$ where V = volume
s = side

In the following table the radii of circles are given. Determine the areas of the circles. Round the answers to the nearest whole number.

	Radius	Area
30.	16.20 mm	
31.	15.60 mm	
32.	0.07 in	
33.	9.28 in	
34.	12.35 mm	

$A = \pi \times R^2$ where A = area
$\pi = 3.14$
R = radius

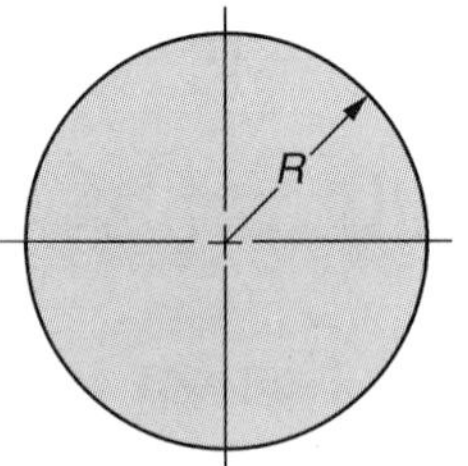

In the following table the diameters of spheres are given. Determine the volumes of the spheres. Round the answers to 1 decimal place where necessary.

	Diameter	Volume
35.	0.65 in	
36.	6.500 mm	
37.	0.75 in	
38.	10.80 mm	
39.	7.060 mm	

$V = \frac{\pi \times D^3}{6}$ where V = volume
$\pi = 3.14$
D = diameter

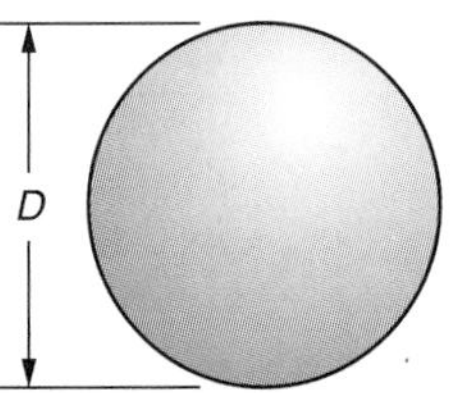

In the following table the radii and heights of cylinders are given. Determine the volumes of the cylinders. Round the answers to the nearest whole number.

	Radius	Height	Volume
40.	5.00 mm	3.20 mm	
41.	1.50 in	2.30 in	
42.	2.25 in	3.00 in	
43.	0.07 in	6.70 in	
44.	7.81 mm	6.72 mm	

$V = \pi \times r^2 \times h$ where V = volume
$\pi = 3.14$
r = radius
h = height

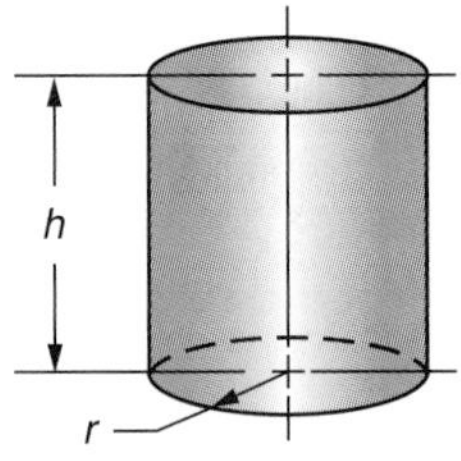

In the following table the diameters and heights of cones are given. Find the volumes of the cones. Round the answers to the nearest whole number.

	Diameter	Height	Volume
45.	3.20 in	4.00 in	
46.	3.00 in	5.00 in	
47.	10.60 mm	13.10 mm	
48.	9.90 mm	6.20 mm	
49.	0.37 in	0.96 in	

$V = 0.2618 \times d^2 \times h$ where V = volume
d = diameter
h = height

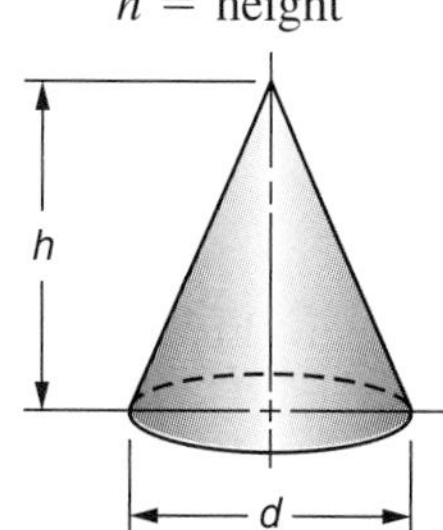

Solve the following problems. Use $\pi = 3.14$. Round answers to the nearest whole number.

50. Find the metal area of this washer. All dimensions are in millimeters. __________

$A = \pi \times R^2$

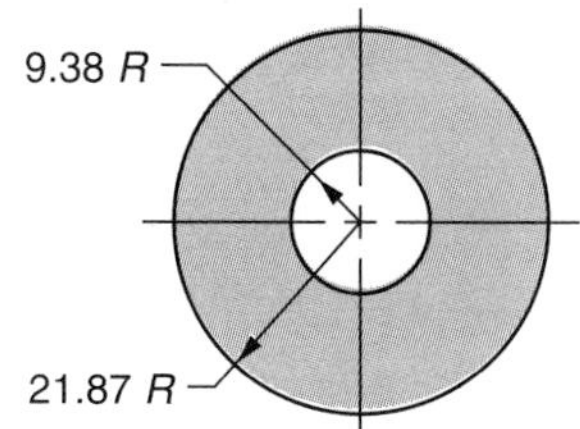

51. Find the metal area of this spacer. All dimensions are in millimeters. __________

Area of square $= s^2$
Area of circle $= \pi \times R^2$

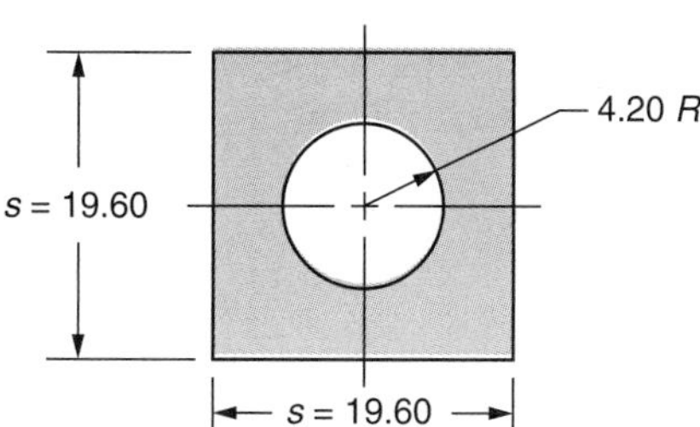

52. Find the area of this plate. All dimensions are in millimeters. __________

➤ **Hint:** The broken lines indicate one method of solution.

$A = s^2$

53. Find the metal volume of this bushing. All dimensions are in inches. __________

$V = \pi \times R^2 \times H$

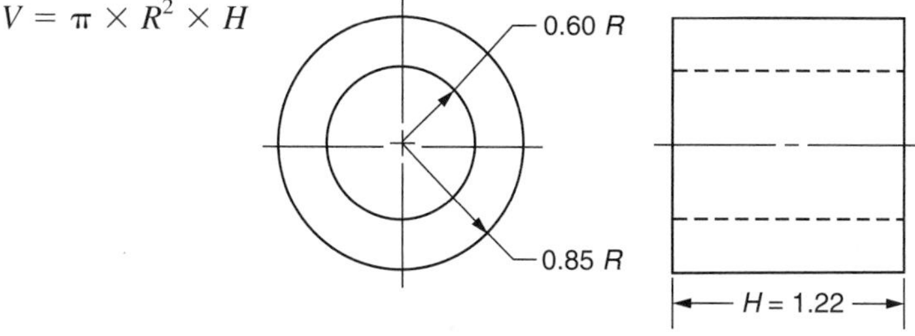

54. Find the volume of this pin. All dimensions are in inches. ____________

Volume of cylinder = $\pi \times R^2 \times H$

Volume of cone = $0.2618 \times D^2 \times H$

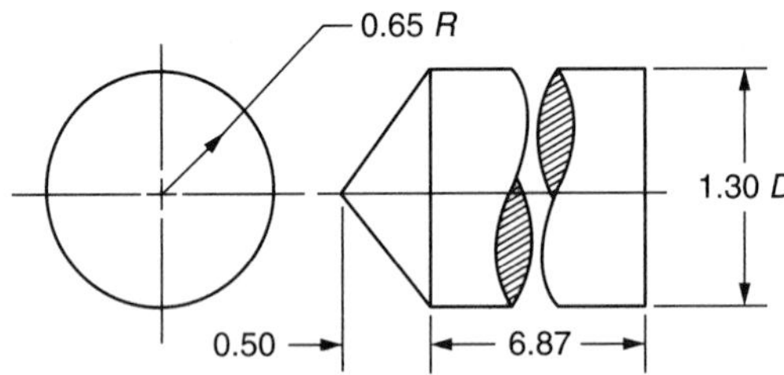

55. A materials estimator finds the weight of aluminum needed for the casting shown. Aluminum weighs 0.0975 pound per cubic inch. Find, to the nearer pound, the weight of aluminum required for 15 castings. All measurements are in inches. ____________

➤ **Hint:** The broken line indicates a method of solution.

$V = s^3$

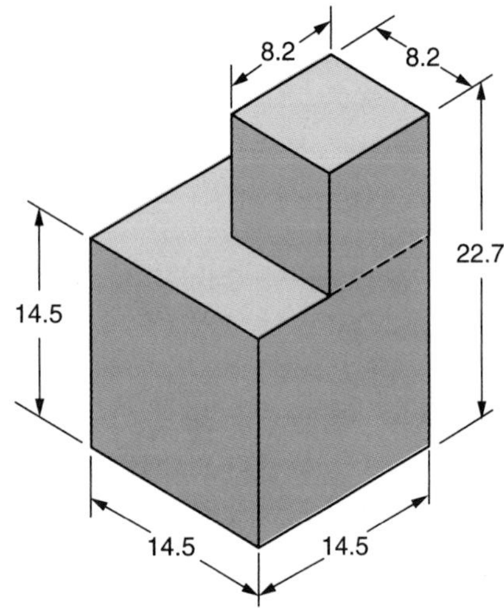

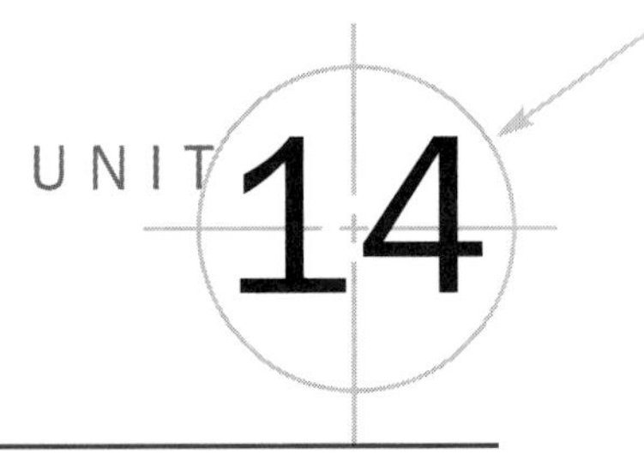

Roots

Objectives ***After studying this unit you should be able to***

- Extract whole number roots.
- Determine the root of any positive number using a calculator.
- Solve problems that involve combinations of roots with other basic arithmetic operations.

The operation of extracting roots of numbers is used to determine lengths of sides and heights of squares and cubes and radii of circular sections when areas and volumes are known. The machinist uses roots in computing distances between various parts of machined pieces from given dimensions.

Description of Roots

A *root* of a number is a quantity that is taken two or more times as an equal factor of the number. The expression $\sqrt[3]{64}$ is called a radical. A *radical* is an indicated root of a number. The symbol $\sqrt{}$ is called a *radical sign* and indicates a root of a number. The digit 3 is called the index. An *index* indicates the number of times that a root is to be taken as an equal factor to produce the given number. The index is written smaller than the number, above, and to the left of the radical sign. The given number 64 is called a radicand. A *radicand* is the number under the radical sign whose root is to be found.

index ⟶ $\sqrt[3]{64} = 4$ ⟵ root

radical sign ↑ ↑ radicand

The index 2 is omitted for an indicated square root. For example, the square root of 9 is written $\sqrt{9}$. The expression $\sqrt{9}$ means to find the number that can be multiplied by itself and equal 9. Since $3 \times 3 = 9$, 3 is the square root of 9.

Examples Find the indicated roots.

1. $\sqrt{36}$ Since $6 \times 6 = 36$, a square root of 36 is 6. Ans
2. $\sqrt{144}$ Since $12 \times 12 = 144$, a square root of 144 is 12. Ans
3. $\sqrt[3]{8}$ Since $2 \times 2 \times 2 = 8$, a cube root of 8 is 2. Ans
4. $\sqrt[3]{125}$ Since $5 \times 5 \times 5 = 125$, a cube root of 125 is 5. Ans
5. $\sqrt[4]{81}$ Since $3 \times 3 \times 3 \times 3 = 81$, a fourth root of 81 is 3. Ans

Roots must be extracted in determining unknown dimensions represented in certain formulas.

Example 1 Compute the length of the side of the square shown in Figure 14-1. This square has an area of 25 square inches.

Since $A = s^2$, the length of a side of the square equals the square root of the area.

$s = \sqrt{A}$

$s = \sqrt{25 \text{ sq in}}$

$s = \sqrt{5 \text{ in} \times 5 \text{ in}}$

$s = 5$ inches Ans

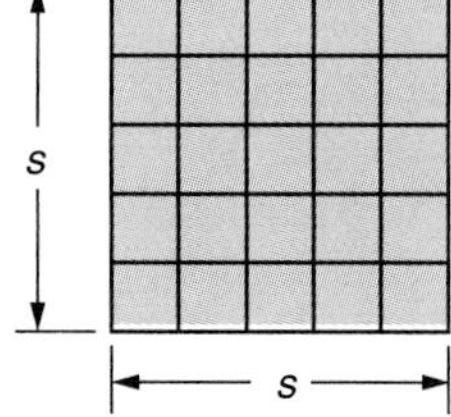

Figure 14-1

Example 2 Compute the length of the side of the cube shown in Figure 14-2. The volume of this cube equals 64 cubic inches.

$s = \sqrt[3]{V}$

$s = \sqrt[3]{64 \text{ cu in}}$

$s = \sqrt[3]{4 \text{ in} \times 4 \text{ in} \times 4 \text{ in}}$

$s = 4$ inches Ans

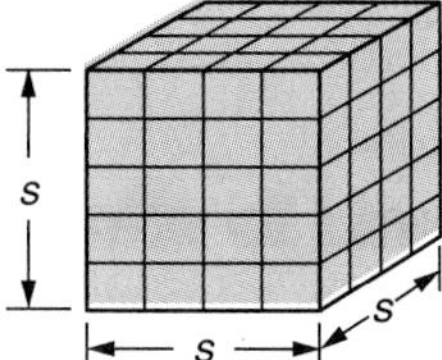

Figure 14-2

Roots of Fractions

In this example, only the root of the numerator is taken.

$$\frac{\sqrt{16}}{25} = \frac{\sqrt{4 \times 4}}{25} = \frac{4}{25} \quad \text{Ans}$$

In this example, only the root of the denominator is taken.

$$\frac{16}{\sqrt{25}} = \frac{16}{\sqrt{5 \times 5}} = \frac{16}{5} = 3\frac{1}{5} \quad \text{Ans}$$

A radical sign that encloses a fraction indicates that the roots of both the numerator and denominator are to be taken. The same answer is obtained by extracting both roots first and dividing second as by dividing first and extracting the root second.

Example Find $\sqrt{\frac{36}{9}}$.

Method 1: Extract both roots then divide.

$$\sqrt{\frac{36}{9}} = \frac{\sqrt{36}}{\sqrt{9}} = \frac{6}{3} = 2 \quad \text{Ans}$$

Method 2: Divide then extract the root.

$$\sqrt{\frac{36}{9}} = \sqrt{4} = 2 \quad \text{Ans}$$

Expressions Enclosed within the Radical Symbol

The radical symbol is a grouping symbol. An expression consisting of operations within the radical symbol is done using the order of operations.

Procedure To solve problems that involve operations within the radical symbol

- Perform the operations within the radical symbol first using the order of operations.
- Then find the root.

Examples Find the indicated roots.

1. $\sqrt{3 \times 12} = \sqrt{36} = \sqrt{6 \times 6} = 6$ Ans
2. $\sqrt{5 + 59} = \sqrt{64} = \sqrt{8 \times 8} = 8$ Ans
3. $\sqrt{128 - 7} = \sqrt{121} = \sqrt{11 \times 11} = 11$ Ans

Problems involving formulas may involve operations within a radical symbol.

Example Compute the length of the chord of the circular segment shown in Figure 14-3. All dimensions are in inches.

$C = 2 \times \sqrt{H \times (2 \times R - H)}$

where C = length of chord

H = height of segment

R = radius of circle

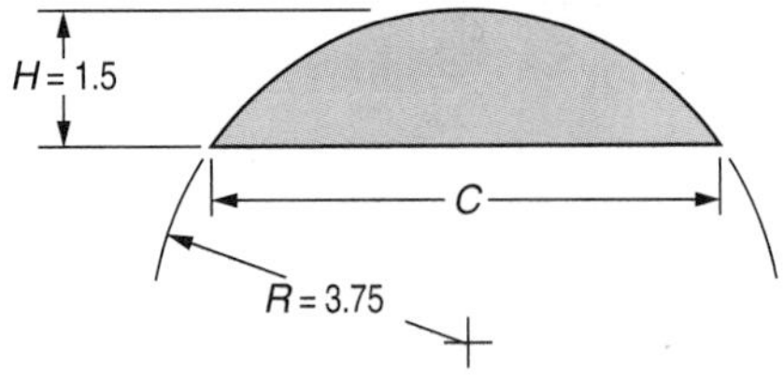

Figure 14-3

$C = 2 \times \sqrt{H \times (2 \times R - H)}$

$C = 2 \times \sqrt{1.5 \times (2 \times 3.75 - 1.5)}$

$C = 2 \times \sqrt{1.5 \times 6}$

$C = 2 \times \sqrt{9}$

$C = 2 \times 3$

$C = 6$

Length of chord = 6 inches Ans

Roots That Are Not Whole Numbers

The root examples and exercises have all consisted of numbers that have whole number roots. These roots are relatively easy to determine by observation.

Most numbers do not have whole number roots. For example, $\sqrt{259} = 16.0935$ (rounded to 4 decimal places) and $\sqrt[3]{17.86} = 2.6139$ (rounded to 4 decimal places). The root of any positive number can easily be computed with a calculator. Calculator solutions to root expressions are given at the end of Unit 16 on pages 82 and 83.

Fractional Exponents

Fractional exponents can be used to indicate roots.

- $a^{1/n} = \sqrt[n]{a}$
- $a^{m/n} = (\sqrt[n]{a})^m = \sqrt[n]{a^m}$

Examples Find the indicated roots.

1. $49^{1/2} = \sqrt{49} = 7$ Ans
2. $125^{1/3} = \sqrt[3]{125} = 5$ Ans
3. $8^{2/3} = \sqrt[3]{8^2} = \sqrt[3]{64} = 4$ Ans

Application

Radicals That Are Whole Numbers

The following problems have either whole number roots or numerators and denominators that have whole number roots. Determine these roots.

1. $\sqrt[3]{216}$ __________
2. $\sqrt{\frac{4}{9}}$ __________
3. $\frac{\sqrt{4}}{9}$ __________
4. $\frac{25}{\sqrt{36}}$ __________
5. $\sqrt{\frac{3}{4} \times \frac{3}{4}}$ __________
6. $\sqrt{0.5 \times 18}$ __________
7. $\sqrt{56.7 + 87.3}$ __________
8. $\sqrt{16.4 - 7.4}$ __________
9. $\sqrt[3]{\frac{428.8}{6.7}}$ __________

The following problems have whole number square roots. Solve for the missing values in the tables.

10. The areas of squares are given in the following table. Determine the lengths of the sides.

$$s = \sqrt{A}$$

	Area (A)	Side (s)
a.	225 mm^2	
b.	121 mm^2	
c.	64 mm^2	
d.	81 sq in	
e.	49 sq in	

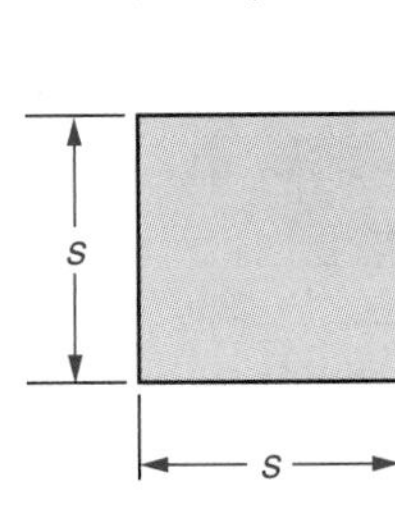

11. The volumes of cubes are given in the following table. Determine the lengths of the sides.

$$s = \sqrt[3]{V}$$

	Volume (*V*)	Side (*s*)
a.	216 mm^3	
b.	64 cu in	
c.	512 cu in	
d.	1000 mm^3	
e.	1 cu in	

12. The areas of circles are given in this table. Determine the lengths of the radii. Use $\pi = 3.14$.

$$R = \sqrt{\frac{A}{\pi}}$$

	Area (*A*)	Radius (*R*)
a.	50.24 sq in	
b.	12.56 sq in	
c.	314 mm^2	
d.	28.26 sq in	
e.	153.86 mm^3	

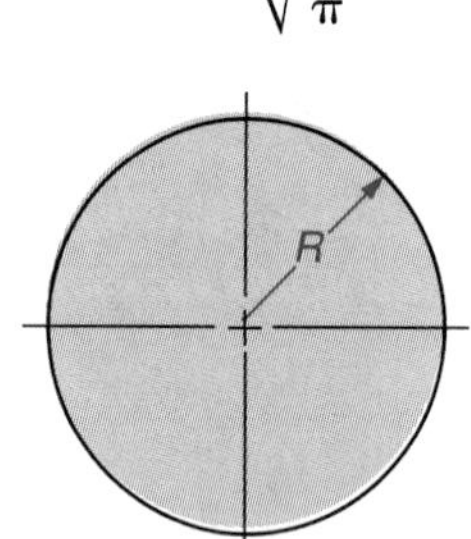

13. The volumes of spheres are given in this table. Determine the lengths of the diameters.

$$D = \sqrt[3]{\frac{V}{0.5236}}$$

	Volume (*V*)	Diameter (*D*)
a.	14.1372 cu in	
b.	113.0976 mm^3	
c.	4.1888 cu in	
d.	0.5236 cu in	
e.	523.6 mm^3	

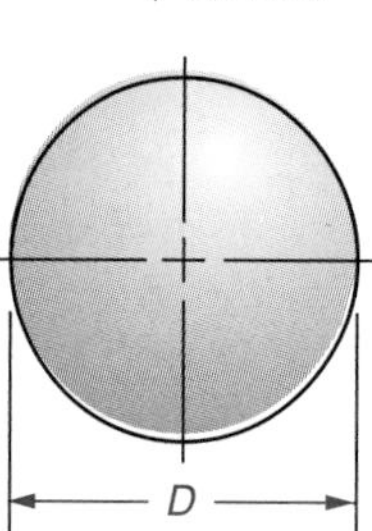

Radicals That Are Not Whole Numbers

The following problems have square roots that are not whole numbers. They require calculator computations. Refer to pages xx and xx for calculator root solutions. Compute these roots to the indicated number of decimal places.

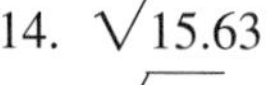

14. $\sqrt{15.63}$ (3 places) __________
15. $\sqrt{391}$ (2 places) __________
16. $\sqrt{\frac{3}{5}}$ (3 places) __________
17. $\sqrt{3\frac{1}{2}}$ (3 places) __________
18. $\sqrt{0.07 \times 28}$ (2 places) __________
19. $\sqrt{15.82 + 3.71}$ (2 places) __________
20. $\sqrt{178.5 - 163.7}$ (3 places) __________
21. $\sqrt{\frac{0.441}{60}}$ (4 places) __________

Determine the following numbers with fractional exponents.

22. $25^{1/2}$ __________
23. $36^{1/2}$ __________
24. $121^{1/2}$ __________
25. $196^{1/2}$ __________
26. $25^{3/2}$ __________
27. $32^{4/5}$ __________
28. $28^{3/4}$ (3 decimal places) __________
29. $75^{5/3}$ (3 decimal places) __________

The following problems have roots that are not whole numbers. Solve for the missing values in the tables.

30. The volumes of cylinders and their heights are given in the following table. Find the lengths of the radii to 2 decimal places. Use $\pi = 3.14$.

	Volume (V)	Height (H)	Radius (R)
a.	249.896 mm^3	7.00 mm	
b.	132.634 mm^3	12.00 mm	
c.	14.00 cu in	29.00 in	
d.	10.00 cu in	28.00 in	

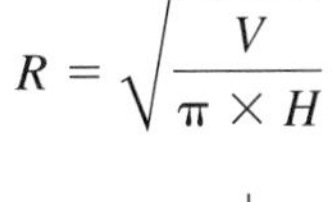

$$R = \sqrt{\frac{V}{\pi \times H}}$$

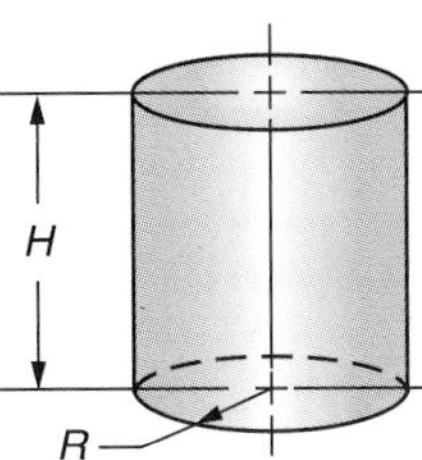

31. The volumes of cones and their heights are given in the following table. Compute the lengths of the diameters to 2 decimal places.

	Volume (V)	Height (H)	Diameter (D)
a.	116.328 mm^3	8.00 mm	
b.	19.388 cu in	2.00 in	
c.	1257.6 mm^3	10.00 mm	
d.	15 cu in	50.00 in	

$$D = \sqrt{\frac{V}{0.262 \times H}}$$

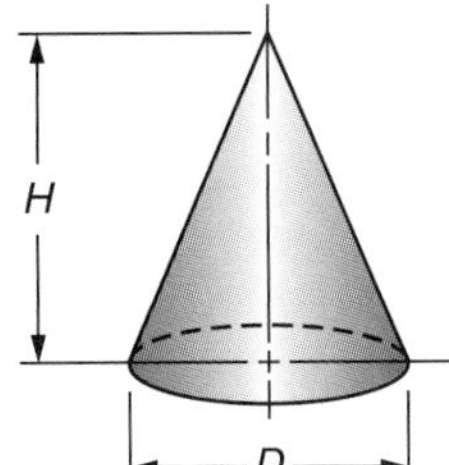

Solve the following problems.

32. The pitch of broach teeth depends upon the length of cut, the depth of cut, and the material being broached.

Minimum Pitch $= 3 \times \sqrt{L \times d \times F}$ where L = length of cut
d = depth of cut
F = a factor related to the type of material being broached

Find the minimum pitch, to 3 decimal places, for broaching cast iron where $L = 0.825''$, $d = 0.007''$, and $F = 5$. __________

33. The dimensions of keys and keyways are determined in relation to the diameter of the shafts with which they are used.

$$D = \sqrt{\frac{L \times T}{0.3}}$$

where D = Shaft diameter
L = key length
T = key thickness

What is the shaft diameter that would be used with a key where L = 2.70" and T = 0.25"? ____________

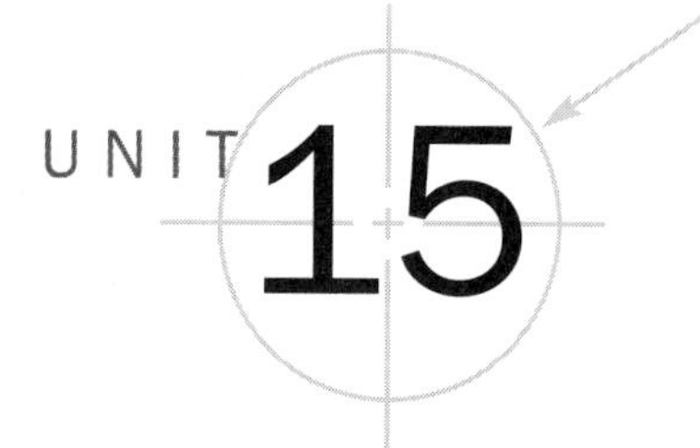

Table of Decimal Equivalents and Combined Operations of Decimal Fractions

Objectives ***After studying this unit you should be able to***

- Write decimal or fraction equivalents using a decimal equivalent table.
- Determine nearer fraction equivalents of decimals by using the decimal equivalent table.
- Solve problems consisting of combinations of operations by applying the order of operations.

Generally, fractional engineering drawing dimensions are given in multiples of 64ths of an inch. A machinist is often required to express these fractional dimensions as decimal equivalents for machine settings. When laying out parts such as castings that have ample stock allowances, it is sometimes convenient to use a fractional steel scale and to express decimal dimensions to the nearer equivalent fractions. The amount of computation and the chances of error can be reduced by using the decimal equivalent table.

Table of Decimal Equivalents

Using a decimal equivalent table saves time and reduces the chance of error. Decimal equivalent tables are widely used in the manufacturing industry. They are posted as large wall charts in work areas and are carried as pocket-size cards. Skilled workers memorize many of the equivalents after using decimal equivalent tables.

The decimals listed in the table are given to six places. For actual on-the-job uses, a decimal is rounded to the degree of precision required for a particular application.

DECIMAL EQUIVALENT TABLE			
1/64 ---- 0.015625	17/64 --- 0.256625	33/64 --- 0.515625	49/64 --- 0.765625
1/32 ---------- 0.03125	9/32 --------- 0.28125	17/32 --------- 0.53125	25/32 --------- 0.78125
3/64 ---- 0.046875	19/64 --- 0.296875	35/64 --- 0.546875	51/64 --- 0.796875
1/16 ------------ 0.0625	5/16 ----------- 0.3125	9/16 ----------- 0.5625	13/16 ----------- 0.8125
5/64 ---- 0.078125	21/64 --- 0.328125	37/64 --- 0.578125	53/64 --- 0.828125
3/32 ---------- 0.09375	11/32 --------- 0.34375	19/32 --------- 0.59375	27/32 -------- 0.84375
7/64 ---- 0.109375	23/64 --- 0.359375	39/64 --- 0.609375	55/64 --- 0.859375
1/8 -------------------- 0.125	3/8 ---------------------- 0.375	5/8 ---------------------- 0.625	7/8 ---------------------- 0.875
9/64 ---- 0.140625	25/64 --- 0.390625	41/64 --- 0.640625	57/64 --- 0.890625
5/32 ---------- 0.15625	13/32 --------- 0.40625	21/32 --------- 0.65625	29/32 --------- 0.90625
11/64 ---- 0.171875	27/64 --- 0.421875	43/64 --- 0.671875	59/64 --- 0.921875
3/16 ------------ 0.1875	7/16 ----------- 0.4375	11/16 ----------- 0.6875	15/16 ----------- 0.9375
13/64 ---- 0.203125	29/64 --- 0.453125	45/64 --- 0.703125	61/64 --- 0.953125
7/32 ---------- 0.21875	15/32 --------- 0.46875	23/32 --------- 0.71875	31/32 --------- 0.96875
15/64 ---- 0.234375	31/64 --- 0.484375	47/64 --- 0.734375	63/64 --- 0.984375
1/4 ---------------------- 0.25	1/2 ------------------------- 0.5	3/4 ----------------------- 0.75	1 ---------------------------- 1.

The following examples illustrate the use of the decimal equivalent table.

Example 1 Find the decimal equivalent of $\frac{23''}{32}$.

The decimal equivalent is shown directly to the right of the common fraction.

$$\frac{23''}{32} = 0.71875'' \quad \text{Ans}$$

Example 2 Find the fractional equivalent of 0.3125".

The fractional equivalent is shown directly to the left of the decimal fraction.

$$0.3125'' = \frac{5''}{16} \quad \text{Ans}$$

Example 3 Find the nearer fractional equivalents of the decimal dimensions given on the casting shown in Figure 15-1. All dimensions are in inches.

Compute dimension A. The decimal 0.757 lies between 0.750 and 0.765625. The difference between 0.757 and 0.750 is 0.007. The difference between 0.757 and 0.765625 is 0.008625. Since 0.007 is less than 0.008625, the 0.750 value is closer to 0.757.

The nearer fractional equivalent of 0.750" is $\frac{3''}{4}$. Ans

Compute dimension B. The decimal 0.978 lies between 0.96875 and 0.984375. The difference between 0.978 and 0.96875 is 0.00925. The difference between 0.978 and 0.984375 is 0.006375. Since 0.006375 is less than 0.00925, the 0.984375 value is closer to 0.978.

The nearer fractional equivalent of 0.984375" is $\frac{63''}{64}$. Ans

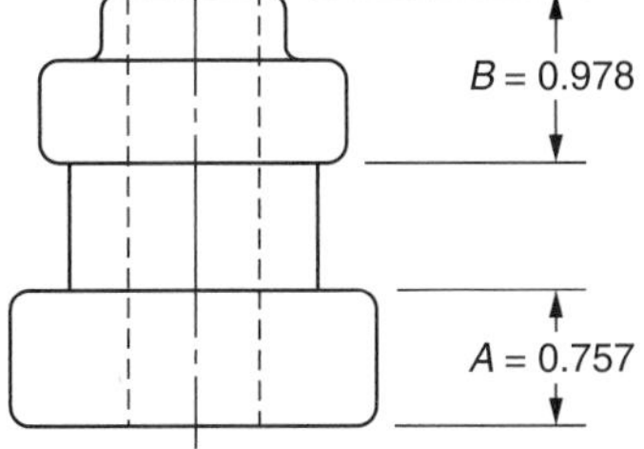

Figure 15-1

Combined Operations of Decimal Fractions

In the process of completing a job, a machinist must determine stock sizes, cutter sizes, feeds and speeds, and roughing allowances as well as cutting dimensions. Usually most and sometimes all of the fundamental operations of mathematics must be used for computations in the manufacture of a part.

Determination of powers and roots must also be considered in the order of operations. The following procedure incorporates all six fundamental operations. Study the following order of operations.

Order of Operations for Combined Operations of Addition, Subtraction, Multiplication, Division, Powers, and Roots

Procedure

- **Do all the work in parentheses first.** Parentheses are used to group numbers. In a problem expressed in fractional form, two or more numbers in the dividend (numerator) and/or divisor (denominator) should be considered as being enclosed in parentheses. For example, $\frac{4.87 + 0.34}{9.75 - 8.12}$ should be considered as $(4.87 + 0.34) \div (9.75 - 8.12)$. If an expression contains parentheses within parentheses or brackets, such as $[5.6 \times (7 - 0.09) + 8.8]$, do the work within the innermost parentheses first.
- **Do powers and roots next.** The operations are performed in the order in which they occur from left to right. If a root consists of two or more operations within the radical sign, perform all operations within the radical sign, then extract the root.
- **Do multiplication and division next.** The operations are performed in the order in which they occur from left to right.
- **Do addition and subtraction last.** The operations are performed in the order in which they occur from left to right.

Some people use the memory aid "**P**lease **E**xcuse **M**y **D**ear **A**unt **S**ally" to help them remember the order of operations. The **P** in "Please" stands for parentheses, the **E** for exponents (or raising to a power) and roots, **M** and **D** for multiplication and division, and the **A** and **S** for addition and subtraction.

Example 1 Find the value of $7.875 + 3.2 \times 4.3 - 2.73$.

Multiply. $7.875 + \underbrace{3.2 \times 4.3} - 2.73$

Add. $\underbrace{7.875 + 13.76} - 2.73$

Subtract. $21.635 - 2.73 = 18.905$ Ans

Example 2 Find the value of $(27.34 - 4.82) \div (2.41 \times 1.78 + 7.89)$. Round the answer to 2 decimal places.

Perform operations within parentheses.

Subtract. $\underbrace{(27.34 - 4.82)} \div (2.41 \times 1.78 + 7.89)$

Multiply. $22.52 \div (\underbrace{2.41 \times 1.78} + 7.89)$

Add. $22.52 \div \underbrace{(4.2898 + 7.89)}$

Divide. $22.52 \div 12.1798 = 1.84896 = 1.85$ Ans

Example 3 Find the value of $\frac{13.79 + (27.6 \times 0.3)^2}{\sqrt{23.04} + 0.875 - 3.76}$.

Round the answer to 3 decimal places.

Grouping symbol operations is done first. Consider the numerator and the denominator as if each were within parentheses. All of the operations are performed in the numerator and in the denominator before the division is performed.

$$\frac{13.79 + (27.6 \times 0.3)^2}{\sqrt{23.04} + 0.875 - 3.76}$$

Rewrite.	$\left[13.79 + (27.6 \times 0.3)^2\right] \div \left(\sqrt{23.04} + 0.875 - 3.76\right)$
Multiply.	$\left[13.79 + \underbrace{(27.6 \times 0.3)}^2\right] \div \left(\sqrt{23.04} + 0.875 - 3.76\right)$
Square.	$\left[13.79 + \underbrace{(8.28)^2}\right] \div \left(\sqrt{23.04} + 0.875 - 3.76\right)$
Extract the square root.	$[13.79 + 68.5584] \div \left(\underbrace{\sqrt{23.04}} + 0.875 - 3.76\right)$
Add.	$\underbrace{[13.79 + 68.5584]} \div (4.8 + 0.875 - 3.76)$
Add.	$82.3484 \div (\underbrace{4.8 + 0.875} - 3.76)$
Subtract.	$82.3484 \div \underbrace{(5.675 - 3.76)}$
Divide.	$82.3484 \div 1.915 = 43.00178$
	$= 43.002$ Ans

Example 4 Blanks in the shape of regular pentagons (5-sided figures) are punched from strip stock as shown in Figure 15-2. Determine the width of strip stock required, using the given dimensions and the formula for dimension R. Round the answer to 3 decimal places. All dimensions are in inches.

$$\text{Width} = R + 0.980 + 2 \times 0.125 \quad \text{where } R = \sqrt{r^2 + s^2 \div 4}$$

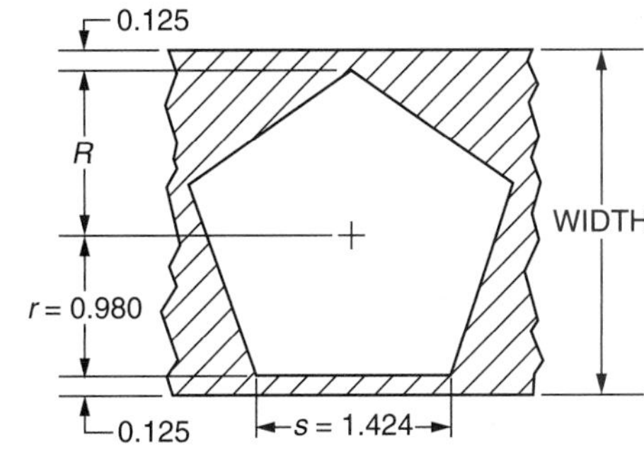

Figure 15-2

$$\sqrt{0.980^2 + 1.424^2 \div 4} + 0.980 + 2 \times 0.125$$

Substitute the given values.

Square.	$\sqrt{\underbrace{0.980^2} + 1.424^2 \div 4} + 0.980 + 2 \times 0.125$
Square.	$\sqrt{0.9604 + \underbrace{1.424^2} \div 4} + 0.980 + 2 \times 0.125$
Divide.	$\sqrt{0.9604 + \underbrace{2.027776 \div 4}} + 0.980 + 2 \times 0.125$
Add.	$\underbrace{\sqrt{0.9604 + 0.506944}} + 0.980 + 2 \times 0.125$
Extract the square root.	$\underbrace{\sqrt{1.467344}} + 0.980 + 2 \times 0.125$
Multiply.	$1.211 + 0.980 + \underbrace{2 \times 0.125}$
Add.	$1.211 + 0.980 + 0.250 = 2.441$
	Width = 2.441 inches Ans

➤ **Note:** In solving expressions that consist of numerous multiplication and power operations, it is often necessary to carry out the work to two or three more decimal places than the number of decimal places required in the answer.

Application

Using the Decimal Equivalent Table

Find the fraction or decimal equivalents of these numbers using the decimal equivalent table.

1. $\frac{25}{32}$ ______________
2. $\frac{7}{32}$ ______________
3. $\frac{11}{32}$ ______________
4. $\frac{13}{16}$ ______________
5. $\frac{5}{64}$ ______________
6. 0.671875 ______________
7. 0.3125 ______________
8. 0.28125 ______________
9. 0.203125 ______________

Find the nearer fraction equivalents of these decimals using the decimal equivalent table.

10. 0.541 ______________
11. 0.762 ______________
12. 0.465 ______________
13. 0.498 ______________
14. 0.209 ______________
15. 0.805 ______________

Combined Operations of Decimal Fractions

Solve these examples of combined operations. Round the answers to 2 decimal places where necessary.

16. $0.5231 + 10.375 \div 4.32 \times 0.521$ ______________
17. $81.07 \div 12.1 + 2 \times 3.7$ ______________
18. $\frac{56.050}{3.8} \times 0.875 - 3.92$ ______________
19. $(24.78 - 19.32) \times 4.6$ ______________
20. $(14.6 \div 4 - 1.76)^2 \times 4.5$ ______________
21. $27.16 \div \sqrt{1.76 + 12.32}$ ______________
22. $(\sqrt{3.98 + 0.87 \times 3.9})^2$ ______________
23. $(3.29 \times 1.7)^2 \div (3.82 - 0.86)$ ______________
24. $0.25 \times \left(\frac{\sqrt{64} \times 3.87}{8.32 \times 5.13}\right) + 18.3^2$ ______________
25. $18.32 - \sqrt{\frac{7.86 \times 13.5}{3.5^2 - 0.52}} \times 0.7$ ______________

Solve the following problems by using combined operations.

26. Figure 15-3 shows the three-wire method of checking screw threads. With proper diameter wires and a micrometer, very accurate pitch diameter measurements can be made. Using the formula given, determine the micrometer dimension over wires of the American (National) Standard threads in the following table. Round the answer to 4 decimal places.

$$M = D - (1.5155 \times P) + (3 \times W)$$

	Major Diameter D (inches)	Pitch P (inches)	Wire Diameter W (inches)	Dimension Over Wires M (inches)
a.	0.8750	0.1250	0.0900	
b.	0.2500	0.0500	0.0350	
c.	0.6250	0.1000	0.0700	
d.	1.3750	0.16667	0.1500	
e.	2.5000	0.2500	0.1500	

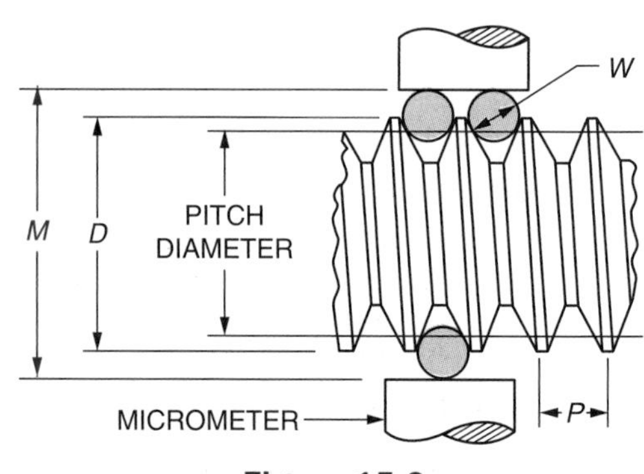

Figure 15-3

27. A bronze bushing with a diameter of 22.225 millimeters is to be pressed into a mounting plate. The assembly print calls for a bored hole in the plate to be 0.038 millimeter less in diameter than the bushing diameter. The hole diameter in the plate checks 22.103 millimeters. By how much must the diameter of the plate hole be increased in order to meet the print specification? __________

28. A stamped sheet steel plate is shown in Figure 15-4. Compute dimensions A–F to 3 decimal places. All dimensions are in inches.

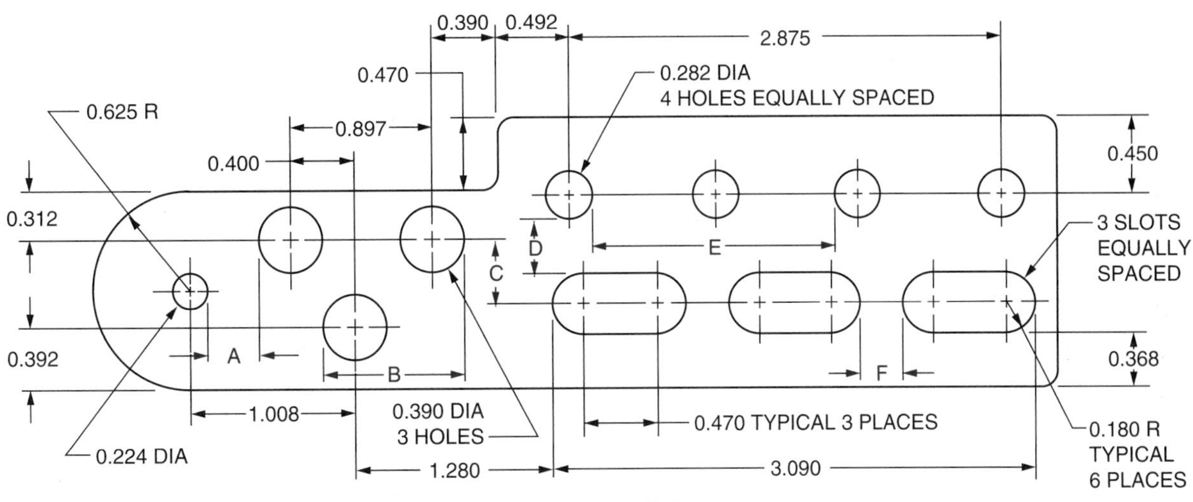

Figure 15-4

A = __________ C = __________ E = __________

B = __________ D = __________ F = __________

29. A flat is to be milled in three pieces of round stock each of a different diameter. The length of the flat is determined by the diameter of the stock and the depth of cut. The table gives the required length of flat and the stock diameter for each piece. Determine the depth of cut for each piece to 2 decimal places using this formula.

$$C = \frac{D}{2} - 0.5 \times \sqrt{4 \times \left(\frac{D}{2}\right)^2 - F^2}$$

	Diameter *D*	Length of Flat *F*	Depth of Cut *C*
a.	34.80 mm	30.50 mm	
b.	55.90 mm	40.60 mm	
c.	91.40 mm	43.40 mm	

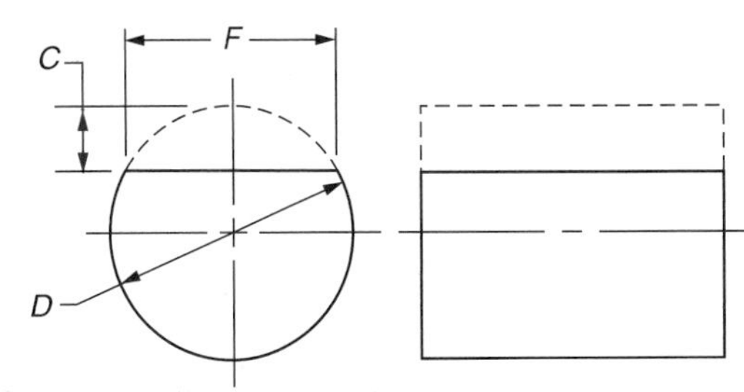

30. A slot is machined in a circular plate with a 41.36-millimeter diameter. Two milling cuts, one 6.30 millimeters deep and the other 3.15 millimeters, are made. A grinding operation then removes 0.40 millimeter. What is the distance from the center of the plate to the bottom of the slot? All dimensions are in millimeters. ______

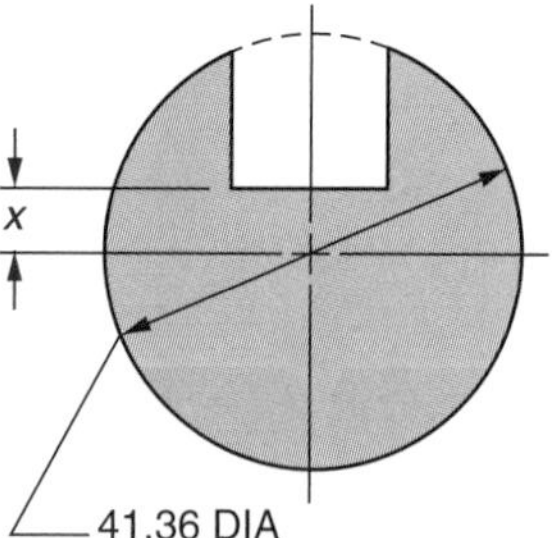

31. A 60° slot has been machined in a fixture. The slot is checked by placing a pin in the slot and indicating the distance between the top of the fixture and the top of the pin as shown. Compute distance H to 3 decimal places by using this formula. All dimensions are in inches. ______

$$H = 1.5 \times D - 0.866 \times W$$

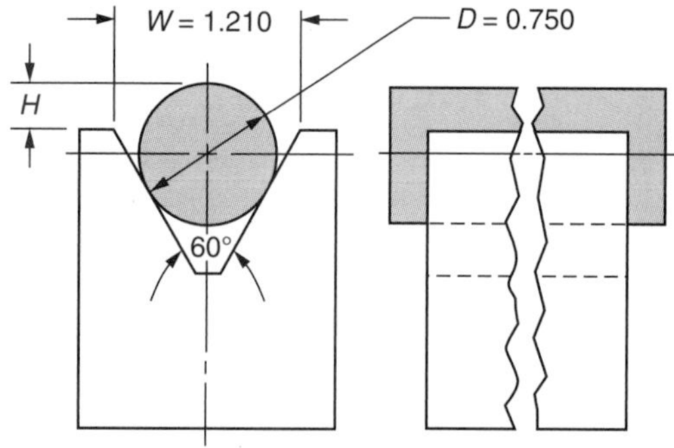

UNIT 16

Computing with a Calculator: Decimals

Objectives ***After studying this unit you should be able to***

- Perform individual operations of addition, subtraction, multiplication, division, powers, and roots with decimals using a calculator.
- Perform combinations of operations with decimals using a calculator.

Decimals

The decimal point key ([·]) is used when entering decimal values in a calculator. When entering a decimal fraction in a calculator, the decimal point key is pressed at the position of the decimal point in the number. For example, to enter the number 0.732, first press [·] and then enter 732. To enter the number 567.409, enter 567 [·] 409.

In calculator examples and illustrations of operations with decimals in this text, the decimal key [·] will *not* be shown to indicate the entering of a decimal point. Wherever the decimal point occurs in a number, it is understood that the decimal point key [·] is pressed.

Decimals with Basic Operations of Addition, Subtraction, Multiplication, and Division

Example 1 Add. 19.37 + 123.9 + 7.04

19.37 [+] 123.9 [+] 7.04 [=] 150.31 Ans

Example 2 Subtract. 2876.78 − 405.052

2876.78 [−] 405.052 [=] 2471.728 Ans

Example 3 Multiply. 427.935 × 0.875 × 93.400 (round answer to 1 decimal place)

427.935 [×] .875 [×] 93.4 [=] 34972.988

34,973.0 Ans

Notice that the two zeros following the 4 are not entered. The final zero or zeros to the right of the decimal point may be omitted.

Notice that the zero to the left of the decimal point is not entered. The leading zero is omitted.

Example 4 Divide. 813.7621 ÷ 6.466 (round answer to 3 decimal places)

813.7621 [÷] 6.466 [=] 125.85247

125.852 Ans

Powers

Expressions involving powers and roots are readily computed with a scientific calculator. The square key is used to raise a number to the second power (to square a number).

To square a number, enter the number, press the square key ([x^2]), and press [=] or [ENTER].

Example To calculate 28.75^2, enter 28.75, press [x^2], and press [=].

28.75 [x^2] [=] 826.5625 Ans

The universal power key, [x^y] or [∧], depending on the calculator used, raises a number to a power. To raise a number to a power using the universal power key, do the following:

Enter the number to be raised to a power.

Press the universal power key, [x^y] or [∧].

Enter the power.

Press the [=] or [ENTER] key.

Examples

1. Calculate 15.72^3. Enter 15.72, press [x^y] or [∧], enter 3, and press [=].
   15.72 [x^y] (or [∧]) 3 [=] 3884.7012 Ans
2. Calculate 0.95^7
   .95 [x^y] (or [∧]) 7 [=] 0.6983373 Ans

Roots

To obtain the square root of any positive number, the square root key ([√]) is used.

Procedure To obtain the square root of a positive number

- Press the square root key ([√]), enter the number, and press [=] or [ENTER].

Example 1 Calculate $\sqrt{27.038}$.

Press [√], enter 27.038, and press [=].

[√] 27.038 [=] 5.199807689 Ans

If the square root key is an alternate function, press [2nd] first.

Example 2 Calculate $\sqrt{27.038}$.

[2nd] [√] 27.038 [=] 5.199807689 Ans

The root of any positive number can be computed with a calculator. Generally, roots are alternate functions.

On calculators with an [x^y] key, press [SHIFT] [x^y] to take the root.

Example Calculate $\sqrt[5]{475.19}$.

5 [SHIFT] [x^y] 475.19 [=] 3.430626662 Ans

Procedure To use the alternate function

- Press [2nd] (or [SHIFT]) directly before pressing the root key ([$\sqrt[x]{\ }$]).

Example Calculate $\sqrt[5]{475.19}$.

5 [2nd] (or [SHIFT]) [$\sqrt[x]{\ }$] 475.19 [=] 3.430626662 Ans

Some calculators automatically insert a left parenthesis when the √ key is pressed. If your calculator does this, then you need to place a right parenthesis to show you are finished taking the root.

Example Evaluate. $\sqrt{67.24} - 5$

Incorrect: [√] 67.24 [−] 5 [=] 7.889233169

What you see on the calculator: √ (67.24 − 5
7.889233169

Correct: [√] 67.24 [)] [−] 5 [=] 3.2, 3.2 Ans

What you see on the calculator: √ (67.24) − 5
3.2

Practice Exercises, Individual Basic Operations

Evaluate the following expressions. The expressions are basic arithmetic operations including powers and roots. Remember to check your answers by doing each problem twice. The solutions to the problems directly follow the practice exercises. Compare your answers to the given solutions. Round each answer to the indicated number of decimal places.

1. $276.84 + 312.094$ (2 places)
2. $16.09 + 0.311 + 5.516$ (1 place)
3. $6704.568 - 4989.07$ (2 places)
4. $0.9244 - 0.0822$ (3 places)
5. 43.4967×6.0913 (4 places)
6. $8.503 \times 0.779 \times 13.248$ (3 places)
7. $54.419 \div 6.7$ (1 place)
8. $0.9316 \div 0.0877$ (4 places)
9. 36.22^2 (2 places)
10. 7.063^5 (1 place)
11. $\sqrt{28.73721}$ (4 places)
12. $\sqrt[5]{1,068.470}$ (3 places)

Solutions to Individual Basic Operations

1. 276.84 [+] 312.094 [=] 588.934, 588.93 Ans
2. 16.09 [+] .311 [+] 5.516 [=] 21.917, 21.9 Ans
3. 6704.568 [−] 4989.07 [=] 1715.498, 1,715.50 Ans
4. .9244 [−] .0822 [=] 0.8422, 0.842 Ans
5. 43.4967 [×] 6.0913 [=] 264.9514487, 264.9514 Ans
6. 8.503 [×] .779 [×] 13.248 [=] 87.752593, 87.753 Ans
7. 54.419 [÷] 6.7 [=] 8.1222388, 8.1 Ans
8. .9316 [÷] .0877 [=] 10.622577, 10.6226 Ans
9. 36.22 [x^2] [=] 1311.8884, 1311.89 Ans
10. 7.063 [x^y] (or [∧]) 5 [=] 17577.052, 17,577.1 Ans
11. [$\sqrt{\ }$] 28.7321 [=] 5.3607098, 5.3607 Ans
12. 5 [2nd] (or [SHIFT]) [$\sqrt[x]{\ }$] 1068.47 [=] 4.03415394, 4.034 Ans

Combined Operations

The expressions are solved by entering numbers and operations into the calculator in the same order as the expressions are written.

Examples

1. Evaluate. $30.75 + 15 \div 4.02$ (round answer to 2 decimal places)

   30.75 [+] 15 [÷] 4.02 [=] 34.481343, 34.48 Ans

2. Evaluate. $51.073 - \frac{4}{0.091} + 33.151 \times 2.707$ (round answer to 2 decimal places)

   51.073 [−] 4 [÷] .091 [+] 33.151 [×] 2.707 [=] 96.856713, 96.86 Ans

3. Evaluate. $46.23 + (5 + 6.92) \times (56.07 - 38.5)$

   As previously discussed in the order of operations, operations enclosed within parentheses are performed first. Calculators having algebraic logic perform the operations within parentheses before performing other operations in combined operations expressions. If an expression contains parentheses, enter the expression into the calculator in the order in which it is written. The parentheses keys [(] and [)] must be used.

   46.23 [+] (5 [+] 6.92 [)] [×] [(] 56.07 [−] 38.51 [)] [=] 255.6644 Ans

4. Evaluate. $\frac{13.463 + 9.864 \times 6.921}{4.373 + 2.446}$ (round answer to 3 decimal places)

Recall that for problems expressed in fractional form, the fraction bar is also used as a grouping symbol. The numerator and denominator are each considered as being enclosed in parentheses.

$$(13.463 + 9.864 \times 6.921) \div (4.373 + 2.446)$$

[(] 13.463 [+] 9.864 [×] 6.921 [)] [÷] [(] 4.373 [+] 2.446 [)] [=]
11.985884, 11.986 Ans

The expression may also be evaluated by using the [=] key to simplify the numerator without having to enclose the entire numerator in parentheses. However, parentheses must be used to enclose the denominator.

13.463 [+] 9.864 [×] 6.921 [=] [÷] [(] 4.373 [+] 2.446 [)] [=] 11.985884, 11.986 Ans

5. Evaluate. $\dfrac{100.32 - (16.87 + 13)}{111.36 - 78.47}$ (round answer to 2 decimal places)

$$\frac{100.32 - (16.87 + 13)}{111.36 - 78.47} = (100.32 - (16.87 + 13)) \div (111.36 - 78.47)$$

Observe these parentheses

To be sure that the complete numerator is evaluated before dividing by the denominator, enclose the complete numerator within parentheses. This is an example of an expression containing parentheses within parentheses.

[(] 100.32 [−] [(] 16.87 [+] 13 [)] [)] [÷] [(] 111.36 [−] 78.47 [)] [=] 2.1419884, 2.14 Ans

Using the [=] key to simplify the numerator:

100.32 [−] [(] 16.87 [+] 13 [)] [=] [÷] [(] 111.36 [−] 78.47 [)] [=] 2.1419884, 2.14 Ans

6. Evaluate. $\dfrac{873.03 + 12.12^3 \times 41}{\sqrt{16.43} - 266.76 \div 107.88}$ (round answer to 2 decimal places)

[(] 873.03 [+] 12.12 [x^y] 3 [×] 41 [)] [÷] [(] [√] (or [2nd] [√]) 16.43 [−] 266.76 [÷] 107.88 [)] [=] 46732.658, 46,732.66 Ans

Using the [=] key to simplify the numerator:

873.03 [+] 12.12 [x^y] 3 [×] 41 [=] [÷] [(] [√] (or [2nd] [√]) 16.43 [−] 266.76 [÷] 107.88 [)] [=]
46732.658, 46,732.66 Ans

7. Evaluate. $[4.73 + (0.24 - 5.16)^2] + \sqrt{12.45}$ (round answer to 3 decimal places)

Use the parenthesis keys when there are brackets or braces.

[(] 4.73 [+] [(] .24 [−] 5.16 [)] [x^2] [)] [+] [√] 12.45 [=] 32.46485575, 32.465 Ans

Practice Exercises, Combined Operations

Evaluate the following combined operations expressions. Remember to check your answers by doing each problem twice. The solutions to the problems directly follow the practice exercises. Compare your answers to the given solutions. Round each answer to the indicated number of decimal places.

1. $503.97 - 487.09 \times 0.777 + 65.14$ (2 places)
2. $27.028 + \dfrac{5}{6.331} - 5.875 \times 1.088$ (3 places)
3. $23.073 \times (0.046 + 5.934 - 3.049) - 17.071$ (3 places)
4. $30.180 \times (0.531 + 12.939 - 2.056) - 60.709$ (3 places)
5. $\dfrac{643.72 - 18.192 \times 0.783}{470.07 - 88.33}$ (2 places)

6. $\dfrac{793.32 - 2.67 \times 0.55}{107.9 + 88.93}$ (1 place)
7. $2{,}446 + 8.917^3 \times 5.095$ (3 places)
8. $679.07 + (36 + 19.973 - 0.887)^2 \times 2.05$ (1 place)
9. $43.71 - \sqrt{256.33 - 107} + 17.59$ (2 places)
10. $\dfrac{\sqrt[5]{14.773} + 93.977 \times \sqrt[3]{282.608}}{3.033}$ (3 places)
11. $\dfrac{1{,}202.03\sqrt[3]{706.8} - 44.317}{(14.03 \times 0.54 - 2.08)^2} - 2.63$ (1 place)

Solutions to Practice Exercises, Combined Operations

1. 503.97 [−] 487.09 [×] .777 [+] 65.14 [=] 190.64107, 190.64 Ans
2. 27.028 [+] 5 [÷] 6.331 [−] 5.875 [×] 1.088 [=] 21.425765, 21.426 Ans
3. 23.073 [×] [(] .046 [+] 5.934 [−] 3.049 [)] [−] 17.071 [=] 50.555963, 50.556 Ans
4. 30.180 [×] [(] .531 [+] 12.939 [−] 2.056 [)] [−] 60.709 [=] 283.76552, 283.766 Ans
5. [(] 634.72 [−] 18.192 [×] .783 [)] [÷] [(] 470.07 [−] 88.33 [)] [=] 1.6489644, 1.65 Ans
   or 643.72 [−] 18.192 [×] .783 [=] [÷] [(] 470.07 [−] 88.33 [)] [=] 1.6489644, 1.65 Ans
6. [(] 793.32 [−] 2.67 [×] .55 [)] [÷] [(] 107.9 [+] 88.93 [)] [=] 4.0230224, 4.0 Ans
   or 793.32 [−] 2.67 [×] .55 [=] [÷] [(] 107.9 [+] 88.93 [)] [=] 4.0230224, 4.0 Ans
7. 2446 [+] 8.917 [x^y] (or [∧]) 3 [×] 5.095 [=] 6058.4387, 6,058.438 Ans
8. 679.07 [+] [(] 36 [+] 19.973 [−] .887 [)] [x^2] [×] 2.05 [=] 6899.7282, 6,899.7 Ans
9. 43.71 [−] [√] [(] 256.33 − 107 [)] [+] 17.59 [=] 49.079935, 49.08 Ans
10. [(] 5 [2nd] (or [SHIFT]) [$\sqrt[x]{\ }$] 14.773 [+] 93.977 [×] 3 [2nd] (or [SHIFT]) [$\sqrt[x]{\ }$] 282.608 [)] [÷] 3.033 [=] 203.89927, 203.899 Ans
11. [(] 1202.03 [×] 3 [2nd] (or [SHIFT]) [$\sqrt[x]{\ }$] [(] 706.8 [−] 44.317 [)] [)] [÷] [(] 14.03 [×] .54 [−] 2.08 [)] [x^2] [−] 2.63 [=] 344.25205, 344.3 Ans

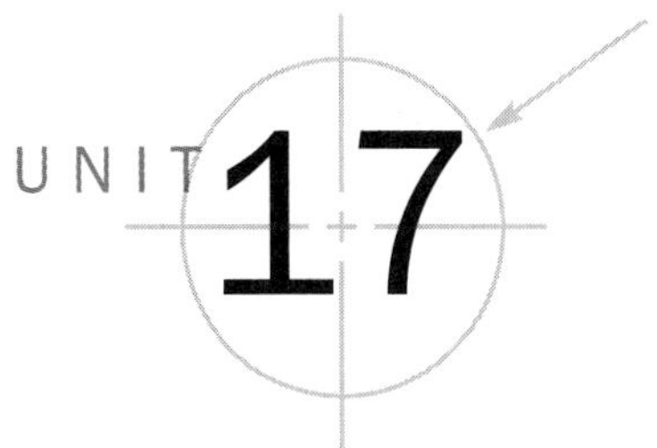

Achievement Review—Section One

Objective *You should be able to solve the exercises and problems in this Achievement Review by applying the principles and methods covered in Units 1–16.*

1. Express each of the following fractions as equivalent fractions as indicated.

   a. $\dfrac{3}{8} = \dfrac{?}{32}$ ________ c. $\dfrac{1}{4} = \dfrac{?}{64}$ ________

   b. $\dfrac{7}{10} = \dfrac{?}{100}$ ________ d. $\dfrac{9}{16} = \dfrac{?}{128}$ ________

2. Express each of the following mixed numbers as improper fractions.

a. $3\frac{1}{5}$ ______ d. $13\frac{3}{8}$ ______

b. $2\frac{9}{10}$ ______ e. $6\frac{9}{32}$ ______

c. $5\frac{3}{4}$ ______

3. Express each of the following improper fractions as mixed numbers.

a. $\frac{5}{2}$ ______ d. $\frac{115}{32}$ ______

b. $\frac{21}{5}$ ______ e. $\frac{329}{64}$ ______

c. $\frac{75}{4}$ ______

4. Express each of the following fractions as a fraction in lowest terms.

a. $\frac{8}{16}$ ______ d. $\frac{18}{64}$ ______

b. $\frac{12}{100}$ ______ e. $\frac{28}{128}$ ______

c. $\frac{30}{32}$ ______

5. Express the fractions in each of the following sets as equivalent fractions having the least common denominator.

a. $\frac{1}{4}, \frac{3}{16}, \frac{9}{32}$ ______ c. $\frac{7}{10}, \frac{3}{4}, \frac{9}{25}, \frac{13}{20}$ ______

b. $\frac{7}{16}, \frac{5}{32}, \frac{9}{64}$ ______

6. Add or subtract each of the following values. Express the answers in lowest terms.

a. $\frac{1}{8} + \frac{5}{8}$ ______ f. $\frac{11}{16} - \frac{7}{16}$ ______

b. $\frac{7}{16} + \frac{15}{16}$ ______ g. $\frac{17}{20} - \frac{3}{5}$ ______

c. $\frac{5}{8} + \frac{13}{32}$ ______ h. $\frac{49}{64} - \frac{3}{8}$ ______

d. $3\frac{7}{10} + \frac{49}{100}$ ______ i. $6 - \frac{11}{16}$ ______

e. $\frac{9}{32} + \frac{1}{4} + \frac{21}{64}$ ______ j. $13\frac{1}{8} - 9\frac{7}{32}$ ______

7. Multiply or divide each of the following values. Express the answers in lowest terms.

a. $\frac{1}{2} \times \frac{5}{8}$ ______ d. $3\frac{1}{10} \times 8\frac{1}{4}$ ______

b. $\frac{3}{4} \times \frac{4}{5} \times \frac{2}{3}$ ______ e. $\frac{3}{16} \times 20 \times 5\frac{1}{2}$ ______

c. $5\frac{7}{32} \times \frac{3}{8}$ ______ f. $\frac{3}{10} \div \frac{2}{5}$ ______

g. $\frac{14}{15} \div \frac{7}{25}$ __________

h. $16 \div \frac{1}{3}$ __________

i. $2\frac{17}{32} \div \frac{9}{24}$ __________

j. $2\frac{29}{32} \div 8\frac{3}{4}$ __________

8. Perform each of the indicated combined operations.

a. $\frac{3}{4} + \frac{5}{16} - \frac{3}{8}$ __________

b. $20\frac{1}{2} + 3\frac{3}{8} \times \frac{1}{8}$ __________

c. $\frac{3}{5} \times \frac{7}{8} + 3\frac{3}{4}$ __________

d. $\left(18 - 5\frac{3}{4}\right) \div \frac{1}{2} + 3\frac{7}{8}$ __________

e. $\dfrac{16 - 4\frac{1}{2}}{\frac{1}{2} + 2\frac{1}{8}}$ __________

f. $\dfrac{5\frac{1}{4} \times \frac{1}{2}}{6 \div 3\frac{3}{4}}$ __________

9. How many complete pieces can be blanked from a strip of aluminum 72 inches long if each stamping requires $1\frac{3}{8}$ inches of material plus an allowance of $\frac{3}{4}$ inch at one end of the strip? __________

10. How many inches of bar stock are needed to make 30 spacers each $1\frac{3}{16}$ inches long? Allow $\frac{1}{8}$ inch waste for each spacer. __________

11. A shaft is turned at 200 revolutions per minute with a tool feed of $\frac{1}{32}$ inch per revolution. How many minutes does it take to cut a distance of 50 inches along the shaft? __________

12. A shop order calls for 1800 steel pins each $1\frac{5}{8}$ inches long. If $\frac{3}{16}$ inch is allowed for cutting off and facing each pin, how many complete 10-foot lengths of stock are needed for the order? __________

13. Compute dimensions A, B, C, D, and E of the support bracket shown in Figure 17-1. All dimensions are given in inches.

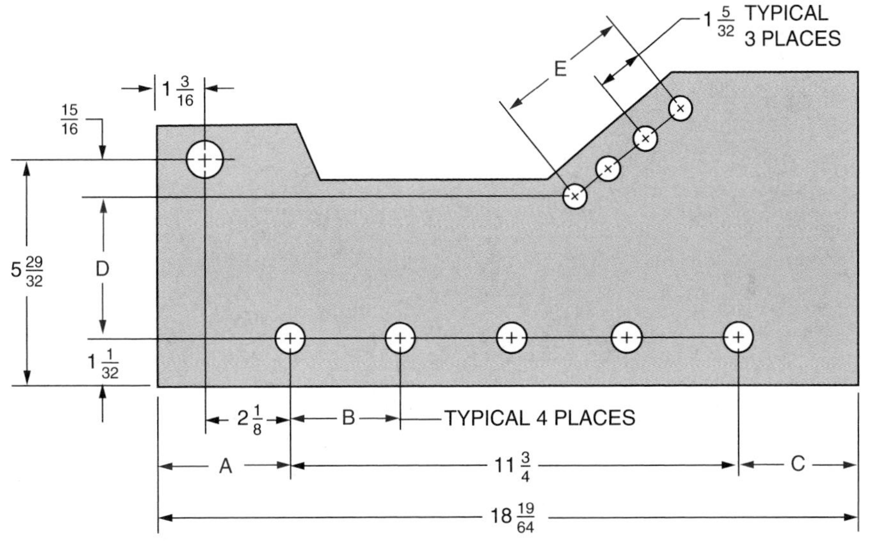

Figure 17-1

A = __________

B = __________

C = __________

D = __________

E = __________

14. Write each of the following numbers as words.

a. 0.6 ____________
b. 0.74 ____________
c. 0.147 ____________
d. 0.0086 ____________
e. 4.208 ____________
f. 16.0419 ____________

15. Write each of the following words as decimal fractions or mixed decimals.

a. three tenths ________
b. twenty-six thousandths ________
c. nine and twenty-six thousandths ________
d. five and eighty-one ten-thousandths ________

16. Round each of the following numbers to the indicated number of decimal places.

a. 0.596 (2 places) ________
b. 5.0463 (3 places) ________
c. 0.80729 (4 places) ________
d. 7.0005 (3 places) ________

17. Express each of the following common fractions as decimal fractions. Where necessary, round the answers to 3 decimal places.

a. $\frac{3}{4}$ ________
b. $\frac{7}{8}$ ________
c. $\frac{2}{3}$ ________
d. $\frac{2}{25}$ ________
e. $\frac{13}{20}$ ________

18. Express each of the following decimal fractions as common fractions in lowest terms.

a. 0.7 ________
b. 0.525 ________
c. 0.007 ________
d. 0.915 ________
e. 0.0075 ________

19. Add or subtract each of the following values.

a. 0.875 + 0.712 ________
b. 5.004 + 0.92 + 0.5034 ________
c. 0.006 + 12.3 + 0.0009 ________
d. 2.99 + 6.015 + 0.1003 ________
e. 23 + 0.0007 + 0.007 + 0.4 ________
f. 0.879 − 0.523 ________
g. 0.1863 − 0.0419 ________
h. 5.400 − 5.399 ________
i. 0.009 − 0.0068 ________
j. 14.001 − 13.999 ________

20. Multiply or divide each of the following values. Round the answer to 4 decimal places where necessary.

a. 0.923 × 0.6 ________
b. 3.63 × 2.30 ________
c. 4.81 × 0.07 ________
d. 0.005 × 0.180 ________
e. 12.123 × 0.001 ________
f. 0.85 ÷ 0.39 ________
g. 0.100 ÷ 0.01 ________
h. 4.016 ÷ 0.03 ________
i. 123 ÷ 0.665 ________
j. 0.0098 ÷ 5.036 ________

21. Raise each of the following values to the indicated powers.

a. 2.6^2 ________

b. 0.50^3 ________

c. 0.006^2 ________

d. $\left(\frac{3}{5}\right)^2$ ________

e. $\left(\frac{20.8}{6.5}\right)^3$ ________

22. Determine the roots of each of the following values as indicated.

a. $\sqrt{49}$ ________

b. $\sqrt[3]{64}$ ________

c. $\sqrt{\frac{36}{81}}$ ________

d. $\sqrt{46.83 + 17.17}$ ________

e. $\sqrt{39.2 \times 1.25}$ ________

23. Determine the square roots of each of the following values to the indicated number of decimal places.

a. $\sqrt{379}$ (2 places) ________

b. $\sqrt{0.8736}$ (3 places) ________

c. $\sqrt{\frac{2}{5}}$ (3 places) ________

d. $\sqrt{93.876 - 47.904}$ (3 places) ________

24. Find the decimal or fraction equivalents of each of the following numbers using the decimal equivalent table.

a. $\frac{5}{8}$ ________

b. $\frac{17}{32}$ ________

c. $\frac{21}{64}$ ________

d. 0.65625 ________

e. 0.671875 ________

25. Determine the nearer fractional equivalents of each of the following decimals using the decimal equivalent table.

a. 0.465 ________

b. 0.769 ________

c. 0.038 ________

d. 0.961 ________

26. Solve each of the following combined operations expressions. Round answers to 2 decimal places.

a. $0.4321 + 10.870 \div 3.43 \times 0.93$ ________

b. $(12.60 \div 3 - 0.98)^2 \times 3.60$ ________

c. $35.98 \div \sqrt{6.35 - 4.81}$ ________

d. $6 \times \left(\frac{\sqrt{81} \times 4.03}{3.30 \times 2.75}\right) - 1.7^2$ ________

27. The basic form of an ISO Metric Thread is shown in Figure 17-2. Given a thread pitch of 1.5 millimeters, compute thread dimensions A, B, C, D, E, and F to 3 decimal places.

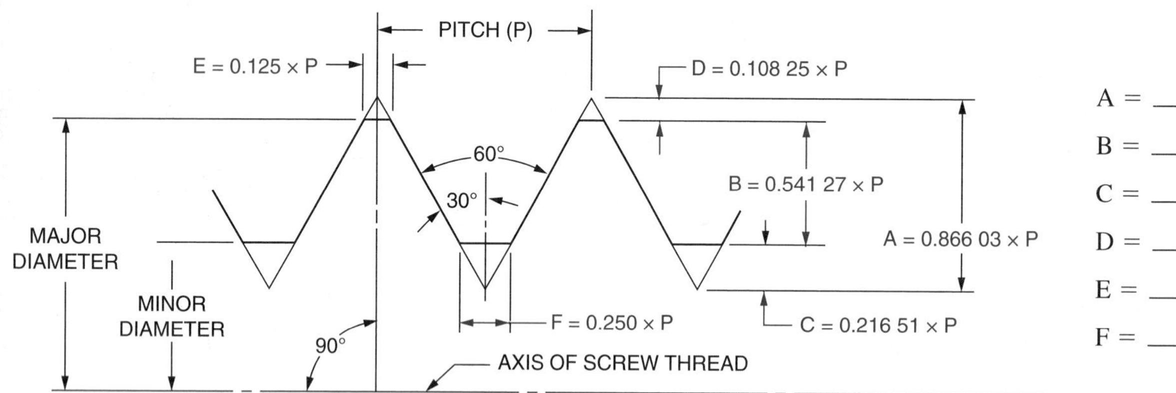

Figure 17-2

A = ______

B = ______

C = ______

D = ______

E = ______

F = ______

28. A combination of gage blocks is selected to provide a total thickness of 0.4573 inch. One block 0.250 inch and one block 0.118 inch thick are selected. What is the required thickness of the remaining blocks? ______

29. A piece of round stock is being turned to a 17.86-millimeter diameter. A machinist measures the diameter of the piece as 18.10 millimeters. What depth of cut should be made to turn the piece to the required diameter? ______

30. A plate 57.20 millimeters thick is to be machined to a thickness of 44.10 millimeters. The plate is to be rough cut with the last cut a finish cut 0.30 millimeter deep. If each rough cut is 3.20 millimeters deep, how many rough cuts are required? ______

31. A shaft is turned in a lathe at 120 revolutions per minute. The cutting tool advances 0.030 inch per revolution. How long is the length of cut along the shaft at the end of 3.50 minutes? ______

Ratio, Proportion, and Percentage

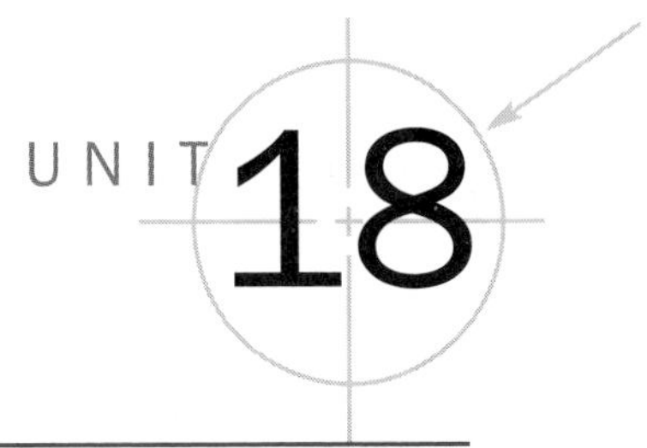

Ratio and Proportion

Objectives ***After studying this unit you should be able to***

- Write comparisons as ratios.
- Express ratios in lowest terms.
- Solve for the unknown term of a proportion.
- Substitute given numerical values for symbols in a proportion and solve for the unknown term.

The ability to solve practical machine shop problems using ratio and proportion is a requirement for the skilled machinist. Ratio and proportion are used for calculating gear and pulley speeds and sizes, for computing thread cutting values on a lathe, for computing taper dimensions, and for determining machine cutting times.

Description of Ratio

Ratio is the comparison of two like quantities.

Examples

1. Two pulleys are shown in Figure 18-1. What is the ratio of the diameter of the small pulley to the diameter of the larger pulley? All dimensions are in inches.

   The ratio is 3 to 5. Ans

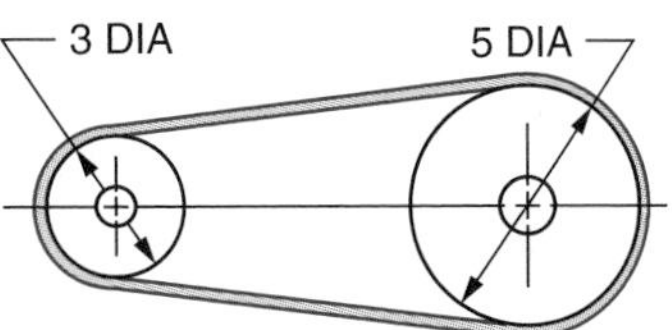

Figure 18-1

2. A triangle with given lengths of 3 meters, 4 meters, and 5 meters for sides a, b, and c is shown in Figure 18-2.

   a. What is the ratio of side a to side b?

      The ratio is 3 to 4. Ans

   b. What is the ratio of side a to side c?

      The ratio is 3 to 5. Ans

   c. What is the ratio of side b to side c?

      The ratio is 4 to 5. Ans

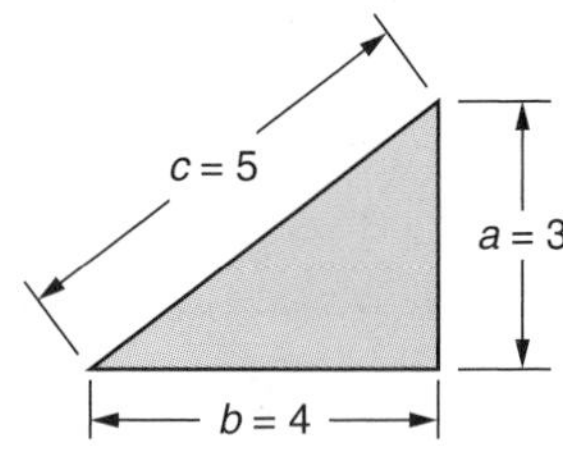

Figure 18-2

The *terms* of a ratio are the two numbers that are compared. **Both terms of a ratio must be expressed in the same units.**

Example Two pieces of bar stock are shown in Figure 18-3. What is the ratio of the short piece to the long piece?

The terms cannot be compared to a ratio until the 2-foot length is expressed as 24 inches.

The ratio is 11 to 24. Ans

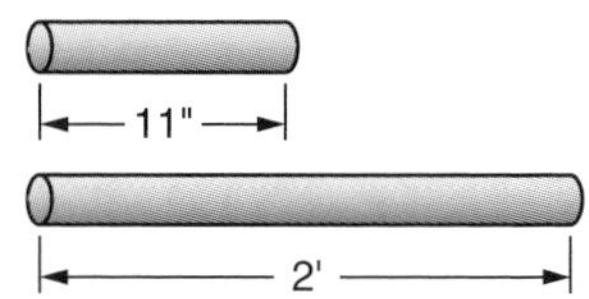

Figure 18-3

It is impossible to express two quantities as ratios if the terms have unlike units that cannot be expressed as like units. Inches and pounds as shown in Figure 18-4 cannot be compared as ratios.

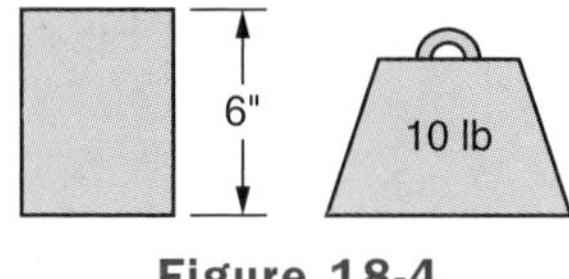

Figure 18-4

Expressing Ratios. Ratios are expressed in the following ways.

- With a colon between the two terms, such as 4:7. The ratio 4:7 is read as 4 to 7.
- With a division sign separating the two numbers, such as $4 \div 7$ or as a fraction, $\frac{4}{7}$.

Order of Terms

The terms of a ratio must be compared in the order in which they are given. The first term is the numerator of a fraction and the second is the denominator.

Examples

1. The ratio 1 to 3 $= 1 \div 3 = \frac{1}{3}$ Ans

2. The ratio 3 to 1 $= 3 \div 1 = \frac{3}{1}$ Ans

3. The ratio $x{:}y = x \div y = \frac{x}{y}$ Ans

4. The ratio $y{:}x = y \div x = \frac{y}{x}$ Ans

Expressing Ratios in Lowest Terms

Generally, a ratio should be expressed in lowest fractional terms.

Examples Each ratio is expressed in lowest terms.

1. $3:9 = \frac{3}{9} = \frac{1}{3}$ Ans
2. $40:15 = \frac{40}{15} = \frac{8}{3}$ Ans
3. $\frac{3}{8}:\frac{9}{16} = \frac{3}{8} \div \frac{9}{16} = \frac{3}{8} \times \frac{16}{9} = \frac{2}{3}$ Ans
4. $10:\frac{5}{6} = 10 \div \frac{5}{6} = \frac{10}{1} \times \frac{6}{5} = \frac{12}{1}$ Ans

Description of Proportions

A *proportion* is an expression that states the equality of two ratios.

Expressing Proportions. Proportions are expressed in the following ways.

- 3:4::6:8, which is read as "3 is to 4 as 6 is to 8."
- 3:4 = 6:8, which is read "the ratio of 3 to 4 equals the ratio of 6 to 8" or "3 is to 4 as 6 is to 8."
- $\frac{3}{4} = \frac{6}{8}$. This equation form is generally the way that proportions are written in practical applications.

A proportion consists of four terms. The first and the fourth term are called *extremes* and the second and third terms are called *means*.

extremes
$a:b = c:d$
means

or

extremes $\frac{a}{b} = \frac{c}{d}$ means

Examples

1. In the proportion 2:3::4:6

   2 and 6 are the extremes; 3 and 4 are the means. Ans

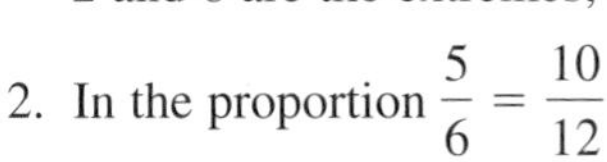

2. In the proportion $\frac{5}{6} = \frac{10}{12}$

   5 and 12 are the extremes; 6 and 10 are the means. Ans

In a proportion the product of the means equals the product of the extremes. If the terms are cross multiplied, their products are equal.

Examples

1. $\frac{3}{4} = \frac{6}{8}$

   Cross multiply, $\frac{3}{4} \times \frac{6}{8}$

   $$3 \times 8 = 4 \times 6$$
   $$24 = 24$$

2. $\frac{a}{b} = \frac{c}{d}$

   Cross multiply, $\frac{a}{b} \times \frac{c}{d}$

   $$a \times d = b \times c$$
   $$ad = bc$$

The method of cross multiplying is used in solving proportions that have an unknown term. You can check your answer by inserting it back in the original proportion.

Examples Solve for the value of x.

1. $\frac{3}{4} = \frac{x}{16}$

$$\frac{3}{4} = \frac{x}{16}$$

Cross multiply.

$$4x = 3(16)$$
$$4x = 48$$

Divide both sides of the equation by 4.

$$\frac{4x}{4} = \frac{48}{4}$$
$$x = 12 \quad \text{Ans}$$

Check.

$$\frac{3}{4} = \frac{x}{16}$$
$$\frac{3}{4} = \frac{12}{16}$$
$$\frac{3}{4} = \frac{3}{4} \quad \text{Ck}$$

2. $\frac{7}{x} = \frac{8}{15}$

$$8x = 7(15)$$
$$8x = 7(15)$$
$$\frac{8x}{8} = \frac{105}{8}$$
$$x = 13\frac{1}{8} \quad \text{Ans}$$

Check.

$$\frac{7}{x} = \frac{8}{15}$$
$$\frac{7}{13\frac{1}{8}} = \frac{8}{15}$$
$$\frac{8}{15} = \frac{8}{15} \quad \text{Ck}$$

3. $\frac{x}{7.5} = \frac{23.4}{20}$

$$20x = 7.5(23.4)$$
$$20x = 175.5$$
$$\frac{20x}{20} = \frac{175.5}{20}$$
$$x = 8.775 \quad \text{Ans}$$

Solving by calculator: 7.5 [×] 23.4 [÷] 20 [=] 8.775 Ans

Check.

$$\frac{x}{7.5} = \frac{23.4}{20}$$
$$\frac{8.775}{7.5} = \frac{23.4}{20} \quad \text{Ck}$$
$$1.17 = 1.17 \quad \text{Ck}$$

Application

Ratios

Express the following ratios in lowest fractional form.

1. 6:21 ________
2. 21:6 ________
3. 2:11 ________
4. 7:21 ________
5. 12":46" ________
6. 3 lb:21 lb ________
7. 13 mi:9 mi ________
8. 156 mm:200 mm ________
9. $\frac{2}{3}:\frac{1}{2}$ ________
10. $\frac{1}{2}:\frac{2}{3}$ ________

Related Ratio Problems

11. Length A in Figure 18-5 is 3 inches and length B is 2.5 feet. Determine the ratio of length A to length B in lowest fractional form. ______

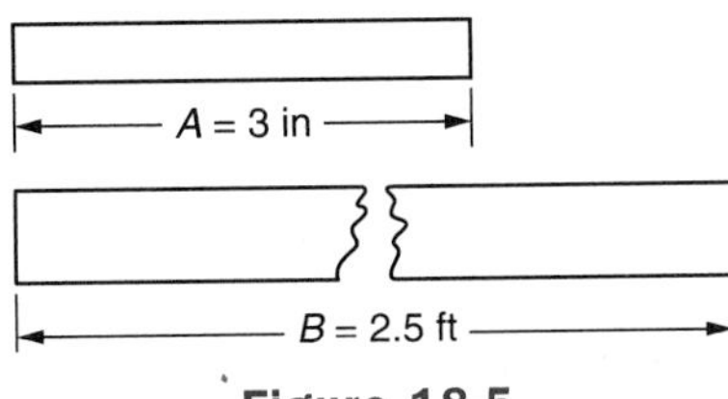

Figure 18-5

12. The diameters of pulleys E, F, G, and H, shown in Figure 18-6, are given in the table. Determine the ratios in lowest fractional form.

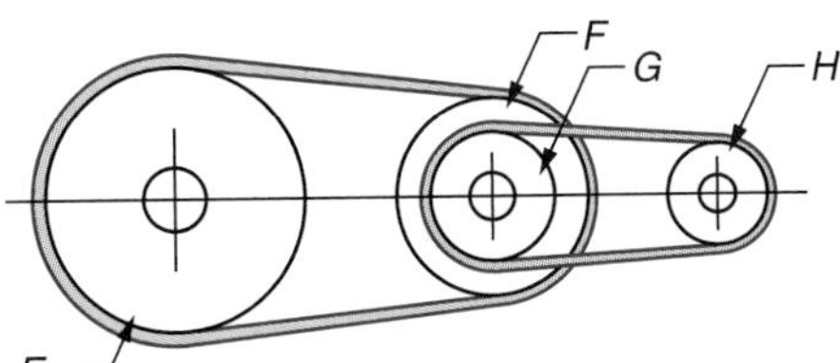

Figure 18-6

	DIAMETERS (inches)				RATIOS							
	E	F	G	H	$\frac{E}{F}$	$\frac{E}{G}$	$\frac{E}{H}$	$\frac{F}{G}$	$\frac{F}{H}$	$\frac{G}{H}$	$\frac{G}{E}$	$\frac{H}{F}$
a.	8	6	4	3								
b.	10	8	5	4								
c.	12	9	6	3								
d.	15	12	10	6								

13. Refer to the hole locations given for the plate in Figure 18-7. Determine the ratios in lowest fractional form. All dimensions are in millimeters.

a. Dimension A to dimension B. ______

b. Dimension A to dimension C. ______

c. Dimension C to dimension D. ______

d. Dimension C to dimension E. ______

e. Dimension D to dimension F. ______

f. Dimension F to dimension B. ______

g. Dimension F to dimension C. ______

h. Dimension E to dimension A. ______

i. Dimension D to dimension B. ______

j. Dimension C to dimension F. ______

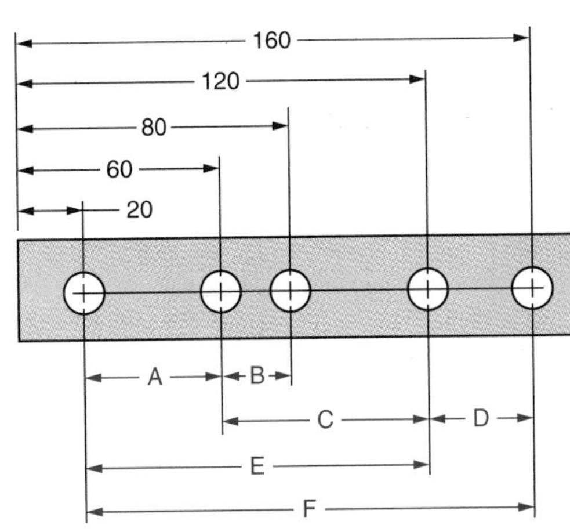

Figure 18-7

14. In Figure 18-8, gear A is turning at 120 revolutions per minute and gear B is turning at 3.6 revolutions per second. Determine the ratio of the speed of gear A to the speed of gear B. ______

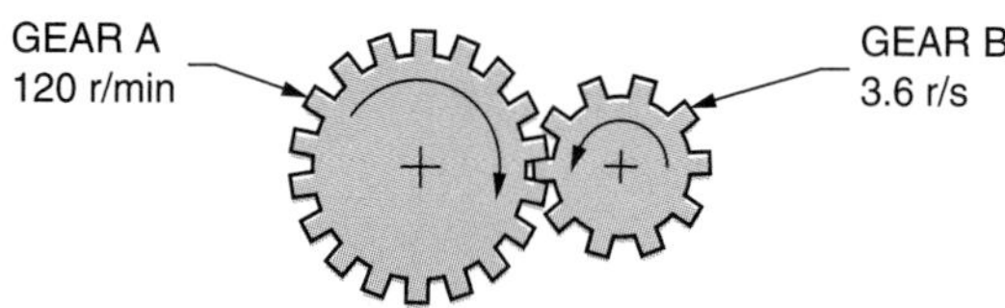

Figure 18-8

Proportions

Solve for the unknown value in each of the following proportions. Check each answer. Round the answers to 3 decimal places where necessary.

15. $\frac{x}{2} = \frac{6}{24}$ ______

16. $\frac{3}{A} = \frac{15}{30}$ ______

17. $\frac{7}{9} = \frac{E}{45}$ ______

18. $\frac{3}{13} = \frac{24}{y}$ ______

19. $\frac{15}{c} = \frac{5}{4}$ ______

20. $\frac{P}{27} = \frac{1}{3}$ ______

21. $\frac{6}{7} = \frac{15}{F}$ ______

22. $\frac{12}{H} = \frac{4}{25}$ ______

23. $\frac{T}{6.6} = \frac{7.5}{22.0}$ ______

24. $\frac{2.4}{3} = \frac{M}{0.8}$ ______

25. $\frac{4}{4.1} = \frac{8}{L}$ ______

26. $\frac{3.4}{y} = \frac{1}{9}$ ______

27. $\frac{24}{5} = \frac{3.2}{A}$ ______

28. $\frac{\frac{3}{8}}{N} = \frac{\frac{1}{2}}{4}$ ______

29. $\frac{3}{\frac{1}{4}} = \frac{5}{F}$ ______

30. $\frac{G}{\frac{1}{4}} = \frac{\frac{7}{8}}{\frac{3}{8}}$ ______

31. $\frac{7}{\frac{1}{8}} = \frac{x}{\frac{9}{16}}$ ______

32. $\frac{4}{R} = \frac{2.5}{12.5}$ ______

33. $\frac{11}{8} = \frac{E}{12}$ ______

34. $\frac{M}{12} = \frac{15}{9}$ ______

35. $\frac{6.08}{H} = \frac{5.87}{12.53}$ ______

36. $\frac{E}{7.53} = \frac{0.36}{1.86}$ ______

Related Proportion Problems

37. The proportion $\frac{A}{B} = \frac{C}{D}$ compares the sides of the two illustrated similar triangles like those in Figure 18-9. Determine the missing values in the table.

	A	*B*	*C*	*D*
a.	18"	4.5"		3"
b.	$6\frac{1}{2}$"	$1\frac{5}{8}$"	$4\frac{1}{2}$"	
c.	87.5 mm		75.0 mm	62.5 mm
d.		25.8 mm	20.6 mm	16.4 mm

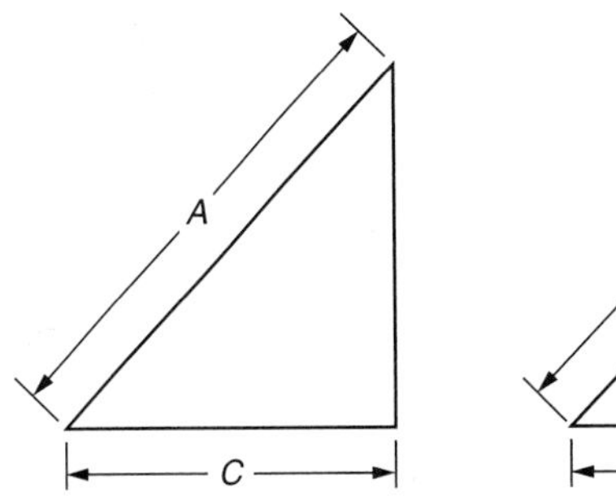

Figure 18-9

38. Where machine parts are doweled in position, it is good practice to extend the pin 1 to $1\frac{1}{2}$ times its diameter into the mating part as shown in Figure 18-10. Use the following proportion to determine the value of each unknown in the table. Round the answers to 3 decimal places where necessary.

$$\frac{N}{1} = \frac{L}{D}$$ where N = the number of times the pin extension is greater than the pin diameter
L = the length of the pin extension
D = the pin diameter

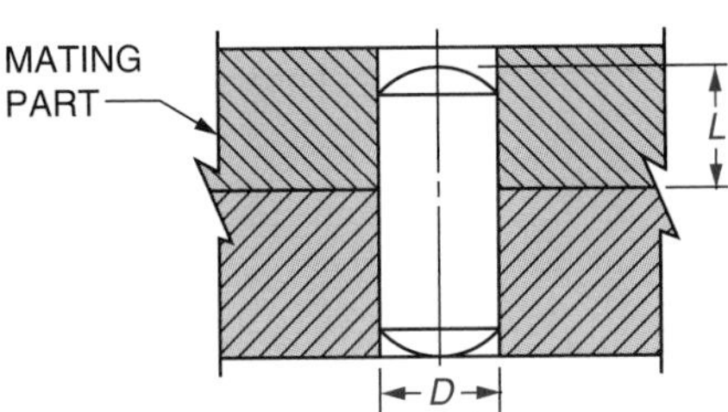

Figure 18-10

	N	*D*	*L*
a.	1.250	7.940 mm	
b.	$1\frac{1}{4}$	$\frac{1}{2}$"	
c.	$1\frac{1}{4}$		$\frac{3}{4}$"
d.	1.375		8.730 mm
e.	1.250	16.120 mm	

	N	*D*	*L*
f.	1.375		1.032"
g.	1.250	0.875"	
h.	1.500	3.680 mm	
i.	1.125		0.281"
j.	1.000	7.500 mm	

39. It is sometimes impractical to make engineering drawings full size. If the part to be drawn is very large or small, a scale drawing is generally made. The scale, which is shown on the drawing, compares the lengths of the lines on the drawing to the dimensions on the part. A scale on a drawing that states $\frac{1}{2}$" = 1" means" the drawing is one-quarter the size of the part. It is expressed as a ratio of 1:4 or $\frac{1}{4}$. A scale drawing that states 2" = 1"

means that the drawing is double the size of the part. It is expressed as a ratio of 2:1 or $\frac{2}{1}$. The actual dimensions of a steel support are given in Figure 18-11. All dimensions are in inches. Using the proportion given, compute the lengths for each unknown in the table. Round the answers to 3 decimal places where necessary.

$$\frac{\text{numerator of scale ratio}}{\text{denominator of scale ratio}} = \frac{\text{drawing length}}{\text{part dimension}}$$

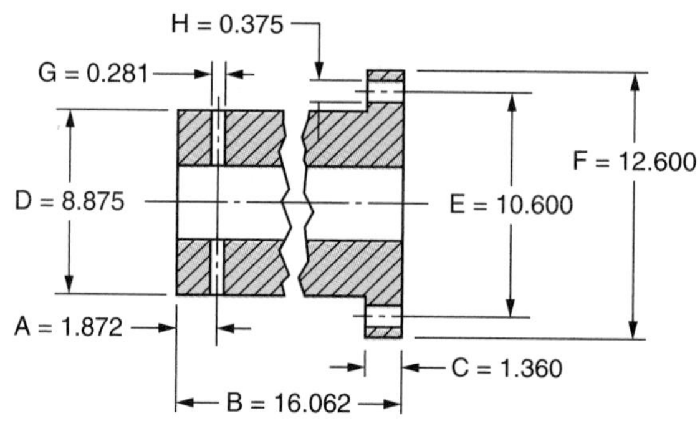

Figure 18-11

	Scale	Drawing Length
a.	$\frac{1}{2}$" = 1"	B =
b.	4" = 1"	G =
c.	$\frac{1}{4}$" = 1"	B =
d.	2" = 1"	C =
e.	$1\frac{1}{2}$" = 1"	A =
f.	$\frac{3}{4}$" = 1"	E =
g.	3" = 1"	H =
h.	$\frac{1}{8}$" = 1"	F =

	Scale	Drawing Length
i.	$\frac{1}{2}$" = 1"	E =
j.	6" = 1"	G =
k.	$\frac{3}{4}$" = 1"	F =
l.	$1\frac{1}{2}$" = 1"	C =
m.	$\frac{1}{2}$" = 1"	F =
n.	3" = 1"	G =
o.	$\frac{1}{4}$" = 1"	B =
p.	2" = 1"	A =

40. Figure 18-12 shows the relationship of gears in a lathe using a simple gear train. The proportion given is used for lathe thread cutting computations using simple gearing. The fixed stud gear and the spindle gear have the same number of teeth. Determine the missing values for each of the following problems.

$$\frac{N_L}{N_C} = \frac{T_S}{T_L}$$

where N_L = number of threads per inch on the lead screw
N_C = number of threads per inch to be cut
T_S = number of teeth on stud gear
T_L = number of teeth on lead screw gear

➤ **Note:** Intermediate gears only change direction.

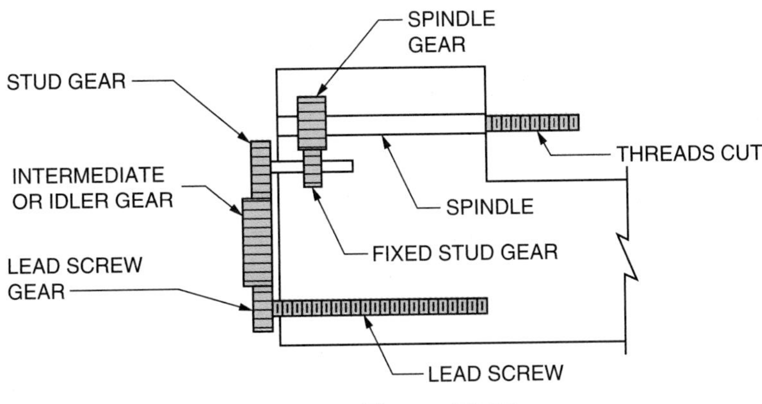

Figure 18-12

a. If $N_L = 4$, $N_C = 8$, and $T_S = 32$, find T_L. __________

b. If $N_L = 7$, $T_S = 35$, and $N_C = 15$, find T_L. __________

c. If $N_C = 10$, $N_L = 6$, and $T_L = 40$, find T_S. __________

d. If $N_L = 8$, $T_L = 42$, and $T_S = 28$, find N_C. __________

41. A template is shown on the left of Figure 18-13. A drafter makes an enlarged drawing of the template as shown on the right. The original length of 1.80 inches on the enlarged drawing is 3.06 inches as shown. Determine the lengths of A, B, C, and D.

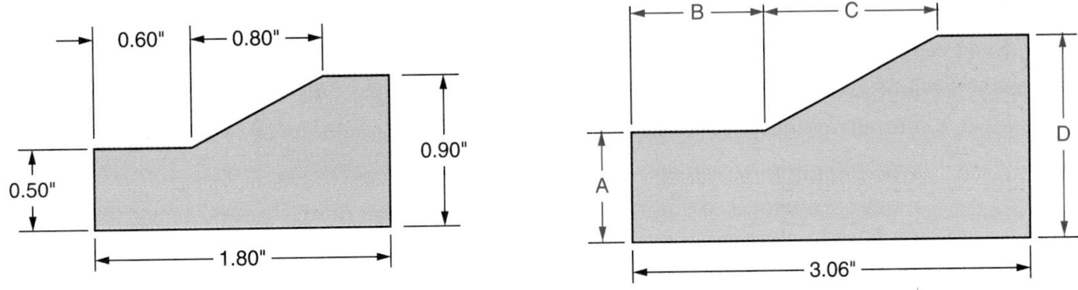

Figure 18-13

A __________ B __________ C __________ D __________

42. A drawing has a scale of $\frac{1}{4}$" = 3'. Determine the value of each missing value in the table.

	Scale Length	Actual Length
a.	$1\frac{1}{8}$"	
b.	$\frac{7}{16}$"	
c.		6' 9"
d.		16' $1\frac{1}{16}$"

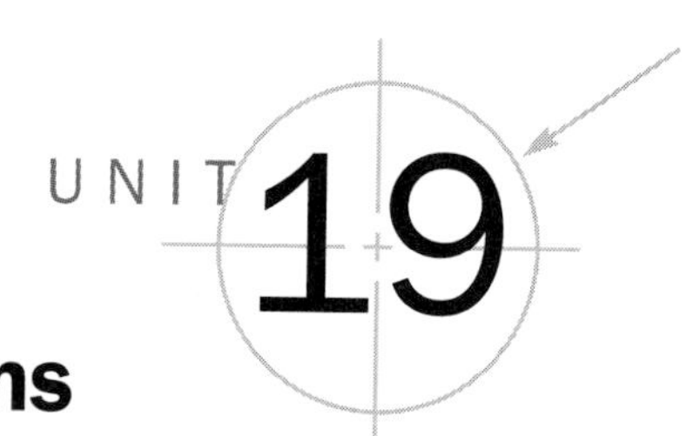

Direct and Inverse Proportions

Objectives ***After studying this unit you should be able to***

- Analyze problems to determine whether quantities are directly or inversely proportional.
- Set up and solve direct and inverse proportions.

Many shop problems are solved by the use of proportions. A machinist may be required to express word statements or other given data as proportions. Generally, three of the four terms of a proportion must be known in order to solve the proportion. When setting up a proportion it is important that the terms be placed in their proper positions.

Direct Proportions

In actual practice, word statements or other data must be expressed as proportions. When a proportion is set up, the terms of the proportion must be placed in their proper positions. A problem that is set up and solved as a proportion must first be analyzed in order to determine where the terms are placed. Depending on the position of the terms, proportions are either direct or inverse.

Two quantities are *directly proportional* if a change in one produces a change in the other in the same direction. If an increase in one produces an increase in the other, or if a decrease in one produces a decrease in the other, the two quantities are directly proportional. The proportions discussed will be those that change at the same rate. An increase or decrease in one quantity produces the same rate of increase or decrease in the other quantity.

When setting up a direct proportion in fractional form, the numerator of the first ratio must correspond to the numerator of the second ratio. The denominator of the first ratio must correspond to the denominator of the second ratio.

Example 1 If a machine produces 120 parts in 2 hours, how many parts are produced in 3 hours?

Analyze the problem. An increase in time (from 2 hours to 3 hours) will produce an increase in the number of pieces produced. Production increases as time increases. The proportion is direct.

Set up the direct proportion. Let x represent the number of parts that are produced in 3 hours. The numerator of the first ratio must correspond to the numerator of the second ratio; 2 hours corresponds to 120 parts. The denominator of the first ratio must correspond to the denominator of the second ratio; 3 hours corresponds to x.

Solve for x.

$$\frac{2 \text{ hours}}{3 \text{ hours}} = \frac{120 \text{ parts}}{x}$$

$$2x = 3(120 \text{ parts})$$

$$2x = 360 \text{ parts}$$

$$x = 160 \text{ parts} \quad \text{Ans}$$

Check.

$$\frac{2 \text{ hours}}{3 \text{ hours}} = \frac{120 \text{ parts}}{x}$$

$$\frac{2 \text{ hours}}{3 \text{ hours}} = \frac{120 \text{ parts}}{180 \text{ parts}}$$

$$\frac{2}{3} = \frac{2}{3} \quad \text{Ck}$$

Example 2 A tapered shaft is one that varies uniformly in diameter along its length. The shaft shown in Figure 19-1 is 15.000 inches long with a 1.200-inch diameter on the large end. A 9.000-inch piece is cut from the shaft. Determine the diameter at the large end of the 9.000-inch piece.

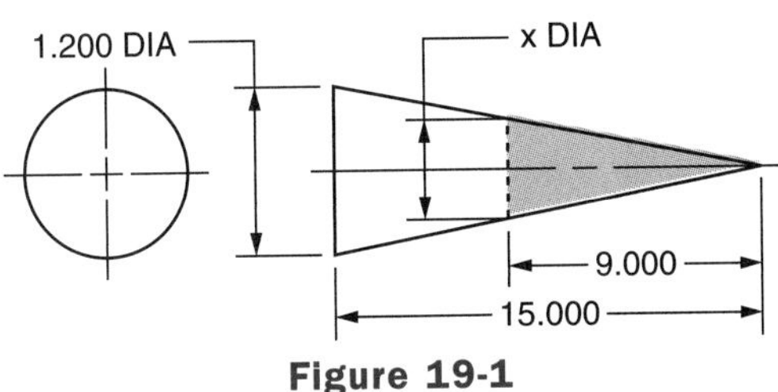

Figure 19-1

Analyze the problem. As the length decreases from 15.000 inches to 9.000 inches, the diameter also decreases at the same rate. The proportion is direct.

Set up the direct proportion. Let x represent the diameter at the large end of the 9.000-inch piece. The numerator of the first ratio must correspond to the numerator of the second ratio; the 15.000-inch piece has a 1.200-inch diameter at the large end. The denominator of the first ratio must correspond to the denominator of the second ratio; the 9.000-inch piece has a diameter of x at the large end.

Solve for x.

$$\frac{\overset{5}{\cancel{15.000\text{ inches}}}}{\underset{3}{\cancel{9.000\text{ inches}}}} = \frac{1.200\text{-inch DIA}}{x\text{ DIA}}$$

$$5x = 3(1.200\text{ inches})$$
$$5x = 3.600\text{ inches}$$
$$x = 0.720\text{ inch} \quad \text{Ans}$$

Check.

$$\frac{15.000\text{ inches}}{9.000\text{ inches}} = \frac{1.200\text{-inch DIA}}{x\text{ DIA}}$$

$$\frac{15.000\text{ inches}}{9.000\text{ inches}} = \frac{1.200\text{-inch DIA}}{0.720\text{-inch DIA}}$$

$$1.67 = 1.67 \quad \text{Ck}$$

Inverse Proportions

Two quantities are *inversely or indirectly proportional* if a change in one produces a change in the other in the opposite direction. If an increase in one produces a decrease in the other, or if a decrease in one produces an increase in the other, the two quantities are inversely proportional. For example, if one quantity increases by 4 times its original value, the other quantity decreases by 4 times its value or is $\frac{1}{4}$ of its original value.

Notice 4 or $\frac{4}{1}$ inverted is $\frac{1}{4}$.

When setting up an inverse proportion in fractional form, the numerator of the first ratio must correspond to the denominator of the second ratio. The denominator of the first ratio must correspond to the numerator of the second ratio.

Example 1 Two gears in mesh are shown in Figure 19-2. The driver gear has 40 teeth and revolves at 360 revolutions per minute. Determine the number of revolutions per minute of a driven gear with 16 teeth.

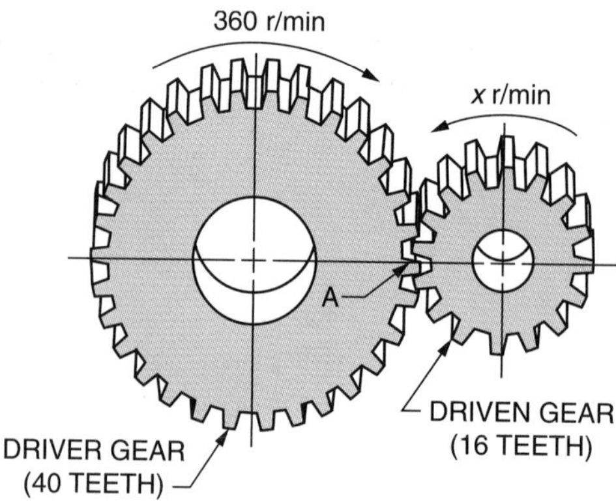

Figure 19-2

Analyze the problem. When the driver gear turns one revolution, 40 teeth pass point A. The same number of teeth on the driven gear must pass point A. Therefore, the driven gear turns more than one revolution for each revolution of the driver gear. The gear with 16 teeth (driven gear) revolves at greater revolutions per minute than the gear with 40 teeth (driver gear). A decrease in the number of teeth produces an increase in revolutions per minute. The proportion is inverse.

Set up the direct proportion. Let x represent the revolutions per minute of the gear with 16 teeth. The numerator of the first ratio must correspond to the denominator of the second ratio; the gear with 40 teeth revolves at 360 r/min. The denominator of the first ratio must correspond to the numerator of the second ratio; the gear with 16 teeth corresponds to x. Note that the product of the first gear's speed and its number of teeth equals the product of the second gear's speed and its number of teeth. Thus, the product 360×40 equals the number of teeth on the driver gear that pass point A in one minute. The corresponding product of the driven gear must also be 360. These are put on diagonals so that the cross-products are equal.

Solve for x.

$$\frac{\overset{5}{\cancel{40}\ \cancel{\text{teeth}}}}{\underset{2}{\cancel{16}\ \cancel{\text{teeth}}}} = \frac{x}{360 \text{ r/min}}$$

$$2x = 1800 \text{ r/min}$$

$$x = 900 \text{ r/min} \quad \text{Ans}$$

Check.

$$\frac{40 \text{ teeth}}{16 \text{ teeth}} = \frac{x}{360 \text{ r/min}}$$

$$\frac{40 \text{ teeth}}{16 \text{ teeth}} = \frac{900 \text{ r/min}}{360 \text{ r/min}}$$

$$2.5 = 2.5 \quad \text{Ck}$$

The driven gear revolves at the rate of 900 revolutions per minute.

Example 2 Five identical machines produce the same parts at the same rate. The 5 machines complete the required number of parts in 2.1 hours. How many hours does it take 3 machines to produce the same number of parts?

Analyze the problem. A decrease in the number of machines (from 5 to 3) requires an increase in time. Time increases as the number of machines decreases; therefore, the proportion is inverse.

Set up the direct proportion. Let x represent the time required by 3 machines to produce the parts.

$$\frac{5 \text{ machines}}{3 \text{ machines}} = \frac{x}{2.1 \text{ hours}}$$

Notice that the numerator of the first ratio corresponds to the denominator of the second ratio; 5 machines corresponds to 2.1 hours. The denominator of the first ratio corresponds to the numerator of the second ratio for a product of 10.5 machine-hours; 3 machines corresponds to x.

Solve for x.

$$\frac{5}{3} = \frac{x}{2.1 \text{ hours}}$$

$$3x = 5(2.1 \text{ hours})$$

$$\frac{3x}{3} = \frac{10.5 \text{ hours}}{3}$$

$$x = 3.5 \text{ hours} \quad \text{Ans}$$

Check.

$$\frac{5}{3} = \frac{x}{2.1 \text{ hours}}$$

$$\frac{5}{3} = \frac{3.5 \text{ hours}}{2.1 \text{ hours}}$$

$$1.\overline{6} = 1.\overline{6} \quad \text{Ck}$$

It will take $3\frac{1}{2}$ hours for 3 machines to produce as many parts as 5 machines did in 2.1 hours (2 hours 6 minutes).

Application

Tapers

Taper is the difference between the diameters at each end of a part. Tapers are expressed as the difference in diameters for a particular length along the centerline of a part.

Note: All dimensions are in millimeters.

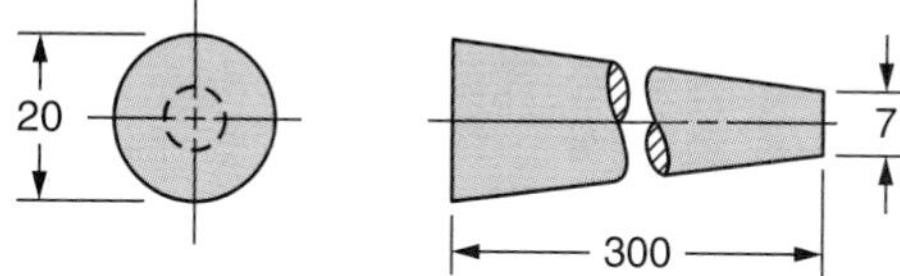

20 mm − 7 mm = 13 mm taper per 300 mm

Figure 19-3

Note: All dimensions are in inches.

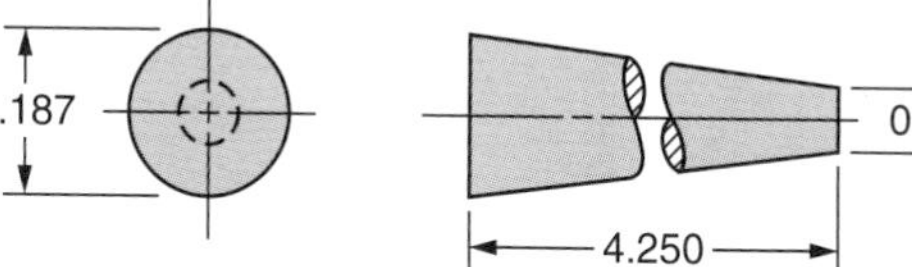

1.187" − 0.885" = 0.302" taper per 4.250"

Figure 19-4

1. A plug gage tapers 3.10 mm along a 38.00 mm length as shown in Figure 19-5. Set up a proportion and determine the amount of taper in the workpiece for each of the following problems. Express the answers to 2 decimal places.

	WORKPIECE THICKNESS	PROPORTION	TAPER IN WORKPIECE
a.	18.40 mm		
b.	31.75 mm		
c.	14.28 mm		
d.	28.58 mm		
e.	23.60 mm		

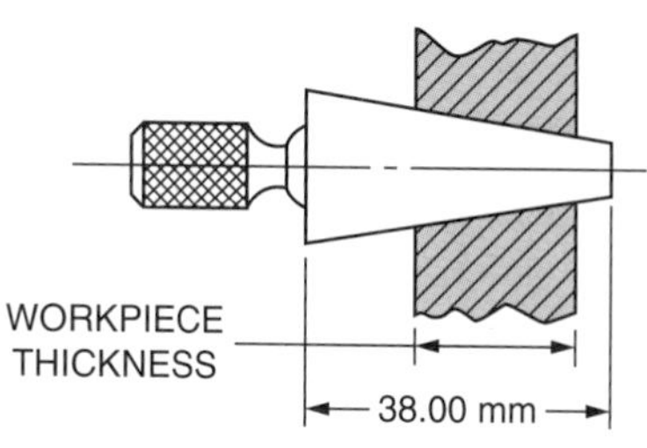

Figure 19-5

2. A reamer tapers 0.130" along a 4.250" length. Set up a proportion and determine length A for each of the following problems. Express the answers to 3 decimal places.

	TAPER IN LENGTH A	PROPORTION	LENGTH A
a.	0.030"		
b.	0.108"		
c.	0.068"		
d.	0.008"		
e.	0.093"		

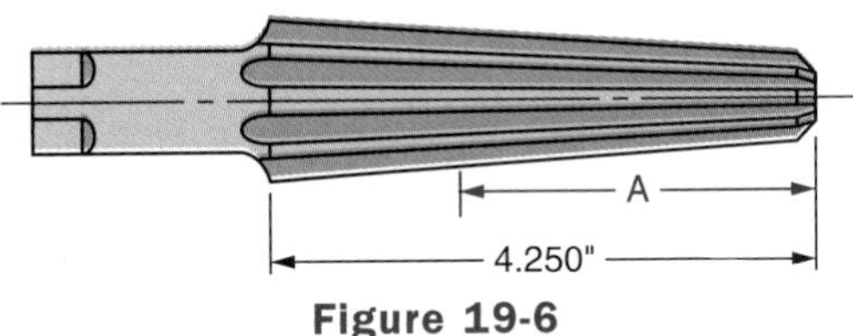

Figure 19-6

3. A micrometer reading is made at dimension D on a tapered shaft. For each of the problems use the dimensions given in the table, compute the taper, set up a proportion, and determine diameter C to 3 decimal places.

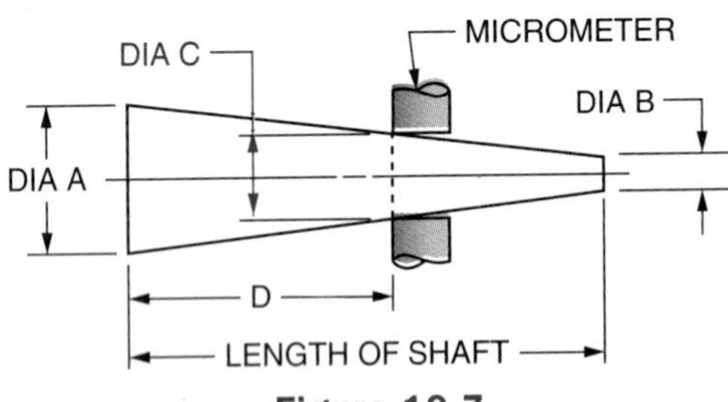

Figure 19-7

	LENGTH OF SHAFT	DIAMETER A	DIAMETER B	DIMENSION D	DIAMETER C
a.	10.200"	1.500"	0.700"	6.500"	
b.	8.750"	1.250"	0.375"	4.875"	
c.	550.000 mm	106.250 mm	62.500 mm	337.500 mm	
d.	147.500 mm	22.500 mm	10.000 mm	112.500 mm	
e.	8.800"	1.325"	0.410"	8.620"	

Proportions

Analyze each of the following problems to determine whether the problem is a direct or inverse proportion. Set up the proportion and solve.

4. A sheet of steel $8\frac{1}{4}$ feet long weighs 325 pounds. A piece $2\frac{1}{2}$ feet long is sheared from the sheet. Determine the weight of the $2\frac{1}{2}$ foot piece to the nearest whole pound. ______
5. If 1350 parts are produced in 6.75 hours, find the number of parts produced in 8.25 hours. ______
6. The production rate for each of 3 machines is the same. Using these 3 machines, 720 parts are produced in 1.6 hours. How many hours will it take 2 of these machines to produce 720 parts? ______
7. Two forgings are made of the same stainless steel alloy. A forging, which weighs 76.00 kilograms, contains 0.38 kilogram of chromium. How many kilograms of chromium does the second forging contain if it weighs 96.00 kilograms? Round the answer to 2 decimal places. ______

Gears and Pulleys

8. A 10.00-inch diameter pulley rotates at 160.0 rpm. A belt connects this 10.00-inch diameter pulley with a 6.5-inch diameter pulley. An 8.00-inch diameter pulley is fixed to the same shaft as the 6.50-inch pulley. A belt connects the 8.00-inch pulley with a 3.50-inch diameter pulley. Determine the revolutions per minute of the 3.50-inch diameter pulley. Round the answer to 1 decimal place. ______

9. Of two gears that mesh, the one that has the greater number of teeth is called the gear, and the one that has the fewer teeth is called the pinion. For each of the problems, set up a proportion, and determine the unknown value, x. Round the answers to 1 decimal place where necessary.

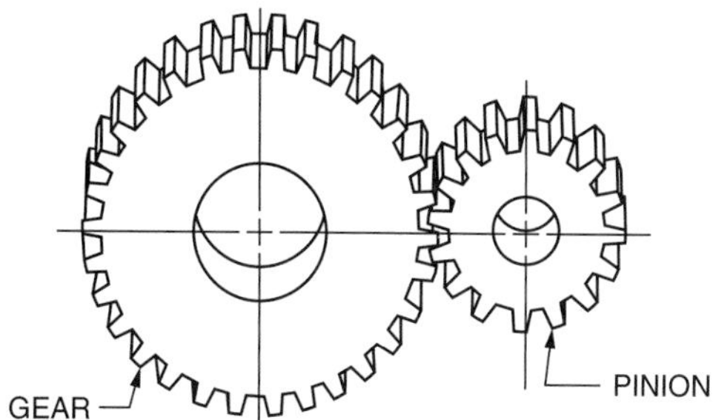

Figure 19-8

	NUMBER OF TEETH ON GEAR	NUMBER OF TEETH ON PINION	SPEED OF GEAR (rpm)	SPEED OF PINION (rpm)
a.	48	20	100.0	$x =$
b.	32	24	$x =$	210.0
c.	35	$x =$	160.0	200.0
d.	$x =$	15	150.0	250.0
e.	54	26	80.0	$x =$

10. Figure 19-9 shows a compound gear train. Gears B and C are keyed to the same shaft; therefore, they turn at the same speed. Gear A and gear C are driving gears. Gear B and gear D are driven gears. Set up a proportion for each problem and determine the unknown values, x, y, and z in the table. Round the answers to 1 decimal place where necessary.

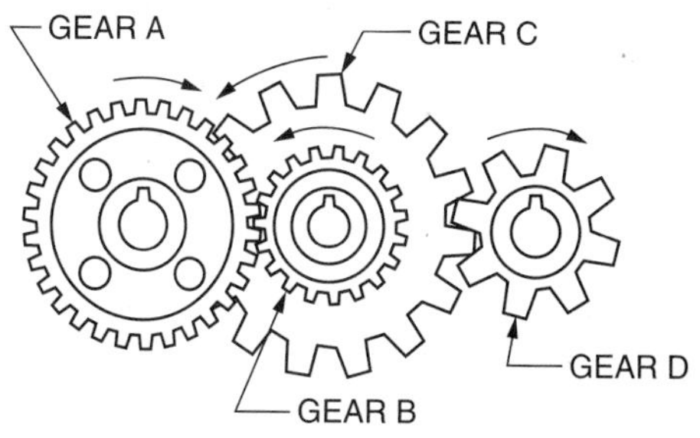

Figure 19-9

	NUMBER OF TEETH				SPEED (rpm)			
	Gear A	Gear B	Gear C	Gear D	Gear A	Gear B	Gear C	Gear D
a.	80	30	50	20	120.0	$x =$	$y =$	$z =$
b.	60	$x =$	45	$y =$	100.0	300.0	$z =$	450.0
c.	$x =$	24	60	36	144.0	$y =$	$z =$	280.0
d.	55	25	$x =$	15	$y =$	$z =$	175.0	350.0

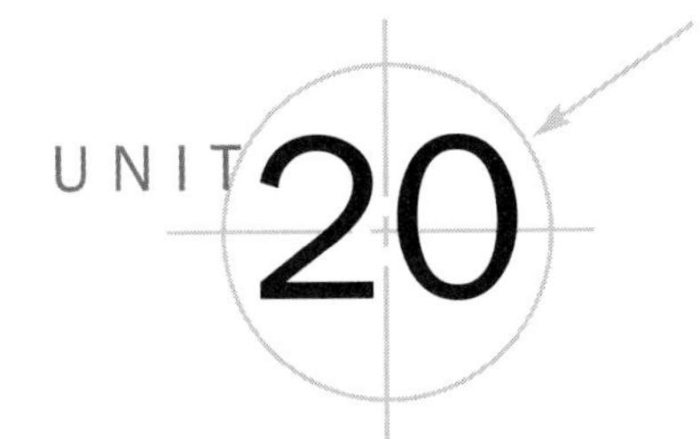

Introduction to Percents

Objectives ***After studying this unit you should be able to***

- Express decimal fractions and common fractions as percents.
- Express percents as decimal fractions and common fractions.

Percents are widely used in both business and nonbusiness fields. Merchandise selling prices and discounts, wage deductions, and equipment depreciation are determined by percentages.

In manufacturing technology percentage concepts have many applications, such as expressing production increases or decreases, power inputs and outputs, quality control product rejections, and material allowances for waste and nonconforming parts.

Definition of Percent

Figure 20-1

The *percent (%)* indicates the number of hundredths of a whole. The square shown to the left is divided into 100 equal parts. The whole (large square) contains 100 small parts, or 100 percent of the small squares. Each small square is one part of 100 parts or $\frac{1}{100}$ of the large square in Figure 20-1. Therefore, each small square is $\frac{1}{100}$ of 100 percent or 1 percent.

1 part of 100 parts

$$\frac{1}{100} = 0.01 = 1\%$$

Figure 20-2

Example What percent of the square in Figure 20-2 is shaded?

The large square is divided into four equal smaller squares. Three of the smaller squares are shaded.

3 parts of 4 parts

$$\frac{3}{4} = 0.75 = 75\% \quad \text{Ans}$$

Expressing Decimal Fractions as Percents

A decimal fraction can be expressed as a percent by moving the decimal point two places to the right and inserting the percent symbol. Moving the decimal point two places to the right is actually multiplying by 100.

Examples

1. Express 0.0152 as a percent.

   Move the decimal point 2 places to the right.

   Insert the percent symbol.

   $0.01{}_{\wedge}52 = 1.52\%$ Ans

2. Express 3.876 as a percent.

   Move the decimal point 2 places to the right. $\quad 3.87\_6 = 387.6\%$ Ans

   Insert the percent symbol.

Expressing Common Fractions and Mixed Numbers as Percents

To express a common fraction as a percent, first express the common fraction as a decimal fraction. Then express the decimal fraction as a percent. If it is necessary to round, the decimal fraction must be two more decimal places than the desired number of places for the percent.

Examples

1. Express $\frac{7}{8}$ as a percent.

   Express $\frac{7}{8}$ as a decimal fraction. $\quad \frac{7}{8} = 0.875$

   Express 0.875 as a percent. $\quad 0.875 = 87.5\%$ Ans

2. Express $5\frac{2}{3}$ as a percent to 1 decimal place.

   Express $5\frac{2}{3}$ as a decimal fraction. $\quad 5\frac{5}{3} = 5.667$

   Express 5.667 as a percent. $\quad 5.667 = 566.7\%$ Ans

   5 [+] 2 [÷] 3 [=] [×] 100 [=] 566.6666667, 566.7% Ans

Expressing Percents as Decimal Fractions

Expressing a percent as a decimal fraction can be done by dropping the percent symbol and moving the decimal point two places to the left. Moving the decimal point two places to the left is actually dividing by 100.

Examples

1. Express $38\frac{16}{21}\%$ as a decimal fraction. Round the answer to 4 decimal places.

   Express $38\frac{16}{21}\%$ as approximately 38.76%

   Drop the percent symbol and move the decimal point 2 places to the left.

   $38\frac{16}{21}\% = 38.76\% = 0.3876$ Ans

38 [+] 16 [÷] 21 [=] [÷] 100 [=] 0.387619048, 0.3876 Ans

or 38 [a b/c] 16 [a b/c] 21 [÷] 100 [=] 0.387619048, 0.3876 Ans

Express each percent as a decimal fraction. Round the answers to 3 decimal places.

1. 0.48% $\quad$ 0.005 Ans
2. $15\frac{3}{4}\%$ $\quad$ 0.158 Ans
3. 7% $\quad$ 0.070 Ans
4. 300% $\quad$ 3.000 Ans

Expressing Percents as Common Fractions

A percent is expressed as a fraction by first finding the equivalent decimal fraction. The decimal fraction is then expressed as a common fraction.

Examples

1. Express 37.5% as a common fraction. $\frac{37.5}{100} = \frac{375}{1000} = \frac{3}{8}$

   Express 37.5% as a decimal fraction. $37.5\% = 0.375$

   Express 0.375 as a common fraction. $0.375 = \frac{375}{1000} = \frac{3}{8}$ Ans

or 375 [a b/c] 1000 [=] 3 ⌋ 8, $\frac{3}{8}$ Ans

Express each percent as a common fraction.

1. 10% $10\% = 0.10 = \frac{10}{100} = \frac{1}{10}$ Ans
2. 0.5% $0.5\% = 0.005 = \frac{5}{1000} = \frac{1}{200}$ Ans
3. $222\frac{1}{2}\%$ $222\frac{1}{2}\% = 222.5\% = 2.225 = 2\frac{225}{1000} = 2\frac{9}{40}$ Ans

Reduce $2\frac{225}{1000}$ 2 [a b/c] 225 [a b/c] 1000 [=] 2 ⌋ 9 ⌋ 40, $2\frac{9}{40}$ Ans

4. $6\frac{1}{4}\%$ $6\frac{1}{4}\% = 6.25\% = 0.0625 = \frac{625}{10{,}000} = \frac{1}{16}$ Ans

Reducing with a calculator: 625 [a b/c] 10000 [ENTER] 1 ⌋16, $\frac{1}{16}$ Ans

Application

Determining Percents

Determine the percent of each figure that is shaded.

1.

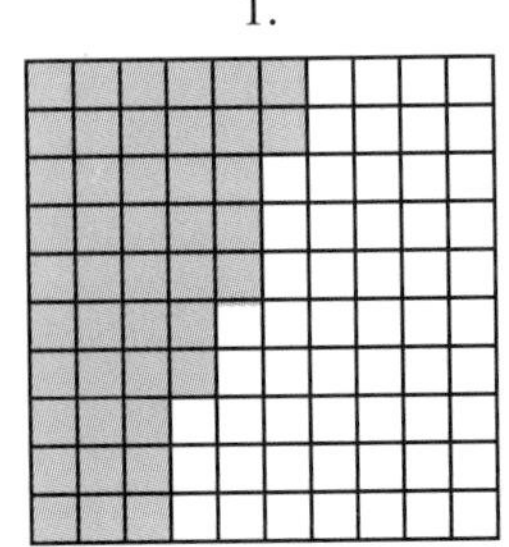

2.

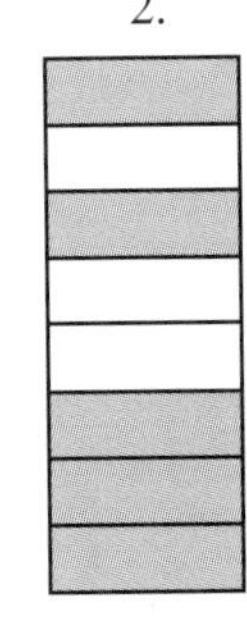

3.

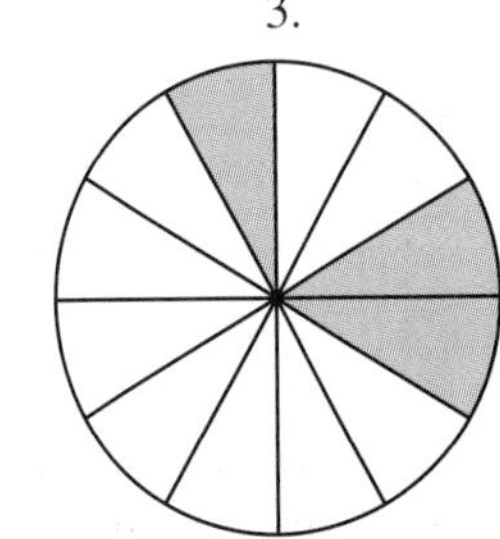

4.

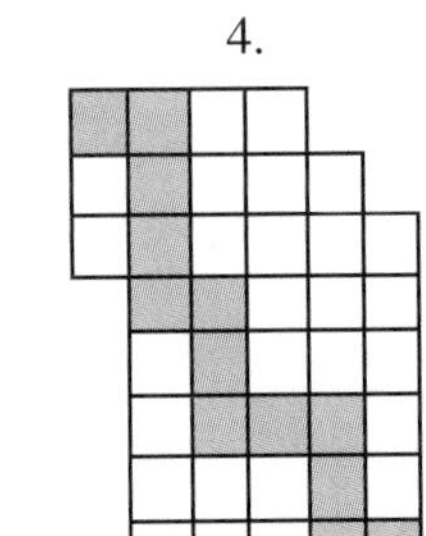

1. ______________
2. ______________
3. ______________
4. ______________

Expressing Decimals and Fractions as Percents

Express each value as a percent.

5. 0.35 ________
6. 0.96 ________
7. 0.04 ________
8. 0.062 ________
9. 0.008 ________
10. 1.33 ________
11. 2.076 ________
12. 0.0639 ________
13. 0.0002 ________
14. 3.005 ________
15. $\frac{1}{4}$ ________
16. $\frac{21}{80}$ ________
17. $\frac{3}{20}$ ________
18. $\frac{37}{50}$ ________
19. $\frac{17}{32}$ ________
20. $\frac{1}{250}$ ________
21. $1\frac{59}{100}$ ________
22. $2\frac{7}{25}$ ________
23. $14\frac{5}{8}$ ________
24. $3\frac{1}{200}$ ________

Expressing Percents as Decimals

Express each percent as a decimal fraction or mixed decimal.

25. 82% ________
26. 19% ________
27. 3% ________
28. 2.6% ________
29. 27.76% ________
30. 103% ________
31. 224.9% ________
32. 0.6% ________
33. 4.73% ________
34. $12\frac{1}{2}\%$ ________
35. $\frac{3}{4}\%$ ________
36. 0.1% ________
37. $2\frac{3}{8}\%$ ________
38. 0.05% ________
39. $37\frac{1}{4}\%$ ________
40. $205\frac{1}{10}\%$ ________

Expressing Percents as Fractions

Express each percent as a common fraction or mixed number.

41. 50% ________
42. 25% ________
43. 62.5% ________
44. 4% ________
45. 16% ________
46. 275% ________
47. 190% ________
48. 0.2% ________
49. 1.8% ________
50. 100.1% ________
51. 0.9% ________
52. 0.05% ________

UNIT 21

Basic Calculations of Percentages, Percents, and Rates

Objectives ***After studying this unit you should be able to***

- Determine the percentage, given the base and rate.
- Determine the percent (rate), given the percentage and base.
- Determine the base, given the rate and percentage.

Types of Simple Percentage Problems

A simple percentage problem has three parts. The parts are the rate, the base, and the percentage. In the problem 10% of \$80 = \$8, the rate is 10%, the base is \$80, and the percentage is \$8. The *rate* is the percent. The *base* is the number of which the rate or percent is taken. It is the whole or a quantity equal to 100%. The *percentage* is the quantity of the percent of the base. In solving problems, the rate, percentage, and base must be identified. Some people like to use the term "amount" for percentage. This is perfectly correct; however, you will often see percentage used in this book.

In solving percentage problems, the words *is* and *of* are often helpful in identifying the three parts. The word *is* generally relates to the rate or percentage and the word *of* generally relates to the base.

The following descriptions may help you recognize the rate, base, and percentage more quickly:

Base: The total, original, or entire amount. The base usually follows the word "of."

Rate: The number with a % sign. Sometimes it is written as a decimal or fraction.

Percentage: The value that remains after the base and rate have been determined. It is a portion of the base. The percentage is often close to the word "is."

Examples

1. What *is* 25% *of* 120? *Is* relates to 25% (the rate) and *of* relates to 120 (the base).
2. What percent *of* 48 *is* 12? *Is* relates to 12 (the percentage) and *of* relates to 48 (the base).
3. 60 *is* 30% *of* what number? *Is* relates to 60 (the percentage) and 30% (the rate). *Of* relates to what number (the base).

There are three types of simple percentage problems. The type used depends on which two quantities are given and which quantity must be found. The three types are as follows:

- Finding the percentage, given the rate (percent) and the base.
  A problem of this type is, "What is 25% of 120?"
  If the rate is less than 100%, the percentage is less than the base.
  If the rate is greater than 100%, the percentage is greater than the base.
- Finding the rate (percent), given the base and the percentage.
  A problem of this type is, "What percent of 48 is 12?"
  If the percentage is less than the base, the rate is less than 100%.
  If the percentage is greater than the base, the rate is greater than 100%.
- Finding the base, given the rate (percent) and the percentage.
  A problem of this type is, "60 is 30% of what number?"

All three types of percentage problems can be solved using the following proportion:

$$\frac{P}{B} = \frac{R}{100}$$

where

B is the base

P is the percentage or part of the base, and

R is the rate or percent.

Practical applications involve numbers that have units or names of quantities called *denominate numbers.* The base and the percentage have the same unit or denomination. For example, if the base unit is expressed in inches, the percentage is expressed in inches. The rate is not a denominate number; it does not have a unit or denomination. Rate is the part to be taken of the whole quantity, the base.

Finding the Percentage, Given the Base and Rate

In some problems, the base and rate are given and the percentage must be found. First, express the rate (percent) as an equivalent decimal fraction. Then solve with the proportion $\frac{P}{B} = \frac{R}{100}$.

Examples

1. What is 15% of 60?

Solution. The rate is $15\% = \frac{15}{100}$, so, $R = 15$.

The base, B, is 60. It is the number of which the rate is taken—the whole or a quantity equal to 100%.

The percentage is to be found. It is the quantity of the percent of the base.

The proportion is $\frac{P}{60} = \frac{15}{100}$.

Now, using cross-products and division.

$$100P = 15 \times 60$$
$$100P = 900$$
$$P = \frac{900}{100} = 9 \quad \text{Ans}$$

2. Find $56\frac{9}{25}$% of \$183.76.

Solution. The rate is $56\frac{9}{25}$%, so $R = 56\frac{9}{25}$.

The base, B, is \$183.76.

The percentage is to be found.

Express R, $56\frac{9}{25}$ as a decimal, $56\frac{9}{25} = 56.36$.

The proportion is $\frac{P}{\$183.76} = \frac{56.36}{100}$.

Again, using cross-products and division.

$$100P = 56.36 \times \$183.76$$
$$100P \approx \$10{,}357$$
$$P = \frac{\$10{,}357}{100} = \$103.57 \quad \text{Ans}$$

56 [+] 9 [÷] 25 [=] [×] 183.76 [÷] 100 [=] 103.567136, \$103.57 Ans

If you use the fraction key [Ab/c], it is not necessary to convert R to a decimal.

56 [Ab/c] 9 [Ab/c] 25 [×] 183.76 [÷] 100 [=] 103.567136, \$103.57 Ans

Finding the Percent (Rate), Given the Base and Percentage

In some problems, the base and percentage are given, and the percent (rate) must be found.

Examples

1. What percent of 12.87 is 9.620? Round the answer to 1 decimal place.

Since a percent of 12.87 is to be taken, the base or whole quantity equal to 100% is 12.87.

The percentage or quantity of the percent of the base is 9.620.

The rate is to be found.

Since the percentage, 9.620, is less than the base, 12.87, the rate must be less than 100%.

The proportion is $\frac{9.620}{12.87} = \frac{R}{100}$.

Cross multiply $9.620 \times 100 = 12.87R$

$$962 = 12.87R$$

Divide $\frac{962}{12.87} = R$

$$R = \frac{962}{12.87} = 74.7474 \approx 74.7\% \quad \text{Ans}$$

9.62 [×] 100 [÷] 12.87 [=] 74.74747475, 74.7% Ans

2. What percent of 9.620 is 12.87? Round the answer to 1 decimal place.

   Notice that although the numbers are the same as in Example 1, the base and percentage are reversed.

   Since a percent of 9.620 is to be taken, the base or whole quantity equal to 100% is 9.620.

   The percentage or quantity of the percent of the base is 12.87.

   Since the percentage, 12.87, is greater than the base, 9.620, the rate must be greater than 100%.

   $B = 9.920$ and $P = 12.87$. Thus, the proportion is $\frac{12.87}{9.620} = \frac{R}{100}$.

$$12.87 \times 100 = 9.620R$$

$$1287 = 9.620R$$

$$\frac{1287}{9.620} = R$$

$$R = \frac{1287}{9.620} = 133.78 \approx 133.8\% \quad \text{Ans (rounded)}$$

12.87 [×] 100 [÷] 9.62 [=] 133.7837838, 133.8% Ans (rounded)

Finding the Base, Given the Percent (Rate) and the Percentage

In some problems, the percent (rate) and the percentage are given, and the base must be found.

Examples

1. 816 is 68% of what number?

   $R = 68$ and $P = 816$. The proportion is $\frac{816}{B} = \frac{68}{100}$.

$$816 \times 100 = 68B$$

$$81{,}600 = 68B$$

$$\frac{81{,}600}{68} = B$$

$$B = \frac{81{,}600}{68} = 1200 \quad \text{Ans}$$

2. \$149.50 is $115\frac{2}{3}$% of what value?

$R = 115.67$ and $P = \$149.50$ and the proportion is $\frac{\$149.50}{B} = \frac{115.67}{100}$.

$$\$149.50 \times 100 = 115.67B$$
$$\$14{,}950 = 115.67B$$
$$\frac{\$14{,}950}{115.67} = B$$
$$B = \frac{\$14{,}950}{115.67} = \$129.25 \quad \text{Ans}$$

149.5 [×] 100 [÷] 115.67 [=] 129.2469958, \$129.25 Ans

or, if you want to leave the percent written as a fraction

149.5 [×] 100 [÷] 115 [Ab/c] 2 [Ab/c] 3 [=] 2nd [F◂▸D] 129.2507205, \$129.25 Ans

Application

Finding Percentage

Find each percentage. Round the answers to 2 decimal places when necessary.

1. 20% of 80 ________
2. 2.15% of 80 ________
3. 60% of 200 ________
4. 15.23% of 150 ________
5. 25% of 312.6 ________
6. 7% of 140.34 ________
7. 156% of 65 ________
8. 0.8% of 214 ________
9. 12.7% of 295 ________
10. 122% of 1.68 ________
11. 140% of 280 ________
12. 1.8% of 1240 ________
13. 39% of 18.3 ________
14. 0.42% of 50 ________
15. 0.03% of 424.6 ________
16. $8\frac{1}{2}$% of 375 ________
17. $\frac{3}{4}$% of 132 ________
18. 296.5% of 81 ________
19. $15\frac{1}{4}$% of $35\frac{1}{4}$ ________
20. $\frac{17}{50}$% of $139\frac{3}{10}$ ________

Finding Percent (Rate)

Find each percent (rate). Round the answers to 2 decimal places when necessary.

21. What percent of 8 is 4? ________
22. What percent of 20.7 is 5.6? ________
23. What percent of 100 is 37? ________
24. What percent of 84.37 is 70.93? ________
25. What percent of 70.93 is 84.37? ________
26. What percent of 258 is 97? ________
27. What percent of 132.7 is 206.3? ________
28. What percent of 19.5 is 5.5? ________
29. What percent of 1.25 is 0.5? ________
30. What percent of 0.5 is 1.25? ________
31. What percent of $6\frac{1}{2}$ is 2? ________
32. What percent of 134 is $156\frac{3}{4}$? ________
33. What percent of $\frac{7}{8}$ is $\frac{3}{8}$? ________
34. What percent of $\frac{3}{8}$ is $\frac{7}{8}$? ________
35. What percent of 3.08 is 4.76? ________

36. What percent of 0.65 is 0.09? ______
37. What percent of $12\frac{1}{4}$ is 3? ______
38. What percent of 312 is 400.9? ______
39. What percent of $\frac{3}{4}$ is $\frac{3}{8}$? ______
40. What percent of $13\frac{4}{5}$ is $6\frac{3}{10}$? ______

Finding Base

Find each base. Round the answers to 2 decimal places when necessary.

41. 15 is 10% of what number? ______
42. 25 is 80% of what number? ______
43. 80 is 25% of what number? ______
44. 3.8 is 95.3% of what number? ______
45. 13.6 is 8% of what number? ______
46. 123.86 is 88.7% of what number? ______
47. 203 is 110% of what number? ______
48. $44\frac{1}{3}$ is 60% of what number? ______
49. $7\frac{1}{2}$ is 180% of what number? ______
50. 10 is $6\frac{1}{4}$% of what number? ______
51. 190.75 is 70.5% of what number? ______
52. 6.6 is 3.3% of what number? ______
53. 88 is 205% of what number? ______
54. 1.3 is 0.9% of what number? ______
55. $\frac{7}{8}$ is 175% of what number? ______
56. $\frac{1}{10}$ is $1\frac{1}{5}$% of what number? ______
57. 9.3 is 238.6% of what number? ______
58. 0.84 is 2.04% of what number? ______
59. $20\frac{1}{2}$ is 71% of what number? ______
60. $\frac{3}{4}$ is 123% of what number? ______

Finding Percentage, Percent, or Base

Find each percentage, percent (rate), or base. Round the answers to 2 decimal places when necessary.

61. What percent of 24 is 18? ______
62. What is 30% of 50? ______
63. What is 123.8% of 12.6? ______
64. 73 is 82% of what number? ______
65. What percent of $10\frac{1}{2}$ is 2? ______
66. ______ is 48% of 94.82.
67. 72.4% of 212.7 is ______.
68. What percent of 228 is 256? ______
69. 51.03 is 88% of what number? ______
70. 36.5 is ______ % of 27.6.
71. $2\frac{1}{4}$% of 150 is ______.
72. ______ is 18% of 120.66.

UNIT 22

Percent Practical Applications

Objectives ***After studying this unit you should be able to***

- Solve simple percentage practical applications in which two of the three parts are given.
- Solve more complex percentage practical applications in which two of the three parts are not directly given.

Identifying Rate, Base, and Percentage in Various Types of Practical Applications

In solving simple problems, generally, there is no difficulty in identifying the rate or percent. A common mistake is to incorrectly identify the percentage and the base. There is sometimes confusion as to whether a value is a percentage or a base; the base and percentage are incorrectly interchanged.

The following statements summarize the information that was given when each of the three types of problems was discussed and solved. A review of the statements should be helpful in identifying the rate, percentage, and base.

- The rate (percent) is the part taken of the whole quantity (base).
- The base is the whole quantity or a quantity that is equal to 100%. It is the quantity of which the rate is taken.
- The percentage is the quantity of the percent that is taken of the base. It is the quantity equal to the percent that is taken of the whole.
- If the rate is 100%, the percentage and the base are the same quantity.
  If the rate is less than 100%, the percentage is less than the base.
  If the rate is greater than 100%, the percentage is greater than the base.
- In practical applications, the percentage and the base have the same unit or denomination. The rate does not have a unit or denomination.
- The word *is* generally relates to the rate or percentage, and the word *of* generally relates to the base.

Finding Percentage in Practical Applications

Example A production run of steel pins is estimated at \$3275. Material cost is estimated as 35% of the total cost. What is the estimated material cost to the nearest dollar?

Think the problem through to determine what is given and what is to be found.

The rate is 35%.

The base, B, is \$3275. It is the total cost or the whole quantity.

The percentage, P, which is the material cost, is to be found.

The proportion is $\dfrac{P}{\$3275} = \dfrac{35}{100}$.

Cross multiply $100P = 35 \times \$3{,}275$

$100P = \$114{,}625$

Divide $P = \dfrac{\$114{,}625}{100} = \$1{,}146.25$ Ans

Finding Percent (Rate) in Practical Applications

Example An inspector rejects 23 out of a total production of 630 electrical switches. What percent of the total production is rejected? Round the answer to 1 decimal place.

Think the problem through to determine what is given and what is to be found.

Since a percent of the total production of 630 switches is to be found, the base or whole quantity equal to 100% is 630 switches.

The proportion is $\dfrac{23 \text{ switches}}{630 \text{ switches}} = \dfrac{R}{100}$.

$$23 \text{ switches} \times 100 = 630 \text{ switches} \times R$$

$$2300 \text{ switches} = 630 \text{ switches} \times R$$

$$\frac{2300 \text{ switches}}{630 \text{ switches}} = R$$

$$R = \frac{2300 \cancel{\text{switches}}}{630 \cancel{\text{switches}}} = 3.651, 3.7\% \quad \text{Ans (rounded)}$$

23 [×] 100 [÷] 630 [=] 3.650793651, 3.7% Ans (rounded)

Finding the Base in Practical Applications

Example A motor is said to be 80% efficient if the output (power delivered) is 80% of the input (power received). How many horsepower does a motor receive if it is 80% efficient with a 6.20-horsepower (hp) output?

Think the problem through to determine what is given and what is to be found. The rate is 80%, so $R = 80$.

Since the output of 6.20 hp is the quantity of the percent of the base, the percentage is 6.20 hp (6.20 hp is 80% of the base).

The base to be found is the input; the whole quantity equal to 100%.

$R = 80$ and $P = 6.20$ hp, so the proportion is $\dfrac{6.20 \text{ hp}}{B} = \dfrac{80}{100}$.

$$6.20 \text{ hp} \times 100 = 80B$$

$$620 \text{ hp} = 80B$$

$$\frac{620 \text{ hp}}{80} = B$$

$$B = \frac{620 \text{ hp}}{80} = 7.75 \text{ hp} \quad \text{Ans}$$

More Complex Percentage Practical Applications

In certain percentage problems, two of the three parts are not directly given. One or more additional operations may be required in setting up and solving a problem. Examples of these types of problems follow.

Examples

1. By replacing high-speed cutters with carbide cutters, a machinist increases production by 35%. Using carbide cutters, 270 pieces per day are produced. How many pieces per day were produced with high-speed steel cutters?

   Think the problem through. The base (100%) is the daily production using high-speed steel cutters. Since the base is increased by 35%, the carbide cutter production of 270 pieces is 100% + 35% or 135% of the base. Therefore, the rate is 135% and the percentage is 270. The base is to be found.

The proportion is $\frac{270 \text{ pieces per day}}{B} = \frac{135}{100}$.

$$270 \text{ pieces per day} \times 100 = 135B$$

$$27{,}000 \text{ pieces per day} = 135B$$

$$\frac{27{,}000 \text{ pieces per day}}{135} = B$$

$$B = \frac{27{,}000 \text{ pieces per day}}{135} = 200 \text{ pieces per day} \quad \text{Ans}$$

2. A mechanic purchases a set of socket wrenches for \$54.94. The purchase price is 33% less than the list price. What is the list price?

   Think the problem through. The base (100%) is the list price. Since the base is decreased by 33%, the purchase price, \$54.94, is 100% − 33% or 67% of the base. Therefore, the rate is 67% and the percentage is \$54.94. The base is to be found.

The proportion is $\frac{\$54.94}{B} = \frac{67}{100}$.

$$\$54.94 \times 100 = 67B$$

$$\$5494 = 67B$$

$$\frac{\$5494}{67} = B$$

$$B = \frac{\$5494}{67} = \$82 \quad \text{Ans}$$

3. An aluminum bar measures 137.168 millimeters before it is heated. When heated, the bar measures 137.195 millimeters. What is the percent increase in length? Express the answer to 2 decimal places.

   Think the problem through. The base (100%) is the bar length before heating, 137.168 millimeters. The increase in length is 137.195 millimeters − 137.168 millimeters or 0.027 millimeter. Therefore, the percentage is 0.027 millimeter, and the base is 137.168 millimeters. The rate (percent) is to be found.

The proportion is $\frac{0.027 \text{ mm}}{137.168 \text{ mm}} = \frac{R}{100}$.

$$0.027 \text{ mm} \times 100 = 137.168 \text{ mm}R$$

$$2.7 \text{ mm} = 137.168 \text{ mm}R$$

$$\frac{2.7 \text{ mm}}{137.168 \text{ mm}} = R$$

$$R = \frac{2.7 \cancel{\text{mm}}}{137.168 \cancel{\text{mm}}} = 0.019968\%, 0.02\% \quad \text{Ans (rounded)}$$

Application

Finding Percentage, Percent, and Base in Practical Applications

Solve the following problems.

1. The total amount of time required to machine a part is 12.5 hours. Milling machine operations take 7.0 hours. What percent of the total time is spent on the milling machine? ____________
2. A casting, when first poured, is 17.875 centimeters long. The casting shrinks 0.188 centimeter as it cools. What is the percent shrinkage? Round the answer to 2 decimal places. ____________
3. A machine operator completes a job in 80% of the estimated time. The estimated time is $8\frac{1}{2}$ hours. How long does the job actually take? ____________
4. A machine is sold for 42% of the original cost. If the original cost is $9,255.00, find the selling price of the used machine. ____________
5. On a production run, 8% of the units manufactured are rejected. If 120 units are rejected how many total units are produced? ____________
6. An engine loses 4.2 horsepower through friction. The power loss is 6% of the total rated horsepower. What is the total horsepower rating? ____________
7. A small manufacturing plant employs 130 persons. On certain days, 16 employees are absent. What percent of the total number of employees are absent? Round the answer to the nearest whole percent. ____________
8. This year's earnings of a company are 140% of last year's earnings. The company earned $910,000 this year. How much did the company earn last year? ____________
9. In three hours 73.50 feet of railing are fabricated. This is 28% of a total order. How many feet of railing were ordered? ____________
10. How many pounds of manganese bronze can be made with 955.0 pounds of copper if the manganese bronze is to contain 58% copper by weight? Round the answer to the nearest whole pound. ____________
11. An alloy of manganese bronze is made up by weight of 58% copper, 40% tin, 1.5% manganese, and 0.5% other materials. How many pounds of each metal are there in 1,250 pounds of alloy? Round the answers to the nearest whole pound.

    Copper: ____________
    Tin: ____________
    Manganese: ____________
    Other: ____________
12. A manufacturer estimates the following percent costs to produce a product: labor, 38%; materials, 45%; overhead, 17%. The total cost of production is $120,000. Determine each of the dollar costs.

    Labor: ____________
    Materials: ____________
    Overhead: ____________
13. An iron casting shrinks $\frac{1}{8}$ inch per foot. What is the percentage shrinkage? Round the answer to the nearest whole percent. ____________
14. A hot brass casting when first poured in a mold is 9.25 inches long. The shrinkage is 1.38%. What is the length of the casting when cooled? Round the answer to 2 decimal places. ____________
15. The following table shows the number of pieces manufactured in three consecutive days. The numbers of defective pieces are shown as rework and scrap for each day. Determine the percents of rework and scrap for each day. Round the answers to 1 decimal place.

Date	Number of Pieces Manufactured	Number of Defective Pieces		% Defective Pieces	
		Rework	Scrap	Rework	Scrap
9/16	1650	44	59		
9/17	1596	29	48		
9/18	1685	52	34		

16. The power output of a machine is equal to the product of the power input and the percent efficiency. Power Output = Power Input × Percent Efficiency. What is the output of a machine with an 8.0-horsepower motor running at full capacity and at 82% efficiency? Round the answer to 1 decimal place. __________

17. Material cost for a job is $1260. The cost is 38.6% of the total cost. What is the total cost? Round the answer to the nearest dollar. __________

18. A machinist's weekly gross income is $745. The following percent deductions are made from the gross income:

    Federal Income Tax, 14.20%

    State Income Tax, 4.50%

    Social Security, 7.60%

    Determine the net income (take-home pay) after the deductions are made. __________

19. In the heat treatment of steel, a rough approximation of temperatures can be made by observing the color of the heated steel. At approximately 1300°F (degrees Fahrenheit) the steel is dark red. What percent increase in temperature from the 1300°F must be made for the heated steel to turn to each of the following colors? Round the answers to the nearest whole percent.

    a. Dull cherry-red at approximately 1470°F __________

    b. Orange-yellow at approximately 2200°F __________

    c. Brilliant white at approximately 2730°F __________

20. The following table lists the percent of carbon by weight for various types of carbon steel tools. Determine the amount of carbon needed to produce 2.60 tons of carbon steel required in the production of each type of tool. Round the answers to the nearest pound.

Type of Machinist's Tool	Percent Carbon	Type of Machinist's Tool	Percent Carbon
1. Twist Drill	1.15	4. Ordinary File	1.25
2. Wrench	0.75	5. Machinist's Hammer	0.95
3. Threading Die	1.05	6. Chuck Jaw	0.85

1. __________
2. __________
3. __________
4. __________
5. __________
6. __________

21. A machine shop has 2600 castings in stock at the beginning of the month. At the end of the first week, 28.0% of the stock is used. At the end of the second week, 50.0% of the stock remaining is used. How many castings remain in stock at the end of the second week? __________

22. It is estimated that 125 meters of channel iron are required for a job. Channel iron is ordered, including an additional 20% allowance for scrap and waste. Actually, 175 meters of channel iron are used for the job. The amount actually used is what percent more than the amount ordered? Round the answer to the nearest whole percent. __________

23. An alloy of red brass is composed of 85% copper, 5% tin, 6% lead, and zinc. Find the number of pounds of zinc required to make 450 pounds of alloy. ______

24. The day shift of a manufacturing firm produces 6% defective pieces out of a total production of 1638 pieces. The night shift produces $4\frac{1}{2}\%$ defective pieces out of a total of 1454 pieces. How many more acceptable pieces are produced by the day shift than by the night shift? ______

25. The following table shows the number of pieces of a product produced each day during one week. Also shown are the number of pieces rejected each day by the quality control department. Find the percent rejection for the week's production. Round the answer to 1 decimal place. ______

	MON.	TUES.	WED.	THUR.	FRI.
Number of Pieces Produced	735	763	786	733	748
Number of Pieces Rejected	36	43	52	47	31

26. A manufacturer estimates that 15,500 pieces per day could be produced with the installation of new machinery. The machines now used produce 11,000 pieces per day. What percent increase in production would be gained by replacing the present machinery with new machinery? Round the answer to the nearest whole percent. ______

27. The average percent defective product of a manufacturing plant is 1.20%. On a particular day 50 pieces were rejected out of a total daily production of 2730 pieces. What is the percent increase of defective pieces for the day above the average percent defective? Round the answer to 2 decimal places. ______

28. Before machining, a steel forging weighs 7.8 pounds. A milling operation removes 1.5 pounds, drilling removes 0.7 pound, and grinding removes 0.5 pound. What percent of the original weight of the forging is the final machined forging? Round the answer to 1 decimal place. ______

29. A machine is 85% efficient and loses 1.3 horsepower through its drivetrain. Determine the horsepower input of the machine. Round the answer to 1 decimal place. ______

30. The cost of one dozen cutters is listed as $525. A multiple discount of 12% and 8% is applied to the purchase. Determine the net (selling) price of the cutters. ______

> **Note:** With multiple discounts, the first discount is subtracted from the list price. The second discount is subtracted from the price computed after the first discount was subtracted.

31. A manufacturer's production this week is 3620 pieces. This is 13.5% greater than last week's production. Find last week's production. Round the answer to the nearest whole piece. ______

32. Two machines are used to produce the same product. One machine has a capability of producing 750 pieces per 8-hour shift. It is operating at 80% of its capability. The second machine has a capability of producing 900 pieces per 8-hour shift. It is operating at 75% of its capability. Find the total number of pieces produced per hour with both machines operating. Round the answer to the nearest whole piece. ______

33. Allowing for scrap, a firm produced 1890 pieces. The number produced is 8% more than the number of pieces required for the order. How many pieces does the order call for? Round the answer to the nearest whole piece. ______

34. A manufacturing company receives $122,000 upon the completion of a job. Total expenses for the job are $110,400. What percent of the job is profit? Round the answer to 1 decimal place. ______

35. Manufacturing costs consist of labor costs, material costs, and overhead. Refer to the following table. What percent of the total manufacturing cost for each of Jobs 1, 2, and 3 is each manufacturing cost? Round the answer to the nearest whole percent.

	MANUFACTURING COSTS		
Job	**Labor Costs**	**Material Costs**	**Overhead Costs**
1	$1890	$ 875	$1240
2	$ 930	$1060	$ 880
3	$2490	$1870	$1600

1. Labor: ______
   Materials: ______
   Overhead: ______
2. Labor: ______
   Materials: ______
   Overhead: ______
3. Labor: ______
   Materials: ______
   Overhead: ______

Achievement Review—Section Two

Objective *You should be able to solve the exercises and problems in this achievement Review by appliying the principles and methods covered in Units 18–22.*

Express these ratios in lowest fractional form.

1. 15:32
2. 46:12
3. 12:46
4. 27 mm:45 mm
5. 21 ft:33 ft
6. 45 in.:27 in.
7. $\frac{1}{4}$ to $\frac{1}{2}$
8. 16 to $\frac{2}{3}$
9. $\frac{1}{4}$ h to 25 min
10. 2 ft to 8 in.

11. The cost and selling price of merchandise are listed in the following table. Determine the cost-to-selling price ratio and the cost-to-profit ratio.

Note: Profit = Selling price − Cost.

	Cost	Selling Price	Ratio of Cost to Selling Price	Ratio of Cost to Profit
a.	$ 60	$ 96		
b.	$105	$180		
c.	$ 18	$ 33		
d.	$204	$440		

12. Bronze is an alloy of copper, zinc, and tin with small amounts of other elements. Two types of bronze castings are listed in the table below with the percent composition of copper, tin, and zinc in each casting. Determine the ratios called for in the table.

	Type of Casting	Percent Composition			Ratios		
		Copper	Tin	Zinc	Copper To Tin	Tin to Zinc	Copper to Zinc
a.	Manganese Bronze	58	1	40			
b.	Hard Bronze	86	10	2			

13. Solve for the unknown value in each of the following proportions and check. Round the answers to 3 decimal places where necessary.

a. $\frac{P}{12.8} = \frac{3}{2}$ __________

b. $\frac{3.6}{0.9} = \frac{E}{2.7}$ __________

c. $\frac{H}{\frac{1}{4}} = \frac{\frac{1}{2}}{\frac{3}{8}}$ __________

d. $\frac{7}{6} = \frac{C}{12}$ __________

e. $\frac{6.5}{M} = \frac{8.2}{41}$ __________

f. $\frac{22.517}{13.503} = \frac{1.297}{x}$ __________

g. $\frac{20.021}{5.773T} = \frac{1.892}{4.518}$ __________

h. $\frac{10.360}{7.890} = \frac{2.015}{N}$ __________

14. Analyze each of the following problems to determine whether the problem is a direct or inverse proportion and solve. Round the answers to 3 decimal places where necessary.
   a. A reamer tapers 0.0975 inch along a 3.2625-inch length. What is the amount of taper along a 2.1250-inch length? __________
   b. A machine produces 2550 parts in 8.5 hours. How many parts are produced by the machine in 10 hours? __________
   c. Of two gears that mesh, one gear with 12 teeth revolves at 420 rpm. What is the revolutions per minute of the other gear, which has 16 teeth? __________

15. Express each value as a percent.

a. 1 ________ b. $1\frac{1}{2}$ ________ c. $2\frac{3}{4}$ ________ d. 0.5 ________

16. Express each value as a percent.

a. 0.72 ________ b. 2.037 ________ c. $\frac{1}{25}$ ________ d. 0.0003 ________

17. Express each percent as a decimal fraction or mixed decimal.

a. 19% ________ b. 0.7% ________ c. $\frac{3}{4}\%$ ________ d. $310\frac{3}{10}\%$ ________

18. Express each percent as a common fraction or mixed number.

a. 30% ________ b. 140% ________ c. 12.5% ________ d. 0.65% ________

19. Find each percentage. Round the answers to 2 decimal places when necessary.

a. 15% of 60 __________

b. 3% of 42.3 __________

c. 72.8% of 120 __________

d. 0.7% of 812 __________

e. 42.6% of 53.76 __________

f. 130% of 212 __________

g. $12\frac{1}{2}\%$ of 32 __________

h. $\frac{1}{4}\%$ of 627.3 __________

20. Find each percent (rate). Round the answers to 2 decimal places when necessary.
   a. What percent of 10 is 2? ______
   b. What percent of 2 is 10? ______
   c. What percent of 88.7 is 21.9? ______
   d. What percent of 275 is 108? ______
   e. What percent of 2.84 is 0.8? ______
   f. What percent of $12\frac{1}{4}$ is 3? ______
   g. What percent of 312 is 400.9? ______
21. Find each base. Round the answers to 2 decimal places when necessary.
   a. 20 is 60% of what number? ______
   b. 4.1 is 24.9% of what number? ______
   c. 340 is 152% of what number? ______
   d. 44.08 is 73.5% of what number? ______
   e. 9.3 is 238.6% of what number? ______
   f. 0.84 is 2.04% of what number? ______
   g. $\frac{3}{4}$ is 123% of what number? ______
22. Find each percentage, percent (rate), or base. Round the answers to 2 decimal places when necessary.
   a. What percent of 24 is 18? ______
   b. What is 30% of 50? ______
   c. What is 123.8% of 12.6? ______
   d. 73 is 82% of what number? ______
   e. What percent of $10\frac{1}{2}$ is 2? ______
   f. __?__ is 48% of 94.82. ______
   g. 72.4% of 212.7 is __?__. ______
   h. What percent of 228 is 256? ______
   i. 51.08 is 88% of what number? ______
   j. 36.5 is __?__% of 27.6. ______
   k. $2\frac{1}{4}$% of 150 is __?__. ______
   l. __?__ is 18% of 120.66. ______
23. The carbon content of machine steel for gages usually ranges from 0.15% to 0.25%. Round the answers for a and b to 2 decimal places.
   a. What is the minimum weight of carbon in 250 kilograms of machine steel? ______
   b. What is the maximum weight of carbon in 250 kilograms of machine steel? ______
24. A piece of machinery is purchased for $8792. In one year the machine depreciates 14.5%. By how many dollars does the machine depreciate in one year? Round the answer to the nearest dollar. ______
25. Engine pistons and cylinder heads are made of an aluminum casting alloy that contains 4% silicon, 1.5% magnesium, and 2% nickel. Round the answers to the nearest tenth kilogram.
   a. How many kilograms of silicon are needed to produce 575 kilograms of alloy? ______
   b. How many kilograms of magnesium are needed to produce 575 kilograms of alloy? ______
   c. How many kilograms of nickel are needed to produce 575 kilograms of alloy? ______
26. Before starting two jobs, a shop has an inventory of eighteen 15.0-foot lengths of flat stock. The first job requires 30% of the inventory. The second job requires 25% of the inventory remaining after the first job. How many feet of flat stock remain in inventory at the end of the second job? Round the answer to the nearest whole foot. ______
27. An alloy of stainless steel contains 73.6% iron, 18% chromium, 8% nickel, 0.1% carbon, and sulfur. How many pounds of sulfur are required to make 5800 pounds of stainless steel? Round the answer to the nearest whole pound. ______
28. Two machines together produce a total of 2015 pieces. Machine A operates for $6\frac{1}{2}$ hours and produces an average of 170 pieces per hour. Machine B operates for 7 hours. What percent of the average hourly production of Machine A is the average hourly production of Machine B? Round the answer to the nearest whole percent. ______

Linear Measurement: Customary (English) and Metric

UNIT 24

Customary (English) Units of Measure

Objectives *After studying this unit you should be able to*

- Express customary lengths as larger or smaller customary linear units.
- Perform arithmetic operations with customary linear units and compound numbers.

The United States uses two systems of weights and measures: the American or U.S. customary system and the International System of Units called the SI metric system.

The American customary system is based on the English system of weights and measures and is sometimes called the "English" system. Throughout this book, American customary units are called "customary" units and SI metric units are called "metric" units.

Both customary and metric systems include all types of units of measure, such as length, area, volume, and capacity. It is important that you have the ability to measure and compute with both customary and metric units.

In the machine trades, linear or length measure is used most often. Throughout this book, linear measure is the primary type of measure presented. However, in Section Six, some fundamentals of area and volume and their applications are presented.

Measurement Definitions

Measurement is the comparison of a quantity with a standard unit. A *linear measurement* is a means of expressing the distance between two points; it is the measurement of lengths. A linear measurement has two parts: a unit of length and a multiplier.

The measurements 3.872 inches and 27.18 millimeters are examples of denominate numbers. A *denominate number* is a number that specifies a unit of measure.

Customary Units of Linear Measure

The yard is the standard unit of linear measure in the customary system. From the yard, other units such as the inch and foot are established. The smallest unit is the inch. Common customary units of length with their symbols are shown in the table below.

CUSTOMARY UNITS OF LINEAR MEASURE
1 yard (yd) = 3 feet (ft)
1 yard (yd) = 36 inches (in)
1 foot (ft) = 12 inches (in)
1 mile (mi) = 1760 yards (yd)
1 mile (mi) = 5280 feet (ft)

In the machine trades, customary linear units other than the inch are seldom used. Customary measure dimensions on engineering drawings are given in inches. Although customary linear units other than the inch are rarely required for on-the-job applications, you should be able to use any units in the system.

Notice that most of the symbols, ft for foot, mi for mile, yd for yard, do not have periods at the end. That is because they are symbols and not abbreviations. The one exception is in. for inch. Many people prefer in. because the period helps you know that they do not mean the word "in."

Expressing Equivalent Units of Measure

When expressing equivalent units of measure, either of two methods can be used. Throughout the unit, examples are given using either of the two methods. Many examples show how both methods are used in expressing equivalent units of measure.

METHOD 1 This is a practical method used for many on-the-job applications. It is useful when simple unit conversions are made. In this method you multiply or divide the given unit of measure by a *conversion factor*.

METHOD 2 This method is called the *unity fraction method*. The unity fraction method eliminates the problem of incorrectly expressing equivalent units of measure. Using this method removes any doubt as to whether to multiply or divide when making a conversion. The unity fraction method is particularly useful in solving problems that involve a number of unit conversions.

This method multiplies the given unit of measure by a fraction equal to one. The unity fraction contains the given unit of measure and its equivalent expressed in the unit of measure to which the given unit is to be converted. The unity fraction is set up in such a way that the original unit cancels out and the unit you are converting to remains. Recall that canceling is the common term used when a numerator and a denominator are divided by a common factor.

Expressing Larger Customary Units of Linear Measure as Smaller Units

Procedure To express a larger unit as a smaller unit of length, either

- Multiply the given length by the number of smaller units contained in one of the larger units (Method 1), or
- Multiply the given length by an appropriate unity fraction (Method 2).

Examples

1. Express 2.28 yards as inches.

METHOD 1

Since 36 inches equal 1 yard, the conversion factor is 36.

Multiply 2.28 by 36. $2.28 \times 36 = 82.08$

$2.28 \text{ yd} \approx 82.1 \text{ inches}$ Ans

METHOD 2

Since 36 inches equal 1 yard, the unity fraction is $\frac{36 \text{ in}}{1 \text{ yd}}$. Multiply 2.28 yd by the unity fraction. $2.28 \cancel{\text{yd}} \times \frac{36 \text{ inches}}{1 \cancel{\text{yd}}} = 2.28 \times 36 \text{ in} \approx 82.1 \text{ in}$ Ans

2. Express $2\frac{1}{2}$ ft as inches.

METHOD 1

Since 12 inches equal 1 foot, the conversion factor is 12.

Multiply $2\frac{1}{2}$ by 12. $2\frac{1}{2} \times 12 = 30$

$2\frac{1}{2} \text{ feet} = 30 \text{ inches}$ Ans

METHOD 2

Since 12 inches equal 1 foot, the unity fraction is $\frac{12 \text{ in}}{1 \text{ ft}}$. Multiply $2\frac{1}{2}$ ft by the unity fraction. $2\frac{1}{2} \cancel{\text{ft}} \times \frac{12 \text{ in}}{1 \cancel{\text{ft}}} = 2\frac{1}{2} \times 12 \text{ in} = 30 \text{ inches}$ Ans

3. How many inches are in 0.25 yard?

METHOD 1

Since 36 inches equal 1 yard, multiply 0.25 by 36. $0.25 \times 36 = 9$

$0.25 \text{ yard} = 9 \text{ inches}$ Ans

METHOD 2

Since 36 inches equal 1 yard, the unity fraction is $\frac{36 \text{ in}}{1 \text{ yd}}$. Multiply 0.25 yd by the unity fraction. $0.25 \cancel{\text{yd}} \times \frac{36 \text{ in}}{1 \cancel{\text{yd}}} = 0.25 \times 36 \text{ in} = 9 \text{ inches}$ Ans

Expressing Smaller Customary Units of Linear Measure as Larger Units

Procedure To express a smaller unit as a larger unit of length, either

- Divide the given length by the number of smaller units contained in one of the larger units (Method 1), or
- Multiply the given length by an appropriate unity fraction (Method 2).

Example 1 Express 67.2 inches as feet.

METHOD 1

Since 12 inches equal 1 foot, divide 67.2 by 12.

$67.2 \div 12 = 5.6$

67.2 inches = 5.6 feet Ans

METHOD 2

Since 12 inches equal 1 foot, the unity fraction is $\frac{1 \text{ ft}}{12 \text{ in}}$. Multiply 67.5 inches by the unity fraction.

$$67.5 \cancel{\text{in}} \times \frac{1 \text{ ft}}{12 \cancel{\text{in}}} = \frac{67.5 \text{ ft}}{12} = 5.6 \text{ feet} \quad \text{Ans}$$

Example 2 How many yards are in 122.4 inches?

METHOD 1

Since 36 inches equal 1 yard, divide 122.4 by 36.

$122.4 \div 36 = 3.4$

122.4 inches = 3.4 yards Ans

METHOD 2

Since 36 inches equal 1 yard, the unity fraction is $\frac{1 \text{ yd}}{36 \text{ in}}$. Multiply 122.4 inches by the unity fraction.

$$122.4 \cancel{\text{in}} \times \frac{1 \text{ yd}}{36 \cancel{\text{in}}} = \frac{122.4 \text{ yd}}{36} = 3.4 \text{ yards} \quad \text{Ans}$$

Arithmetic Operations with Linear Units

In order to add two measurements, they must be in the same units. The addition of unlike units can not be performed unless one of the measurements is converted to the other unit. For example, 9 inches and 4 inches can be added. Both units are in inches. But, 7 inches and 5 feet cannot be added unless 7 inches is expressed in feet or 5 feet is expressed in inches.

Procedure To add or subtract measures in the same units

- Add or subtract the numerical values, or
- Leave the units unchanged.

Example 1 Add 5 inches and 13 inches.

$$\begin{array}{l} 5 \text{ inches} \\ \underline{13 \text{ inches}} \\ 18 \text{ inches} \quad \text{Ans} \end{array}$$

Example 2 Subtract $7\frac{1}{2}$ inches from $21\frac{1}{4}$ inches.

$$\begin{aligned} 21\tfrac{1}{4} \text{ inches} &= 20\tfrac{5}{4} \text{ inches} \\ 7\tfrac{1}{2} \text{ inches} &= 7\tfrac{2}{4} \text{ inches} \\ \hline & 13\tfrac{3}{4} \text{ inches} \quad \text{Ans} \end{aligned}$$

Addition and Subtraction of Compound Numbers

To add or subtract compound numbers, arrange like units in the same column, then add each column. When necessary, simplify the answer.

Example Add 3 feet and 9 inches, 2 feet $8\frac{3}{4}$ inches, and 2 feet $10\frac{1}{2}$ inches.

Arrange line units in the same column

$$\begin{array}{ll} 3 \text{ ft} & 9 \text{ in} \\ 2 \text{ ft} & 8\frac{3}{4} \text{ in} \\ 2 \text{ ft} & 10\frac{1}{2} \text{ in} \\ \hline 7 \text{ ft} & 28\frac{1}{4} \text{ in} \end{array}$$

Add each column

Simplify the sum. Divide $28\frac{1}{4}$ by 12 to express $28\frac{1}{4}$ inches as 2 feet $4\frac{1}{4}$ inches.

$$28\frac{1}{4} \text{ inches} = 2 \text{ feet } 4\frac{1}{4}$$

Add. $7 \text{ feet} + 2 \text{ feet } 4\frac{1}{4} \text{ inches} = 9 \text{ feet } 4\frac{1}{4} \text{ inches}$ Ans

Multiplication of Compound Numbers

To multiply compound numbers, multiply each unit of the compound number by the multiplier. When necessary, simplify the product.

Example Six pieces are to be cut from a piece of bar stock. Each piece is 1 foot $5\frac{3}{4}$ inches long. Allow $\frac{1}{8}$ inch for cutting each piece.

Add the length of each piece and the cut. 1 foot $5\frac{3}{4}$ inches + $\frac{1}{8}$ inch

1 foot $5\frac{7}{8}$ inches

Multiply this sum by 6.

$$\begin{array}{l} 1 \text{ foot } 5\frac{7}{8} \text{ inches} \\ \times 6 \\ \hline 6 \text{ feet } 30\frac{42}{8} \text{ inches} \\ 6 \text{ feet } 35\frac{1}{4} \text{ inches} \end{array}$$

Simplify the product. 8 feet $11\frac{1}{4}$ inches Ans

Division of Compound Numbers

To divide compound numbers, divide each unit of the divisor starting at the left. If a unit is not exactly divisible, express the remainder as the next smaller unit and add it to the given number of smaller units. When necessary, simplify the product.

Example The five holes in the angle iron shown in Figure 24-1 are equally spaced. Determine the distance between two consecutive holes.

Figure 24-1

Since there are four spaces between holes, divide 14 feet 6 inches by 4.	
Divide 14 feet by 4.	14 ft ÷ 4 = 3 ft (quotient) and a 2 ft remainder.
Express the 2-foot remainder as 24 inches.	2 ft = 2 × 12 in = 24 in
Add 24 inches to the 6 inches given in the problem.	24 in + 6 in = 30 in
Divide 30 inches by 4.	30 in ÷ 4 = $7\frac{1}{2}$ in (quotient)
Collect quotients.	3 ft $7\frac{1}{2}$ in Ans

Application

Customary Units of Linear Measure

1. Express each of the following lengths as indicated.

   a. 96 inches as feet ________
   b. 123 inches as feet ________
   c. $3\frac{1}{2}$ feet as inches ________
   d. 0.4 yard as inches ________
   e. $1\frac{1}{4}$ yards as inches ________
   f. 144 inches as yards ________
   g. 75 inches as feet ________
   h. 8 yards as feet ________
   i. 4.2 yards as feet ________
   j. 27 feet as yards ________
   k. 51 feet as yards ________
   l. $\frac{1}{3}$ yard as inches ________
   m. 258 inches as feet ________
   n. $7\frac{2}{3}$ feet as inches ________
   o. 0.20 yard as inches ________
   p. 140.25 feet as yards ________
   q. 333 inches as yards ________
   r. 186 inches as feet ________
   s. $20\frac{2}{3}$ yards as feet ________
   t. 9.25 feet as inches ________

2. A $3\frac{1}{2}$-inch diameter milling cutter revolving at 130.0 revolutions per minute has a cutting speed of 120.0 feet per minute. What is the cutting speed in inches per minute? ________
3. How many complete 6-foot lengths of round stock should be ordered to make 230 pieces each 1.300 inches long? Allow $1\frac{1}{2}$ lengths of stock for cutoff and scrap. ________
4. Pieces each 3.25 inches long are to be cut from lengths of bar stock. Allowing 0.10 inch for cutoff per piece, how many complete pieces can be cut from twelve 8-foot lengths of stock? ________

Arithmetic Operations with Linear Units

5. Add or subtract as indicated. Express each answer in the same unit as given in the exercise.

a. 3 in + 7 in ____________

b. 12 in + 5 in ____________

c. 15 in − 8 in ____________

d. 12 in − 9 in ____________

e. $11\frac{1}{4}$ in + $4\frac{3}{8}$ in ____________

f. $7\frac{3}{4}$ in + $6\frac{5}{8}$ in ____________

g. $7\frac{3}{4}$ in − $4\frac{5}{8}$ in ____________

h. $8\frac{1}{4}$ in − $4\frac{11}{16}$ in ____________

6. Add or subtract as indicated. Express each answer in the same units as given in the exercise.

a. 5 ft 4 in + 2 ft 6 in ____________

b. 3 ft 8 in + 4 ft 3 in ____________

c. 3 ft 7 in − 2 ft 6 in ____________

d. 5 ft 9 in − 2 ft 7 in ____________

e. 7 ft $3\frac{1}{2}$ in + 5 ft $2\frac{3}{8}$ in ____________

f. 2 ft $5\frac{3}{4}$ in + 3 ft $4\frac{7}{16}$ in ____________

g. 7 ft $6\frac{1}{2}$ in − 6 ft $5\frac{3}{4}$ in ____________

h. 3 ft $9\frac{7}{8}$ in − 1 ft $7\frac{5}{16}$ in ____________

Arithmetic Operations with Compound Numbers

7. Perform the indicated operation. Express each answer in the same units as given in the exercise. Regroup the answer when necessary.

a. 3 ft 9 in + 4 ft 7 in ____________

b. 6 ft 5 in − 4 ft 9 in ____________

c. 7 ft 9 in × 3 ____________

d. 17 ft 11 in ÷ 5 ____________

e. 6 ft $3\frac{3}{8}$ in + 4 ft $1\frac{1}{2}$ in + 8 ft $10\frac{3}{4}$ in ____________

f. 10 ft $1\frac{3}{8}$ in − 7 ft $9\frac{7}{16}$ in ____________

g. 12 ft $3\frac{3}{4}$ in × 5.25 ____________

h. 10 ft $6\frac{1}{4}$ in ÷ 4 ____________

UNIT 25 Metric Units of Linear Measure

Objectives ***After studying this unit you should be able to***

- Express metric lengths as larger or smaller metric linear units.
- Express metric length units as customary length units.
- Express customary length units as metric length units.

An advantage of the metric system is that it allows easy and fast computations. Since metric units are based on powers of ten, computations are simplified. To express a metric unit as a smaller or larger unit, all that is required is to move the decimal point a certain number of places to the left or right.

The metric system does not require difficult conversions as with the customary system. For example, it is easier to remember that 1000 meters equal 1 kilometer than to remember that 1760 yards equal 1 mile. The meter is the standard unit of linear measure in the metric system. Other linear metric units are based on the meter.

Metric measure dimensions on engineering drawings are given in millimeters. In the machine trades, metric linear units other than the millimeter are seldom used. However, you should be able to use any units in the system. Some metric units of length with their symbols are shown in this table. Observe that each unit is ten times greater than the unit directly above it.

METRIC UNITS OF LINEAR MEASURE	
1 millimeter (mm) = 0.001 meter (m)	1000 millimeters (mm) = 1 meter (m)
1 centimeter (cm) = 0.01 meter (m)	100 centimeters (cm) = 1 meter (m)
1 decimeter (dm) = 0.1 meter (m)	10 decimeters (dm) = 1 meter (m)
1 meter (m) = 1 meter (m)	1 meter (m) = 1 meter (m)
1 dekameter (dam) = 10 meters (m)	0.1 dekameter (dam) = 1 meter (m)
1 hectometer (hm) = 100 meters (m)	0.01 hectometer (hm) = 1 meter (m)
1 kilometer (km) = 1000 meters (m)	0.001 kilometer (km) = 1 meter (m)

The following metric power of ten prefixes are used throughout the metric system.

milli means one thousandth (0.001)
centi means one hundredth (0.01)
deci means one tenth (0.1)
deka means ten (10)
hecto means hundred (100)
kilo means thousand (1000)

The most frequently used metric units of length are the kilometer (km), meter (m), centimeter (cm), and millimeter (mm). In actual applications, the dekameter (dam) and hectometer (hm) are not used. The decimeter (dm) is seldom used.

Periods are *not* used after the unit symbols. For example, write 1 mm, *not* 1 m.m. or 1mm., when expressing the millimeter as a symbol. Some people spell the unit of measure *metre* so that it is not confused with a meter used as a measuring instrument.

Expressing Equivalent Units Within the Metric System

To express a given unit of length as a larger unit, move the decimal point a certain number of places to the left. To express a given unit of length as a smaller unit, move the decimal point a certain number of places to the right. The procedure of moving decimal points is shown in the following examples. Refer to the table of metric units of linear measure.

Example 1 Express 72 millimeters (mm) as centimeters (cm).

Since a centimeter is the next *larger* unit to a millimeter, move the decimal point 1 place to the *left*. (In moving the decimal point 1 place to the left, you are actually dividing by 10.)

7 2 .

72 mm = 7.2 cm Ans

Example 2 Express 0.96 centimeter (cm) as millimeters (mm).

Since a millimeter is the next *smaller* unit to a centimeter, move the decimal point 1 place to the *right*. (In moving the decimal point 1 place to the right, you are actually multiplying by 10.)

0 . 9 6

0.96 cm = 9.6 mm Ans

Example 3 Express 0.245 meter (m) as millimeters (mm).

Since a millimeter is three units smaller than a meter, move the decimal point 3 places to the right. (In moving the decimal point 3 places to the right, you are actually multiplying by 10^3 or 1000.)

0 . 2 4 5

0.245 m = 245 mm Ans

Example 4 Add. 0.3 meter (m) + 12.6 centimeters (cm) + 76 millimeters (mm). Express the answer in millimeters.

Express each value in millimeters.

0.3 m = 300 mm
12.6 cm = 126 mm
76 mm = 76 mm

Add. 300 ⊞ 126 ⊞ 76 ⊟ 502, 502 mm Ans

Metric-Customary Linear Equivalents (Conversion Factors)

Since both the customary and metric systems are used in this country, it is sometimes necessary to express equivalents between systems. Dimensioning an engineering drawing with both customary and metric dimensions is called *dual dimensioning*. Since dual dimensioning tends to clutter a drawing and introduces additional opportunities for error, many companies do not use the system. Instead, some companies use metric dimensions only, with an inch-millimeter conversion table on or attached to the print. However, certain companies use dual dimensioning; it can be a practical method for industries that have plants in foreign countries. Examples of two types of dual dimensioning are shown in Figure 25-1.

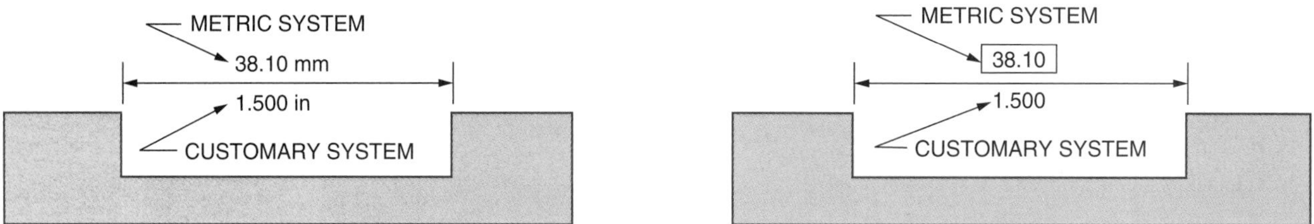

Figure 25-1

The commonly used equivalent factors of linear measure are shown in this table. Equivalent factors are commonly called conversion factors.

METRIC-CUSTOMARY LINEAR EQUIVALENTS (CONVERSION FACTORS)	
Metric to Customary Units	**Customary to Metric Units**
1 millimeter (mm) = 0.03937 inch (in)	1 inch (in) = 25.4 millimeters (mm)
1 centimeter (cm) = 0.3937 inch (in)	1 inch (in) = 2.54 centimeters (cm)
1 meter (m) = 39.37 inches (in)	1 foot (ft) = 0.3048 meter (m)
1 meter (m) = 3.2808 feet (ft)	1 yard (yd) = 0.9144 meter (m)
1 kilometer (km) = 0.6214 mile (mi)	1 mile (mi) = 1.609 kilometers (km)

Metric-customary linear equivalents other than millimeter-inch equivalents are seldom used in the machine trades. However, you should be able to express any unit in one measuring system as a unit in the other system. The relationship between customary decimal inch units and metric millimeter units is shown in Figure 25-2 by comparing these scales.

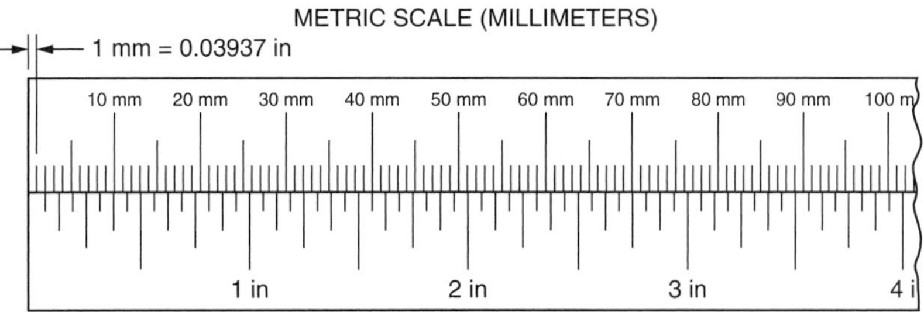

Figure 25-2

Procedure To express a unit in one system as an equivalent unit in the other system, either use unity fractions or multiply the given measurement by the appropriate conversion factor in the Metric-Customary Linear Equivalent Table.

Examples

1. Express 12.700 inches as millimeters.

METHOD 1

Since 1 inch = 25.4 mm, 12.700 × 25.4 mm = 322.58 mm Ans

METHOD 2

Multiply by the unity fraction $\frac{25.4\text{ mm}}{1\text{ in}}$. $12.700\cancel{\text{in}} \times \frac{25.4\text{ mm}}{1\cancel{\text{in}}} = 322.58\text{ mm}$ Ans

2. Express 6.780 centimeters as inches. Round the answer to 3 decimal places.

METHOD 1

Since 1 cm = 0.3937 in, 6.78 × 0.3937 in = 2.669 in Ans

METHOD 2

Multiply by the unity fraction $\frac{1\text{ in}}{2.54\text{ cm}}$. $6.78\cancel{\text{cm}} \times \frac{1\text{ in}}{2.54\cancel{\text{cm}}} = \frac{6.78\text{ in}}{2.54} = 2.669\text{ in}$ Ans

3. The template in Figure 25-3 is dimensioned in millimeters. Determine, in inches, the total length of the template. Round the answer to 3 decimal places.

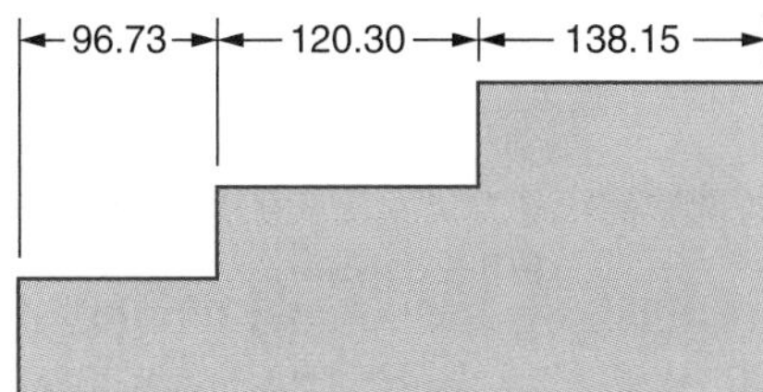

Figure 25-3

Add the dimensions in millimeters as they are given and express the sum in inches.

METHOD 1

Since 1 mm = 0.03937 in, multiply the sum in millimeters by 0.03937.

96.73 [+] 120.3 [+] 138.15 [=] [×] 0.03937 [=] 13.983437, 13.983 in Ans

or [(] 96.73 [+] 120.3 [+] 138.15 [)] [×] 0.03937 [=] 13.983437, 13.983 in Ans

METHOD 2

Multiply by the unity fraction $\frac{1 \text{ in}}{25.4 \cancel{\text{mm}}}$.

$$(96.73 + 120.3 + 138.15)\frac{1 \text{ in}}{25.4 \cancel{\text{mm}}}$$

(96.73 + 120.3 + 138.15) ÷ 25.4 = 13.983465, 13.983 in Ans

4. Express 453.7 millimeters as yards.

There is no conversion factor in the table for millimeters to yards, so we will have to use two conversion factors. First, convert millimeters to inches and then convert inches to yards. The conversion factors are $\frac{0.03937 \text{ in}}{1 \text{ mm}}$ and $\frac{1 \text{ yd}}{36 \text{ in}}$.

$$\frac{453.7 \cancel{\text{mm}}}{1} \times \frac{0.03937 \cancel{\text{in}}}{1 \cancel{\text{mm}}} \times \frac{1 \text{ yd}}{36 \cancel{\text{in}}} = \frac{453.7 \times 0.03937 \text{ yd}}{36} \approx 0.5 \text{ yd.}$$

Another way to solve this problem is to first convert millimeters to meters and then convert meters to yards. The conversion factors in this case are $\frac{1 \text{ m}}{1000 \text{ mm}}$ and $\frac{1 \text{ yd}}{0.9144 \text{ m}}$.

$$\frac{453.7 \cancel{\text{mm}}}{1} \times \frac{1 \cancel{\text{m}}}{1000 \cancel{\text{mm}}} \times \frac{1 \text{ yd}}{0.9144 \cancel{\text{m}}} = \frac{453.7 \text{ yd}}{1000 \times 0.9144} \approx 0.5 \text{ yd.}$$

Application

Metric Units of Linear Measure

1. Express each of the following lengths as indicated.

a. 2.9 centimeters as millimeters ______
b. 15.78 centimeters as millimeters ______
c. 219.75 millimeters as centimeters ______
d. 97.83 millimeters as centimeters ______
e. 0.97 meter as centimeters ______
f. 0.17 meter as millimeters ______
g. 153 millimeters as meters ______
h. 673 centimeters as meters ______
i. 0.93 millimeter as centimeters ______
j. 0.08 centimeter as millimeters ______
k. 0.0086 meter as millimeters ______
l. 1.046 meters as centimeters ______
m. 30.03 centimeters as millimeters ______
n. 876.84 millimeters as centimeters ______
o. 2039 millimeters as meters ______
p. 3.47 centimeters as meters ______
q. 0.049 meter as millimeters ______
r. 7.321 meters as centimeters ______
s. 6.377 centimeters as millimeters ______
t. 0.934 meter as millimeters ______

2. Perform the indicated operations. Express the answer in the indicated unit.

a. 25.73 mm + 7.6 cm = ? mm ______
b. 3.7 m + 98 cm = ? m ______
c. 59.6 cm − 63.7 mm = ? cm ______
d. 184.8 mm − 12.3 cm = ? mm ______
e. 1.06 m − 43.7 cm = ? cm ______
f. 0.793 m − 523.8 mm = ? mm ______
g. 214 mm + 87.6 cm + 0.9 m = ? m ______
h. 0.056 m + 4.93 cm + 57.3 mm = ? mm ______
i. 54.4 mm + 5.05 cm + 204.3 mm = ? mm ______
j. 3.927 m − 812 mm = ? m ______

3. An aluminum slab 0.082 meter thick is machined with three equal cuts; each cut is 10 millimeters deep. Determine the finished thickness of the slab in millimeters. ______

4. A piece of sheet metal is 1.12 meters wide. Strips each 3.4 centimeters wide are cut. Allow 3 millimeters for cutting each strip.

   a. Determine the number of complete strips cut. ______

   b. Determine the width of the waste strip in millimeters. ______

Metric-Customary Linear Equivalents

5. Express each of the following customary units of length as the indicated metric unit of length. Where necessary round the answer to 3 decimal places.

   a. 37.000 millimeters as inches ______
   b. 126.800 millimeters as inches ______
   c. 17.300 centimeters as inches ______
   d. 0.840 centimeter as inches ______
   e. 2.400 meters as inches ______
   f. 0.090 meter as inches ______
   g. 8.000 meters as feet ______
   h. 10.200 meters as feet ______
   i. 736.00 millimeters as inches ______
   j. 34.050 millimeters as inches ______
   k. 56.300 centimeters as inches ______
   l. 2.000 meters as yards ______
   m. 45.000 centimeters as feet ______
   n. 780.000 millimeters as feet ______

6. Express each of the following metric units of length as the indicated customary unit of length. Where necessary, round the answer to 2 decimal places.

   a. 4.000 inches as millimeters ______
   b. 0.360 inch as millimeters ______
   c. 34.00 inches as centimeters ______
   d. 20.85 inches as centimeters ______
   e. 6.00 feet as meters ______
   f. 0.75 foot as meters ______
   g. 3.50 yards as meters ______
   h. 1.30 yards as meters ______
   i. 2.368 inches as millimeters ______
   j. 0.73 inch as centimeters ______
   k. 216.00 inches as meters ______
   l. $\frac{1}{2}$ inch as millimeters ______
   m. $3\frac{1}{4}$ inches as centimeters ______
   n. $75\frac{3}{8}$ inches as meters ______

7. Determine the total length of stock in inches required to make 35 bushings, each 34.2 mm long. Allow $\frac{3}{16}$" waste for each bushing. Round the answer to 1 decimal place. ______

8. The part shown in Figure 25-4 is to be made in a machine shop using decimal-inch machinery and tools. All dimensions are in millimeters. Express each of the dimensions, A–L, in inches to 3 decimal places.

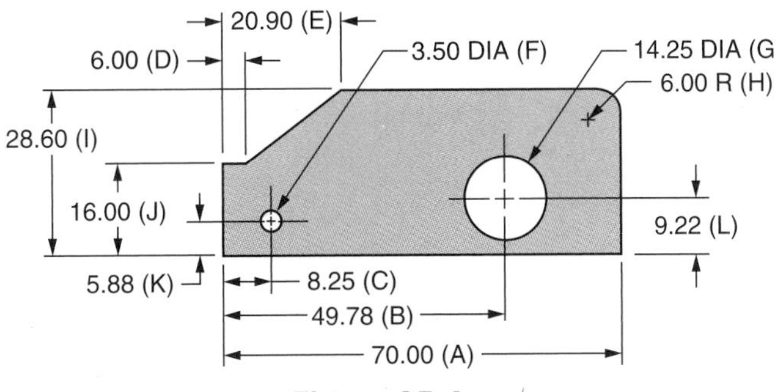

Figure 25-4

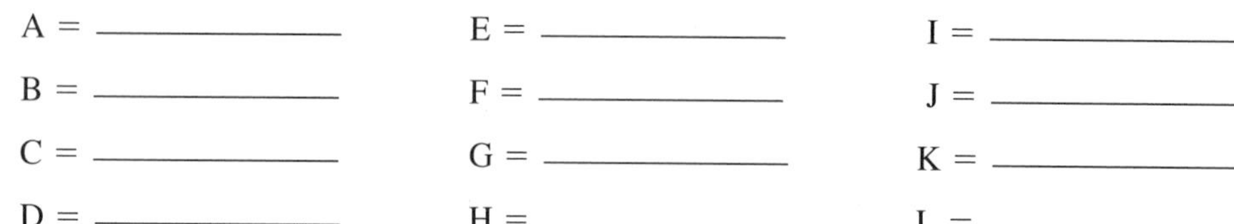

9. The shaft shown in Figure 25-5 is dimensioned in inches. Express each dimension, A–J, in millimeters and round each dimension to 2 decimal places.

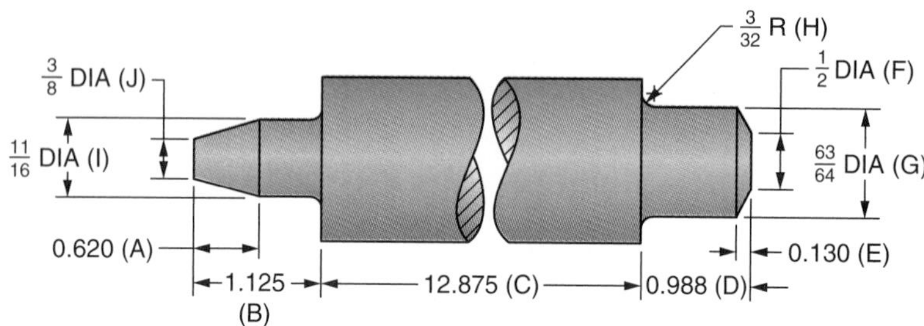

Figure 25-5

A = ____________ E = ____________ H = ____________

B = ____________ F = ____________ I = ____________

C = ____________ G = ____________ J = ____________

D = ____________

UNIT 26

Degree of Precision, Greatest Possible Error, Absolute Error, and Relative Error

Objectives ***After studying this unit you should be able to***

- Determine the degree of precision of any given number.
- Compute the greatest possible error of customary and metric length units.
- Compute absolute error and relative error.

Degree of Precision

The cost of producing a part increases with the degree of precision called for; therefore, no greater degree of precision than is actually required should be specified on a drawing. The degree of precision specified for a particular machining operation dictates the type of machine, the machine setup, and the measuring instrument used for that operation.

The *exact* length of an object cannot be measured. All measurements are approximations. By increasing the number of graduations on a measuring instrument, the degree of precision is increased. Increasing the number of graduations enables the user to get closer to the *exact* length. The precision of a measurement depends on the measuring instrument used. The *degree of precision of a measuring instrument* depends on the smallest graduated unit of the instrument.

No measurement is truly correct, but some are more precise than others. Depending on the need for precision in a measurement, you might measure one object with a tape measure and something else with a micrometer, a caliper, or a dial bore gage. As a general rule, a measuring instrument should measure to a degree 10 times finer than the smallest unit it will be used to measure.

Machinists often work to 0.001-inch or 0.02-millimeter precision. In the manufacture of certain products, very precise measurements to 0.00001 inch or 0.0003 millimeter and 0.000001 inch or 0.00003 millimeter are sometimes required.

Various measuring instruments have different limitations on the degree of precision possible. There is always some "rounding" involved in measuring. If the edge of a part falls between two graduations on the measuring scale, you should round to the nearer one. The accuracy achieved in measurement does not depend only on the limitations of the measuring instrument. Accuracy can also be affected by errors of measurement. Errors can be caused by defects in the measuring instruments and by environmental changes such as differences in temperature. Perhaps the greatest cause of error is the inaccuracy of the person using the measuring instrument.

Limitations of Measuring Instruments

Following are the limitations on the degree of precision or the accuracy possible of some commonly used manufacturing measuring instruments.

Steel rules: $\frac{1}{64}''$ (fractional-inch); 0.01" (decimal-inch); 0.5 mm (metric).

Micrometers: 0.001" (decimal-inch) and 0.0001" (with vernier scale); 0.01 mm (metric) and 0.002 mm (with vernier scale).

Digital micrometers: Accurate to 0.00005" (decimal-inch); 0.0001 mm (metric).

Vernier and dial calipers: 0.001" (decimal-inch); 0.02 mm (metric).

Dial indicators (comparison measurement): Graduations as small as 0.00005" (decimal-inch); 0.002 mm (metric).

Precision gage blocks (comparison measurement): Accurate to 0.000002" (decimal-inch); 0.00006 mm (metric). The degree of precision of measurement is only as precise as the measuring instrument that is used with the blocks.

High amplification comparators (mechanical, optical, pneumatic, electronic): Accurate to 0.000006" (decimal-inch); 0.00001 mm (metric).

Degree of Precision of Numbers

The *degree of precision of a number* depends upon the unit of measurement. The degree of precision of a number increases as the number of decimal places increases.

Suppose a ruler is marked in 0.1 inch. Then, any measurement we read covers a whole range of values. For example, as indicated by the "rounding funnels" in Figure 26-1, a reading of 1.2 inches means that the actual measurement is between 1.15 inches and 1.2499 . . . inches; a reading of 1.3 inches means the actual measurement is between 1.25 inches and 1.3499 . . . inches; and a reading of 1.4 inches means the actual measurement is between 1.35 inches and 1.4499 . . . inches.

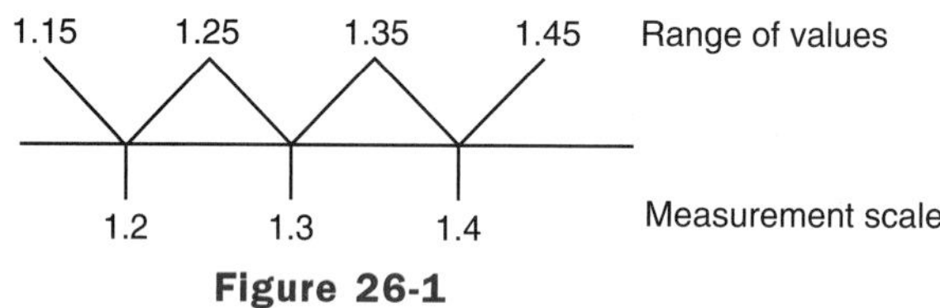

Figure 26-1

Example 1 The degree of precision of 2" is to the nearer inch as shown in Figure 26-2. The range of values includes all numbers equal to or greater than 1.5" or less than 2.5".

Figure 26-2

Figure 26-3

Example 2 The degree of precision of 2.0" is to the nearer 10th of an inch as shown in Figure 26-3. The range of values includes all numbers equal to or greater than 1.95" and less than 2.05".

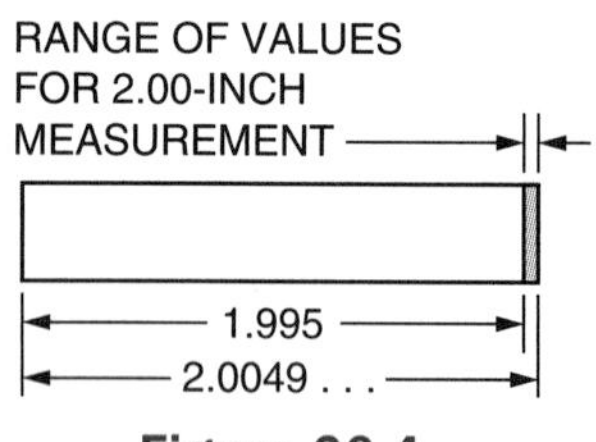

Figure 26-4

Example 3 The degree of precision of 2.00" is to the nearer 100th of an inch as shown in Figure 26-4. The range of values includes all numbers equal to or greater than 1.995" and less than 2.005".

Example 4 The degree of precision of 2.000" is to the nearer 1000th of an inch. The range of values includes all numbers equal to or greater than 1.9995" and less than 2.0005".

Greatest Possible Error

The *greatest possible error* of a measurement is one-half the smallest graduated unit of the measurement used to make the measurement. Therefore, the greatest possible error is equal to $\frac{1}{2}$ or 0.5 of the precision.

Examples

1. A machinist reads a measurement of 36 millimeters on a steel rule. The smallest graduation on the rule used is 1 millimeter; therefore, the precision is 1 millimeter. Since the greatest possible error is one-half of the smallest graduated unit, the greatest possible error is 0.5×1 mm or 0.5 mm. The actual length measured is between 36 mm $-$ 0.5 mm and 36 mm $+$ 0.5 mm or between 35.5 mm and 36.5 mm.
2. A tool and die maker reads a measurement of 0.4754 inch on a vernier scale micrometer. The smallest graduation on the micrometer is 0.0001 inch, therefore the precision is 0.0001 inch. The greatest possible error is $0.5 \times 0.0001"$ or 0.00005". The actual length measured is between 0.4754" $-$ 0.00005" and 0.4754" $+$ 0.00005" or between 0.47535" and 0.47545".
3. A machinist reads a measurement of 114 mm on a steel rule. The smallest graduation on the rule is 0.5 mm.

   Greatest possible error: $\frac{1}{2} \times 0.5 \text{ mm} = 0.5 \times 0.5 \text{ mm} = 0.25 \text{ mm}$

   Smallest possible length: $114 \text{ mm} - 0.25 \text{ mm} = 113.75 \text{ mm}$

   Largest possible length: $114 \text{ mm} + 0.25 \text{ mm} = 114.25 \text{ mm}$

Absolute Error and Relative Error

Absolute error and relative error are commonly used to express the amount of error between an actual or true value and a measured value.

Absolute error is the difference between a true value and a measured value. Since the measured value can be either a smaller or larger value than the true value, subtract the smaller value from the larger value.

$$\text{Absolute Error} = \text{True Value} - \text{Measured Value}$$

or

$$\text{Absolute Error} = \text{Measured Value} - \text{True Value}$$

Relative error is the ratio of the absolute error to the true value. It is expressed as a percent.

$$\text{Relative Error} = \frac{\text{Absolute Error}}{\text{True Value}} \times 100$$

Examples

1. The actual or true value of the diameter of a shaft is 1.7056 inches. The shaft is measured as 1.7040 inches. Compute the absolute and relative error.

   The true value is larger than the measured value, therefore:

   $$\text{Absolute Error} = \text{True Value} - \text{Measured Value}$$

   $$\text{Absolute Error} = 1.7056 \text{ in} - 1.7040 \text{ in} = 0.0016 \text{ in} \quad \text{Ans}$$

   $$\text{Relative Error} = \frac{\text{Absolute Error}}{\text{True Value}} \times 100$$

   $$\text{Relative Error} = \frac{0.0016 \text{ in}}{1.7056 \text{ in}} \times 100 \approx 0.094\% \quad \text{Ans (rounded)}$$

   .0016 [÷] 1.7056 [×] 100 [=] 0.09380863, 0.094% Ans (rounded)

2. An inspector measured a taper angle as 3.01 degrees. The true value of the angle is 2.98 degrees. Compute the absolute and relative error.

   The measured value is larger than the true value, therefore:

   $$\text{Absolute Error} = \text{Measured Value} - \text{True Value}$$

   $$\text{Absolute Error} = 3.01^\circ - 2.98^\circ = 0.03^\circ \quad \text{Ans}$$

   $$\text{Relative Error} = \frac{\text{Absolute Error}}{\text{True Value}} \times 100$$

   $$\text{Relative Error} = \frac{0.03^\circ}{2.98^\circ} \times 100 \approx 1.0\% \quad \text{Ans (rounded)}$$

Application

Degree of Precision

For each measurement find

a. the degree of precision.

b. the value that is equal to or less than the range of values.

c. the value that is greater than the range of values.

1. 4.3" a. ______ b. ______ c. ______
2. 1.62" a. ______ b. ______ c. ______
3. 4.078" a. ______ b. ______ c. ______
4. 6.07" a. ______ b. ______ c. ______
5. 15.885" a. ______ b. ______ c. ______
6. 9.1837" a. ______ b. ______ c. ______
7. 11.003" a. ______ b. ______ c. ______
8. 36.0" a. ______ b. ______ c. ______
9. 7.01" a. ______ b. ______ c. ______
10. 23.00" a. ______ b. ______ c. ______
11. 6.1" a. ______ b. ______ c. ______
12. 14.01070" a. ______ b. ______ c. ______
13. 26.87 mm a. ______ b. ______ c. ______
14. 15.4 mm a. ______ b. ______ c. ______
15. 117.06 mm a. ______ b. ______ c. ______
16. 0.976 mm a. ______ b. ______ c. ______
17. 48.01 mm a. ______ b. ______ c. ______
18. 104.799 mm a. ______ b. ______ c. ______
19. 7.00 mm a. ______ b. ______ c. ______
20. 34.0825 mm a. ______ b. ______ c. ______
21. 8.001 mm a. ______ b. ______ c. ______
22. 14.0000 mm a. ______ b. ______ c. ______

Greatest Possible Error

For each of the exercises in the following tables, the measurement made and the smallest graduation of the measuring instrument are given. Determine the greatest possible error and the smallest and largest possible actual length for each.

	CUSTOMARY SYSTEM				
	Measurement Made (inches)	Smallest Graduation of Measuring Instrument Used (inches)	Greatest Possible Error (inches)	Actual Length: Smallest Possible (inches)	Actual Length: Largest Possible (inches)
23.	5.30	0.05 (steel rule)			
24.	15.68	0.02 (steel rule)			
25.	0.753	0.001 (vernier caliper)			
26.	0.226	0.001 (micrometer)			
27.	0.9369	0.0001 (vernier micrometer)			
28.	$3\frac{5}{8}$	$\frac{1}{64}$ (steel rule)			

	METRIC SYSTEM				
				Actual Length	
	Measurement Made (millimeters)	Smallest Graduation of Measuring Instrument Used (millimeters)	Greatest Possible Error (millimeters)	Smallest Possible (millimeters)	Largest Possible (millimeters)
29.	64	1 (steel rule)			
30.	105	0.5 (steel rule)			
31.	98.5	0.5 (steel rule)			
32.	53.38	0.02 (vernier caliper)			
33.	13.37	0.01 (micrometer)			
34.	12.778	0.002 (vernier micrometer)			

Absolute and Relative Error

For each of the values in the following table, the true value and measured value are given. Determine the absolute and relative error of each. Where necessary, round the answers to 3 decimal places.

	True Value	Measured Value	Absolute Error	Relative Error
35.	38.720 in	38.700 in		
36.	0.530 mm	0.520 mm		
37.	12.700°	12.900°		
38.	0.485 in	0.482 in		
39.	23.860 mm	24.000 mm		
40.	6.056°	6.100°		
41.	1.050 mm	1.020 mm		
42.	0.9347 in	0.9341 in		
43.	1.005°	1.015°		
44.	27.200 in	26.900 in		
45.	18.276 in	18.302 in		
46.	0.983 mm	1.000 mm		

Tolerance, Clearance, and Interference

Objectives ***After studying this unit you should be able to***

- Compute total tolerances and maximum and minimum limits of dimensions.
- Compute maximum and minimum clearances of mating parts.
- Compute maximum and minimum interferences of mating parts.
- Express unilateral tolerances as bilateral tolerances.

Tolerance

Tolerance is the amount of variation permitted on the dimensions or surfaces of manufactured parts. *Limits* are the extreme permissible dimensions of a part. Tolerance is equal to the difference between the maximum and minimum limits of any specified dimension of a part.

Example The maximum limit of a hole diameter is 0.878 inch and the minimum limit is 0.872 inch. Find the tolerance.

The tolerance is 0.878" − 0.872" = 0.006" Ans

A *basic dimension* is the standard size from which the maximum and minimum limits are determined. Usually tolerances are given in such a way as to show the amount of variation and in which direction from the basic dimension these variations can occur. *Unilateral tolerance* means that the total tolerance is taken in one direction from the basic dimension. *Bilateral tolerance* means that the tolerance is divided partly plus (+) or above and partly minus (−) or below the basic dimension. A *mean dimension* is a value that is midway between the maximum and minimum limits. Where bilateral tolerances are used with equal plus and minus tolerances, the mean dimension is equal to the basic dimension.

Example 1 The part shown in Figure 27-1 is dimensioned with a unilateral tolerance. The dimensions are given in inches. The basic dimension is 3.7500". The total tolerance is a minus (−) tolerance. Find the maximum permissible dimension (maximum limit) and minimum permissible dimension (minimum limit).

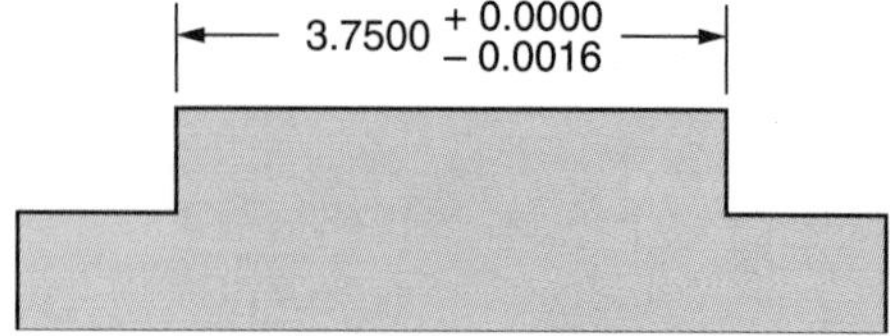

Figure 27-1

Maximum Limit: 3.7500" + 0.0000" = 3.7500" Ans

Minimum Limit: 3.7500" − 0.0016" = 3.7484" Ans

Example 2 The part shown in Figure 27-2 is dimensioned with a bilateral tolerance. The dimensions are given in millimeters. The basic dimension is 62.79 mm. The tolerance is given in two directions, plus (+) and minus (−). Find the maximum limit and the minimum limit.

62.79 ± 0.04

Figure 27-2

Maximum Limit: 62.79 mm + 0.04 mm = 62.83 mm Ans

Minimum Limit: 62.79 mm − 0.04 mm = 62.75 mm Ans

Example 3 What is the mean dimension of a part if the maximum dimension (maximum limit) is 46.35 millimeters and the minimum dimension (minimum limit) is 46.27 millimeters?

Subtract. 46.35 mm − 46.27 mm = 0.08 mm

Divide. 0.08 mm ÷ 2 = 0.04 mm

Subtract. 46.35 mm − 0.04 mm = 46.31 mm Ans

➤ **Note:** The mean dimension is midway between 46.35 mm and 46.27 mm.

46.31 mm +0.04 mm = 46.35 mm (Max. limit)
−0.04 mm = 46.27 mm (Min. limit)

Expressing Unilateral Tolerance as Bilateral Tolerance

In the actual processing of parts, given unilateral tolerances are sometimes changed to bilateral tolerances. A machinist may prefer to work to a mean dimension and take equal plus and minus tolerances while machining a part. The following example shows the procedure for expressing a unilateral tolerance as a bilateral tolerance.

Example The part shown in Figure 27-3 is dimensioned with unilateral tolerances. Express the unilateral tolerance as a bilateral tolerance. Dimensions are given in inches.

Divide the total tolerance by 2. 0.0028" ÷ 2 = 0.0014"

Determine the mean dimension. 1.2590" + 0.0014" = 1.2604"

Show as a bilateral tolerance. 1.2604" ± 0.0014" Ans

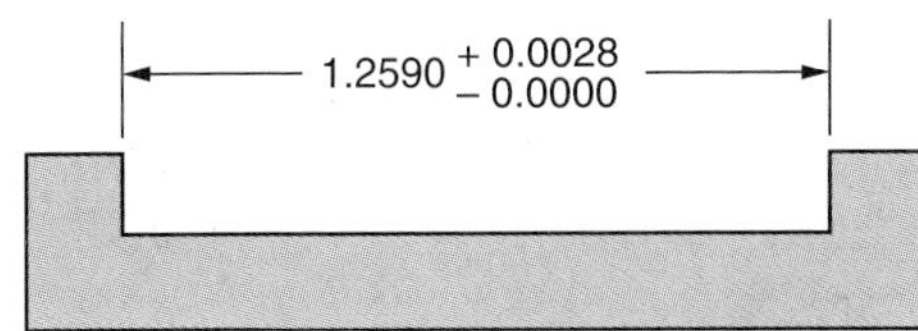

Figure 27-3

Fits of Mating Parts

Fits between mating parts, such as between shafts and holes, have wide application in the manufacturing industry. The tolerances applied to each of the mating parts determines the relative looseness or tightness of fit between parts.

When one part is to move within another there is a *clearance* between the parts. A shaft made to turn in a bushing is an example of a clearance fit. The shaft diameter is less than the bushing hole diameter. When one part is made to be forced into the other there is *interference* between parts. A pin pressed into a hole is an example of an interference fit. The pin diameter is greater than the hole diameter.

Allowance is the intentional difference in the dimensions of mating parts that provides for different classes of fits. Allowance is the minimum clearance or the maximum interference intended between mating parts. Allowance represents the condition of the tightest permissible fit.

Example 1 A mating shaft and a hole with a clearance fit dimensioned with bilateral tolerances is shown in Figure 27-4. All dimensions are in inches. Determine the following:

a. Maximum shaft diameter 0.7502" + 0.0008" = 0.7510" Ans

b. Minimum shaft diameter 0.7502" − 0.0008" = 0.7494" Ans

c. Maximum hole diameter 0.7536" + 0.0008" = 0.7544" Ans

d. Minimum hole diameter 0.7536" − 0.0008" = 0.7528" Ans

e. Maximum clearance equals maximum hole diameter minus minimum shaft diameter 0.7544" − 0.7494" = 0.0050" Ans

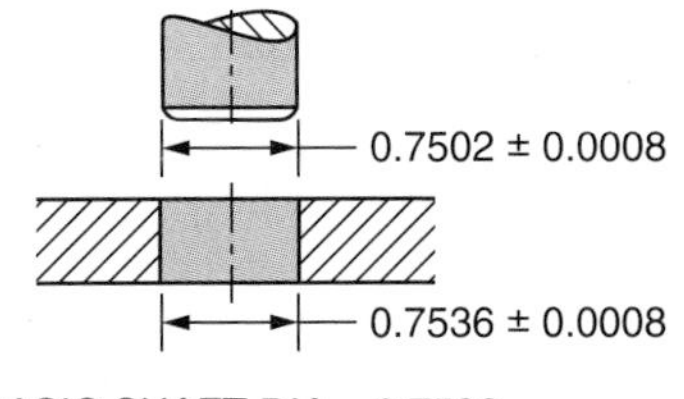

BASIC SHAFT DIA = 0.7502
BASIC HOLE DIA = 0.7536

Figure 27-4

f. Minimum clearance equals minimum hole diameter minus maximum shaft diameter 0.7528" − 0.7510" = 0.0018" Ans

Since allowance is defined as the minimum clearance, the allowance = 0.0018" Ans

This example can be summarized by the following table:

	Basic Dimension	Maximum Diameter (Max. Limit)	Minimum Diameter (Min. Limit)	Maximum Clearance	Minimum Clearance (Allowance)
Shaft	0.7502"	0.7510"	0.7494"	0.0050"	0.0018"
Hole	0.7536"	0.7544"	0.7528"		

Example 2 A pin intended to be pressed into a hole is shown in Figure 27-5. This is an example of an interference fit dimensioned with unilateral tolerances. Dimensions are in millimeters. Determine the following:

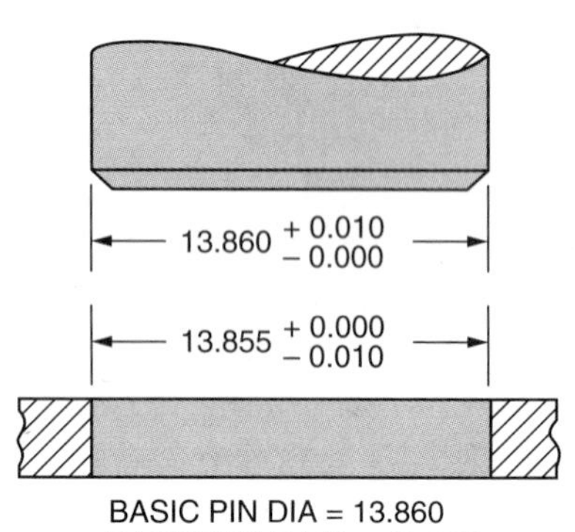

BASIC PIN DIA = 13.860
BASIC HOLE DIA = 13.855

Figure 27-5

a. Maximum pin diameter 13.860 mm + 0.010 mm = 13.870 mm Ans

b. Minimum pin diameter 13.860 mm − 0.000 mm = 13.860 mm Ans

c. Maximum hole diameter 13.855 mm + 0.000 mm = 13.855 mm Ans

d. Minimum hole diameter 13.855 mm − 0.010 mm = 13.845 mm Ans

e. Minimum interference equals minimum pin diameter minus maximum hole diameter 13.860 mm − 13.855 mm = 0.005 mm Ans

f. Maximum interference equals maximum pin diameter minus minimum hole diameter 13.870 mm − 13.845 mm = 0.025 mm Ans

Since allowance is defined as the maximum interference, the allowance = 0.025 mm Ans

This example can be summarized by the following table:

	Basic Dimension	Maximum Diameter (Max. Limit)	Minimum Diameter (Min. Limit)	Maximum Interference (Allowance)	Minimum Interference
Pin	13.860 mm	13.870 mm	13.860 mm	0.025 mm	0.005 mm
Hole	13.855 mm	13.855 mm	13.845 mm		

Application

Tolerance, Maximum and Minimum Limits

Refer to the following tables and determine the tolerance, maximum limit, or minimum limit as required for each problem.

1. Customary System

	Tolerance	Maximum Limit	Minimum Limit
a.		$5\frac{7}{16}$"	$5\frac{13}{32}$"
b.		7'9$\frac{1}{16}$"	7'8$\frac{15}{16}$"
c.	0.03"	16.76"	
d.	0.007"		0.904"
e.		1.7001"	1.6998"
f.	0.004"		10.999"

2. Metric System

	Tolerance	Maximum Limit	Minimum Limit
a.		50.7 mm	49.9 mm
b.		26.8 cm	26.6 cm
c.	0.04 mm		258.03 mm
d.	0.12 mm	79.65 mm	
e.	0.006 cm		12.731 cm
f.		4.01 mm	3.98 mm

Unilateral and Bilateral Tolerance

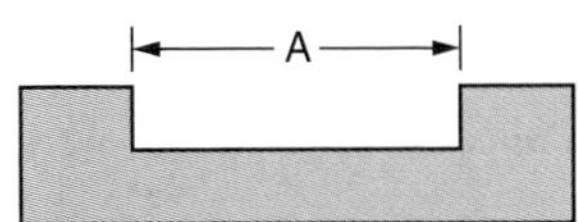

Figure 27-6

3. Refer to Figure 27-6. Dimension A with its tolerance is given in each of the following problems. Determine the maximum dimension (maximum limit) and the minimum dimension (minimum limit) for each.

a. Dimension A = 4.640" $^{+0.003"}_{-0.000"}$
maximum ________ minimum ________

b. Dimension A = 5.927" $^{+0.005"}_{-0.000"}$
maximum ________ minimum ________

c. Dimension A = 2.004" $^{+0.000"}_{-0.004"}$
maximum ________ minimum ________

d. Dimension A = 4.6729" $^{+0.0000"}_{-0.0012"}$
maximum ________ minimum ________

e. Dimension A = 1.0875" $^{+0.0009"}_{-0.0000"}$
maximum ________ minimum ________

f. Dimension A = 28.16 mm $^{+0.00\text{ mm}}_{-0.06\text{ mm}}$
maximum ________ minimum ________

g. Dimension A = 43.94 mm $^{+0.04\text{ mm}}_{-0.00\text{ mm}}$
maximum ________ minimum ________

h. Dimension A = 118.66 mm $^{+0.07\text{ mm}}_{-0.00\text{ mm}}$
maximum ________ minimum ________

i. Dimension A = 73.398 mm $^{+0.000\text{ mm}}_{-0.012\text{ mm}}$
maximum ________ minimum ________

j. Dimension A = 45.106 mm $^{+0.009\text{ mm}}_{-0.000\text{ mm}}$
maximum ________ minimum ________

4. The following dimensions are given with bilateral tolerances. For each value determine the maximum dimension (maximum limit) and the minimum dimension (minimum limit).

a. 2.812" ± 0.006"
maximum ________ minimum ________

b. 3.003" ± 0.004"
maximum ________ minimum ________

c. 3.971" ± 0.010"
maximum ________ minimum ________

d. 4.0562" ± 0.0012"
maximum ________ minimum ________

e. 1.3799" ± 0.0009"
maximum ________ minimum ________

f. 2.0000" ± 0.0007"
maximum ________ minimum ________

g. 43.46 mm ± 0.05 mm
maximum ________ minimum ________

h. 107.07 mm ± 0.08 mm
maximum ________ minimum ________

i. 62.04 mm ± 0.10 mm
maximum ________ minimum ________

j. 10.203 mm ± 0.024 mm
maximum ________ minimum ________

k. 289.005 mm ± 0.007 mm
maximum ________ minimum ________

l. 66.761 mm ± 0.015 mm
maximum ________ minimum ________

5. Express each of the following unilateral tolerances as bilateral tolerances having equal plus and minus values.

a. 0.938" $^{+0.010"}_{-0.000"}$	______	i. 44.30 mm $^{+0.02\text{ mm}}_{-0.00\text{ mm}}$	______
b. 1.686" $^{+0.002"}_{-0.000"}$	______	j. 10.06 mm $^{+0.00\text{ mm}}_{-0.08\text{ mm}}$	______
c. 3.000" $^{+0.000"}_{-0.004"}$	______	k. 64.89 mm $^{+0.06\text{ mm}}_{-0.00\text{ mm}}$	______
d. 0.073" $^{+0.000"}_{-0.008"}$	______	l. 37.988 mm $^{+0.056\text{ mm}}_{-0.000\text{ mm}}$	______
e. 4.1873" $^{+0.0014"}_{-0.0000"}$	______	m. 125.00 mm $^{+0.000\text{ mm}}_{-0.017\text{ mm}}$	______
f. 1.0021" $^{+0.0000"}_{-0.0074"}$	______	n. 43.091 mm $^{+0.000\text{ mm}}_{-0.026\text{ mm}}$	______
g. 1.0010" $^{+0.0000"}_{-0.0008"}$	______	o. 98.879 mm $^{+0.009\text{ mm}}_{-0.000\text{ mm}}$	______
h. 8.4649" $^{+0.0022"}_{-0.0000"}$	______		

Fits of Mating Parts

The following problems require computations with both clearance fits and interference fits between mating parts. Find the missing values in the following tables.

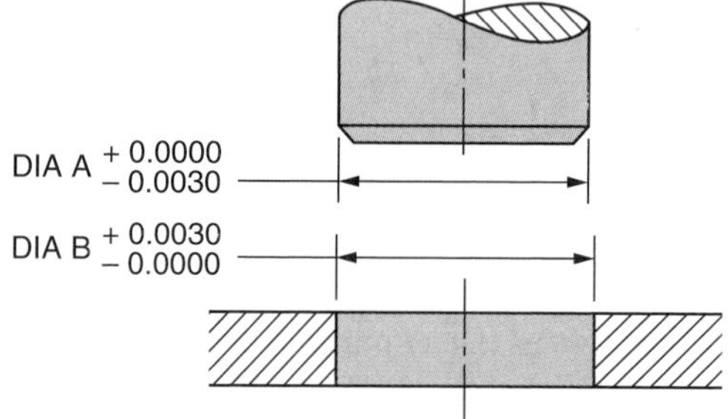

Figure 27-7

6. Refer to Figure 27-7 to determine the values in the table. The answer to the first problem is given. Allowance is equal to the minimum clearance. All dimensions are in inches.

		Basic Dimension	Maximum Diameter (Max. Limit)	Minimum Diameter (Min. Limit)	Maximum Clearance	Minimum Clearance (Allowance)
a.	DIA A	1.4580	1.4580	1.4550	0.0090	0.0030
	DIA B	1.4610	1.4640	1.4610		
b.	DIA A	0.6345				
	DIA B	0.6365				
c.	DIA A	2.1053				
	DIA B	2.1078				

7. Refer to Figure 27-8 to determine the values in the table. Allowance is equal to the maximum interference. All dimensions are in millimeters.

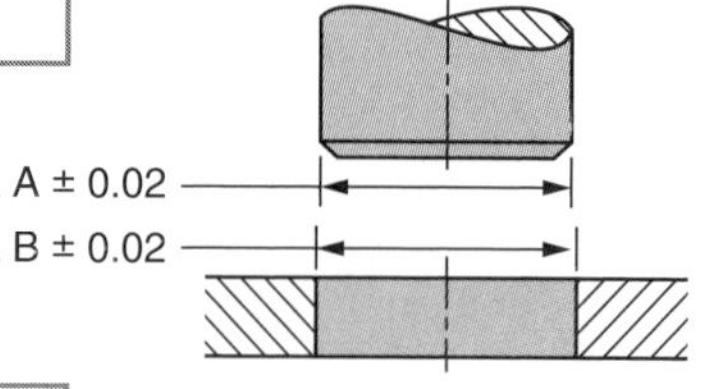

Figure 27-8

		Basic Dimension	Maximum Diameter (Max. Limit)	Minimum Diameter (Min. Limit)	Maximum Interference (Allowance)	Minimum Interference
a.	DIA A	20.73				
	DIA B	20.68				
b.	DIA A	32.07				
	DIA B	32.01				
c.	DIA A	12.72				
	DIA B	12.65				

8. Refer to Figure 27-9 to determine the values in the table. Allowance is equal to minimum clearance. All dimensions are in inches.

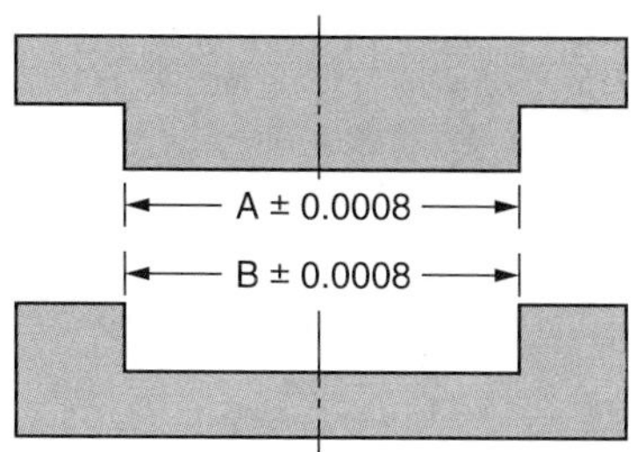

Figure 27-9

		Basic Dimension	Maximum Dimension (Max. Limit)	Minimum Dimension (Min. Limit)	Maximum Clearance	Minimum Clearance (Allowance)
a.	DIM A	0.9995				
	DIM B	1.0020				
b.	DIM A	2.0334				
	DIM B	2.0360				
c.	DIM A	1.4392				
	DIM B	1.4412				

9. Refer to Figure 27-10 to determine the values in the table. Allowance is equal to the maximum interference. All dimensions are in millimeters.

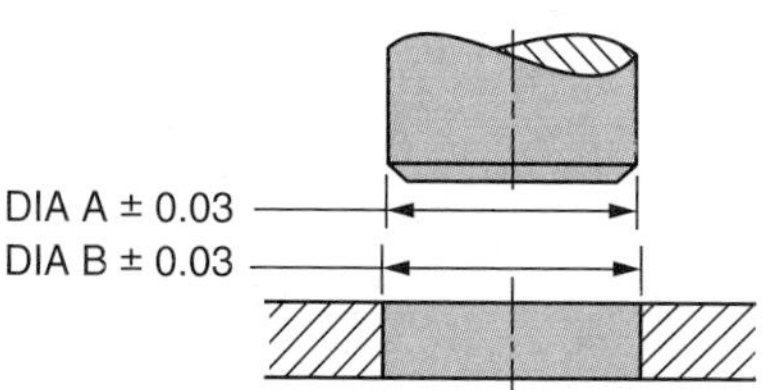

Figure 27-10

		Basic Dimension	Maximum Diameter (Max. Limit)	Minimum Diameter (Min. Limit)	Maximum Interference (Allowance)	Minimum Interference
a.	DIA A	87.58				
	DIA B	87.50				
b.	DIA A	9.94				
	DIA B	9.85				
c.	DIA A	130.03				
	DIA B	129.96				

Related Problems

10. Spacers are manufactured to the mean dimension and tolerance shown in Figure 27-11. An inspector measures 10 spacers and records the following thicknesses:

0.372"	0.379"	0.370"	0.377"	0.371"
0.376"	0.375"	0.373"	0.378"	0.380"

Which spacers are defective (above the maximum limit or below the minimum limit)? All dimensions are in inches.

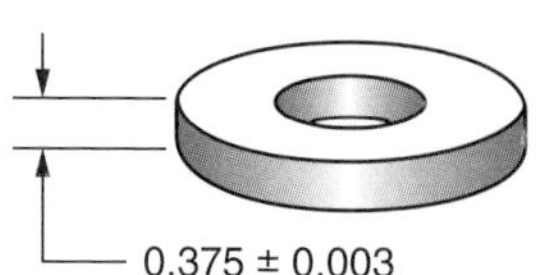

Figure 27-11

11. A tool-and-die maker grinds a pin to an 18.25-millimeter diameter as shown in Figure 27-12. The pin is to be pressed (an interference fit) into a hole. The minimum interference permitted is 0.03 millimeter. The maximum interference permitted is 0.07 millimeter. Determine the mean diameter of the hole. All dimensions are in millimeters.

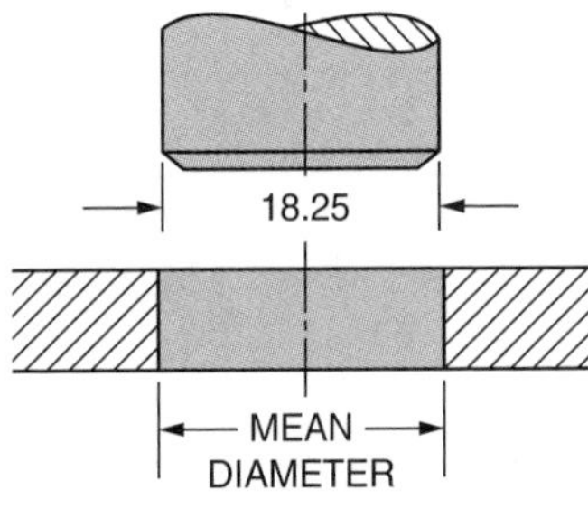

Figure 27-12 __________

12. A piece is to be cut to the dimensions and tolerances shown in Figure 27-13. Determine the maximum permissible value of length A. All dimensions are in inches.

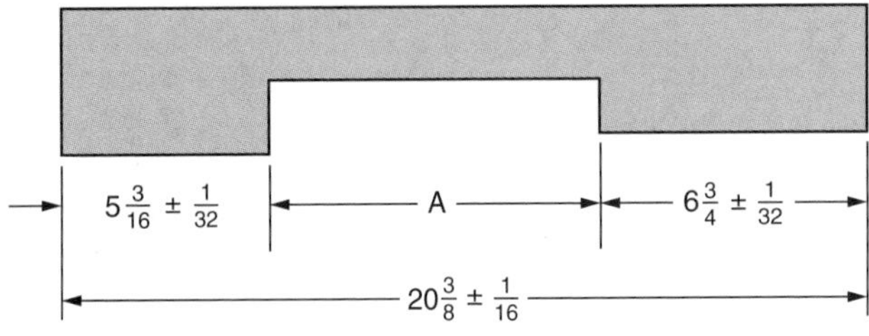

Figure 27-13 __________

13. Determine the maximum and minimum permissible wall thickness of the steel sleeve shown in Figure 27-14. All dimensions are in millimeters.

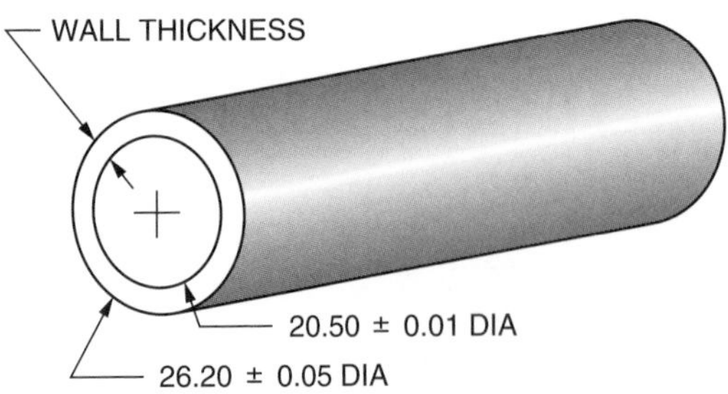

Figure 27-14

maximum __________

minimum __________

14. Mating parts are shown in Figure 27-15. The pins in the top piece fit into the holes in the bottom piece. All dimensions are in inches. Determine the following:

a. The mean pin diameters __________

b. The mean hole diameters __________

c. The maximum dimension A __________

d. The minimum dimension A __________

e. The maximum dimension B __________

f. The minimum dimension B __________

g. The maximum total clearance between dimension C and dimension D __________

h. The minimum total clearance between dimension C and dimension D __________

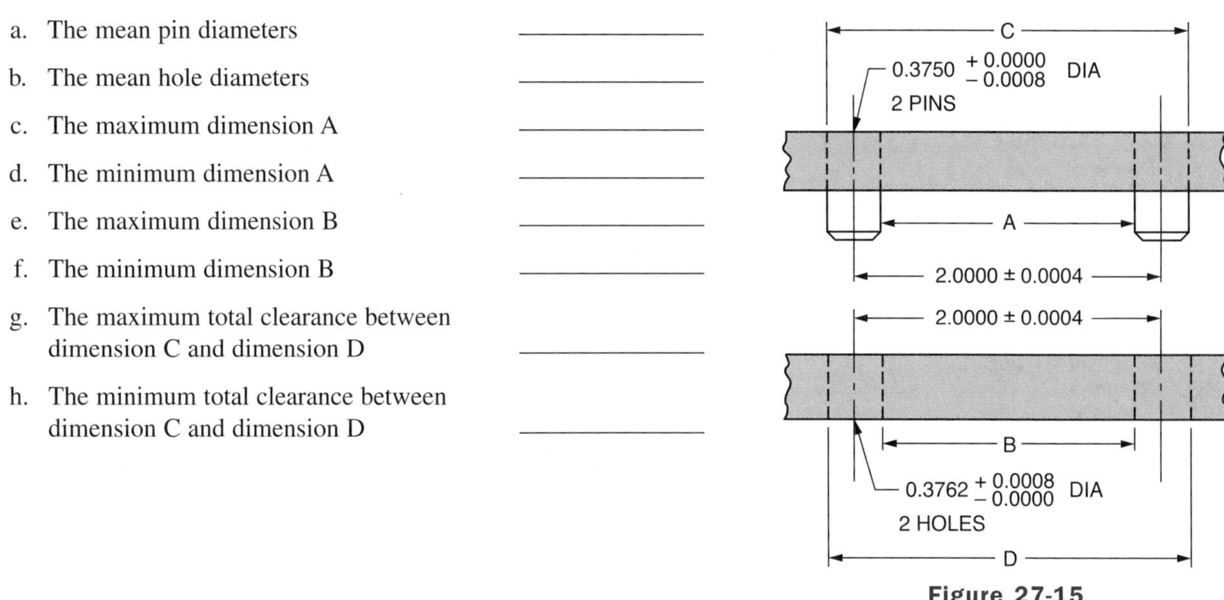

Figure 27-15

15. Figure 27-16 gives the locations with tolerances of 6 holes that are to be drilled in a length of angle iron. A machinist drills the holes then checks them for proper locations from edge A. The actual locations of the drilled holes are shown in Figure 27-17. Which holes are drilled out of tolerance (located incorrectly)?

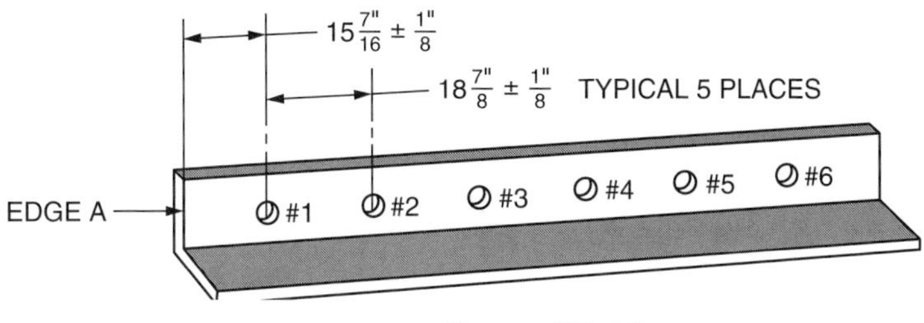

Figure 27-16

Figure 27-16 shows specifications for locations of holes.

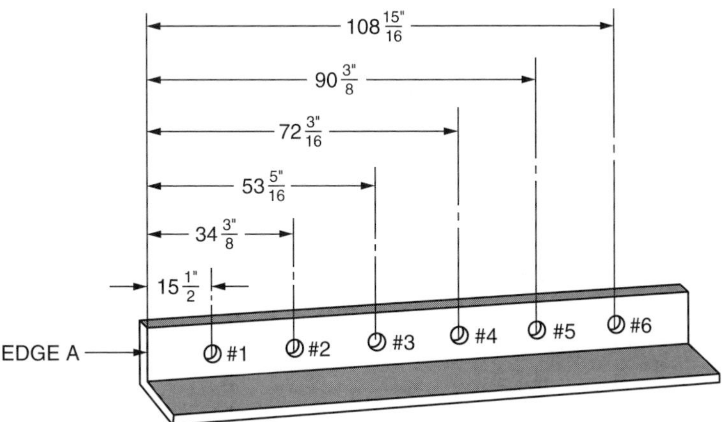

Figure 27-17

Figure 27-17 shows actual locations of drilled holes.

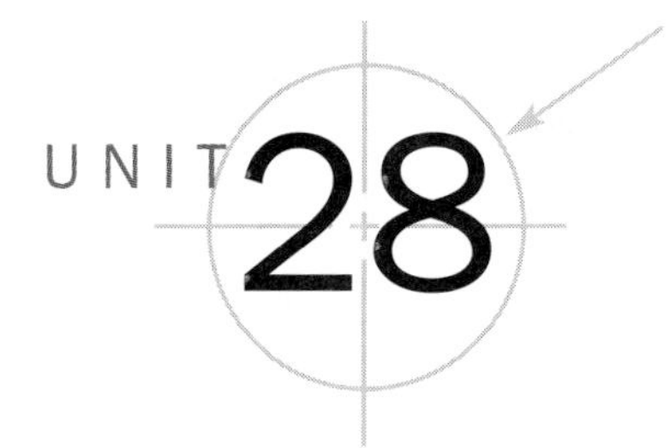

Customary and Metric Steel Rules

Objectives *After studying this unit you should be able to*

- Read measurements on fractional-inch and decimal-inch steel rules.
- Measure lengths using fractional-inch and decimal-inch scales.
- Read measurements on metric steel rules.
- Measure lengths using metric scales.

Steel rules are widely used for machine shop applications that do not require a high degree of precision. The steel rule is often the most practical measuring instrument to use for checking dimensions where stock allowances for finishing are provided. Steel rules are also used for locating roughing cuts on machined pieces and for determining the approximate locations of parts for machine setups. Steel rules used in the machine shop are generally six inches long, although rules anywhere from a fraction of an inch to several inches in length are also used.

Correct Procedure in the Use of Steel Rules

The end of a rule receives more wear than the rest of the rule. Therefore, the end should not be used as a reference point unless it is used with a knee (a straight block) as shown in Figure 28-1.

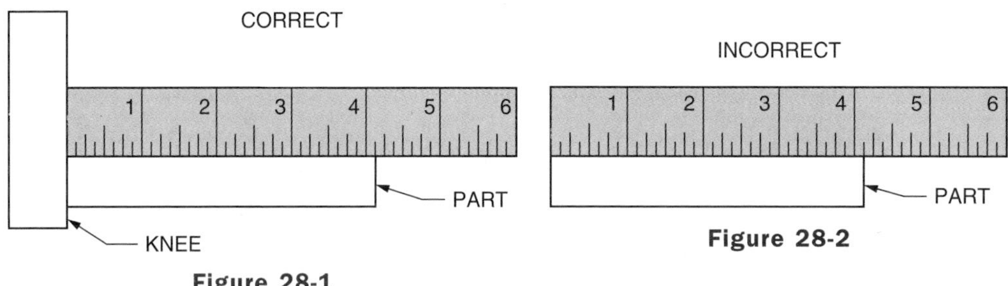

Figure 28-1

Figure 28-2

If a knee is not used, the 1-inch graduation of customary measure rules should be used as a reference point as shown in Figure 28-3. The 1 inch must be subtracted from the measurement obtained. For metric measure rules, use the 10-millimeter graduation as the reference point as in Figure 28-4. The 10 millimeters must be subtracted from the measurement obtained.

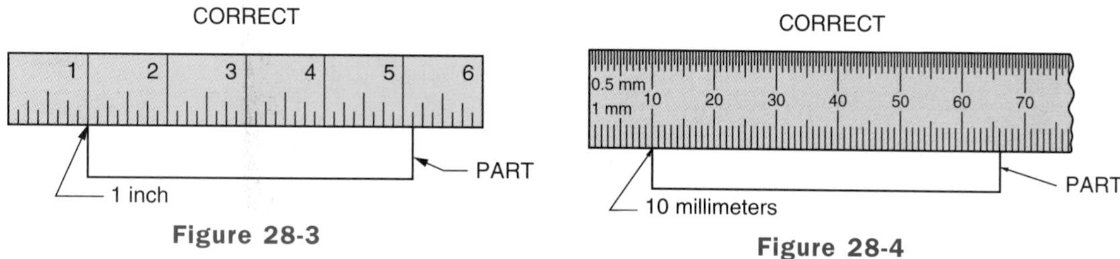

Figure 28-3

Figure 28-4

The scale edge of the rule should be put on the part to be measured. Following the correct procedure shown in Figure 28-5 eliminates parallax error (error caused by the scale and the part being in different planes) like that in Figure 28-6.

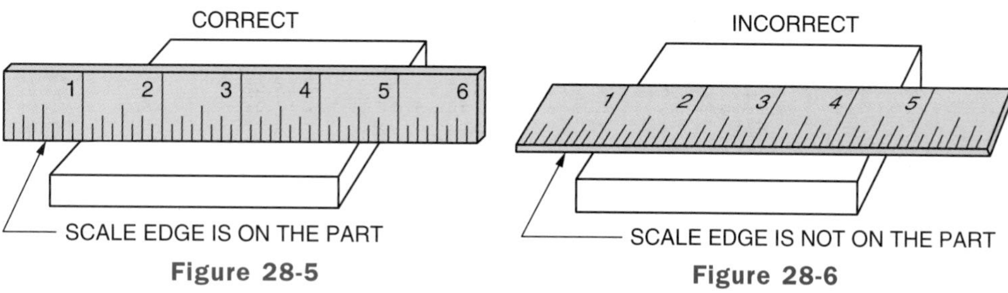

Figure 28-5

Figure 28-6

Reading Fractional-Inch Rules

The smallest division of fractional rules is $\frac{1}{64}$ inch. An enlarged fractional-inch rule is shown. The top scale in Figure 28-7 is graduated in 64ths of an inch and the bottom scale in 32nds of an inch. The staggered graduations are halves, quarters, eighths, sixteenths, and thirty-seconds of an inch.

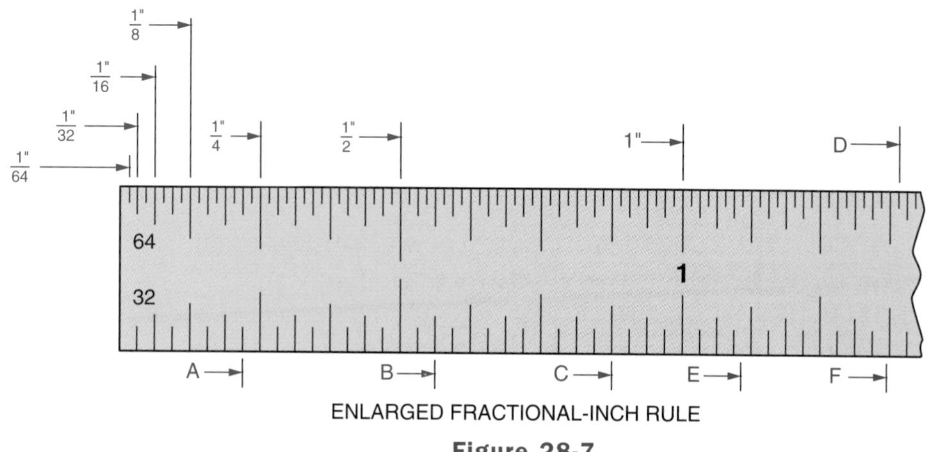

Figure 28-7

Measurements can be read by noting the last complete inch unit and counting the number of fractional units past the inch unit. Generally, a short-cut method of reading measurements is used as illustrated by the following examples.

Examples Read the following measurements on the enlarged fractional-inch rule shown in Figure 28-7.

1. Length A: Subtract one $\frac{1}{32}"$ graduation from $\frac{1}{4}"$.

   $\frac{1"}{4} - \frac{1}{32}" = \frac{8}{32}" - \frac{1}{32}" = \frac{7}{32}"$ Ans

2. Length B: Add one $\frac{1}{16}"$ graduation to $\frac{1}{2}"$.

   $\frac{1"}{2} + \frac{1}{16}" = \frac{8}{16}" + \frac{1}{16}" = \frac{9}{16}"$ Ans

3. Length C: Subtract one $\frac{1"}{8}$ graduation from 1".

   $1" - \frac{1"}{8} = \frac{8"}{8} - \frac{1"}{8} = \frac{7"}{8}$ Ans

4. Length D: Add one $\frac{1}{64}"$ graduation to $1\frac{3"}{8}$.

   $1\frac{3"}{8} + \frac{1}{64}" = 1\frac{24"}{64} + \frac{1}{64}" = 1\frac{25"}{64}$ Ans

Often the edge of an object being measured does not fall exactly on a rule graduation. In these cases, read the measurement to the nearer rule graduation.

Examples Read the following measurements, to the nearer graduation, on the enlarged fractional-inch rule shown in Figure 28-7.

1. Length E: Since the measurement is nearer to $1\frac{3}{32}"$ than $1\frac{1"}{8}$,

   Length E is read as $1\frac{3}{32}"$. Ans

2. Length F: Since the measurement is nearer to $1\frac{3"}{8}$ than $1\frac{11"}{32}$,

   Length F is read as $1\frac{3"}{8}$. Ans

Reading Decimal-Inch Rules

An enlarged decimal-inch rule is shown in Figure 28-8. The top scale is graduated in hundredths of an inch (0.01"). The bottom scale is graduated in fiftieths of an inch (0.02"). The staggered graduations are halves, tenths, and fiftieths of an inch.

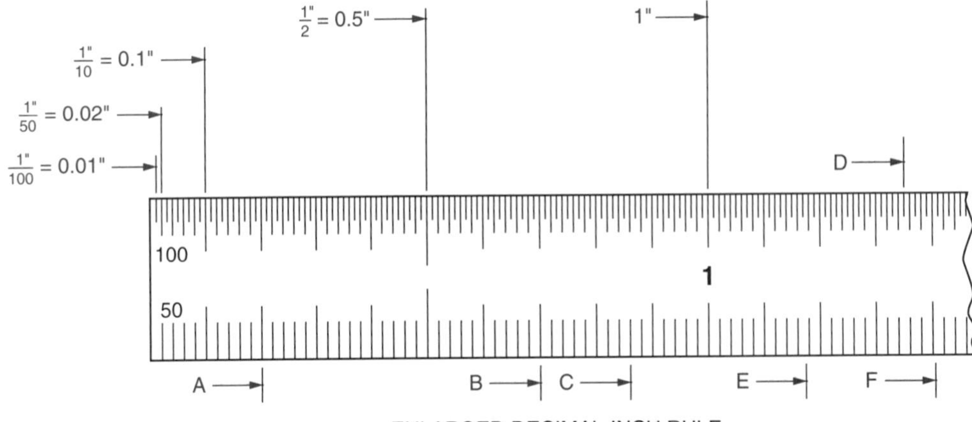

ENLARGED DECIMAL-INCH RULE

Figure 28-8

Examples Read the following measurements on the enlarged decimal-inch rule shown in Figure 28-8.

1. Length A: Count two 0.1" graduations.

   2 × 0.1" = 0.2" Ans

2. Length B: Add two 0.1" graduations to the 0.5".

   0.5" + 0.2" = 0.7" Ans

3. Length C: Add three 0.02" graduations to 0.8".

   0.8" + 0.06" = 0.86" Ans

4. Length D: Add 1", plus three 0.1" graduations, plus five 0.01" graduations.

   1" + 0.3" + 0.05" = 1.35" Ans

5. Length E: Since the measurement is nearer to 1.18" than 1.16",

   Length E is read as 1.18". Ans

6. Length F: Since the measurement is nearer 1.40" than 1.42",

   Length F is read as 1.40". Ans

Reading a Metric Rule

An enlarged metric rule is shown in Figure 28-9. The top scale is graduated in half millimeters (0.5 mm). The bottom scale is graduated in millimeters (1 mm).

The following examples show the method of reading measurements with a metric rule with 0.5-mm and 1-mm scales.

Examples Read the following measurements on the enlarged metric rule shown.

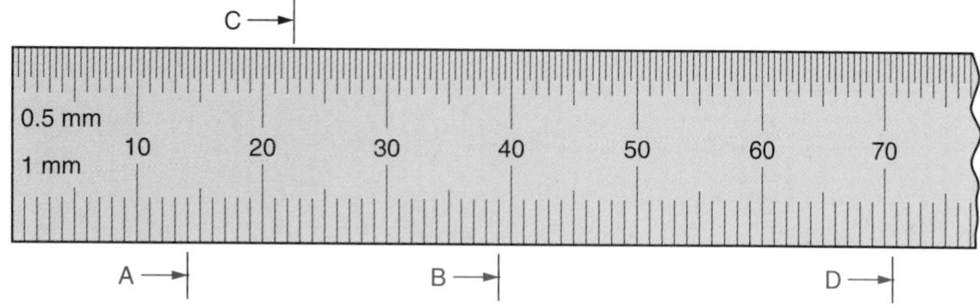

ENLARGED METRIC RULE (1 mm and 0.5 mm)

Figure 28-9

1. Length A: Add four 1-mm graduations to 10 mm.

   10 mm + 4 mm = 14 mm Ans

2. Length B: Subtract one 1-mm graduation from 40 mm.

   40 mm − 1 mm = 39 mm Ans

3. Length C: Add 20 mm, plus two 1-mm graduations, plus one 0.5-mm graduation.

   20 mm + 2 mm + 0.5 mm = 22.5 mm Ans

4. Length D: Since the measurement is nearer 71 mm than 70 mm,

   Length D is read as 71 mm. Ans

Application

Fractional-Inch Steel Rules

1. Read measurements a–p on the enlarged fractional-inch rule shown in Figure 28-10.

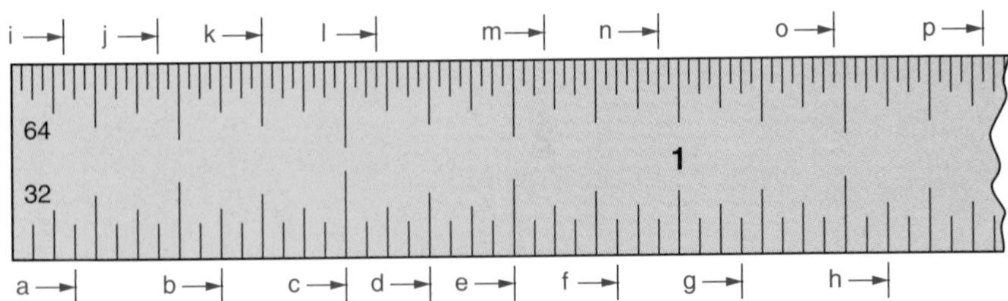

Figure 28-10

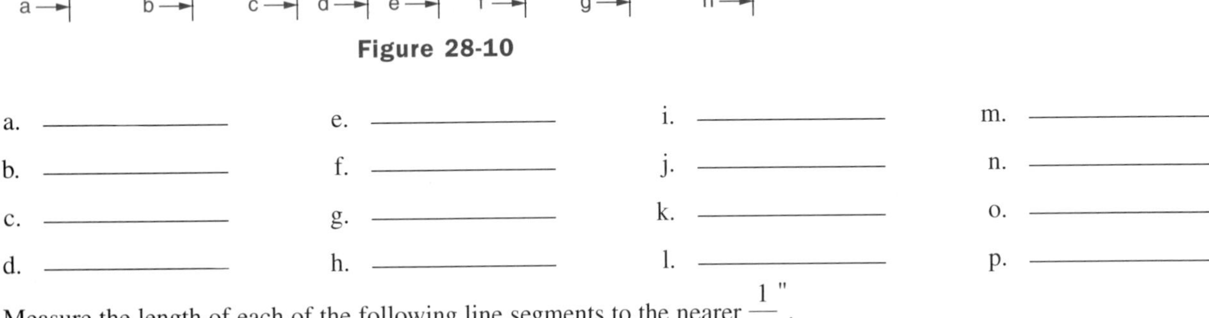

2. Measure the length of each of the following line segments to the nearer $\frac{1}{16}$".

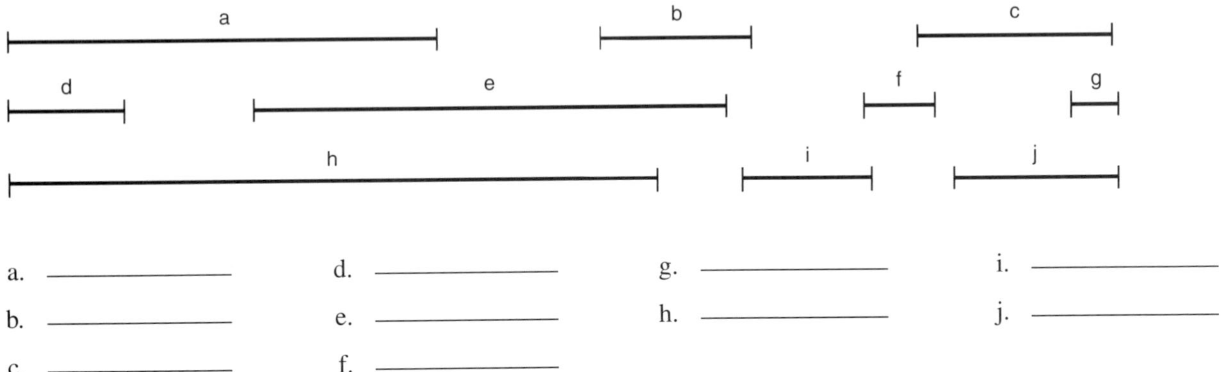

3. Measure the lengths of dimensions a–n of the template shown in Figure 28-11 to the nearer $\frac{1}{32}$".

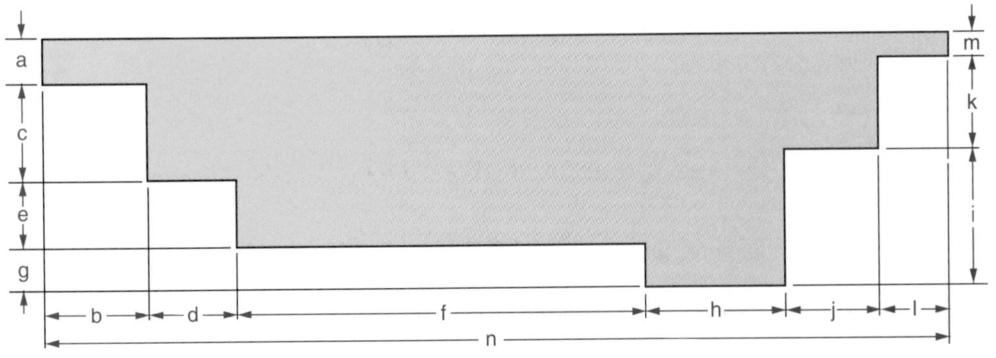

Figure 28-11

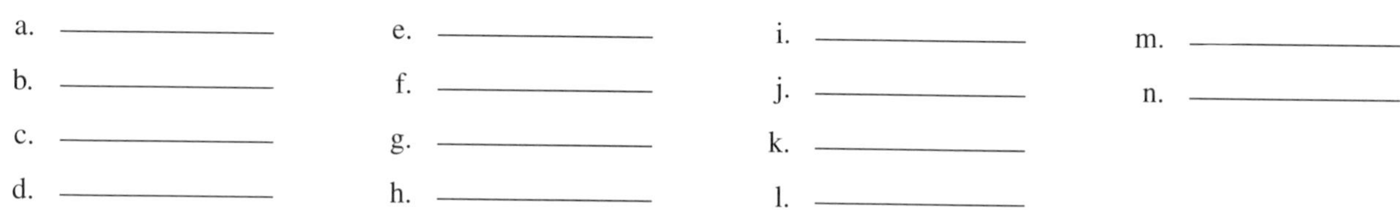

4. Measure the length of each of the following line segments to the nearer $\frac{1}{64}$".

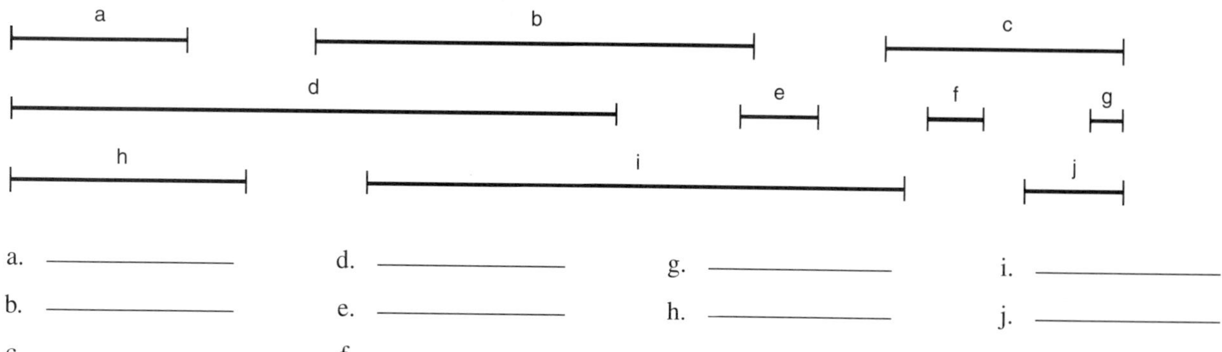

a. ______ d. ______ g. ______ i. ______
b. ______ e. ______ h. ______ j. ______
c. ______ f. ______

Decimal-Inch Steel Rules

5. Read measurements a–p on the enlarged decimal-inch rule shown in Figure 28-12.

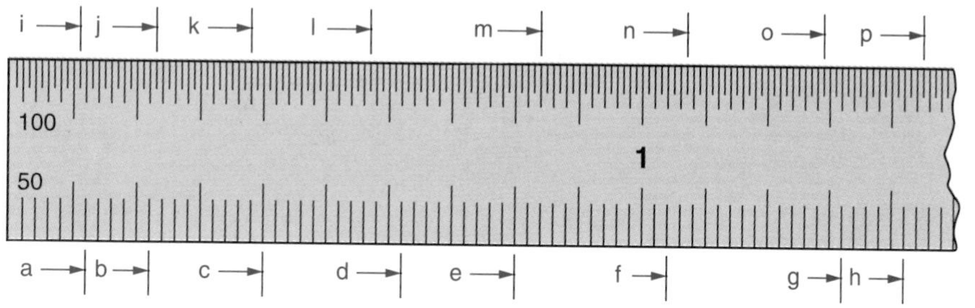

Figure 28-12

a. ______ e. ______ i. ______ m. ______
b. ______ f. ______ j. ______ n. ______
c. ______ g. ______ k. ______ o. ______
d. ______ h. ______ l. ______ p. ______

6. Measure the length of each of the following line segments to the nearer fiftieth of an inch (0.02").

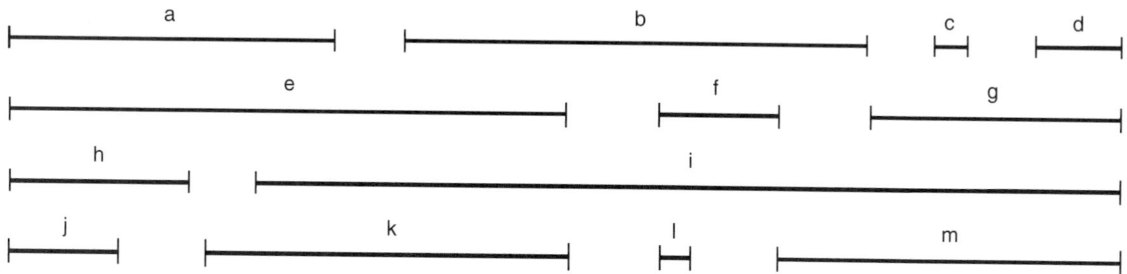

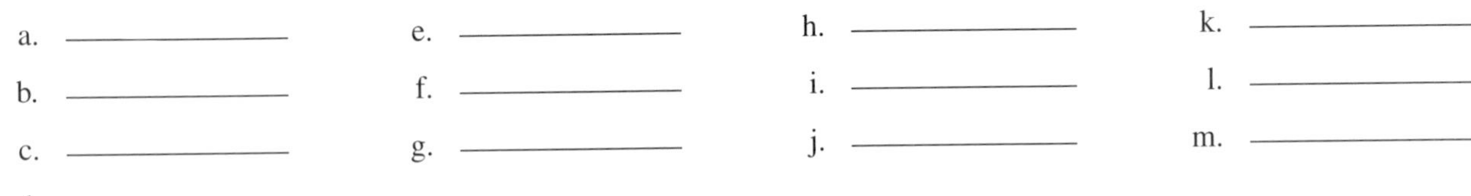

7. Measure the diameters of the holes in the plate shown in Figure 28-13 to the nearer fiftieth of an inch (0.02").

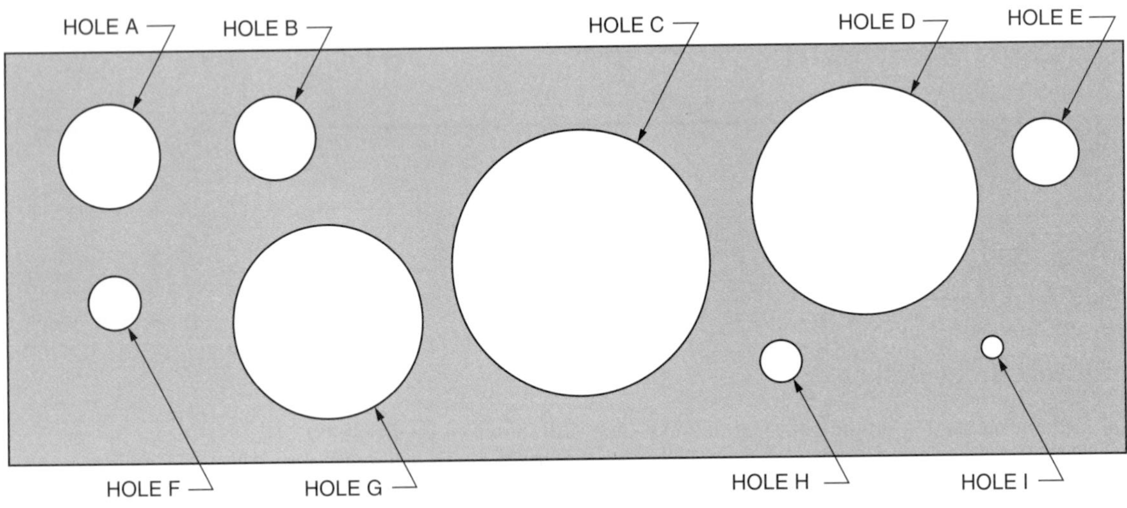

Figure 28-13

➤ **Note:** Measure to the inside of the hole line thickness.

A = ______ D = ______ F = ______ H = ______

B = ______ E = ______ G = ______ I = ______

C = ______

Metric Steel Rules

8. Read measurements a–p on the enlarged metric rule with 1-millimeter and 0.5-millimeter graduations shown.

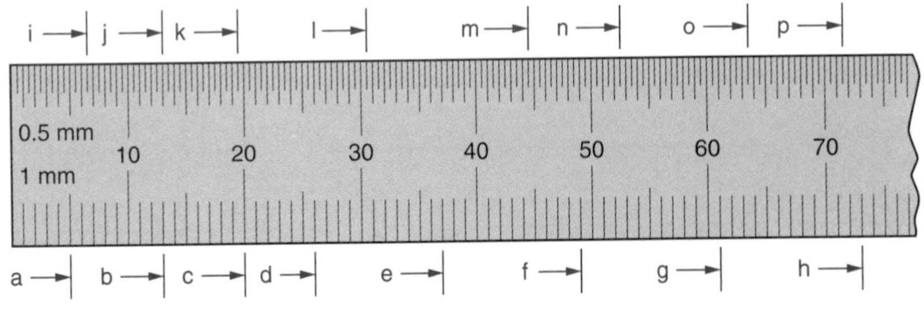

Figure 28-14

a. ______ e. ______ i. ______ m. ______
b. ______ f. ______ j. ______ n. ______
c. ______ g. ______ k. ______ o. ______
d. ______ h. ______ l. ______ p. ______

9. Measure the length of each of the following line segments to the nearer whole millimeter.

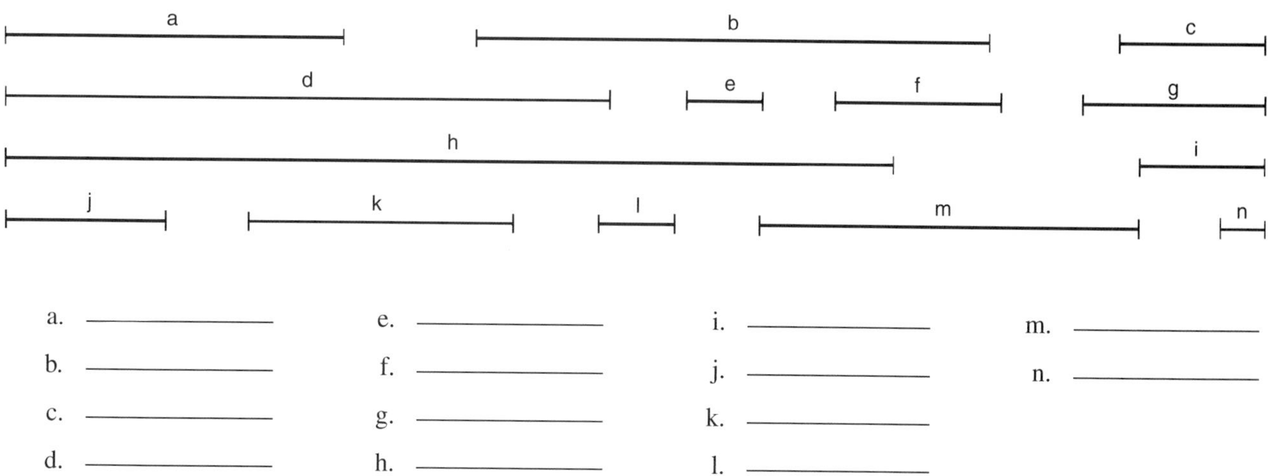

a. ______ e. ______ i. ______ m. ______
b. ______ f. ______ j. ______ n. ______
c. ______ g. ______ k. ______
d. ______ h. ______ l. ______

10. Measure dimensions a–k on the pattern shown in Figure 28-15 to the nearer whole millimeter.

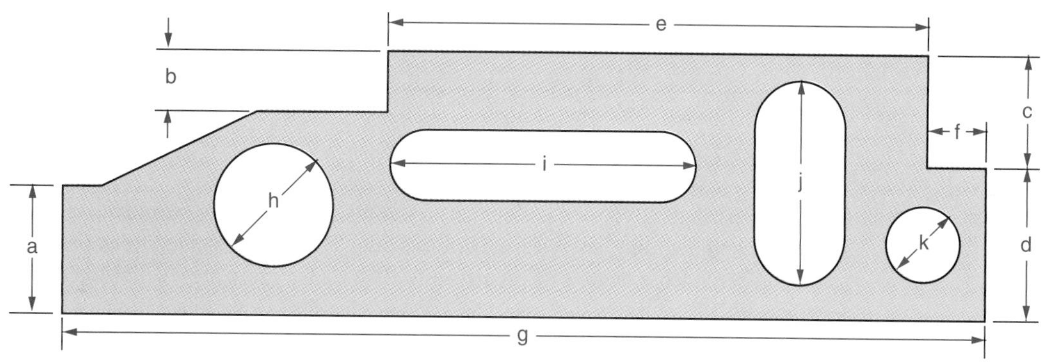

Figure 28-15

a. ______ d. ______ g. ______ j. ______
b. ______ e. ______ h. ______ k. ______
c. ______ f. ______ i. ______

Customary Vernier Calipers and Height Gages

Objectives ***After studying this unit you should be able to***

- Read measurements set on a decimal-inch vernier caliper.
- Set given measurements on a decimal-inch vernier caliper.
- Read measurements set on a decimal-inch vernier height gage.
- Set given measurements on a decimal-inch vernier height gage.

Digital calipers and height gages are widely used. Measurements are read directly on a five-digit LCD readout display with instant inch/millimeter conversion.

Vernier calipers and height gages have been largely replaced by digital instruments. However, many conventional (non-digital) customary vernier calipers and height gages that have been used for years are still in use. Therefore, these customary measuring instruments are retained in this edition of the book, but metric vernier calipers and height gages have been eliminated.

Decimal-inch vernier calipers are used in machine shop applications when the degree of precision to thousandths of an inch is adequate. They are used for measuring lengths of parts, distances between holes, and both inside and outside diameters of cylinders.

Vernier height gages are widely used on surface plates and on machine tables. The height gage with an indicator attachment is used for checking locations of surfaces and holes. The height gage with a scriber attachment is used to mark reference lines, locations, and stock allowances on castings and forgings.

Decimal-Inch Vernier Caliper

The basic parts of a vernier caliper are a main scale, which is similar to a steel rule with a fixed jaw, and a sliding jaw with a vernier scale. The vernier scale slides parallel to the main scale and provides a degree of precision to 0.001". Calipers are available in a wide range of lengths with different types of jaws and scale graduations. A vernier caliper, which is commonly used in machine shops, is shown in Figure 29-1.

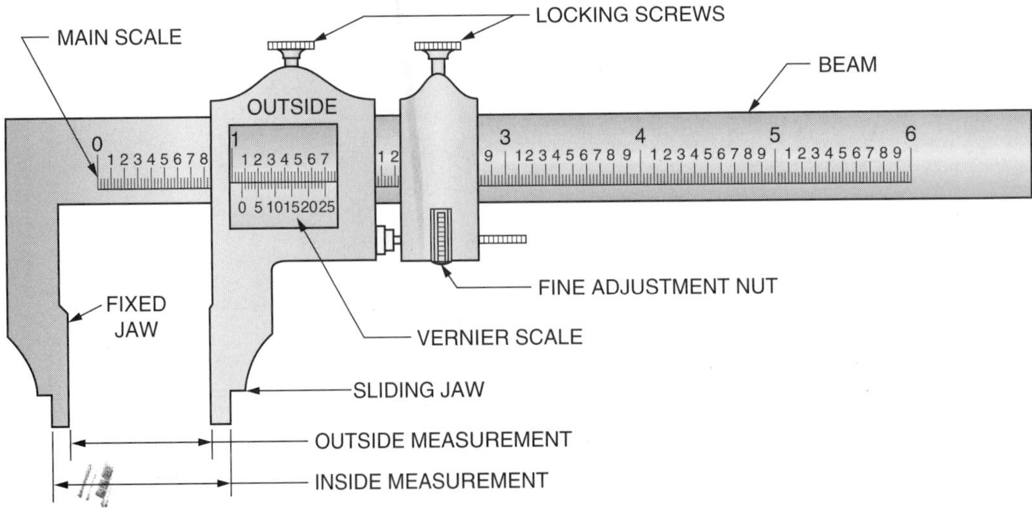

Figure 29-1

The main scale is divided into inches and the inches are divided into 10 divisions each equal to 0.1". The 0.1" divisions are divided into 4 parts each equal to 0.025". The vernier scale consists of 25 divisions.

A vernier scale is shown in Figure 29-2. The vernier scale has 25 divisions in a length equal to a length on the main scale that has 24 divisions. The difference between a main scale division and a vernier division is $\frac{1}{25}$ of 0.025" or 0.001".

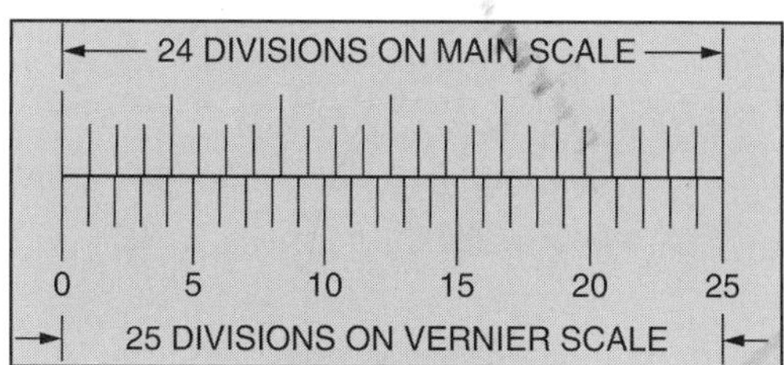

Figure 29-2

Reading and Setting a Measurement on a Decimal-Inch Vernier Caliper

A measurement is read by adding the thousandths reading on the vernier scale to the reading from the main scale.

Procedure To read a measurement on a decimal-inch vernier caliper

- Read the number of 1" graduations, 0.1" graduations, and 0.025" graduations on the main scale that are left of the zero graduation on the vernier scale.
- On the vernier scale, find the graduation that most closely coincides with a graduation on the main scale. Add this vernier reading, which indicates the number of 0.001" graduations to the main scale reading.

Setting a given measurement is the reverse procedure of reading a measurement on the vernier caliper.

Example 1 Read the measurement set on the vernier caliper in Figure 29-3.

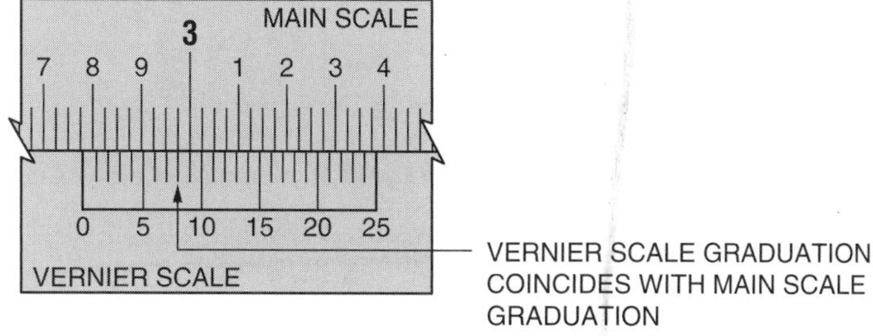

Figure 29-3

In reference to the zero division on the vernier scale, read two 1" divisions, seven 0.1" divisions, and three 0.025" divisions on the main scale.

$$(2" + 0.7" + 0.075" = 2.775")$$

Observe which vernier scale graduation most closely coincides with a main scale graduation. The 8 vernier scale graduation coincides; therefore, 0.008" is added to 2.775".

Measurement: 2.775" + 0.008" = 2.783" Ans

Example 2 Set 1.237" on a vernier caliper.

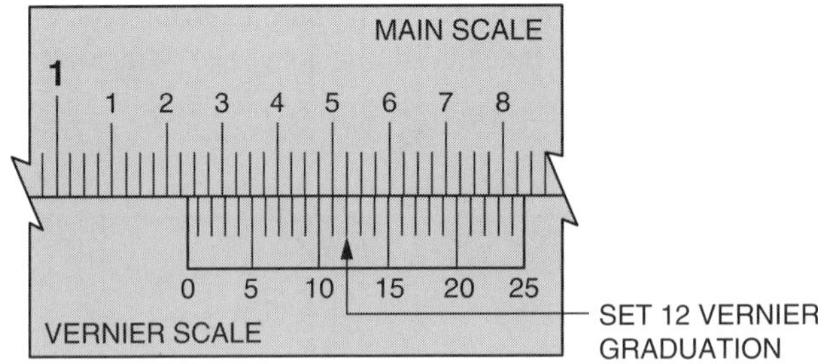

Figure 29-4

Move the vernier zero graduation to 1" + 0.2" + 0.025" on the main scale.

An additional 0.012" (1.237" − 1.225") is set by adjusting the sliding jaw until the 12 graduation on the vernier scale coincides with a graduation on the main scale.

The 1.237-inch setting is shown in Figure 29-4.

The accuracy of measurement obtainable with a vernier caliper depends on the user's ability to align the caliper with the part that is being measured and the user's "feel" when measuring. The line of measurement must be parallel to the beam of the caliper and lie in the same plane as the caliper. Care must be used to prevent a caliper setting that is too loose or too tight.

The front side of the customary vernier caliper (25 divisions) is used for outside measurements as shown in Figure 29-5. The reverse or back side is used for inside measurements as shown in Figure 29-6.

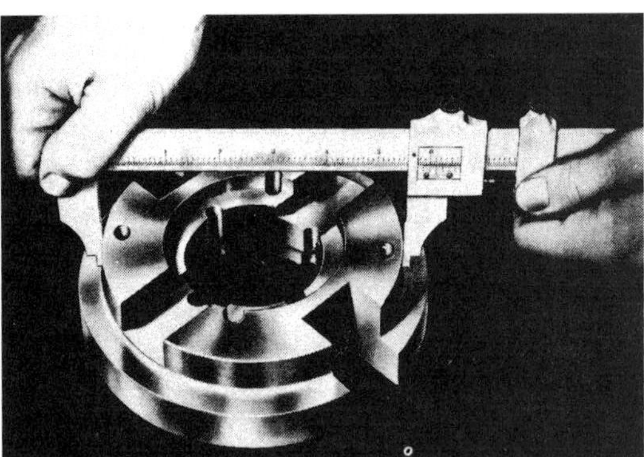

Figure 29-5 Measure an outside diameter The measurement is read on the front side of the caliper. (*Courtesy of L.S. Starrett Company*)

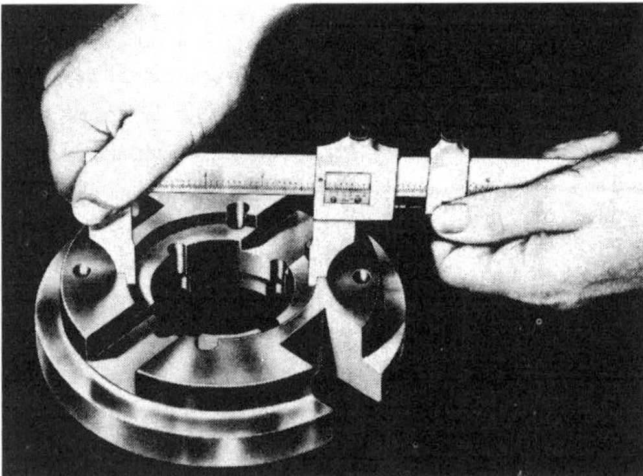

Figure 29-6 Measure an inside diameter The measurement is read on the back side of the caliper. (*Courtesy of L.S. Starrett Company*)

Decimal-Inch Vernier Height Gage

The vernier height gage and vernier caliper are similar in operation. The height gage also has a sliding jaw; the fixed jaw is the surface plate with which the height gage is usually used. The gage can be used with a scriber, a depth gage attachment, or an indicator. The indicator is the most widely used and, generally, the most accurate attachment. The parts of a vernier height gage are shown in Figure 29-7.

Measurements on the vernier height gage are read and set using the same procedure as with the vernier caliper.

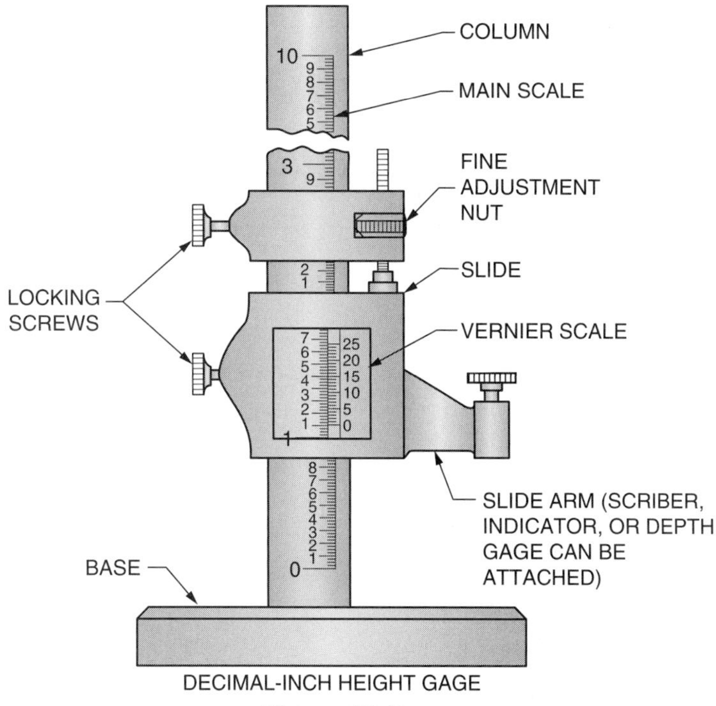

Figure 29-7

Example 1 Read the measurement set on the vernier height gage in Figure 29-8.

In reference to the zero division on the vernier scale read 5", four 0.1" divisions, and two 0.025" divisions on the main scale.
(5" + 0.4" + 0.050" = 5.450")
Observe which vernier scale graduation most closely coincides with the main scale graduation. The 21 vernier scale graduation coincides; therefore, 0.021" is added to 5.450".

Measurement = 5.450" + 0.021" = 5.471" Ans

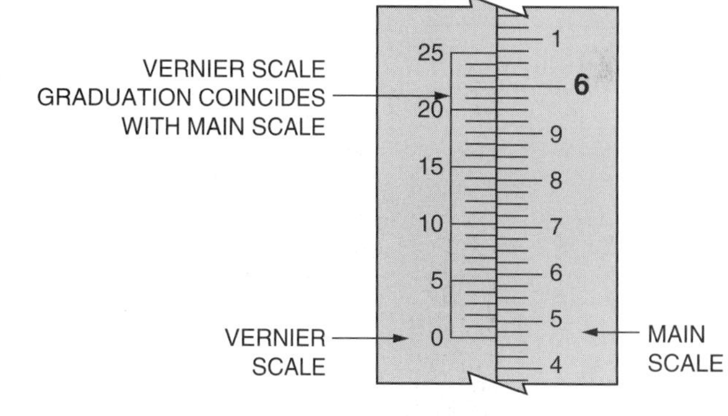

Figure 29-8

Example 2 Set 8.398" on a vernier height gage.

Move the vernier zero graduation to
8" + 0.3" + 0.075" = 8.375".
Lock the upper stide in place.

An additional 0.023" (8.398" − 8.375") is set by turning the fine adjustment screw until the 23 graduation on the vernier scale coincides with a graduation on the main scale.

The 8.398-inch setting is shown in Figure 29-9.

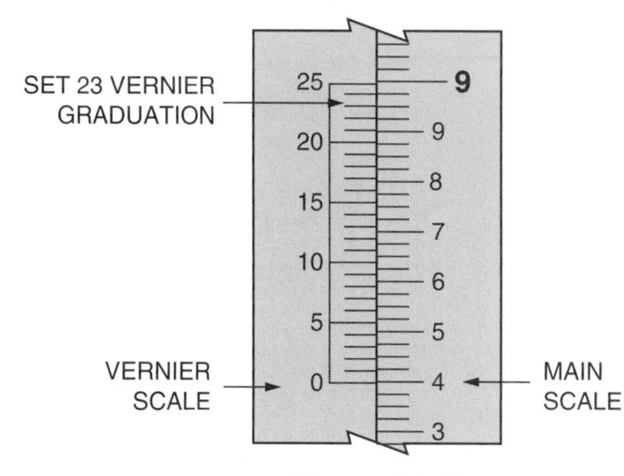

Figure 29-9

Application

Decimal-Inch Vernier Caliper

1. Read the decimal-inch vernier caliper measurements a–h for the following settings.

a. ______________

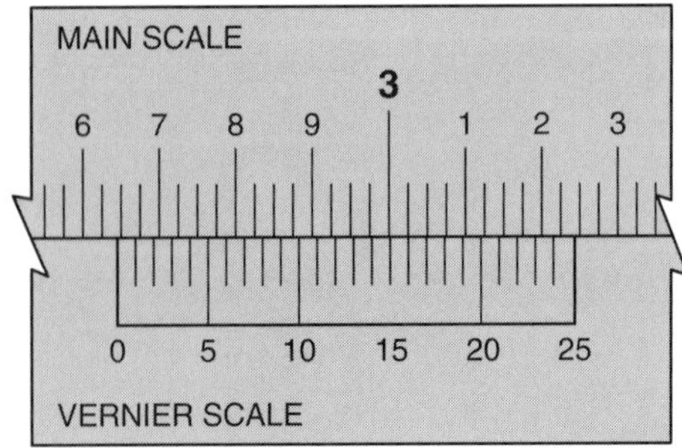

b. ______________

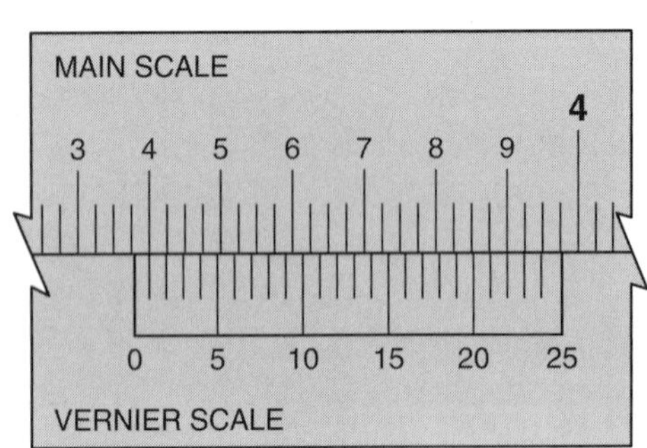

c. ______________

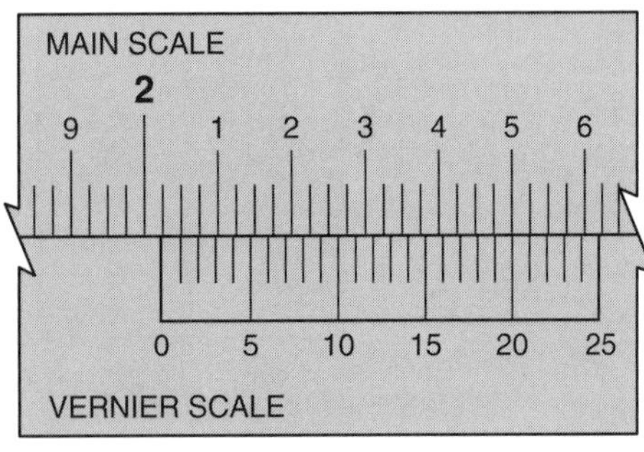

d. ______________

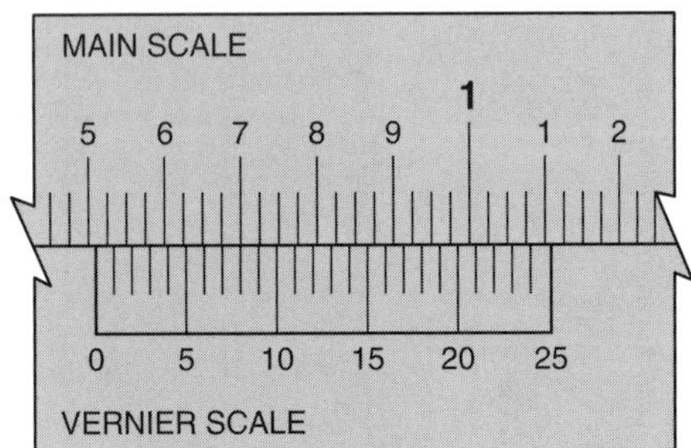

e. ______________

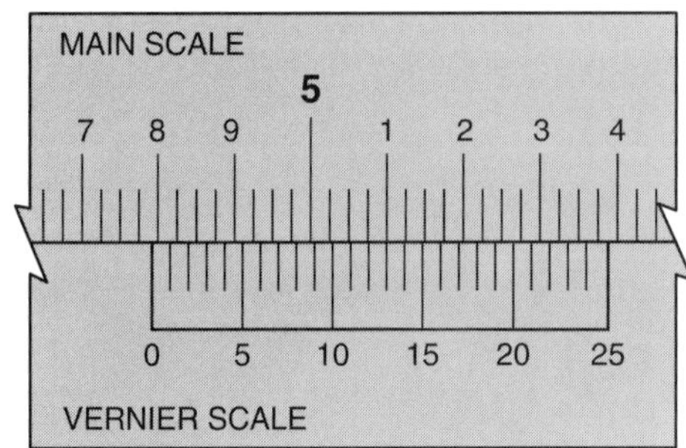

f. ______________

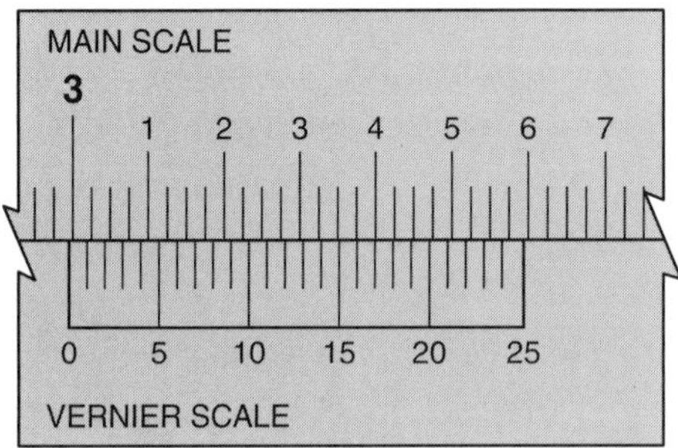

g. ______________

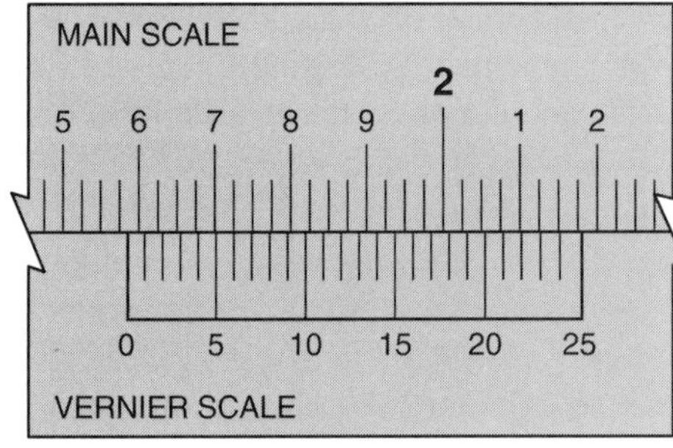

h. ______________

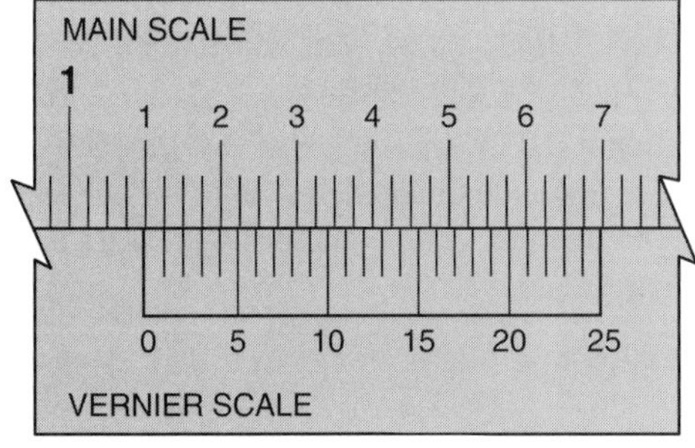

2. The following tables give the position of the zero graduation on the vernier scale in reference to the main scale and the vernier scale graduation that coincides with a main scale graduation. Determine the vernier caliper settings. The answer to the first problem is given.

	Zero Vernier Graduation Lies Between These Main Scale Graduations (inches)	Vernier Graduation That Coincides with a Main Scale Graduation	Vernier Caliper Setting (inches)
a.	1.875–1.900	19	1.894
b.	3.025–3.050	21	
c.	0.050–0.075	6	
d.	5.775–5.800	11	
e.	1.225–1.250	7	
f.	0.075–0.100	16	
g.	3.000–3.025	4	
h.	2.650–2.675	9	
i.	1.000–1.025	13	
j.	5.975–6.000	18	
k.	2.825–2.850	8	
l.	4.950–4.975	1	

	Zero Vernier Graduation Lies Between These Main Scale Graduations (inches)	Vernier Graduation That Coincides with a Main Scale Graduation	Vernier Caliper Setting (inches)
m.	0.000–0.025	5	
n.	0.825–0.850	17	
o.	3.550–3.575	23	
p.	5.075–5.100	20	
q.	3.325–3.350	15	
r.	2.075–2.100	6	
s.	4.400–4.425	10	
t.	1.025–1.050	13	
u.	0.675–0.700	18	
v.	0.050–0.075	2	
w.	3.000–3.025	21	
x.	2.925–2.950	22	

3. Refer to the following sentence and to the following given vernier caliper settings to find values A, B, and C. "The zero vernier scale graduation lies between A and B on the main scale and the vernier graduation C coincides with the main scale graduation." The answer to the first problem is given.

	Vernier Caliper Setting (inches)	A (inches)	B (inches)	C
a.	3.242	3.225	3.250	17
b.	2.877			
c.	4.839			
d.	0.611			
e.	4.369			
f.	0.084			
g.	7.857			

	Vernier Caliper Setting (inches)	A (inches)	B (inches)	C
h.	1.646			
i.	4.034			
j.	0.022			
k.	3.333			
l.	5.999			
m.	0.278			
n.	0.965			

4. The distance between the centers of two holes can be checked with a vernier caliper. The position of the caliper in measuring the inside distance between two holes is shown in Figure 29-10. To determine the setting on the caliper, subtract the radius of each hole (one-half the diameter) from the center distance. The following problems give the hole diameters and the distances between centers. For each determine (1) the main scale setting and (2) the vernier scale setting. All dimensions are in inches.

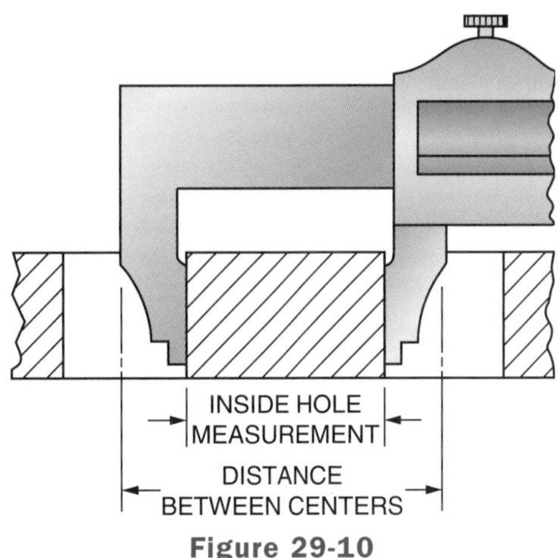

Figure 29-10

a.

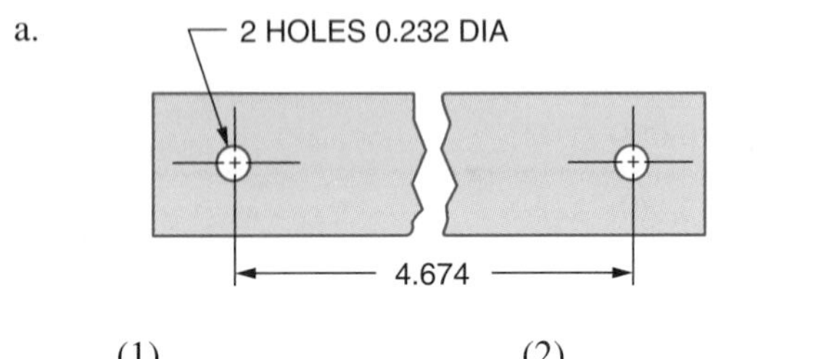

(1) ______________ (2) ______________

b. 2 HOLES 0.186 DIA

4.358

(1) ______________ (2) ______________

c.

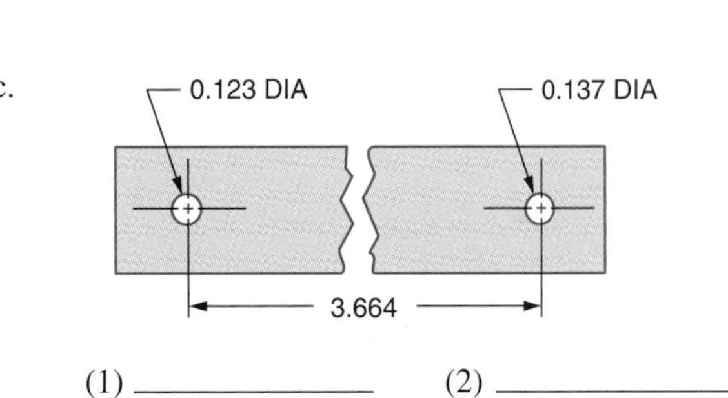

(1) ______________ (2) ______________

d. **Note:** Hole tolerances are shown. Maximum and minimum vernier scale settings are required.

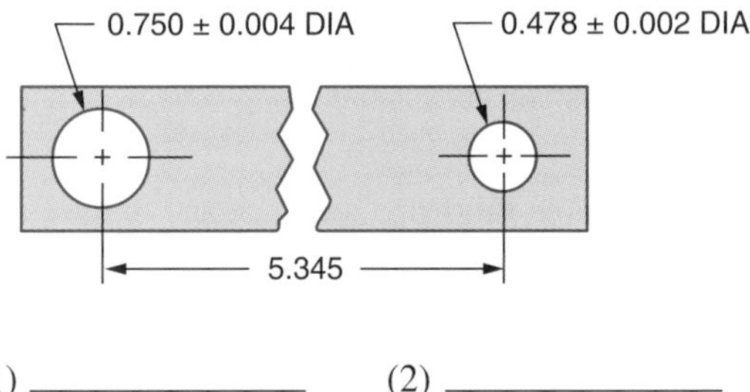

(1) ______________ (2) ______________

e. **Note:** Hole tolerances and center distance tolerances are shown. Maximum and minimum vernier scale settings are required.

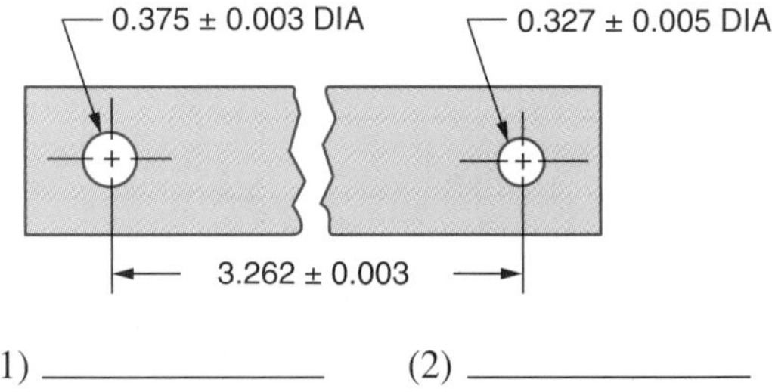

(1) ______________ (2) ______________

Decimal-Inch Height Gage

5. Read height gage measurement a–h for the following settings.

a. ________ b. ________ c. ________ d. ________

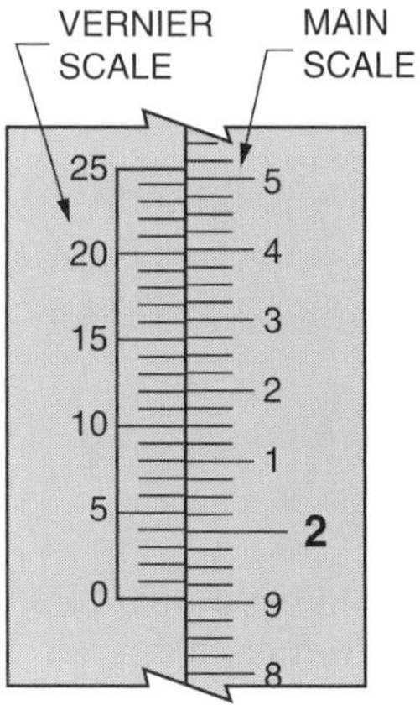

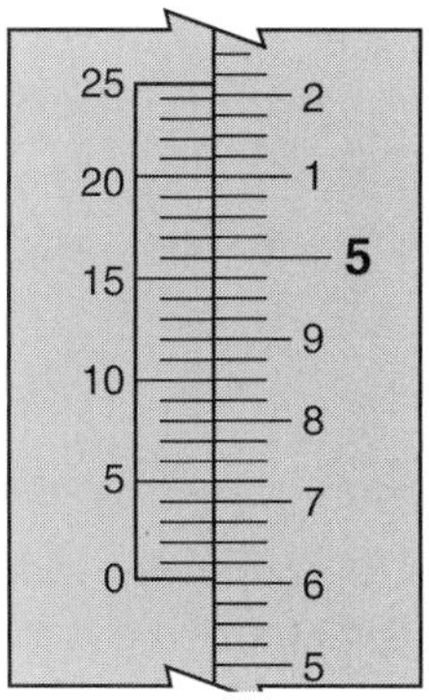

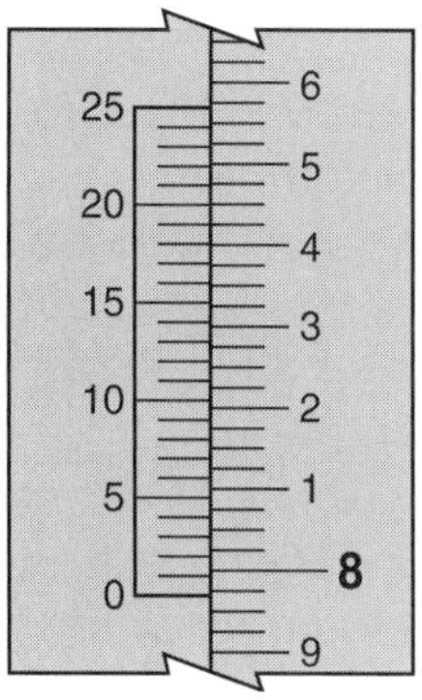

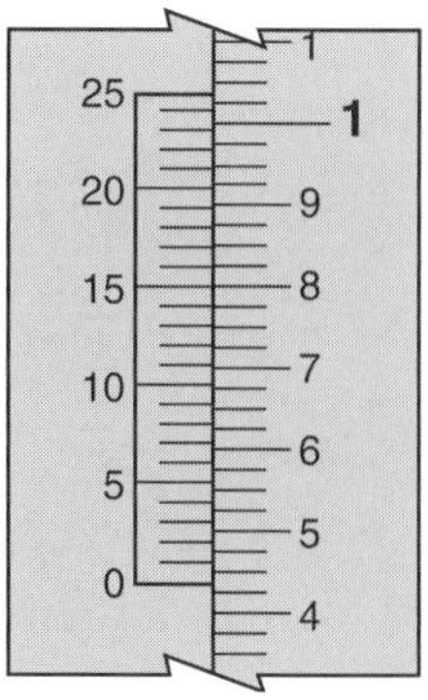

e. ________ f. ________ g. ________ h. ________

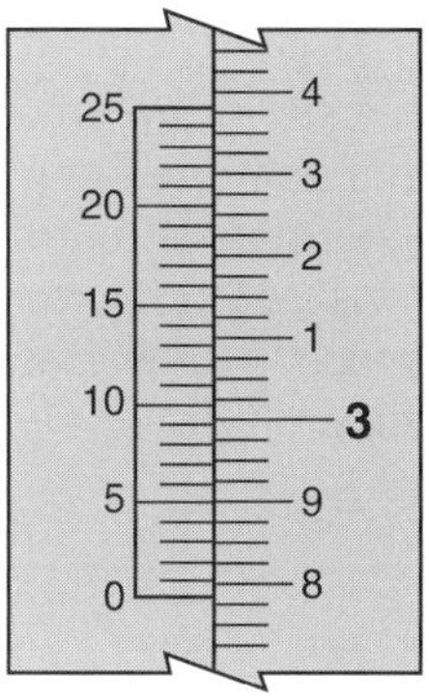

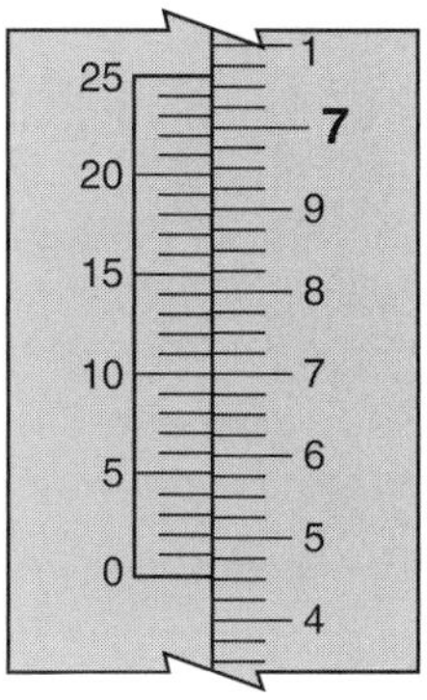

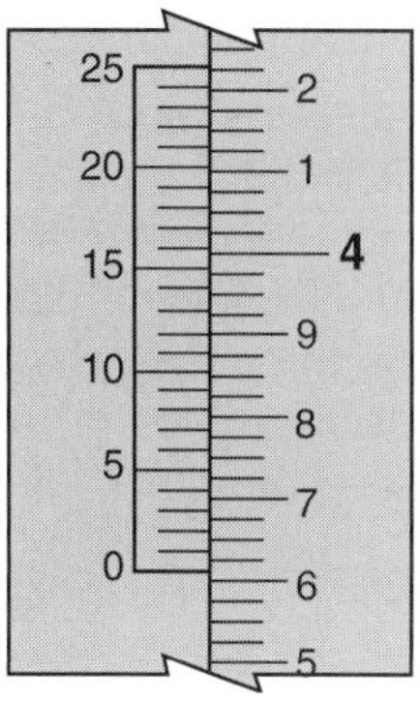

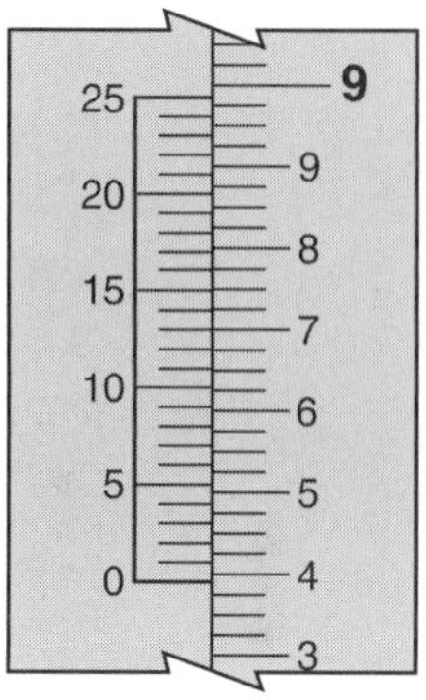

6. The hole locations of the block in Figure 29-11 are checked by placing the block on a surface plate and indicating the bottom of each hole using a height gage with an indicator attachment. Determine the height gage settings from the bottom of the part to the bottom of the holes. Assume that the actual hole diameters and locations are the same as the given dimensions. The setting for the first problem is given.

Hole Number	Hole Diameter (inches)	Given Locations to Centers of Holes (inches)	Height Gage Settings	
			Main Scale Setting (inches)	Vernier Scale Setting (inches)
1	0.376	A = 0.640	0.450–0.475	2
2	0.258	B = 1.008		
3	0.188	C = 0.514		
4	0.496	D = 0.312		
5	0.127	E = 0.810		

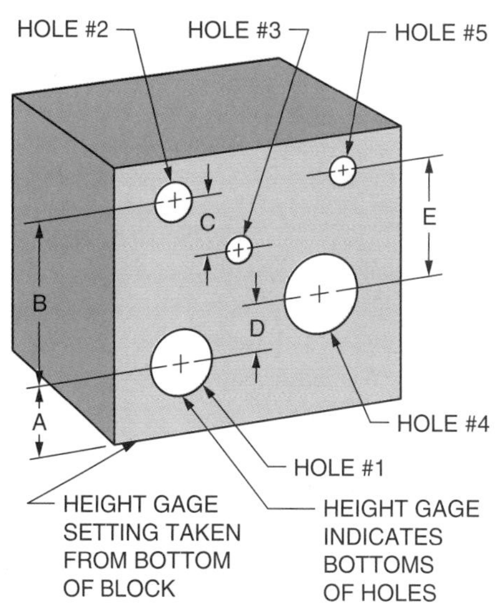

Figure 29-11

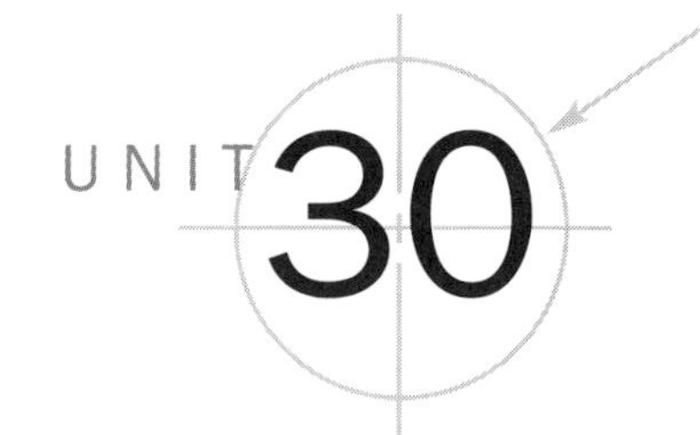

Customary Micrometers

Objectives ***After studying this unit you should be able to***

- Read settings from the barrel and thimble scales of a 0.001-inch micrometer.
- Set given dimensions on the scales of 0.001-inch and 0.0001-inch micrometers.
- Read settings from the barrel, thimble, and vernier scales of 0.0001-inch micrometers.

As with digital calipers, digital micrometers are widely used. Measurements are read directly on a five-digit LCD readout display with inch/millimeter selection.

Also, as with calipers, many conventional (non-digital) customary micrometers that have been used for years are still in use. Therefore, these customary micrometers are retained in this edition of the book, but conventional (non-digital) metric micrometers have been eliminated.

Micrometers are basic measuring instruments used by machinists in the processing and checking of parts. Micrometers are available in a wide range of sizes and types. Outside micrometers are used to measure dimensions between parallel surfaces of parts and outside diameters of cylinders. Other types, such as depth micrometers, screw thread micrometers, disc and blade micrometers, bench micrometers, and inside micrometers, also have wide application in the machine shop. A few of the many types of non-digital micrometers are shown in Figure 30-1.

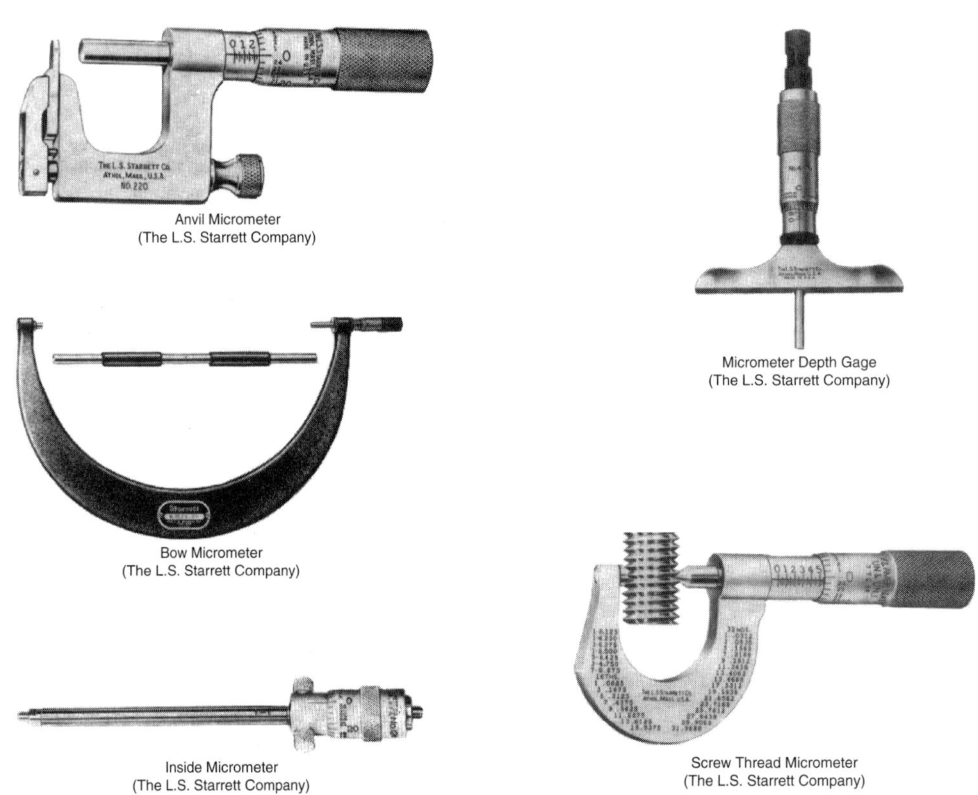

Anvil Micrometer
(The L.S. Starrett Company)

Micrometer Depth Gage
(The L.S. Starrett Company)

Bow Micrometer
(The L.S. Starrett Company)

Inside Micrometer
(The L.S. Starrett Company)

Screw Thread Micrometer
(The L.S. Starrett Company)

Figure 30-1

The 0.001-Inch Micrometer

A 0.001-inch outside micrometer is shown in Figures 30-2 and 30-3 with its principal parts labeled.

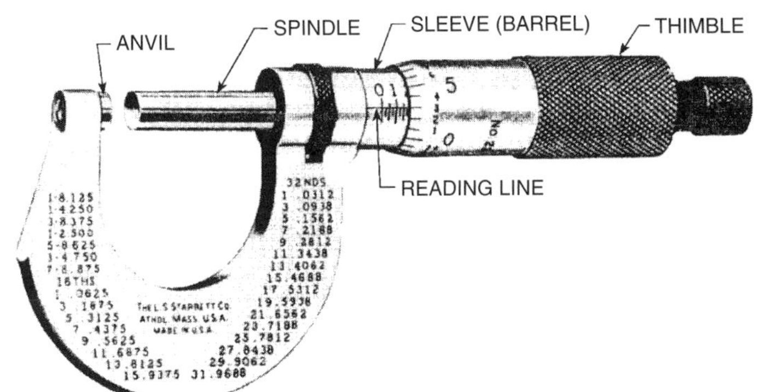

Figure 30-2

BARREL SCALE
THIMBLE SCALE
0.001" DIVISION
0 1 2 3
0
0.025" DIVISION (DISTANCE MOVED IN ONE REVOLUTION OF THIMBLE)
0.100" DIVISION

Figure 30-3

The part to be measured is placed between the anvil and the spindle. The barrel of a micrometer consists of a scale, which is one inch long. The one-inch length is divided into ten divisions each equal to 0.100 inch. The 0.100-inch divisions are further divided into four divisions each equal to 0.025 inch.

The thimble has a scale that is divided into twenty-five parts. One revolution of the thimble moves 0.025 inch on the barrel scale. Therefore, a movement of one graduation on the thimble equals $\frac{1}{25}$ of 0.025 inch or 0.001 inch along the barrel.

Reading and Setting a 0.001-Inch Micrometer

A micrometer is read by observing the position of the bevel edge of the thimble in reference to the scale on the barrel. Observe the greatest 0.100-inch division and the number of 0.025-inch divisions on the barrel scale. To this barrel reading, add the number of the 0.001-inch divisions on the thimble that coincide with the horizontal line (reading line) on the barrel scale.

Procedure To read a 0.001-inch micrometer

- Observe the greatest 0.100-inch division on the barrel scale.
- Observe the number of 0.025-inch divisions on the barrel scale.
- Add the thimble scale reading (0.001-inch division) that coincides with the horizontal line on the barrel scale.

Example 1 Read the micrometer setting shown in Figure 30-4.

Observe the greatest 0.100-inch division on the barrel scale.
(three 0.100" = 0.300")

Observe the number of 0.025-inch divisions between the 0.300-inch mark and the thimble. (two 0.025" = 0.050")

Add the thimble scale reading that coincides with the horizontal line on the barrel scale. (eight 0.001" = 0.008")

Micrometer reading: 0.300" + 0.050" + 0.008" = 0.358" Ans

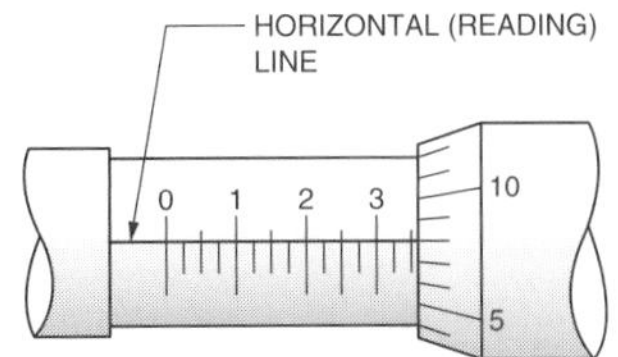

Figure 30-4

Example 2 Read the micrometer setting shown in Figure 30-5.

On the barrel scale, two 0.100" = 0.200".

On the barrel scale, zero 0.025" = 0".

On the thimble scale, twenty-three 0.001" = 0.023".

Micrometer reading: 0.200" + 0.023" = 0.223" Ans

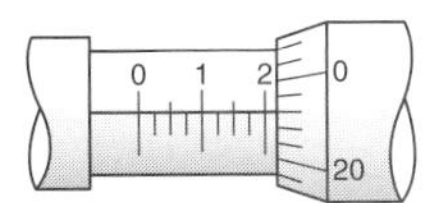

Figure 30-5

Procedure To set a 0.001-inch micrometer to a given dimension

- Turn the thimble until the barrel scale indicates the required number of 0.100-inch divisions plus the necessary number of 0.025-inch divisions.
- Turn the thimble until the thimble scale indicates the required additional 0.001-inch divisions.

Example 1 Set 0.949 inch on a micrometer.

Figure 30-6

Turn the thimble to nine 0.100-inch divisions plus one 0.025-inch division on the barrel scale. ($9 \times 0.100" + 0.025" = 0.925"$)

Turn the thimble an additional twenty-four 0.001-inch thimble scale divisions. ($0.949" - 0.925" = 0.024"$)

The 0.949-inch setting is shown in Figure 30-6.

Example 2 Set 0.520 inch on a micrometer.

Figure 30-7

Turn the thimble to five 0.100-inch divisions on the barrel scale. ($5 \times 0.100" = 0.500"$)

Turn the thimble an additional twenty 0.001-inch divisions. ($0.520" - 0.500" = 0.020"$)

The 0.520-inch setting is shown in Figure 30-7.

The Vernier (0.0001-Inch) Micrometer

The addition of a vernier scale on the barrel of a 0.001-inch micrometer increases the degree of precision of the instrument to 0.0001 inch. The barrel scale and thimble scale of a vernier micrometer are identical to that of a 0.001-inch micrometer. Figure 30-8 shows the relative positions of the barrel scale, thimble scale, and vernier scale of a 0.0001-inch vernier micrometer.

The vernier scale consists of ten divisions. Ten vernier divisions on the circumference of the barrel are equal in length to nine divisions of the thimble scale. The difference between one vernier division and one thimble division is 0.0001-inch. A flattened view of a vernier and a thimble scale is shown in Figure 30-9.

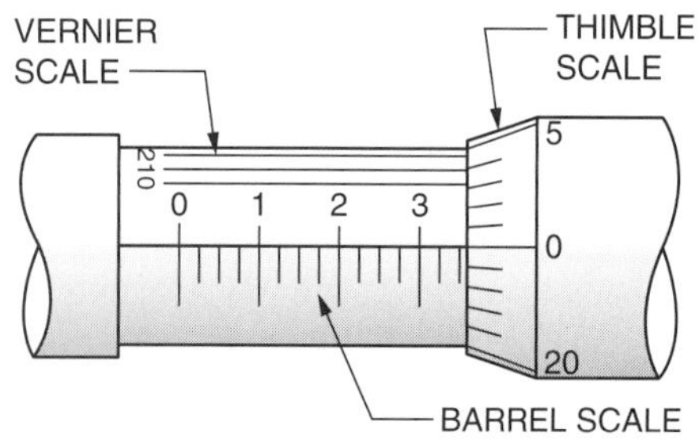

Figure 30-8

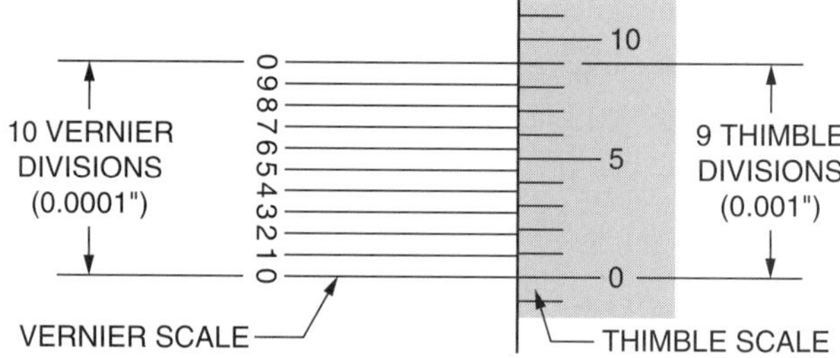

Figure 30-9

Reading and Setting the Vernier (0.0001-Inch) Micrometer

Reading a vernier micrometer is the same as reading a 0.001-inch micrometer except for the addition of reading the vernier scale. A particular vernier graduation coincides with a thimble scale graduation. This vernier graduation gives the number of 0.0001-inch divisions that are added to the barrel and thimble scale readings.

Example 1 Read the vernier micrometer setting shown in the flattened view in Figure 30-10.

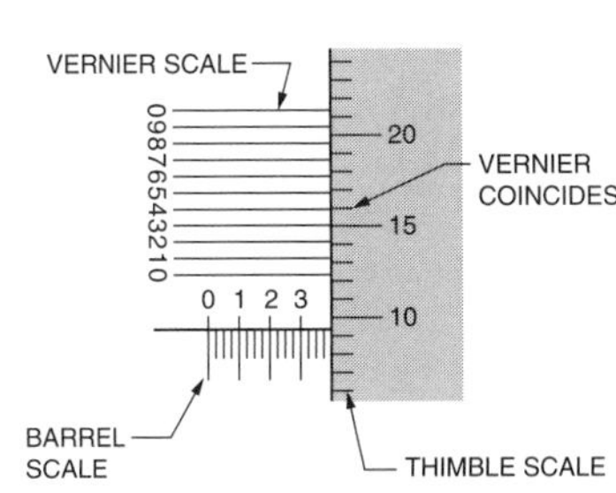

Figure 30-10

Read the barrel scale reading. Three 0.100" divisions plus three 0.025" divisions = 0.375".

Read the thimble scale. The reading is between the 0.009" and 0.010" divisions, therefore, the thimble reading is 0.009".

Read the vernier scale. The 0.0004" division of the vernier scale coincides with a thimble division.

Vernier micrometer reading:
0.375" + 0.009" + 0.0004" = 0.3844" Ans

Example 2 Read the vernier micrometer setting shown in this flattened view in Figure 30-11.

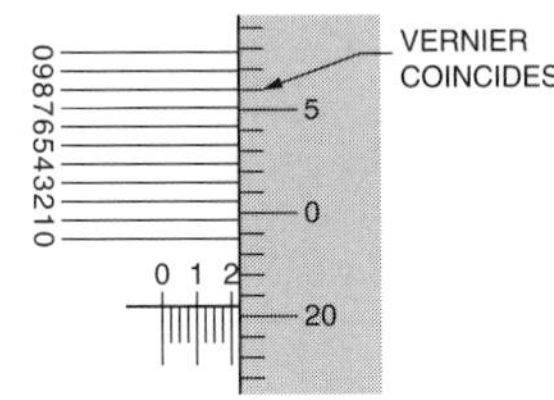

Figure 30-11

On the barrel scale read 0.200".

On the thimble scale read 0.020".

On the vernier scale read 0.0008".

Vernier micrometer reading:
0.200" + 0.020" + 0.0008" = 0.2208" Ans

Setting a vernier (0.0001-inch) micrometer is the same as setting a 0.001-inch micrometer except for the addition of setting the vernier scale.

Example Set 0.2336 inch on a vernier micrometer.

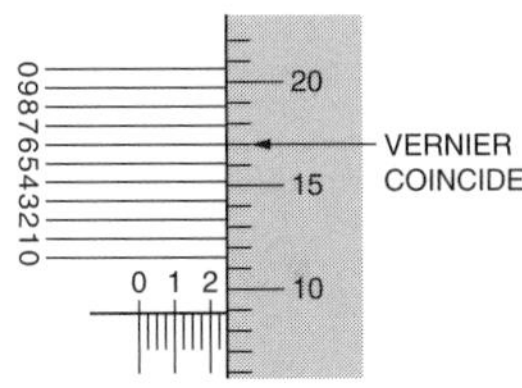

Figure 30-12

Turn the thimble to two 0.100-inch divisions plus one 0.025-inch division on the barrel scale. (2 × 0.100" + 0.025" = 0.225")

Turn the thimble an additional eight 0.001-inch divisions. (0.2336" − 0.225" = 0.0086")

Turn the thimble carefully until a graduation on the thimble scale coincides with the 0.0006-inch division on the vernier scale. (0.2336" − 0.233" = 0.0006")

The 0.2336-inch setting is shown in Figure 30-12.

Application

0.001-Inch Micrometer

Read the settings on the following 0.001-inch micrometer scales.

1. 

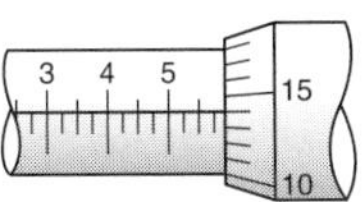

2. ______________

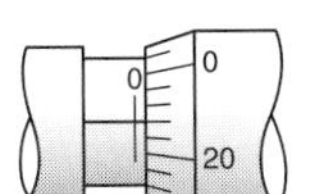

3. ______________

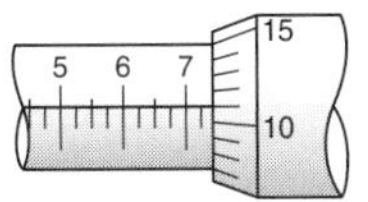

4. ______________

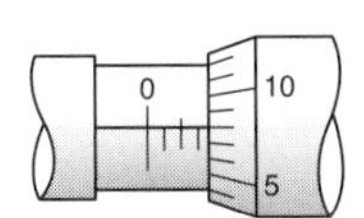

5. ______________

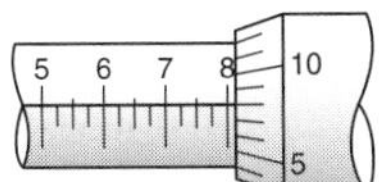

6. 

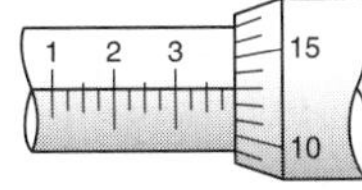

7. ______________

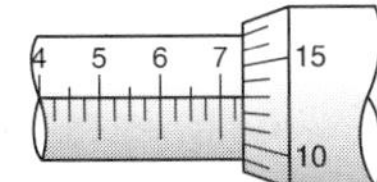

8. 

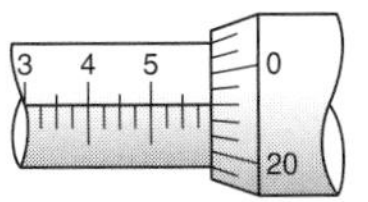

9. ______________

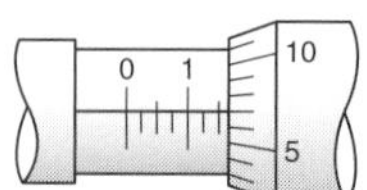

10. ______________

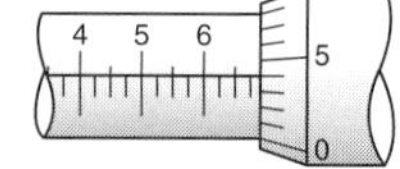

11. ______________

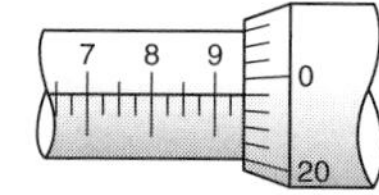

12. ______________

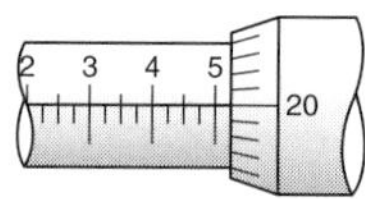

Given the following barrel scale and thimble scale settings of a 0.001-inch micrometer, determine the readings in the tables. The answer to the first problem is given.

	Barrel Scale Setting Is Between: (inches)	Thimble Scale Setting (inches)	Micrometer Reading (inches)
13.	0.425–0.450	0.016	0.441
14.	0.075–0.100	0.009	
15.	0.150–0.175	0.003	
16.	0.875–0.900	0.012	
17.	0.400–0.425	0.024	

	Barrel Scale Setting Is Between: (inches)	Thimble Scale Setting (inches)	Micrometer Reading (inches)
18.	0.000–0.025	0.023	
19.	0.025–0.050	0.013	
20.	0.750–0.775	0.017	
21.	0.975–1.000	0.008	
22.	0.625–0.650	0.016	

Given the following 0.001-inch micrometer readings, determine the barrel scale and thimble scale settings. The answer to the first problem is given.

	Micrometer Reading (inches)	Barrel Scale Setting Is Between: (inches)	Thimble Scale Setting (inches)
23.	0.387	0.375–0.400	0.012
24.	0.841		
25.	0.973		
26.	0.002		
27.	0.079		

	Micrometer Reading (inches)	Barrel Scale Setting Is Between: (inches)	Thimble Scale Setting (inches)
28.	0.998		
29.	0.038		
30.	0.281		
31.	0.427		
32.	0.666		

The Vernier (0.0001-Inch) Micrometer

Read the settings on the following 0.0001-inch micrometer scales. The vernier, thimble, and barrel scales are shown in flattened views.

33. ______________

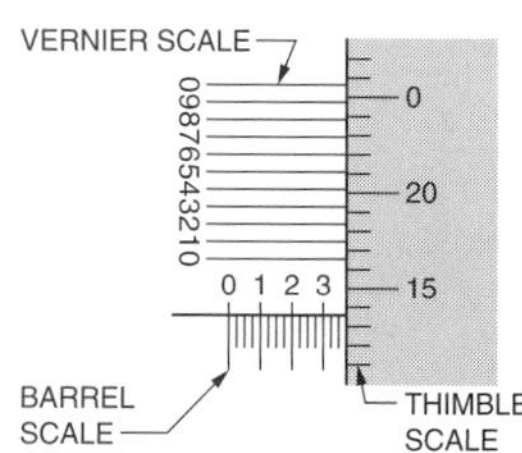

34. ______________

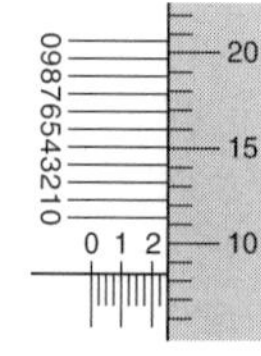

35. ______________

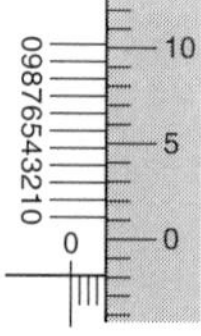

36. ______________ 37. ______________ 38. ______________

39. ______________ 40. ______________ 41. ______________

42. ______________ 43. ______________ 44. ______________

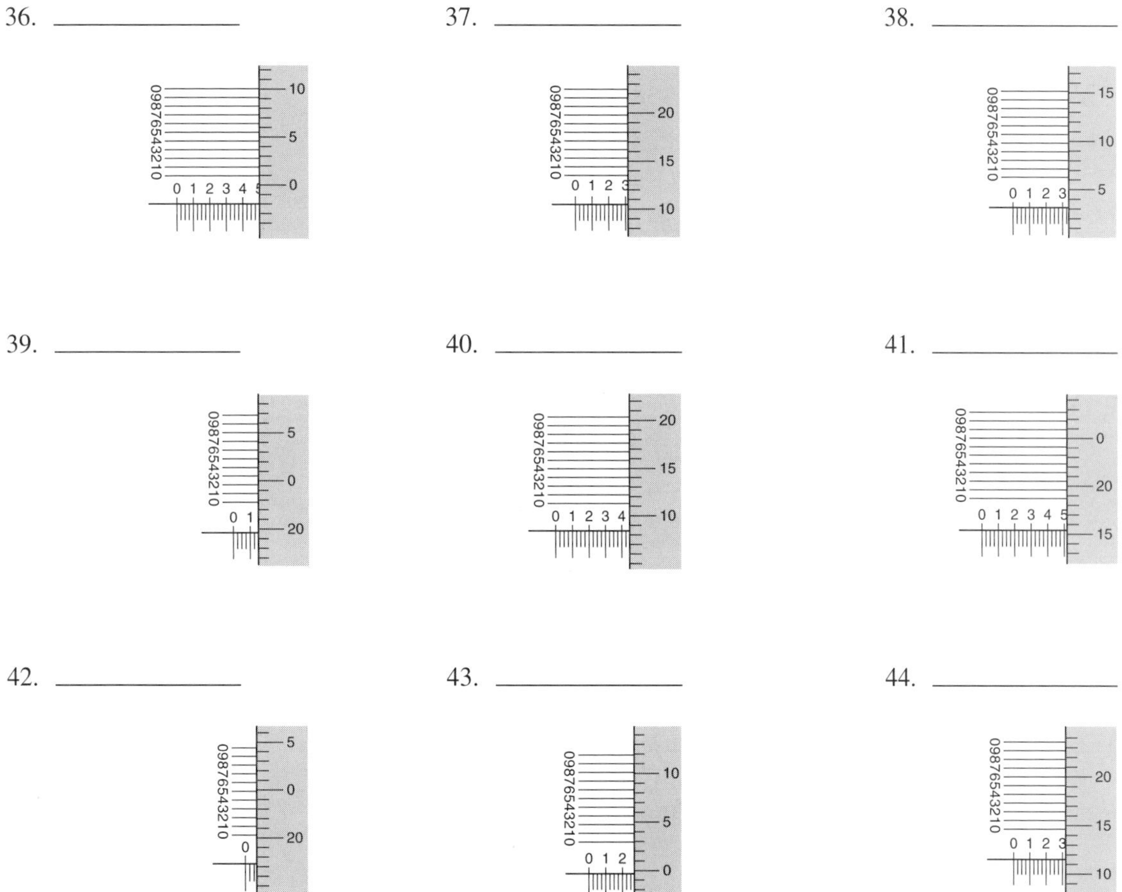

Given the following barrel scale, thimble scale, and vernier scale settings of a 0.0001-inch micrometer, determine the micrometer readings in these tables. The answer to the first problem is given.

	Barrel Scale Setting Is Between: (inches)	Thimble Scale Setting Is Between: (inches)	Vernier Scale Setting (inches)	Micrometer Reading (inches)
45.	0.375–0.400	0.017–0.018	0.0008	0.3928
46.	0.125–0.150	0.008–0.009	0.0003	
47.	0.950–0.975	0.021–0.022	0.0007	
48.	0.075–0.100	0.011–0.012	0.0005	
49.	0.300–0.325	0.000–0.001	0.0004	
50.	0.625–0.650	0.021–0.022	0.0002	
51.	0.000–0.025	0.000–0.001	0.0009	
52.	0.275–0.300	0.020–0.021	0.0007	
53.	0.850–0.875	0.009–0.010	0.0004	
54.	0.225–0.250	0.014–0.015	0.0008	

Given the following 0.0001-inch micrometer readings, determine the barrel scale, thimble scale, and vernier scale settings. The answer to the first problem is given.

	Micrometer Reading (inches)	Barrel Scale Setting Is Between: (inches)	Thimble Scale Setting Is Between: (inches)	Vernier Scale Setting (inches)
55.	0.7846	0.775–0.800	0.009–0.010	0.0006
56.	0.1035			
57.	0.0083			
58.	0.9898			
59.	0.3001			
60.	0.0012			
61.	0.8008			
62.	0.3135			
63.	0.9894			
64.	0.0479			

Customary and Metric Gage Blocks

Objectives ***After studying this unit you should be able to***

- Determine proper gage block combinations for specified customary or metric system dimensions.

Gage blocks are used in machine shops as standards for checking and setting (calibration) of micrometers, calipers, dial indicators, and other measuring instruments. Other applications of gage blocks are for layout, machine setups, and surface plate inspection.

Description of Gage Blocks

Gage blocks like those in Figure 31-1 are square- or rectangular-shaped hardened steel blocks that are manufactured to a high degree of accuracy, flatness, and parallelism. Gage blocks, when properly used, provide millionths of an inch accuracy with millionths of an inch precision.

By *wringing* blocks (slipping blocks one over the other using light pressure), a combination of the proper blocks can be achieved to provide a desired length. Wringing the blocks produces a very thin air gap that is similar to liquid film in holding the blocks together. There are a variety of both customary unit and metric gage block sets available. These tables list the thicknesses of blocks of a frequently used customary gage block set and the thicknesses of blocks of a commonly used metric gage block set. To reduce the possibility of error, it is customary to use the fewest number of blocks possible to achieve the stack.

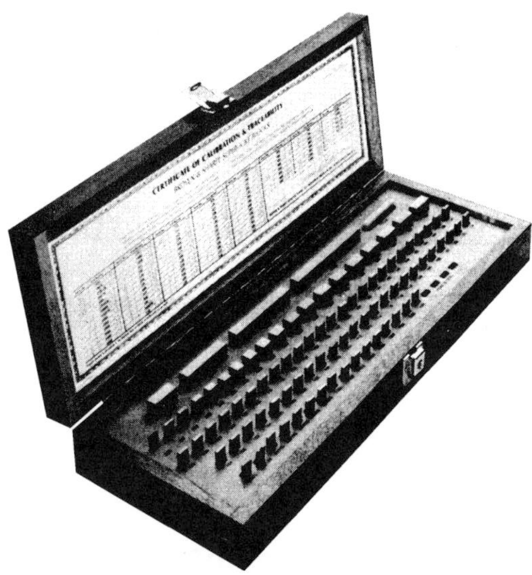

Figure 31-1 A complete set of gage blocks
(*Courtesy of Brown & Sharpe Mfg. Co.*)

BLOCK THICKNESSES OF A CUSTOMARY GAGE BLOCK SET*

9 Blocks 0.0001" Series

0.1001	0.1002	0.1003	0.1004	0.1005	0.1006	0.1007	0.1008	0.1009

49 Blocks 0.001" Series

0.101	0.102	0.103	0.104	0.105	0.106	0.107	0.108	0.109
0.110	0.111	0.112	0.113	0.114	0.115	0.116	0.117	0.118
0.119	0.120	0.121	0.122	0.123	0.124	0.125	0.126	0.127
0.128	0.129	0.130	0.131	0.132	0.133	0.134	0.135	0.136
0.137	0.138	0.139	0.140	0.141	0.142	0.143	0.144	0.145
0.146	0.147	0.148	0.149					

19 Blocks 0.050" Series

0.050	0.100	0.150	0.200	0.250	0.300	0.350	0.400	0.450
0.500	0.550	0.600	0.650	0.700	0.750	0.800	0.850	0.900
0.950								

4 Blocks 1.000" Series

1.000	2.000	3.000	4.000

*All thicknesses are in inches

BLOCK THICKNESSES OF A METRIC GAGE BLOCK SET*								
9 Blocks 0.001 mm Series								
1.001	1.002	1.003	1.004	1.005	1.006	1.007	1.008	1.009
9 Blocks 0.01 mm Series								
1.01	1.02	1.03	1.04	1.05	1.06	1.07	1.08	1.09
9 Blocks 0.1 mm Series								
1.1	1.2	1.3	1.4	1.5	1.6	1.7	1.8	1.9
9 Blocks 1 mm Series								
1	2	3	4	5	6	7	8	9
9 Blocks 10 mm Series								
10	20	30	40	50	60	70	80	90

*All thicknesses are in millimeters

Determining Gage Block Combinations

Usually there is more than one combination of blocks that will give a desired length. The most efficient procedure for determining block combinations is to eliminate digits of the desired measurement from right to left. This procedure saves time, minimizes the number of blocks, and reduces the chances of error. The following examples show how to apply the procedure in determining block combinations.

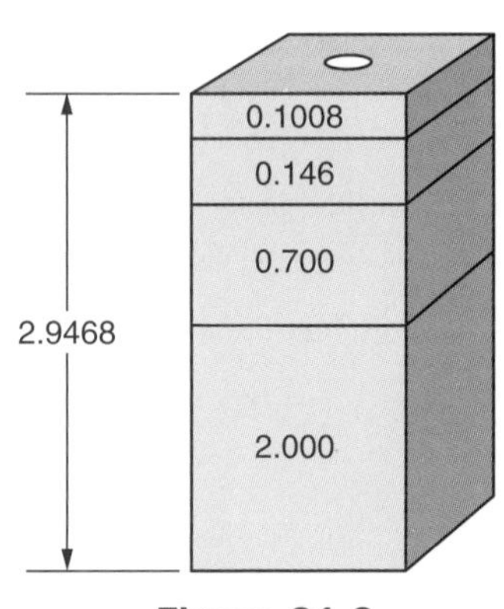

Figure 31-2

Example 1 Determine a combination of gage blocks for 2.9468 inches. Refer to the gage block sizes given in the Table of Block Thicknesses of a Customary Gage Block Set. All dimensions are in inches.

Choose the block that eliminates the last digit to the right, the 8. Choose the 0.1008" block. Subtract. (2.9468" − 0.1008" = 2.846")

Eliminate the last digit, 6, of 2.846". Choose the 0.146" block that eliminates the 4 as well as the 6. Subtract. (2.846" − 0.146" = 2.700")

Eliminate the last non-zero digit, 7, of 2.700". Choose the 0.700" block. Subtract. (2.700" − 0.700" = 2.000")

The 2.000" block completes the required dimension as shown in Figure 31-2.

Check. Add the blocks chosen.
0.1008" + 0.146" + 0.700" + 2.000" = 2.9468"

Example 2 Determine a combination of gage blocks for 10.2843 inches. Refer to the gage block sizes given in the Table of Block Thicknesses for a Customary Gage Block Set. All dimensions are in inches.

Eliminate the 3. Choose the 0.1003" block.
Subtract. (10.2843" − 0.1003" = 10.184")

Eliminate the 4. Choose the 0.134" block.
Subtract. (10.184" − 0.134" = 10.050")

Eliminate the 5. Choose the 0.050" block.
Subtract. (10.050" − 0.050" = 10.000")

The 1.000", 2.000", 3.000", and 4.000" blocks complete the required dimensions as shown in Figure 31-3.

Check. (0.1003" + 0.134" + 0.050" + 1.000" + 2.000" + 3.000" + 4.000" = 10.2843")

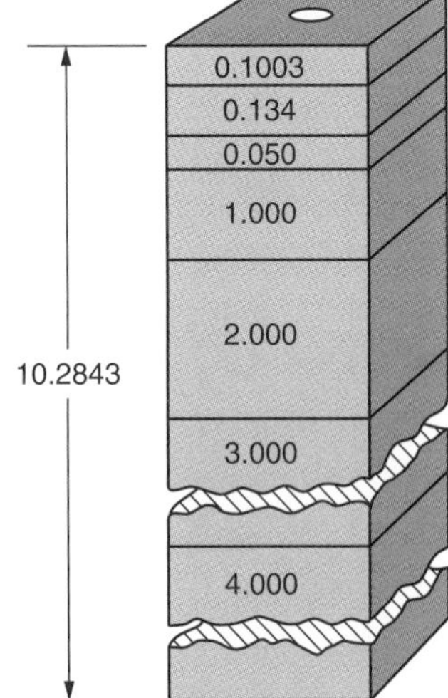

Figure 31-3

Example 3 Determine a combination of gage blocks for 157.372 millimeters. Refer to the gage block sizes given in the Table of Block Thicknesses for a Metric Gage Block Set. All dimensions are in millimeters.

Eliminate the 2. Choose the 1.002 mm block.
Subtract. (157.372 mm − 1.002 mm = 156.37 mm)

Eliminate the 7. Choose the 1.07 mm block.
Subtract. (156.37 mm − 1.07 mm = 155.3 mm)

Eliminate the 3. Choose the 1.3 mm block.
Subtract. (155.3 mm − 1.3 mm = 154 mm)

Eliminate the 4. Choose the 4 block.
Subtract. (154 mm − 4 mm = 150 mm)

The 60 and 90 block complete the required dimension as shown in Figure 31-4.

Check. (1.002 mm + 1.07 mm + 1.3 mm + 4 mm + 60 mm + 90 mm = 157.372 mm)

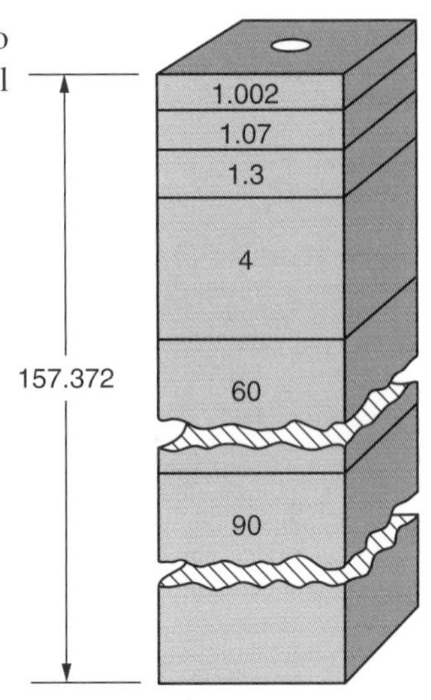

Figure 31-4

Application

Customary Gage Blocks

Using the Table of Block Thicknesses for a Customary Gage Block Set, determine a combination of gage blocks for each of the following dimensions.

➤ **Note:** Usually more than one combination of blocks will give the desired dimension.

1. 3.8638" ______
2. 1.8702" ______
3. 3.1222" ______
4. 0.6333" ______
5. 0.2759" ______
6. 5.8002" ______
7. 7.973" ______
8. 0.9999" ______
9. 10.250" ______
10. 9.050" ______
11. 4.8757" ______
12. 1.0001" ______
13. 0.2621" ______
14. 2.7311" ______
15. 5.090" ______
16. 6.0807" ______
17. 2.9789" ______
18. 0.6754" ______
19. 7.7777" ______
20. 10.0101" ______
21. 9.4346" ______
22. 4.8208" ______
23. 6.003" ______
24. 10.0021" ______
25. 0.6998" ______

Metric Gage Blocks

Using the Table of Block Thicknesses for a Metric Gage Block Set, determine a combination of gage blocks for each of the following dimensions.

➤ **Note:** Usually more than one combination of blocks will give the desired dimension.

26. 43.285 mm	________	34. 157.08 mm	________	42. 41.87 mm	________
27. 14.073 mm	________	35. 13.86 mm	________	43. 2.007 mm	________
28. 34.356 mm	________	36. 28.727 mm	________	44. 107.23 mm	________
29. 156.09 mm	________	37. 6.071 mm	________	45. 193.03 mm	________
30. 213.9 mm	________	38. 85.111 mm	________	46. 73.061 mm	________
31. 43.707 mm	________	39. 39.099 mm	________	47. 10.804 mm	________
32. 9.999 mm	________	40. 134.44 mm	________	48. 149.007 mm	________
33. 76.46 mm	________	41. 67.005 mm	________	49. 55.555 mm	________

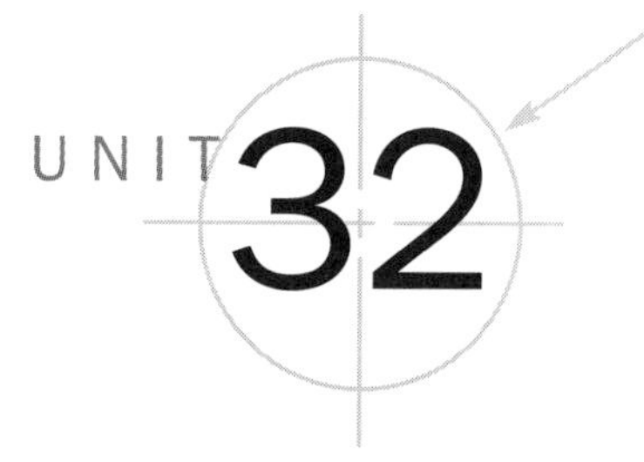

Achievement Review—Section Three

Objectives *You should be able to solve the exercises and problems in this Achievement Review by applying the principles and methods covered in Units 24–31.*

1. Express each of the following lengths as indicated.
   a. 81 inches as feet ________
   b. $6\frac{1}{4}$ feet as inches ________
   c. 9.6 yards as feet ________
   d. 2.7 centimeters as millimeters ________
   e. 0.8 meter as millimeters ________
   f. 218 millimeters as centimeters ________
2. Holes are to be drilled in the length of angle iron as shown in Figure 32-1. What is the distance between 2 consecutive holes?

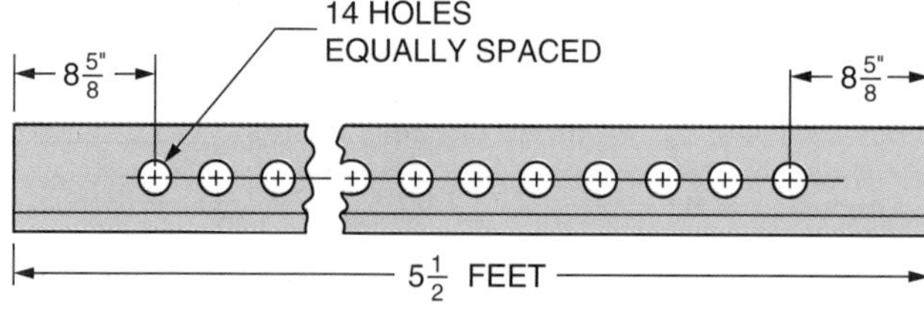

Figure 32-1

3. How many complete 3-meter lengths of tubing are required to make 250 pieces each 54 millimeters long? Allow a total one-half length of tubing for cutoff and scrap.

4. Express each of the following lengths as indicated. When necessary, round the answer to 3 decimal places.

   a. 47 millimeters as inches ________

   b. 5.5 meters as feet ________

   c. 16.8 centimeters as inches ________

   d. 4.75 inches as millimeters ________

   e. 31 inches as centimeters ________

   f. 4.5 feet as meters ________

5. For each of the exercises in the following table, the measurement made and the smallest graduation of the measuring instrument is given. Determine the greatest possible error and the smallest and largest possible actual length measure for each.

	Measurement Made	Smallest Graduation of Measuring Instrument Used	Greatest Possible Error	ACTUAL LENGTH	
				Smallest Possible	Largest Possible
a.	4.28"	0.02" (steel rule)			
b.	0.8367"	0.0001" (vernier micrometer)			
c.	46.16 mm	0.02 mm (vernier caliper)			
d.	16.45 mm	0.01 mm (micrometer)			

6. Compute the Absolute Error and Relative Error of each of the values in the following table. Where necessary, round the answers to 3 decimal places.

	True Value	Measured Value
a.	5.963 in	5.960 in
b.	0.392 mm	0.388 in
c.	7.123°	7.200°

	True Value	Measured Value
d.	0.1070 in	0.0990 in
e.	0.8639 in	0.8634 in
f.	0.713°	0.706°

a. Absolute Error ________
   Relative Error ________

b. Absolute Error ________
   Relative Error ________

c. Absolute Error ________
   Relative Error ________

d. Absolute Error ________
   Relative Error ________

e. Absolute Error ________
   Relative Error ________

f. Absolute Error ________
   Relative Error ________

7. The following dimensions with tolerances are given. Determine the maximum dimension (maximum limit) and the minimum dimension (minimum limit) for each.

   a. 1.714" ± 0.005"
   maximum ________
   minimum ________

   b. $4.0688"\ {}^{+0.0000"}_{-0.0012"}$
   maximum ________
   minimum ________

c. 5.9047" $^{+0.0008"}_{-0.0000"}$
maximum ______________
minimum ______________

d. 64.91 mm ± 0.08 mm
maximum ______________
minimum ______________

e. 173.003 mm $^{+0.000\text{ mm}}_{-0.013\text{ mm}}$
maximum ______________
minimum ______________

8. Express each of the following unilateral tolerances as bilateral tolerances having equal plus and minus values.

a. 0.876" $^{+0.006"}_{-0.000"}$ ______________

b. 5.2619" $^{+0.0000"}_{-0.0012"}$ ______________

c. 37.53 mm $^{+0.00\text{ mm}}_{-0.03\text{ mm}}$ ______________

d. 78.909 mm $^{+0.009\text{ mm}}_{-0.000\text{ mm}}$ ______________

9. The following problems require computations with both clearance fits and interference fits between mating parts. Determine the clearance or interference values as indicated. All dimensions are given in inches.

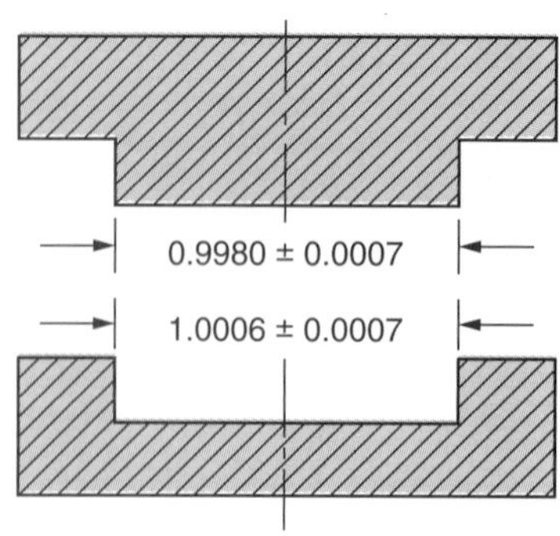

a. Find the maximum clearance. ______________

b. Find the minimum clearance. ______________

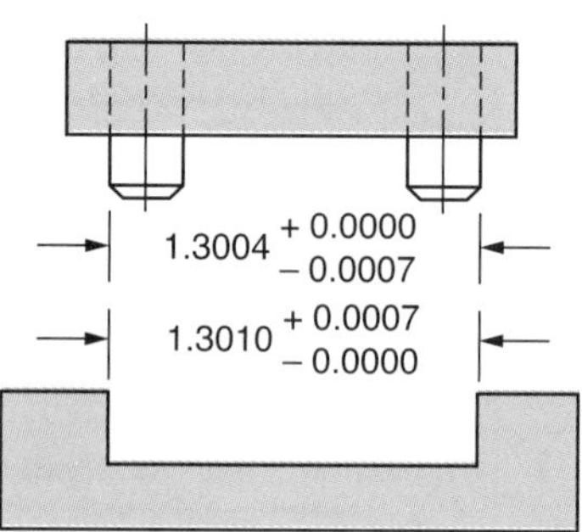

c. Find the maximum clearance. ______________

d. Find the minimum clearance. ______________

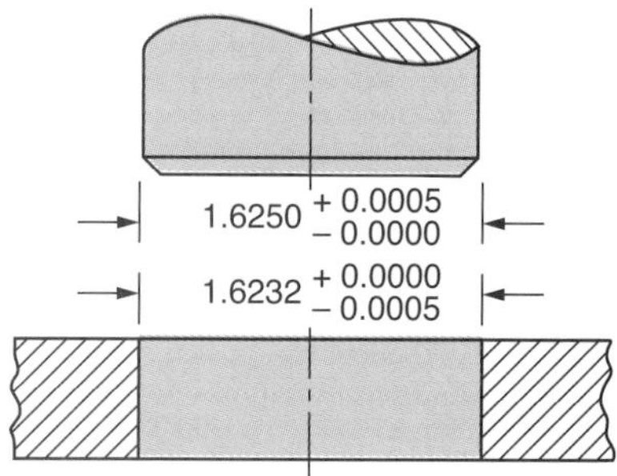

e. Find the maximum interference (allowance). ______________

f. Find the minimum interference. ______________

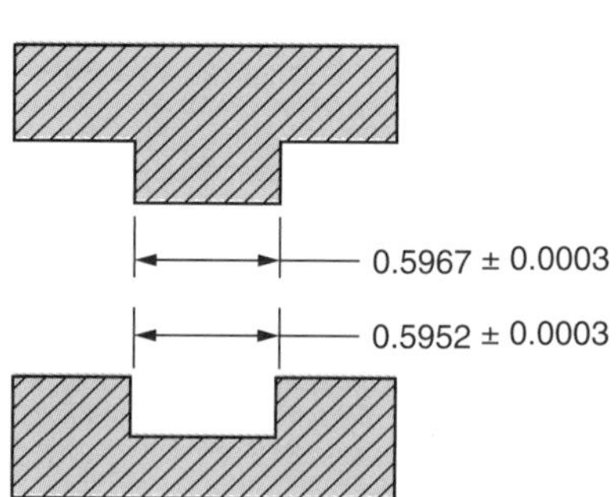

g. Find the maximum interference. (allowance). ______________

h. Find the minimum interference. ______________

10. Determine the minimum permissible length of distance A of the part shown in Figure 32-2. All dimensions are in millimeters.

____

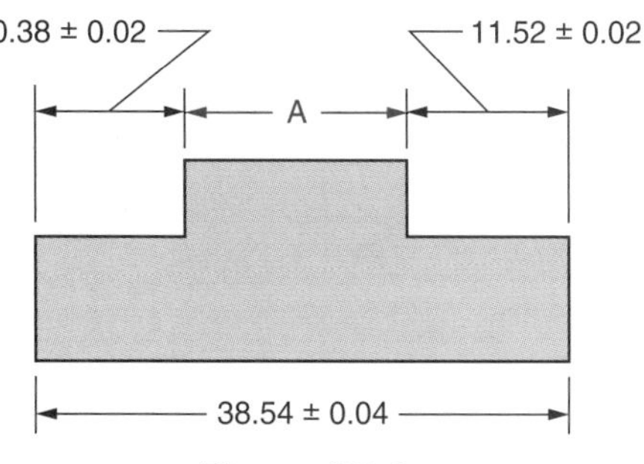

Figure 32-2

11. Read measurements a–p on the enlarged 32nds and 64ths graduated fractional rule shown in Figure 32-3.

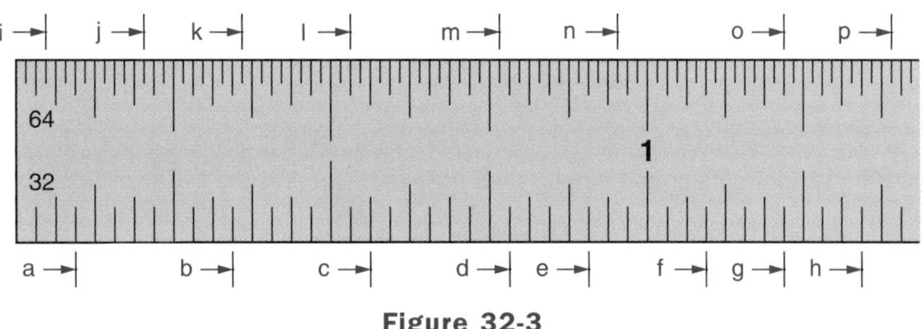

Figure 32-3

a.	_______	e.	_______	i.	_______	m.	_______
b.	_______	f.	_______	j.	_______	n.	_______
c.	_______	g.	_______	k.	_______	o.	_______
d.	_______	h.	_______	l.	_______	p.	_______

12. Read measurements a–p on the enlarged 50th and 100ths graduated decimal-inch rule shown in Figure 32-4.

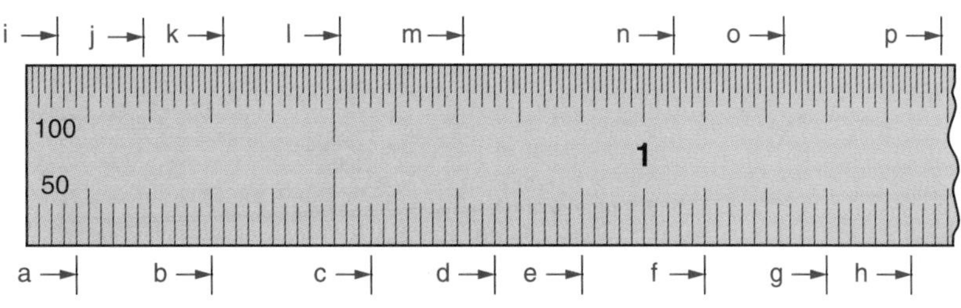

Figure 32-4

a.	_______	e.	_______	i.	_______	m.	_______
b.	_______	f.	_______	j.	_______	n.	_______
c.	_______	g.	_______	k.	_______	o.	_______
d.	_______	h.	_______	l.	_______	p.	_______

13. Read measurements a–p on the enlarged 1 millimeter and 0.5 millimeter graduated metric rule shown in Figure 32-5.

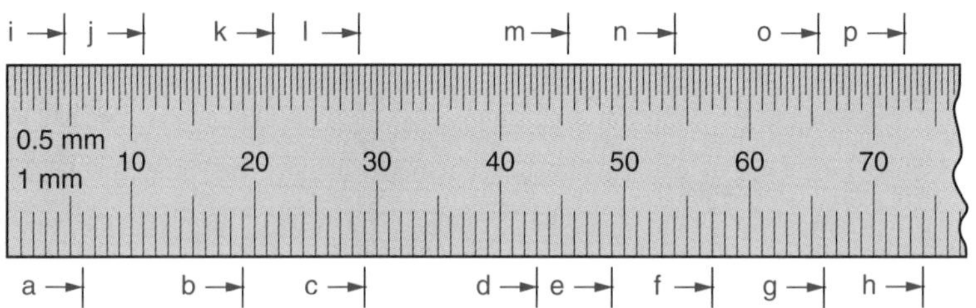

Figure 32-5

a. ______ e. ______ i. ______ m. ______

b. ______ f. ______ j. ______ n. ______

c. ______ g. ______ k. ______ o. ______

d. ______ h. ______ l. ______ p. ______

14. Read the vernier caliper and height gage measurements for the following decimal-inch settings.

a. ______ b. ______ c. ______

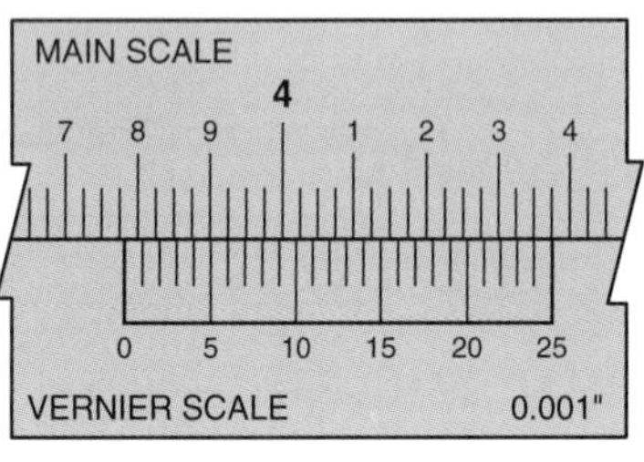

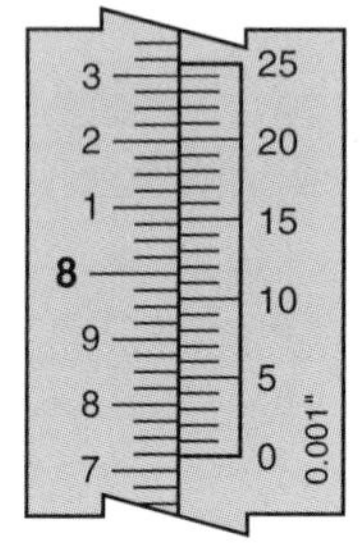

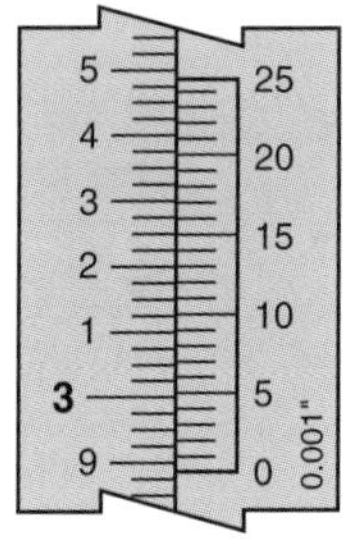

15. Read the settings on the following micrometer scales.

a. 0.001 Decimal-Inch Micrometer

(1) ______ (3) ______

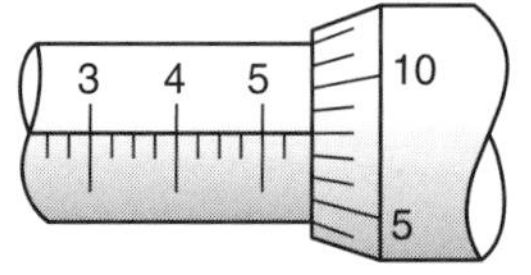

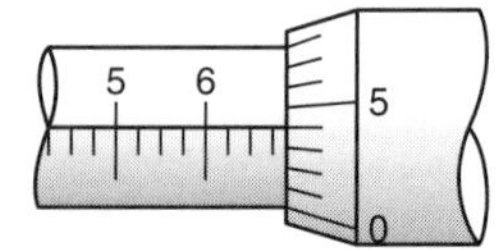

(2) ______ (4) ______

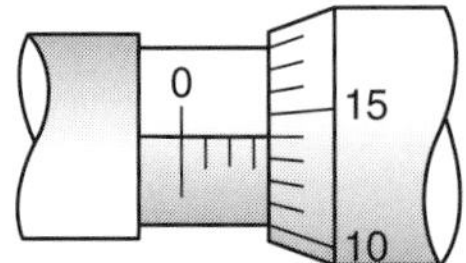

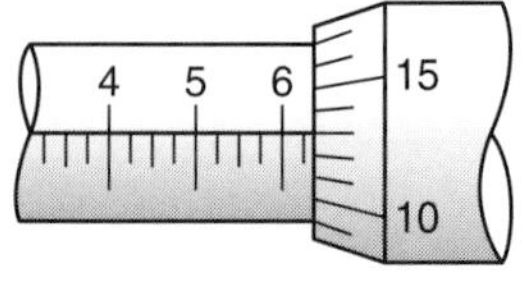

b. 0.0001 Decimal-Inch Vernier Micrometer

 (1) ______  (2) ______  (3) ______  (4) ______

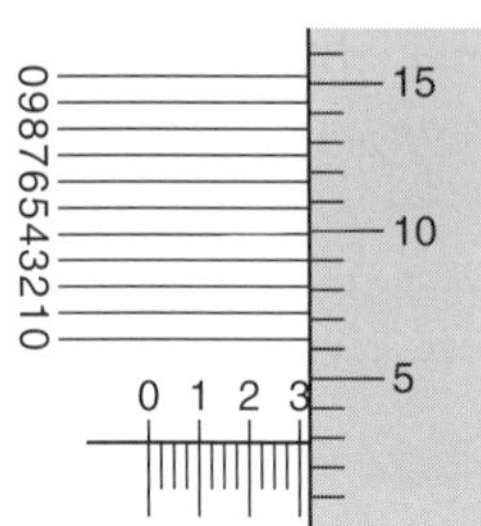

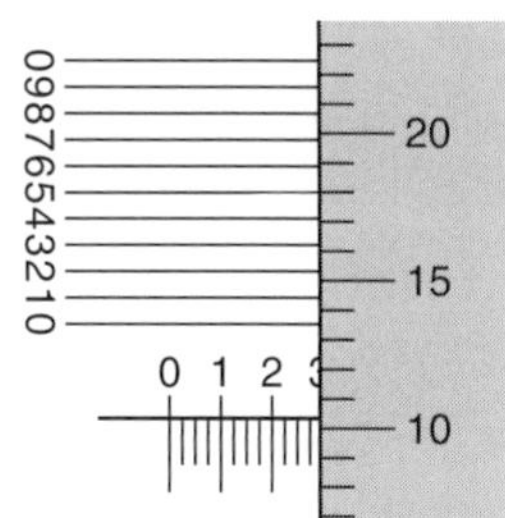

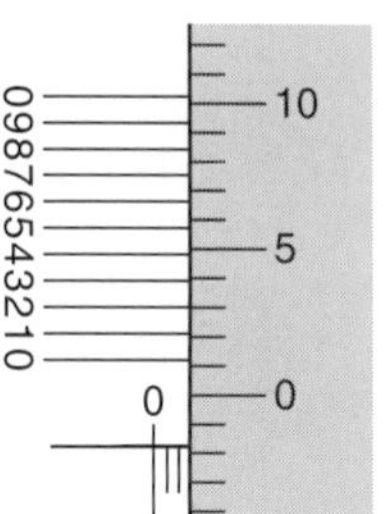

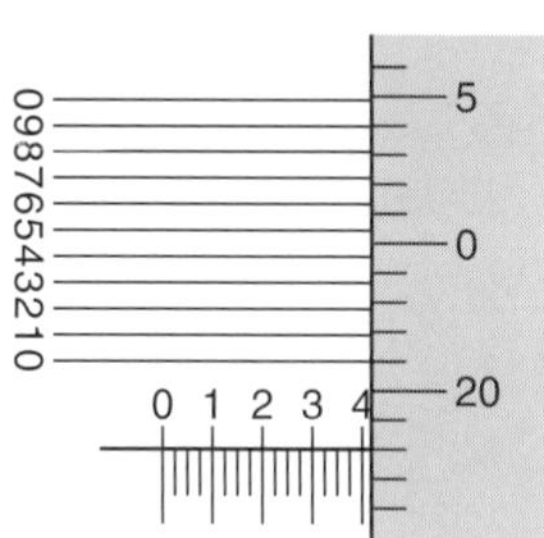

16. Using the Table of Block Thicknesses for a Customary Gage Block Set found in unit 31, determine a combination of gage blocks for each of the following dimensions.

➤ **Note:** Usually more than one combination of blocks will give the desired dimension.

a. 0.3784" ______ e. 3.0901" ______

b. 2.5486" ______ f. 0.2009" ______

c. 1.7062" ______ g. 7.8895" ______

d. 5.6467" ______ h. 8.0014" ______

17. Using the Table of Block Thicknesses for a Metric Gage Block Set found in unit 31, determine a combination of gage blocks for each of the following dimensions.

➤ **Note:** Usually more than one combination of blocks will give the desired dimension.

a. 67.53 mm ______ e. 66.066 mm ______

b. 125.22 mm ______ f. 43.304 mm ______

c. 85.092 mm ______ g. 99.998 mm ______

d. 13.274 mm ______ h. 107.071 mm ______

SECTION 4

Fundamentals of Algebra

Symbolism and Algebraic Expressions

Objectives ***After studying this unit you should be able to***

- Express word statements as algebraic expressions.
- Express diagram dimensions as algebraic expressions.
- Evaluate algebraic expressions by substituting numbers for symbols.

Algebra is a branch of mathematics in which letters are used to represent numbers. By the use of letters, general rules called *formulas* can be stated mathematically. Algebra is an extension of arithmetic; therefore, the rules and procedures that apply to arithmetic also apply to algebra. Many problems that are difficult or impossible to solve by arithmetic can be solved by algebra.

The basic principles of algebra discussed in this text are intended to provide a practical background for machine shop applications. A knowledge of algebraic fundamentals is essential in the use of trade handbooks and for the solutions of many geometric and trigonometric problems.

Symbolism

Symbols are the language of algebra. Both arithmetic numbers and literal numbers are used in algebra. *Arithmetic numbers* are numbers that have definite numerical values, such as 4, 5.17, and $\frac{7}{8}$. *Literal numbers* are letters that represent arithmetic numbers, such as *a, x, V,* and *P*. Depending on how it is used, a literal number can represent one particular arithmetic number, a wide range of numerical values, or all numerical values.

Customarily the multiplication sign (×) is not used in algebra because it can be misinterpreted as the letter *x*. When a literal number is multiplied by a numerical value, or when two or more literal numbers are multiplied, no sign of operation is required.

Examples

1. 5 times a is written $5a$
2. 17 times c is written $17c$
3. V times P is written VP
4. 6 times a times b times c is written $6abc$

Parentheses ( ) are often used in place of the multiplication sign (×) when numerical values are multiplied; 3×4 is written $3(4)$; $18 \times 3.4 \times 5^2$ is written $18(3.4)(5^2)$.

A raised dot · is also used by some people as a multiplication sign. Here, 3×4 is written $3 \cdot 4$ and 6×7.25 as $6 \cdot 7.25$. It is better to use parentheses if there is any chance that the raised dot might be confused for a decimal point.

An *algebraic expression* is a word statement put into mathematical form by using literal numbers, arithmetic numbers, and signs of operation. The following are examples of algebraic expressions.

Example 1 As shown in Figure 33-1, a dimension is increased by 0.5 inch. How long is the increased dimension? All dimensions are in inches.

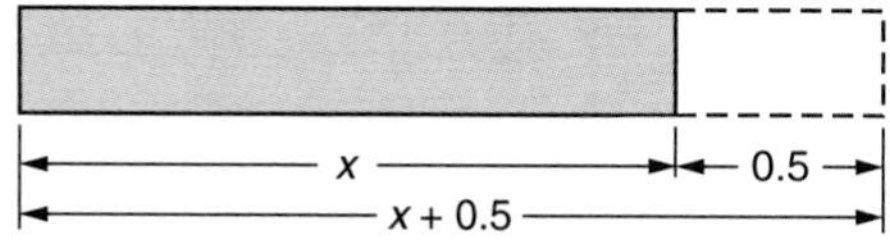

Figure 33-1

If x is the original dimension, the increased dimension is $x + 0.5''$. Ans

Example 2 The production rate of a new machine is 4 times as great as an old machine. Write an algebraic expression for the production rate of the new machine.

If the old machine produced y parts per hour, the new machine produces $4y$ parts per hour. Ans

Example 3 As shown in Figure 33-2, a drill rod is cut into 3 equal pieces. How long is each piece? (Disregard waste)

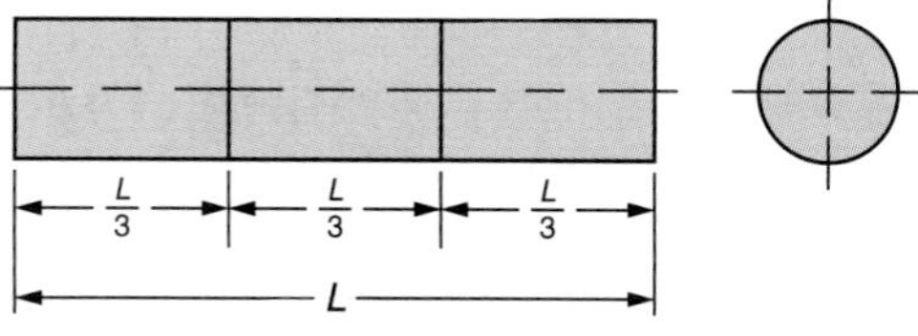

Figure 33-2

If L is the length of the drill rod, the length of each piece is $\frac{L}{3}$. Ans

Example 4 In the step block shown in Figure 33-3, dimension B equals $\frac{3}{4}$ of dimension A and dimension C is twice dimension A. Find the total height of the block.

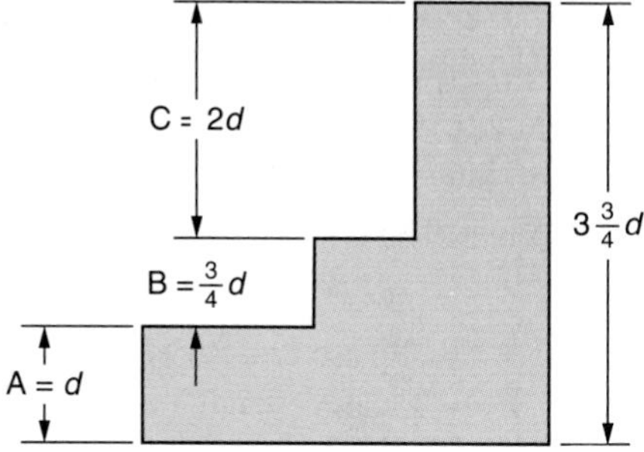

Figure 33-3

If d is the length of dimension A, dimension B is $\frac{3}{4}d$ and dimension C is $2d$. The total height is $d + \frac{3}{4}d + 2d$ or $3\frac{3}{4}d$. Ans

➤ **Note:** If no arithmetic number appears before a literal number, it is assumed that the value is the same as if a one (1) appeared before the letter, $d = 1d$.

Example 5 A plate with 8 drilled holes is shown in Figure 33-4. The distance from the left edge of the plate to hole 1 and the distance from the right edge of the plate to hole 8 are each represented by a. The distances between holes 1 and 2, holes 2 and 3, and holes 3 and 4 are each represented by b. The distances between holes 4 and 5, holes 5 and 6, holes 6 and 7, and holes 7 and 8 are each represented by c. Find the total length of the plate. All dimensions are in millimeters.

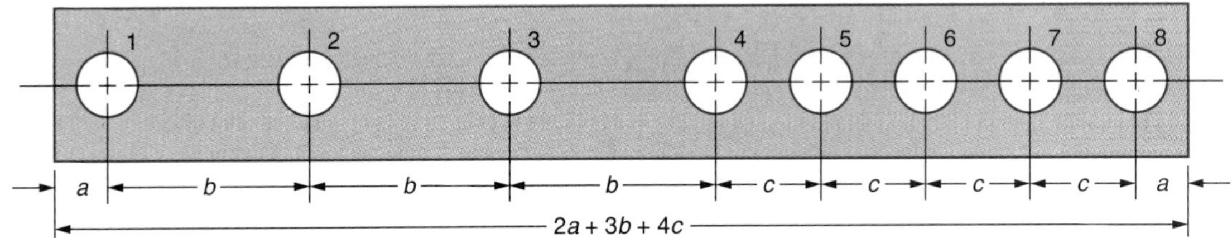

Figure 33-4

The total length of the plate is $a + b + b + b + c + c + c + c + a$, or $2a + 3b + 4c$. Ans

➤ **Note:** Only like literal numbers may be arithmetically added.

Evaluation of Algebraic Expressions

Certain problems in this text involve the use of formulas. Some problems require substituting numerical values for letter values. The problems are solved by applying the order of operations of arithmetic as presented in Section 1, Unit 15.

Order of Operations for Combined Operations of Addition, Subtraction, Multiplication, Division, Powers, and Roots

- **Do all the work in parentheses first.** Parentheses are used to group numbers. In a problem expressed in fractional form, two or more numbers in the dividend (numerator) and/or divisor (denominator) may be considered as being enclosed in parentheses. For example, $\frac{4.87 + 0.34}{9.75 - 8.12}$ may be considered as $(4.87 + 0.34) \div (9.75 - 8.12)$. If an algebraic expression contains parentheses within parentheses or brackets, such as $[5.6 \times (7 - 0.09) + 8.8]$, do the work within the innermost parentheses first.
- **Do powers and roots next.** The operations are performed in the order in which they occur from left to right. If a root consists of two or more operations within the radical sign, perform all the operations within the radical sign, then extract the root.
- **Do multiplication and division next.** The operations are performed in the order in which they occur from left to right.
- **Do addition and subtraction last.** The operations are performed in the order in which they occur from left to right.

Again, you can use the memory aid "**P**lease **E**xcuse **M**y **D**ear **A**unt **S**ally" to help remember the order of operations. The **P** in "Please" stands for parentheses, the **E** for exponents (or raising to a power) and roots, the **M** and **D** for multiplication and division, and the **A** and **S** for addition and subtraction.

Example 1 The formula for finding the perimeter of a rectangle is given. Find the perimeter of the rectangle shown in Figure 33-5. All dimensions are in millimeters.

$P = 2L + 2W$	$P = 2L + 2W$ where
$P = 2(50\text{ mm}) + 2(30\text{ mm})$	P = perimeter
$P = 100\text{ mm} + 60\text{ mm}$	L = length
$P = 160\text{ mm}$ Ans	W = width

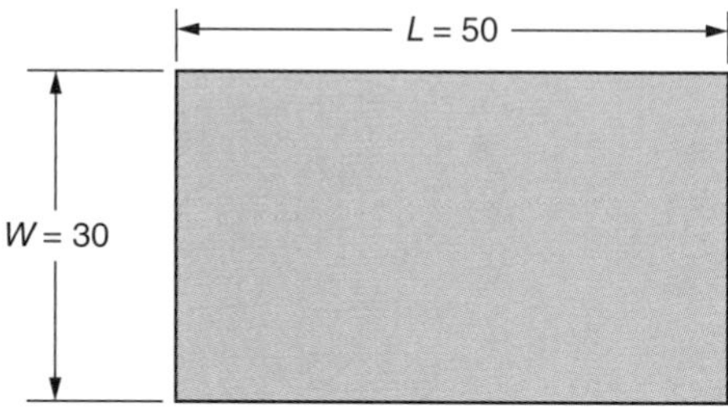

Figure 33-5

Example 2 The formula for finding the area of a ring is given. Find the area of the ring shown in Figure 33-6. All dimensions are in inches. Round the answer to 2 decimal places.

$A = \pi R^2 - \pi r^2$ where A = area
R = outside radius
r = inside radius

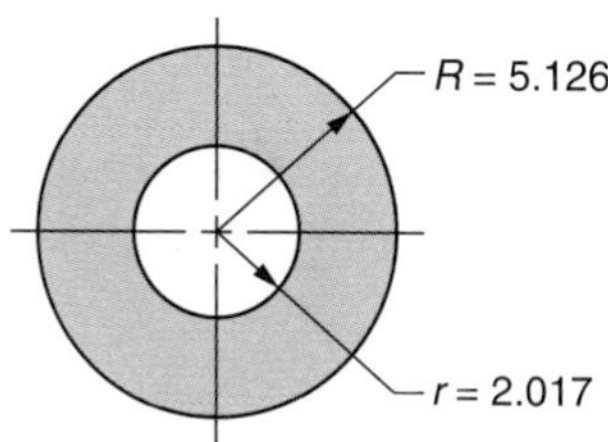

Figure 33-6

The symbol π (pi) represents a constant value used in mathematical relationships involving circles. It is described in Unit 49 on page 324

Scientific calculators have a pi key, [π]. Depressing the pi key, [π], enters the value of pi to 10 digits (3.141592654) on most calculators. On some calculators, [π] is an alternate function. You may have to press [SHIFT] or [2nd] first.

$$A = \pi R^2 - \pi r^2$$
$$A = \pi(5.126\text{ in})^2 - \pi(2.017\text{ in})^2$$

[π] [×] 5.126 [x^2] [−] [π] [×] 2.017 [x^2] [=] 69.76719217

$A \approx 69.77$ sq in Ans (rounded)

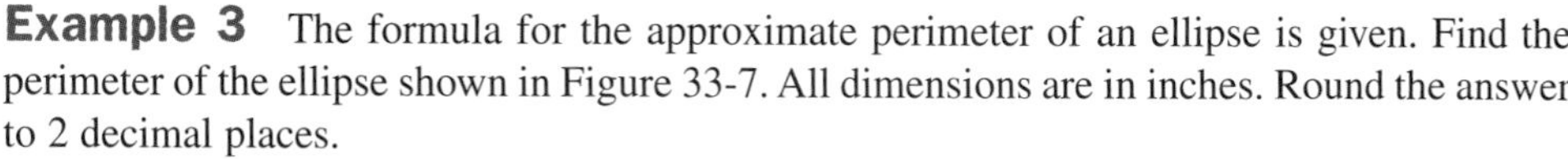

Example 3 The formula for the approximate perimeter of an ellipse is given. Find the perimeter of the ellipse shown in Figure 33-7. All dimensions are in inches. Round the answer to 2 decimal places.

$P = \pi\sqrt{2(a^2 + b^2)}$ where P = perimeter
a = 0.5 (major axis)
b = 0.5 (minor axis)

$P = \pi\sqrt{2(8.50^2 + 6.42^2)}$
[π] [×] [√] [(] 2 [×] [(] 8.5 [x^2] [+] 6.42 [x^2] [)] [)]
[=] 47.32585933

$P \approx 47.33$ in Ans (rounded)

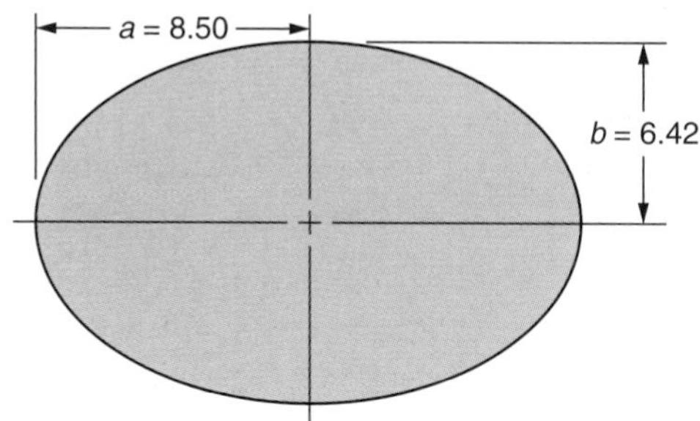

Figure 33-7

Example 4 Find the value of $\dfrac{3(2b + 3dy)}{4(7d - bd)}$ when $b = 6$, $d = 4$, and $y = 2$.

$$\frac{3[2(6) + 3(4)(2)]}{4[7(4) - 6(4)]} = \frac{3(12 + 24)}{4(28 - 24)} = \frac{3(36)}{4(4)} = \frac{108}{16} = 6.75 \quad \text{Ans}$$

Example 5 Find the value of $3m[4p + 5(x - m) + p]^2$ when $m = 2$, $p = 3$, and $x = 8$.

$$\begin{aligned} 3(2)[4(3) + 5(8 - 2) + 3]^2 &= 6[12 + 5(6) + 3]^2 \\ &= 6(12 + 30 + 3)^2 \\ &= 6(45)^2 \\ &= 6(2025) = 12{,}150 \quad \text{Ans} \end{aligned}$$

Example 6 Find the value of $\dfrac{6a}{b} + \dfrac{abc}{20}(a^3 - 12b)$ when $a = 5$, $b = 10$, and $c = 8$.

$$\begin{aligned} \frac{6(5)}{10} + \frac{5(10)(8)}{20}[5^3 - 12(10)] &= \frac{30}{10} + \frac{400}{20}(125 - 120) \\ &= 3 + 20(5) \\ &= 3 + 100 = 103 \quad \text{Ans} \end{aligned}$$

Application

Algebraic Expressions

Express each of the following problems as an algebraic expression.

1. The product of 6 and x increased by y. ________
2. The sum of a and 12. ________
3. Subtract b from 21. ________
4. Subtract 21 from b. ________
5. Divide r by s. ________
6. Twice L minus one-half P. ________
7. The product of x and y divided by the square of m. ________

8. In the part shown in Figure 33-8, all dimensions are in inches.
   a. What is the total length of this part? ________
   b. What is the length from point A to point B? ________

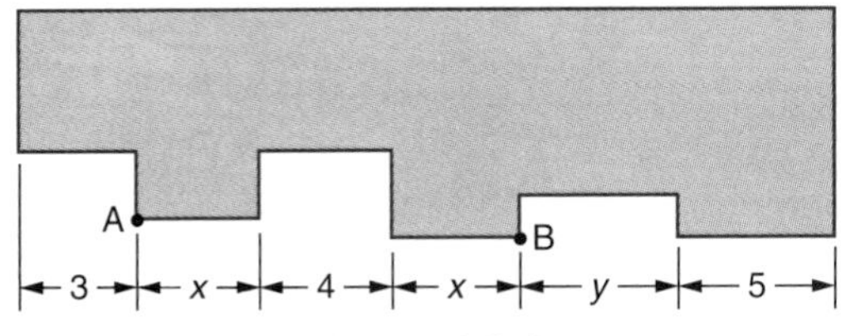

Figure 33-8

9. Find the distance between the indicated points of Figure 33-9.
   a. Point A to point B ________
   b. Point F to point C ________
   c. Point B to point C ________
   d. Point D to point E ________

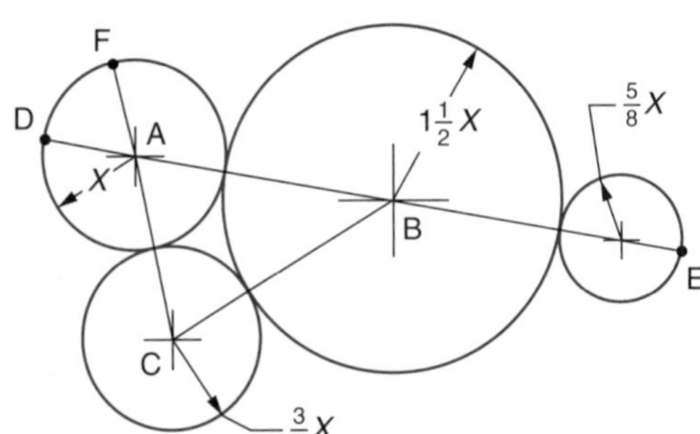

Figure 33-9

10. What are the lengths of the following dimensions in Figure 33-10? All dimensions are in millimeters.

 a. Dimension A __________

 b. Dimension B __________

 c. Dimension C __________

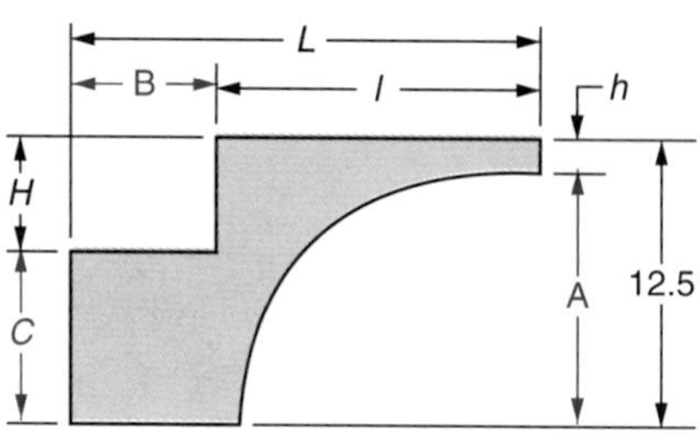

Figure 33-10

11. Stock is removed from a block in two operations. The original thickness of the block is represented by n. The thickness removed by the milling operation is represented by p and the thickness removed by the grinding operation is represented by t. What is the final thickness of the block?

12. Given: In Figure 33-11, s is the length of a side of a hexagon, r is the radius of the inside circle, and R is the radius of the outside circle.

 a. What is the length of r if r equals the product of 0.866 and the length of a side of the hexagon? __________

 b. What is the length of R if R equals the product of 1.155 and the radius of the inside circle? __________

 c. What is the area of the hexagon if the area equals the product of 2.598 and the square of the radius of the outside circle? __________

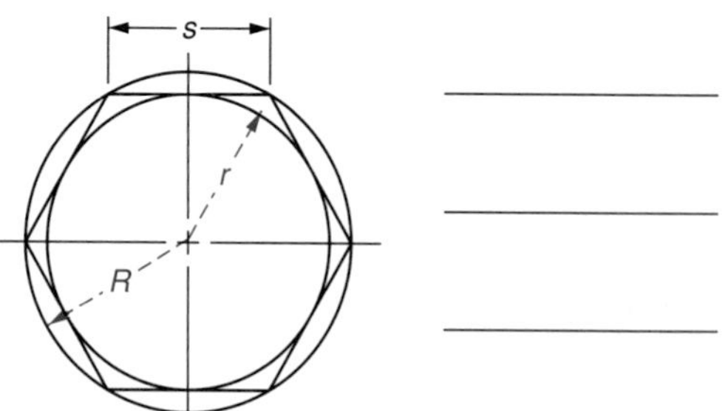

Figure 33-11

Evaluation of Algebraic Expressions

Substitute the given numbers for letters and find the values of the following expressions.

13. If $a = 5$ and $c = 3$, find

 a. $5a + 3c^2$ __________

 b. $5c + a$ __________

 c. $\dfrac{10c}{a}$ __________

 d. $\dfrac{a + c}{a - c}$ __________

 e. $\dfrac{a + 5c}{ac + a}$ __________

 f. $\dfrac{5c + a}{5c - a}$ __________

14. If $b = 8$, $d = 4$, and $e = 2$, find

 a. $\dfrac{b}{d} + e - 3$ __________

 b. $bd(3 + 4d - b)$ __________

 c. $5b - (bd + 3)$ __________

 d. $3e(b - e) - d\left(\dfrac{b}{2}\right)$ __________

 e. $\dfrac{12d}{e} - [3b - (d + e) + 4]$ __________

 f. $\dfrac{b + d + e}{d - e} + bd^2 - de^3$ __________

15. If $x = 12$ and $y = 6$, find

 a. $2xy + 7$ __________

 b. $3x - 2y + xy$ __________

 c. $\dfrac{5xy - 2y}{8x - xy}$ __________

 d. $\dfrac{4x - 4y}{3}$ __________

 e. $6x - 3y + xy$ __________

 f. $x^2 - y^2$ __________

16. If $m = 5, p = 4$, and $r = 3$, find

a. $m + mp^2 - r^2$ __________

b. $(p + 2)^2 (m - r)^2$ __________

c. $\dfrac{(pr)^2}{2} - pr + m^3$ __________

d. $\dfrac{p^3 + 3p - 12}{m^2 + 15}$ __________

e. $\dfrac{r^3}{3p - 9} + m^2(mp - 6r)^2$ __________

f. $\dfrac{m^3 - p^2 + 5}{r^2} + \dfrac{p + 2m}{m + p + r}$ __________

For problems 17–28, round the answers to 1 decimal place.

17. All dimensions in Figure 33-12 are in inches.

a. Find the area (A) of this square.

$A = \dfrac{1}{2}d^2$ __________

b. Find the side (S) of this square.

$S = 0.7071d$ __________

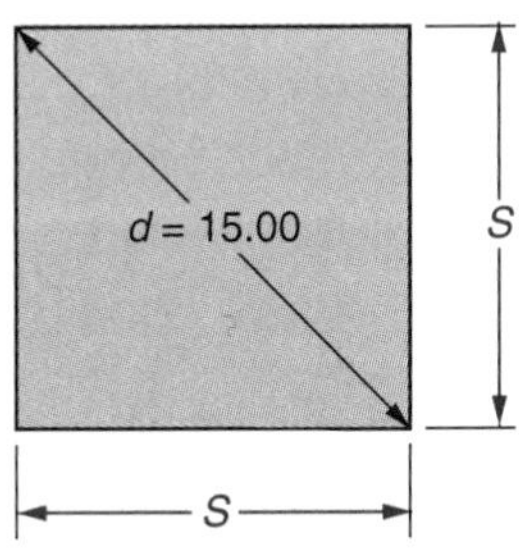

Figure 33-12

18. All dimensions in Figure 33-13 are in millimeters.

a. Find the length of this arc (l).

$l = \dfrac{\pi R\alpha}{180°}$ __________

b. Find the area of this sector (A).

$A = \dfrac{1}{2}Rl$ __________

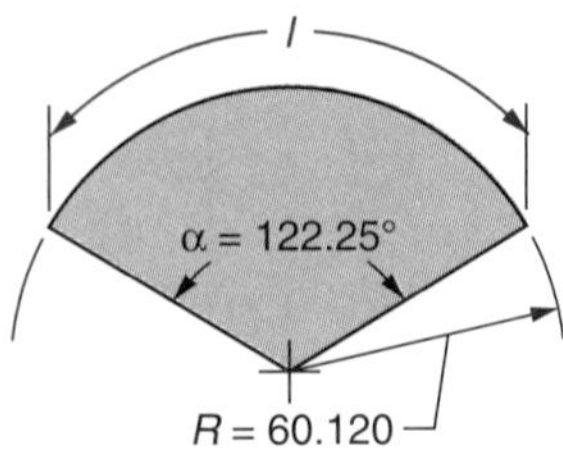

Figure 33-13

19. All dimensions in Figure 33-14 are in inches. Refer to the triangle shown.

a. Find S when $S = \dfrac{1}{2}(a + b + c)$.

b. Find the area (A) when

$A = \sqrt{S(S - a)(S - b)(S - c)}$. __________

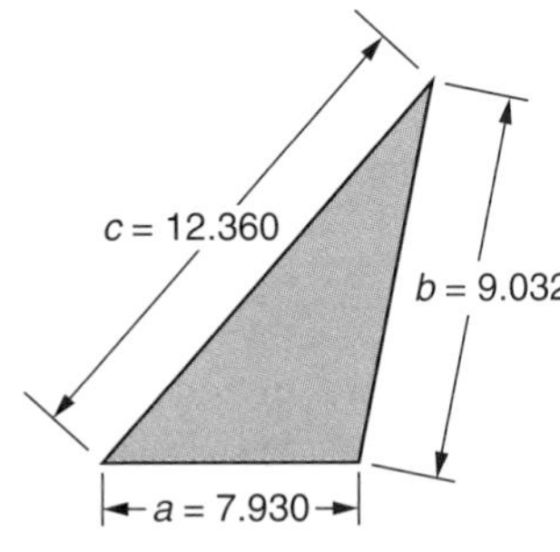

Figure 33-14

20. All dimensions in Figure 33-15 are in millimeters.

a. Find the radius of this circle.

$r = \dfrac{c^2 + 4h^2}{8h}$ __________

b. Find the length of the arc (l).

$l = 0.0175r\alpha$ __________

Figure 33-15

21. All dimensions are in inches. Find the shaded area of Figure 33-16.

$$\text{Area} = \frac{(H + h)b + ch + aH}{2}$$ ____________

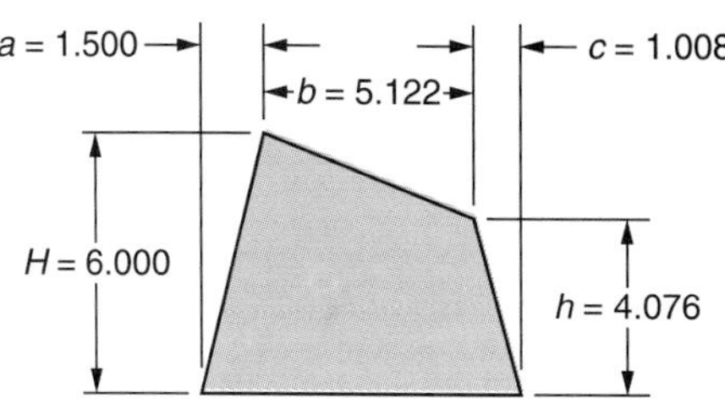

Figure 33-16

22. All dimensions in Figure 33-17 are in inches. Find the length of belt on the pulleys.

Length of belt =

$$2C + \frac{11D + 11d}{7} + \frac{(D - d)^2}{4C}$$ ____________

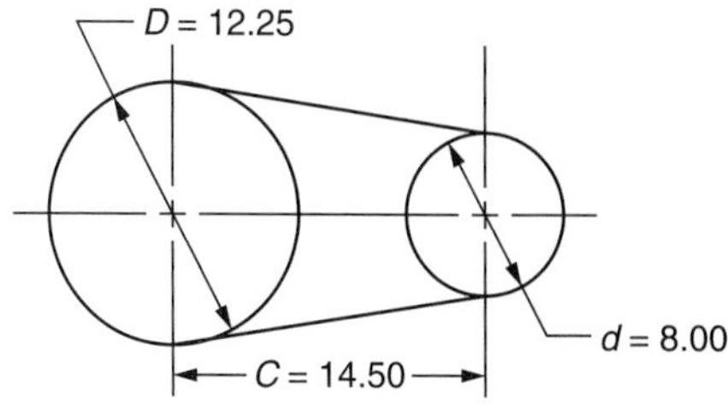

Figure 33-17

23. All dimensions are in millimeters. Find the shaded area of Figure 33-18.

$\text{Area} = dt + 2a(s + n)$ ____________

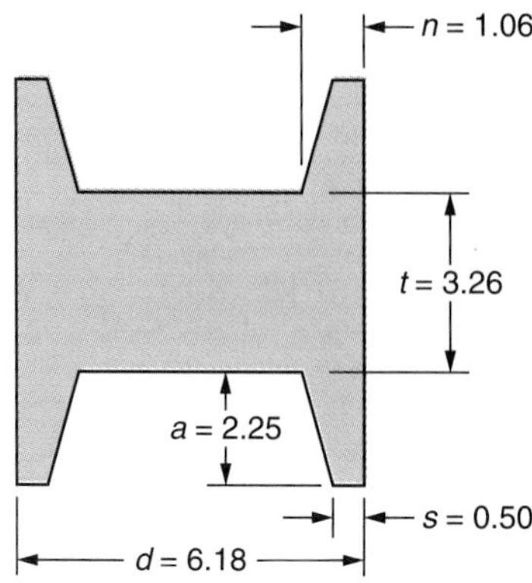

Figure 33-18

24. All dimensions are in inches. Find the shaded area of Figure 33-19.

$\text{Area} = \pi(ab - cd)$ ____________

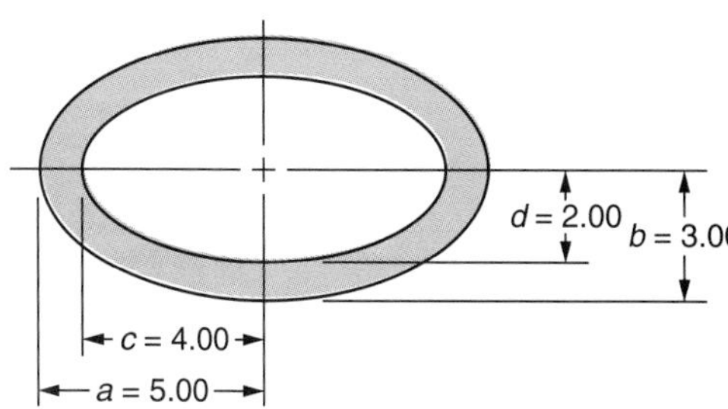

Figure 33-19

25. All dimensions are in inches. Find the shaded area of Figure 33-20.

$$\text{Area} = \frac{\pi(R^2 - r^2)}{2}$$ ____________

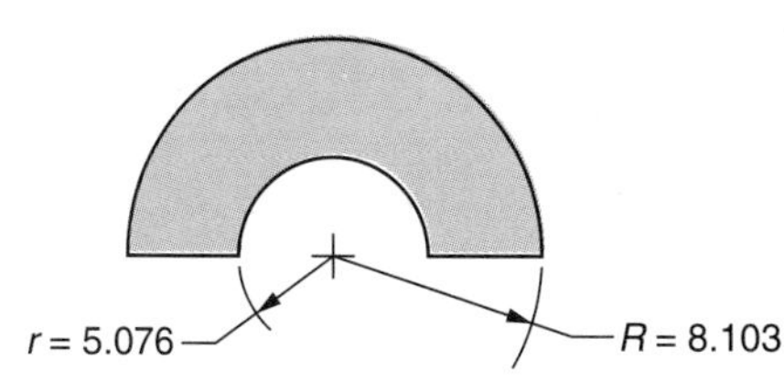

Figure 33-20

26. All dimensions are in centimeters. Find the shaded area of Figure 33-21.

    Area $= t[b + 2(a - t)]$ ____________

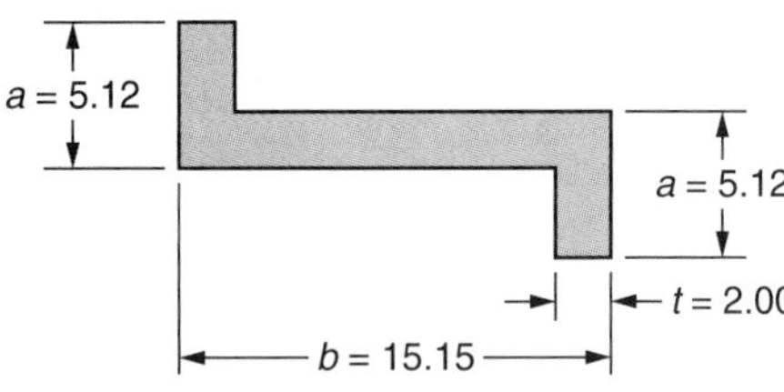

Figure 33-21

27. All dimensions in Figure 33-22 are in inches.

    a. Find the slant height (S).

       $S = \sqrt{(R - r)^2 + h^2}$ ____________

    b. Find the volume.

       Volume $= 1.05h(R^2 + Rr + r^2)$ ____________

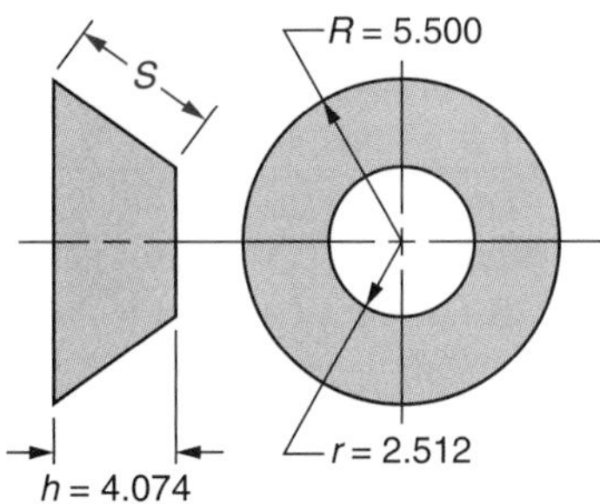

Figure 33-22

28. All dimensions in Figure 33-23 are in inches. Find the volume.

    Volume $= \dfrac{(2a + c)bh}{6}$ ____________

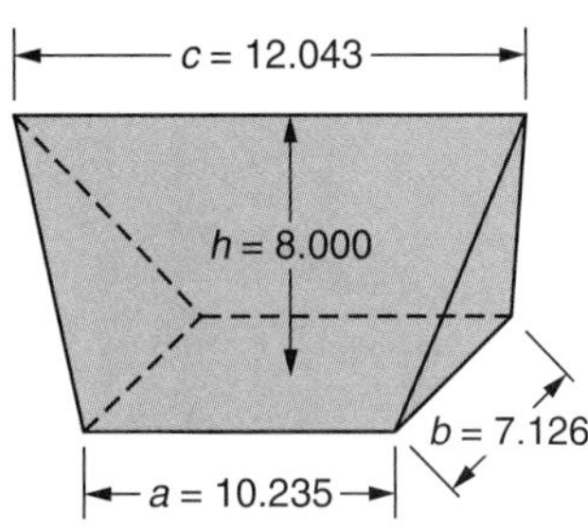

Figure 33-23

UNIT 34

Signed Numbers

Objectives ***After studying this unit you should be able to***

- Compare signed numbers according to size and direction using the number scale.
- Determine absolute values of signed numbers.
- Perform basic operations of addition, subtraction, multiplication, division, powers, and roots using signed numbers.
- Solve expressions that involve combined operations of signed numbers.

Signed numbers are required for solving problems in mechanics and trigonometry. Positive and negative numbers express direction, such as machine table movement from a reference point. Signed numbers are particularly useful in programming machining operations for numerical control.

Meaning of Signed Numbers

Plus and minus signs, which you have worked with so far in this book, have been *signs of operation.* These are signs used in arithmetic, with the plus sign (+) indicating the operation of addition and the minus sign (−) indicating the operation of subtraction.

In algebra, plus and minus signs are used to indicate both operation and direction from a reference point or zero. A *positive number* is indicated either with no sign or with a plus sign (+) preceding the number. For example, +7 or 7 is a positive number that is 7 units greater than zero. A *negative number* is indicated with a minus sign (−) preceding the number. For example, −7 is a negative number that is 7 units less than zero. Positive and negative numbers are called *signed numbers* or *directed numbers.*

The Number Scale

A number scale like the one in Figure 34-1 shows the relationship of positive and negative numbers. It shows both distance and direction between numbers. Considering a number as a starting point and counting to a number to the right represents positive (+) direction with numbers increasing in value. Counting to the left represents negative (−) direction with numbers decreasing in value. The number 0 does not have a sign because 0 is neither positive nor negative.

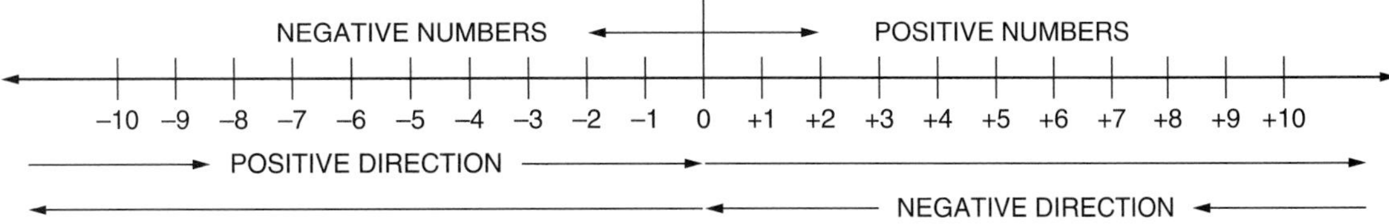

Figure 34-1

Examples

1. Starting at 0 and counting to the right to +5 represents 5 units in a positive (+) direction; +5 is 5 units greater than 0.
2. Starting at 0 and counting to the left to −5 represents 5 units in a negative (−) direction; −5 is 5 units less than 0.
3. Starting at −2 and counting to the right to +6 represents 8 units in a positive (+) direction; +6 is 8 units greater than −2.
4. Starting at +6 and counting to the left to −2 represents 8 units in a negative (−) direction; −2 is 8 units less than a +6.
5. Starting at −3 and counting to the left to −10 represents 7 units in a (−) direction; −10 is 7 units less than −3.
6. Starting at −9 and counting to the right to 0 represents 9 units in a (+) direction; 0 is 9 units greater than −9.

Operations Using Signed Numbers

In order to solve problems in algebra, you must be able to perform the basic operations using signed numbers. The following procedures and examples show how to perform operations of addition, subtraction, multiplication, division, powers, and roots with signed numbers.

Absolute Value

The procedures for performing certain operations of signed numbers are based on an understanding of absolute value.

The *absolute value* of a number is the number without regard to its sign. For example, the absolute value of +4 is 4, the absolute value of −4 is also 4. Therefore, the absolute value of +4 and −4 is the same value, 4. The absolute value of a number is indicated by placing the number between a pair of vertical bars, $|+4| = 4$, $|-4| = 4$. The absolute value of 0 is 0, that is, $|0| = 0$.

The absolute value of −20 is 15 greater than the absolute value of +5; 20 is 15 greater than 5.

Addition of Signed Numbers

Procedure To add two or more positive numbers

- Add the numbers as in arithmetic.
  Positive numbers do not require a positive sign as a prefix.

Examples Add the following numbers.

1. $\begin{array}{r} +3 \\ +5 \\ \hline +8 \end{array}$ Ans
2. $\begin{array}{r} 15 \\ 7 \\ \hline 22 \end{array}$ Ans
3. $2 + 9 + 13 = 24$ Ans
4. $+12 + (+15) = +27$ Ans

Procedure To add two or more negative numbers

- Add the absolute values of the numbers.
- Prefix a minus sign to the sum.

Examples Add the following numbers.

1. $\begin{array}{r} -5 \\ -2 \\ \hline -7 \end{array}$ Ans
2. $\begin{array}{r} -13 \\ -\ 4 \\ -15 \\ \hline -32 \end{array}$ Ans
3. $-6 + (-5) = -11$ Ans
4. $-8 + (-10) + (-4) + (-3) = -25$ Ans

Procedure To add a positive and a negative number

- Subtract the smaller absolute value from the larger absolute value.
- Prefix the sign of the number having the larger absolute value to the difference.

Examples Add the following numbers.

1. $\begin{array}{r} +5 \\ -3 \\ \hline +2 \end{array}$ Ans
2. $\begin{array}{r} -5 \\ +3 \\ \hline -2 \end{array}$ Ans
3. $\begin{array}{r} -17 \\ +17 \\ \hline 0 \end{array}$ Ans
4. $+12 + (-8) = +4$ Ans
5. $-12 + (+8) = -4$ Ans

Procedure To add more than two positive and negative numbers

- Add all the positive numbers.
- Add all the negative numbers.
- Add their sums following the procedure for adding signed numbers.

Examples Add the following numbers.

1. $-2 + 4 + (-10) + 5 = 9 + (-12) = -3$ Ans
2. $8 + 7 + (-6) + 4 + (-3) + (-5) + 10 = 29 + (-14) = 15$ Ans
3. $4 + (-6) + 12 + 3 + (-7) + 1 + (-5) + (-2) = 20 + (-20) = 0$ Ans

Enter a negative number in a calculator by first pressing the negative sign key [(-)], then enter the absolute value of the number.

Examples

1. Add. $-25.873 + (-138.029)$

   [(-)] 25.873 [+] [(-)] 138.029 [=] −163.902 Ans

2. Add. $-6.053 + (-0.072) + (-15.763) + (-0.009)$

   [(-)] 6.053 [+] [(-)] .072 [+] [(-)] 15.763 [+] [(-)] .009

   [=] −21.897 Ans

Subtraction of Signed Numbers

Procedure To subtract signed numbers

- Change the sign of the number subtracted (subtrahend) to the opposite sign.
- Follow the procedure for addition of signed numbers.

➤ **Note:** When the sign of the subtrahend is changed, the problem becomes one in addition. Therefore, subtracting a negative number is the same as adding a positive number. Subtracting a positive number is the same as adding a negative number.

Examples

1. Subtract 5 from 8. $8 - (+5) = 8 + (-5) = 3$ Ans
2. Subtract 8 from 5. $5 - (+8) = 5 + (-8) = -3$ Ans
3. Subtract −5 from 8. $8 - (-5) = 8 + (+5) = 13$ Ans
4. Subtract −5 from −8. $-8 - (-5) = -8 + (+5) = -3$ Ans
5. $-3 - (+7) = -3 + (-7) = -10$ Ans
6. $0 - (-14) = 0 + (+14) = 14$ Ans
7. $0 - (+14) = 0 + (-14) = -14$ Ans
8. $-14 - (-14) = -14 + (+14) = 0$ Ans

Examples

1. Subtract. $-163.94 - (-150.65)$

   [(-)] 163.94 [−] [(-)] 150.65 [=] −13.29 Ans

2. Subtract. $-27.55 - (-8.64 + 0.74) - (-53.41)$

   [(-)] 27.55 [−] [(] [(-)] 8.64 [+] .74 [)] [−] [(-)] 53.41 [=] 33.76 Ans

Multiplication of Signed Numbers

Procedure To multiply two or more signed numbers

- Multiply the absolute values of the numbers.
- Count the number of negative signs.

  If there is an odd number of negative signs, the product is negative.

  If there is an even number of negative signs, the product is positive.

  If all numbers are positive, the product is positive.

It is not necessary to count the number of positive values in an expression consisting of both positive and negative numbers. Count only the number of negative values to determine the sign of the product.

Examples

1. $4(-3) = -12$ Ans (There is one negative sign. Since one is an odd number, the product is negative.)
2. $-4(-3) = +12$ Ans (There are two negative signs. Since two is an even number, the product is positive.)
3. $(-2)(-4)(-3)(-1)(-2)(-1) = +48$ Ans (6 negatives, even number, positive product)
4. $(-2)(-4)(-3)(-1)(-2) = -48$ Ans (5 negatives, odd number, negative product)
5. $(2)(4)(3)(1)(2) = +48$ Ans (all positives, positive product)
6. $(2)(-4)(-3)(1)(-2) = -48$ Ans (3 negatives, odd number, negative product)
7. $(-2)(4)(-3)(-1)(-2) = +48$ Ans (4 negatives, even number, positive product)

➤ **Note:** The product of any number or numbers and $0 = 0$; for example, $0(9) = 0$; $0(-9) = 0$; $8(-6)(0)(6) = 0$.

Example Multiply. $(-8.61)(3.04)(-1.85)(-4.03)(0.162)$. Round the answer to 1 decimal place.

[(−)] 8.61 [×] 3.04 [×] [(−)] 1.85 [×] [(−)] 4.03 [×] .162 =
-31.61320475, -31.6 Ans (rounded)

Division of Signed Numbers

Procedure To divide two signed numbers

- Divide the absolute values of the numbers.
- Determine the sign of the quotient.

  If both numbers have the same sign (both negative or both positive) the quotient is positive.

  If the two numbers have unlike signs (one positive and one negative) the quotient is negative.

Examples

1. $\frac{-8}{-2} = +4$ Ans
2. $\frac{8}{2} = +4$ Ans
3. $15 \div 3 = +5$ Ans
4. $-3\overline{)-15} = +5$ Ans
5. $\frac{-30}{5} = -6$ Ans
6. $\frac{30}{-5} = -6$ Ans
7. $-21 \div 3 = -7$ Ans
8. $-3\overline{)21} = -7$ Ans

Divide. 31.875 [÷] (−56.625). Round the answer to 3 decimal places.
31.875 [÷] [(−)] 56.625 [=] -0.562913907, -0.563 Ans (rounded)

➤ **Note:** Zero divided by any number equals zero. For example, $0 \div (+3) = 0$, $0 \div (-3) = 0$. Dividing by zero is undefined. For example, $+3 \div 0$ and $-3 \div 0$ are not defined.

Powers of Signed Numbers

Procedure To raise numbers with positive exponents to a power

- Apply the procedure for multiplying signed numbers to raising signed numbers to powers.

Examples

1. $3^2 = +9$ Ans
2. $3^3 = +27$ Ans
3. $2^4 = +16$ Ans
4. $2^5 = +32$ Ans
5. $(-3)^2 = (-3)(-3) = +9$ Ans
6. $(-3)^3 = (-3)(-3)(-3) = -27$ Ans
7. $(-2)^4 = (-2)(-2)(-2)(-2) = +16$ Ans
8. $(-2)^5 = (-2)(-2)(-2)(-2)(-2) = -32$ Ans

➤ **Note:**

- A positive number raised to any power is positive.
- A negative number raised to an even power is positive.
- A negative number raised to an odd power is negative.
- Use parentheses to enclose a negative number if it is to be raised to a power.

As presented in Unit 16, the universal power key, [x^y], or caret key [^] raises any *positive* number to a power.

Solve. 2.073^5. Round the answer to 2 decimal places.
2.073 [x^y] (or [^]) 5 [=] 38.28216674, 38.28 Ans (rounded)

The universal power key can also be used to raise a *negative* number to a power. The negative number to be raised must be enclosed within parentheses. Notice that this was done in Examples 5 through 8 above. The reason is that a negative number, such as -7, is considered to be $-1 \cdot 7$, so $-7^2 = -1 \cdot 7^2 = -1 \cdot 49 = -49$. But, $(-7)^2 = (-7)(-7) = +49 = 49$.

Examples Round the answers to 1 decimal place.

1. Solve. $(-3.874)^4$
   [(] [(−)] 3.874 [)] [x^y] (or [^]) 4 [=] 225.236342, 225.2 Ans (rounded)

➤ **Note:** −3.874 must be enclosed within parentheses.

2. Solve. $(-3.874)^5$
   [(] [(−)] 3.874 [)] [x^y] (or [^]) 5 [=] −872.565589, −872.6 Ans (rounded)

A number with a negative exponent is equal to the reciprocal of the number (the number inverted) with a positive exponent. If x represents a number and n represents an exponent then $\frac{x^{-n}}{1} = \frac{1}{x^n}$.

Procedure To raise numbers with negative exponents to a power

- Invert the number.
- Change the negative exponent to a positive exponent.

Examples

1. $3^{-2} = \frac{3^{-2}}{1} = \frac{1}{3^2} = \frac{1}{9}$ or 0.111 Ans (rounded)
2. $2^{-3} = \frac{2^{-3}}{1} = \frac{1}{2^3} = \frac{1}{8}$ or 0.125 Ans
3. $-4^{-3} = \frac{-4^{-3}}{1} = \frac{1}{-4^3} = \frac{1}{-64}$ or -0.016 Ans (rounded)

A *negative* exponent is entered with the negative key [(−)]. The rest of the procedure is the same as used with positive exponents.

Examples Round the answers to 3 decimal places.

1. Calculate. 3.162^{-3}
   3.162 [x^y] (or [^]) [(-)] 3 [=] 0.0316311078, 0.032 Ans (rounded)
2. Calculate. $(-3.162)^{-3}$
   [(] [(-)] 3.162 [)] [x^y] (or [^]) [(-)] 3 [=] −0.031631108,
   −0.032 Ans (rounded)

Roots of Signed Numbers

When either a positive number or a negative number is squared, a positive number results. For example, $3^2 = 9$ and $(-3)^2 = 9$. Therefore, every positive number has two square roots, one positive root and one negative root. The square roots of 9 are +3 and −3. The expression $\sqrt{9}$ is used to indicate the positive or *principal square root,* +3 or 3. The expression $-\sqrt{9}$ is used to indicate the negative square root, −3. The expression $\pm\sqrt{9}$ indicates both the positive and negative square roots, ±3. The principal cube root of 8 is 2, $\sqrt[3]{8} = 2$. The principal cube root of −8 is −2, $\sqrt[3]{-8} = -2$. In this book, only principal roots are to be determined or used in problem solving.

Examples

1. $\sqrt{36} = \sqrt{(6)(6)} = 6$ Ans
2. $\sqrt[4]{16} = \sqrt[4]{(2)(2)(2)(2)} = 2$ Ans
3. $\sqrt[3]{-27} = \sqrt{(-3)(-3)(-3)} = -3$ Ans
4. $\sqrt[5]{32} = \sqrt[5]{(2)(2)(2)(2)(2)} = 2$ Ans
5. $\sqrt[3]{\frac{-8}{27}} = \sqrt[3]{\frac{(-2)(-2)(-2)}{(3)(3)(3)}} = \frac{-2}{3}$ Ans

As presented in Unit 16, generally, roots are alternate functions.

Solve. $\sqrt[4]{562.824}$.

4 [2nd] (or [SHIFT]) [$\sqrt[x]{\ }$] 562.824 [=] 4.870719863 Ans

The following example shows the procedure for calculating roots of *negative* numbers.

Solve. $\sqrt[5]{-85.376}$.

5 [2nd] (or [SHIFT]) [$\sqrt[x]{\ }$] [(-)] 85.376 [=] −2.433700665 Ans

The square root of a negative number has no solution in the real number system. For example, $\sqrt{-4}$ has no solution; $\sqrt{-4}$ is not equal to $\sqrt{(-2)(-2)}$ and is not equal to $\sqrt{(+2)(+2)}$. Any even root (even index) of a negative number has no solution in the real number system. For example, $\sqrt[4]{-16}$ and $\sqrt[6]{-64}$ have no solution. When you use a calculator to take an even root of a negative number, it will give an error message to show that there is no solution in the real number system. For example, some calculators will give a message like `ERROR` or `DOMAIN Error`.

Expressing Numbers with Fractional Exponents as Radicals

Procedure To simplify numbers with fractional exponents

- Write the numerator of the fractional exponent as the power of the radicand.
- Write the denominator of the fractional exponent as the root index of the radicand.
- Simplify.

Examples

1. $25^{1/2} = \sqrt[2]{25^1} = \sqrt{25} = \sqrt{(5)(5)} = 5$ Ans
2. $8^{1/3} = \sqrt[3]{8^1} = \sqrt[3]{(2)(2)(2)} = 2$ Ans
3. $8^{2/3} = \sqrt[3]{8^2} = \sqrt[3]{64} = \sqrt[3]{(4)(4)(4)} = 4$ Ans
4. $36^{-1/2} = \dfrac{1}{36^{1/2}} = \dfrac{1}{\sqrt{36}} = \dfrac{1}{\sqrt{(6)(6)}} = \dfrac{1}{6}$ Ans

Use the universal power key, [x^y], or caret key [^] with fractional exponents. Enclose the fractional exponent in parentheses.

Examples

1. Solve. $8.732^{2/3}$

   8.732 [x^y] (or [^]) [(] 2 [ab/c] 3 [)] [=] 4.240422706 Ans

   or 8.732 [x^y] (or [^]) [(] 2 [÷] 3 [)] [=] 4.240422706 Ans
2. Solve. $8.732^{-2/3}$

   8.732 [x^y] (or [^]) [(] [(-)] 2 [ab/c] 3 [)] [=] 0.235825546 Ans

   or 8.732 [x^y] (or [^]) [(] [(-)] 2 [÷] 3 [)] [=] 0.235825546 Ans

Combined Operations of Signed Numbers

Expressions consisting of two or more operations of signed numbers are solved using the same order of operations as in arithmetic.

Example Compute the value of $50 + (-2)[6 + (-2)^3(4)]$.

$$\begin{aligned}50 + (-2)[6 + (-2)^3(4)] &= 50 + (-2)[6 + (-8)(4)]\\ &= 50 + (-2)[6 + (-32)]\\ &= 50 + (-2)(-26)\\ &= 50 + 52 = 102 \quad \text{Ans}\end{aligned}$$

Examples

1. Solve. $\sqrt{38.44} - (-3)[8.2 - (5.6)^3(-7)]$

   [√] 38.44 [−] [(-)] 3 [×] [(] 8.2 [−] 5.6 [x^y] (or [^]) 3 [×] [(-)] 7 [)] [=]

   3718.736 Ans
2. Solve.

   $18.32 - (-4.52) + \dfrac{\sqrt[4]{93.724 - 6.023}}{-1.236^3}$. Round the answer to 2 decimal places.

   18.32 [−] [(-)] 4.52 [+] 4 [2nd] (or [SHIFT]) [$\sqrt[x]{\ }$] [(] 93.724

   [−] 6.023 [)] [÷] [(-)] 1.236 [x^y] (or [^]) 3 [=]

   21.21932578, 21.22 Ans (rounded)

Application

The Number Scale

1. Refer to the number scale in Figure 34-2 and give the direction (+ or −) and the number of units counted going from the first to the second number.

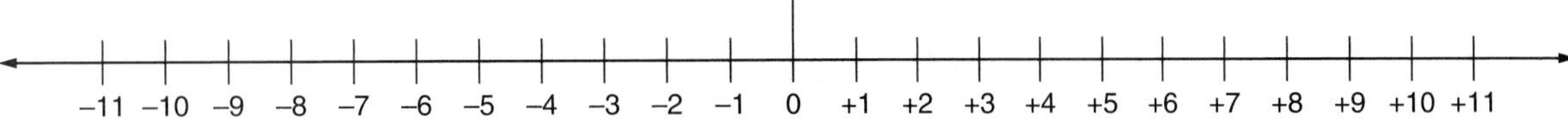

Figure 34-2

a. -11 to -2 ______ g. $+10$ to -10 ______ m. -7.5 to $+10$ ______

b. -8 to -3 ______ h. $+10$ to 0 ______ n. $+10$ to -7.5 ______

c. -6 to 0 ______ i. $+4$ to $+7$ ______ o. -10.8 to -4.3 ______

d. -2 to -8 ______ j. $+9$ to $+1$ ______ p. -2.3 to -0.8 ______

e. $+2$ to -8 ______ k. $+11$ to 0 ______ q. $+7\frac{1}{2}$ to $2\frac{1}{4}$ ______

f. $+3$ to $+10$ ______ l. 0 to -6 ______ r. $+5\frac{3}{4}$ to 0 ______

Comparing Signed Numbers

2. Select the greater of the two signed values and indicate the number of units by which it is greater.

a. $+5, -14$ ______ d. $+8, +13$ ______ g. $+14.3, +23$ ______

b. $+7, -3$ ______ e. $+20, -22$ ______ h. $-1.8, +1.8$ ______

c. $-8, -1$ ______ f. $-16, -4$ ______ i. $+17.6, -21.9$ ______

3. List the following signed numbers in order of increasing value starting with the smallest number.

a. $+17, -1, +2, 0, -18, +4, -25$ ______

b. $-5, +5, 0, +13, +27, -21, -2, -19$ ______

c. $+10, -10, -7, +7, 0, +25, -25, +14$ ______

d. $0, 15, -3.6, -2.5, -14.9, +17, +0.3$ ______

e. $-16, +14\frac{1}{8}, -13\frac{7}{8}, +6, -3\frac{5}{8}$ ______

Absolute Value

4. Express each of the following pairs of signed numbers as absolute values and subtract the smaller absolute value from the larger absolute value.

a. $+23, -14$ ______ c. $-6, +6$ ______ e. $-16, +16$ ______

b. $-17, +9$ ______ d. $+25, +13$ ______ f. $-33.7, -29.7$ ______

➤ **Note:** For problems 5 through 35 that follow, round the answers to 3 decimal places wherever necessary.

Addition of Signed Numbers

5. Add the following signed numbers as indicated.

a. $+15 + (+8)$ ______

b. $7 + (+18) + 5$ ______

c. $0 + (+25)$ ______

d. $-8 + (-15)$ ______

e. $-18 + (-4) + (-11)$ ______

f. $+12 + (-5)$ ______

g. $+18 + (-26)$ ______

h. $-20 + (+19)$ ______

i. $-23 + 17$ ______

j. $-25 + 3$ ______

k. $-9\frac{1}{4} + \left(-3\frac{3}{4}\right)$ ______

l. $18\frac{5}{8} + \left(-21\frac{3}{4}\right)$ ______

m. $-13 + \left(-\frac{5}{16}\right)$ ______

n. $-4.25 + (-7) + (-3.22)$ ______

o. $18.07 + (-17.64)$ ______

p. $16 + (-4) + (-11)$ ______

q. $-53.07 + (-6.37) + 19.82$ ______

r. $30.88 + (-0.95) + 1.32$ ______

s. $-12.77 + (-9) + (-7.61) + 0.48$ ______

t. $2.53 + 16.09 + (-54.05) + 21.37$ ______

Subtraction of Signed Numbers

6. Subtract the following signed numbers as indicated.

a. $-10 - (-4)$ ________
b. $+5 - (-13)$ ________
c. $-22 - (-14)$ ________
d. $+17 - (+6)$ ________
e. $+40 - (+40)$ ________
f. $-40 - (-40)$ ________
g. $-40 - (+40)$ ________
h. $0 - (-12)$ ________
i. $-52 - (-8)$ ________
j. $16.5 - (+14.3)$ ________
k. $-18.4 - (-14.3)$ ________
l. $-50.2 - (+51)$ ________
m. $+50.2 - (-51)$ ________
n. $0.03 - (+0.06)$ ________
o. $-10\frac{1}{2} - \left(-7\frac{1}{4}\right)$ ________
p. $5\frac{7}{8} - \left(-4\frac{1}{8}\right)$ ________
q. $(6 + 10) - (-7 + 9)$ ________
r. $(-14 + 5) - (2 - 10)$ ________
s. $(7.23 - 6.81) - (-10.73)$ ________
t. $[-8.76 + (-5.83)] - [12.06 - (-0.97)]$ ________

Multiplication of Signed Numbers

7. Multiply the following signed numbers as indicated.

a. $(-4)(6)$ ________
b. $(-4)(-6)$ ________
c. $(+10)(-3)$ ________
d. $(-10)(-3)$ ________
e. $(-5)(7)$ ________
f. $(-2)(-14)$ ________
g. $0(-16)$ ________
h. $(6.5)(-5)$ ________
i. $(-3.2)(-0.1)$ ________
j. $(-0.06)(-0.60)$ ________
k. $\left(1\frac{1}{2}\right)\left(-\frac{3}{4}\right)$ ________
l. $\frac{1}{4}(0)$ ________
m. $(-2)(-2)(-2)$ ________
n. $(-2)(+2)(+2)$ ________
o. $(8)(-4)(3)(0)(-1)$ ________
p. $(-3.86)(-2.1)(27.85)(-32.56)$ ________
q. $(8)(-2.65)(0.5)(-1)$ ________
r. $(-6.3)(-0.35)(2)(-1)(0.05)$ ________
s. $(-4.03)(-0.25)(-3)(-0.127)$ ________
t. $(-0.03)(-100)(-0.10)$ ________

Division of Signed Numbers

8. Divide the following signed numbers as indicated.

a. $-10 \div (-5)$ ________
b. $-10 \div (+2.5)$ ________
c. $+18 \div (+9)$ ________
d. $-21 \div 3$ ________
e. $-30 \div (-6)$ ________
f. $+48 \div (-6)$ ________
g. $-35 \div 7$ ________
h. $\frac{-16}{-4}$ ________
i. $\frac{0}{-10}$ ________
j. $\frac{-48}{-8}$ ________

k. $-\frac{1}{2} \div \left(-\frac{1}{2}\right)$ ____________

l. $\frac{-60}{-0.5} - 6 \div \frac{3}{4}$ ____________

m. $\frac{-10}{-2.5} + \frac{1}{3} \div \left(-\frac{2}{3}\right)$ ____________

n. $\frac{-17.92}{3.28}$ ____________

o. $0.562 \div (-0.821)$ ____________

p. $-29.96 \div 5.35$ ____________

q. $-4.125 \div (-0.75)$ ____________

r. $-41.87 \div 7.9$ ____________

s. $-20.47 \div 0.537$ ____________

t. $-44.876 \div (-7.836)$ ____________

Powers of Signed Numbers

9. Raise the following signed numbers to the indicated powers.

a. $(-2)^2$ ____________

b. 2^3 ____________

c. $(-2)^3$ ____________

d. $(-4)^3$ ____________

e. $(-2)^4$ ____________

f. $(-2)^5$ ____________

g. $(-6)^2$ ____________

h. $(-5)^3$ ____________

i. $(-2)^6$ ____________

j. $(-1.6)^2$ ____________

k. $(-0.4)^3$ ____________

l. 0.93^6 ____________

m. $(-1.58)^2$ ____________

n. $(-0.85)^3$ ____________

o. 0.73^3 ____________

p. $\left(-\frac{2}{3}\right)^3$ ____________

q. $(-1.038)^{-5}$ ____________

r. 17.66^{-2} ____________

s. $(-0.83)^{-3}$ ____________

t. $(-6.087)^{-4}$ ____________

Roots of Signed Numbers

10. Determine the indicated root of the following signed numbers.

a. $\sqrt[3]{64}$ ____________

b. $\sqrt[3]{-64}$ ____________

c. $\sqrt[3]{-27}$ ____________

d. $\sqrt[3]{-1000}$ ____________

e. $\sqrt[3]{1000}$ ____________

f. $\sqrt[5]{-32}$ ____________

g. $\sqrt[3]{125}$ ____________

h. $\sqrt[5]{+32}$ ____________

i. $\sqrt[3]{+1}$ ____________

j. $\sqrt[3]{-1}$ ____________

k. $\sqrt[7]{-1}$ ____________

l. $\sqrt[3]{216}$ ____________

m. $\sqrt[3]{\frac{-8}{-64}}$ ____________

n. $\sqrt[3]{\frac{+8}{-27}}$ ____________

o. $\sqrt[4]{\frac{+1}{+16}}$ ____________

p. $\sqrt[3]{\frac{+27}{-125}}$ ____________

q. $\sqrt[3]{-236.539}$ ____________

r. $\sqrt[5]{-86.009}$ ____________

s. $\sqrt[3]{\frac{-97.326}{123.592}}$ ____________

t. $\sqrt[3]{\frac{-89.096}{-17.323}}$ ____________

Expressing Numbers with Fractional Exponents as Radicals

11. Determine the value of the following:

a. $9^{1/2}$ ____________

b. $81^{1/2}$ ____________

c. $8^{1/3}$ ____________

d. $64^{1/3}$ ____________

e. $-8^{1/3}$ ____________

f. $16^{1/4}$ ____________

g. $-125^{1/3}$ ____________

h. $125^{1/3}$ ____________

i. $273.19^{2/3}$ ____________

j. $41.673^{-1/2}$ ____________

k. $8.007^{2/3}$ ____________

l. $67.725^{-2/3}$ ____________

Combined Operations of Signed Numbers

Solve each of the following problems using the proper order of operations.

12. $19 - (3)(-2) + (-5)^2$ ________
13. $4 - 5(8 - 10)$ ________
14. $-2(4 + 2) + 3(5 - 7)$ ________
15. $5 - 3(8 - 6) - [1 + (-6)]$ ________
16. $\dfrac{2(-1)(-3) - (6)(5)}{3(7) - 9}$ ________
17. $(-3)^3 + 3^3 - (-6)(3) - \left(\dfrac{-6}{2}\right)$ ________
18. $5^2 + \sqrt[3]{-8} + (-4)(0)(-3)$ ________
19. $[4^2 + (2)(5)(-3)]^2 + 2(-3)^3$ ________
20. $(-2)^3 + \sqrt{16} - (5)(3)(8)$ ________
21. $\dfrac{2(-5)^2}{2(5)} - \dfrac{(-4)^3}{18 + (-2)}$ ________
22. $(-2.87)^3 + \sqrt{15.93} - (5.63)(4)(-5.26)^3$ ________
23. $\dfrac{2(-5.16)^2}{3.07(4.98)} - \dfrac{(-4.66)^3}{18.37 + (-2.02)}$ ________
24. $(-2.46)^3 + \sqrt[3]{(-3.86)(-10.42) - (-6.16)}$ ________
25. $10.78^{-2} + [43.28 + (9)(-0.563)]^{-3}$ ________

Substitute the given numbers for letters in the following expressions and solve.

26. Find $6xy + 5 - xy$ when $x = -2$ and $y = 7$. ________
27. Find $\dfrac{-3ab - 2bc}{abc - 35}$ when $a = -3$, $b = 10$, and $c = -4$. ________
28. Find $(x - y)(3x - 2y)$ when $x = -5$ and $y = -7$. ________
29. Find $\dfrac{d^3 + 4f - fh}{h^2 - (2 + d)}$ when $d = -2$, $f = -4$, and $h = 4$. ________
30. Find $\dfrac{x^2}{n} - \dfrac{21 + y^3}{xy}$ when $n = 5$, $x = -5$, and $y = -1$. ________
31. Find $\sqrt{6(ab - 6)} - (b)^3$ when $a = -6$ and $b = -2$. ________
32. $\dfrac{x^2}{n} - \dfrac{21 + y^3}{xy}$; $n = 5.31$, $x = -5.67$, $y = -1.87$ ________
33. $\sqrt{6(ab - 6)} - (c)^3$; $a = -6.07$, $b = -2.91$, $c = 1.56$ ________
34. $5\sqrt[3]{e} + (ef - d) - (d)^3$; $d = -10.55$, $e = 8.26$, $f = -7.09$ ________
35. $\dfrac{\sqrt[4]{(mpt + pt + 19)}}{t^2 + 2p - 7}$; $m = 2$, $p = -2.93$, $t = -5.86$ ________

Algebraic Operations of Addition, Subtraction, and Multiplication

Objectives ***After studying this unit you should be able to***

- Perform the basic algebraic operations of addition, subtraction, and multiplication.
- Express decimal numbers in scientific notation form and multiply and divide using scientific notation.

A knowledge of basic algebraic operations is essential in order to solve equations. For certain applications, formulas given in machine trade handbooks cannot be used directly as given, but must be rearranged. Formulas are rearranged by using the principles of algebraic operations.

Definitions

It is important to understand the following definitions in order to apply the procedures that are required for solving problems involving basic operations.

A *term* of an algebraic expression is that part of the expression that is separated from the rest by a plus or a minus sign. For example, $4x + \frac{ab}{2x} - 12 + 3ab^2x - 8a\sqrt{b}$ is an expression that consists of five terms: $4x$, $\frac{ab}{2x}$, 12, $3ab^2x$, and $8a\sqrt{b}$.

A *factor* is one of two or more literal and/or numerical values of a term that are multiplied. For example, 4 and x are each factors of $4x$; 3, a, b^2, and, x are each factors of $3ab^2x$; 8, a, and $\sqrt{b}$ are each factors of $8a\sqrt{b}$.

➤ **Note:** It is absolutely necessary that you distinguish between factors and terms.

A *numerical coefficient* is the number factor of a term. The letter factors of a term are the *literal factors*. For example, in the term $5x$, 5 is the numerical coefficient; x is the literal factor. In the term $\frac{1}{3}ab^2c^3$, $\frac{1}{3}$ is the numerical coefficient; a, b^2, and c^2 are the literal factors.

Like terms are terms that have identical literal factors including exponents. The numerical coefficients do not have to be the same. For example, $6x$ and $13x$ are like terms; $15ab^2c^3$, $3.2ab^2c^3$, and $\frac{1}{8}ab^2c^3$ are like terms.

Unlike terms are terms that have different literal factors or exponents. For example, $12x$ and $12y$ are unlike terms because they have different literal factors. The terms $15xy$, $3x^2y$, and $4x^2y^2$ are unlike terms. Although the literal factors are x and y in each of the terms, these literal factors are raised to different powers.

Addition

Only like terms can be added. The addition of unlike terms can only be indicated. As in arithmetic, like things can be added, but unlike things cannot be added. For example, 4 inches + 5 inches = 9 inches. Both values are inches; therefore, they can be added. But 4 inches + 5 pounds cannot be added because they are unlike things.

Procedure To add like terms

- Add the numerical coefficients applying the procedure for addition of signed numbers. If a term does not have a numerical coefficient, the coefficient 1 is understood: $x = 1x$, $abc = 1abc$, $n^2rs^3 = 1n^2rs^3$.
- Leave the literal factors unchanged.

Examples Add the following like terms.

1. $3x$; $12x$; $15x$ Ans
2. x; $-14x$; $-13x$ Ans
3. $-5xy^2$; $+5xy^2$; 0 Ans
4. $6x^2y^3$; $-13x^2y^3$; $-7x^2y^3$ Ans
5. $2(a + b)$; $-3(a + b)$; $7(a + b)$; $6(a + b)$ Ans

Procedure To add unlike terms

- The addition of unlike terms can only be indicated.

Examples Add the following unlike terms.

1. 15; x; $15 + x$ Ans
2. $7x$; $8y$; $7x + 8y$ Ans
3. $3x$; $-7x^2$; $3x + (-7x^2)$ Ans
4. $8a$; $-6b$; $2c$; $8a + (-6b) + 2c$ Ans

Procedure

To add two or more expressions that consist of two or more terms

- Group the like terms in the same column.
- Add like terms and indicate the addition of unlike terms.

Examples Add the following expressions.

1. $12x - 2xy + 6x^2y^3$ and $-4x - 7xy + 5x^2y^3$

Group like terms in the same column.

$$\begin{array}{rrr} 12x & -\ 2xy & +\ 6x^2y^3 \\ -4x & -\ 7xy & +\ 5x^2y^3 \\ \hline \end{array}$$

Add like terms. $8x - 9xy + 11x^2y^3$ Ans

2. $6a - 7b$ and $18b - 3ab + a$ and $-14a + ab^2 - 5ab$

Group like terms.

$$\begin{array}{rrrr} 6a & -\ 7b & & \\ a & +\ 18b & -\ 3ab & \\ -14a & & -\ 5ab & +\ ab^2 \\ \hline -7a & +\ 11b & -\ 8ab & +\ ab^2 \end{array}$$

Add like terms and indicate the addition of unlike terms. $-7a + 11b - 8ab + ab^2$ Ans

Subtraction

As in addition, only like terms can be subtracted. The subtraction of unlike terms can only be indicated. The same principles apply in arithmetic. For example, 8 feet −3 feet = 5 feet, but 8 feet −3 ounces cannot be subtracted because they are unlike things.

Procedure To subtract like terms

- Subtract the numerical coefficients applying the procedure for subtraction of signed numbers.
- Leave the literal factors unchanged.

Examples Subtract the following like terms as indicated.

1. $18ab - 7ab = 11ab$ Ans
2. $bx^2y^3 - 13bx^2y^3 = -12bx^2y^3$ Ans
3. $-5x^2y - 8x^2y = -13x^2y$ Ans
4. $-24dmr - (-24dmr) = 0$ Ans

Procedure To subtract unlike terms

- The subtraction of unlike terms can only be indicated.

Examples Subtract the following unlike terms as indicated.

1. $3x^2 - (+2x) = 3x^2 - 2x$ Ans
2. $-13abc - (+8abc^2) = -13abc - 8abc^2$ Ans
3. $-2xy - (-7y) = -2xy + 7y$ Ans

Procedure To subtract expressions that consist of two or more terms

- Group like terms in the same column.
- Subtract like terms and indicate the subtractions of the unlike terms.

➤ **Note:** Each term of the subtrahend is subtracted following the procedure for subtraction of signed numbers.

Examples Subtract the following expressions as indicated.

1. Subtract. $7a + 3b - 3d$ from $8a - 7b + 5d$

   Group like terms in the same column. Change the sign of each term in the subtrahend and follow the procedure for addition of signed numbers.

$$\begin{array}{rcr} 8a - 7b + 5d & = & 8a - 7b + 5d \\ \underline{-(7a + 3b - 3d)} & = & \underline{+(-7a - 3b + 3d)} \\ & & a - 10b + 8d \end{array}$$ Ans

2. Subtract as indicated: $(3x^2 + 5x - 12xy) - (7x^2 - x - 3x^3 + 6y)$

$$\begin{array}{rcr} 3x^2 + 5x - 12xy \qquad\qquad & = & 3x^2 + 5x - 12xy \qquad\qquad \\ \underline{-(7x^2 - x \qquad\quad - 3x^3 + 6y)} & = & \underline{+(-7x^2 + x \qquad\quad + 3x^3 - 6y)} \\ & & -4x^2 + 6x - 12xy + 3x^3 - 6y \end{array}$$ Ans

Multiplication

It was shown that unlike terms cannot be added or subtracted. In multiplication, unlike terms can be multiplied. For example, x^2 can be multiplied by x^4. The term x^2 means $(x)(x)$. The term x^4 means $(x)(x)(x)(x)$.

$$(x^2)(x^4) = (x)(x)(x)(x)(x)(x) = x^{2+4} = x^6$$

Also, $2x$ can be multiplied by $4y$. The product is $8xy$.

Procedure To multiply two or more terms

- Multiply the numerical coefficients following the procedure for multiplication of signed numbers.
- Add the exponents of the same literal factors.
- Show the product as a combination of all numerical and literal factors.

Examples Multiply as indicated.

1. Multiply. $(-3x^2)(6x^4)$

   Multiply numerical coefficients $(-3)(6) = -18$

   Add exponents of like literal factors $(x^2)(x^4) = x^{2+4} = x^6$

   Show product as combination of all numerical and literal factors.

   $(-3x^2)(6x^4) = -18x^6$ Ans
2. $(3a^2b^3)(7ab^3) = (3)(7)(a^{2+1})(b^{3+3}) = 21a^3b^6$ Ans
3. $(-4a)(-7b^2c^2)(-2ac^3d^3) = (-4)(-7)(-2)(a^{1+1})(b^2)(c^{2+3})d^3 = -56a^2b^2c^5d^3$ Ans

Procedure To multiply expressions that consist of more than one term within an expression

- Multiply each term of one expression by each term of the other expression.
- Combine like terms.

Before applying the procedure to algebraic expressions, an example is given to show that the procedure is consistent with arithmetic.

Examples in Arithmetic:

Multiply. $3(4 + 2)$

From arithmetic: $3(4 + 2) = 3(6) = 18$ Ans

From algebra:
Multiply each term of one expression by each term of the other expression. $3(4 + 2) = 3(4) + 3(2) = 12 + 6 = 18$ Ans

Combine like terms.

Examples in Algebra:

1. $3a(6 + 2a^2) = (3a)(6) + 3a(2a^2) = 18a + 6a^3$ Ans
2. $-5x^2y^5(3xy - 4x^3y^2 + 5y) = -5x^2y(3xy) - 5x^2y(-4x^3y^2) - 5x^2y(5y)$

   $= -15x^3y^2 + 20x^5y^3 - 25x^2y^2$ Ans
3. Multiply. $(3c + 5d^2)(4d^2 - 2c)$

   This is an example in which both expressions have two terms. The solution illustrates a shortcut of the distributive property called the FOIL method.

FOIL Method

Find the sum of the products of

1. the **F**irst terms: **F**
2. the **O**uter terms: **O**
3. the **I**nner terms: **I**
4. the **L**ast terms: **L**

Then combine like terms.

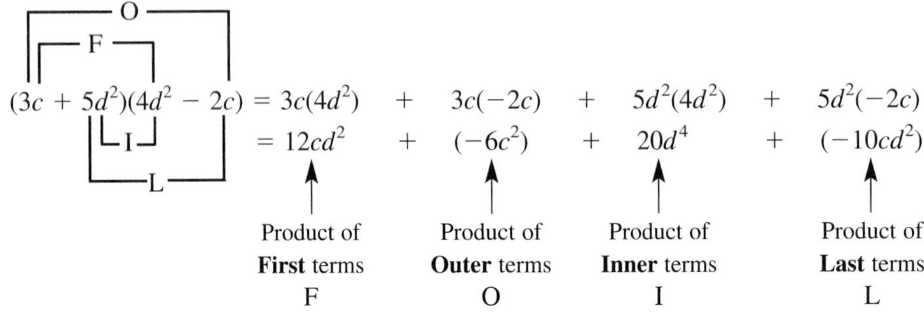

Combine like terms.

COMBINE

$$= 12cd^2 + (-6c)^2 + 20d^4 + (-10cd^2)$$
$$= 2cd^2 + (-6c^2) + 20d^4$$
$$= 2cd^2 - 6c^2 + 20d^4 \quad \text{Ans}$$

Application

Addition of Single Terms

Add the terms in the following expressions.

1. $18y + y$ ________
2. $15xy + 7xy$ ________
3. $-15xy + (-7xy)$ ________
4. $22m^2 + (-m^2)$ ________
5. $-5x^2y + 5x^2y$ ________
6. $4c^3 + 0$ ________
7. $-9pt + (-pt)$ ________
8. $0.4x + (-0.8x)$ ________
9. $8.3a^2b + 6.9a^2b$ ________
10. $-0.04y + 0.07y$ ________
11. $\frac{1}{2}xy + \frac{3}{4}xy$ ________
12. $2\frac{3}{4}c^2d + \left(-3\frac{1}{8}c^2d\right)$ ________
13. $-2.06gh^3 + (-0.85gh^3)$ ________
14. $-50.6abc + 50.5abc$ ________
15. $4P + (-6P) + P + 12P$ ________
16. $-0.3dt^2 + (-1.7dt^2) + (-dt^2)$ ________
17. $5P + 2P^2$ ________
18. $-a^3 + 2a^2$ ________
19. $7ab^2 + (-2a^2b) + (-a^2b^2)$ ________
20. $(-xyz) + x^2yz + (-xy^2z) + 5xyz^2$ ________
21. $\frac{1}{4}xy + \frac{7}{8}xy + xy + (-4xy)$ ________
22. $20.06D + (-19.97D) + (-0.7D)$ ________
23. $6M + 0.6M + 0.06M + 0.006M$ ________
24. $-3xy^2 + 8xy^2 + 7.8xy^2$ ________
25. $5T + 2T^2 + (-8T)$ ________
26. $2x^2 + 5ax^2 + (-7x^2)$ ________
27. $15ax^2 + 3a^2x + (-15ax^2) + (-10a^2x)$ ________
28. $-8abc + 8ab^2c + (-8abc^2) + (-8ab^2c)$ ________

29. The machined plate distances shown in Figure 35-1 are dimensioned, in millimeters, in terms of x. Determine dimensions A–G.

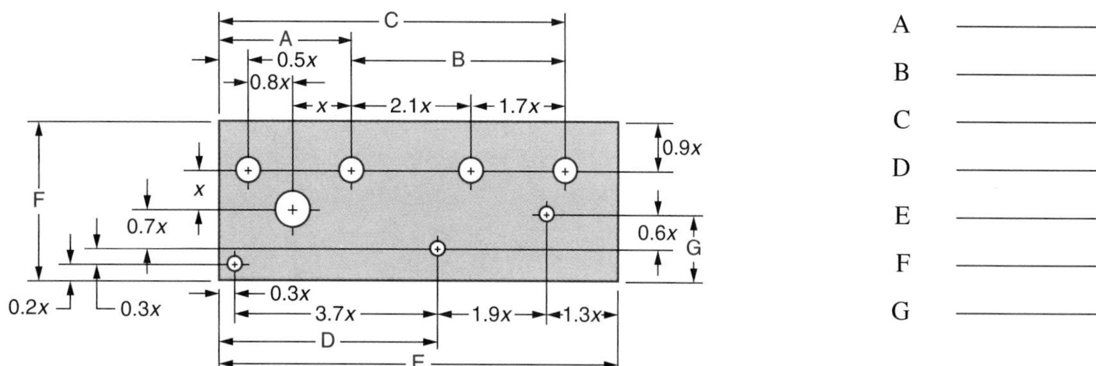

Figure 35-1

A ______
B ______
C ______
D ______
E ______
F ______
G ______

Addition of Expressions with Two or More Terms

Add the following expressions.

30. $\begin{array}{r} -5x + 7xy - 8y \\ -9x - 12xy + 13y \\ \hline \end{array}$

31. $\begin{array}{r} 3a - 11d - 8m \\ -a + 11d - 3m \\ \hline \end{array}$

32. $\begin{array}{r} -6ab - 5a^2b^2 - 3a^2b \\ -5ab + 14a^2b^2 - 12a^3b \\ -9ab - 7a^2b^2 + a^3b \\ ab \qquad - 2a^3b \\ \hline \end{array}$

33. $(3xy^2 + x^2y - x^2y^2), (2x^2y + x^2y^2)$ ______

34. $(10a - 5b), (-12a - 7b), (11a + b)$ ______

35. $(x^3 + 5), (3x - 7x^2 + 7), (x - 3x^3)$ ______

36. $(b^4 + 4b^3c - 2b^2c), (4b^3c - 7bc)$ ______

37. $(x^2 - 4xy), (4xy - y^2), (-x^2 + y^2)$ ______

38. $(1.3M - 3N), (-8M + 0.5N), (20M + 0.4N)$ ______

39. $(c + 3.6cd - 4.9d), (-1.4c + 8.6d)$ ______

Subtraction of Single Terms

Subtract the following terms as indicated.

40. $7xy^2 - (-18xy^2)$ ______
41. $3xy - xy$ ______
42. $-3xy - xy$ ______
43. $-3xy - (-xy)$ ______
44. $9ab - (-9ab)$ ______
45. $-5a^2 - (5a^2)$ ______
46. $0.7a^2b^2 - 2.3a^2b^2$ ______
47. $0 - (-12mn^3)$ ______
48. $-8mn^3 - 0$ ______
49. $\frac{7}{8}x^2 - \left(-\frac{3}{8}x^2\right)$ ______
50. $13a - 9a^2$ ______
51. $-13a - (-7a^2)$ ______
52. $0.2xy - 0.9xy^2$ ______
53. $-ax^2 - ax^2$ ______
54. $\frac{1}{2}dt - \left(-\frac{3}{8}dt\right)$ ______
55. $\frac{1}{2}d^2t^2 - \left(-\frac{1}{2}d^2t^2\right)$ ______
56. $21 - 3x$ ______
57. $3x - 21$ ______
58. $-3.2d - 6.4d$ ______
59. $-1.4xy - (-1.4xy)$ ______

Subtraction of Expressions with Two or More Terms

Subtract the following expressions as indicated.

60. $(2a^2 - 3a) - (7a^2 - 10a)$ ________
61. $(4x^2 + 8xy) - (3x^2 + 5xy)$ ________
62. $(9b^2 + 1) - (9b^2 - 1)$ ________
63. $(9b^2 - 1) - (9b^2 - 1)$ ________
64. $(xy^2 - x^2y^2 + x^2y^2) - 0$ ________
65. $(2a^3 - 0.3a^2) - (-a^3 + a^2 - a)$ ________
66. $(5x + 3xy - 7y) - (3y^2 - x^2y)$ ________
67. $(-d^2 - dt + dt^2) - (-4 + dt)$ ________
68. $(15L - 12H) - (-12L + 6H - 4)$ ________
69. $(9.08e + 14.76f) - (e - f - 10.03)$ ________

Multiplication of Single Terms

Multiply the following terms as indicated.

70. $(-5b^2c)(3b^3)$ ________
71. $(x)(x^2)$ ________
72. $(-3a^2)(-5a^4)$ ________
73. $(8ab^2c)(7a^3bc^2)$ ________
74. $(-x^3y^3)(5a^3b)$ ________
75. $(-3xy)(0)$ ________
76. $(7ab^4)(3a^4b)$ ________
77. $(-3d^5r^4)(-d^3)$ ________
78. $(-3d^5r^4)(-d^3)(-1)$ ________
79. $(0.3x^2y^4)(0.7x^5)$ ________
80. $\left(\frac{1}{4}a^3\right)\left(\frac{3}{8}a^2\right)$ ________
81. $(-5x)(0)(-5x)$ ________
82. $(m^2t)(st^2)$ ________
83. $(-1.6bc)(2.1)$ ________
84. $(abc^3)(c^3d)$ ________
85. $(2x^6y^6)(-x^2)$ ________
86. $\left(-\frac{2}{3}mt\right)\left(t^4\right)$ ________
87. $(7ab^3)(-7a^3b)$ ________
88. $(-0.3a^3b^2)(-4b^3)$ ________
89. $(-x^2y)(-xy)(-x)$ ________
90. $(d^4m^2)(-1)(-m^3)$ ________

Multiplication of Expressions with Two or More Terms

Multiply the following expressions as indicated and combine like terms where possible.

91. $-5xy(2xy^2 - 3x^4)$ ________
92. $3a^2(-a^2 + a^3b)$ ________
93. $-2a^3b^2(4ab^3 - b^2 - 2)$ ________
94. $xy^2(x^2 + y^3 + xy)$ ________
95. $-4(dt + t^2 - 1)$ ________
96. $(m^2t^3s^4)(-m^4s^2 + m - s^5)$ ________
97. $(3x + 7)(x^2 + 9)$ ________
98. $(7x^2 - y^3)(-2x^3 + y^2)$ ________
99. $(5ax^3 + bx)(2a^2x^3 + b^2x)$ ________
100. $(-3a^2b^3 + 5xy^2)(4a^2b^3 - 5xy)$ ________

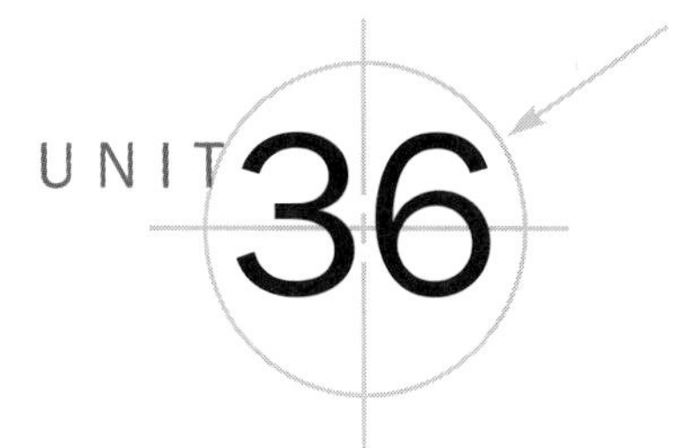

Algebraic Operations of Division, Powers, and Roots

Objectives ***After studying this unit you should be able to***

- Perform the basic algebraic operations of division, powers, and roots.
- Remove parentheses that are preceded by a plus or minus sign.
- Simplify algebraic expressions that involve combined operations.
- Write decimal numbers as scientific notation.
- Compute expressions using scientific notation.

Division

As with multiplication, unlike terms can be divided. For example, x^4 can be divided by x.

$$\frac{x^4}{x} = \frac{(x)(x)(x)(x)}{x} = x^{4-1} = x^3 \quad \text{Ans}$$

Procedure To divide two terms

- Divide the numerical coefficients following the procedure for division of signed numbers.
- Subtract the exponents of the literal factors of the divisor from the exponents of the same literal factors of the dividend.
- Combine numerical and literal factors.

This division procedure is consistent with arithmetic.

Example in Arithmetic:

Divide. $\frac{2^5}{2^2}$

From arithmetic: $\frac{2^5}{2^2} = \frac{(2)(2)(2)(2)(2)}{(2)(2)} = (2)(2)(2) = 8$ Ans

From algebra: $\frac{2^5}{2^2} = 2^{5-2} = 2^3 = 8$ Ans

Examples in Algebra:

1. Divide $-16x^3$ by $8x$.

   Divide the numerical coefficients following the procedure for signed numbers. $\quad -16 \div 8 = -2$

   Subtract the exponents of the literal factors in the divisor from the exponents of the same literal factors in the dividend. $\quad x^3 \div x = x^{3-1} = x^3$

   Combine the numerical and literal factors. $\quad \frac{-16x^3}{8x} = -2x^2$ Ans

2. $\dfrac{-30a^3b^5c^2}{-5a^2b^3} = \left(\dfrac{-30}{-5}\right)\left(a^{3-2}\right)\left(b^{5-3}\right)\left(c^2\right) = 6ab^2c^2$ Ans

In arithmetic, any number except 0 divided by itself equals 1. For example, $4 \div 4 = 1$. Applying the division procedure $4 \div 4 = 4^{1-1} = 4^0$. Therefore, $4^0 = 1$. Any number except 0 raised to the zero power equals 1.

Example 1 $\dfrac{5^3}{5^3} = 5^{3-3} = 5^0 = 1$ Ans

Example 2 $\dfrac{a^3b^2c}{a^3b^2c} = (a^{3-3})(b^{2-2})(c^{1-1}) = a^0b^0c^0 = (1)(1)(1) = 1$ Ans

Procedure To divide when the divisor consists of one term and the dividend consists of more than one term

- Divide each term of the dividend by the divisor following the procedure shown above.
- Combine terms.

Example in Arithmetic:

Divide. $\dfrac{6+8}{2}$

From arithmetic: $\dfrac{6+8}{2} = \dfrac{14}{2} = 7$ Ans

From algebra: $\dfrac{6+8}{2} = \dfrac{6}{2} + \dfrac{8}{2} = 3 + 4 = 7$ Ans

Example in Algebra:

Divide. $\dfrac{-20xy^2 + 15x^2y^3 + 35x^3y}{-5xy}$

$$\frac{-20xy^2 + 15x^2y^3 + 35x^3y}{-5xy} = \frac{-20xy^2}{-5xy} + \frac{15x^2y^3}{-5xy} + \frac{35x^3y}{-5xy}$$

$$= 4y - 3xy^2 - 7x^2 \quad \text{Ans}$$

Powers

Procedure To raise a single term to a power

- Raise the numerical coefficients to the indicated power following the procedure for powers of signed numbers.
- Multiply each of the literal factor exponents by the exponent of the power to which it is raised.
- Combine numerical and literal factors.

This power procedure is consistent with arithmetic.

Example in Arithmetic:

Raise to the indicated power. $(2^2)^3$

From arithmetic: $(2^2)^3 = (4)^3 = (4)(4)(4) = 64$ Ans

From algebra: $(2^2)^3 = 2^{2(3)} = 2^6 = (2)(2)(2)(2)(2)(2) = 64$ Ans

Examples in Algebra:

1. Raise to the indicated power. $(5x^3)^2$

   Raise the numerical coefficient to the indicated power following the procedure for powers of signed numbers. $5^2 = 25$

   Multiply each literal factor exponent by the exponent of the power to which it is to be raised. $(x^3)^2 = x^{3(2)} = x^6$

   Combine numerical and literal factors. $(5x^3)^2 = 25x^6$ Ans

➤ **Note:** $(x^3)^2$ is not the same as x^3x^2.
$(x^3)^2 = (x^3)(x^3) = (x)(x)(x)(x)(x)(x) = x^6$
$x^3x^2 = (x)(x)(x)(x)(x) = x^5$

2. $(-3a^2b^4c)^3 = (-3)^3a^{2(3)}b^{4(3)}c^{1(3)} = -27a^6b^{12}c^3$ Ans

3. $\left[-\frac{1}{2}x^3(yd^2)^3r^4\right]^2 = \left[-\frac{1}{2}x^3y^3d^6r^4\right]^2 = \frac{1}{4}x^6y^6d^{12}r^8$ Ans

Procedure To raise two or more terms to a power

- Apply the procedure for multiplying expressions that consist of more than one term.

Example Raise to the indicated power.

Solve. $(2x + 5y^3)^2$

$(2x + 5y^3)^2 = (2x + 5y^3)(2x + 5y^3)$

F	O	I	L
Step 1	*Step* 2	*Step* 3	*Step* 4

$= 2x(2x) + 2x(5y^3) + 5y^3(2x) + 5y^3(5y^3)$

$= 4x^2 + 10xy^3 + 10xy^3 + 25y^6 = 4x^2 + 20xy^3 + 25y^6$ Ans

(*Combine* $10xy^3 + 10xy^3$)

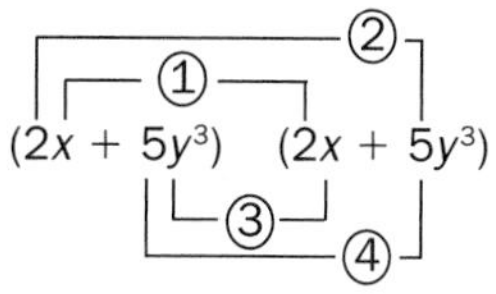

Roots

Procedure To extract the root of a term

- Determine the root of the numerical coefficient following the procedure for roots of signed numbers.
- The roots of the literal factors are determined by dividing the exponent of each literal factor by the index of the root.
- Combine the numerical and literal factors.

  This procedure for extracting roots is consistent with arithmetic.

Example in Arithmetic:

Find the indicated root. $\sqrt{2^6}$

From arithmetic: $\sqrt{2^6} = \sqrt{(2)(2)(2)(2)(2)(2)} = \sqrt{64} = 8$ Ans

From algebra: $\sqrt{2^6} = 2^{6 \div 2} = 2^3 = (2)(2)(2) = 8$ Ans

Examples in Algebra:

1. $\sqrt{25a^6b^4c^8} = \sqrt{25}(a^{6\div 2})(b^{4\div 2})(c^{8\div 2}) = 5a^3b^2c^4$ Ans
2. $\sqrt[3]{-27d^3x^9y^2} = \sqrt[3]{-27}(d^{3\div 3})(x^{9\div 3})\sqrt[3]{y^2} = -3dx^3\sqrt[3]{y^2}$ Ans
3. $\sqrt[4]{\frac{16}{81}d^8t^{12}y^2} = \sqrt[4]{\frac{16}{81}}(d^{8\div 4})(t^{12\div 4})(y^{2\div 4}) = \frac{2}{3}d^2t^3y^{1/2} = \frac{2}{3}d^2t^3\sqrt{y}$ Ans

➤ **Note:** Roots of expressions that consist of two or more terms cannot be extracted by this procedure. For example, $\sqrt{x^2 + y^2}$ consists of two terms and does *not* equal $\sqrt{x^2} + \sqrt{y^2}$. The mistake of considering the expressions equal is commonly made by students and must be avoided. This fact is consistent with arithmetic as shown.

$\sqrt{3^2 + 4^2} = \sqrt{9 + 16} = \sqrt{25} = 5$, but $\sqrt{3^2} + \sqrt{4^2} = 3 + 4 = 7$.

5 does *not* equal 7; therefore $\sqrt{3^2 + 4^2} \neq \sqrt{3^2} + \sqrt{4^2}$.

Removal of Parentheses

In certain expressions, terms are enclosed within parentheses, which are preceded by a plus or minus sign. In order to combine like terms, it is necessary to first remove parentheses.

Procedure To remove parentheses preceded by a plus sign

- Remove the parentheses without changing the signs of any terms within the parentheses.
- Combine like terms.

Example $5a + (4b + 7a - 3d) = 5a + 4b + 7a - 3d$
$= 12a + 4b - 3d$ Ans

Procedure To remove parentheses preceded by a minus sign

- Remove the parentheses and change the sign of each term within the parentheses.
- Combine like terms.

Example $-(7a^2 + b - 3) + 12 - (-b + 5) = -7a^2 - b + 3 + 12 + b - 5$
$= -7a^2 + 10$ Ans

Combined Operations

Procedure To solve expressions, consisting of two or more different operations

- Apply the proper order of operations. The order of operations as presented in Units 15 and 33 is repeated as follows:

Order of Operations

- First, do all operations within grouping symbols. Grouping symbols are parentheses ( ), brackets [ ], braces { }, and absolute value signs | |.
- Second, do powers and roots.
- Next, do multiplication and division operations in order from left to right.
- Last, do addition and subtraction operations in order from left to right.

Once again, you can use the memory aid "**P**lease **E**xcuse **M**y **D**ear **A**unt **S**ally" to help remember the order of operations. The **P** in "Please" stands for parentheses, the **E** for exponents (or raising to a power) and roots, the **M** and **D** for multiplication and division, and the **A** and **S** for addition and subtraction.

Examples

1. $10x - 3x(2 + x - 4x^2) = 10x - 6x - 3x^2 + 12x^3$
   $= 4x - 3x^2 + 12x^3$ Ans
2. $15a^6b^3 + (2a^2b)^3 - \dfrac{a^7(b^3)^2}{ab^3} = 15a^6b^3 + 8a^6b^3 - \dfrac{a^7b^6}{ab^3}$
   $= 15a^6b^3 + 8a^6b^3 - a^6b^3 = 22a^6b^3$ Ans
3. $-4a[15 - 3(2a + ab) + a] - 2a^2b = -4a(15 - 6a - 3ab + a) - 2a^2b$
   $= -60a + 24a^2 + 12a^2b - 4a^2 - 2a^2b$
   $= -60a + 20a^2 + 10a^2b$ Ans

Scientific Notation

In scientific applications and certain technical fields, computations with very large and very small numbers are required. The numbers in their regular or standard form are inconvenient to read, to write, and to use in computations. For example, copper expands 0.00000900 per unit of length per degree Fahrenheit. Scientific notation simplifies reading, writing, and computing with large and small numbers.

In scientific notation, a number is written as a whole number or a decimal with an absolute value between 1 and 10 multiplied by 10 with a suitable exponent. For example, a value of 325,000 is written in scientific notation as 3.25×10^5.

The effect of multiplying a number by 10 is to shift the position of the decimal point. Changing a number from the standard decimal form to scientific notation involves counting the number of decimal places the decimal point must be shifted.

Expressing Decimal (Standard Form) Numbers in Scientific Notation

A positive or negative number whose absolute value is 10 or greater has a positive exponent when expressed in scientific notation.

Examples

1. Rewrite 146,000 using scientific notation.
   a. Write the number as a value between 1 and 10: 1.46
   b. To determine the exponent of 10, count the number of places the decimal point is shifted: 1 46000. The decimal point is shifted 5 places. The exponent of 10 is 5: 10^5
   c. Multiply 1.46×10^5
      $146{,}000 = 1.46 \times 10^5$ Ans
2. Express 63,150,000 using scientific notation.
   6 3,150,000. $= 6.315 \times 10^7$ Ans
   Shift 7 places
3. Express −97.856 using scientific notation.
   −9 7.856 $= -9.7856 \times 10^1$ Ans
   Shift 1 place

A positive or negative number whose absolute value is less than 1 has a negative exponent when expressed in scientific notation.

Examples

1. Rewrite 0.0289 using scientific notation.

   $0.02\,89 = 2.89 \times 10^{-2}$ Ans

   Shift 2 places. Observe that the decimal point is shifted to the right, resulting in a negative exponent.

2. Rewrite 0.0000318 using scientific notation.

   $0.00003\,18 = 3.18 \times 10^{-5}$ Ans

   Shift 5 places

3. Rewrite 0.859 using scientific notation.

   $0.8\,59 = 8.59 \times 10^{-1}$ Ans

   Shift 1 place

Expressing Scientific Notation as Decimal (Standard Form) Numbers

To express a number given in scientific notation as a decimal number, shift the decimal point in the reverse direction and attach required zeros. Move the decimal point according to the exponent of 10. With positive exponents the decimal point is moved to the right; with negative exponents it is moved to the left.

Examples Express the following values in decimal form.

1. $4.3 \times 10^3 = 4{,}300.$ Ans

   Shift 3 places. Attach required zeros.

2. $8.907 \times 10^5 = 8\,90{,}700.$ Ans

   Shift 5 places. Attach required zeros.

3. $3.8 \times 10^{-4} = 0.0003\,8$ Ans

   Shift 4 places. Attach required zeros.

Multiplication and Division Using Scientific Notation

Scientific notation is used primarily for multiplication and division operations. The procedures presented in Unit 34 for the algebraic operations of multiplication and division are applied to operations involving scientific notation.

Examples Compute the following expressions.

1. $(2.8 \times 10^3) \times (3.5 \times 10^5)$

   a. Multiply the decimals: $2.8 \times 3.5 = 9.8$

   b. The product of the 10's equals 10 raised to a power that is the sum of the exponents:

      $10^3 \times 10^5 = 10^{3+5} = 10^8$

c. Combine both parts (9.8 and 10^8) as a product:

$(2.8 \times 10^3) \times (3.5 \times 10^5) = 9.8 \times 10^8$ Ans

2. $340{,}000 \times 7{,}040{,}000$

Rewrite the numbers in scientific notation and solve:

$340{,}000 \times 7{,}040{,}000 = (3.4 \times 10^5) \times (7.04 \times 10^6) = 23.936 \times 10^{11}$.

Notice that the decimal part is greater than 10. Rewrite the decimal part and solve: $23.936 = 2.3936 \times 10^1$.

$(2.3936 \times 10^1) \times 10^{11} = 2.3936 \times 10^{12}$ Ans

3. $-840{,}000 \div 0.0006$

$$\begin{aligned}-840{,}000 \div 0.0006 &= (-8.4 \times 10^5) \div (6 \times 10^{-4})\\ &= (-8.4 \div 6) \times (10^5 \div 10^{-4})\\ &= -1.4 \times 10^{5-(-4)}\\ &= -1.4 \times 10^9 \quad \text{Ans}\end{aligned}$$

4. $\dfrac{(3.4 \times 10^{-8}) \times (7.9 \times 10^5)}{(-2 \times 10^6)}$

$$\begin{aligned}3.4 \times 7.9 \div -2 &= -13.43\\ 10^{-8} \times 10^5 \div 10^6 &= 10^{-8+5-6} = 10^{-9}\\ -13.43 \times 10^{-9} &= (-1.343 \times 10^1) \times 10^{-9}\\ (-1.343 \times 10^1) \times 10^{-9} &= -1.343 \times 10^{1-9}\\ &= -1.343 \times 10^{-8} \quad \text{Ans}\end{aligned}$$

With 10-digit calculators, the number shown in the calculator display is limited to 10 digits. Calculations with answers that are greater than 9,999,999,999 or less than 0.000000001 are automatically expressed in scientific notation.

Examples

1. 80000000 [×] 400000 [=] 3.2^{13} (Answer displayed as 3.2^{13} or, on a SHARP, as 3.2×10^{13}.)

The display shows the number (mantissa) and the exponent of 10; it does *not* necessarily show the 10. The displayed answer of 3.2^{13} does *not* mean that 3.2 is raised to the thirteenth power. The display 3.2^{13} means 3.2×10^{13}; $80{,}000{,}000 \times 400{,}000 = 3.2 \times 10^{13}$.

2. .0000007 [×] .000002 [=] 1.4^{-12} (Answer is displayed as 1.4^{-12})

The display 1.4^{-12} means 1.4×10^{-12}; $0.0000007 \times 0.000002 = 1.4 \times 10^{-12}$.

Numbers in scientific notation can be directly entered in a calculator. For calculations whose answer does *not exceed* the number of digits in the calculator display, the answer is displayed in standard decimal form.

The answer is displayed in decimal (standard) form on certain calculators with the exponent entry key, [EE], or exponent key, [EXP]. Exponent entry is often an alternate [2nd] function as shown.

Example Solve. $(3.86 \times 10^3) \times (4.53 \times 10^4)$

3.86 [2nd] [EE] 3 [×] 4.53 [2nd] [EE] 4 [=] 174858000 Ans

or 3.86 [EXP] 3 [×] 4.53 [EXP] 4 [=] 174858000 Ans

The answer is displayed in standard form.

For calculations with answers that *exceed* the number of digits in the calculator display, the answer is displayed in scientific notation. Both calculators with the [EE] key or [EXP] key display the answer in scientific notation.

Examples Solve. $\dfrac{(-1.96 \times 10^7) \times (2.73 \times 10^5)}{8.09 \times 10^{-4}}$

1. Using the [EE] key:

   [(-)] 1.96 [2nd] [EE] 7 [×] 2.73 [2nd] [EE] 5 [÷] 8.09 [2nd] [EE] [(-)] 4 [=]

   $-6.614091471 \times 10^{15}$, $-6.614091471 \times 10^{15}$ Ans

2. Using the [EXP] key:

   [(-)] 1.96 [EXP] 7 [×] 2.73 [EXP] 5 [÷] 8.09 [EXP] [(-)] 4 [=]

   $-6.614091471 \times 10^{15}$, $-6.614091471 \times 10^{15}$ Ans

Some calculators can be set so that all results are displayed in scientific notation. On some calculators, this can be set by pressing [2nd] and then `SCI/ENG.` When these keys are pressed, the calculator displays something like that shown in Figure 36-1. The default setting for the calculator is `FLO` (Float). Press the [▶] until `SCI` is underlined and then press [ENTER]. To return to the default display, press [2nd] `SCI/ENG` and press the [◀] until `FLO` is underlined and press [ENTER].

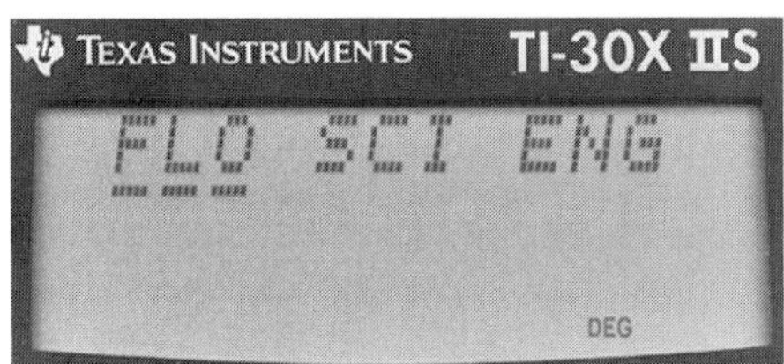

Figure 36-1

On some calculators, you first press [MODE] and then change to scientific notation as described above.

Application

Division of Single Terms

Divide the following terms as indicated.

1. $\dfrac{4x^2}{2x}$ ________
2. $\dfrac{-16a^4b^5}{4ab^3}$ ________
3. $\dfrac{FS^2}{-FS^2}$ ________
4. $\dfrac{-FS^2}{-FS^2}$ ________
5. $0 \div 14mn$ ________
6. $(-42a^5d^2) \div (-6a^2d^2)$ ________
7. $(-3.6H^2P) \div (0.6HP)$ ________
8. $DM^2 \div (-1)$ ________
9. $3.7ab \div ab$ ________
10. $0.8PV^2 \div (-0.2V)$ ________
11. $1\frac{1}{4}c^2d^3 \div \frac{1}{4}cd^2$ ________
12. $\left(-\frac{1}{3}x^3y^3\right) \div \frac{1}{9}x^3$ ________
13. $-6g^3h^2 \div \left(-\frac{3}{4}gh\right)$ ________
14. $-24x^2y^5 \div (-0.5x^2y^4)$ ________
15. $x^2y^3z^4 \div xy^3z^4$ ________
16. $18a^2bc^2y \div (-a^2)$ ________
17. $0.25P^2V \div 0.0625$ ________
18. $-0.08xy \div 0.02y$ ________
19. $-\frac{3}{4}FS^3 \div (-3S)$ ________
20. $-9.6x^2yz \div (-1.2x)$ ________

Division of Expressions with Two or More Terms in the Dividend

Divide the following expressions as indicated.

21. $(8x^3 + 12x^2) \div x$
22. $(12x^3y^3 - 8x^2y^2) \div 4xy$
23. $(9x^6y^3 - 6x^2y^5) \div (-3xy^2)$
24. $(2x - 4y) \div 4$
25. $(15a^2 + 25a^5) \div (-a)$
26. $(-18a^2b^7 - 12a^5c^5) \div (-6a^2b^5)$
27. $(14cd - 35c^2d - 7) \div (-7)$
28. $(0.8x^5y^6 + 0.2x^4y^7) \div (2x^2y^4)$
29. $(-0.9a^2x - 0.3ax^2 + 0.6) \div (-0.3)$
30. $(5y^2 - 25xy^2 - 10y^4) \div 5y^2$
31. $\left(\frac{1}{2}a^2c - \frac{3}{4}a^3c^2 - ac^3\right) \div \frac{1}{8}ac$
32. $(-2.5e^2f - 0.5ef^2 + e^2f^2) \div 0.5f$

Powers of Single Terms

Raise the following terms to indicated powers.

33. $(3ab)^2$
34. $(-4xy)^3$
35. $(2x^2y)^3$
36. $(4a^4b^3)^2$
37. $(-3c^3d^2e^4)^3$
38. $(2MS^2)^2$
39. $(-7x^4y^5)^2$
40. $(-3N^2P^2T^3)^4$
41. $(a^3bc^2)^3$
42. $(-2a^2bc^3)^3$
43. $(-x^4y^5z)^3$
44. $(8C^3FH^2)^2$
45. $(0.4x^3y)^3$
46. $(-0.5c^2d^3e)^3$
47. $(4.3M^2N^2P)^2$
48. $\left(\frac{3}{4}abc^3\right)^3$
49. $[-8(a^2b^3)^2c]^3$
50. $[-3x^2(y^2)^2z^3]^3$
51. $[0.6d^3(ef^2)^3]^2$
52. $[(-2x^2y)^2(xy^2)^2]^3$

Powers of Expressions of Two or More Terms

Raise the following terms to the indicated powers and combine like terms where possible.

53. $(3x^2 - 5y^3)^2$
54. $(a^4 + b^3)^2$
55. $(5t^2 - 6x)^2$
56. $(a^2b^3 + ab^3)^2$
57. $(0.4d^2t^3 - 0.2t)^2$
58. $(-0.2x^2y - y^4)^2$
59. $\left(\frac{2}{3}c^2d + \frac{3}{4}cd^2\right)^2$
60. $[(x^2)^3 - (y^3)^2]^2$
61. $[(-a^4b)^2 + (x^2y)^3]^2$

Roots

Determine the roots of the following terms.

62. $\sqrt{16c^2d^6}$
63. $\sqrt{m^6n^4s^2}$
64. $\sqrt[3]{64x^3y^9}$
65. $\sqrt{81x^8y^6}$
66. $\sqrt[3]{p^9t^6w^3}$
67. $\sqrt[3]{-27x^6y^{12}}$
68. $\sqrt{0.25h^4y^2}$
69. $\sqrt{0.16a^8c^2f^6}$
70. $\sqrt{\frac{4}{9}a^2b^4c^6}$
71. $\sqrt{\frac{1}{16}x^2y^2}$

72. $\sqrt[3]{\frac{8}{27}m^6n^3}$ ______

73. $\sqrt[3]{-64d^6t^9}$ ______

74. $\sqrt[4]{16x^4y^8}$ ______

75. $\sqrt[5]{32h^{10}}$ ______

76. $\sqrt{25ab^2}$ ______

77. $\sqrt[3]{64a^3c}$ ______

78. $\sqrt[3]{-\frac{1}{64}x^3y^6z^2}$ ______

79. $\sqrt{\frac{9}{16}a^2bc^2}$ ______

80. $\sqrt[3]{27d^3e^6f^2}$ ______

81. $\sqrt[5]{-32a^5b^3}$ ______

Removal of Parentheses

Simplify.

82. $6a + (3a - 2a^2 + a^3)$ ______

83. $9b - (15b^2 - c + d)$ ______

84. $15 + (x^2 - 10)$ ______

85. $-(ab + a^2b - a)$ ______

86. $-10c^3 - (-8c^3 - d + 12)$ ______

87. $-(16 + xy - x) + (-x)$ ______

88. $-25a^2b - (-2a^2b - a + b^2)$ ______

89. $15 - (r^2 + r) + (r^2 - 14)$ ______

90. $-(a^2 + b^2) + (a^2 + b^2)$ ______

91. $-(3x + xy - 6) + 18 + (x + xy)$ ______

92. $20 + (cd - c^2d + d) + 14 - (cd + d)$ ______

93. $20 - (cd - c^2d + d) - 14 + (cd + d)$ ______

Combined Operations

Simplify the following expressions.

94. $15 - 2(3xy)^2 + x^2y^2 - 8$ ______

95. $5(a^2 - b) + a^2 - b$ ______

96. $(2 - c^2)(2 + c^2) + 2c$ ______

97. $\frac{ab}{a} - \left(\frac{-a^2b}{a^2} - \frac{a^3b}{a^3}\right)$ ______

98. $\frac{4 - 8x + 16x^2}{2} + \frac{3x^4}{x^2}$ ______

99. $\frac{16xy^8}{2xy^2} - (y^2)^3 + 15$ ______

100. $\frac{\sqrt{25x^2}}{-5}(3xy^3) - (-10)$ ______

101. $\sqrt{\frac{64d^6}{9}} \div d^2$ ______

102. $\frac{12x^6 + 6x^4y}{(2x)^2} - (16x^4y^2)^{1/2}$ ______

103. $-5a(-8 + (ab^2)^3 - 12)$ ______

104. $5a[-6 + (ab^2)^3 - 10]$ ______

105. $(10f^6 + 12f^4h) \div \sqrt{4f^4}$ ______

Rewriting Numbers in Scientific Notation

Rewrite the following standard form numbers in scientific notation.

106. 625 ______

107. 80,000 ______

108. 1,320,000 ______

109. 976,000 ______

110. 0.0073 ______

111. 0.015 ______

112. 0.00004 ______

113. 0.2 ______

114. 39 ______

115. 0.00039 ______

116. 175,000 ______

117. 0.00175 ______

Rewriting Scientific Notation Values

Rewrite the following scientific notation values in standard decimal form.

118. 3×10^3 ________
119. 1.6×10^5 ________
120. 8.5×10^2 ________
121. 5.09×10^6 ________
122. 4.7×10^{-1} ________
123. 6.32×10^{-5} ________
124. 1.05×10^{-3} ________
125. 3.123×10^{-6} ________
126. 7.312×10^4 ________
127. 7.321×10^{-4} ________
128. 2.09×10^6 ________
129. 2.09×10^{-2} ________

Multiplying and Dividing in Scientific Notation

The following problems are given in scientific notation. Solve and leave answers in scientific notation. Round the answers (mantissas) to 2 decimal places.

130. $(2.50 \times 10^3) \times (5.10 \times 10^5)$ ________
131. $(3.10 \times 10^{-3}) \times (5.20 \times 10^{-4})$ ________
132. $(-7.60 \times 10^4) \times (1.90 \times 10^5)$ ________
133. $(2.43 \times 10^{-6}) \div (7.60 \times 10^3)$ ________
134. $(8.51 \times 10^7) \div (6.30 \times 10^{-5})$ ________
135. $\dfrac{(1.25 \times 10^4) \times (6.30 \times 10^5)}{(7.83 \times 10^3)}$ ________
136. $\dfrac{(8.76 \times 10^{-5}) \times (1.05 \times 10^9)}{(6.37 \times 10^3)}$ ________
137. $\dfrac{(5.50 \times 10^4) \times (-6.00 \times 10^6)}{(6.92 \times 10^{-3})}$ ________
138. $\dfrac{(8.46 \times 10^{-5})}{(3.90 \times 10^7) \times (6.77 \times 10^{-3})}$ ________

The following problems are given in decimal (standard) form. Calculate and give answers in scientific notation. Round the answers (mantissas) to 2 decimal places.

139. $1510 \times 30{,}500$ ________
140. 0.000300×0.00210 ________
141. $-56{,}100 \times 781{,}000$ ________
142. $61{,}770 \times 53{,}100$ ________
143. $0.0000821 \div -315$ ________
144. $\dfrac{-0.00623 \times 742{,}000}{651{,}000}$ ________
145. $\dfrac{65{,}300 \times 517{,}000}{0.00786}$ ________
146. $\dfrac{-0.000829}{405{,}000 \times 0.00312}$ ________
147. $\dfrac{518{,}000 \times 0.00612}{37{,}400 \times 0.0000830}$ ________

148. The amount of expansion of metal when heated is computed as follows:

Expansion = original length × linear expansion per unit of length per degree Fahrenheit × temperature change.

Calculate the amount of expansion for the metals shown in the table. Give the answers in decimal (standard) form to 3 decimal places.

	Metal	Original Length of Metal	Linear Expansion Per Unit Length Per Degree Fahrenheit	Original Temperature	Temperature to Which Heated
a.	Aluminum	6.7520 in	1.244×10^{-5}	68.0°F	225.0°F
b.	Copper	35.750 ft	9.000×10^{-6}	35.0°F	97.0°F
c.	Carbon Steel	3.0950 in	6.330×10^{-6}	84.0°F	743.0°F

a. ________
b. ________
c. ________

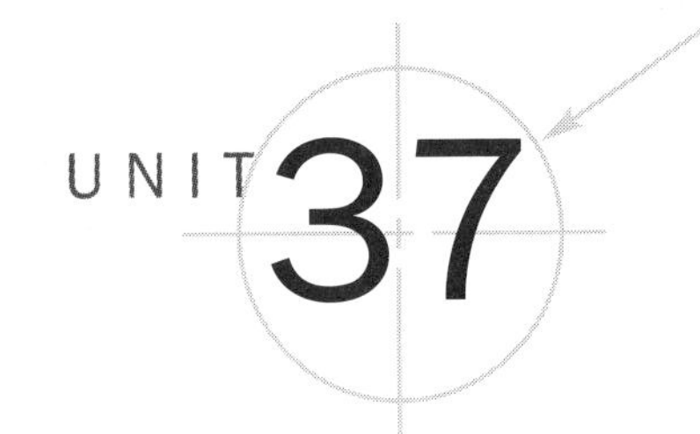

Introduction to Equations

Objectives ***After studying this unit you should be able to***

- Express word problems as equations.
- Express problems given in graphic form as equations.
- Solve simple equations using logical reasoning.

It is essential that the skilled machine technician understand equations and their applications. The solution of equations is required to compute problems using trade handbook formulas. Often machine shop problems are solved using a combination of equations with elements of geometry and trigonometry.

Expression of Equality

An *equation* is a mathematical statement of equality between two or more quantities and always contains the equal sign (=). The value of all quantities on the left side of the equal sign equals the value of all quantities on the right side of the equal sign. A *formula* is a particular type of an equation that states a mathematical rule.

The following are examples of simple equations:

$$7 + 2 = 5 + 4 \qquad \frac{12}{3} + 2 \times 5 = 18 - 4$$

$$3 \times 5\frac{1''}{2} = 16\frac{1''}{2} \qquad 360° = 5 \times 80° - 40°$$

$$a + b = c + d \qquad \frac{xy}{2} = x + y$$

Because it expresses the equality of the quantities on the left and on the right of the equal sign, an equation is a balanced mathematical statement. An equation may be considered similar to a balanced scale as illustrated in Figure 37-1(A). The total weight on the left side of the scale equals the total weight on the right side; therefore, the scale balances.

$$3 \text{ pounds} + 5 \text{ pounds} + 2 \text{ pounds} = 4 \text{ pounds} + 6 \text{ pounds}$$
$$10 \text{ pounds} = 10 \text{ pounds}$$

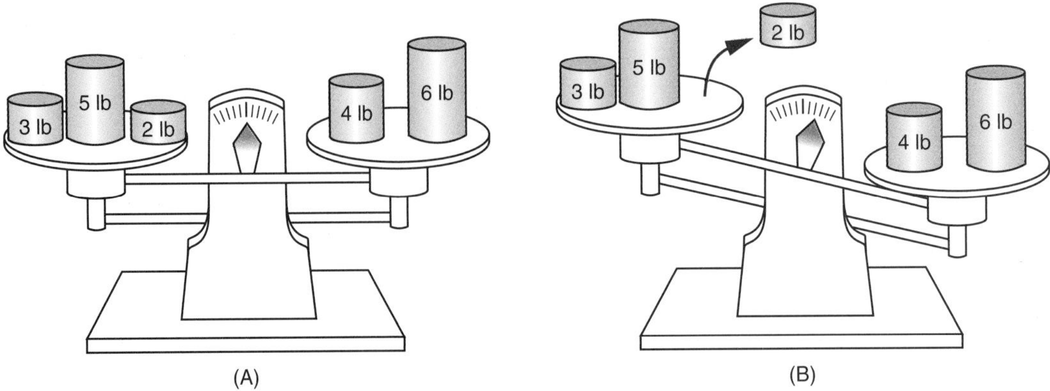

Figure 37-1

When the 2-pound weight is removed from the scale, the scale is no longer in balance as illustrated in Figure 37-1(B).

$$3 \text{ pounds} + 5 \text{ pounds} \neq 4 \text{ pounds} + 6 \text{ pounds}$$
$$8 \text{ pounds} \neq 10 \text{ pounds}$$

The Unknown Quantity

In general, an equation is used to determine the numerical value of an unknown quantity. Although any letter or symbol can be used to represent the unknown quantity, the letter x is commonly used.

The first letter of the unknown quantity is often used to represent a quantity. Some common letter designations are

L	to represent length	P	to represent pressure
A	to represent area	F	to represent feed of cutter
t	to represent time	W	to represent weight
D	to represent diameter	h	to represent height

Writing Equations from Word Statements

An equation asks a question. It asks for the value of the unknown, which makes the left side of the equation equal to the right side. The question asked may not be in equation form; instead it may be expressed in words.

It is important to develop the ability to express word statements as mathematical symbols, or equations. A problem must be fully understood before it can be written as an equation.

Whether the word problem is simple or complex, a definite logical procedure should be followed to analyze the problem. A few or all of the following steps may be required, depending on the complexity of the particular problem.

- Carefully read the entire problem, several times if necessary.
- Break the problem down into simpler parts.
- It is sometimes helpful to draw a simple picture as an aid in visualizing the various parts of the problem.
- Identify and list the unknowns. Give each unknown a letter name, such as x.
- Decide where the equal sign should be, and group the parts of the problem on the proper side of the equal sign.
- Check. Are the statements on the left equal to the statements on the right of the equal sign?
- After writing the equation, check it against the original problem, step-by-step. Does the equation state mathematically what the problem states in words?

The following examples illustrate the method of writing equations from given word statements. After each equation is written the value of the unknown quantity is obtained. No specific procedures are given at this time in solving for the unknowns. The unknown quantity values are determined by logical reasoning.

Example 1 What weight must be added to a 12-pound weight so that it will be in balance with a 20-pound weight?

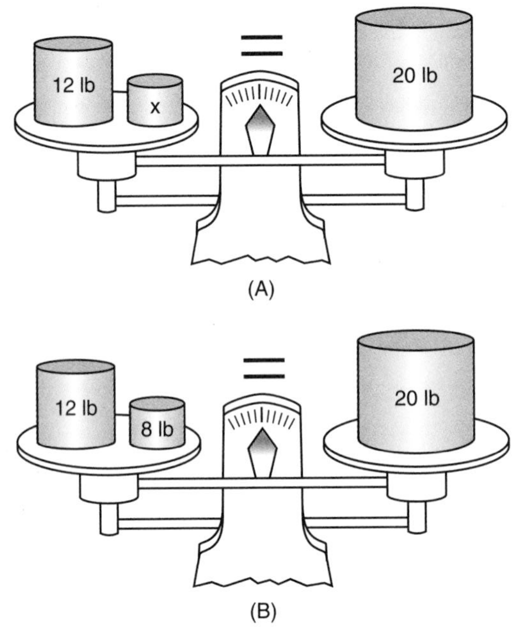

Figure 37-2

Ask the question: 12 pounds + what weight = 20 pounds?

To help visualize the problem, a picture is shown in Figure 37-2(A).

Identify the unknowns.

Let x represent the unknown weight.

Write the equation. $12 \text{ lb} + x = 20 \text{ lb}$

Ask the question: What number added to 12 pounds equals 20 pounds? Since 8 pounds added to 12 pounds equals 20 pounds, $x = 8 \text{ lb}$ Ans

Check the answer by substituting 8 pounds for x in the original equation.

$$12 \text{ lb} + 8 \text{ lb} = 20 \text{ lb}$$
$$20 \text{ lb} = 20 \text{ lb} \quad \text{Ck}$$

The equation is balanced as shown in Figure 37-2(B) since the left side of the equation equals the right side.

Example 2 A $9\frac{1}{2}$-inch piece is cut from a 12-inch length of bar stock. Find the length of the unused piece. Make no allowance for thickness of the cut.

Ask the question: What number subtracted from 12 inches = $9\frac{1}{2}$ inches?

A picture of the problem is shown in Figure 37-3. All dimensions are in inches. Let x represent the number of inches cut off.

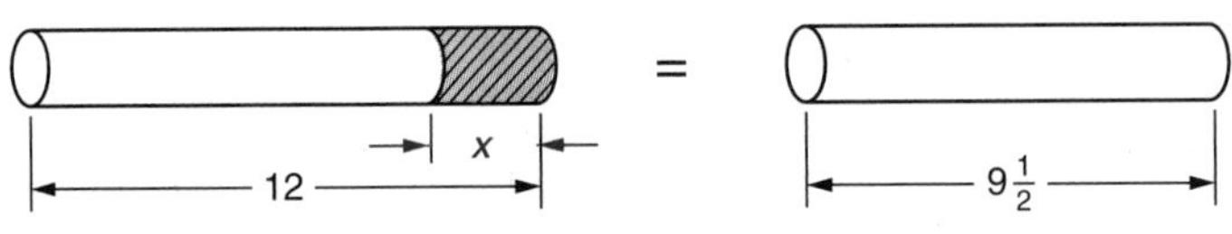

Figure 37-3

Express the problem as an equation. $12'' - x = 9\frac{1}{2}''$

Since $2\frac{1}{2}$ inches subtracted from 12 inches is equal to $9\frac{1}{2}$ inches, as shown in Figure 37-4, $x = 2\frac{1}{2}''$ Ans

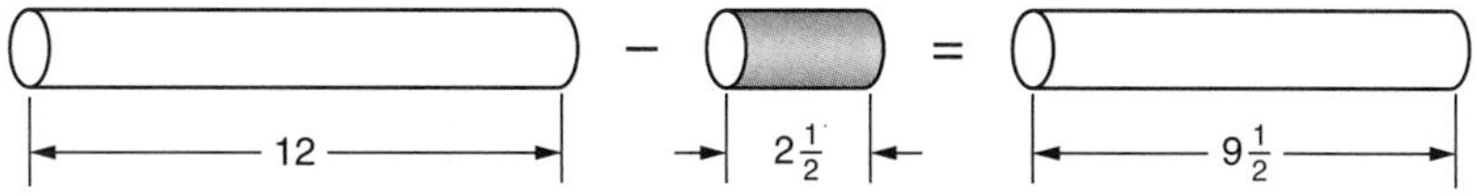

12 − $2\frac{1}{2}$ = $9\frac{1}{2}$

Figure 37-4

Check the answer by substituting $2\frac{1}{2}$ inches for x in the original equation.

$$12'' - 2\frac{1}{2}'' = 9\frac{1}{2}''$$

$$9\frac{1}{2}'' = 9\frac{1}{2}'' \quad \text{Ck}$$

The equation is balanced.

Example 3 The sum of two angles equals 90°. One angle is twice as large as the other. What is the size of the smaller angle?

An angle + an angle twice as large = 90°.

A picture of the problem is shown in Figure 37-5.

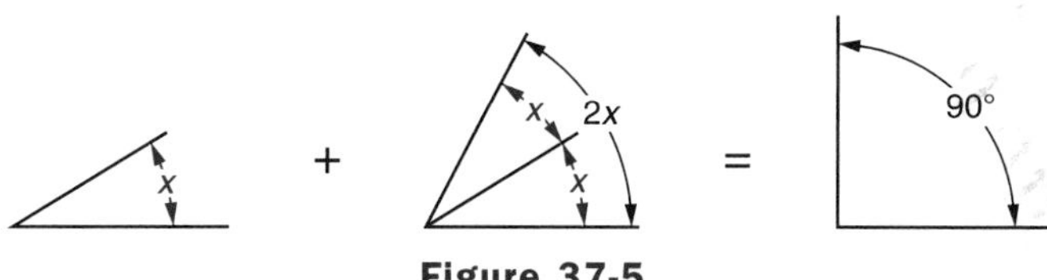

Figure 37-5

Let x represent the smaller angle. Let $2x$ represent the larger angle.

Express the problem as an equation. $x + 2x = 90°$ or $3x = 90°$

Ask the question: What number multiplied by 3 = 90°?

Since 3 multiplied by 30° = 90°, $x = 30°$

The smaller angle is $x = 30°$ Ans

The larger angle is $2x = 60°$ Ans

Figure 37-6 shows the sizes of both angles and also shows the solution.

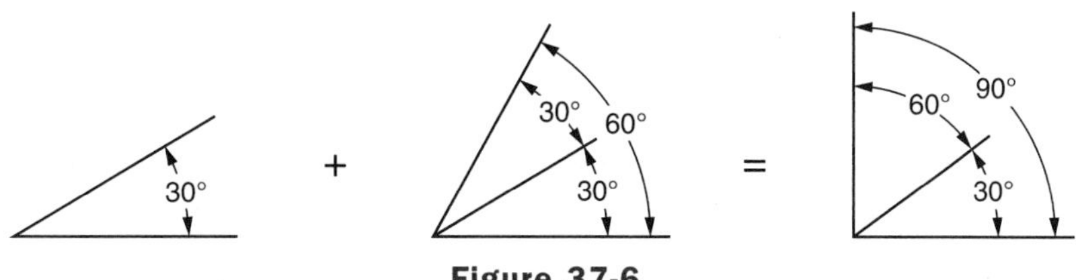

Figure 37-6

Check the answer by substituting 30° for x in the original equation.

$$30° + 2(30°) = 90°$$
$$90° = 90° \quad \text{Ck}$$

The equation is balanced.

Example 4 Three gage blocks are used to tilt a sine plate. The total height of the three blocks is 2.75 inches. The bottom block is 4 times as thick as the middle block. The middle block is twice as thick as the top block. How thick is each block?

Convert the problem from word form to equation form.

Let x represent the thickness of the thinnest block, the top block.

The middle block is twice as thick as the top block, or $2x$.

The bottom block is four times as thick as the middle block, or $(4)(2x) = 8x$.

The sum of the three blocks = 2.75".

Therefore, $x + 2x + 8x = 2.75"$, or $11x = 2.75"$.

A picture of the problem is shown in Figure 37-7. All dimensions are in inches.

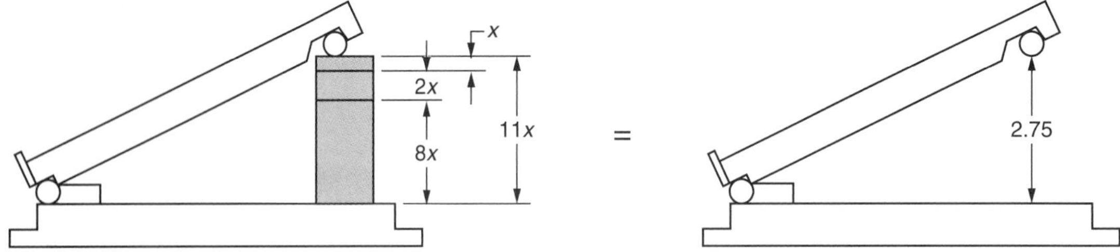

Figure 37-7

Ask the question: What number multiplied by 11 = 2.75"?
Since 11 × 0.25" = 2.75", $x = 0.25"$.

The top block is x or 0.25". Ans

The middle block is $2x$ or 2(0.25") = 0.50". Ans

The bottom block is $8x$ or 8(0.25") = 2.00". Ans

The thickness of each block is shown in Figure 37-8. All dimensions are in inches.

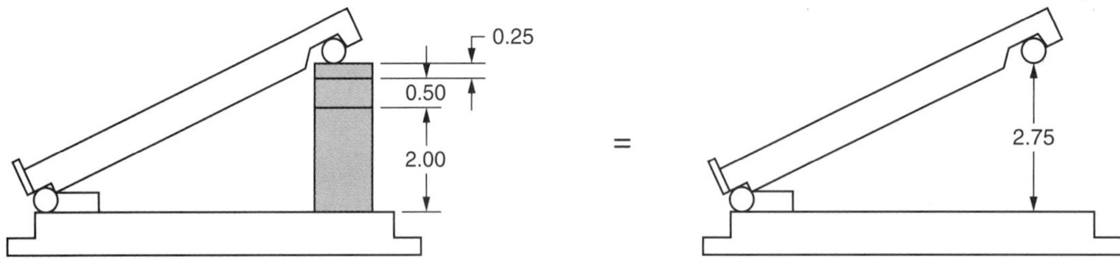

Figure 37-8

Check the answer by substituting 0.25" for x in the original equation:

$$x + 2x + 4(2x) = 2.75"$$

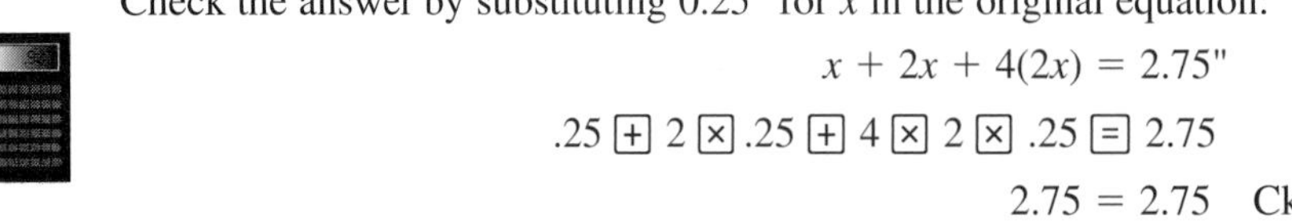

$$2.75 = 2.75 \quad \text{Ck}$$

The equation is balanced.

In many cases the problems to be solved in actual machine shop applications will be more difficult than the preceding examples. It is essential, therefore, to be able to use the procedure shown to analyze the problem, determine the unknowns, and set up the equation. If the solution to a problem is a rounded value, the check may result in a very small difference between both sides.

Checking the Equation

In the final step in each of the preceding examples, the value found for the unknown was substituted in the original equation to prove that it was the correct value. If an equation is properly written and if both sides of the equation are equal, the equation is balanced and the solution is correct.

All work in a machine shop should be checked and rechecked to prevent errors. It is important that you check your computations. When working with equations on the job, checking your work is essential. Errors in computation can often be costly in terms of time, labor, and materials.

Application

Express each of the following word problems as equations. Let the unknown number equal x and by logical reasoning solve for the value of the unknown. Check the equation by comparing it to the word problem. Does the equation state mathematically what the problem states in words? Check whether the equation is balanced by substituting the value of the unknown in the equation.

1. A number plus 20 equals 32. Find the number. ______
2. A number less 7 equals 15. Find the number. ______
3. Five times a number equals 55. Find the number. ______
4. A number divided by 4 equals 9. Find the number. ______
5. Thirty-two divided by a number equals 8. Find the number. ______
6. A number plus twice the number equals 36. Find the number. ______
7. Five times a number minus the number equals 48. Find the number. ______
8. Seven times a number plus eight times the number equals 60. Find the number. ______
9. Sixty divided by the product of 3 and a number equals 4. Find the number. ______
10. A piece of bar stock 32 inches long is cut into two unequal lengths. One piece is 3 times as long as the other. How long is each piece? ______
11. Three blocks are used to tilt a sine plate. The total height of the three blocks is 4.5 inches. The first block is 3 times as thick as the second block. The second block is twice as thick as the third block. How thick is each block? ______
12. Five holes are drilled in a steel plate on a bolt circle as shown in Figure 37-9. There are 300° between hole 1 and hole 5. The number of degrees between any two consecutive holes doubles in going from hole 1 to hole 5. Find the number of degrees between the indicated holes.

    a. 1 and 2 ______
    b. 2 and 3 ______
    c. 3 and 4 ______
    d. 4 and 5 ______

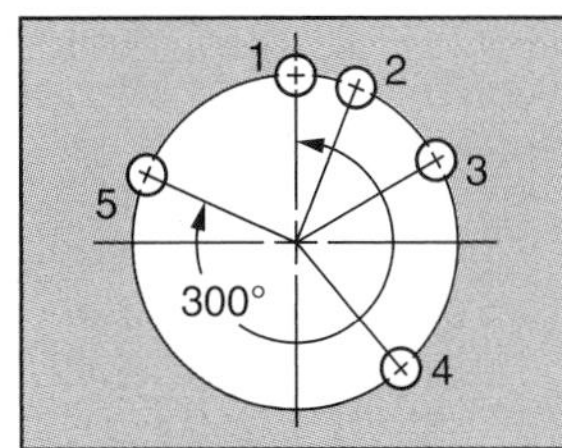

Figure 37-9

13. The total amount of stock milled off an aluminum casting in two cuts is 8.58 millimeters. The roughing cut is 6.35 millimeters greater than the finish cut. What is the depth of the finish cut? ______

In each of the following problems, refer to the corresponding figure. Write an equation, solve for x, and check.

14. All dimensions are in inches.

______ $x =$ ______

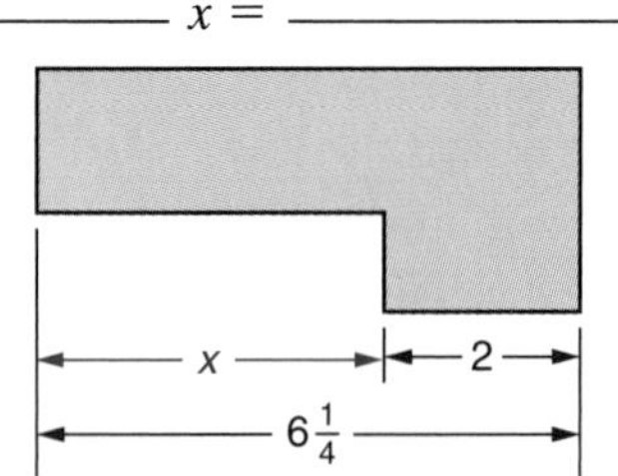

15. All dimensions are in millimeters.

______ $x =$ ______

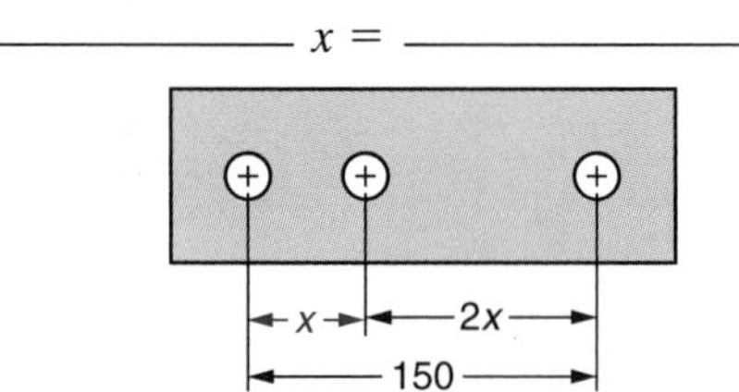

16. All dimensions are in inches.

______ $x =$ ______

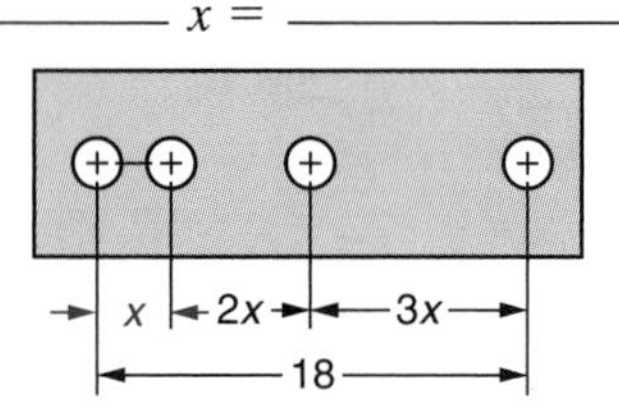

17. ______ $x =$ ______

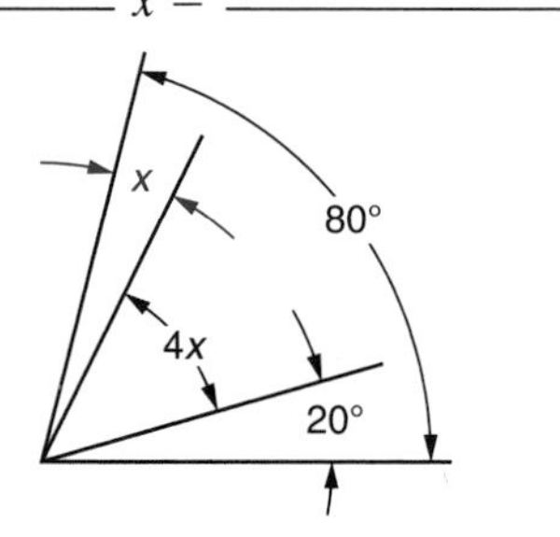

18. All dimensions are in inches.

______ $x =$ ______

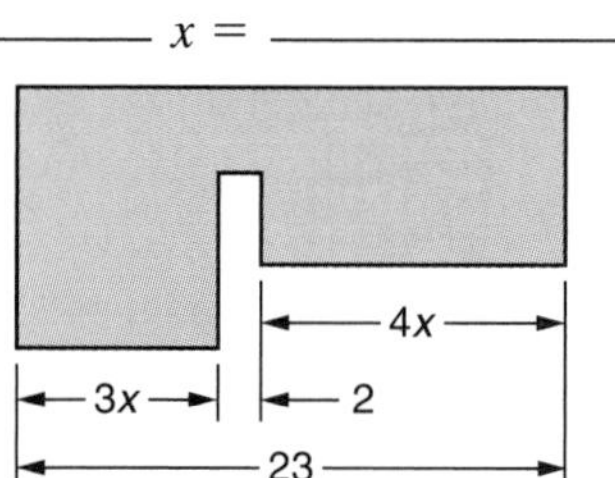

19. ______ $x =$ ______

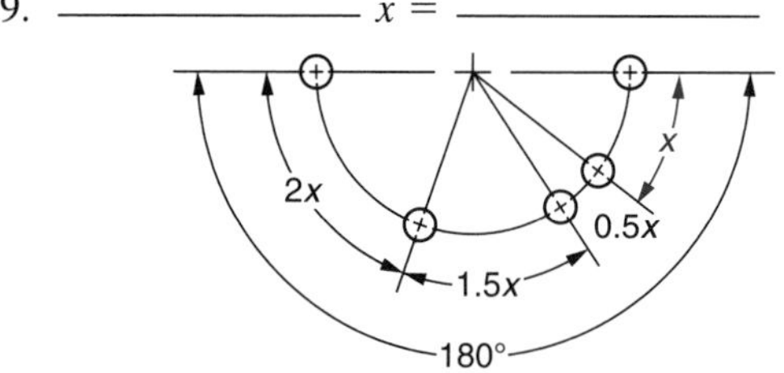

20. All dimensions are in millimeters.

______ $x =$ ______

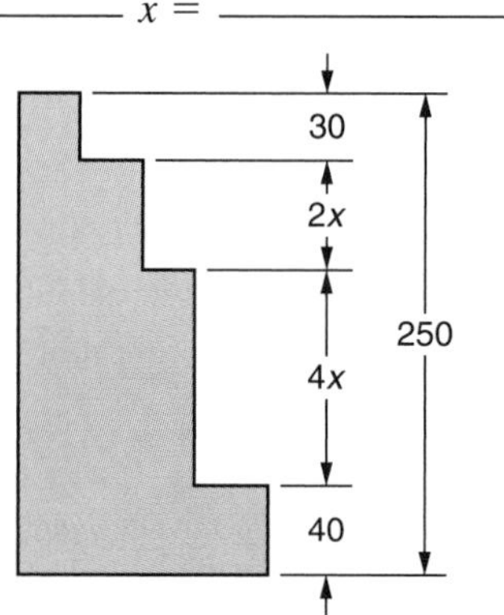

21. All dimensions are in inches.

______ $x =$ ______

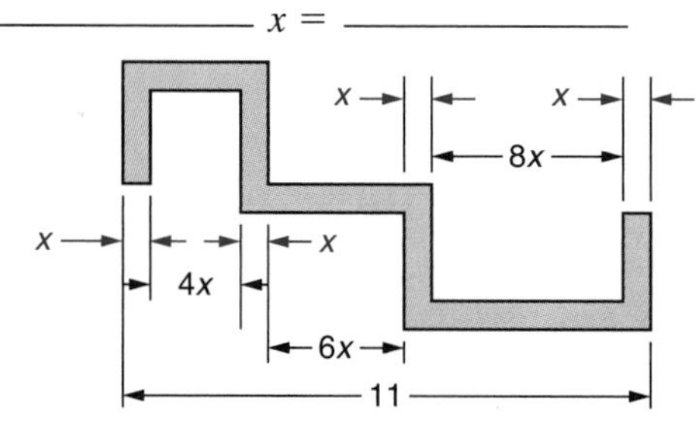

For each of the following problems, refer to the given figure, solve for the unknowns, and check.

22. Find the distances between the indicated holes. All dimensions are in millimeters.

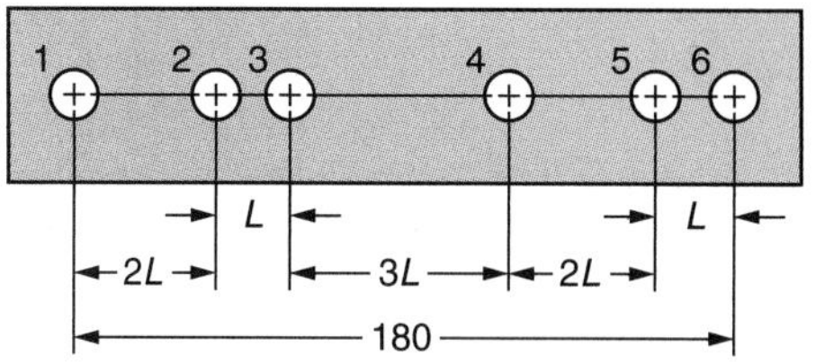

a. Hole 1 to Hole 2 __________

b. Hole 2 to Hole 3 __________

c. Hole 3 to Hole 4 __________

d. Hole 4 to Hole 5 __________

e. Hole 5 to Hole 6 __________

f. Hole 2 to Hole 4 __________

g. Hole 3 to Hole 6 __________

23. Find the distances between the indicated points. All dimensions are in inches.

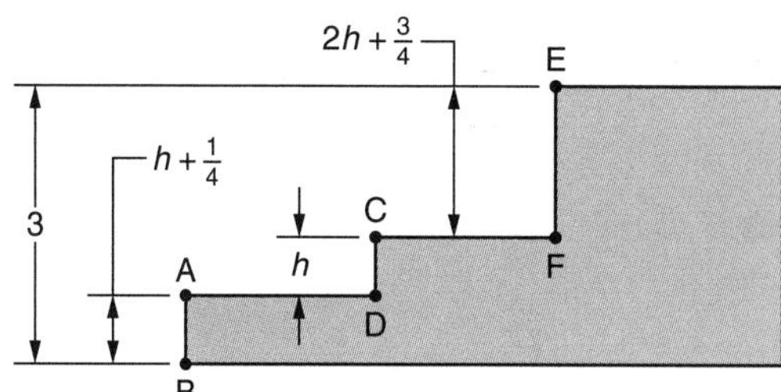

a. A and B __________

b. C and D __________

c. E and F __________

24. Find the value of each of the four angles.

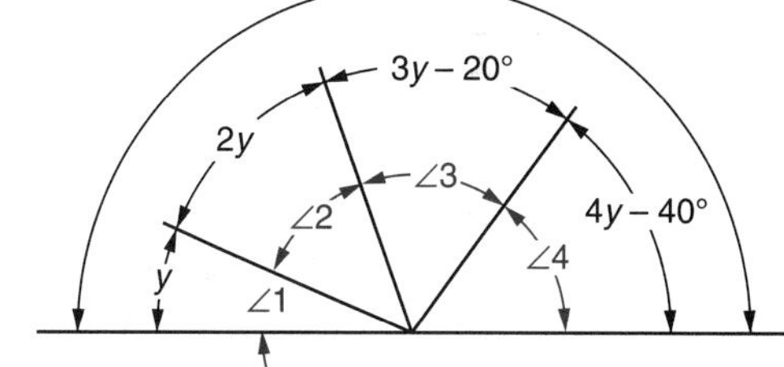

a. $\angle 1$ __________

b. $\angle 2$ __________

c. $\angle 3$ __________

d. $\angle 4$ __________

Solve for the unknown values in the following equations.

25. $x + 9x = 30$ __________

26. $x + 3 = 12$ __________

27. $2y + 5y + 3y = 70$ __________

28. $32 = 17 + y$ __________

29. $18 - a = 12$ __________

30. $b - 13 = 80$ __________

31. $3b + 5b - 2b = 96$ __________

32. $3(5a) = \frac{6(30)}{2}$ __________

33. $\frac{1}{2}x = 42$ __________

34. $\frac{x}{4} = 15$ __________

35. $\frac{27}{x} = 9$ __________

36. $\frac{d}{6} + 4 = 9$ __________

37. $0.75x - 0.5x = \frac{18 + 30}{4}$ __________

38. $6(2.5x) + 5x = 80$ __________

39. $\frac{2y + 4y + 6y}{3} = 80$ __________

40. $27 - (3)(6) = b + 3$ __________

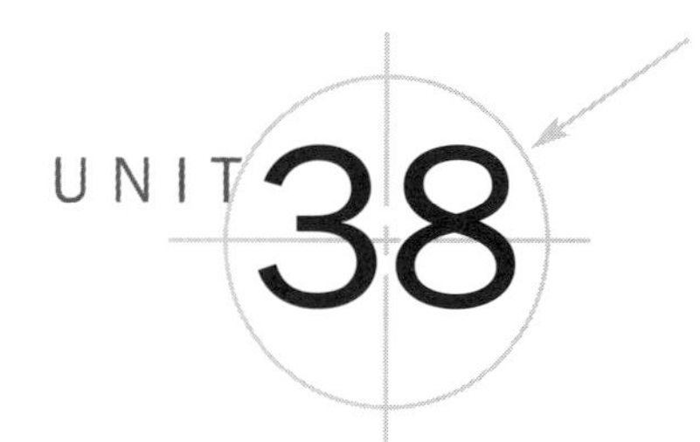

Solution of Equations by the Subtraction, Addition, and Division Principles of Equality

Objectives *After studying this unit you should be able to*

- Solve equations using the subtraction principle of equality.
- Solve equations using the addition principle of equality.
- Solve equations using the division principle of equality.
- Solve equations using transposition.

Principles of Equality

In actual practice, equations cannot usually be solved by inspection or common sense. There are specific procedures for solving equations using the fundamental principles of equality. The principles of equality that will be presented in this unit are those of subtraction, addition, and division. Multiplication, root, and power principles are presented in Unit 39.

Solution of Equations by the Subtraction Principle of Equality

The subtraction principle of equality states that if the same number is subtracted from both sides of an equation, the sides remain equal, and the equation remains balanced. The subtraction principle is used to solve an equation in which a number is added to the unknown, such as $x + 15 = 20$.

The values on each side of an equation are equal and an equation is balanced. If the same value is subtracted from both sides, the equation remains balanced. The equation 8 pounds + 4 pounds = 12 pounds is pictured in Figure 38-1(A). If 4 pounds are removed from the left side only, the scale is not in balance as shown in Figure 38-1(B). If 4 pounds are removed from both the left and right sides, the scale remains in balance as in Figure 38-1(C).

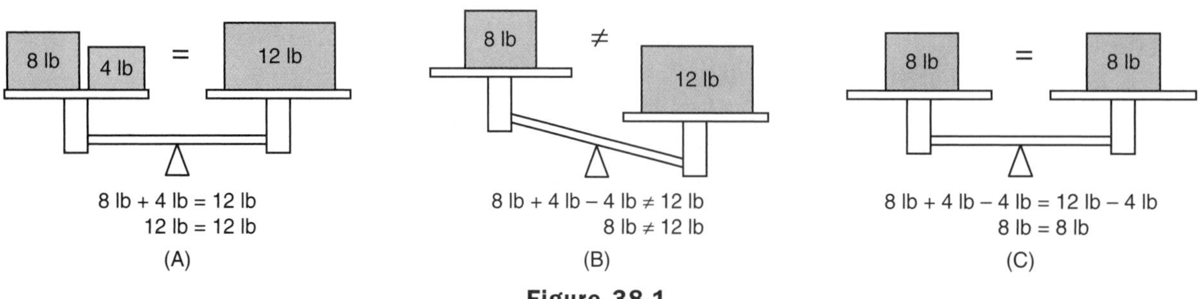

Figure 38-1

Procedure To solve an equation in which a number is added to the unknown

- Subtract the number that is added to the unknown from both sides of the equation.
- Check.

Examples

1. $x + 4 = 9$. Solve for x.

   In the equation, the number 4 is added to x as in Figure 38-2(A). To solve, subtract 4 from both sides of the equation.

$$\begin{array}{r} x + 4 = 9 \\ \underline{\quad -4 = -4} \\ x = 5 \quad \text{Ans} \end{array}$$

   Check.

$$\begin{array}{r} x + 4 = 9 \\ 5 + 4 = 9 \\ 9 = 9 \quad \text{Ck} \end{array}$$

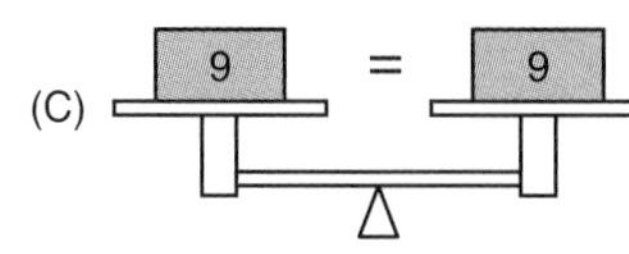

Figure 38-2

2. In the part shown in Figure 38-3, determine dimension y. All dimensions are in inches.

   Write an equation.

$$5.5'' + y = 17''$$

   Subtract 5.5" from both sides.

$$\begin{array}{r} 5.5'' + y = 17'' \\ \underline{-5.5'' \qquad = -5.5''} \\ y = 11.5'' \quad \text{Ans} \end{array}$$

   Check.

$$\begin{array}{r} 5.5'' + y = 17'' \\ 5.5'' + 11.5'' = 17'' \\ 17'' = 17'' \quad \text{Ck} \end{array}$$

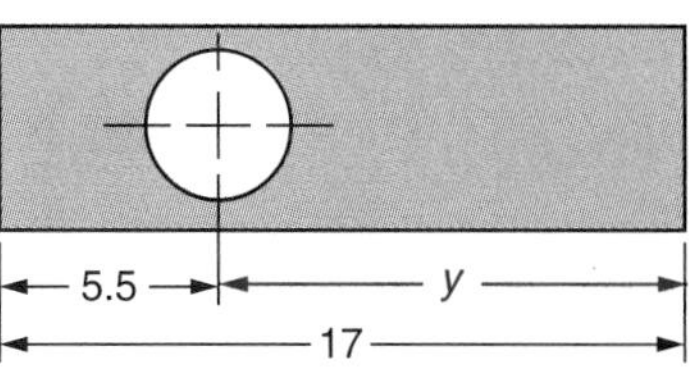

Figure 38-3

3. $-39 = P + 18$. Solve for P.

$$\begin{array}{r} -39 = P + 18 \\ \underline{-18 = \quad -18} \\ -57 = P \quad \text{Ans} \end{array}$$

   Check.

$$\begin{array}{r} -39 = P + 18 \\ -39 = -57 + 18 \\ -39 = -39 \quad \text{Ck} \end{array}$$

4. $W + 4\frac{3}{4} = 12$. Solve for W.

$$\begin{array}{r} \underline{-4\frac{3}{4} = -4\frac{3}{4}} \\ W = 7\frac{1}{4} \quad \text{Ans} \end{array}$$

   Check.

$$\begin{array}{r} W + 4\frac{3}{4} = 12 \\ 7\frac{1}{4} + 4\frac{3}{4} = 12 \\ 12 = 12 \quad \text{Ck} \end{array}$$

Transposition

With your instructor's permission, an alternate method of solving certain equations may be used. The alternate method is called transposition. *Transposition* or transposing a term means that a term is moved from one side of an equation to the opposite side with the sign changed.

Transposition is not a mathematical process although it is based on the addition and subtraction principles of equality. Transposition should only be used after the principles of equality are fully understood and applied.

Transposition is a quick and convenient means of solving equations in which a term is added to or subtracted from the unknown. The purpose of using transposition is the same as that of using the addition and subtraction principles of equality. Both methods involve getting the unknown term to stand alone on one side of the equation in order to determine the value of the unknown.

The following example is solved by applying the subtraction principle of equality and transposition. Notice that when applying the subtraction principle of equality, a term is eliminated on one side of the equation and appears on the other side with the sign changed.

Example 1 Solve for x.

$$x + 15 = 25$$

Method 1: The Subtraction Principle of Equality

$x + 15 = 25$
$\quad - 15 = -15$
$x = 25 - 15$
$x = 10$ Ans

← Observe that $+15$ is eliminated from the left side of the equation and appears as -15 on the right side.

Method 2: Transpostion

$x + 15 = 25$
$x + 15 = 25 - 15$
$x = 25 - 15$
$x = 10$ Ans

← Observe that this expression is identical to the expression obtained when applying the subtraction principle of equality.

The following examples are solved by transposition.

Examples

1. $y + 10.7 = 18$. Solve for y.

   Move $+10.7$ from the left side of the equation to the right side and change to -10.7.

   $y + 10.7 = 18$
   $y = 18 - 10.7$
   $y = 7.3$ Ans

2. $T + 6\frac{1}{8} = -19$. Solve for T.

   Move $+6\frac{1}{8}$ from the left side of the equation to the right side and change to $-6\frac{1}{8}$.

   $T + 6\frac{1}{8} = -19$
   $T = -19 - 6\frac{1}{8}$
   $T = -25\frac{1}{8}$ Ans

Solution of Equations by the Addition Principle of Equality

The addition principle of equality states that if the same number is added to both sides of an equation, the sides remain equal and the equation remains balanced. The addition principle is used to solve an equation in which a number is subtracted from the unknown, such as $x - 17 = 30$.

Procedure To solve an equation in which a number is subtracted from the unknown

- Add the number that is subtracted from the unknown to both sides of the equation.
- Check.

Examples

1. $x - 6 = 15$. Solve for x.

   In the equation, the number 6 is subtracted from the x.

   To solve, add 6 to both sides of the equation.

   $x - 6 = 15$
   $\quad + 6 = +6$
   $x = 21$ Ans

   Check. $x - 6 = 15$
   $21 - 6 = 15$
   $15 = 15$ Ck

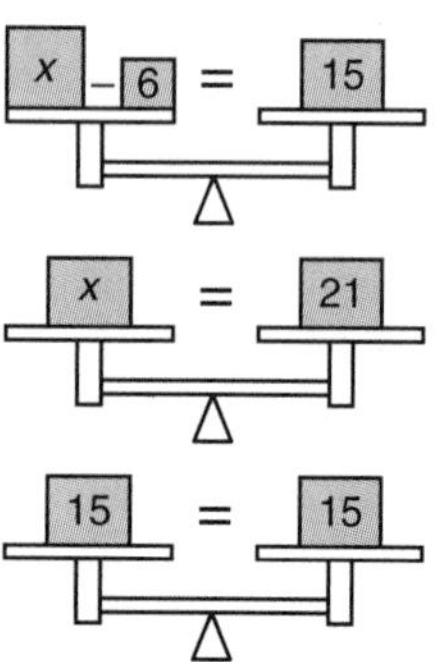

Figure 38-4

2. A 7-inch piece is cut from the height of a block as shown in Figure 38-5. The remaining block is 10 inches high. What is the height of the original block? All dimensions are in inches. Make no allowance for thickness of cut.

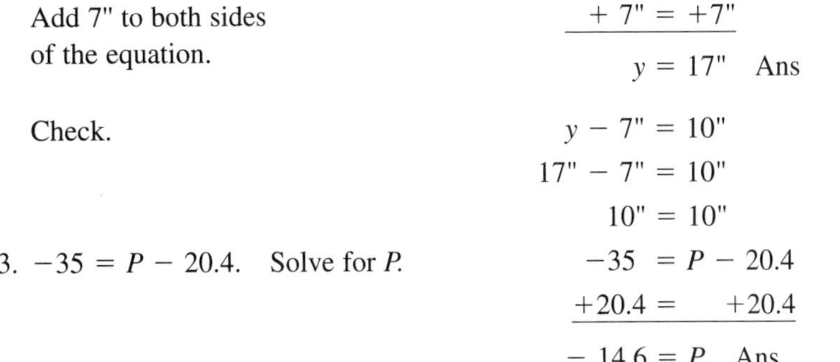

Let y = the height of the original block.

Write an equation. $y - 7'' = 10''$

Add 7" to both sides of the equation.

$$\begin{aligned} y - 7'' &= 10'' \\ +7'' &= +7'' \\ \hline y &= 17'' \quad \text{Ans} \end{aligned}$$

Check.

$$\begin{aligned} y - 7'' &= 10'' \\ 17'' - 7'' &= 10'' \\ 10'' &= 10'' \end{aligned}$$

Figure 38-5

3. $-35 = P - 20.4$. Solve for P.

$$\begin{aligned} -35 &= P - 20.4 \\ +20.4 &= \quad +20.4 \\ \hline -14.6 &= P \quad \text{Ans} \end{aligned}$$

Check.

$$\begin{aligned} -35 &= P - 20.4 \\ -35 &= -14.6 - 20.4 \\ -35 &= -35 \quad \text{Ck} \end{aligned}$$

The following examples are solved by transposition.

Examples

1. $x - 4 = 19$. Solve for x.

Move -4 from the left side of the equation to the right and change to $+4$.

$$\begin{aligned} x - 4 &= 19 \\ x &= 19 + 4 \\ x &= 23 \quad \text{Ans} \end{aligned}$$

2. $y - 16.9 = 30$. Solve for y.

Move -16.9 from the left side of the equation to the right and change to $+16.9$.

$$\begin{aligned} y - 16.9 &= 30 \\ y &= 30 + 16.9 \\ y &= 46.9 \quad \text{Ans} \end{aligned}$$

Solution of Equations by the Division Principle of Equality

The division principle of equality states that if both sides of an equation are divided by the same number, the sides remain equal and the equation remains balanced. The division principle is used to solve an equation in which a number is multiplied by the unknown, such as $3x = 18$.

Procedure To solve an equation in which the unknown is multiplied by a number

- Divide both sides of the equation by the number that multiplies the unknown.
- Check.

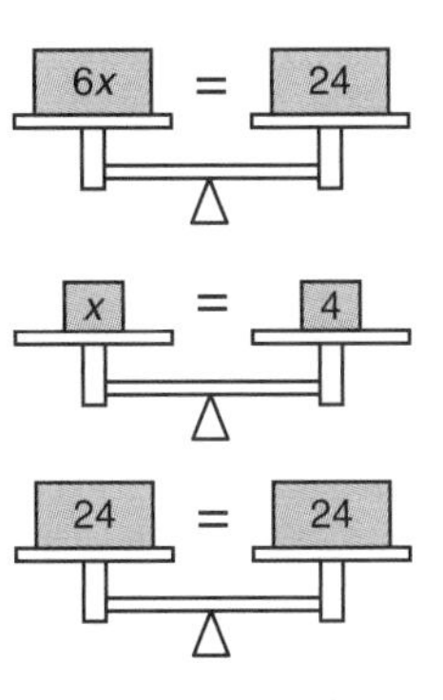

Figure 38-6

Examples

1. $6x = 24$. Solve for x.

In the equation, x is multiplied by 6. $6x = 24$

To solve, divide both sides of the equation by 6.

$$\frac{6x}{6} = \frac{24}{6}$$

$$x = 4 \quad \text{Ans}$$

Check.

$$6x = 24$$
$$6(4) = 24$$
$$24 = 24 \quad \text{Ck}$$

2. A part is shown in Figure 38-7. Solve for y.

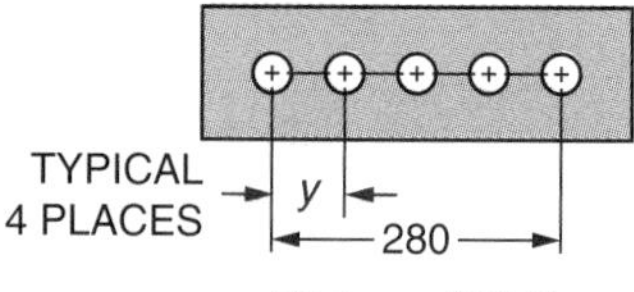

Figure 38-7

All dimensions are in millimeters.

Write an equation. $4y = 280 \text{ mm}$

Divide both sides of the equation by 4. $\frac{4y}{4} = \frac{280}{4} \text{ mm}$

$y = 70 \text{ mm}$ Ans

Check.

$$4y = 280$$
$$4(70 \text{ mm}) = 280 \text{ mm}$$
$$280 \text{ mm} = 280 \text{ mm} \quad \text{Ck}$$

3. $-14.4 = 3.2F$. Solve for F.

$$-14.4 = 3.2F$$
$$\frac{-14.4}{3.2} = \frac{3.2F}{3.2}$$
$$-4.5 = F \quad \text{Ans}$$

Check.

$$-14.4 = 3.2$$
$$-14.4 = 3.2(-4.5)$$
$$-14.4 = -14.4 \quad \text{Ck}$$

4. $7\frac{1}{4}A = 21\frac{3}{4}$. Solve for A.

$$7\frac{1}{4}A = 21\frac{3}{4}$$
$$\frac{7\frac{1}{4}A}{7\frac{1}{4}} = \frac{21\frac{3}{4}}{7\frac{1}{4}}$$
$$A = 3 \quad \text{Ans}$$

Check.

$$7\frac{1}{4}A = 21\frac{3}{4}$$
$$7\frac{1}{4}(3) = 21\frac{3}{4}$$
$$21\frac{3}{4} = 21\frac{3}{4} \quad \text{Ck}$$

Application

Solution by the Subtraction Principle of Equality

Solve each of the following equations using the subtraction principle of equality. Check each answer.

1. $P + 15 = 22$ ______
2. $x + 18 = 27$ ______
3. $M + 24 = 43$ ______
4. $y + 48 = 82$ ______
5. $13 = T + 9$ ______
6. $37 = D + 2$ ______
7. $62 = a + 19$ ______
8. $y + 16 = 15$ ______
9. $y + 30 = -23$ ______
10. $x + 63 = 17$ ______
11. $10 + R = 53$ ______
12. $51 = 48 + E$ ______
13. $-36 = 14 + x$ ______
14. $H + 7.6 = 14.7$ ______
15. $22.5 = L + 3.7$ ______
16. $-36.2 = y + 6.2$ ______
17. $T + 9.07 = 9.07$ ______
18. $H + 3\frac{1}{4} = 6\frac{1}{2}$ ______
19. $-\frac{7}{8} = x + \frac{3}{4}$ ______
20. $20\frac{3}{16} = A + 17\frac{1}{8}$ ______

21. $39\frac{5}{8} = y + 40\frac{7}{8}$ __________

22. $1\frac{7}{16} = W + \frac{9}{16}$ __________

23. $x + 13\frac{1}{8} = -10$ __________

24. $0.023 = 1.009 + H$ __________

Write an equation for each of the following problems, solve for the unknown, and check.

25. All dimensions are in inches.
 Find x. ________ $x =$ ________

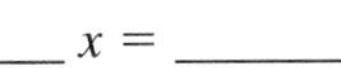

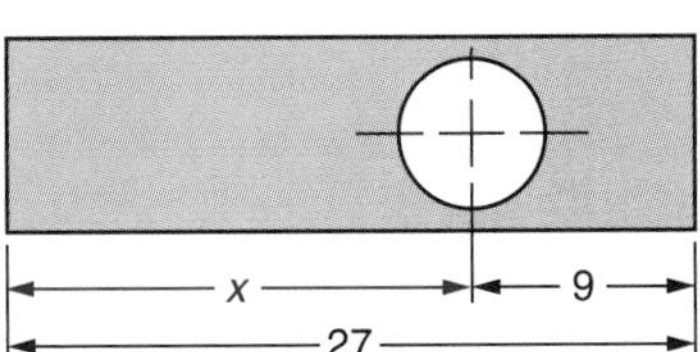

26. All dimensions are in millimeters.
 Find y. ________ $y =$ ________

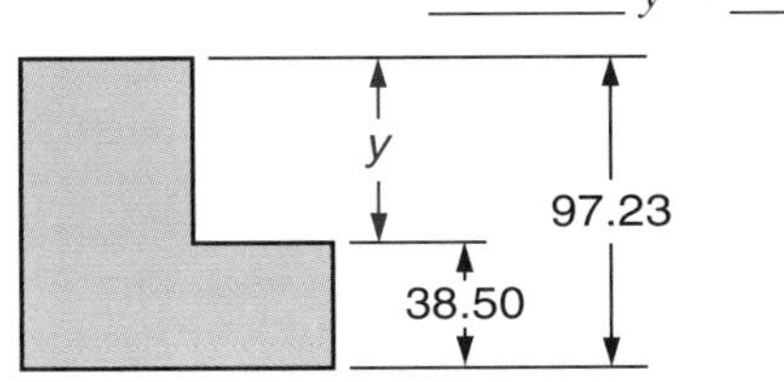

27. All dimensions are in inches.
 Find r. ________ $r =$ ________

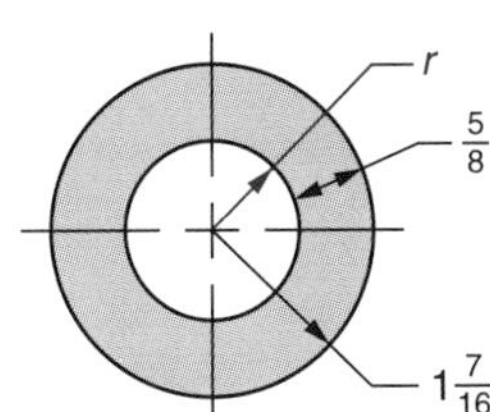

28. All dimensions are in inches.
 Find T. ________ $T =$ ________

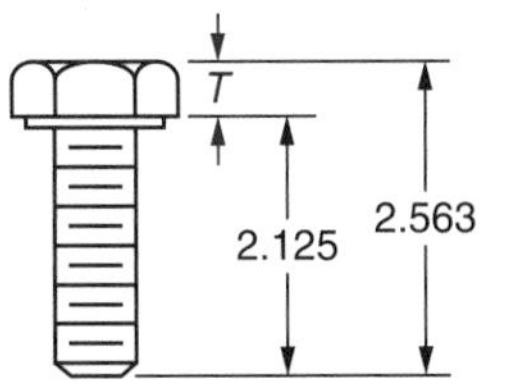

29. All dimensions are in millimeters.
 Find x. ________ $x =$ ________

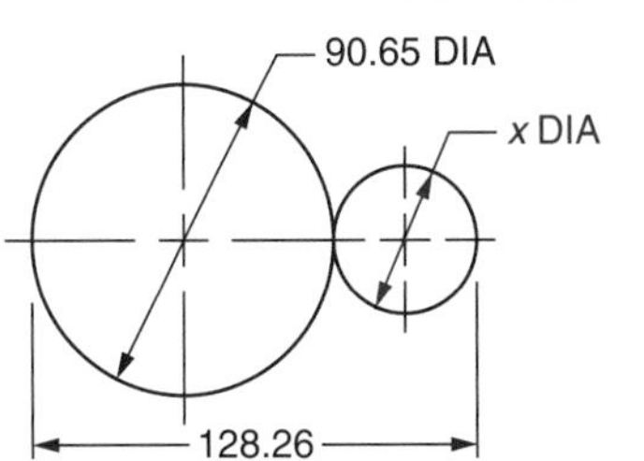

30. All dimensions are in inches.
 Find H. ________ $H =$ ________

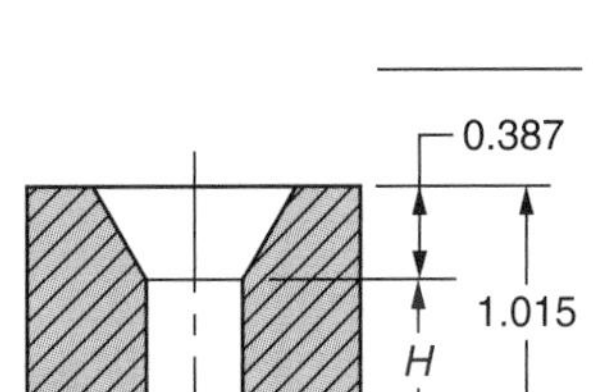
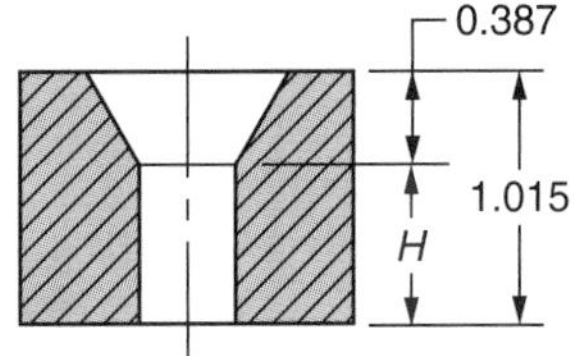

31. The height of 2 gage blocks is 0.8508 inch. One block is 0.750 inch thick. What is the thickness of the other block? __________

32. Three holes are drilled on a horizontal line in a housing. The center distance between the first hole and the second hole is 193.75 millimeters and the center distance between the first hole and the third hole is 278.12 millimeters. What is the distance between the second hole and the third hole? __________

33. A metal bar is $7\frac{5}{8}$ inches long. If $\frac{9}{32}$ inch is cut off one end, how long is the bar after the cut? __________

34. A shaft rotates in a bearing that is 0.3968 inch in diameter. The total clearance between the shaft and bearing is 0.0008 inch. What is the diameter of the shaft? __________

For each of the following problems, substitute the given values in the formula and solve for the unknown. Check.

35. One of the formulas used in computing spur gear dimensions is $D_O = D + 2a$. Determine D when $a = 0.1429$ inch and $D_O = 4.7144$ inches. __________

36. A formula used to compute the dimensions of a ring is $D = d + 2T$. Determine d when $D = 52.0$ millimeters and $T = 9.40$ millimeters. __________

37. A formula used in relation to the depth of a gear tooth is $WD = a + d$. Determine d when $WD = 0.3082$ inch and $a = 0.1429$ inch. ________

38. A sheet metal formula used in computing the size of a stretch-out is $L.S. = 4s + W$. Determine W when $s = 3$ inches and $L.S. = 12\frac{1}{8}$ inches. ________

Solution by the Addition Principle of Equality

Solve each of the following equations using the addition principle of equality. Check each answer.

39. $25 = d - 9$
40. $T - 12 = 34$
41. $x - 9 = -19$
42. $B - 4 = 9$
43. $P - 48 = 87$
44. $y - 23 = -20$
45. $16 = M - 12$
46. $-40 = E - 21$
47. $47 = R - 36$
48. $h - 8 = 12$
49. $39 = F - 39$
50. $W - 18 = 33$
51. $N - 2.4 = 6.9$
52. $A - 0.8 = 0.3$
53. $x - 10.09 = -13.78$
54. $5.07 = r - 3.07$
55. $-30.003 = x - 29.998$
56. $91.96 = L - 13.74$
57. $x - 8.12 = -13.01$
58. $D - \frac{1}{2} = \frac{1}{2}$
59. $y - \frac{7}{8} = -\frac{3}{8}$
60. $15\frac{5}{8} = H - 2\frac{7}{8}$
61. $-46\frac{3}{32} = x - 29\frac{15}{16}$
62. $C - 5\frac{7}{16} = -5\frac{7}{16}$
63. $W - 10.0039 = 8.0481$
64. $-14\frac{15}{32} = y - 14\frac{7}{16}$

Write an equation for each of the following problems, solve for the unknown, and check.

65. The point of a conical workpiece has been faced off to a $1\frac{11}{16}$ inch length of the tapered portion. If $6\frac{3}{4}$ inches of the original length was removed, what was the original length x of the tapered portion? ________

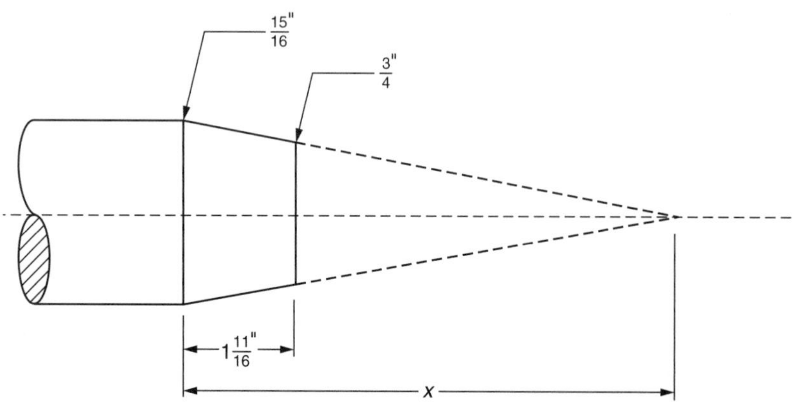

66. The bushing shown has a body diameter of 44.45 millimeters, which is 14.29 millimeters less than the head diameter. What is the size of the head diameter? All dimensions are in millimeters.

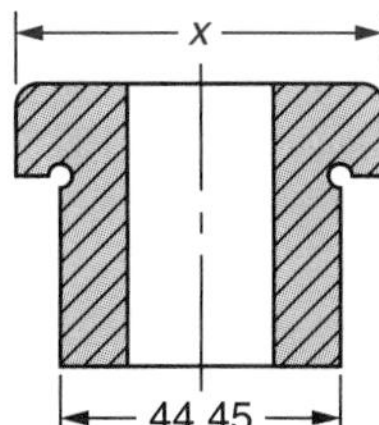

67. The flute length of the reamer shown is $1\frac{1}{8}$ inches, which is $3\frac{3}{8}$ inches less than the shank length. How long is the shank? All dimensions are in inches.

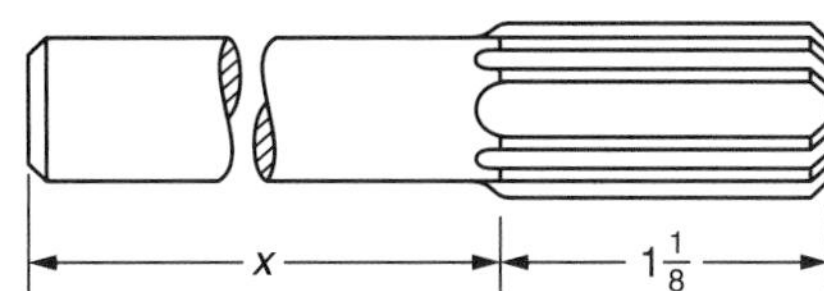

68. A hole is countersunk as shown to a depth of 0.250 inch. The depth of the countersink is 1.650 inches less than the depth of the 0.625-inch hole. Find depth x. All dimensions are in inches.

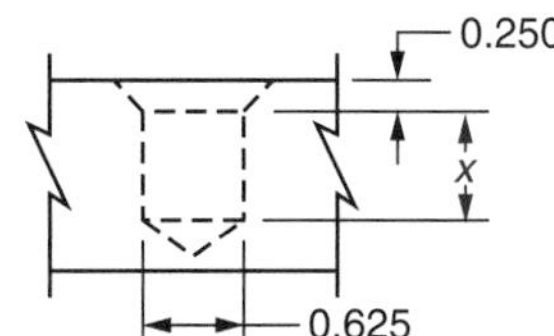

For each of the following problems, substitute the given values in the formula and solve for the unknown. Check each answer.

69. The total taper of a shaft equals the diameter of the large end minus the diameter of the small end, $T = D - d$. Determine D when $T = 22.5$ millimeters and $d = 30.8$ millimeters.

70. Using the spur gear formula, $D_R = D - 2d$, compute the pitch diameter (D) when the root diameter (D_R) = 3.0118 inches and the dedendum (d) = 0.1608 inch.

71. Using a sheet metal formula, $W = L.S. - 4S$, determine the length size ($L.S.$) when $W = 382$ millimeters and $S = 112$ millimeters.

72. The formula $d = D - \frac{1}{N}$ is used to determine the size of a hole into which threads will be tapped. If $N = 10$ threads per inch and $d = 0.65$ inch, what is D, the outside diameter of the threading tool?

Solution by the Division Principle of Equality

Solve each of the following equations using the division principle of equality. Check each answer.

73. $5A = 115$
74. $4D = 32$
75. $7x = -21$
76. $15M = 75$
77. $54 = 9P$
78. $-27 = 3y$
79. $54 = 6x$
80. $10y = 0.80$
81. $18T = 41.4$
82. $12x = -54$

83. $-5C = 0$ ______________

84. $7.1E = 21.3$ ______________

85. $0.6L = 12$ ______________

86. $-2.7x = 23.76$ ______________

87. $0.1y = -0.18$ ______________

88. $13.2W = 0$ ______________

89. $-x = -19.75$ ______________

90. $0.125P = 1.500$ ______________

91. $\frac{1}{4}D = 8$ ______________

92. $24 = \frac{3}{8}B$ ______________

93. $-\frac{1}{2}y = 36$ ______________

94. $1\frac{5}{8}L = 9\frac{3}{4}$ ______________

95. $-48\frac{3}{8} = 10\frac{3}{4}x$ ______________

96. $-\frac{7}{16} = -\frac{7}{16}y$ ______________

97. $50.98W = 10.196$ ______________

98. $0.0621 = 0.027t$ ______________

Write an equation for each of the following problems, solve for the unknown.

99. The depth, D, of an American Standard thread is given by the formula $0.6495D = P$, where P is the pitch. Compute the depth of the thread shown in the figure. ______________

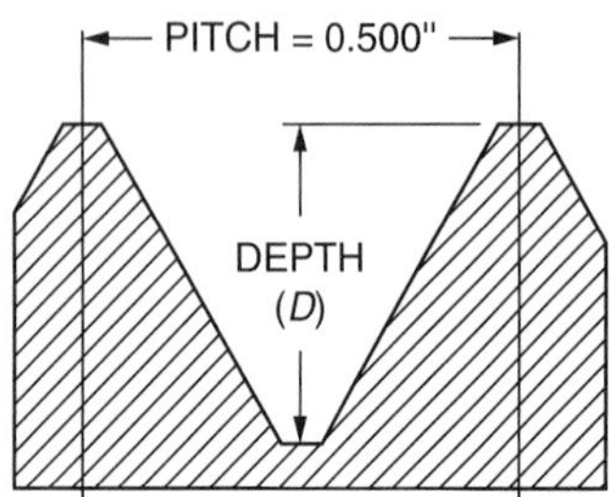

100. All dimensions are in millimeters.
Find x. ________ $x =$ ________

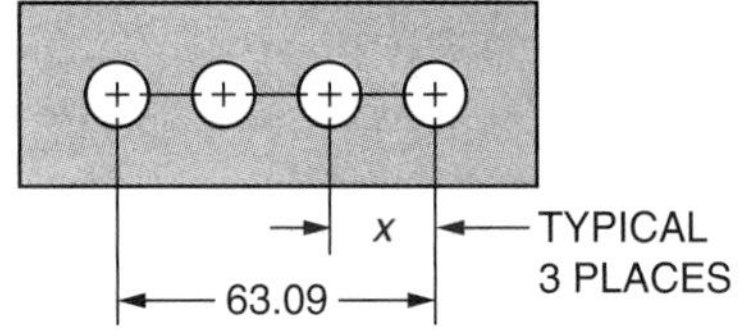

101. Find x. ________ $x =$ ________

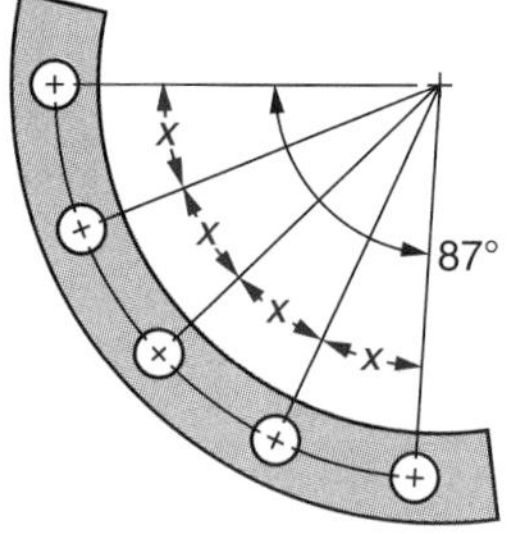

102. The feed of a drill is the depth of material that the drill penetrates in one revolution. The total depth of penetration equals the product of the number of revolutions and the feed. Compute the feed of a drill that cuts to a depth of 3.300 inches while turning 500.0 revolutions. ______________

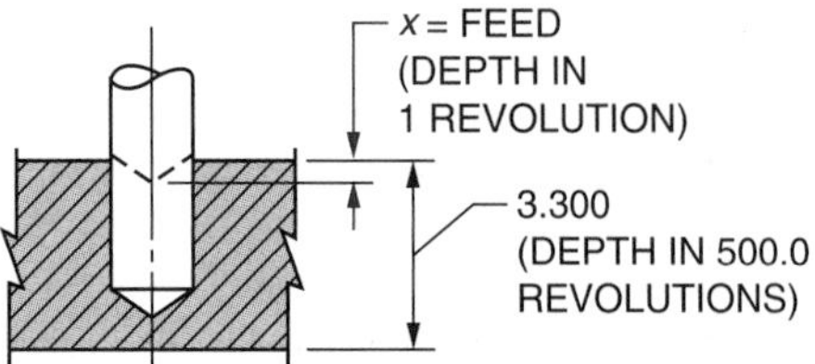

For each of the following problems, substitute the given values in the formula and solve for the unknown. Check each answer. Round the answers to 2 decimal places.

103. The circumference of a circle (C) equals π (approximately 3.1416) times the diameter (d) of the circle, $C = \pi d$. Determine d when $C = 392.50$ millimeters. ________

104. The depth (d) of a sharp V-thread is equal to 0.866 times the pitch (p), or $d = 0.866p$. Determine p when $d = 0.125$ inch. ________

105. The length of cut (L) in inches of a workpiece in a lathe is equal to the product of the cutting time (T) in minutes, the tool feed (F) in inches per revolution, and the number of revolutions per minute (N) of the workpiece: $L = TFN$. Determine N when $L = 9.50$ inches, $T = 3.00$ minutes, and $F = 0.050$ inch per revolution. ________

106. The length of cut, L, in inches, of a workpiece in a lathe is equal to the product of the cutting time, in minutes, T; the tool feed, F, in inches per revolution; and N, the number of revolutions per minute of the workpiece: $L = TFN$. Determine N when $L = 18$ inches, $T = 2.5$ minutes, and $F = 0.050$ inch per revolution. ________

Solve each of the following equations using either the addition, subtraction, or division principle of equality. Check each answer.

107. $T - 19 = -5$ ________

108. $-0.006x = 4.938$ ________

109. $9.37R = 103.07$ ________

110. $-22 = x - 31$ ________

111. $C + 34 = 12$ ________

112. $x + 6 = -13$ ________

113. $E - 29.8936 = 18.3059$ ________

114. $A - 16.37 = 9.03$ ________

115. $78.09 = x + 61.95$ ________

116. $F + 0.007 = 1.006$ ________

117. $-\frac{3}{16} = -1\frac{1}{16}y$ ________

118. $-0.66x = 4.752$ ________

119. $P - 0.20 = 0.07$ ________

120. $G - 59\frac{7}{8} = 48\frac{13}{16}$ ________

121. $-x = 19$ ________

122. $0 = 7H$ ________

123. $-14.067 = 3.034 + x$ ________

124. $20.863 = D + 25.942$ ________

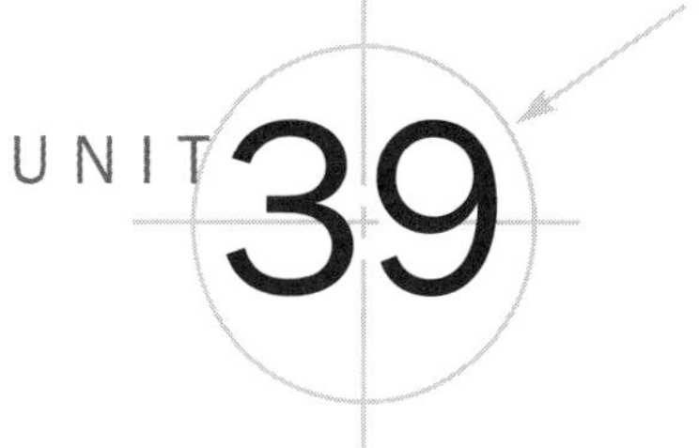

Solution of Equations by the Multiplication, Root, and Power Principles of Equality

Objectives ***After studying this unit you should be able to***

- Solve equations using the multiplication principle of equality.
- Solve equations using the root principle of equality.
- Solve equations using the power principle of equality.

Solution of Equations by the Multiplication Principle of Equality

The multiplication principle of equality states that if both sides of an equation are multiplied by the same number, the sides remain equal, and the equation remains balanced.

The multiplication principle is used to solve an equation in which the unknown is divided by a number, such as $\frac{x}{4} = 10$.

Procedure To solve an equation in which the unknown is divided by a number

- Multiply both sides of the equation by the number that divides the unknown.
- Check.

Examples

1. $\frac{x}{3} = 7$. Solve for x.

$$\frac{x}{3} = 7$$

To solve, multiply both sides of the equation by 3.

$$3\left(\frac{x}{3}\right) = 3(7)$$

$$x = 21 \quad \text{Ans}$$

Check.

$$\frac{x}{3} = 7$$

$$\frac{21}{3} = 7$$

$$7 = 7 \quad \text{Ck}$$

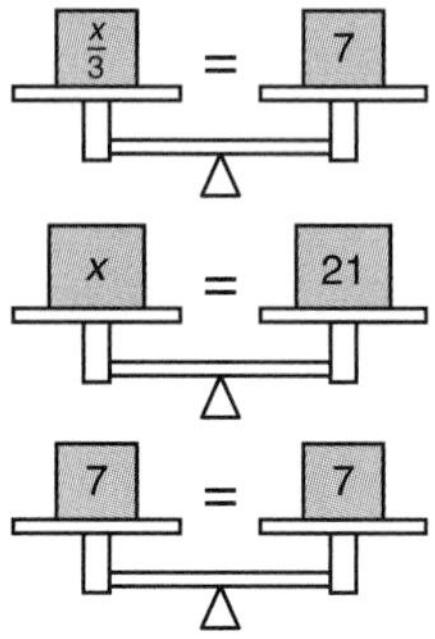

Figure 39-1

2. The length of bar stock shown in Figure 39-2 is cut into 5 equal pieces. Each piece is 4.5 inches long. Find y, the length of the bar before it was cut. All dimensions are in inches. Make no allowance for thickness of cuts.

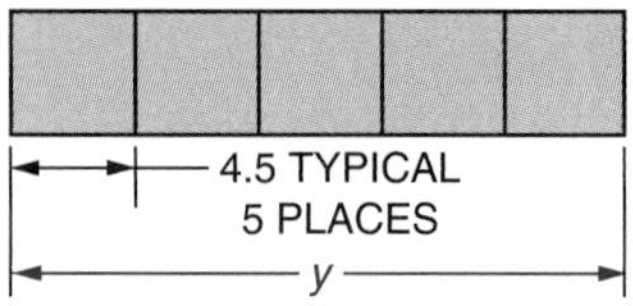

Figure 39-2

Write the equation.

$$\frac{y}{5} = 4.5''$$

Multiply both sides of the equation by 5.

$$5\left(\frac{y}{5}\right) = 5(4.5)''$$

$$y = 22.5'' \quad \text{Ans}$$

Check.

$$\frac{y}{5} = 4.5''$$

$$\frac{22.5''}{5} = 4.5''$$

$$4.5'' = 4.5'' \quad \text{Ck}$$

3. $6\frac{1}{8} = \frac{F}{-5}$. Solve for F.

$$6\frac{1}{8} = \frac{F}{-5}$$

$$-5\left(6\frac{1}{8}\right) = -5\left(\frac{F}{-5}\right)$$

$$-30\frac{5}{8} = F \quad \text{Ans}$$

Check.

$$6\frac{1}{8} = \frac{F}{-5}$$

$$6\frac{1}{8} = \frac{-30\frac{5}{8}}{-5}$$

$$6\frac{1}{8} = 6\frac{1}{8} \quad \text{Ck}$$

Solution of Equations by the Root Principle of Equality

The root principle of equality states that if the same root of both sides of an equation is taken, the sides remain equal, and the equation remains balanced.

The root principle is used to solve an equation that contains an unknown that is raised to a power, such as $x^2 = 36$.

Procedure To solve an equation in which an unknown is raised to a power

- Extract the root of both sides of the equation that leaves the unknown with an exponent of one.
- Check.

Examples

1. $x^2 = 9$ Solve for x in Figure 39-3.

   To solve, extract the square root of both sides of the equation.

   $$x^2 = 9$$
   $$\sqrt{x^2} = \sqrt{9}$$
   $$x = 3 \quad \text{Ans}$$

   Check.

   $$x^2 = 9$$
   $$3^2 = 9$$
   $$9 = 9 \quad \text{Ck}$$

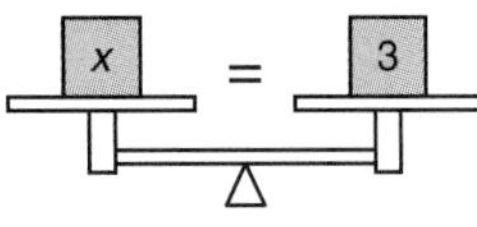

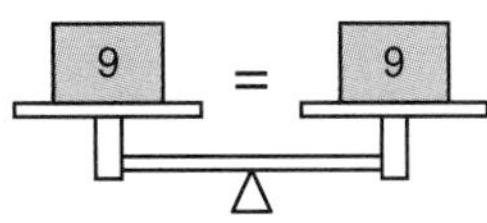

Figure 39-3

2. The area of a square piece of sheet steel shown in Figure 39-4 equals 16 square feet. What is the length of each side (s)?

   Write an equation.

   $$s^2 = 16 \text{ sq ft}$$

   Extract the square root of both sides of the equation.

   $$\sqrt{s^2} = \sqrt{16 \text{ sq ft}}$$
   $$s = 4 \text{ ft} \quad \text{Ans}$$

   Check.

   $$s^2 = 16 \text{ sq ft}$$
   $$(4 \text{ ft})^2 = 16 \text{ sq ft}$$
   $$16 \text{ sq ft} = 16 \text{ sq ft} \quad \text{Ck}$$

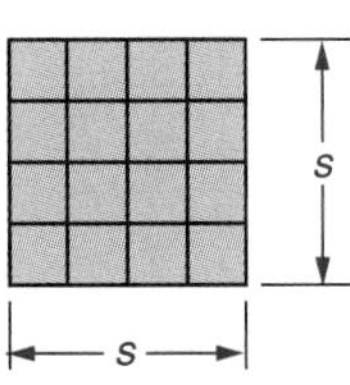

Figure 39-4

3. Solve for T.

   $$T^3 = -64$$
   $$\sqrt[3]{T^3} = \sqrt[3]{-64}$$
   $$T = -4 \quad \text{Ans}$$

   Check.

   $$T^3 = -64$$
   $$(-4)^3 = -64$$
   $$-64 = -64 \quad \text{Ck}$$

4. Solve for V.

   $$V^2 = \frac{9}{64}$$
   $$\sqrt{V^2} = \sqrt{\frac{9}{64}}$$
   $$V = \frac{3}{8} \quad \text{Ans}$$

   Check.

   $$V^2 = \frac{9}{64}$$
   $$\left(\frac{3}{8}\right)^2 = \frac{9}{64}$$
   $$\frac{9}{64} = \frac{9}{64} \quad \text{Ck}$$

Solution of Equations by the Power Principle of Equality

The power principle of equality states that if both sides of an equation are raised to the same power, the sides remain equal and the equation remains balanced. The power principle is used to solve an equation that contains a root of the unknown, such as $\sqrt{x} = 8$.

Procedure To solve an equation that contains a root of the unknown

- Raise both sides of the equation to the power that leaves the unknown with an exponent of one.
- Check.

Examples

1. $\sqrt{x} = 8$. Solve for x in Figure 39.5.

   In the equation, x is expressed as a root.

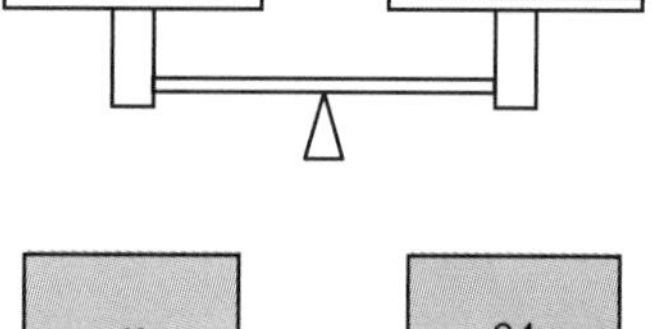

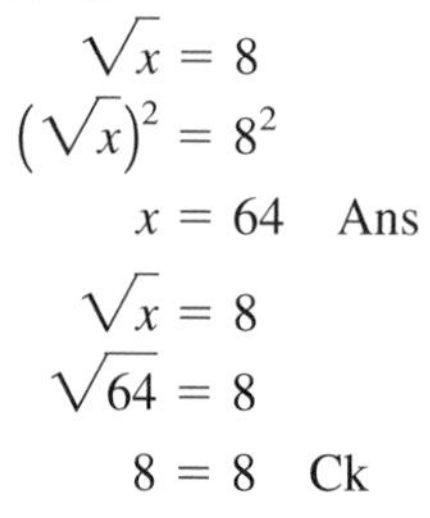

$$\sqrt{x} = 8$$

To solve, square both sides of the equation.

$$(\sqrt{x})^2 = 8^2$$

$$x = 64 \quad \text{Ans}$$

Check.

$$\sqrt{x} = 8$$

$$\sqrt{64} = 8$$

$$8 = 8 \quad \text{Ck}$$

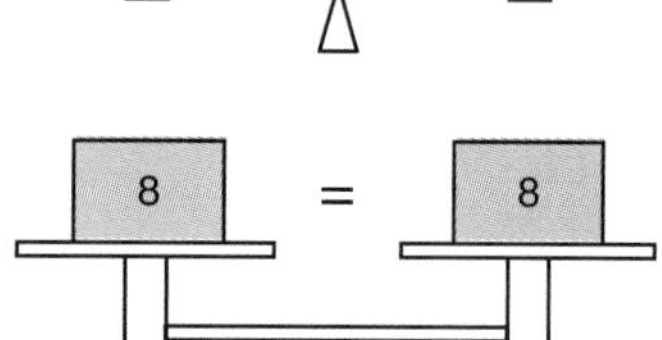

Figure 39-5

2. The length of a side of the cube shown in Figure 39-6 equals 2.8620 inches. The cube root of the volume equals the length of a side. Find the volume of the cube. Round the answer to 3 decimal places.

   Let V = the volume of the cube.

Write the equation. $\sqrt[3]{V} = 2.8620 \text{ in}$

Cube both sides of the equation. $(\sqrt[3]{V})^3 = (2.8620 \text{ in})^3$

$$V = 23.443 \text{ cu in} \quad \text{Ans}$$

2.8620 2.8620 2.8620

Figure 39-6

Check.

$$\sqrt[3]{V} = 2.8620 \text{ in}$$

$$\sqrt[3]{23.443 \text{ cu in}} = 2.8620 \text{ in}$$

$$2.8620 \text{ in} = 2.8620 \text{ in} \quad \text{Ck}$$

2.862 [x^y] (or [^]) 3 [=] 23.44276793, 23.443 cu in Ans (rounded)

Check. 3 [SHIFT] (or [2nd]) [$\sqrt[x]{\ }$] 23.443 [=] 2.862 (rounded)

2.862 = 2.862

Application

Solution by the Multiplication Principle of Equality

Solve each of the following equations using the multiplication principle of equality. Check each answer.

1. $\dfrac{P}{5} = 6$
2. $\dfrac{M}{12} = 5$
3. $D \div 9 = 7$
4. $3 = L \div 8$
5. $3 = W \div 9$
6. $\dfrac{N}{12} = -2$

7. $\frac{C}{14} = 0$

8. $\frac{x}{-10} = 9$

9. $\frac{F}{4.3} = 5$

10. $\frac{A}{-0.5} = 24$

11. $S \div (7.8) = 3$

12. $x \div (-0.3) = 16$

13. $-20 = \frac{y}{0.3}$

14. $\frac{T}{-1.8} = 2.4$

15. $0 = H \div (-3.8)$

16. $M \div 9.5 = -12$

17. $1.04 = \frac{H}{0.06}$

18. $\frac{B}{\frac{1}{2}} = 7$

19. $V \div 1\frac{1}{4} = 3$

20. $\frac{x}{\frac{3}{8}} = -\frac{1}{4}$

21. $D \div \left(-\frac{1}{16}\right) = -32$

22. $4 = y \div \left(-\frac{7}{8}\right)$

23. $\frac{1}{2} = \frac{T}{1\frac{1}{2}}$

24. $H \div (-2) = 7\frac{9}{16}$

Write an equation for each of the following problems, solve for the unknown, and check.

25. All dimensions are in millimeters.
Find x. ________ x = ________

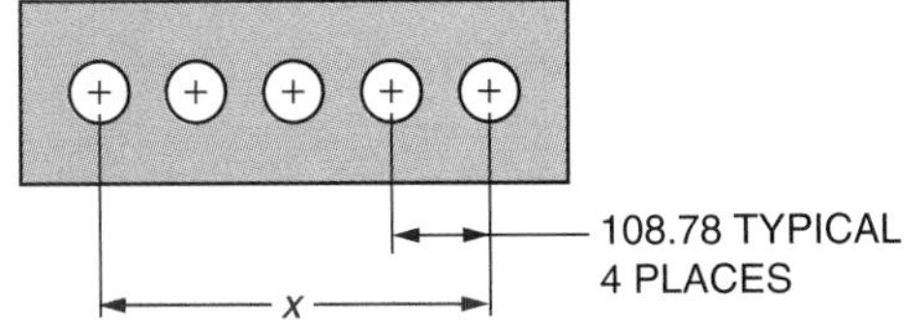

26. Find x. ________ x = ________

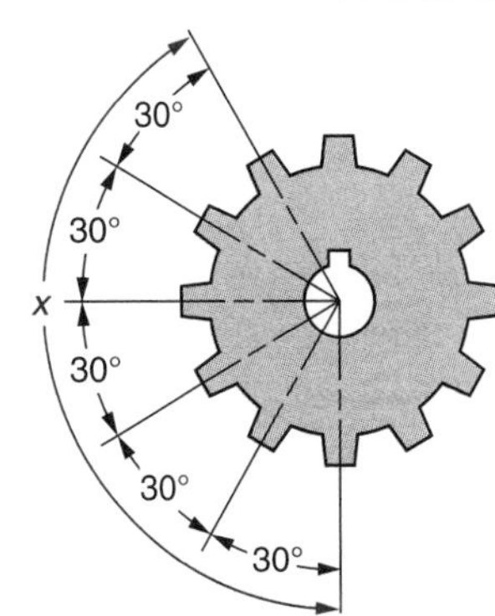

27. A 10-inch sine plate is tilted at an angle of 45° as shown. The gage block height divided by 10 equals 0.70711 inch. Compute the height of the gage blocks. All dimensions are in inches. ________

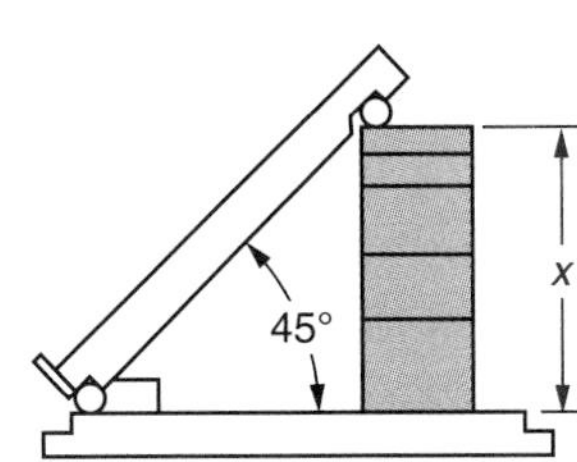

28. The width of a rectangular sheet of metal shown is equal to the area of the sheet divided by its length. Compute the area of a sheet that is $3\frac{1}{4}$ feet wide and $5\frac{1}{2}$ feet long. __________

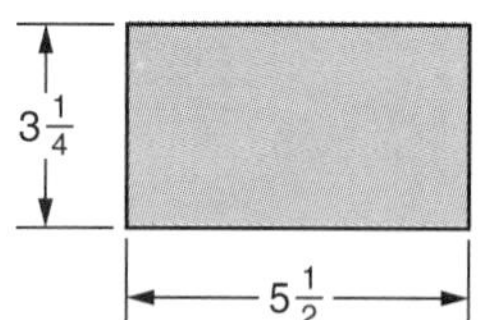

29. The depth of an American Standard thread shown divided by 0.6495 is equal to the pitch. Compute the depth of a thread with a 0.0500-inch pitch. All dimensions are in inches. Round the answer to 3 decimal places. __________

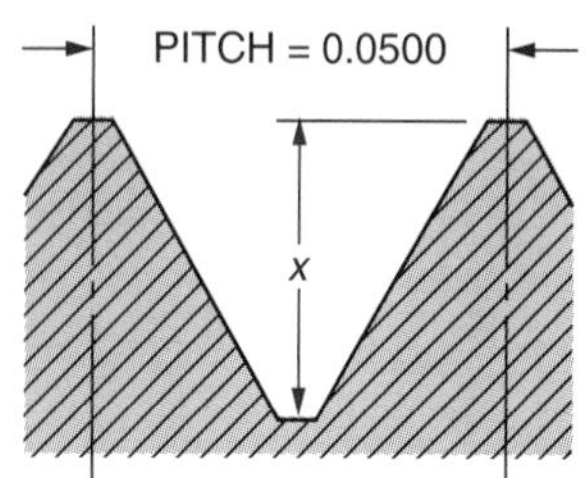

For each of the following problems, substitute the given values in the formula and solve for the unknown. Check each answer.

30. In mechanical energy applications, force (F) in pounds equals work (W) in foot-pounds divided by distance (D) in feet, $F = \frac{W}{D}$. Determine W when $F = 150.0$ pounds and $D = 7.500$ feet. __________
31. The diameter (D) of a circle equals the circle circumference (C) divided by 3.1416, $D = \frac{C}{3.1416}$. Determine C when $D = 52.14$ millimeters. Round the answer to 1 decimal place. __________
32. The pitch (P) of a spur gear equals the number of gear teeth (N) divided by the pitch diameter (D), $P = \frac{N}{D}$. Determine N when $P = 5$ teeth per inch and $D = 5.6000$ inches. __________

Solution by the Root Principle of Equality

Solve each of the following equations using the root principle of equality. Round the answers to 3 decimal places where necessary.

33. $S^2 = 16$ __________
34. $P^2 = 81$ __________
35. $81 = M^2$ __________
36. $49 = B^2$ __________
37. $D^3 = 64$ __________
38. $x^3 = -64$ __________
39. $144 = F^2$ __________
40. $-64 = y^3$ __________
41. $10{,}000 = L^2$ __________
42. $-125 = x^3$ __________
43. $\frac{9}{25} = W^2$ __________
44. $C^2 = \frac{1}{16}$ __________
45. $P^2 = \frac{16}{49}$ __________
46. $M^3 = \frac{1}{64}$ __________
47. $-\frac{1}{8} = y^3$ __________
48. $D^3 = \frac{64}{27}$ __________

49. $E^2 = 0.04$ ________
50. $0.64 = H^2$ ________
51. $W^2 = 2.753$ ________
52. $0.0017 = R^2$ ________
53. $N^3 = 0.123$ ________
54. $-0.123 = x^3$ ________
55. $7.843 = F^4$ ________
56. $T^2 = 7.056$ ________

Write an equation for each of the following problems, solve for the unknown, and check.

57. The area of a square equals the length of a side squared, $A = s^2$. For each area of a square given, compute the length of a side. Round the answers to 3 decimal places where necessary.

a. 36 square inches ________ $s =$ ________

b. $\frac{25}{64}$ square foot ________ $s =$ ________

c. 1.44 square meters ________ $s =$ ________

d. 64.700 square meters ________ $s =$ ________

e. 0.049 square foot ________ $s =$ ________

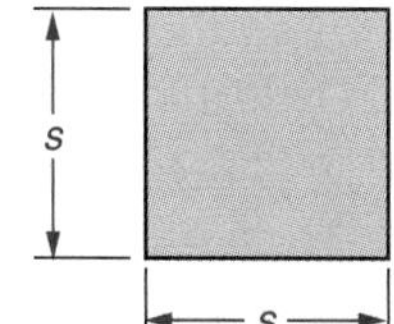

58. The volume of a cube equals the length of a side cubed, $V = s^3$. For each volume of a cube given, compute the length of a side. Round the answers to 3 decimal places where necessary.

a. 125 cubic inches ________ $s =$ ________

b. $\frac{27}{216}$ cubic foot ________ $s =$ ________

c. 0.642 cubic meter ________ $s =$ ________

d. 92.76 cubic millimeters ________ $s =$ ________

e. 0.026 cubic foot ________ $s =$ ________

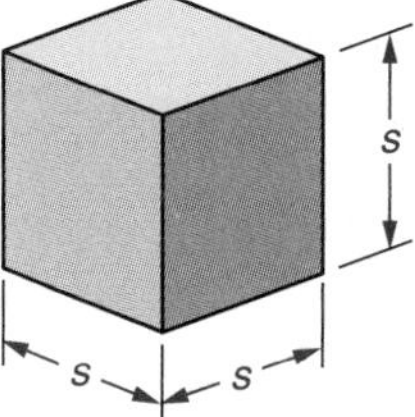

Solution by the Power Principle of Equality

Solve each of the following equations using the power principle of equality. Check all answers. Round the answers to 3 decimal places where necessary.

59. $\sqrt{C} = 6$ ________
60. $\sqrt{T} = 12$ ________
61. $\sqrt{P} = 1.2$ ________
62. $0.8 = \sqrt{M}$ ________
63. $0.82 = \sqrt{F}$ ________
64. $\sqrt[3]{V} = 3$ ________
65. $\sqrt[3]{H} = 1.7$ ________
66. $\sqrt[3]{x} = -4$ ________
67. $\sqrt{A} = 0$ ________
68. $\sqrt[5]{N} = 1$ ________
69. $-2 = \sqrt[5]{y}$ ________
70. $0.3 = \sqrt[4]{D}$ ________
71. $\sqrt[3]{x} = -0.6$ ________
72. $\sqrt[4]{P} = 0.1$ ________
73. $0.1 = \sqrt[3]{B}$ ________
74. $\frac{1}{4} = \sqrt{A}$ ________
75. $\sqrt[4]{F} = \frac{1}{4}$ ________
76. $-\frac{3}{5} = \sqrt[3]{y}$ ________

77. $\frac{5}{8} = \sqrt{H}$ ________
78. $\sqrt{P} = 1.256$ ________
79. $\sqrt[3]{B} = 2.868$ ________
80. $\sqrt[5]{x} = -1.090$ ________
81. $0.7832 = \sqrt[3]{y}$ ________
82. $0.364 = \sqrt[3]{y}$ ________

Write an equation for each of the following problems, solve for the unknown, and check. Round the answers to 3 decimal places where necessary.

83. The length of a side of a square equals the square root of the area, $s = \sqrt{A}$. For each side of a square given, compute the area.

a. 3.4" ________ $A =$ ________

b. 0.75' ________ $A =$ ________

c. 0.652 m ________ $A =$ ________

d. 2.162 mm ________ $A =$ ________

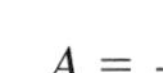

e. 1.290" ________ $A =$ ________

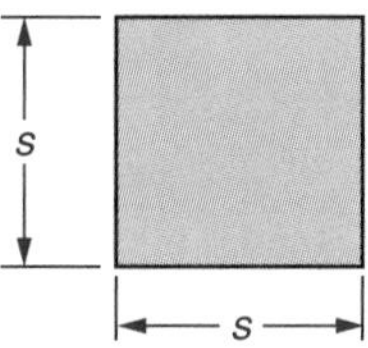

84. The length of a side of a cube equals the cube root of the volume, $s = \sqrt[3]{V}$. For each side of a cube given, compute the volume. Round the answers to 2 decimal places where necessary.

a. 3.300" ________ $V =$ ________

b. 0.900' ________ $V =$ ________

c. 0.62 m ________ $V =$ ________

d. 4.073 mm ________ $V =$ ________

e. 1.281" ________ $V =$ ________

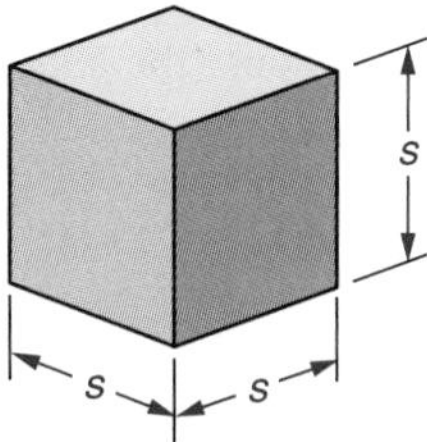

Solve each of the following equations using either the multiplication, root, or power principle of equality. Check each answer.

85. $L^3 = -125$ ________
86. $T^5 = 0$ ________
87. $\frac{E}{-2} = -18$ ________
88. $13 = y \div (-5)$ ________
89. $-0.1 = \sqrt[3]{y}$ ________
90. $\sqrt[4]{M} = 3$ ________
91. $\frac{y}{-0.1} = -0.01$ ________
92. $\frac{R}{12.6} = 0.002$ ________
93. $\sqrt{R} = \frac{3}{8}$ ________
94. $\sqrt[3]{V} = \frac{2}{3}$ ________
95. $G^3 = \frac{64}{125}$ ________
96. $x^3 = \frac{-64}{125}$ ________
97. $\sqrt[3]{x} = -2.9631$ ________
98. $\sqrt[5]{x} = 0.797$ ________
99. $y^3 = 0.0393$ ________
100. $-2.127 = y^5$ ________
101. $\frac{M}{0.009} = 100$ ________
102. $x \div 6.004 = -0.2125$ ________

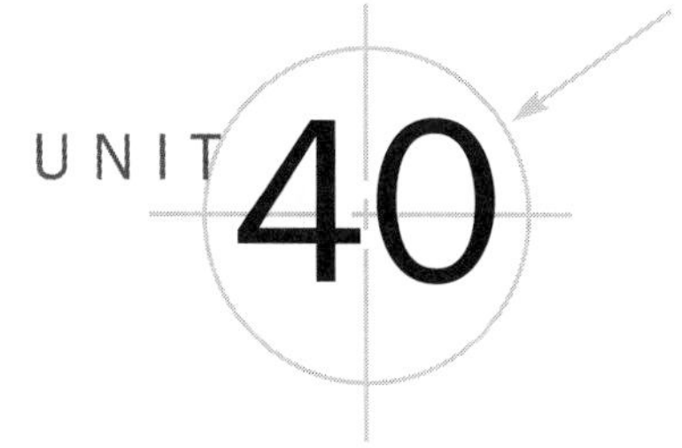

Solution of Equations Consisting of Combined Operations and Rearrangement of Formulas

Objectives ***After studying this unit you should be able to***

- Solve equations involving several operations.
- Rearrange formulas in terms of any letter value.
- Substitute values in formulas and solve for unknowns.
- Solve for the unknown term of a proportion.

Often in actual occupational applications, the formulas used result in complex equations. These equations require the use of two or more principles of equality for their solutions. For example,

$$0.13x - 4.73(x + 6.35) = 5.06x - 2.87$$

requires a definite procedure in determining the value of x. Use of proper procedure results in the unknown standing alone on one side of the equation with its value on the other.

Procedure for Solving Equations Consisting of Combined Operations

It is essential that the steps used in solving an equation be taken in the following order. Some or all of these steps may be used depending upon the particular equation.

- Remove parentheses.
- Combine like terms on each side of the equation.
- Apply the addition and subtraction principles of equality to get all unknown terms on one side of the equation and all known terms on the other side.
- Combine like terms.
- Apply the multiplication and division principles of equality.
- Apply the power and root principles of equality.

➤ **Note:** Always solve for a positive unknown. A positive unknown may equal a negative value, but a negative unknown is not a solution. For example, $x = -10$ is correct, but $-x = 10$ is incorrect. When solving equations where the unknown remains a negative value, multiply both sides of the equation by -1. Multiplying a negative unknown by -1 results in a positive unknown. For example, multiplying both sides of $-x = 10$ by -1 gives $(-1)(-x) = (-1)(10)$ with the result $x = -10$.

Examples The following are examples of equations consisting of combined operations.

1. $5x + 7 = 22$. Solve for x.

   The operations involved are multiplication and addition. Follow the procedure for solving equations consisting of combined operations.

Apply the subtraction principle. Subtract 7 from both sides of the equation.

$$5x + 7 = 22$$
$$5x + 7 - 7 = 22 - 7$$
$$5x = 15$$

Apply the division principle. Divide both sides of the equation by 5.

$$\frac{5x}{5} = \frac{15}{5}$$
$$x = 3 \quad \text{Ans}$$

Check.

$$5x + 7 = 22$$
$$5(3) + 7 = 22$$
$$15 + 7 = 22$$
$$22 = 22 \quad \text{Ck}$$

2. $6x + 4x = 3x - 5x + 19 + 5$. Solve for x.

Combine like terms on each side of the equation.

$$6x + 4x = 3x - 5x + 19 + 5$$
$$10x = -2x + 24$$

Apply the addition principle. Add $2x$ to both sides of the equation.

$$10x + 2x = -2x + 24 + 2x$$
$$12x = 24$$

Apply the division principle. Divide both sides of the equation by 12.

$$\frac{12x}{12} = \frac{24}{12}$$
$$x = 24 \div 12$$
$$x = 2 \quad \text{Ans}$$

Check.

$$6x + 4x = 3x - 5x + 19 + 5$$
$$6(2) + 4(2) = 3(2) - 5(2) + 19 + 5$$
$$12 + 8 = 6 - 10 + 19 + 5$$
$$20 = 20 \quad \text{Ck}$$

3. $9x + 7(x + 3) = 25$. Solve for x.

$$9x + 7(x + 3) = 25$$

Remove parentheses.

$$9x + 7x + 21 = 25$$

Combine like terms.

$$16x + 21 = 25$$

Apply the subtraction principle. Subtract 21 from both sides of the equation.

$$16x + 21 - 21 = +25 - 21$$
$$16x = 4$$

Apply the division principle. Divide both sides of the equation by 16.

$$\frac{16x}{16} = \frac{4}{16}$$
$$x = \frac{1}{4} \quad \text{Ans}$$

Check.

$$9x + 7(x + 3) = 25$$
$$9\left(\frac{1}{4}\right) + 7\left(\frac{1}{4} + 3\right) = 25$$
$$2\frac{1}{4} + 22\frac{3}{4} = 25$$
$$25 = 25 \quad \text{Ck}$$

4. $-x = 14$. Solve for x.

$$-x = 14$$

Apply the multiplication principle. Multiply both sides of the equation by -1.

$$(-1)(-x) = (-1)(14)$$
$$x = -14 \quad \text{Ans}$$

Check.

$$-x = 14$$
$$-(-14) = 14$$
$$14 = 14 \quad \text{Ck}$$

5. $\dfrac{x^2}{4} - 32 = -23$. Solve for x.

$$\frac{x^2}{4} - 32 = -23$$

Apply the addition principle. Add 32 to both sides of the equation.

$$\frac{x^2}{4} - 32 + 32 = -23 + 32$$
$$\frac{x^2}{4} = 9$$

Apply the multiplication principle. Multiply both sides of the equation by 4.

$$4\left(\frac{x^2}{4}\right) = 4(9)$$
$$x^2 = 36$$

Apply the root principle. Extract the square root of both sides of the equation.

$$\sqrt{x^2} = \sqrt{36}$$
$$x = 6 \quad \text{Ans}$$

Check.

$$\frac{x^2}{4} - 32 = -23$$
$$\frac{6^2}{4} - 32 = -23$$
$$9 - 32 = -23$$
$$-23 = -23 \quad \text{Ck}$$

6. $6\sqrt[3]{x} = 4(\sqrt[3]{x} + 1.5)$. Solve for x.

$$6\sqrt[3]{x} = 4(\sqrt[3]{x} + 1.5)$$

Remove parentheses.

$$6\sqrt[3]{x} = 4\sqrt[3]{x} + 6$$

Apply the subtraction principle. Subtract $4\sqrt[3]{x}$ from both sides of the equation.

$$6\sqrt[3]{x} - 4\sqrt[3]{x} = 4\sqrt[3]{x} + 6 - 4\sqrt[3]{x}$$
$$2\sqrt[3]{x} = 6$$

Apply the division principle. Divide both sides of the equation by 2.

$$\frac{2\sqrt[3]{x}}{2} = \frac{6}{2}$$
$$\sqrt[3]{x} = 3$$

Apply the power principle. Raise both sides of the equation to the third power.

$$(\sqrt[3]{x})^3 = 3^3$$
$$x = 27 \quad \text{Ans}$$

Check.

$$6\sqrt[3]{27} = 4(\sqrt[3]{27} + 1.5)$$
$$6(3) = 4(3 + 1.5)$$
$$18 = 4(4.5)$$
$$18 = 18 \quad \text{Ck}$$

Substituting Values and Solving Formulas

Manufacturing applications often require solving formulas in which all but one numerical value for letter values is known. The unknown letter value can appear anywhere within the formula. To determine the numerical value of the unknown, write the original formula, substitute the known number values for their respective letter values, and simplify. Then follow the procedure given for solving equations consisting of combined operations.

Example An open belt pulley system is shown in Figure 40-1. The number of inches between the pulley centers is represented by x. The larger pulley diameter (D) is 6.25 inches and the smaller pulley diameter (d) is 4.25 inches. The belt length is 56.0 inches. Find the distance between pulley centers using this formula found in a trade handbook. Round the answer to 1 decimal place.

$$L = 3.14(0.5D + 0.5d) + 2x$$

where L = belt length
D = the diameter of the larger pulley
d = the diameter of the smaller pulley
x = the distance between pulley centers

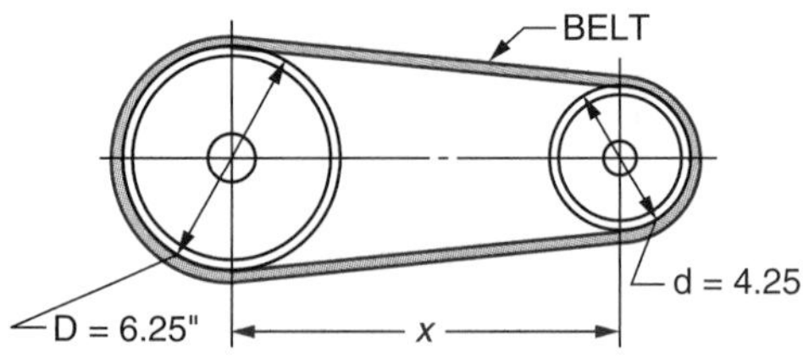

Figure 40-1

Write the formula. $L = 3.14(0.5D + 0.5d) + 2x$

Substitute the known numerical values for their respective letter values and simplify.

$56.0 \text{ in} = 3.14[0.5(6.25 \text{ in}) + 0.5(4.25 \text{ in})] + 2x$

3.14 [×] [(] .5 [×] 6.25 [+] .5 [×] 4.25 [)] [=] 16.485

$56.0 \text{ in} = 16.485 \text{ in} + 2x$

Apply the subtraction principle. Subtract 16.485 inches from both sides.

$56.0 \text{ in} = 16.485 \text{ in} + 2x$

$56.0 \text{ in} - 16.485 \text{ in} = 16.485 \text{ in} + 2x - 16.485 \text{ in}$

$39.515 \text{ in} = 2x$

Apply the division principle. Divide both sides by 2.

$$\frac{39.515 \text{ in}}{2} = \frac{2x}{2}$$

$19.7575 \text{ in} = x$, $x = 19.8 \text{ in}$ Ans (rounded)

Check. $L = 3.14(0.5D + 0.5d) + 2x$

56 = 3.14 [×] [(] .5 [×] 6.25 [+] .5 [×] 4.25 [)] [+] 2 [×] 19.7575 [=] 56

$56 \text{ in} = 56 \text{ in}$

Rearranging Formulas

A formula that is used to find a particular value must sometimes be rearranged to solve for another value. Consider the letter to be solved for as the unknown term and the other letters in the formula as the known values. The formula must be rearranged so that the unknown term is on one side of the equation and all other values are on the other side. A formula is rearranged by using the same procedure that is used for solving equations consisting of combined operations.

Problems are often solved more efficiently by first rearranging formulas than by directly substituting values in the original formula and solving for the unknown. This is particularly true in solving more complex formulas that involve many operations. Also, it is sometimes necessary to solve for the same unknown after a formula has been rearranged using different known values. Since the formula has been rearranged in terms of the specific unknown, solutions are more readily computed.

First rearranging formulas and then substituting known values enables you to solve for the unknown using a calculator for continuous operations. This is illustrated in Example 4 on page 249.

Examples Given the following formulas, rearrange and solve for the designated letter.

1. $A = bh$. Solve for h.

$$A = bh$$

Apply the division principle. Divide both sides of the equation by b.

$$\frac{A}{b} = \frac{bh}{b}$$

$$\frac{A}{b} = h \quad \text{Ans}$$

2. In the figure shown in Figure 40-2, $L = a + b$. Solve for a.

$$L = a + b$$

Apply subtraction principle. Subtract b from both sides of the equation.

$$L - b = a + b - b$$
$$L - b = a \quad \text{Ans}$$

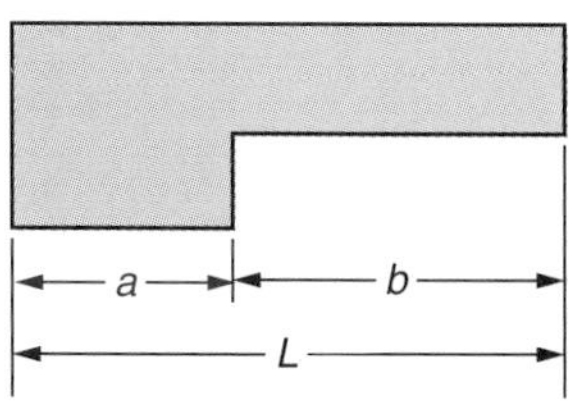

Figure 40-2

3. A screw thread is checked using a micrometer and 3 wires as shown in Figure 40-3. The measurement is checked using the following formula. Solve the formula for W.

$$M = D - 1.5155P = 3W$$

where M = measurement over the wires
D = major diameter
P = pitch
W = wire size

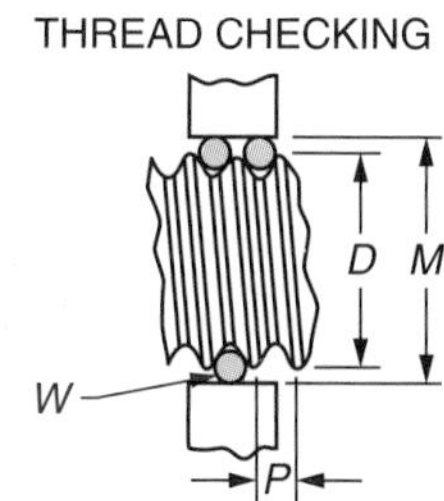

Figure 40-3

Apply subtraction principle. Subtract D from both sides of the equation.

$$M = D - 1.5155P + 3W$$
$$M - D = D - 1.5155P + 3W - D$$
$$M - D = -1.5155P + 3W$$

Apply addition principle. Add $1.5155P$ to both sides of the equation.

$$M - D + 1.5155P = -1.5155P + 3W + 1.5155P$$
$$M - D + 1.5155P = 3W$$

Apply division principle. Divide both sides of the equation by 3.

$$\frac{M - D + 1.5155P}{3} = \frac{3W}{3}$$
$$\frac{M - D + 1.5155P}{3} = W \quad \text{Ans}$$

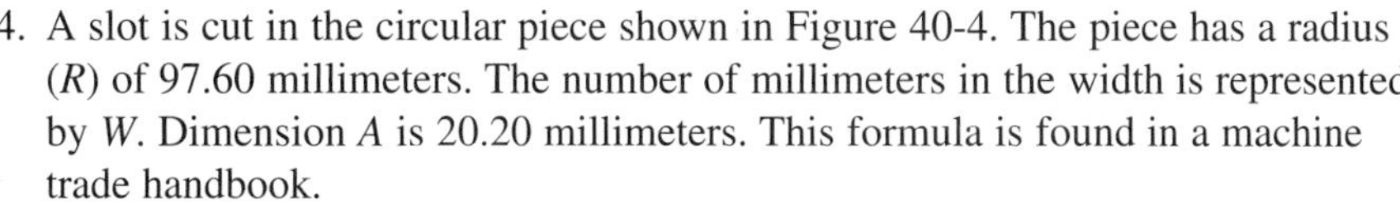

4. A slot is cut in the circular piece shown in Figure 40-4. The piece has a radius (R) of 97.60 millimeters. The number of millimeters in the width is represented by W. Dimension A is 20.20 millimeters. This formula is found in a machine trade handbook.

$$A = R - \sqrt{R^2 - 0.2500W^2}$$

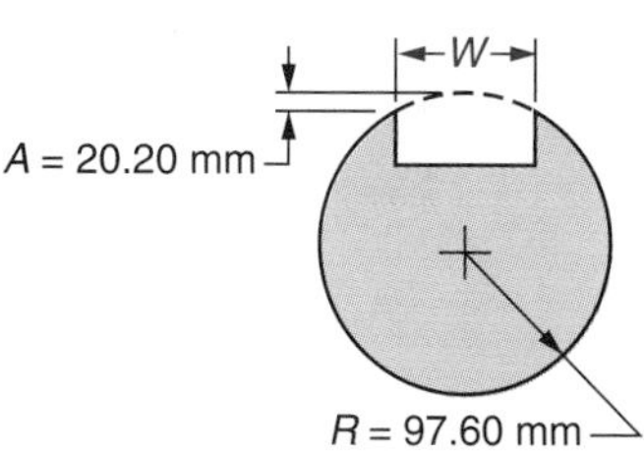

Figure 40-4

Solve for W.

Apply the subtraction principle. Subtract R from both sides of the equation.

$$A = R - \sqrt{R^2 - 0.2500W^2}$$
$$\underline{-R = -R}$$
$$A - R = -\sqrt{R^2 - 0.2500W^2}$$

Apply the power principle. Square both sides of the equation.

$$(A - R)^2 = (-\sqrt{R^2 - 0.2500W^2})^2$$
$$(A - R)^2 = R^2 - 0.2500W^2$$

Apply the subtraction principle. Subtract R^2 from both sides of the equation.

$$(A - R)^2 = R^2 - 0.2500W^2$$
$$\underline{-R^2 = -R^2}$$
$$(A - R)^2 - R^2 = -0.2500W^2$$

Apply the division principle. Divide both sides of the equation by -0.2500.

$$\frac{(A - R)^2 - R^2}{-0.2500} = \frac{-0.2500W^2}{-0.2500}$$

Simplify.

$$\frac{(A - R)^2 - R^2}{-0.2500} = W^2$$

Apply the root principle. Take the square root of both sides.

$$\sqrt{\frac{(A - R)^2 - R^2}{-0.2500}} = \sqrt{W^2}$$

$$\sqrt{\frac{(A - R)^2 - R^2}{-0.2500}} = W$$

Substitute the given numerical values for their respective letter values and find dimension W.

$$W = \sqrt{\frac{(20.20 - 97.60)^2 - 97.60^2}{-0.2500}}$$

$W =$ [√] [(] [(] [(] 20.2 [−] 97.6 [)] [x²] [−] 97.6 [x²] [)] [÷] [(−)] .25 [)] [=] 118.911732

$W = 118.9$ mm Ans (rounded)

Check. Substitute numerical values in the *original* formula.

20.20 = 97.6 [−] [√] [(] 97.6 [x²] [−] .25 [×] 118.911732 [x²] [)] [=] 20.20000001

20.20 = 20.20

Proportions

As we saw in Unit 18, in a proportion the product of the means equals the product of the extremes. If the terms are cross multiplied, their products are equal.

Cross multiplying is used to solve proportions that have an unknown term. Since a proportion is an equation, the principles used for solving equations are applied in determining the value of the unknown after the terms have been cross multiplied.

Examples

1. Solve the proportion $\frac{a}{b} = \frac{c}{x}$ for the value of x.

$$\frac{a}{b} = \frac{c}{x}$$

Cross multiply.

$$ax = bc$$

Apply the division principle of equality. Divide both sides of the equation by a.

$$\frac{ax}{a} = \frac{bc}{a}$$

$$x = \frac{bc}{a} \quad \text{Ans}$$

Check.

$$\frac{a}{b} = \frac{c}{x}$$

$$\frac{a}{b} = \frac{c}{\frac{bc}{a}} \text{ or } \frac{c}{1} \times \frac{a}{bc}$$

$$\frac{a}{b} = \frac{a}{b} \quad \text{Ck}$$

2. Solve the proportion $\frac{4}{R} = \frac{2R}{12.5}$ for R.

$$\frac{4}{R} = \frac{2R}{12.5}$$

Cross multiply.

$$2R^2 = (4)(12.5)$$
$$2R^2 = 50$$

Apply the division principle of equality. Divide both sides of the equation by 2.

$$\frac{2R^2}{2} = \frac{50}{2}$$
$$R^2 = 25$$

Apply the root principle. Take the square root of both sides.

$$\sqrt{R^2} = \sqrt{25}$$
$$R = 5 \quad \text{Ans}$$

Check.

$$\frac{4}{R} = \frac{2R}{12.5}$$
$$\frac{4}{5} = \frac{2(5)}{12.5}$$
$$\frac{4}{5} = \frac{10}{12.5}$$
$$0.8 = 0.8 \quad \text{Ck}$$

Application

Equations Consisting of Combined Operations

Solve for the unknown and check each of the following combined operations equations. Round answers to 2 decimal places where necessary.

1. $5x - 33 = 12$ ______
2. $10M + 5 + 4M = 89$ ______
3. $8E - 14 = 2E + 28$ ______
4. $4B - 7 = B + 21$ ______
5. $7T - 14 = 0$ ______
6. $6N + 4 = 84 + N$ ______
7. $2.5A + 8 = 15 - 4.5$ ______
8. $12 - (-x + 8) = 18$ ______
9. $3H + (2 - H) = 20$ ______
10. $12 = -(2 + C) - (4 + 2C)$ ______
11. $-5(R + 6) = 10(R - 2)$ ______
12. $0.29E = 9.39 - 0.01E$ ______
13. $7.2F + 5(F - 8.1) = 0.6F + 15.18$ ______
14. $\frac{P}{7} + 8 = 6.3$ ______
15. $\frac{1}{4}W + (W - 8) = \frac{3}{4}$ ______
16. $\frac{1}{8}D - 3(D - 7) = 5\frac{1}{8}D - 3$ ______
17. $0.58y = 18.3 - 0.02y$ ______
18. $2H^2 - 20 = (H + 4)(H - 4)$ ______
19. $4A^2 + 3A + 36 = 8A^2 + 3A$ ______
20. $x(4 + x) + 20 = x^2 - (x - 5)$ ______
21. $\left(\frac{b}{2}\right)^3 + 34 = 42$ ______
22. $3F^3 + F(F + 8) = 8F + F^2 + 81$ ______
23. $9 + y^2 = (y - 4)(y - 1)$ ______
24. $\frac{1}{4}(2B - 12) + B^2 = \frac{1}{2}B + 22$ ______
25. $-4(y - 1.5) = 2\sqrt{y} - 4y$ ______
26. $14\sqrt{x} = 6(\sqrt{x} + 8) + 16$ ______
27. $8.12P^2 + 6.83P + 5.05$
    $= 16.7P^2 + 6.83P$ ______
28. $7.3\sqrt{x} = 3(\sqrt{x} + 8.06) - 4.59$ ______
29. $\sqrt{B^2} - 2.53B = -2.53(B - 3.95)$ ______
30. $(2y)^3 - 2.80(5.89 + 3y)$
    $= 23.87 - 8.40y$ ______

Substituting Values and Solving Formulas

The following formulas are used in the machine trades. Substitute the given values in each formula and solve for the unknown. Round the answers to 3 decimal places where necessary.

31. $F = 2.380P + 0.250$
    Given: $F = 2.375$.
    Solve for P. __________
32. $a = 3H \div 8$
    Given: $a = 0.1760$.
    Solve for H. __________
33. $H.P. = 0.000016MN$
    Given: $H.P. = 22, N = 50.8$.
    Solve for M. __________
34. $N = 0.707DP_n$
    Given: $N = 24, P_n = 8$.
    Solve for D. __________
35. $S = T - \dfrac{1.732}{N}$
    Given: $S = 0.4134, N = 20$.
    Solve for T. __________
36. $a = \dfrac{D_2 - D_1}{2}$
    Given: $a = 0.250, D_1 = 0.875$.
    Solve for D_2. __________
37. $S = \dfrac{0.290W}{t^2}$
    Given: $S = 1000, t = 0.750$.
    Solve for W. __________
38. $W = St(0.55d^2 - 0.25d)$
    Given: $W = 1150, d = 0.750$.
    Solve for St. __________
39. $M = E - 0.866P + 3W$
    Given: $M = 3.3700, E = 3.2000, P = 0.125$.
    Solve for W. __________
40. $S = \dfrac{L_1}{L_2}\left[\dfrac{1}{2}(D_1 - D_2)\right]$
    Given: $S = \dfrac{1}{4}, L_1 = 16, L_2 = 4, D_2 = 2\dfrac{1}{2}$.
    Solve for D_1. __________
41. $C = \dfrac{\pi DN}{12}$
    Given: $C = 210, D = 6, \pi = 3.1416$.
    Solve for N. __________
42. $D_o = \dfrac{P_c(N + 2)}{\pi}$
    Given: $D_o = 4.3750, \pi = 3.1416, P_c = 0.3927$.
    Solve for N. __________
43. $S = \sqrt{\dfrac{d^2}{4} = h^2}$
    Given: $S = 12.700, d = 6$.
    Solve for h. __________
44. $C = 2\sqrt{h(2r - h)}$
    Given: $C = 7.600, h = 3.750$.
    Solve for r. __________

Rearranging Formulas

The following formulas are used in machine trade calculations. Rearrange the formulas in terms of the designated letters.

45. The dimensions shown can be found using these two formulas.

    (1) $A = ab$ (2) $d = \sqrt{a^2 + b^2}$

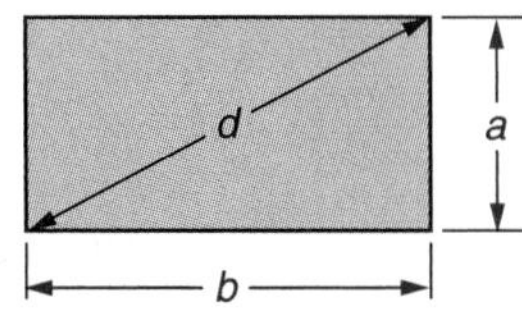

   a. Solve formula (1) for a. __________
   b. Solve formula (1) for b. __________
   c. Solve formula (2) for a. __________
   d. Solve formula (2) for b. __________

46. The radii shown in this figure can be found using these two formulas.

    (1) $R = 1.155r$ (2) $A = 2.598R^2$

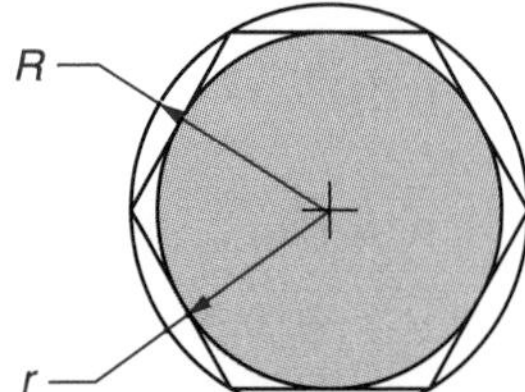

   a. Solve formula (1) for r. __________
   b. Solve formula (2) for R. __________

47. The dimensions shown can be found using these two formulas.

    (1) $FW = \sqrt{D_o^2 - D^2}$ (2) $D_o = 2C - d + 2a$

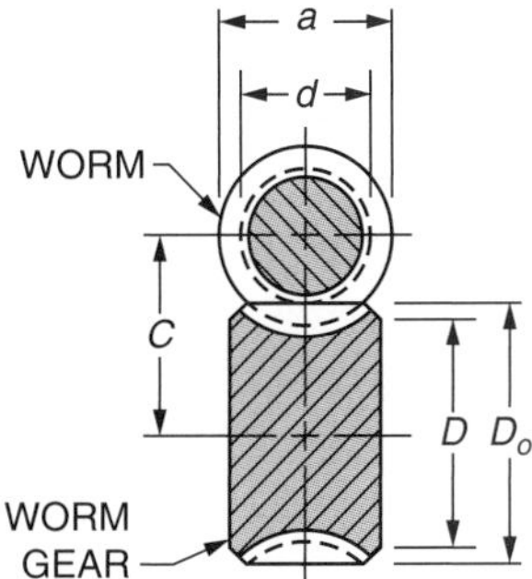

   a. Solve formula (1) for D_o. ________
   b. Solve formula (1) for D. ________
   c. Solve formula (2) for d. ________
   d. Solve formula (2) for a. ________

48. $A = \pi(R^2 - r^2)$

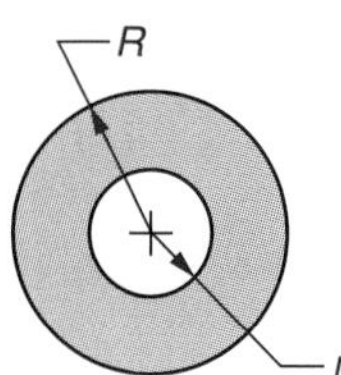

   a. Solve for R. ________
   b. Solve for r. ________

49. $M = D - 1.5155P + 3W$

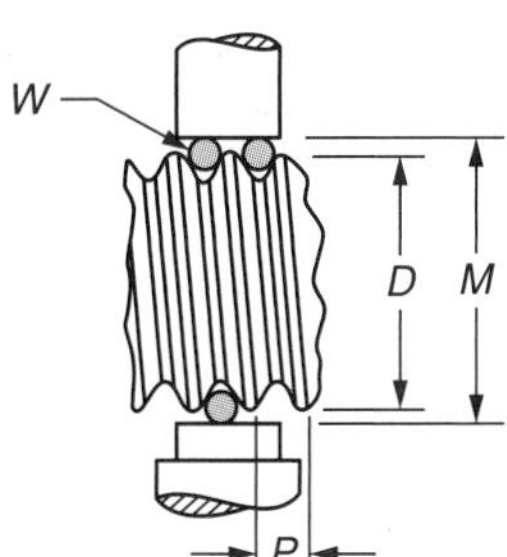

   a. Solve for D. ________
   b. Solve for P. ________
   c. Solve for W. ________

50. $\angle A + \angle B + \angle C = 180°$

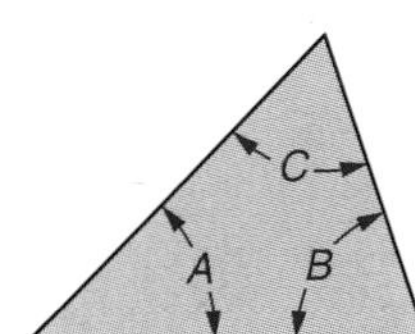

   a. Solve for $\angle A$. ________
   b. Solve for $\angle B$. ________
   c. Solve for $\angle C$. ________

51. $L = 3.14(0.5D + 0.5d) + 2x$

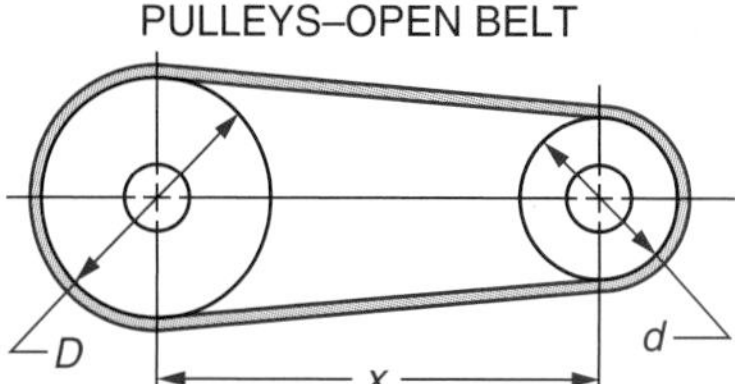

   a. Solve for D. ________
   b. Solve for d. ________
   c. Solve for x. ________

52. $By(F - 1) = Cx$

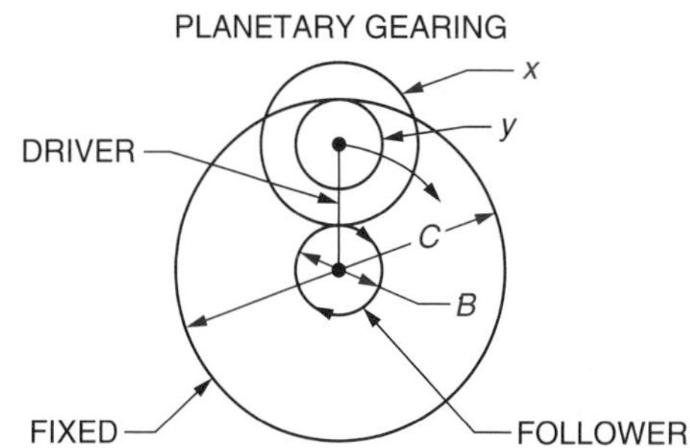

   a. Solve for x. ________
   b. Solve for B. ________
   c. Solve for C. ________

53. $Ca = S(C - F)$

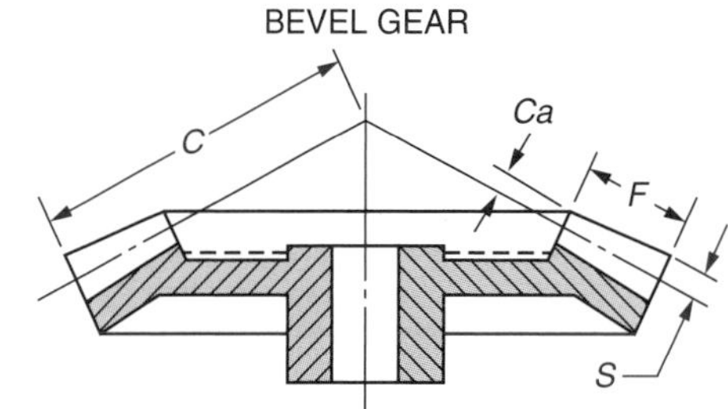

   a. Solve for S. ________
   b. Solve for C. ________

For problems 54 and 55, rearrange each formula for the designated letter and solve.

54. The horsepower of an electric motor is found with this formula.

$$hp = \frac{6.2832T(\text{rpm})}{33{,}000} \quad \text{where} \quad hp = \text{horsepower}$$

T = torque in pound-feet (lb-ft)

rpm = revolutions per minute

Solve for T when $hp = 1.50$ and $rpm = 2250$. Round the answer to 2 decimal places. __________

55. A tapered pin is shown.

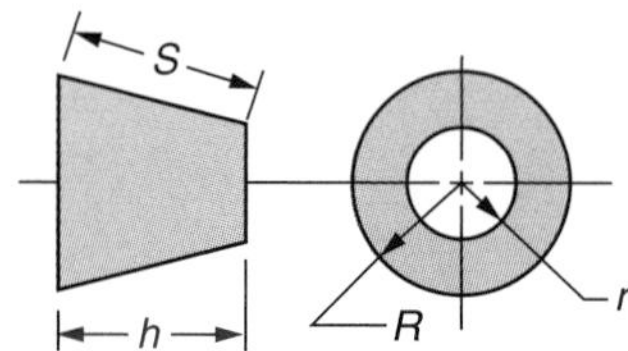

a. Solve for h when $R = 2.38$ cm, $r = 1.46$ cm, and $V = 69.5$ cm^3. Round the answer to 2 decimal places.

$$V = 1.05(R + Rr + r^2)h$$ __________

b. Solve for h when $S = 0.875$ in, $R = 0.420$ in, and $r = 0.200$ in. Round the answer to 3 decimal places.

$$S = \sqrt{(R - r)^2 + h^2}$$ __________

Proportions

Solve for the unknown value in each of the following proportions. Check each answer. Round the answers to 3 decimal places where necessary.

56. $\frac{10}{x} = \frac{5x}{8}$ __________

57. $\frac{A}{12.5} = \frac{4.761}{5A}$ __________

58. $\frac{11}{8} = \frac{E + 3}{12}$ __________

59. $\frac{M - 5}{12} = \frac{15}{8}$ __________

60. $\frac{5 - B}{4.38} = \frac{B + 9.2}{11.71}$ __________

61. $\frac{E - 15}{E + 7.53} = \frac{0.36}{1.86}$ __________

62. $\frac{6.08}{3H^2 - 12} = \frac{5.87}{12.53}$ __________

63. $\frac{12.8}{5 - 3C^2} = \frac{7.43}{-31.29}$ __________

64. $\frac{8.62M + 23.30}{12.36} = \frac{7.62M + 0.05}{0.86}$ __________

65. $\frac{P^2 - 186.73}{5.65P} = \frac{-23.30P}{3.04}$ __________

66. $\frac{B + 1}{B - 2} = \frac{B + 3}{B - 5}$ __________

67. $\frac{C + 4}{C - 3} = \frac{C - 1}{C + 6}$ __________

68. $\frac{N^3 - 4}{3} = \frac{N^3}{6}$ __________

69. $\frac{2A^3 + 7A - 5}{7} = \frac{A^3 + 5A + 8}{5}$ __________

Applications of Formulas to Cutting Speed, Revolutions per Minute, and Cutting Time

Objectives ***After studying this unit you should be able to***

- Solve cutting speed, revolutions per minute, and cutting time problems by substitution in given formulas.
- Solve production time and cutting feed problems by rearranging and combining formulas.

In order to perform cutting operations efficiently, a machine must be run at the proper cutting speed. Proper cutting speed is largely determined by the type of material that is being cut, the feed and depth of cut, the cutting tool, and the machine characteristics. The machinist must be able to determine proper cutting speeds by using trade handbook data and formulas.

Cutting Speed Using Customary Units of Measure

Cutting speeds or surface speeds for lathes, drills, milling cutters, and grinding wheels are computed using the same formula. On the lathe, the workpiece revolves. On drill presses, milling machines, and grinders, the tool revolves. Speeds are computed in reference to the tool rather than the workpiece. The speed of a revolving object equals the product of the circumference times the number of revolutions per minute made by the object. Generally, diameters are expressed in inches and cutting speeds are in feet per minute (fpm). In order to express inches per minute as feet per minute, it is necessary to divide by 12.

➤ **Note:** $\pi = 3.1416$ rounded to 4 decimal places. When solving problems with a calculator use the [π] key. On certain calculators, you must use the [2nd] key to get π.

$$C = \frac{3.1416DN}{12}$$

or $$C = \frac{\pi DN}{12}$$

where C = cutting speed in feet per minute (fpm)
D = diameter in inches
N = revolutions per minute (rpm)

Lathe

The *cutting speed* of a lathe is the number of feet that the revolving workpiece travels past the cutting edge of the tool in one minute.

Example A steel shaft 2.500 inches in diameter is turned in a lathe at 184.0 rpm. Determine the cutting speed to 1 decimal place.

$$C = \frac{\pi DN}{12} = \frac{\pi(2.500)(184.0)}{12}$$

[π] [×] 2.5 [×] 184 [÷] 12 [=]

120.4277184, 120.4 fpm Ans (rounded)

Milling Machine, Drill Press, and Grinder

The *cutting speed* or surface speed of a drill press, milling machine, and grinder is the number of feet that a point on the circumference of the tool travels in 1 minute.

Example 1 A 10-inch diameter grinding wheel runs at 1910 rpm. Determine the surface speed to the nearer whole number.

$$C = \frac{3.1416DN}{12} = \frac{3.1416(10)(1910)}{12} \approx$$

5000.38, 5000 fpm Ans (rounded)

Example 2 Determine the cutting speed to the nearer whole number of a $3\frac{1}{2}$-inch diameter milling cutter revolving at 120 rpm.

$$C = \frac{\pi DN}{12} = \frac{\pi(3.5)(120)}{12}$$

[π] [×] 3.5 [×] 120 [÷] 12 [=]

109.9557429, 110 fpm Ans (rounded)

Revolutions per Minute Using Customary Units of Measure

The cutting speed formula is rearranged in terms of N in order to determine the revolutions per minute of a workpiece or tool.

$$C = \frac{3.1416DN}{12}$$

$$12C = 3.1416DN$$

$$\frac{12C}{3.1416D} = N \text{ or } N = \frac{12C}{3.1416D}$$

Lathe

Example An aluminum cylinder with a 6.000-inch outside diameter is turned in a lathe at a cutting speed of 225 feet per minute. Determine the revolutions per minute to the nearest whole revolution.

$$N = \frac{12C}{3.1416D} = \frac{12(225)}{3.1416(6)} \approx 143.239, \ 143 \text{ rpm} \quad \text{Ans (rounded)}$$

Milling Machine, Drill Press, and Grinder

Example 1 A $\frac{1}{2}$-inch diameter twist drill has a cutting speed of 60.0 feet per minute. Determine the revolutions per minute to the nearest whole revolution.

$$N = \frac{12C}{3.1416D} = \frac{12(60)}{3.1416(0.5)} \approx 458.365, \ 458 \text{ rpm} \quad \text{Ans (rounded)}$$

Example 2 A 6.00-inch diameter grinding wheel operates at a cutting speed of 6000 feet per minute. Determine the revolutions per minute to the nearest whole revolution.

$$N = \frac{12C}{\pi D} = \frac{12(6000)}{\pi(6.00)}$$

12 [×] 6000 [÷] [(] [π] [×] 6 [)] [=]

3819.7186, 3820 rpm Ans (rounded)

Cutting Time Using Customary Units of Measure

The same formula is used to compute cutting times for machines that have a revolving workpiece, such as the lathe, as is used for machines that have a revolving tool, such as the milling machine, drill press, and grinder. Cutting time is determined by the length or depth to be cut in inches, the revolutions per minute of the revolving workpiece or revolving tool, and the tool feed in inches for each revolution of the workpiece or tool.

$$T = \frac{L}{FN}$$

where T = cutting time per cut in minutes
L = length of cut in inches
F = tool feed in inches per revolution
N = speed of revolving workpiece or tool in revolutions per minute

Lathe

Example 1 How many minutes are required to take one cut 22.00 inches in length on a steel shaft when the lathe feed is 0.050 inch per revolution and the shaft turns 152 rpm? Round the answer to 1 decimal place.

$$T = \frac{L}{FN} = \frac{22.00}{0.050(152)} \approx 2.895, 2.9 \text{ min}$$ Ans (rounded)

Example 2 A 3.250-inch diameter cast iron sleeve that is 20.00 inches long is turned in a lathe to a 2.450-inch diameter. Roughing cuts are each made to a 0.125-inch depth of cut. One finish cut using a 0.025-inch depth of cut is made. The feed is 0.100 inch per revolution for roughing and 0.030 inch for finishing. Roughing cuts are made at 150 rpm and the finish cut at 200 rpm. What is the total cutting time required?

Compute the total depth of cut.

$$\frac{3.250'' - 2.450''}{2} = \frac{0.800''}{2} = 0.400''$$

Compute the number of roughing cuts required.

$$\frac{0.400'' - 0.025''}{0.125''} = 3$$

Compute the time required for one roughing cut.

$$T = \frac{20.00}{(0.100)(150)} = 1.33 \text{ min}$$

Compute the total time for roughing.

$3 \times 1.33 \text{ min} = 4.0 \text{ min}$

Compute the time required for finishing.

$$T = \frac{20.00}{(0.030)(200)} = 3.3 \text{ min}$$

Compute the total cutting time.

$4.0 \text{ min} + 3.3 \text{ min.} = 7.3 \text{ min}$ Ans

Milling Machine, Drill Press, and Grinder

Example 1 Determine the cutting time required to drill through a workpiece that is 3.600 inches thick with a drill revolving 300 rpm and a feed of 0.025 inch per revolution.

$$T = \frac{L}{FN} = \frac{3.600}{0.025(300)} = 0.48 \text{ min} \quad \text{Ans}$$

Example 2 A milling machine cutter makes 460 rpm with a table feed of 0.020 inch per revolution. Four cuts are required to mill a slot in an aluminum plate 28.68 inches long. Compute the total cutting time. Round the answer to 1 decimal place.

$$\text{Total Cutting Time} = \frac{L}{FN} \times 4 = \frac{28.68}{0.020(460)} \times 4$$

28.68 [÷] [(] .02 [×] 460 [)] [×] 4 [=]
12.46956, 12.5 min Ans (rounded)

Cutting Speed Using Metric Units of Measure

Diameters are expressed in millimeters. Cutting speeds are expressed in meters per minute. The symbol for meters per minute is m/min. In order to express speed in millimeters per minute as meters per minute, it is necessary to divide by 1000 or to move the decimal point 3 places to the left.

$$C = \frac{3.1416DN}{1000}$$

where C = cutting speed in meters per minute
D = diameter in millimeters
N = revolutions per minute

Example A medium-steel shaft is cut in a lathe using a high-speed tool. The shaft has a diameter of 55 millimeters and is turning at 260 revolutions per minute. Determine the cutting speed to the nearest whole number.

$$C = \frac{3.1416DN}{1000} = \frac{3.1416(55)(260)}{1000} \approx 45 \text{ m/min} \quad \text{Ans (rounded)}$$

Revolutions per Minute Using Metric Units of Measure

In the metric system, the symbol for revolutions per minute is r/min. The cutting speed formula is rearranged in terms of N in order to determine the revolutions per minute of a workpiece or tool.

$$C = \frac{3.1416DN}{1000}$$

$$1000C = 3.1416DN$$

$$\frac{1000C}{3.1416D} = N \text{ or } N = \frac{1000C}{3.1416D}$$

Example A high-speed steel milling cutter with a 45 millimeter diameter and a cutting speed of 12 meters per minute is used for a roughing operation on an annealed chromium-nickel steel workpiece. Determine the revolutions per minute to the nearest whole number.

$$N = \frac{1000C}{3.1416D} = \frac{1000(12)}{3.1416(45)}$$

1000 [×] 12 [÷] [(] [π] [×] 45 [)] [=]
84.88264, 85 r/min Ans (rounded)

Cutting Time Using Metric Units of Measure

Cutting time is determined by the length or depth to be cut in millimeters, the revolutions per minute of the revolving workpiece or revolving tool, and the tool feed in millimeters for each revolution of the workpiece or tool.

$$T = \frac{L}{FN}$$

where T = cutting time per cut in minutes
L = length of cut in millimeters
F = tool feed in millimeters per revolution
N = r/min of revolving workpiece or tool

Example An 88-millimeter diameter cast iron cylinder is turned in a lathe at 260 revolutions per minute. Each length of cut is 700 millimeters and 5 cuts are required. A carbide tool is fed into the workpiece at 0.40 millimeter per revolution. What is the total cutting time? Round the answer to the nearest minute.

Calculate the time required for one cut.

$$T = \frac{L}{FN} = \frac{700}{0.40(260)} \approx 6.73 \text{ min}$$

Calculate the total cutting time.

5(6.73 min) = 33.65 min, 34 min Ans (rounded)

Using Data from a Cutting Speed Table

Tables of cutting speeds have been developed and are used in determining machine spindle speed (revolutions per minute) settings. The tables take into consideration the material to be cut and the tool material.

In addition to the material being cut and the type of tool used, other factors must be taken into consideration. Variables are considered, such as the depth and width of cut, the design of the cutting tool, the rate of feed, the coolant used, and the finish required.

Because cutting speed depends upon many factors, data given in cutting speed tables should be considered as recommended values. Generally it is not possible to set machines to an exact calculated spindle speed. Therefore, a simplified spindle speed formula is used in computing revolutions per minute. In the simplified formula, 3.1416 is rounded to 3.

$$N = \frac{12C}{3D} = \frac{4C}{D}$$

Comprehensive detailed cutting speed tables are available that list cutting speeds for specific materials based on material alloy composition, hardness, and condition. Some tables list cutting speeds separately for rough and finish cuts.

Selected materials are listed in the following table with their respective cutting speeds using high-speed steel and carbide tools.

Material	CUTTING SPEEDS: FEET PER MINUTE (fpm)						
	Turning		*Milling*		*Drilling*	*Reaming*	
	High-Speed Steel Tool	Carbide Tool	High-Speed Steel Tool	Carbide Tool	High-Speed Steel Tool	High-Speed Steel Tool	Carbide Tool
Carbon Steel (1020), BHN 175–225	100	350	70–130	200–400	70	40	175
Alloy Steel (4320), BHN 220–275	70	300	50–100	225–450	60	40	150
Malleable Cast Iron (32510), BHN 110–160	200	600	130–225	400–800	130	90	240
Stainless Steel (305), BHN 225–275	60	200	50–80	175–275	40	25	100
Aluminum (5052)	600	1200	500–800	1000–1800	250	250	700
Brass, annealed	300	650	250–450	500–900	160	160	320
Manganese Bronze, cold drawn	250	550	200–350	450–650	140	120	275
Beryllium Copper, annealed	100	200	80–140	180–275	60	50	180

Revolutions per minute are generally computed using table cutting speeds with the simplified revolutions per minute formula. Where a range of cutting speed table values is listed, use the average of the low and high speeds given. For example, the cutting speed for milling the alloy steel shown in the table with a high-speed steel cutter is listed as 50–100 feet per minute. Use the average cutting speed of

$$75 \text{ feet per minute } \left(\frac{50 + 100}{2} = 75\right).$$

After revolutions per minute are calculated, generally, the machine spindle speed is set to the closest spindle speed below the calculated revolutions per minute. The spindle speed may then be increased or decreased depending on the performance of the operation.

Example 1 Calculate the revolutions per minute required to turn a 3.500-inch diameter piece of stainless steel using a carbide toolbit. Express the answer to the nearest revolution per minute.

Refer to the table of cutting speeds. The recommended cutting speed is 200 feet per minute.

$$N = \frac{4C}{D} = \frac{4(200)}{3.500} \approx 228.571, 229 \text{ rpm} \qquad \text{Ans (rounded)}$$

Example 2 A carbon steel plate is milled using a 2.75-inch diameter high-speed steel cutter. Compute, to the nearest whole number, the revolutions per minute.

Refer to the table of cutting speeds. The recommended cutting speed is 100 feet per minute.

$$\left(\frac{70 + 130}{2} = 100\right) \quad N = \frac{4C}{D} = \frac{4(100)}{2.75} \approx 145.455, 145 \text{ rpm} \qquad \text{Ans (rounded)}$$

Application

Cutting Speeds

Given the workpiece or tool diameters and the revolutions per minute, determine the cutting speeds in the following tables to the nearest whole number. Use $C = \frac{3.1416DN}{12}$ or $\frac{\pi DN}{12}$ for customary units and $C = \frac{3.1416DN}{1000}$ or $\frac{\pi DN}{1000}$ for metric units.

	Workpiece or Tool Diameter	Revolutions per Minute	Cutting Speed (fpm)
1.	0.475"	460	
2.	2.750"	50	
3.	4.000"	86	
4.	0.850"	175	
5.	1.750"	218	

	Workpiece or Tool Diameter	Revolutions per Minute	Cutting Speeds (m/min)
6.	190.00 mm	59	
7.	53.98 mm	764	
8.	3.25 mm	1525	
9.	133.35 mm	254	
10.	6.35 mm	4584	

Revolutions per Minute

Given the cutting speed and the tool or workpiece diameter, determine the revolutions per minute in the following tables to the nearest whole number. Use $N = \frac{12C}{3.1416D}$ or $\frac{12C}{\pi D}$ for customary units and $N = \frac{1000C}{3.1416D}$ or $\frac{1000C}{\pi D}$ for metric units.

	Cutting Speed (fpm)	Workpiece or Tool Diameter	Revolutions per Minute
11.	70	2.400"	
12.	120	0.750"	
13.	90	8.000"	
14.	180	8.000"	
15.	200	0.375"	

	Cutting Speed (m/min)	Workpiece or Tool Diameter	Revolutions per Minute
16.	130	25.50 mm	
17.	100	66.70 mm	
18.	30	6.35 mm	
19.	180	15.80 mm	
20.	150	114.30 mm	

Cutting Time

Given the number of cuts, the length of cut, the revolutions per minute of the workpiece or tool, and the tool feed, determine the total cutting time in the table to 1 decimal place. Use $T = \frac{L}{FN}$.

	Number of Cuts	Feed (per revolution)	Length of Cut	Revolutions per Minute	Tool Cutting Time (Minutes)
21.	1	0.002"	20"	2100	
22.	1	0.12 mm	925 mm	610	
23.	4	0.008"	8"	350	

Cutting Speed and Surface Speed Problems

Compute the following problems. Express the answers to the nearer whole number. Use $C = \frac{3.1416DN}{12}$ or $\frac{\pi DN}{12}$ for customary units and $C = \frac{3.1416DN}{1000}$ or $\frac{\pi DN}{1000}$ for metric units.

24. A $3\frac{1}{2}$-inch diameter high-speed steel cutter, running at 55 rpm, is used to rough mill a steel casting. What is the cutting speed? ______
25. A 50-millimeter diameter carbon steel drill running at 286 r/min is used to drill an aluminum plate. Find the cutting speed. ______
26. What is the surface speed of a 16-inch diameter surface grinder wheel running at 1194 rpm? ______
27. A medium-steel shaft is cut in a lathe using a high-speed steel tool. Determine the cutting speed if the shaft is 2.125 inches in diameter, and is turning at 275 rpm. ______
28. A finishing cut is taken on a brass workpiece using a 100-millimeter diameter carbon steel milling cutter. What is the cutting speed when the cutter is run at 86 r/min? ______

Revolutions per Minute Problems

Compute the following problems. Express the answers to the nearest whole number. Use $N = \frac{12C}{3.1416D}$ or $\frac{12C}{\pi D}$ for customary units and $N = \frac{1000C}{3.1416D}$ or $\frac{1000C}{\pi D}$ for metric units.

29. Grooves are cut in a stainless steel plate using a 3.750-inch diameter carbide milling cutter with a cutting speed of 180 feet per minute. Determine the revolutions per minute. ______
30. An annealed cast iron housing is drilled with a cutting speed of 20 meters per minute using a 22-millimeter diameter carbon steel drill. Find the revolutions per minute. ______
31. A grinding operation is performed using a 150-millimeter diameter wheel with a cutting speed of 1800 meters per minute. Determine the revolutions per minute. ______
32. Determine the revolutions per minute of an aluminum alloy rod 1.250 inches in diameter with a cutting speed of 550 feet per minute. ______
33. A high-speed steel milling cutter with a 1.750-inch diameter and a cutting speed of 40 feet per minute is used for a roughing operation on an annealed chromium-nickel steel workpiece. Find the revolutions per minute. ______

Cutting Time Problems

Compute the following problems. Express the answers to 1 decimal place. Use

$$T = \frac{L}{FN}$$

34. Cast iron, $3\frac{1}{4}$ inches in diameter, is turned in a lathe at 270 rpm. Each length of cut is 27.00 inches and five cuts are required. A carbide tool is fed into the work at 0.015 inch per revolution. What is the total cutting time? ______
35. A slot 812.00 millimeters long is cut into a carbon steel baseplate with a feed of 0.80 millimeter per revolution. Find the cutting time using a 75-millimeter diameter carbide milling cutter running at 640 r/min. ______
36. Fifteen 3.20 millimeter diameter holes each 57.15 millimeters deep are drilled in an aluminum workpiece. The high-speed steel drill runs at 9200 r/min with a feed of 0.05 millimeter per revolution. Determine the total cutting time. ______

37. Thirty 2-inch diameter stainless steel shafts are turned in a lathe at 250 rpm. Two cuts each 14.5 inches long are required using a feed of 0.020 inch per revolution. Setup and handling time averages 3 minutes per piece. Calculate the total production time. ______

38. Seven brass plates 9.00 inches wide and 21.00 inches long are machined with a milling cutter along the length of the plates. The entire top face of each plate is milled. The width of each cut allowing for overlap is $2\frac{1}{4}$ inches. Using a feed of 0.020 inch per revolution and 525 rpm, determine the total cutting time. ______

Complex Problems

The solution of the following problems requires more than one formula and the rearrangement of formulas.

$$\text{Use } C = \frac{3.1416DN}{12} \text{ or } \frac{\pi DN}{12} \text{ for customary units and}$$

$$C = \frac{3.1416DN}{1000} \text{ or } \frac{\pi DN}{1000} \text{ for metric units.}$$

$$N = \frac{12C}{3.1416D} \text{ or } \frac{12C}{\pi D} \text{ for customary units and}$$

$$N = \frac{1000C}{3.1416D} \text{ or } \frac{1000C}{\pi D} \text{ for metric units.}$$

$$T = \frac{L}{FN}$$

39. A 3.000-inch diameter cylinder is turned for an 11.300-inch length of cut. The cutting speed is 300 feet per minute and the cutting time is 1.02 minutes. Calculate the tool feed in inches per revolution. Round the answer to 3 decimal places. ______

40. A combination drilling and countersinking operation on bronze round stock is performed on an automatic screw machine. The length of cut per piece is $1\frac{3}{4}$ inches. The total cutting time for 2300 pieces is $6\frac{1}{2}$ hours running at 1600 rpm. What is the tool feed in inches per revolution? Round the answer to 3 decimal places. ______

41. Steel shafts, $1\frac{1}{4}$ inches in diameter, are turned on an automatic machine. One finishing operation is required for a 16.5-inch length of cut. The tool feed is 0.015 inch per revolution using a cutting speed of 200 feet per minute. Determine the number of hours of cutting time required for 1500 shafts. Round the answer to the nearest hour. ______

42. A carbide milling cutter is used for machining a 560.00-millimeter length of stainless steel. The cutting time is 11.95 minutes, the cutting speed is 60.000 meters per minute, and the feed is 0.250 millimeter per revolution. What is the diameter of the carbide milling cutter? Round the answer to 1 decimal place. ______

43. Aluminum baseplates are produced that are $1\frac{5}{8}$-inches thick. Six $\frac{1}{4}$-inch diameter holes are drilled in each plate using a feed of 0.004 inch per revolution and a cutting speed of 300 feet per minute. Setup and handling time is estimated at 0.5 minute per piece. What is the total number of hours required to produce 850 aluminum baseplates? Round the answer to 1 decimal place.

Cutting Speed Table

Refer to cutting speed table on page 260. Use the table values and the simplified revolutions per minute formula, $N = \frac{4C}{D}$. Compute the revolutions per minute to the nearer revolution for each problem in the following table.

	Material Machined	Cutting Operation	Tool Material	Tool or Workpiece Diameter (inches)	Speed (rpm)
44.	Aluminum (5052)	Milling	High-Speed Steel	3.500	
45.	Stainless Steel (305), BHN 225–275	Turning	Carbide	5.200	
46.	Alloy Steel (4320), BHN 220–275	Reaming	High-Speed Steel	0.480	
47.	Manganese Bronze, cold drawn	Drilling	High-Speed Steel	0.375	
48.	Brass, annealed	Milling	High-Speed Steel	4.000	
49.	Carbon Steel (1020), BHN 175–225	Turning	Carbide	6.100	
50.	Beryllium Copper, annealed	Drilling	High-Speed Steel	1.100	
51.	Malleable Cast Iron (32510), BHN 110–160	Milling	Carbide	3.000	
52.	Alloy Steel (4320), BHN 220–275	Milling	Carbide	2.500	
53.	Aluminum (5052)	Turning	High-Speed Steel	5.800	
54.	Carbon Steel (1020), BHN 175–225	Milling	Carbide	4.500	
55.	Brass, annealed	Turning	High-Speed Steel	2.750	
56.	Stainless Steel (305), BHN 225–275	Reaming	Carbide	0.620	
57.	Malleable Cast Iron (32510), BHN 110–160	Turning	Carbide	7.000	
58.	Carbon Steel (1020), BHN 175–225	Drilling	High-Speed Steel	0.375	

UNIT 42

Applications of Formulas to Spur Gears

Objectives ***After studying this unit you should be able to***

- Identify the proper gear formula to use depending on the unknown and the given data.
- Compute gear part dimensions by substituting known values directly into formulas.

- Compute gear part dimensions by rearranging given formulas in terms of the unknowns.
- Compute gear part dimensions by the application of two or more formulas in order to determine an unknown.

Gears have wide application in machine technology. They are basic to the design and operation of machinery. Most machine shops are equipped to cut gears, and some shops specialize in gear design and manufacture. It is essential that the machinist and drafter have an understanding of gear parts and the ability to determine gear dimensions by the use of trade handbook formulas.

Description of Gears

Gears are used for transmitting power by rotary motion between shafts. Gears are designed to prevent slippage and to ensure positive motion while maintaining a high degree of accuracy of the speed ratios between driving and driven gears. The shape of the gear tooth is of primary importance in providing a smooth transmission of motion. The shape of most gear teeth is an *involute curve*. This curve is formed by the path of a point on a straight line as it rolls along a circle. *Spur gears* are gears that are in mesh between parallel shafts. Of two gears in mesh, the smaller gear is called the *pinion* and the larger gear is called the *gear*.

Spur Gear Definitions

Spur gears and the terms that apply to these gears are shown in Figures 42-1 and 42-2. It is essential to study the figures and gear terms before computing gear problems by the use of formulas.

Pitch Circles are the imaginary circles of two meshing gears that make contact with each other. The circles are the basis of gear design and gear calculations.

Pitch Diameter is the diameter of the pitch circle.

Root Circle is a circle that coincides with the bottoms of the tooth spaces.

Root Diameter is the diameter of the root circle.

Outside Diameter is the diameter measured to the tops of the gear teeth.

Addendum is the height of the tooth above the pitch circle.

Dedendum is the depth of the tooth space below the pitch circle.

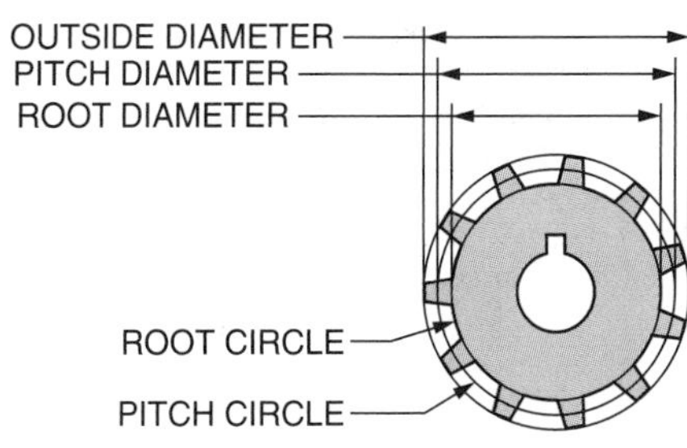

Figure 42-1

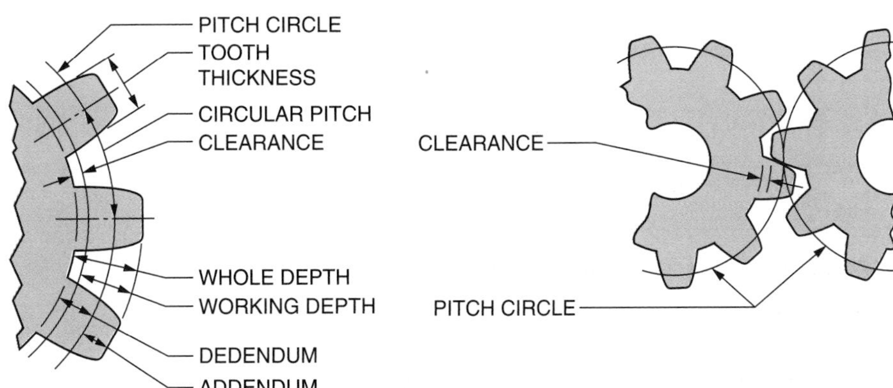

Figure 42-2

Whole Depth is the total depth of the tooth space. It is equal to the addendum plus the dedendum.

Working Depth is the total depth of mating teeth when two gears are in mesh. It is equal to twice the addendum.

Clearance is the distance between the top of a tooth and the bottom of the mating tooth space of two gears in mesh. It is equal to the whole depth minus the working depth.

Tooth Thickness (Circular) is the length of the arc, on the pitch circle, between the two sides of a tooth.

Circular Pitch is the length of the arc measured on the pitch circle between the centers of two adjacent teeth. It is equal to the circumference of the pitch circle divided by the number of teeth on the gear.

Diametral Pitch (Pitch) is the ratio of the number of gear teeth to the number of inches of pitch diameter. It is equal to the number of gear teeth for each inch of pitch diameter.

When the pitch of a gear is mentioned, the reference is to diametral pitch, rather than circular pitch. For example, if a gear has 28 teeth and a pitch diameter of 4 inches, it has a pitch (diametral pitch) of $\frac{28}{4}$ or 7. It has 7 teeth per inch of pitch diameter, and it is called a 7-pitch gear. It will only mesh with other 7-pitch gears. Gears must have the same pitch in order to mesh.

Gearing–Diametral Pitch System

The diametral pitch system is the system of gear design that is generally applied to decimal-inch dimensional gears. The following table lists the symbols and formulas used in the diametral pitch system.

DECIMAL-INCH SPUR GEARS (American National Standard)		
Term	*Symbol*	*Formulas*
Pitch (diametral pitch)	P	$P = \frac{N}{D}$ $P = \frac{3.1416}{P_C}$
Circular Pitch	P_C	$P_C = \frac{3.1416D}{N}$ $P_C = \frac{3.1416}{P}$
Pitch Diameter	D	$D = \frac{N}{P}$ $D = \frac{NP_C}{3.1416}$
Outside Diameter	D_O	$D_O = \frac{N + 2}{P}$ $D_O = \frac{P_C(N + 2)}{3.1416}$ $D_O = D + 2a$

DECIMAL-INCH SPUR GEARS (*Continued*)		
Term	*Symbol*	*Formulas*
Root Diameter	D_R	$D_R = D - 2d$
Addendum	a	$a = \frac{1}{P}$* $a = 0.3183P_C$*
Dedendum	d	$d = \frac{1.157}{P}$* $d = 0.3683P_C$*
Whole Depth	WD	$WD = \frac{2.157}{P}$* $WD = 0.6866P_C$* $WD = a + d$*
Working Depth	W_D	$W_D = \frac{2.000}{P}$* $W_D = 0.6366P_C$*
Clearance	c	$c = \frac{0.157}{P}$* $c = 0.050P_C$*
Tooth Thickness	T	$T = \frac{1.5708}{P}$
Number of Teeth	N	$N = PD$ $N = \frac{3.1416D}{P_C}$

**Note:* Formulas for $14\frac{1}{2}$-degree Involute and Composite Full-Depth Teeth and 20-degree Involute Full-Depth Teeth.

Gear Calculations

Most gear calculations are made by identifying the proper formula, which is given in terms of the unknown, and substituting the known dimensions. It is sometimes necessary to rearrange a formula in terms of a particular unknown. The solution of a problem may require the substitution of values in two or more formulas.

Examples Refer to the Decimal-Inch Spur Gears Table on pages 266 and 267.

1. Determine the pitch diameter of a 5-pitch gear that has 28 teeth.

Identify the formula whose parts consist of pitch diameter, pitch, and number of teeth. $D = \frac{N}{P}$

Solve. $D = \frac{28}{5}$

$D = 5.6000$ inches Ans

2. Determine the outside diameter of a gear that has 16 teeth and a circular pitch of 0.7854 inch.

Identify the formula whose parts consist of outside diameter, number of teeth, and circular pitch.

$$D_O = \frac{P_C(N + 2)}{3.1416}$$

Solve.

$$D_O = \frac{0.7854(16 + 2)}{3.1416}$$

.7854 [×] [(] 16 [+] 2 [)] [÷] 3.1416 [=] 4.50001523

$D_O = 4.5000$ inches Ans (rounded)

3. Determine the circular pitch of a gear with a whole depth dimension of 0.3081inch.

Identify the formula whose parts consist of circular pitch and whole depth.

$$WD = 0.6866P_C$$

The formula must be rearranged in terms of circular pitch.

$$P_C = \frac{WD}{0.6866}$$

Solve.

$$P_C = \frac{0.3081}{0.6866}$$

$P_C \approx 0.4487$ inch Ans (rounded)

4. Determine the addendum of a gear that has an outside diameter of 3.0000 inches and a pitch diameter of 2.7500 inches.

Identify the formula whose parts consist of addendum, outside diameter, and pitch diameter.

$$D_O = D + 2a$$

The formula must be rearranged in terms of the addendum.

$$D_O - D = 2a$$

$$a = \frac{D_O - D}{2}$$

Solve.

$$a = \frac{3.0000 - 2.7500}{2}$$

$a = 0.1250$ inch Ans

5. Determine the working depth of a gear that has 46 teeth and a pitch diameter of 11.5000 inches.

There is no single formula in the table that consists of working depth, number of teeth, and pitch diameter. Therefore, it is necessary to substitute in two formulas in order to solve the problem.

Observe $W_D = \frac{2.0000}{P}$.

$$P = \frac{N}{D}$$

The pitch must be found first.

$$P = \frac{46}{11.5000}$$

$$P = 4$$

Solve for W_D.

$$W_D = \frac{2.000}{P}$$

$$W_D = \frac{2.000}{4}$$

$W_D = 0.5000$ inch Ans

Gearing–Metric Module System

The *module system* of gear design is generally the system that is used with metric system units of measure. The *module* of a gear equals the pitch diameter divided by the number of teeth. In the metric system, the module of a gear means the pitch diameter in millimeters is divided by the number of teeth. Module is an actual dimension in millimeters, not a ratio as with diametral pitch. For example, if a gear has 20 teeth and a 50-millimeter pitch diameter, the module is 2.5 millimeters (50 mm ÷ 20). A module of 2.5 millimeters means that there are 2.5 millimeters of pitch diameter per tooth.

A partial list of a standard series of modules (in millimeters) is listed as follows:

1	2	3	4	6	9
1.25	2.25	3.25	4.5	6.5	10
1.5	2.5	3.5	5	7	11
1.75	2.75	3.75	5.5	8	12

The relation between module and various gear parts using a metric module system is shown in the following table.

METRIC SPUR GEARS		
Term	*Symbol*	*Formulas*
Module	m	$m = \frac{D}{N}$
Circular Pitch	P_C	$P_C = \frac{m}{0.3183}$
Pitch Diameter	D	$D = mN$
Outside Diameter	D_O	$D_O = m(N + 2)$
Addendum	a	$a = m$
Dedendum	d	$d = 1.157m$* $d = 1.167m$**
Whole Depth	WD	$WD = 2.157m$* $WD = 2.167m$**
Working Depth	W_D	$W_D = 2m$
Clearance	c	$c = 0.157m$* $c = 0.1667m$**
Tooth Thickness	T	$T = 1.5708m$

*When clearance = 0.157 × module
**When clearance = 0.1667 × module

Examples Refer to the Metric Spur Gears Table.

1. Determine the circular pitch of a 6 millimeter module gear.

   Circular Pitch = *Module* ÷ 0.3183

   Circular Pitch = 6 mm ÷ 0.3183 ≈ 18.850 mm Ans (rounded)

2. Determine the outside diameter of a 3.5 millimeter module gear with 20 teeth.

   Outside Diameter = *Module* × (*Number of teeth* + 2)

   Outside Diameter = 3.5 mm × (20 + 2) = 3.5 mm × 22 = 77.000 mm Ans

3. Compute the dedendum of a 4.5 millimeter module gear designed with a clearance of 0.157 × module.

   When *Clearance* = 0.157 × *Module,* the *Dedendum* = 1.157 × *Module*

   Dedendum = 1.157 × 4.5 mm ≈ 5.207 mm Ans (rounded)

4. Compute the whole depth of 7 millimeter module gear designed with a clearance of 0.1667 × module.

   When *Clearance* = 0.1667 × *Module,* the *Whole Depth* = 2.167 × *Module*

   Whole Depth = 2.167 × 7 mm = 15.169 mm Ans

Application

Gearing–Diametral Pitch System

Refer to the Decimal-Inch Spur Gears Table on pages 266 and 267 for each of the following gearing problems.

	GIVEN VALUES	FIND	ANSWER
1.	Circular Pitch = 1.5708"	Pitch	
2.	Pitch = 10	Circular Pitch	
3.	Pitch Diameter = 5.2000" Number of Teeth = 26	Circular Pitch	
4.	Pitch Diameter = 12.5714" Number of Teeth = 44	Pitch	
5.	Pitch = 7 Number of Teeth = 26	Pitch Diameter	
6.	Circular Pitch = 0.3142" Number of Teeth = 12	Pitch Diameter	
7.	Pitch = 18	Circular Pitch	
8.	Pitch Diameter = 0.7273" Number of Teeth = 16	Pitch	
9.	Pitch = 12 Pitch Diameter = 1.1667"	Number of Teeth	
10.	Circular Pitch = 0.6283" Pitch Diameter = 8.4000"	Number of Teeth	
11.	Number of Teeth = 56 Pitch = 8	Outside Diameter	
12.	Pitch = 14	Addendum	
13.	Pitch Diameter = 1.3333" Dedendum = 0.0643"	Root Diameter	
14.	Pitch = 3.4	Whole Depth	
15.	Circular Pitch = 0.2856"	Working Depth	
16.	Circular Pitch = 1.4650"	Clearance	

	GIVEN VALUES	FIND	ANSWER
17.	Pitch = 20	Tooth Thickness	
18.	Pitch Diameter = 3.5000" Addendum = 0.1818"	Outside Diameter	
19.	Circular Pitch = 0.0954"	Dedendum	
20.	Circular Pitch = 0.8976"	Addendum	
21.	Addendum = 0.1429" Dedendum = 0.1653"	Whole Depth	
22.	Pitch = 4	Dedendum	
23.	Circular Pitch = 0.3076"	Whole Depth	
24.	Pitch = 17	Clearance	
25.	Pitch = 9	Working Depth	

Refer to the Decimal-Inch Spur Gears Table on pages 266 and 267. The formula in terms of the unknown is not given. Choose the formula that consists of the given parts, rearrange in terms of the unknown, and solve.

	GIVEN VALUES	FIND	ANSWER
26.	Addendum = 0.0857"	Circular Pitch	
27.	Addendum = 0.0666"	Pitch	
28.	Addendum = 0.2000" Outside Diameter = 4.8000"	Pitch Diameter	
29.	Outside Diameter = 2.7144" Number of Teeth = 17	Pitch	
30.	Outside Diameter = 4.3750" Circular Pitch = 0.3927"	Number of Teeth	
31.	Working Depth = 0.0769"	Pitch	
32.	Working Depth = 0.4500"	Circular Pitch	
33.	Outside Diameter = 4.7144" Pitch Diameter = 4.4286"	Addendum	

Refer to the Decimal-Inch Spur Gears Table on pages 266 and 267. No single formula is given that consists of the given parts and the unknown. Two or more formulas, some in rearranged form, must be used in solving these problems.

	GIVEN VALUES	FIND	ANSWER
34.	Number of Teeth = 72 Pitch Diameter = 6.0000"	Addendum	
35.	Number of Teeth = 44 Pitch Diameter = 3.6667"	Dedendum	
36.	Number of Teeth = 10 Pitch Diameter = 2.5000"	Whole Depth	

	GIVEN VALUES	FIND	ANSWER
37.	Number of Teeth = 90 Pitch Diameter = 12.8571"	Working Depth	
38.	Pitch Diameter = 1.0625" Pitch = 16	Outside Diameter	
39.	Pitch Diameter = 2.9167" Pitch = 12	Root Diameter	
40.	Number of Teeth = 29 Pitch Diameter = 2.0714"	Root Diameter	
41.	Number of Teeth = 75 Pitch Diameter = 6.8182"	Clearance	
42.	Addendum = 0.1429"	Tooth Thickness	
43.	Pitch Diameter = 1.0455" Addendum = 0.0455"	Number of Teeth	

Backlash is the amount that a tooth space is greater than the engaging tooth on the pitch circles of two gears. Determine the average backlash of each of the following using this formula.

$$\text{Average backlash} = \frac{0.030}{P}$$

44. A 7-pitch gear ________
45. A 20-pitch gear ________
46. A 3.5-pitch gear ________
47. A gear with a whole depth of 0.2696" ________
48. A gear with a working depth of 0.1176" ________
49. A gear with a pitch diameter of 4.800" and 24 teeth ________

PITCH CIRCLE
PITCH CIRCLE
BACKLASH

The *center distance* of a pinion and a gear is the distance between the centers of the pitch circles. Determine the center distance of each of the following using this formula.

$$\text{Center distance} = \frac{\text{pitch diameter of gear} + \text{pitch diameter of pinion}}{2}$$

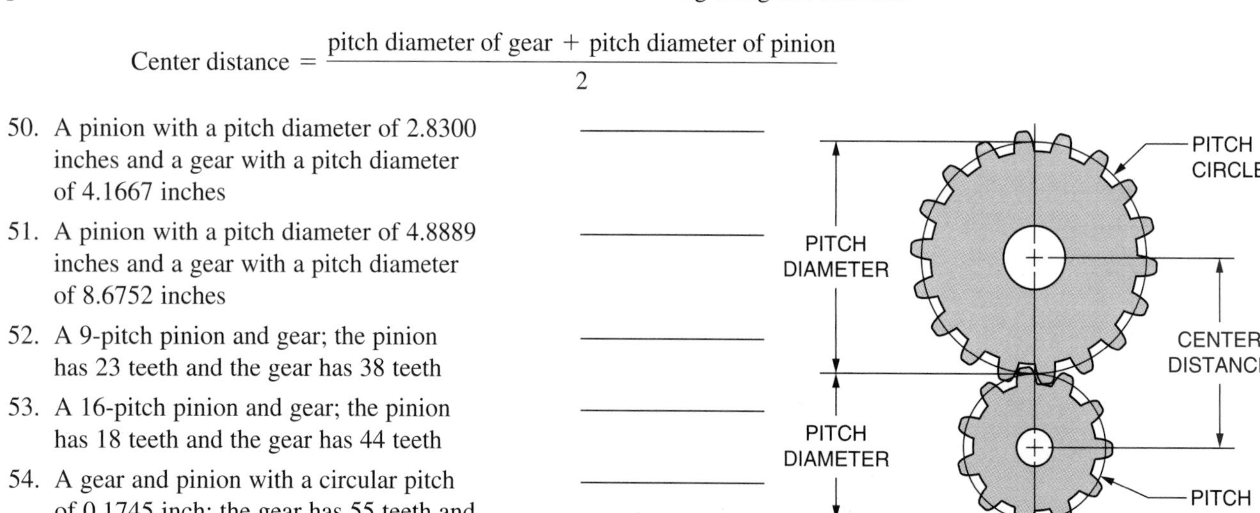

50. A pinion with a pitch diameter of 2.8300 inches and a gear with a pitch diameter of 4.1667 inches ________
51. A pinion with a pitch diameter of 4.8889 inches and a gear with a pitch diameter of 8.6752 inches ________
52. A 9-pitch pinion and gear; the pinion has 23 teeth and the gear has 38 teeth ________
53. A 16-pitch pinion and gear; the pinion has 18 teeth and the gear has 44 teeth ________
54. A gear and pinion with a circular pitch of 0.1745 inch; the gear has 55 teeth and the pinion has 37 teeth ________

Gearing–Metric Module System

Refer to the Metric Spur Gears Table on page 269 and determine the values in the following table.

	Module	Number of Teeth	a. Pitch Diameter	b. Circular Pitch	c. Outside Diameter	d. Addendum	e. Working Depth	f. Tooth Thickness
55.	6.5 mm	18						
56.	9 mm	24						
57.	2.5 mm	10						
58.	3.75 mm	16						
59.	10 mm	26						

Solve the following metric module system gearing problems. Certain problems require rearranging the data given in the Metric Spur Gears Table on page 269.

60. Compute the whole depth of a 5-millimeter module gear designed with a clearance of 0.157 × module. ______
61. What is the number of teeth on a 4-millimeter module gear with a pitch diameter of 120 millimeters? ______
62. What is the module of a gear that has a working depth of 13 millimeters? ______
63. Compute the dedendum of a 7-millimeter module gear designed with a clearance of 0.1667 × module. ______
64. What is the module of a gear with 38 teeth and an outside diameter of 220 millimeters? ______

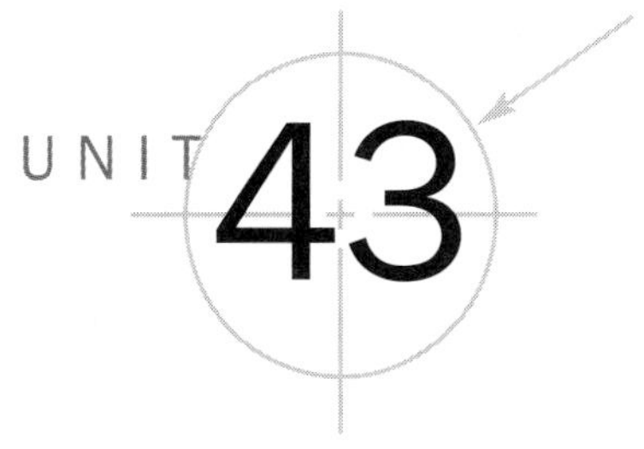

Achievement Review— Section Four

Objective ***You should be able to solve the exercises and problems in this Achievement Review by applying the principles and methods covered in Units 33–42.***

1. Express each of the following problems as an algebraic expression.
   a. The sum of x and y reduced by c. ______
   b. The product of a and b divided by d. ______
   c. Twice M minus the square of P. ______
2. Substitute the given numbers for letters and find the value for each of the following expressions. Round the answers to 3 decimal places where necessary.
   a. Find $(5a + 6b) \div 4b$ when $a = 4$ and $b = 5$. ______
   b. Find $3xy - (2x + y)$ when $x = 6$ and $y = 3$. ______

c. Find $e^2 + (2f)^2 - \left(\frac{m}{f}\right)^2$ when e = 5.125, f = 3.062, and m = 6.127. ________

d. Find $\sqrt{hr + 5p\left(\frac{5h}{p} + h\right)}$ when h = 10.26, p = 8.00, and r = 6.59. ________

3. Perform the operation or operations as indicated for each of the following exercises. Round the answers to 3 decimal places where necessary.

a. $-25 + (-12)$ ________

b. $24 + (-8)$ ________

c. $-1.8 - (12.6)$ ________

d. $18(-4)$ ________

e. $(-0.3)(-2.6)$ ________

f. $-18 \div 3$ ________

g. $-12.8 \div (-0.4)$ ________

h. $(-6)^2$ ________

i. $(-5)^3$ ________

j. $\sqrt[3]{-27}$ ________

k. $\left(-\frac{1}{4}\right)^2$ ________

l. $\sqrt[3]{\frac{27.063}{8.920}}$ ________

m. $(-4.02)^2 + \sqrt[3]{8.96} - (3.86)(-5.66)$ ________

n. $[4(-10.66)(0.37)] \div (12 + 18.95)$ ________

4. The following expressions consist of literal terms. Perform the indicated operations.

a. $-8P + 21P$ ________

b. $-0.05H^2 - 1.13H^2$ ________

c. $12d + 8d^2 - 7d + 14d^2$ ________

d. $(5a - 2a^2) - (6a - 4a^2)$ ________

e. $(-10x)(9x^2y)$ ________

f. $(-5.9e^2f^2)(-f^2)$ ________

g. $(16x^2 - 4x^3) \div 0.5x$ ________

h. $(0.6f^2g^3 - fg - 2f^2) \div 0.2f$ ________

i. $(6x^2 - y^3)(-3x^3 + y^2)$ ________

j. $(-4a^3bc^2)^3$ ________

k. $[(xy^2)^3 - (x^2y^2)]^2$ ________

l. $\sqrt[3]{-27a^6b^3c^9}$ ________

m. $\sqrt{\frac{25}{64}e^4g^2d}$ ________

n. $-6(x^2 + y - 2x)$ ________

o. $36 - (m^3 + m) + (m^3 - 12)$ ________

p. $-3a(2a)^2 + \sqrt{36a^4}$ ________

q. $9y - x[-8 + (xy)^2 - y] + 12x$ ________

r. $b(b + 7m)^2 - b(b - 7m)^2$ ________

5. Solve for the unknown in each of the following equations using one of the six principles of equality. Check each answer. Round the answers to 3 decimal places where necessary.

a. $x + 12 = 33$ ________

b. $y - 15 = 23$ ________

c. $-32 = B - 46$ ________

d. $H + 11.7 = 43.9$ ________

e. $14.3 = x + 53.6$ ________

f. $R - 7.8 = -9.2$ ________

g. $7y = -84$ ________

h. $1.3E = 7.54$ ________

i. $11.22 = \frac{L}{6.6}$ ________

j. $\frac{x}{-0.8} = 8.48$ ________

k. $6.75 = -20.25x$ ________

l. $\frac{3}{4}x = 2\frac{1}{4}$ ________

m. $s^2 = 81$ ________

n. $x^3 = \frac{-8}{27}$ ________

o. $\sqrt{M} = 12.892$ ________

p. $\sqrt[3]{V} = 5.873$ ________

q. $-0.0284 = y^3$ ________

r. $\sqrt[3]{B} = \dfrac{3.866}{4.023}$ ________

6. Solve for the unknown and check each of the following combined operations equations. Round the answers to 3 decimal places where necessary.

a. $32 - (-P + 18) = 45$ ________

b. $10(M - 4) = -5(M - 4)$ ________

c. $7.1E + 3(E - 6) = 0.5E + 22.8$ ________

d. $\dfrac{H}{4} + 7.8 = 13.6$ ________

e. $\dfrac{1}{4}F + 6\left(F - \dfrac{1}{2}\right) = 15\dfrac{3}{4}$ ________

f. $12x^2 - 53 = (x - 3)(x + 3)$ ________

g. $-8.53(G - 3.67) = 5.7(-18.36)$ ________

h. $(T - 7.8)(T - 8) = T^2 + 0.3$ ________

i. $59.66\sqrt{x} = 8.71(1.07\sqrt{x} + 55.32)$ ________

j. $H + 5.023\sqrt{H} = -9.777\sqrt{H} + H$ ________

7. In each of the following formulas, substitute given numerical values for letter values and solve for the unknown and check. Round the answers to 3 decimal places where necessary.

a. $N = 0.707DP_n$. Solve for D when $N = 36$ and $P_n = 12$. ________

b. $I = \dfrac{nE}{R + nr}$. Solve for E when $I = 0.3$, $n = 5$, $R = 6$, and $r = 4$. ________

c. $S = \dfrac{0.290W}{t^2}$. Solve for W when $S = 600$ and $t = 0.375$. ________

d. $c = \sqrt{a^2 + b^2}$. Solve for b when $a = 8.053$ and $c = 10.096$. ________

e. $V = 1.570h(R^2 + r^2)$. Solve for R when $V = 105.823$, $h = 5.897$, and $r = 2.023$. ________

8. Rearrange each of the following formulas in terms of the designated letter.

a. $E = I(R + r)$. Solve for r. ________

b. $D_O = 2C - d + 2a$. Solve for a. ________

c. $M = D - 1.5155P + 3W$. Solve for W. ________

d. $r = \sqrt{x^2 + y^2}$. Solve for y. ________

e. $L = 3.14(0.5D + 0.5d) + 2x$. Solve for x. ________

f. $HP = \dfrac{D^2N}{2.5}$. Solve for D. ________

9. Solve for the unknown value in each of the following proportions and check. Round the answers to 3 decimal places where necessary.

a. $\dfrac{P}{12.8} = \dfrac{3}{2}$ ________

b. $\dfrac{3.6}{0.9} = \dfrac{E}{2.7}$ ________

c. $\dfrac{H-1}{\frac{1}{4}} = \dfrac{\frac{1}{2}}{\frac{3}{8}}$ __________

d. $\dfrac{7}{6} = \dfrac{C+7}{12}$ __________

e. $\dfrac{10.360}{7.890} = \dfrac{2.015}{N-2.515}$ __________

f. $\dfrac{10+D}{D} = \dfrac{5}{9}$ __________

g. $\dfrac{G+5}{G+6} = \dfrac{G}{G+2}$ __________

h. $\dfrac{R-4}{R+2} = \dfrac{R-6}{R+8}$ __________

i. $\dfrac{n^3+6}{10} = \dfrac{n^3-8}{8}$ __________

j. $\dfrac{T^3+3T-0.7}{3} = \dfrac{T^3+5T+9.1}{5}$ __________

10. The following problems are given in scientific notation. Solve and leave answers in scientific notation. Round the answers (mantissas) to 2 decimal places.

a. $(3.76 \times 10^4) \times (2.87 \times 10^3)$ __________

b. $(8.63 \times 10^7) \div (5.77 \times 10^{-5})$ __________

c. $\dfrac{(9.76 \times 10^{-5}) \times (1.77 \times 10^9)}{(5.87 \times 10^3)}$ __________

d. $\dfrac{(9.09 \times 10^{-5})}{(4.72 \times 10^6) \times (6.15 \times 10^{-3})}$ __________

11. The following problems are given in decimal (standard) form. Calculate and give answers in scientific notation. Round the answers (mantissas) to 2 decimal places.

a. 0.00021×0.00039 __________

b. $1{,}476{,}000 \times -0.0000373$ __________

c. $\dfrac{0.0000287 \times 216{,}000{,}000}{0.00981}$ __________

d. $\dfrac{0.00503 \times 0.000406}{416{,}000 \times 0.00392}$ __________

12. Refer to the Decimal-Inch Spur Gears Table on pages 266 and 267 for each of the following. The formula in terms of the unknown may not be given.

	Given Values	Find	Answer
a.	Circular Pitch = 0.5712" Number of Teeth = 18	Pitch Diameter	
b.	Pitch = 1.574"	Addendum	
c.	Dedendum = 0.052"	Pitch	
d.	Clearance = 0.0281"	Circular Pitch	
e.	Pitch Diameter = 4.32" Number of Teeth = 36	Tooth Thickness	
f.	Pitch Diameter = 1.0346" Dedendum = 0.0855"	Number of Teeth	

13. Solve the following cutting speed and gear problems.

    a. A steel shaft 2.300 inches in diameter is turned in a lathe at 250 rpm. Determine the cutting speed to the nearer whole number.

    $$C = \frac{3.1416DN}{12}$$ __________

    b. Determine the revolutions per minute to the nearer whole number of an aluminum cylinder 40.00 millimeters in diameter with a cutting speed of 160 meters per minute.

    $$N = \frac{1000C}{3.1416D}$$ __________

    c. Twenty 5.50-millimeter diameter holes each 58.00 millimeters deep are drilled in a workpiece. The drill turns at 6500 r/min with a feed of 0.05 millimeter per revolution. Determine the total cutting time to the nearest hundredth minute.

    $$T = \frac{L}{FN}$$ __________

    d. What is the pitch of a gear with 44 teeth and a pitch diameter of 12.5714 inches?

    $$P = \frac{N}{D}$$ __________

    e. Determine the whole depth to 4 decimal places of a gear with 20 teeth and a pitch diameter of 5.0000 inches.

    $$P = \frac{N}{D} \text{ and } WD = \frac{2.157}{P}$$ __________

    f. Determine the center distance of a pinion with a pitch diameter of 6.4445 inches and a gear with a pitch diameter of 10.7533 inches. __________

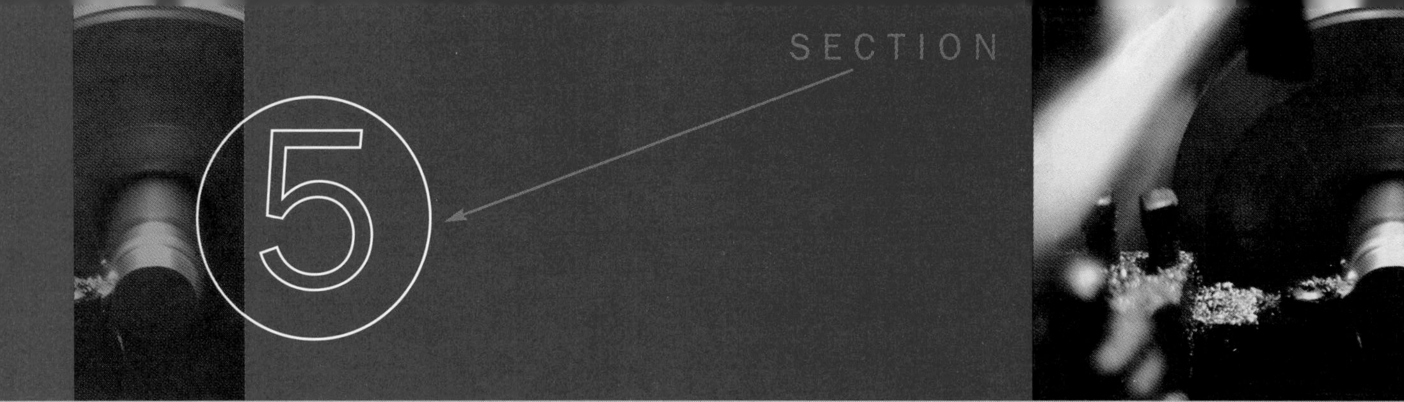

Fundamentals of Plane Geometry

Lines and Angular Measure

Objectives ***After studying this unit you should be able to***

- Add, subtract, multiply, and divide angles in terms of degrees, minutes, and seconds.
- Express decimal degrees as degrees, minutes, and seconds.
- Express degrees, minutes, and seconds as decimal degrees.

The fundamental principles of geometry generally applied to machine shop problems are those used to make the calculations required for machining parts from engineering drawings. An engineering drawing is an example of applied geometry.

Plane Geometry

Plane geometry is the branch of mathematics that deals with points, lines, and various figures that are made of combinations of points and lines. The figures lie on a flat surface, or *plane.* Examples of plane geometry are the views of a part as shown on an engineering drawing.

Since geometry is fundamental to machine technology, it is essential to understand the definitions and terms of geometry. It is equally important to be able to apply the geometric principles in problem solving. The methods and procedures used in problem solving are the same as those required for the planning, making, and checking of machined parts.

Procedure To solve a geometry problem

- Study the figure.
- Relate it to the principle or principles that are needed for the solution.
- Base all conclusions on fact: given information and geometric principles.
- Do not assume that something is true because of its appearance or because of the way it is drawn.

➤ **Note:** The same requirements are applied in reading engineering drawings.

Axioms and Postulates

In the study of geometry, certain basic statements called *axioms* or *postulates* are assumed to be true without requiring proof. Axioms or postulates may be compared to the rules of a game. Some axioms or postulates are listed. Others will be given as they are required for problem solving.

- Quantities equal to the same quantity, or to equal quantities, are equal to each other. A quantity may be substituted for an equal quantity.
- If equals are added to or subtracted from equals, the sums or remainders are equal.
- If equals are multiplied or divided by equals, the products or quotients are equal.
- The whole is equal to the sum of its parts.
- Only one straight line can be drawn between two given points.
- Through a given point, only one line can be drawn parallel to a given line.
- Two different or distinct straight lines can intersect at only one point.

Points and Lines

A *point* has no size or form; it has only location. A point is shown as a dot. Each point is usually named by a capital letter as shown.

$A \bullet$ $B \bullet$

$C \bullet$

$D \bullet$

A *line* extends without end in two directions. A line has no width. It has an infinite number of points. A *ray* starts with an end point and continues indefinitely. Sometimes a ray is referred to as a *half-line*.

A *line,* as it is used in this book, always means a straight line. A line other than a straight line, such as a curved line, is identified. Formally, arrowheads are used in drawing a line to show that there are no end points. In this book, arrowheads are *not* used to identify lines.

A *line segment* is that part of a line that lies between two definite points. Line segments are often named by placing a bar over the end point letters. For example, segment *AB* may be shown as $\overline{AB}$. In this book, segments are shown *without* a bar. Segment *AB* is shown as *AB*. In this book, *no* distinction is made between naming a line and a line segment.

Parallel lines do not meet regardless of how far they are extended. They are the same distance apart (equidistant) at all points. The symbol ∥ means parallel. In Figure 44-1, line *AB* is parallel to line *CD*; therefore, *AB* and *CD* are equidistant (distance *x*) at all points.

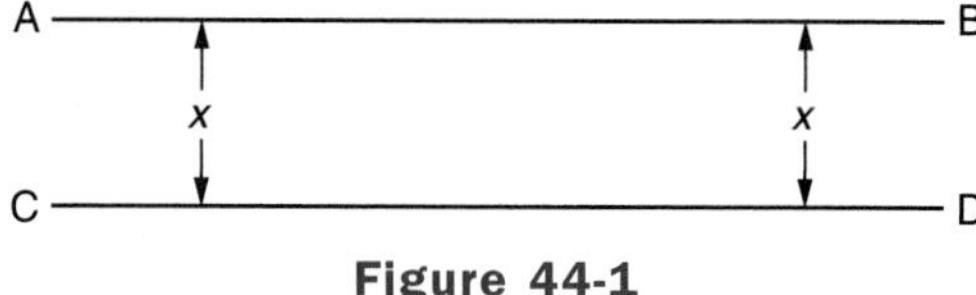

Figure 44-1

Perpendicular lines meet or intersect at a right, or 90°, angle. The symbol ⊥ means perpendicular. Figure 44-2 shows three examples of perpendicular lines.

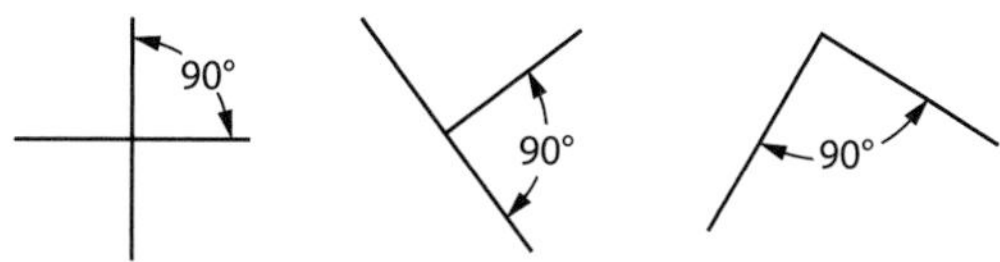

Figure 44-2

Oblique lines meet or intersect at an angle other than 90°. They are neither parallel nor perpendicular. Three examples of oblique lines are shown in Figure 44-3.

Figure 44-3

Angles

An *angle* is a figure that consists of two lines that meet at a point called the *vertex*. The symbol ∠ means angle. The size of an angle is determined by the number of degrees one side of the angle is rotated from the other. The length of the side does *not* determine the size of the angle. For example, in Figure 44-4, ∠1 is equal to ∠2. The rotation of side *AC* from side *AB* is equal to the rotation of side *DF* from side *DE* although the lengths of the sides are not equal.

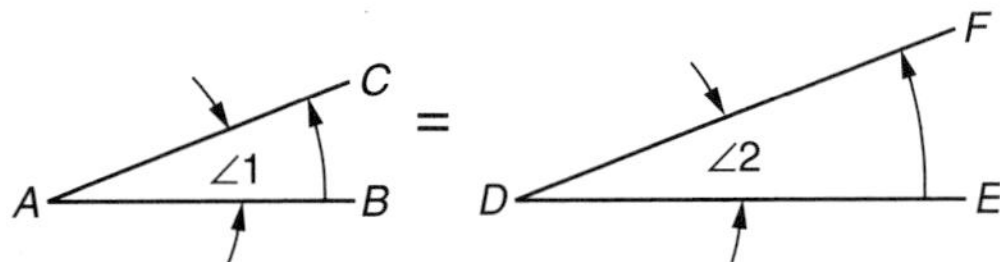

Figure 44-4

Units of Angular Measure

The degree is the basic unit of angular measure. The symbol for degree is °. A radius that is rotated one revolution makes a complete circle or 360°.

A circle may be thought of as a ray with a fixed end point. The ray is rotated. One rotation makes a complete circle of 360° as shown in Figure 44-5.

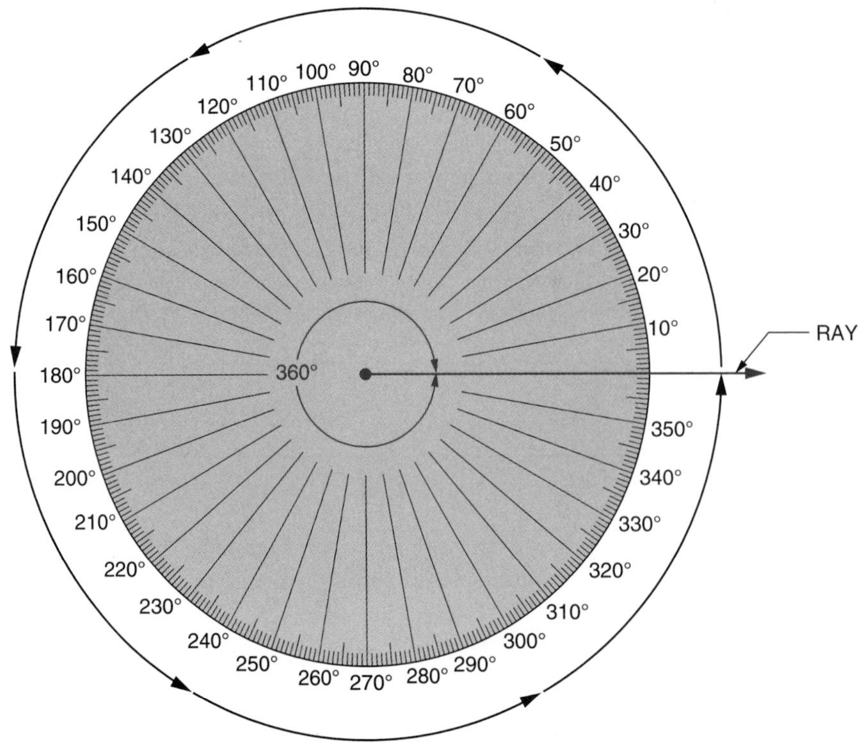

Figure 44-5

The degree of precision required in computing and measuring angles depends on how the angle is used. Some manufactured parts are designed and processed to a very high degree of precision.

In metric calculations, the decimal degree is generally the preferred unit of measurement. In the customary system, angular measure is expressed in these ways.

- As decimal degrees, such as 6.5 degrees and 108.274 degrees.
- As fractional degrees, such as $12\frac{1}{4}$ degrees and $53\frac{1}{10}$ degrees. Fractional degrees are seldom used.
- As degrees, minutes, and seconds, such as 37 degrees, 18 minutes and 123 degrees, 46 minutes, 53 seconds. In the customary system, angular measure is generally expressed this way.

Decimal and fractional degrees are added, subtracted, multiplied, and divided in the same way as any other numbers.

Units of Angular Measure in Degrees, Minutes, and Seconds

A degree is divided into 60 equal parts called *minutes.* The symbol for minute is '. A minute is divided into 60 equal parts called *seconds.* The symbol for second is ". The relationship between degrees, minutes, and seconds is shown in the following chart.

1 Circle = 360 Degrees (°)	1 Degree (°) = $\frac{1}{360}$ of a Circle
1 Degree (°) = 60 Minutes (')	1 Minutes (') = $\frac{1}{60}$ of a Degree (°)
1 Minute (') = 60 Seconds (")	1 Second (") = $\frac{1}{60}$ of a Minute (')

Expressing Decimal Degrees as Degrees, Minutes, and Seconds

The measure of an angle given in the form of decimal degrees, such as 47.1938°, must often be expressed as degrees, minutes, and seconds.

Procedure To express decimal degrees as degrees, minutes, and seconds

- Multiply the decimal part of the degrees by 60' in order to obtain minutes.
- If the number of minutes obtained is not a whole number, multiply the decimal part of the minutes by 60" in order to obtain seconds. Round to the nearer whole second if necessary.
- Combine degrees, minutes, and seconds.

Example Express 47.1938° as degrees, minutes and seconds.

Multiply 0.1938 by 60' to obtain minutes. 60' × 0.1938 = 11.6280'

Multiply 0.6280 by 60" to obtain seconds. Round to the nearer whole second. 60" × 0.6280 = 37.68", 38"

Combine degrees, minutes, and seconds. 47° + 11' + 38" = 47°11'38" Ans

There are two basic procedures used in converting decimal degrees to degrees, minutes, and seconds. Depending on the make and model of your calculator, one of the two procedures should apply.

Procedure 1 Enter decimal degrees, press [=], press [SHIFT], press [←].

Example Convert 23.3075° to degrees, minutes, and seconds.

23.3075 [=] [SHIFT] [←] 23°18°27.

➤ **Note:** Minutes are not displayed with the symbol ' and seconds are not displayed with the symbol ". 23°18'27" Ans

Procedure 2 The scroll or cursor key, ◀ is used with the degrees, minutes, seconds key, [° ' "], in converting to degrees, minutes, and seconds.

Enter decimal degrees, press [° ' "], press ◀, press [ENTER], press [ENTER].

Example Convert 23.3075° to degrees, minutes, and seconds.

23.3075 [° ' "] ◀ [ENTER] [ENTER] 23°18'27" Ans

Expressing Degrees, Minutes, and Seconds as Decimal Degrees

Often an angle given in degrees and minutes is to be expressed as decimal degrees. This is often the case when computations involve metric system units of measure.

Procedure To express degrees and minutes as decimal degrees

- Divide the minutes by 60 to obtain the decimal degree.
- Combine whole degrees and the decimal degree. Round the answer to 2 decimal places.

Example Express 76°29' as decimal degrees.

Divide 29' by 60 to obtain decimal degree. $29' \div 60 = 0.48°$

Combine whole degrees with the decimal degree. $76° + 0.48° = 76.48°$ Ans

When working with customary and metric units of measure, it may be necessary to express angles given in degrees, minutes, and seconds as angles in decimal degrees.

Procedure To express degrees, minutes, and seconds as decimal degrees

- Divide the seconds by 60 in order to obtain the decimal minute.
- Add the decimal minute to the given number of minutes.
- Divide the sum of the minutes by 60 in order to obtain the decimal degree.
- Add the decimal degree to the given number of degrees. Round the answer to 4 decimal places.

Example Express 23°18'44" as decimal degrees.

Divide 44" by 60 to obtain the decimal minute.	44" ÷ 60 = 0.7333'
Add the decimal minute (0.7333') to the given minutes (18').	18' + 0.7333' = 18.7333'
Divide the sum of the minutes (18.7333') by 60 to obtain the decimal degree.	18.7333' ÷ 60 = 0.3122°
Add the decimal degree (0.3122°) to the given degrees (23°).	23° + 0.3122° = 23.3122° Ans

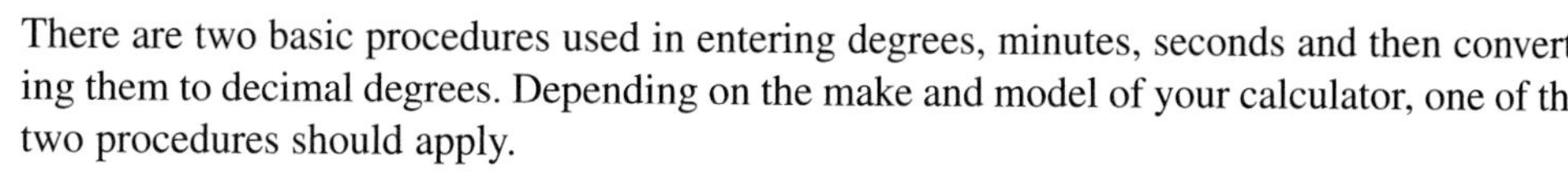

There are two basic procedures used in entering degrees, minutes, seconds and then converting them to decimal degrees. Depending on the make and model of your calculator, one of the two procedures should apply.

The degrees, minutes, seconds key, [° ' "], is used for both procedures.

Procedure 1 Enter degrees, press [° ' "], enter minutes, press [° ' "], enter seconds, press [° ' "], press [=], press [SHIFT], press [←].

➤ **Note:** ← is the alternate function of the primary function key [° ' "].

Example Convert 53°47'25" to decimal degrees.

53 [° ' "] 47 [° ' "] 25 [° ' "] [=] [SHIFT] [←] → 53.79027778 Ans

Degrees, minutes, seconds are entered — Conversion to decimal degrees

Procedure 2 The scroll or cursor key ▶ is used with [° ' "] in order to enter minutes and seconds. It is pressed twice with seconds.

Enter degrees, press [° ' "], enter minutes, press [° ' "], press ▶, enter seconds, press [° ' "], press ▶, press ▶, press [ENTER], press [ENTER].

Example Convert 53°47'25" to decimal degrees.

53 [° ' "] 47 [° ' "] ▶ 25 [° ' "] ▶ ▶ [ENTER]

[ENTER] 53.79027778° Ans

Conversion to decimal degree

Arithmetic Operations on Angular Measure in Degrees, Minutes, and Seconds

The division of degrees into minutes and seconds permit very precise computations and measurements. In machining operations, dimensions at times are computed to seconds in order to ensure the proper functioning of parts.

When computing with degrees, minutes, and seconds, it is sometimes necessary to exchange units. When exchanging units, keep in mind that 1 degree equals 60 minutes and 1 minute equals 60 seconds. These examples illustrate adding, subtracting, multiplying, and dividing angles in degrees, minutes, and seconds.

Arithmetic operations with degrees, minutes, and seconds computed with a calculator may require entering an angle as degrees, minutes, and seconds, converting to decimal degrees, and converting back to degrees, minutes, and seconds. Calculator applications for each of the arithmetic operations are shown.

Adding Angles Expressed in Degrees, Minutes, and Seconds

Example 1 Determine the size of ∠1 shown in Figure 44-6.

∠1 = 15°18' + 63°37'

    15°18'
  + 63°37'
    78°55' Ans

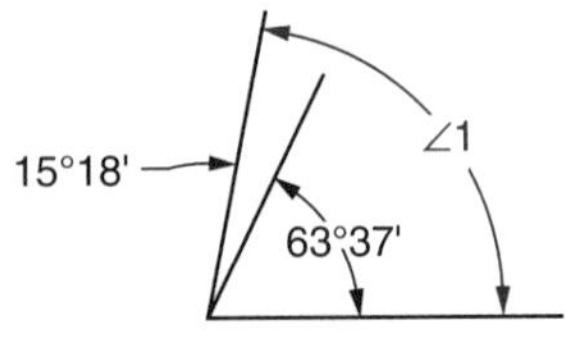

Figure 44-6

Example 2 Determine the size of ∠2 shown in Figure 44-7.

∠2 = 43°37' + 82°54'

    43°37'
  + 82°54'
    125°91' = 126°31' Ans

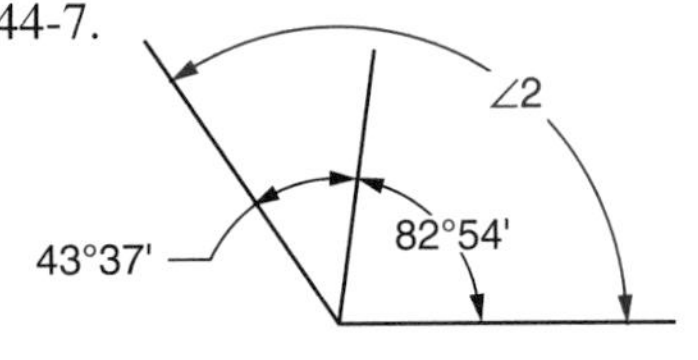

Figure 44-7

➤ **Note:** Express 91' as degrees and minutes. 91' = 60' + 31' = 1°31'. Add, 125° + 1°31' = 126°31' Ans

Example 3 Determine ∠3 shown in Figure 44-8.

∠3 = 78°43'27" + 29°38'52"

    78°43'27"
  + 29°38'52"
    107°81'79" = 107°82'19" = 108°22'19" Ans

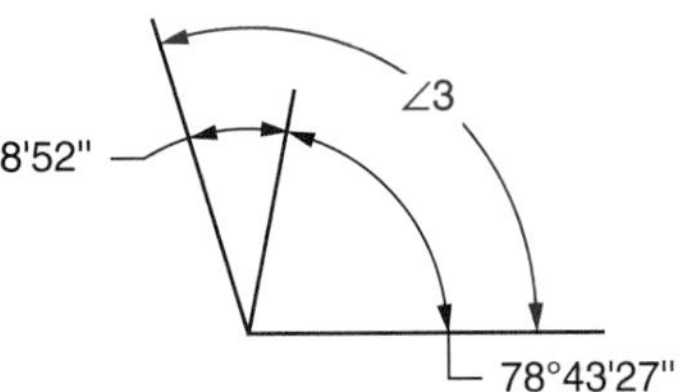

Figure 44-8

➤ **Note:** 79" = 60" + 19" = 1'19" therefore, 107°81'79" = 107°82'19". 82' = 60' + 22' = 1°22' therefore, 107°82'19" = 108°22'19".

Example 4 Determine ∠3 shown in Figure 44-8.

∠3 = 78°43'27" + 29°38'52"

78 [° ' "] 43 [° ' "] 27 [° ' "] [+] 29 [° ' "] 38 [° ' "] 52 [° ' "] [=] 108°22°19, 108°22'19" Ans

or 78 [° ' "] 43 [° ' "] ▶ 27 [° ' "] ▶ ▶ [+]
29° [° ' "] 38 [° ' "] ▶ 52 [° ' "] ▶ ▶ [ENTER]
[° ' "] ◀ [ENTER] [ENTER] 108°22'19" Ans

Conversion from decimal degrees to degrees, minutes, seconds

Subtracting Angles Expressed in Degrees, Minutes, and Seconds

Example 1 Determine ∠1 shown in Figure 44-9.

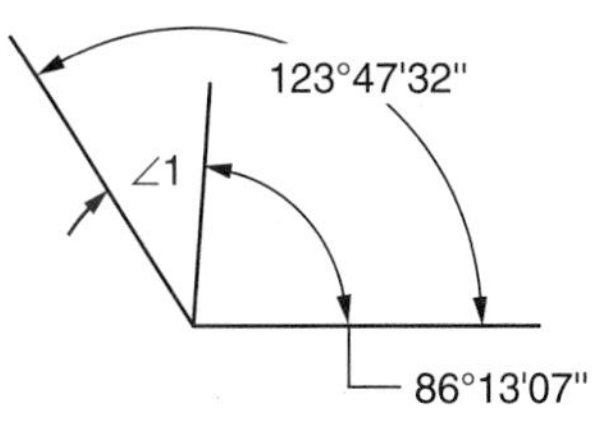

Figure 44-9

∠1 = 123°47'32" − 86°13'07"

$$\begin{array}{r} 123°47'32" \\ -\ \ 86°13'07" \\ \hline 37°34'25" \end{array}\quad \text{Ans}$$

Example 2 Determine ∠2 shown in Figure 44-10.

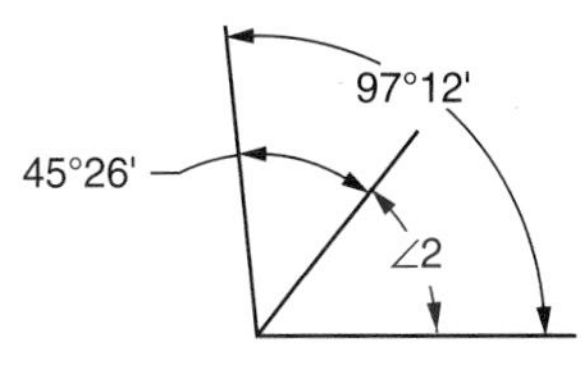

Figure 44-10

∠2 = 97°12' − 45°26'

$$\begin{array}{r} 97°12' = 96°72' \\ -\ 45°26' = 45°26' \\ \hline 51°46' \end{array}\quad \text{Ans}$$

➤ **Note:** Since 26' cannot be subtracted from 12', 1° is exchanged for 60'. 97°12' = 96° + 1° + 12' = 96° + 60' + 12' = 96°72'.

Example 3 Determine ∠3 shown in Figure 44-11.

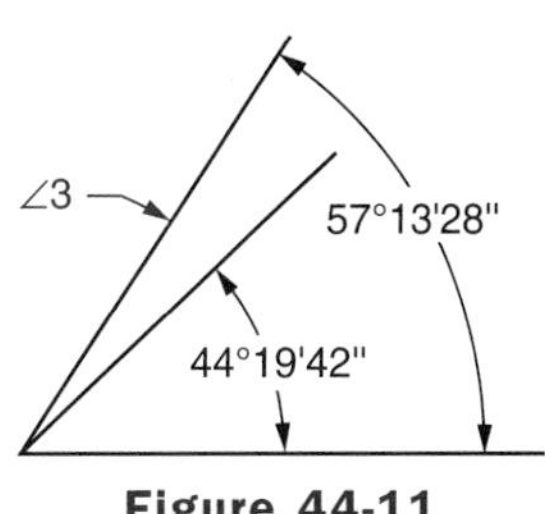

Figure 44-11

∠3 = 57°13'28" − 44°19'42"

$$\begin{array}{r} 57°13'28" = 56°73'28" = \ \ 56°72'88" \\ -44°19'42" = 44°19'42" = -44°19'42" \\ \hline 12°53'46" \end{array}\quad \text{Ans}$$

➤ **Note:** Since 19' cannot be subtracted from 13', and 42" cannot be subtracted from 28", 1° is exchanged for 60' and 1' is exchanged for 60".
57°13'28" = 56° + 1° + 13' + 28" = 56°73'28" = 56°72' + 1' + 28"
= 56°72'88".

Example 4 Determine ∠3 shown in Figure 44-11.

∠3 = 57°13'28" − 44°19'42"

57 [° ' "] 13 [° ' "] 28 [° ' "] [−] 44 [° ' "] 19 [° ' "] 42 [° ' "] [=] 12°53°46,

12°53'46" Ans

or 57 [° ' "] 13 [° ' "] [▶] 28 [° ' "] [▶] [▶] [−] 44 [° ' "]

19 [° ' "] [▶] 42 [° ' "] [▶] [▶] [ENTER]

[° ' "] [◀] [ENTER] [ENTER] 12°53'46" Ans

Conversion from decimal degrees to degrees, minutes, seconds

Multiplying Angles Expressed in Degrees, Minutes, and Seconds

Example 1 Five holes are drilled on a circle as shown in Figure 44-12. The angular measure between two consecutive holes is 32°18'. Determine the angular measure, ∠1, between hole 1 and hole 5.

∠1 = 4(32°18')

32°18'
× 4
128°72' = 129°12' Ans

➤ Note: 72' = 1° 12'

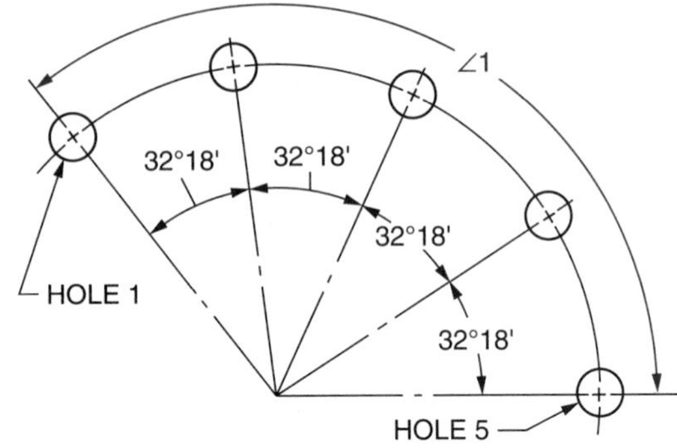

Figure 44-12

Example 2 Determine the size of ∠2 shown in Figure 44-13 when x = 41°27'42".

∠2 = 5x = 5(41°27'42")

41° 27' 42"
× 5
205°135'210" = 205°138'30" = 207°18'30" Ans

➤ Note: 210" = 3'30" and 138' = 2°18'.

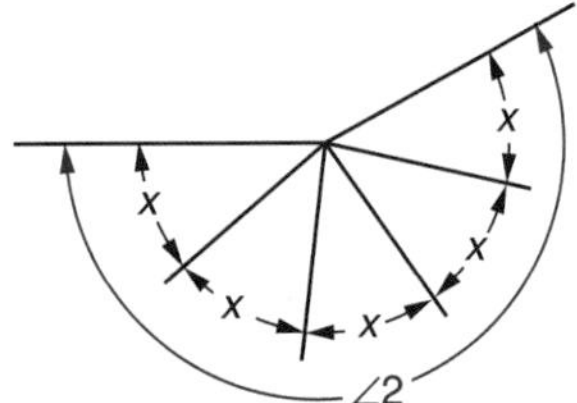

Figure 44-13

Example 3 Determine ∠2 in Figure 44-13 when x = 41°27'42".

∠2 = 5x = 5(41°27'42")

5 [×] 41 [° ' "] 27 [° ' "] 42 [° ' "] [=] 207°18°30, 207°18'30" Ans

or 5 [×] 41 [° ' "] 27 [° ' "] [▶] 42 [° ' "] [▶] [▶] [ENTER]

[° ' "] [◀] [ENTER] [ENTER] 207°18'30" Ans

Conversion from decimal degrees to degrees, minutes, seconds

Dividing Angles Expressed in Degrees, Minutes, and Seconds

Example 1 ∠1 and ∠2 are the same size. Determine the size of ∠1 and ∠2 shown in Figure 44-14.

∠1 = ∠2 = 104°58' ÷ 2

$$2\overline{)104°58'}\quad 52°29'$$

Ans: 52°29'

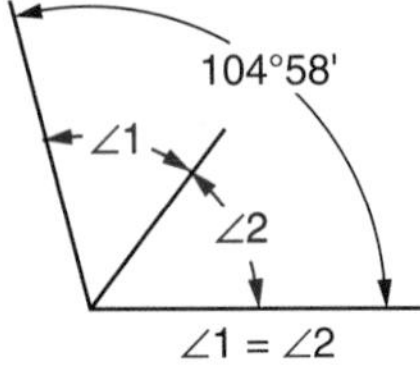

Figure 44-14

Example 2 Determine the size of ∠1, ∠2, and ∠3 shown in Figure 44-15 if each is the same size.

∠1 = ∠2 = ∠3 = 128°37'21" ÷ 3

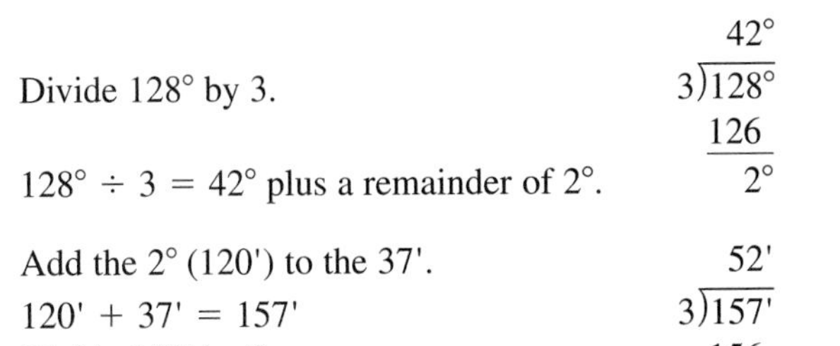

Divide 128° by 3. $3\overline{)128°}$ = 42°; 126; 2°

128° ÷ 3 = 42° plus a remainder of 2°.

Add the 2° (120') to the 37'.

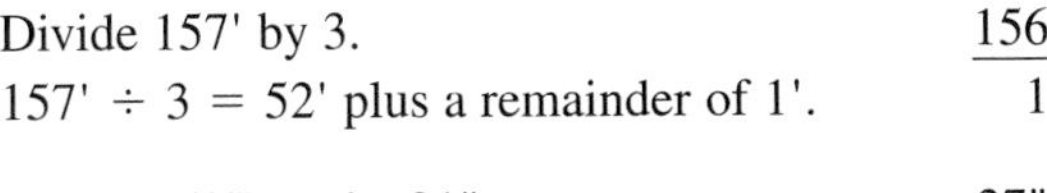

120' + 37' = 157'

Divide 157' by 3. $3\overline{)157'}$ = 52'; 156; 1'

157' ÷ 3 = 52' plus a remainder of 1'.

Add 1' (60") to the 21".

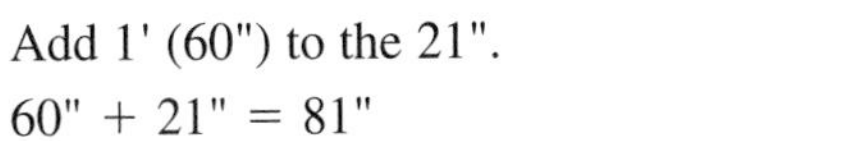

60" + 21" = 81" $3\overline{)81"}$ = 27"

Divide 81" by 3.

81" ÷ 3 = 27"

Combine. 42° + 52' + 27" = 42°52'27" Ans

128°37'21"

∠1 = ∠2 = ∠3

Figure 44-15

Example 3 Determine ∠1, ∠2, and ∠3 in Figure 44-15.

∠1 = ∠2 = ∠3 = 128°37'21" ÷ 3

128 [° ' "] 37 [° ' "] 21 [° ' "] [÷] 3 [=] 42°52°27, 42°52'27" Ans

or 128 [° ' "] 37 [° ' "] ▶ 21 [° ' "] ▶ ▶ [ENTER] [÷] 3

[° ' "] ◀ [ENTER] [ENTER] 42°52'27" Ans

Conversion from decimal degrees to degrees, minutes, seconds

Application

Definitions and Terms

1. Refer to Figure 44-16 and identify each of the following as parallel, perpendicular, or oblique lines.

   a. Line AB and line CD __________

   b. Line AB and EF __________

   c. Line CD and GH __________

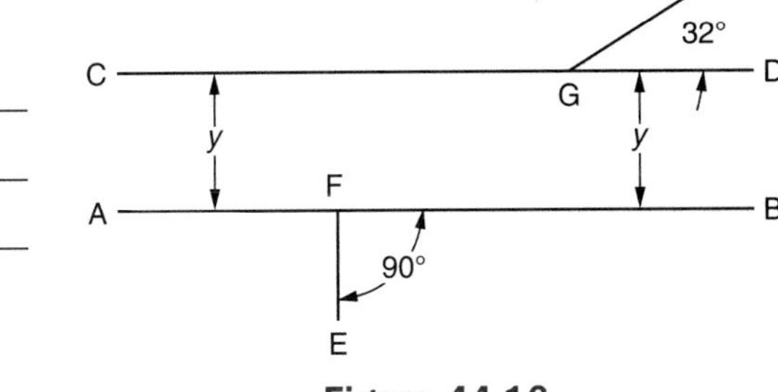

Figure 44-16

2. a. How many degrees are in a circle? __________

   b. How many minutes are in 1 degree? __________

   c. How many seconds are in 1 minute? __________

   d. How many seconds are in 1 degree? __________

   e. How many minutes are in a circle? __________

3. Write the symbols for the following words.

a. parallel ______________ d. minute ______________

b. perpendicular ______________ e. second ______________

c. degree ______________

Expressing Decimal Degrees as Degrees and Minutes

Express the following decimal degrees as degrees and minutes. When necessary, round the answer to the nearest whole minute.

4. 13.50° ______________ 9. 93.15° ______________
5. 67.85° ______________ 10. 81.08° ______________
6. 48.10° ______________ 11. 6.47° ______________
7. 117.70° ______________ 12. 125.91° ______________
8. 18.60° ______________ 13. 77.67° ______________

Expressing Decimal Degrees as Degrees, Minutes, and Seconds

Express the following decimal degrees as degrees, minutes, and seconds. When necessary, round the answer to the nearest whole second.

14. 52.1380° ______________ 19. 103.0090° ______________
15. 212.0710° ______________ 20. 37.9365° ______________
16. 7.9250° ______________ 21. 89.9056° ______________
17. 44.4440° ______________ 22. 182.0692° ______________
18. 73.9330° ______________ 23. 19.8973° ______________

Expressing Degrees and Minutes as Decimal Degrees

Express the following degrees and minutes as decimal degrees. Round the answer to 2 decimal places.

24. 22°40' ______________ 29. 56°48' ______________
25. 107°45' ______________ 30. 87°37' ______________
26. 6°10' ______________ 31. 2°19' ______________
27. 87°16' ______________ 32. 32°08' ______________
28. 122°07' ______________ 33. 79°59' ______________

Expressing Degrees, Minutes, and Seconds as Decimal Degrees

Express the following degrees, minutes, and seconds as decimal degrees. Round the answer to 4 decimal places.

34. 28°18'30" ______________ 38. 176°27'18" ______________
35. 57°08'45" ______________ 39. 2°07'13" ______________
36. 130°50'10" ______________ 40. 19°49'59" ______________
37. 98°20'25" ______________ 41. 61°12'06" ______________

Adding Angles Expressed in Degrees, Minutes, and Seconds

42. Determine ∠1. ____________

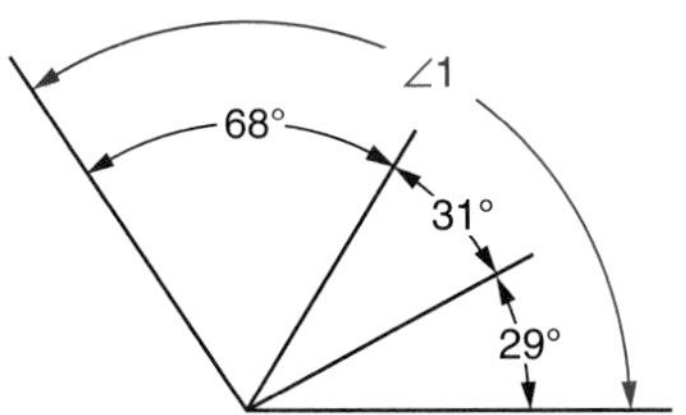

46. Determine ∠5. ____________

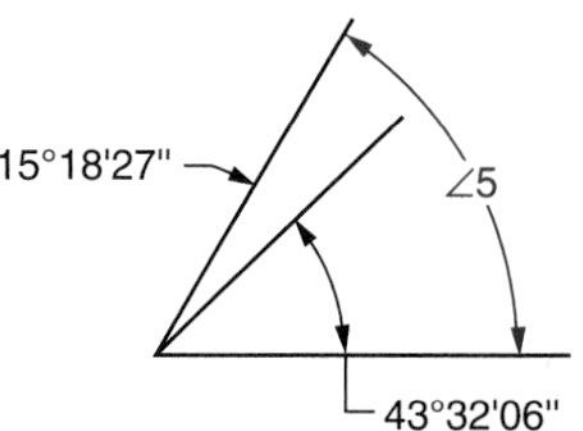

43. Determine ∠2. ____________

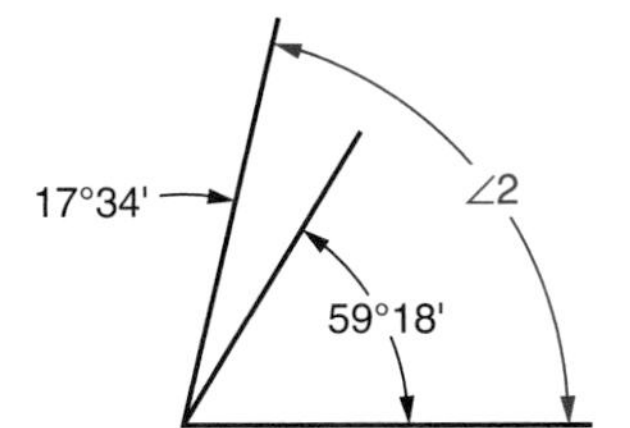

47. Determine ∠6. ____________

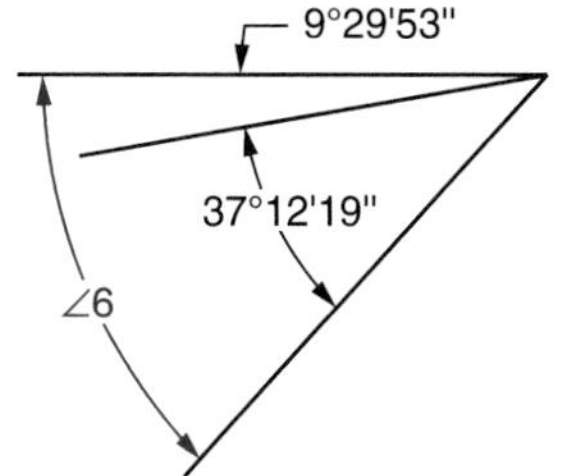

44. Determine ∠3. ____________

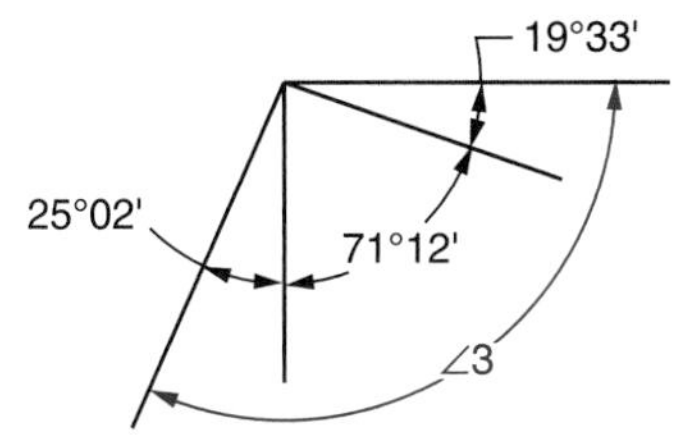

48. Determine ∠7 + ∠8 + ∠9. ____________

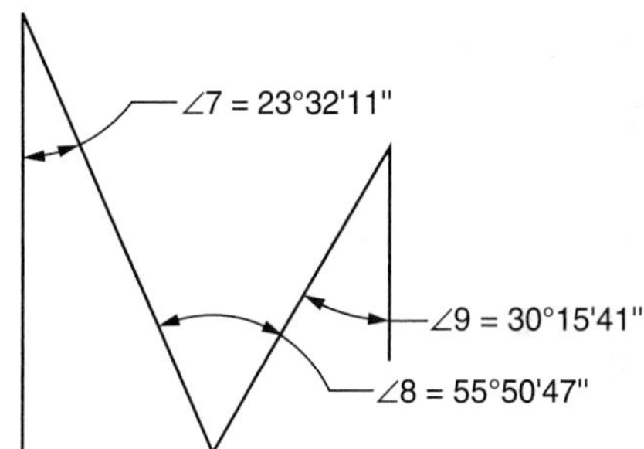

45. Determine ∠1 + ∠2 + ∠3. ____________

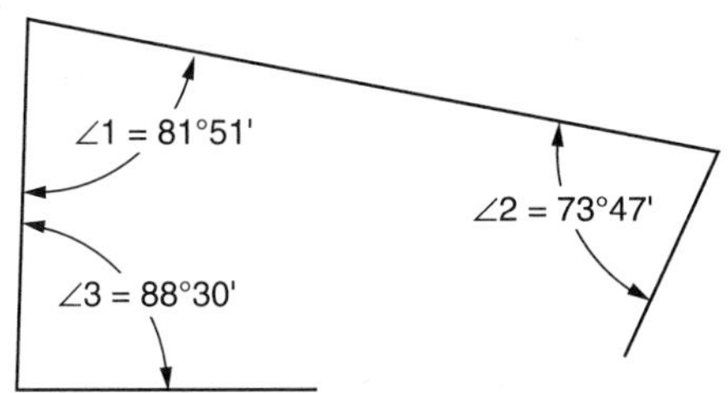

49. Determine ∠1 + ∠2 + ∠3 + ∠4 + ∠5. ____________

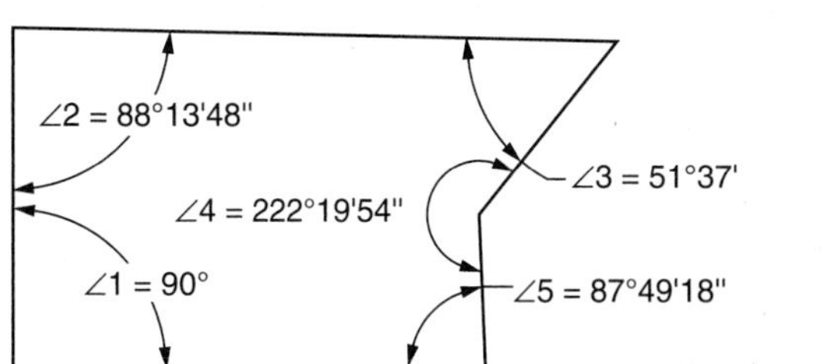

Subtracting Angles Expressed in Degrees, Minutes, and Seconds

Subtract the angles in each of the following exercises.

50. 114° − 89° ____________
51. 92°35' − 76°26' ____________
52. 63°23' − 32°58' ____________
53. 122°36'17" − 13°15'08" ____________
54. 49°34'12" − 19°13'42" ____________

55. Determine ∠1. ______________

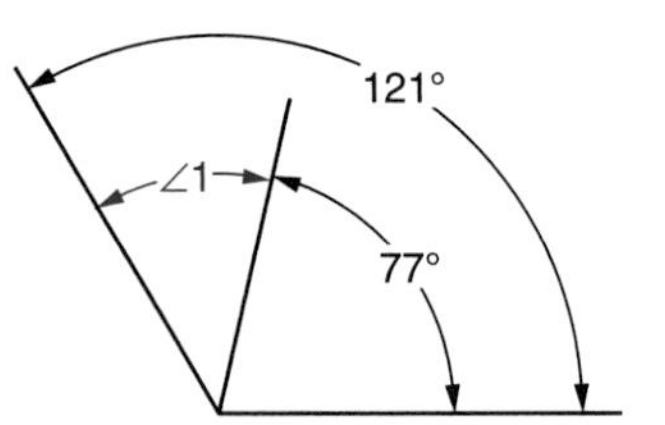

57. Determine ∠2. ______________

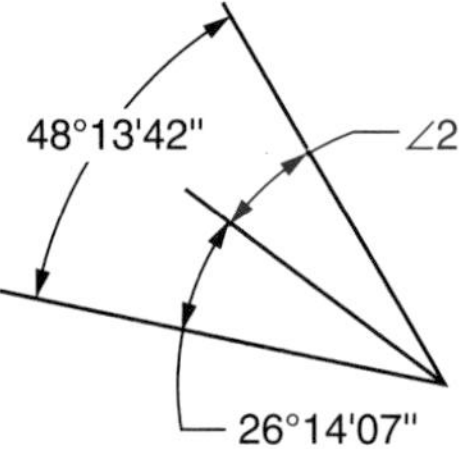

56. Determine ∠3 − ∠2. ______________

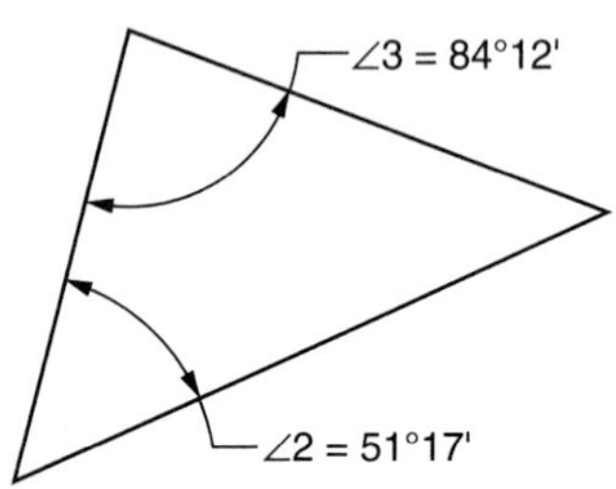

58. Determine ∠1 − ∠2. ______________

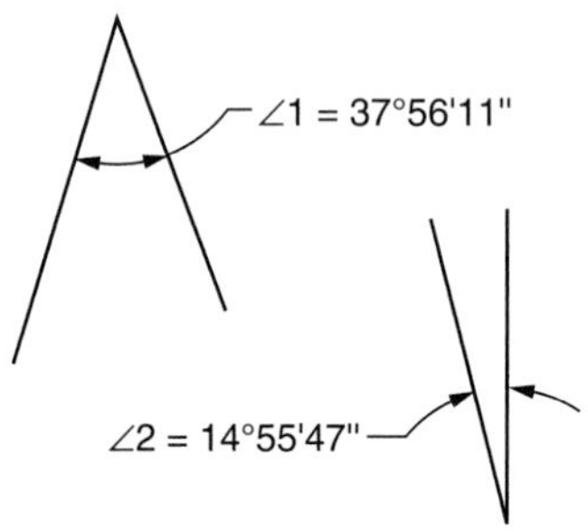

59. In the figure shown ∠6 = 720° − (∠1 + ∠2 + ∠3 + ∠4 + ∠5). Determine ∠6. ______________

➤ **Note:** 720° = 719°59'60"

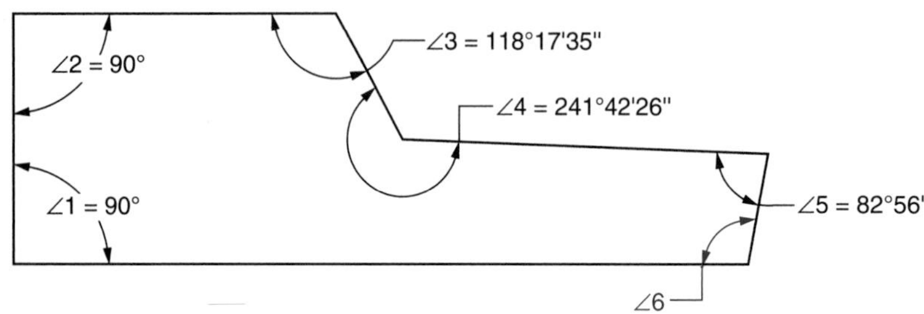

Multiplying Angles Expressed in Degrees, Minutes, and Seconds

Multiply the angles in each of the following exercises.

60. 7(15°) ______________
61. 3(29°19') ______________
62. 2(43°43') ______________
63. 5(22°10'13") ______________
64. 8(43°23'28") ______________

65. In the figure shown, ∠1 = ∠2 = 42°. Determine ∠3. ______________

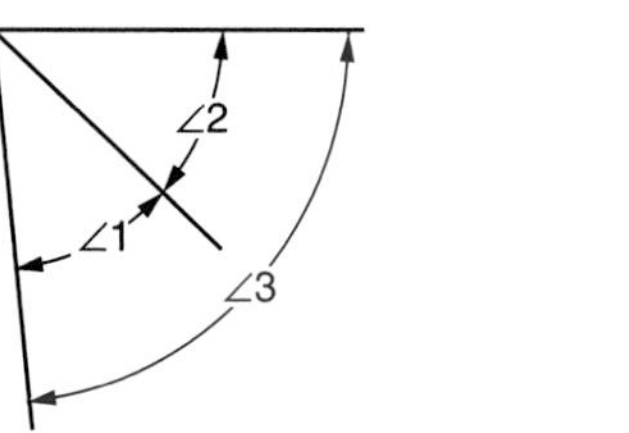

66. If x = 39°14', find ∠4. ______________

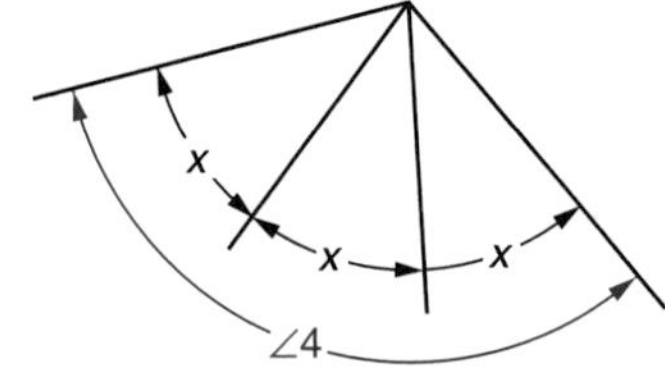

67. In the figure shown, $\angle 1 = \angle 2 = \angle 3 = \angle 4 = \angle 5 = 54°03'$.
Determine $\angle 6$. ____________

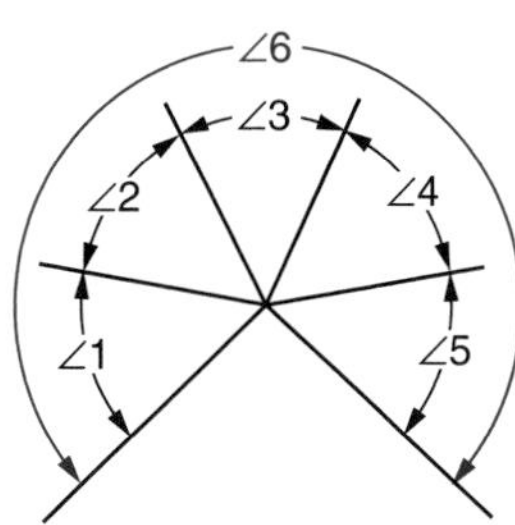

Dividing Angles Expressed in Degrees, Minutes, and Seconds

Divide the angles in each of the following exercises.

68. $94° \div 2$ ____________
69. $87° \div 2$ ____________
70. $105°20' \div 4$ ____________
71. If $\angle 1 = \angle 2$, find $\angle 1$. ____________

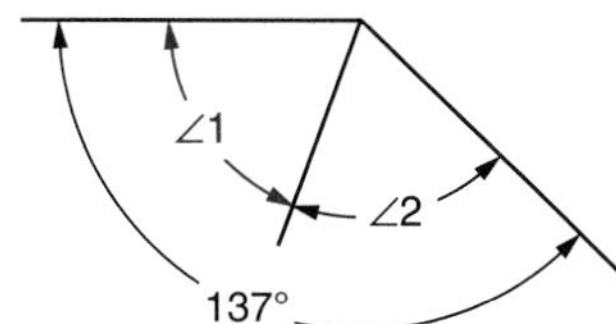

72. Determine x. ____________

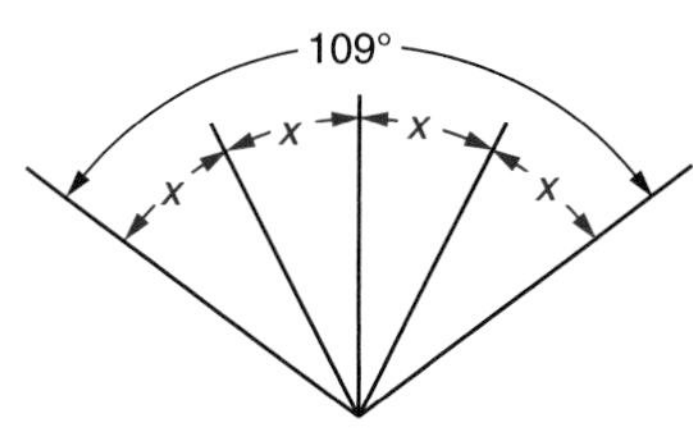

73. Determine y. ____________

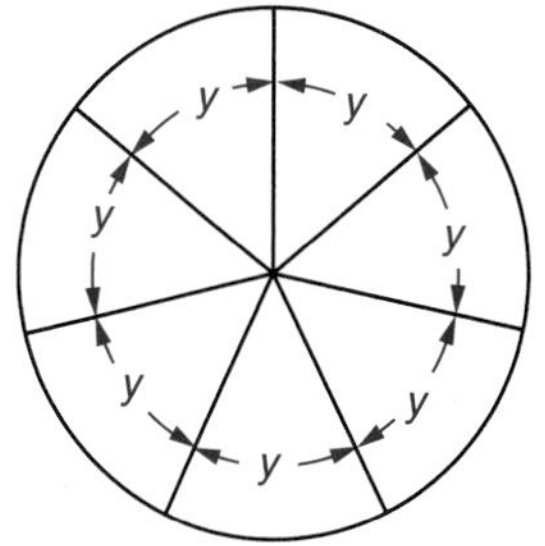

74. If $\angle 1 = \angle 2 = \angle 3 = \angle 4$, find $\angle 1$. ____________

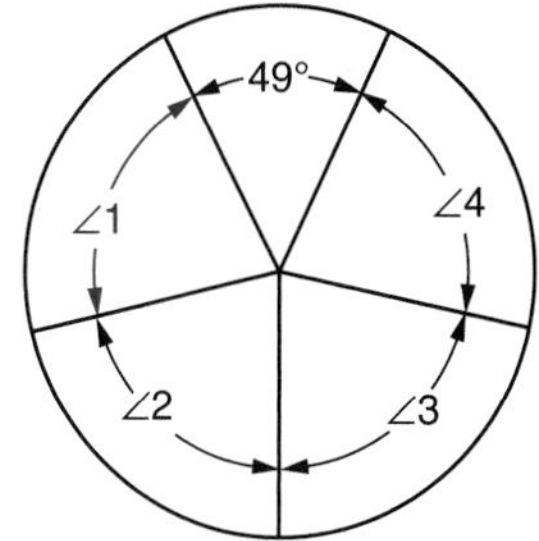

75. The sum of the angles in the figure shown equals 1440°.

If $\angle 1 = \angle 2 = \angle 3 = \angle 4 = \angle 5 = \angle 6$ and $\angle 7 = \angle 8 = \angle 9 = \angle 10 = 118°14'23''$, find $\angle 1$. ____________

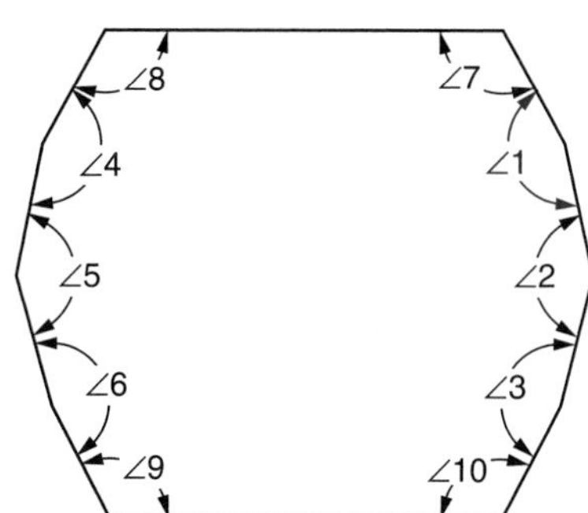

UNIT 45

Protractors—Simple Semicircular and Vernier

Objectives ***After studying this unit you should be able to***

- Measure angles with a simple protractor.
- Lay out angles with a simple protractor.
- Read settings on a vernier bevel protractor.
- Compute complements and supplements of angles.

Protractors are used for measuring, drawing, and laying out angles. Various types of protractors are available, such as the simple semicircular protractor, swinging blade protractor, and bevel protractor. The type of protractor used depends on its application and the degree of precision required. Protractors have wide occupational use, particularly in the metal and woodworking trades.

Simple Semicircular Protractor

A simple semicircular protractor like the one in Figure 45-1 has two scales, each graduated from 0° to 180° so that it can be read from either the left or right side. The vertex of the angle to be measured or drawn is located at the center of the base of the protractor.

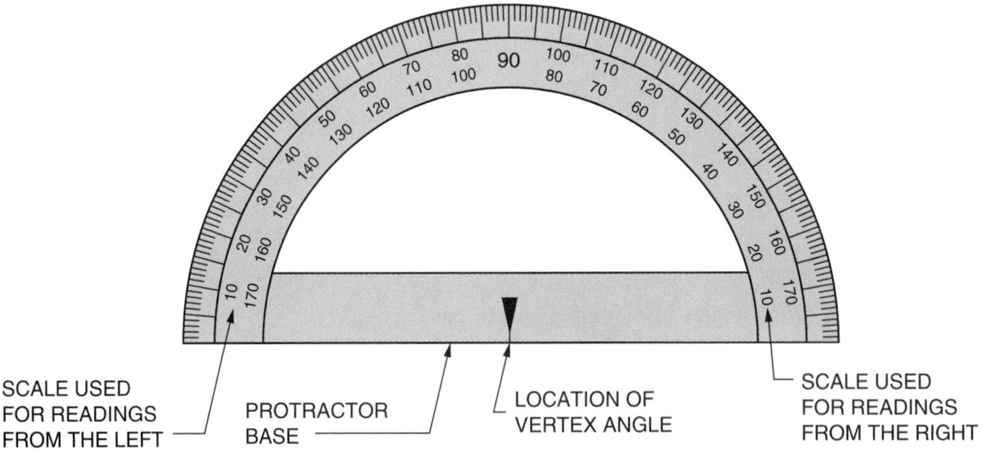

Figure 45-1

Procedure To lay out (construct) a given angle

- Draw a baseline.
- On the baseline mark a point as the vertex.
- Place the protractor base on the baseline with the protractor center on the vertex.
- If the angle rotates from the right, choose the scale that has a zero degree reading on the right side of the protractor. If the angle rotates from the left, choose the scale that has a zero degree reading on the left side of the protractor. At the scale reading for the angle being drawn, mark a point.
- Remove the protractor and connect the two points.

Example Lay out an angle of 105°.

Draw baseline AB.

On AB mark point O as the vertex.

Place the protractor base on AB with the protractor center on point O as shown in Figure 45-2.

The angle is rotated from the right. The inside scale has a zero degree reading on the right side of the protractor. Use the inside scale and mark a point (point P) at the scale reading of 105°.

Remove the protractor and connect points P and O.

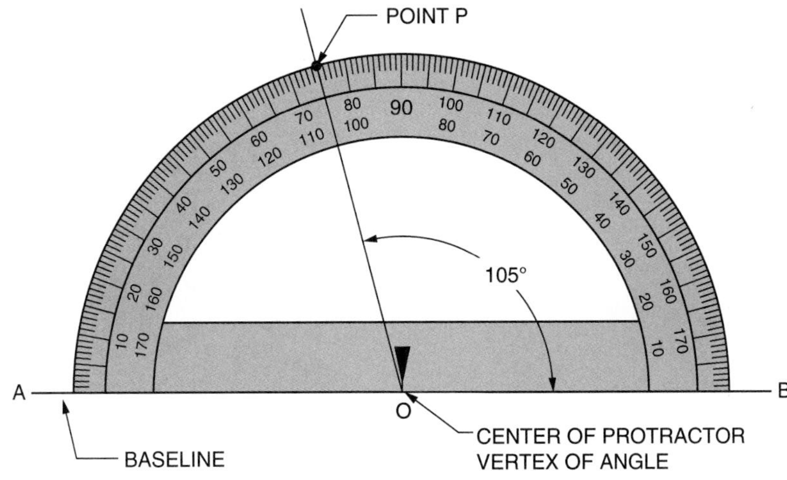

Figure 45-2

Procedure To measure a given angle

- Place the protractor base on one side of the angle with the protractor center on the angle vertex.
- If the angle rotates from the right, choose the scale that has the zero degree reading on the right side of the protractor. If the angle rotates from the left, choose the scale that has the zero degree reading on the left side of the protractor. Read the measurement where the side crosses the protractor scale.

Example 1 Measure ∠1.

Extend the sides OA and OB of ∠1 as shown by the dashed line in Figure 45-3.

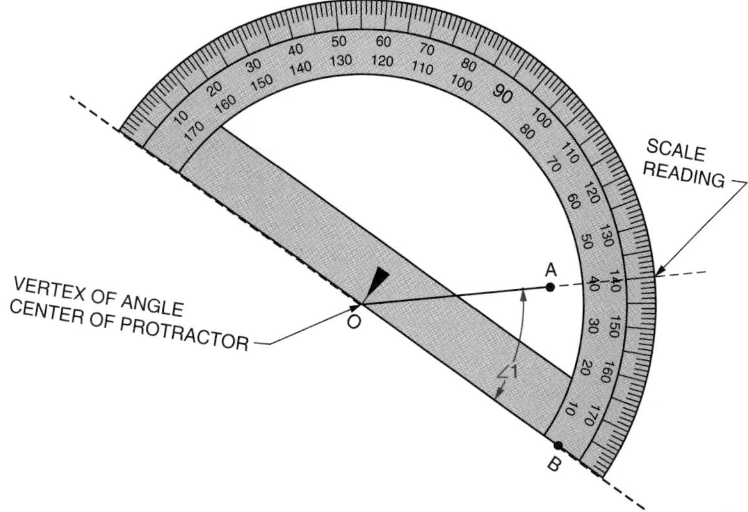

Figure 45-3

Place the protractor base on side OB with the protractor center on the angle vertex, point O.

Angle 1 is rotated from the right. The angle measurement is read from the inside scale since the inside scale has a zero degree (0°) reading on the right side of the protractor base. Read the measurement where the extension of side OA crosses the protractor scale. Angle 1 = 40° Ans

Example 2 Measure ∠2 as shown in Figure 45-4.

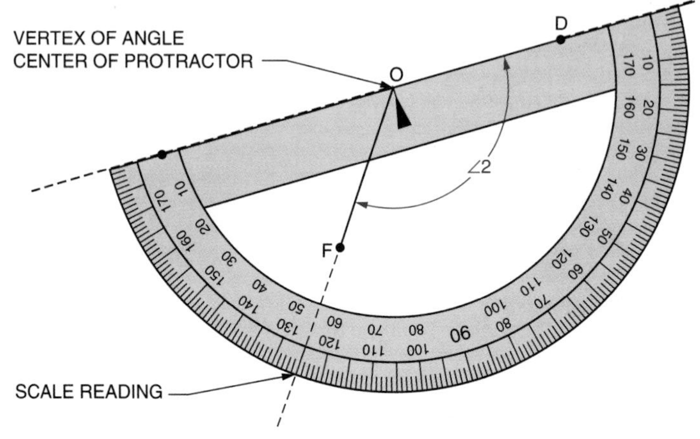

Figure 45-4

Extend the sides OD and OF of ∠2 as shown.

The protractor is positioned upside down. Place the protractor base on the side OD with the protractor center on the angle vertex point O.

Angle 2 is rotated from the right. The angle measurement is read from the outside scale since the outside scale has a zero degree (0°) reading on the right side of the protractor base. Read the measurement where the extension of side OF crosses the protractor scale. Angle 2 = 125° Ans

Bevel Protractor with Vernier Scale

The bevel protractor is the most widely used vernier protractor in the machine shop. A vernier bevel protractor is shown in Figure 45-5.

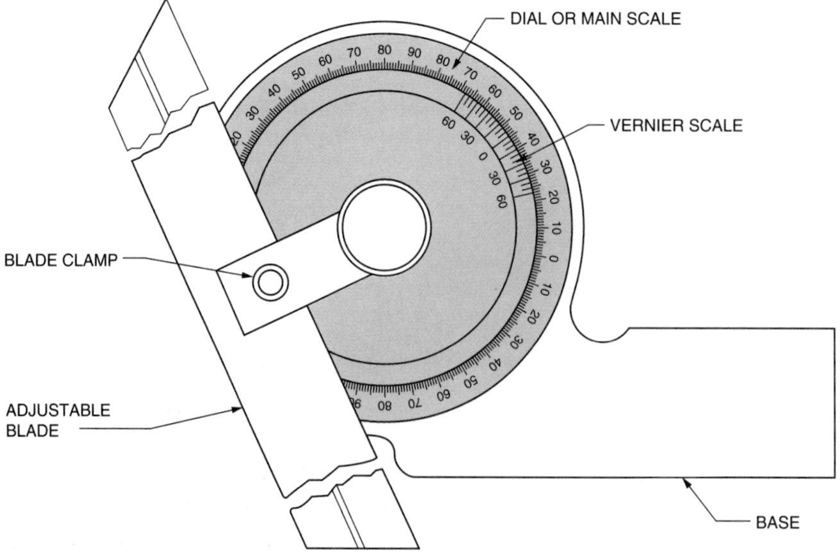

Figure 45-5

A vernier bevel protractor consists of a fixed dial or main scale. The main scale is divided into four sections, each from 0° to 90°. The vernier scale rotates within the main scale. A blade that can be adjusted to required positions is rotated to a desired angle.

The vernier scale permits accurate readings to $\frac{1}{12}$ degree or 5 minutes. The vernier scale is divided into 24 units, with 12 units on each side of zero. The divisions on the vernier scale are in minutes. Each division is equal to 5 minutes.

The left vernier scale is used when the vernier zero is to the left of the dial zero. The right vernier scale is used when the vernier zero is to the right of the dial zero.

Example Read the setting on the vernier protractor shown in Figure 45-6.

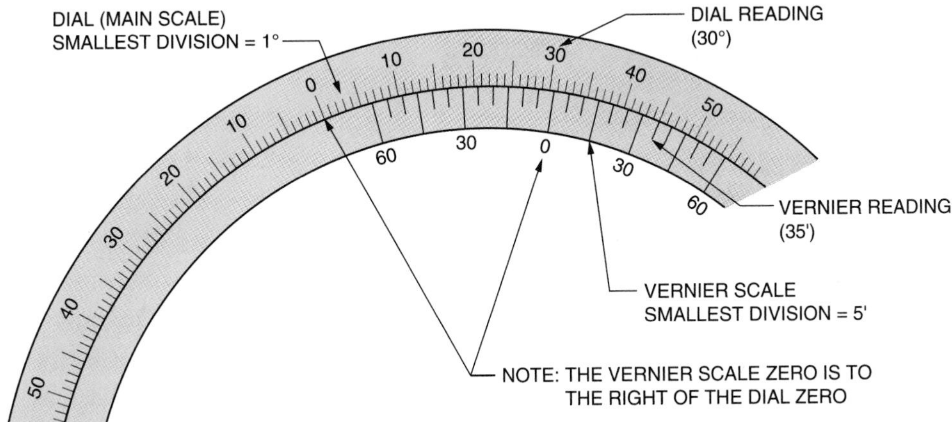

Figure 45-6

The zero mark on the vernier scale is just to the right of the 30° division of the dial scale. The vernier zero is to the right of the dial zero; therefore, the right vernier scale is read. The 35' vernier graduation coincides with a dial graduation. The protractor reading is 30°35'. Ans

Complements and Supplements of Scale Readings

When using the bevel protractor, the machinist must determine whether the desired angle of the part being measured is the actual reading on the protractor or the complement or the supplement of the protractor reading. Particular caution must be taken when measuring angles close to 45° and 90°.

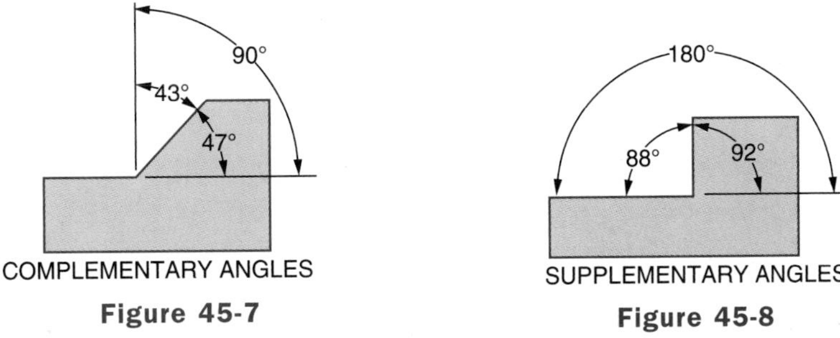

Figure 45-7 **Figure 45-8**

Two angles are *complementary* when their sum is 90°. For example, in Figure 45-7 $43° + 47° = 90°$. Therefore, 43° is the complement of 47° and 47° is the complement of 43°.

Two angles are *supplementary* when their sum is 180°. For example, in Figure 45-8 $92° + 88° = 180°$. Therefore, 92° is the supplement of 88° and 88° is the supplement of 92°.

Application

Simple Protractor

1. Write the values of angles A–J on the protractor scale shown.

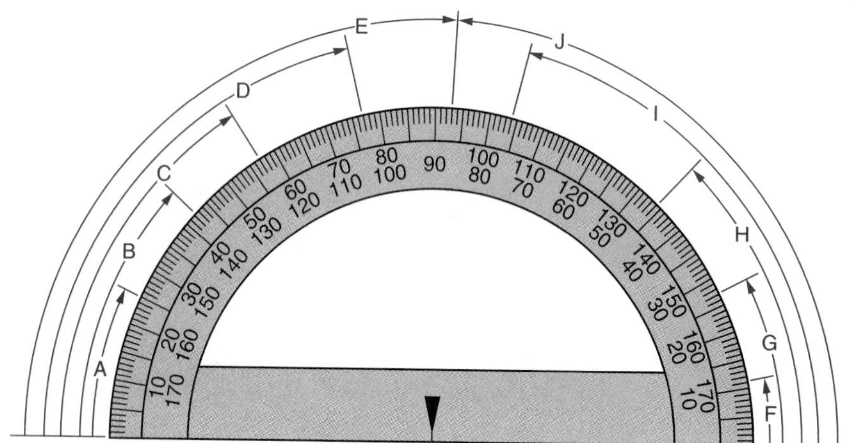

A = ______
B = ______
C = ______
D = ______
E = ______
F = ______
G = ______
H = ______
I = ______
J = ______

2. Using a protractor, lay out the following angles.

a. 19°	c. 80°	e. 4°	g. 97°	i. 150°
b. 65°	d. 12°	f. 98°	h. 123°	j. 166°

3. Lay out a 3-sided closed figure (triangle) of any size containing angles of 47° and 105°. Measure the third angle. How many degrees are contained in the third angle? ______
4. Lay out a 4-sided figure (quadrilateral) of any size containing angles of 89°, 69°, and 124°. Measure the fourth angle. How many degrees are contained in the fourth angle? ______
5. Measure each of the following angles, ∠1–∠14, to the nearer degree. Extend the sides of the angles if necessary.

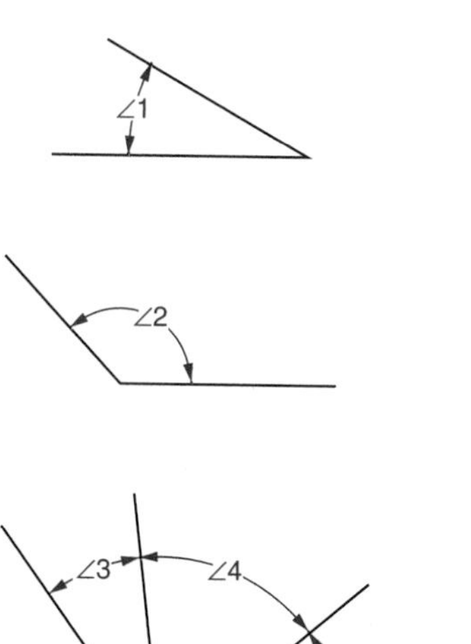

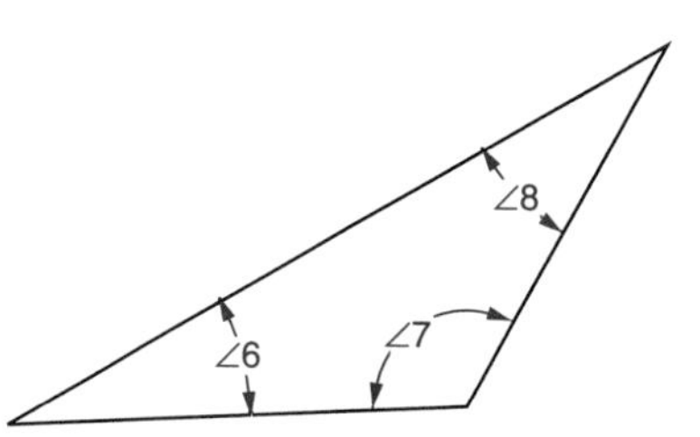

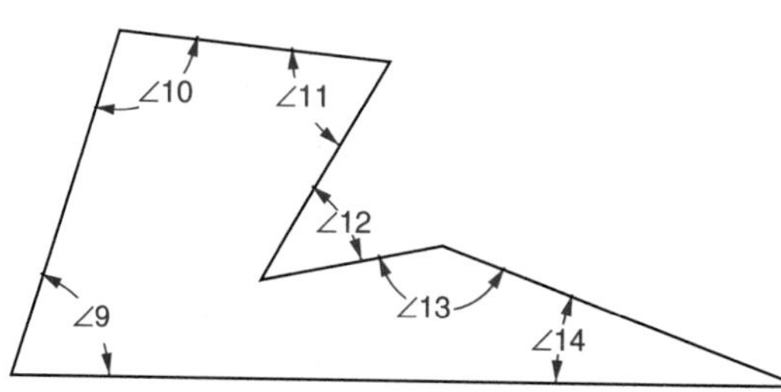

∠1 = ______
∠2 = ______
∠3 = ______
∠4 = ______
∠5 = ______
∠6 = ______
∠7 = ______
∠8 = ______
∠9 = ______
∠10 = ______
∠11 = ______
∠12 = ______
∠13 = ______
∠14 = ______

Vernier Protractor

Write the values of the settings on the following vernier protractor scales.

6.

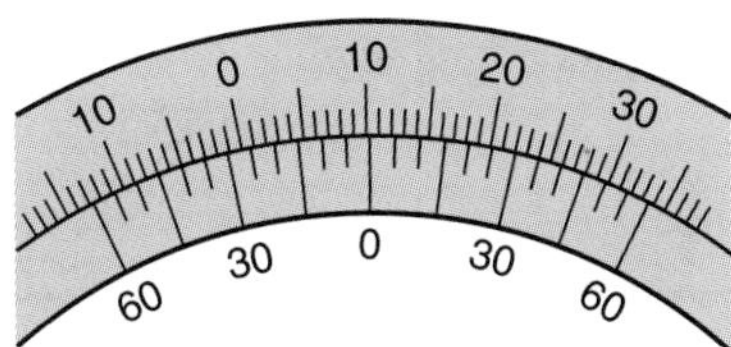

7.

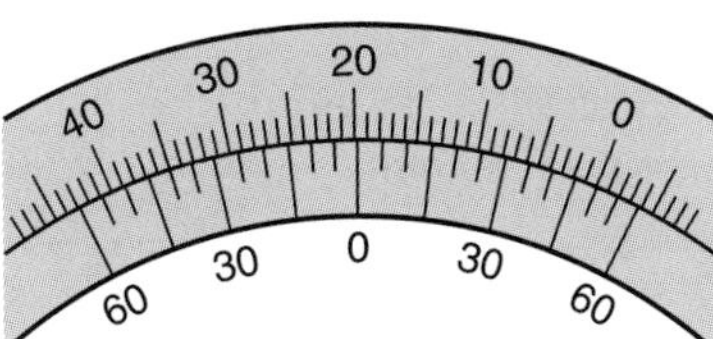

8.

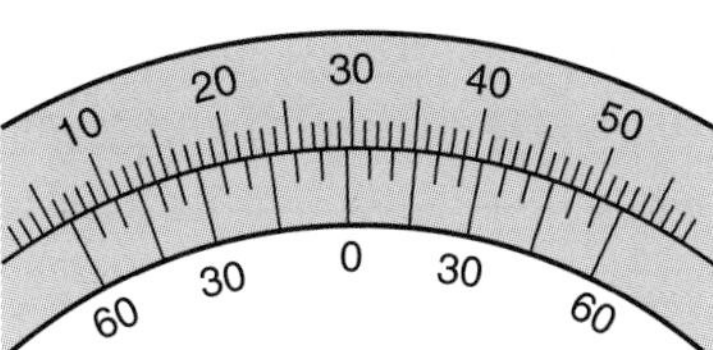

9.

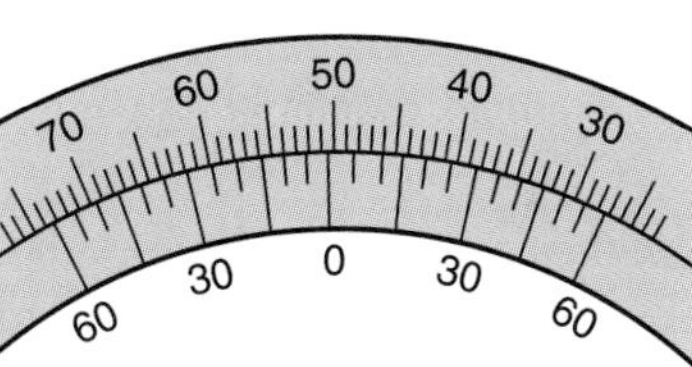

10.

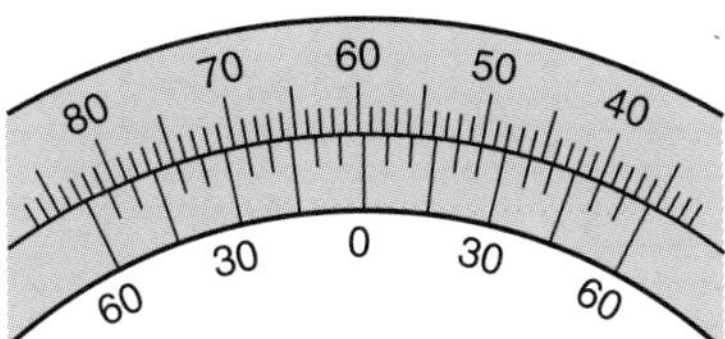

11.

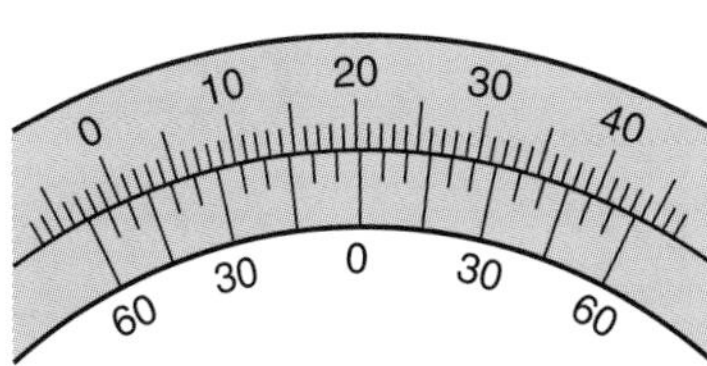

12.

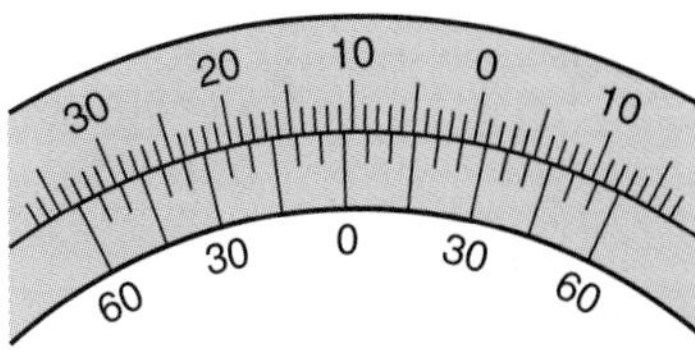

13.

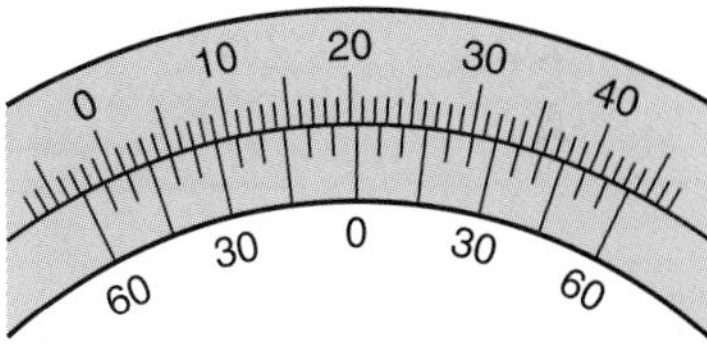

14.

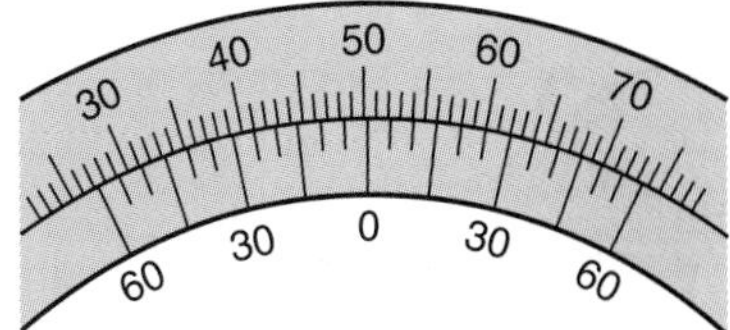

Complementary and Supplementary Angles

15. Write the complements of the following angles.

a. 43° __________
b. 76° __________
c. 17° __________
d. 5° __________
e. 67°49' __________
f. 45°19' __________
g. 21°43' __________
h. 78°19'27" __________
i. 59°0'59" __________

16. Write the supplements of the following angles.

a. 13° __________
b. 65° __________
c. 91° __________
d. 179°59' __________
e. 0°49' __________
f. 89°59' __________
g. 2°43'20" __________
h. 68°21'29" __________
i. 133°32'08" __________

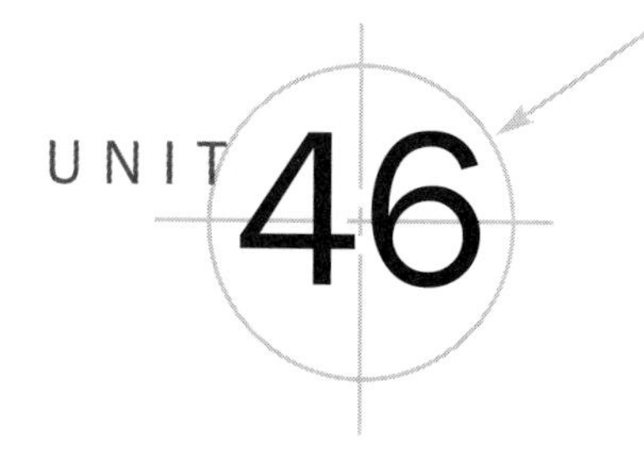

Types of Angles and Angular Geometric Principles

Objectives ***After studying this unit you should be able to***

- Identify different types of angles.
- Determine unknown angles in geometric figures using the principles of opposite, alternate interior, corresponding, parallel, and perpendicular angles.

Solving a practical application may require working with a number of different angles. To avoid confusion, angles must be properly named and their types identified. Determination of required unknown angular and linear dimensions is often based on the knowledge and understanding of angular geometric principles and their practical applications.

Naming Angles

Angles are named by a number, a letter, or three letters. When an angle is named with three letters, the vertex must be the middle letter. For example, the angle shown in Figure 46-1 can be called ∠1, ∠C, ∠ACB, or ∠BCA.

Figure 46-1

In cases where a point is the vertex of more than one angle, a single letter cannot be used to name an angle. For example, in Figure 46-2, point E is the vertex of three different angles; the single letter E cannot be used in naming the angle.

∠1 is called ∠GEH or ∠HEG.

∠2 is called ∠FEG or ∠GEF.

∠3 is called ∠FEH or ∠HEF.

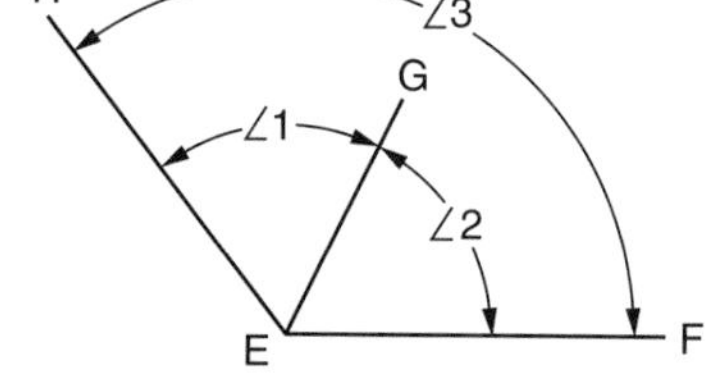

Figure 46-2

Types of Angles

An *acute angle* is an angle that is less than 90°. Angle 1 in Figure 46-3 is acute.

A *right angle* is an angle of 90°. Angle A in Figure 46-4 is a right angle. Right angles are often marked with a small square.

An *obtuse angle* is an angle greater than 90° but less than 180°. Angle ABC in Figure 46-5 is an obtuse angle.

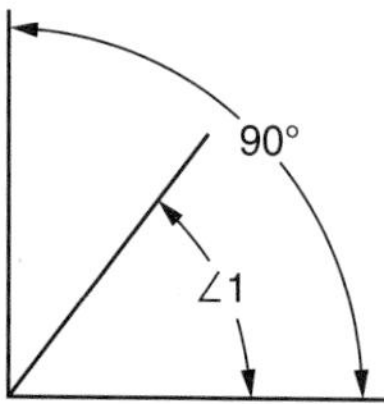

Figure 46-3

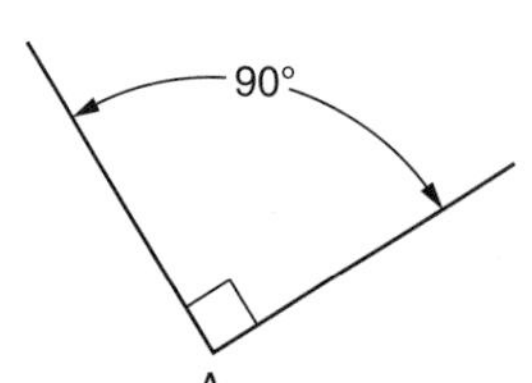

Figure 46-4

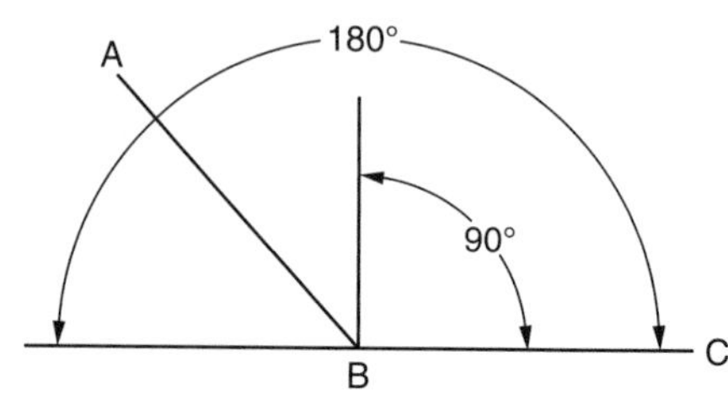

Figure 46-5

A *straight angle* is an angle of 180°. A straight line is a straight angle. Line EFG in Figure 46-6 is a straight angle.

A *reflex angle* is an angle greater than 180° and less than 360°. Angle 3 in Figure 46-7 is a reflex angle.

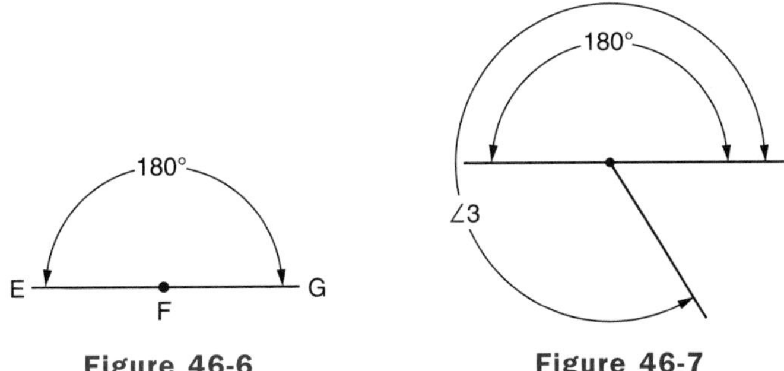

Figure 46-6

Figure 46-7

Adjacent Angles

Two angles are *adjacent* if they have a common side and a common vertex. Angle 1 and angle 2 shown in Figure 46-8 are adjacent since they both contain the common side *BC* and the common vertex B. Angle 4 and angle 5 shown in Figure 46-9 are not adjacent. The angles do not have a common vertex.

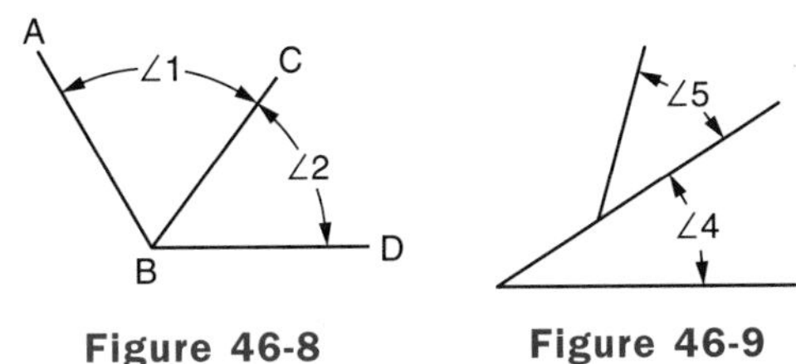

Figure 46-8

Figure 46-9

Angles Formed by a Transversal

A *transversal* is a line that intersects (cuts) two or more lines. Line EF in Figure 46-10 is a transversal since it cuts lines AB and CD.

Alternate interior angles are pairs of interior angles on opposite sides of the transversal. The angles have different vertices. For example, in Figure 46-10, angles 3 and 5 are alternate interior angles. Angles 4 and 6 are also alternate interior angles.

Corresponding angles are pairs of angles, one interior and one exterior with both angles on the same side of the transversal. The angles have different vertices. For example, in Figure 46-10, angles 1 and 5, 2 and 6, 3 and 7, and 4 and 8 are pairs of corresponding angles.

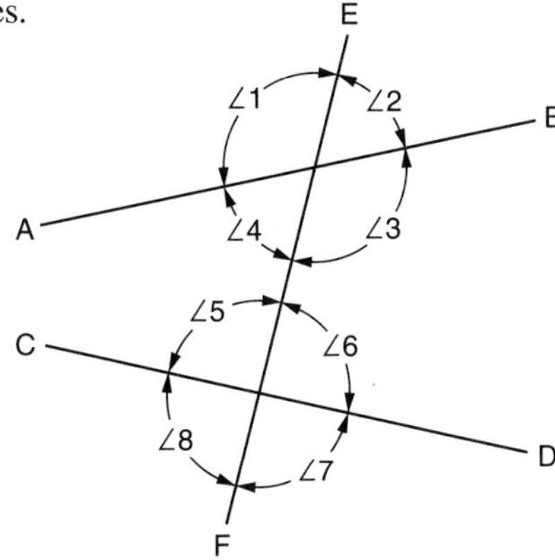

Figure 46-10

Geometric Principles

In this book, geometric postulates, theorems, and corollaries are grouped together and are called geometric principles. *Geometric principles* are statements of truth that are used as geometric rules. The principles will not be proved, but they will be used as the basis for problem solving.

➤ **Note:** Angles that are referred to or shown as equal are angles of equal measure (m). For example, ∠A = ∠B means m∠A = m∠B. Also, line segments that are referred to or shown as equal are segments of equal length. For example, AB = CD means length AB = length CD.

➤ **Principle 1**

If two lines intersect, the opposite or vertical angles are equal.

Given: AB intersects CD in Figure 46-11.

Conclusion: ∠1 = ∠3 and ∠2 = ∠4.

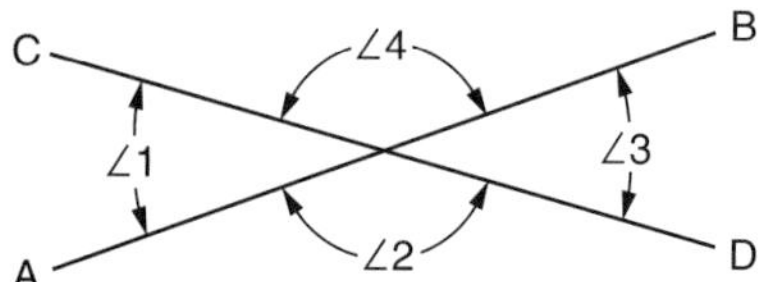

Figure 46-11

➤ **Principle 2**

If two parallel lines are intersected by a transversal, the alternate interior angles are equal.

Given: AB ∥ CD as shown in Figure 46-12.

Conclusion: ∠3 = ∠5 and ∠4 = ∠6.

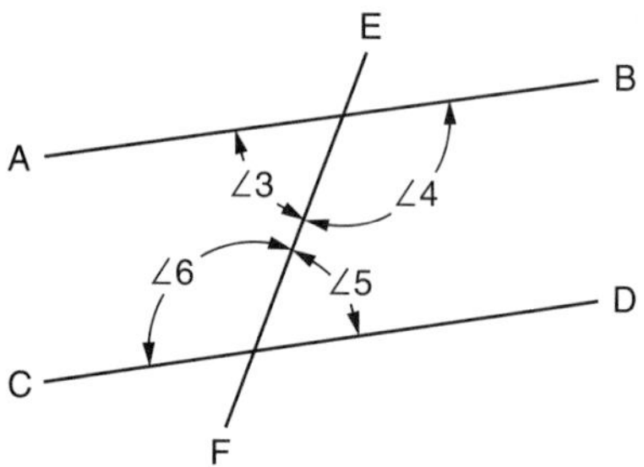

Figure 46-12

➤ **If two lines are intersected by a transversal and a pair of alternate interior angles are equal, the lines are parallel.**

Given: ∠1 = ∠2 as in Figure 46-13.

Conclusion: AB ∥ CD.

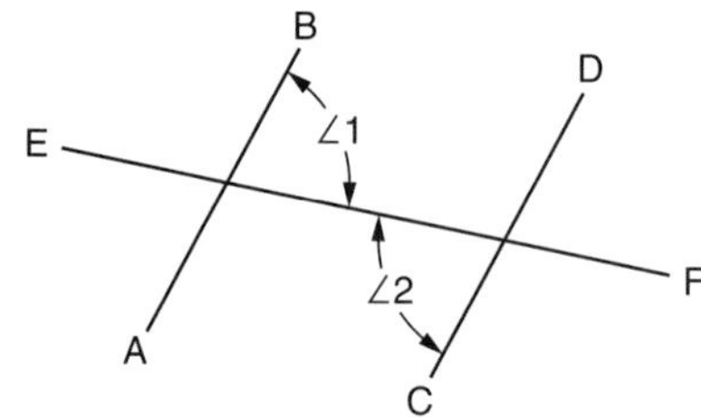

Figure 46-13

➤ **Principle 3**

If two parallel lines are intersected by a transversal, the corresponding angles are equal.

Given: In Figure 46-14, AB ∥ CD.

Conclusion: ∠1 = ∠5, ∠2 = ∠6, ∠3 = ∠7, and ∠4 = ∠8.

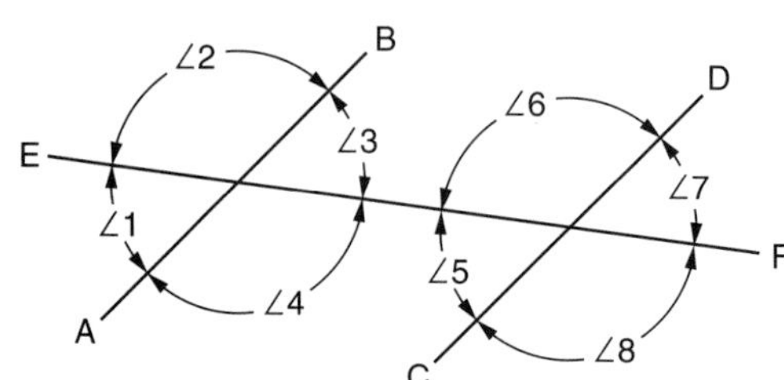

Figure 46-14

➤ **If two lines are intersected by a transversal and a pair of corresponding angles are equal, the lines are parallel.**

Given: In Figure 46-15, $\angle 1 = \angle 2$.

Conclusion: AB ∥ CD.

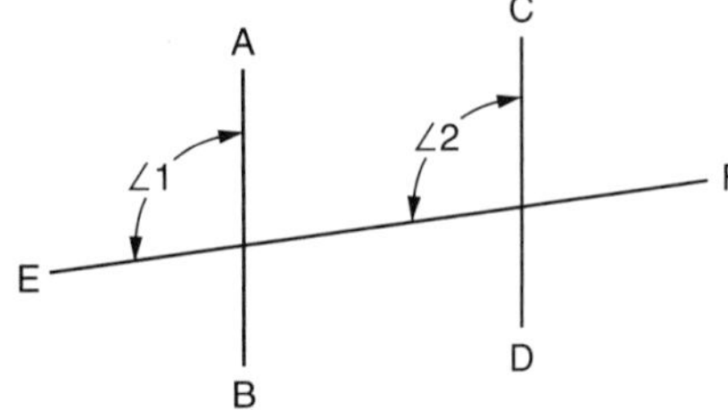

Figure 46-15

➤ **Principle 4**
Two angles are either equal or supplementary if their corresponding sides are parallel.

Given: AB ∥ FG and BC ∥ DE as in Figure 46-16.

Conclusion: $\angle 1 = \angle 3$ and $\angle 1$ and $\angle 2$ are supplementary ($\angle 1 + \angle 2 = 180°$).

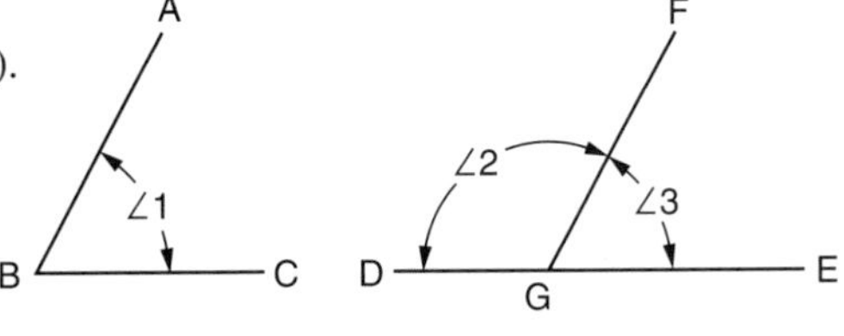

Figure 46-16

➤ **Principle 5**
Two angles are either equal or supplementary if their corresponding sides are perpendicular.

Given: In Figure 46-17, AB ⊥ DH and BC ⊥ EF.

Conclusion: $\angle 1 = \angle 2$; $\angle 1$ and $\angle 3$ are supplementary ($\angle 1 + \angle 3 = 180°$).

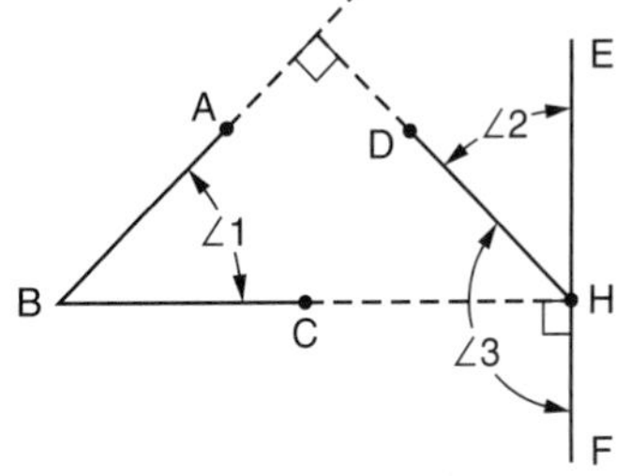

Figure 46-17

The following example illustrates the method of solving angular measure problems. Values of angles are determined by applying angular geometric principles and the fact that a straight angle (straight line) contains 180°.

Example Given: In Figure 46-18, AB ∥ CD, EF ∥ GH, $\angle 1 = 115°$, and $\angle 2 = 82°$. Determine the values of $\angle 3$ through $\angle 9$.

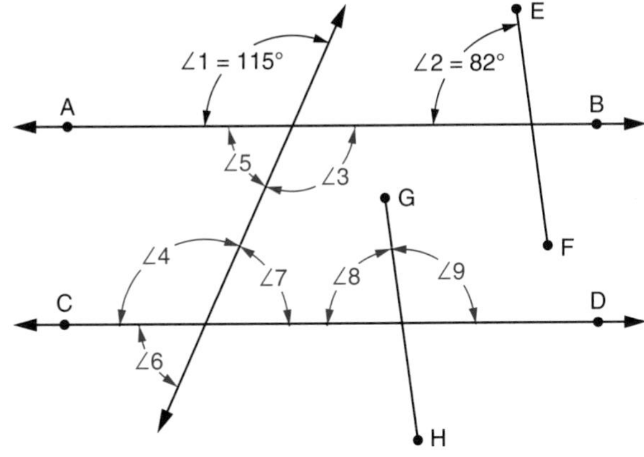

Figure 46-18

Solve for $\angle 3$. *Apply Principle 1.*
If two lines intersect, the opposite or vertical angles are equal.

$$\angle 3 = \angle 1 = 115^\circ \quad \text{Ans}$$

Solve for $\angle 4$. *Apply either Principle 2 or Principle 3.*

Applying Principle 2: If two parallel lines are intersected by a transversal, the alternate interior angles are equal.

$$\angle 4 = \angle 3 = 115^\circ \quad \text{Ans}$$

or,

Applying Principle 3: If two parallel lines are intersected by a transversal, the corresponding angles are equal.

$$\angle 4 = \angle 1 = 115^\circ \quad \text{Ans}$$

Solve for $\angle 5$. Since a straight angle (straight line) contains 180°, $\angle 5$ and $\angle 1$ are supplementary.

$$\angle 5 = 180^\circ - \angle 1 = 180^\circ - 115^\circ = 65^\circ \quad \text{Ans}$$

Solve for $\angle 6$. *Apply Principle 3.*
If two parallel lines are intersected by a transversal, the corresponding angles are equal.

$$\angle 6 = \angle 5 = 65^\circ \quad \text{Ans}$$

Solve for $\angle 7$. *Apply Principle 1.*
If two lines intersect, the opposite or vertical angles are equal.

$$\angle 7 = \angle 6 = 65^\circ \quad \text{Ans}$$

Solve for $\angle 8$. *Apply Principle 4.*
Two angles are either equal or supplementary if their corresponding sides are parallel.

$$\angle 8 = \angle 2 = 82^\circ \quad \text{Ans}$$

Solve for $\angle 9$. Since a straight angle (straight line) contains 180°, $\angle 8$ and $\angle 9$ are supplementary.

$$\angle 9 = 180^\circ - \angle 8 = 180^\circ - 82^\circ = 98^\circ \quad \text{Ans}$$

Application

Naming Angles

1. Name each of the following angles in three additional ways.

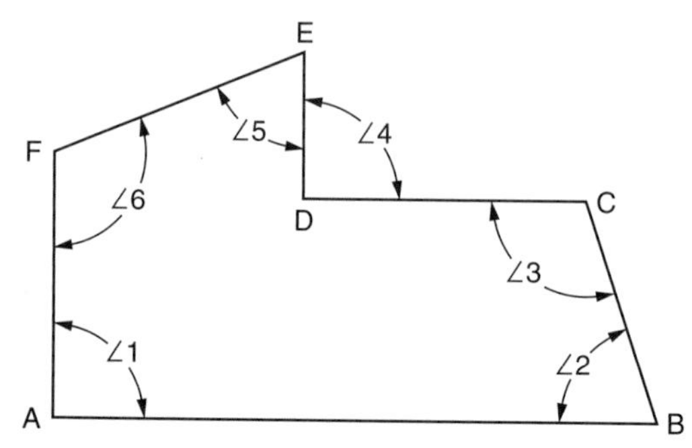

a. $\angle 1$ __________

b. $\angle 2$ __________

c. $\angle C$ __________

d. $\angle D$ __________

e. $\angle E$ __________

f. $\angle F$ __________

2. Name each of the following angles in two additional ways.

a. ∠1 ______________

b. ∠CBF ______________

c. ∠3 ______________

d. ∠ECB ______________

e. ∠5 ______________

f. ∠BCD ______________

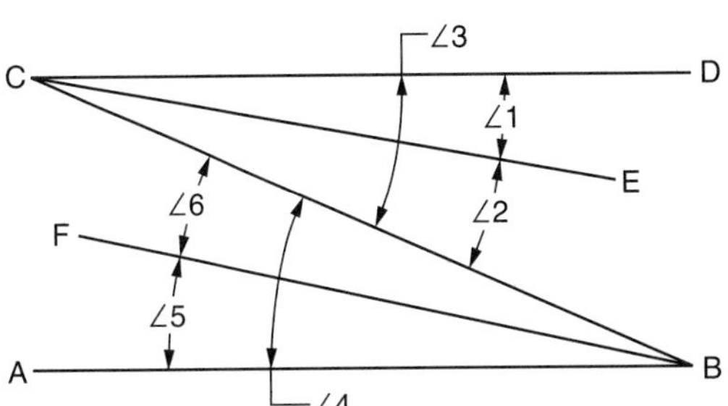

Types of Angles

3. Identify each of the following angles as acute, obtuse, right, straight, or reflex.

a. ∠BAF ______________

b. ∠ABF ______________

c. ∠CBF ______________

d. ∠DCA ______________

e. ∠BFA ______________

f. ∠BFD ______________

g. ∠ABC ______________

h. ∠BCD ______________

i. ∠AED ______________

j. ∠1 ______________

k. ∠DFA ______________

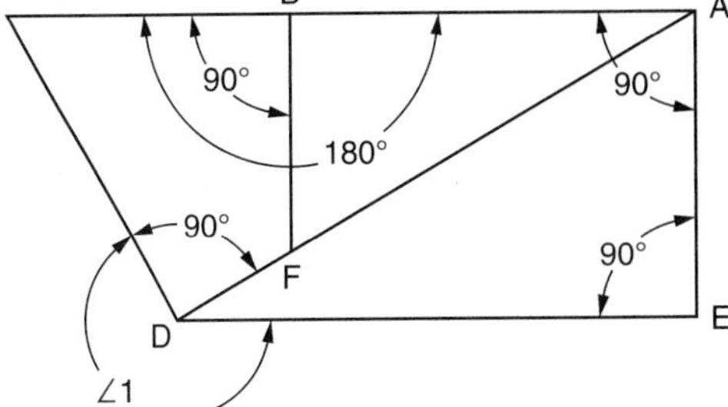

4. Name all pairs of adjacent angles shown in the figure.

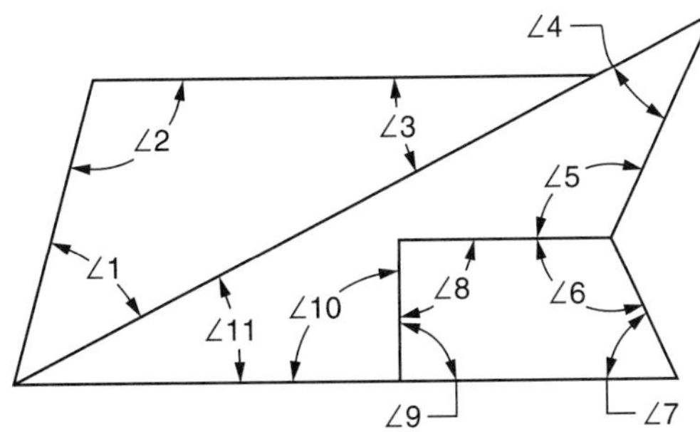

5. Alternate interior angles and corresponding angles are shown in the figure.

a. Name all pairs of alternate interior angles. ______________

b. Name all pairs of corresponding angles. ______________

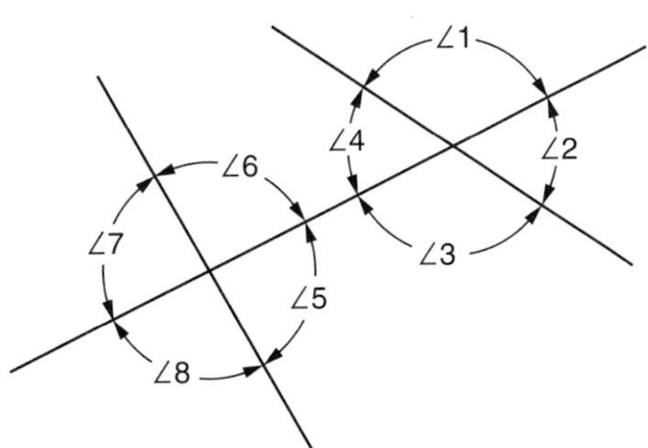

Applications of Geometric Principles

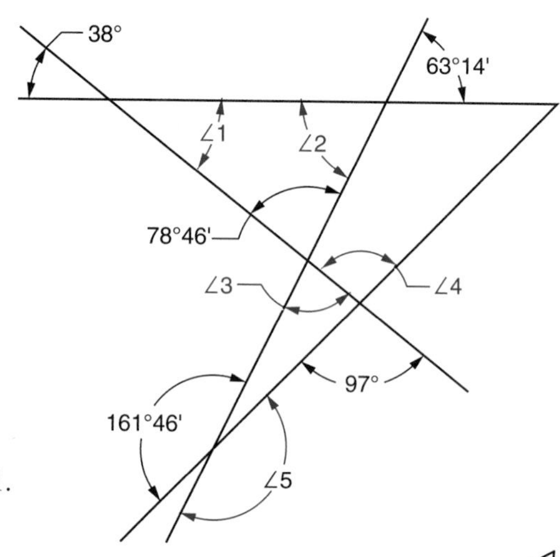

Solve the following problems.

6. Determine the values of ∠1 through ∠5.

∠1 = ______
∠2 = ______
∠3 = ______
∠4 = ______
∠5 = ______

7. Determine the values of ∠2, ∠3, and ∠4 for these given values of ∠1.

a. ∠1 = 32°
∠2 = ______ ∠4 = ______
∠3 = ______

b. ∠1 = 35°19'
∠2 = ______ ∠4 = ______
∠3 = ______

8. Given: AB ∥ CD. Determine the values of ∠2 through ∠8 for these given values of ∠1.

a. ∠1 = 68°
∠2 = ______ ∠6 = ______
∠3 = ______ ∠7 = ______
∠4 = ______ ∠8 = ______
∠5 = ______

b. ∠1 = 52°55'
∠2 = ______ ∠6 = ______
∠3 = ______ ∠7 = ______
∠4 = ______ ∠8 = ______
∠5 = ______

9. Given: Hole centerlines EF ∥ GH and MP ∥ KL. Determine the values of ∠1 through ∠15 for these values of ∠16.

a. ∠16 = 71°
∠1 = ______ ∠6 = ______ ∠11 = ______
∠2 = ______ ∠7 = ______ ∠12 = ______
∠3 = ______ ∠8 = ______ ∠13 = ______
∠4 = ______ ∠9 = ______ ∠14 = ______
∠5 = ______ ∠10 = ______ ∠15 = ______

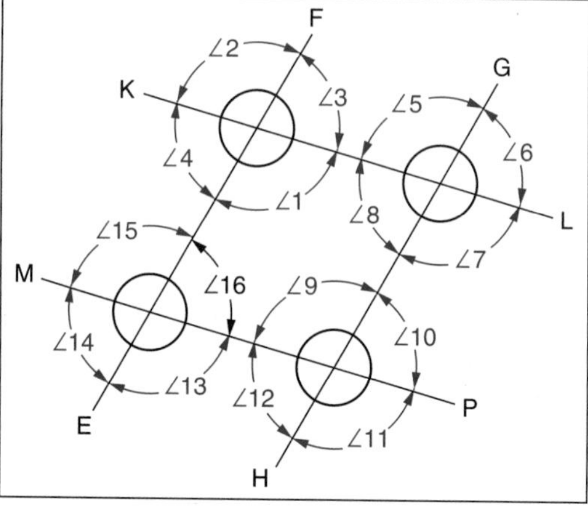

b. ∠16 = 86°52'
∠1 = ______ ∠6 = ______ ∠11 = ______
∠2 = ______ ∠7 = ______ ∠12 = ______
∠3 = ______ ∠8 = ______ ∠13 = ______
∠4 = ______ ∠9 = ______ ∠14 = ______
∠5 = ______ ∠10 = ______ ∠15 = ______

10. Given: Hole centerlines AB ∥ CD and EF ∥ GH. Determine the values of ∠1 through ∠22 for these given values of ∠23, ∠24, and ∠25.

    a. ∠23 = 97°, ∠24 = 34°, and ∠25 = 102°

    b. ∠23 = 112°23', ∠24 = 27°53', and ∠25 = 95°18'

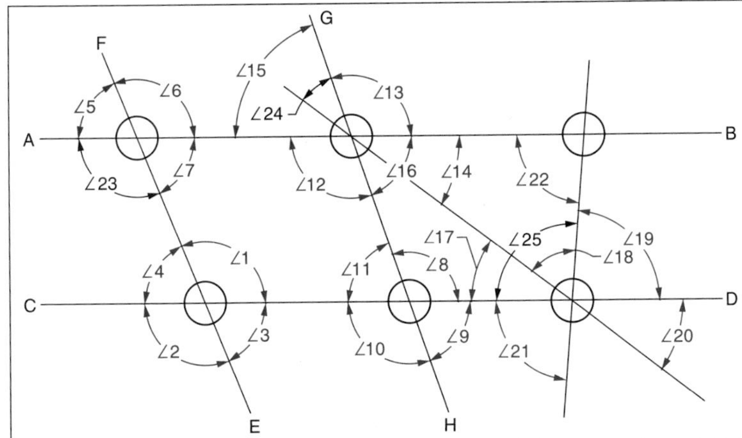

a. ∠1 = ________ ∠7 = ________ ∠13 = ________ ∠19 = ________

∠2 = ________ ∠8 = ________ ∠14 = ________ ∠20 = ________

∠3 = ________ ∠9 = ________ ∠15 = ________ ∠21 = ________

∠4 = ________ ∠10 = ________ ∠16 = ________ ∠22 = ________

∠5 = ________ ∠11 = ________ ∠17 = ________

∠6 = ________ ∠12 = ________ ∠18 = ________

b. ∠1 = ________ ∠7 = ________ ∠13 = ________ ∠19 = ________

∠2 = ________ ∠8 = ________ ∠14 = ________ ∠20 = ________

∠3 = ________ ∠9 = ________ ∠15 = ________ ∠21 = ________

∠4 = ________ ∠10 = ________ ∠16 = ________ ∠22 = ________

∠5 = ________ ∠11 = ________ ∠17 = ________

∠6 = ________ ∠12 = ________ ∠18 = ________

11. Given: AB ∥ CD, AC ∥ ED. Determine the value of ∠2 and ∠3 for these values of ∠1.

    a. ∠1 = 67° ∠2 = ________ ∠3 = ________

    b. ∠1 = 74°12' ∠2 = ________ ∠3 = ________

E D B ∠2 ∠3 ∠1 A C

12. Given: FH ∥ GS ∥ KM and FG ∥ HK. Determine the values of ∠1, ∠2, and ∠3 for these values of ∠4.

    a. ∠4 = 116°

    ∠1 = ________ ∠3 = ________

    ∠2 = ________

    b. ∠4 = 107°43'

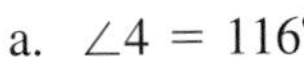

    ∠1 = ________ ∠3 = ________

    ∠2 = ________

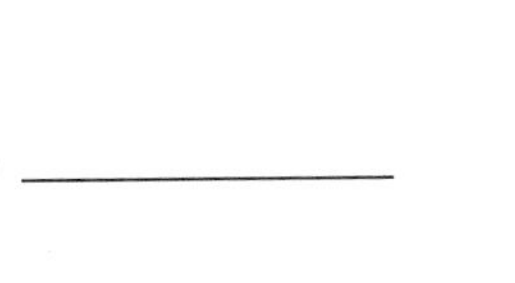

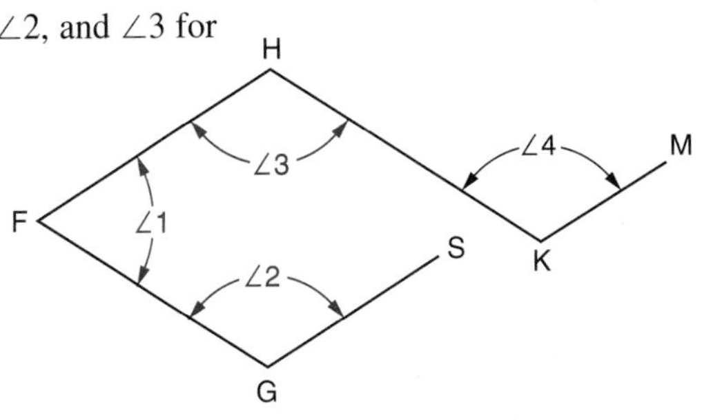

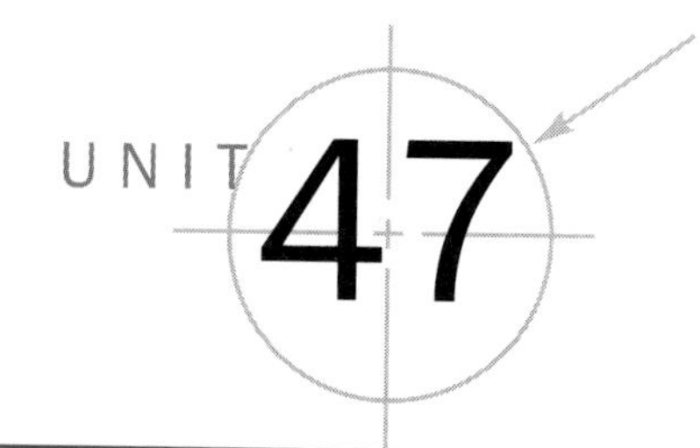

Introduction to Triangles

Objectives ***After studying this unit you should be able to***

- Identify different types of triangles.
- Determine unknown angles based on the principle that all triangles contain 180°.
- Identify corresponding parts of triangles.

A *polygon* is a closed plane figure formed by three or more straight line segments. A *triangle* is a three-sided polygon; it is the simplest kind of polygon. The symbol △ means triangle. Triangles are widely applied in engineering and manufacturing. The triangle is a rigid figure that is the basic figure in many designs. Machine technicians and drafters require a knowledge of triangles in laying out work.

Types of Triangles

A *scalene triangle* has three unequal sides. It also has three unequal angles. In Figure 47-1, triangle ABC is scalene. Sides AB, AC, and BC are unequal and angles A, B, and C are unequal.

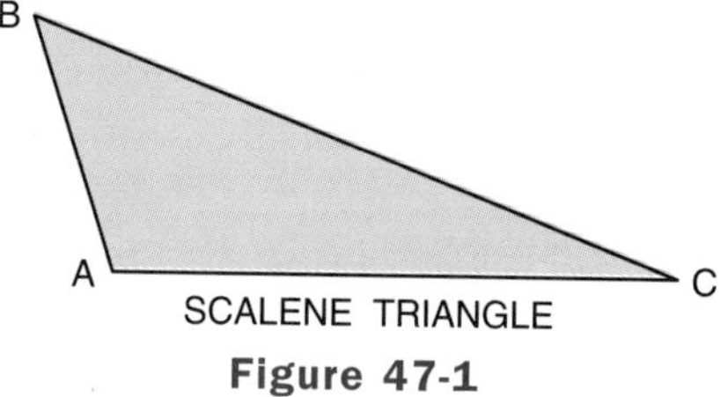

Figure 47-1

An *isosceles triangle* has two equal sides. The equal sides are called *legs.* The third side is called the *base.* An isosceles triangle also has two equal base angles. *Base angles* are the angles that are opposite the legs. Figure 47-2 shows isosceles triangle RST, with side RT = side ST and ∠R = ∠S.

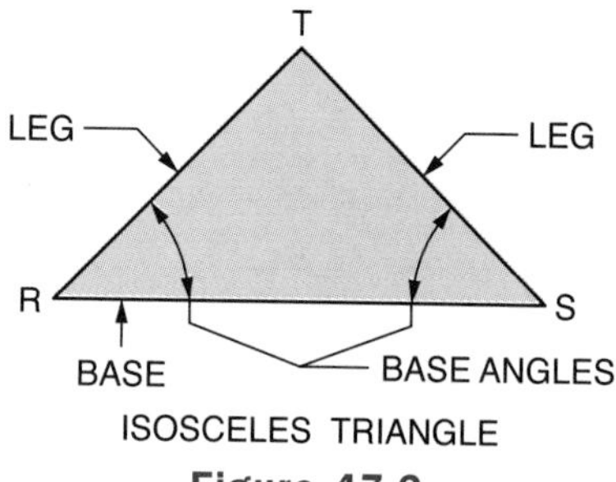

Figure 47-2

An *equilateral triangle* has three equal sides. It also has three equal angles. In Figure 47-3 is equilateral triangle DEF, with sides DE = DF = EF and ∠D = ∠E = ∠F. Because an equilateral triangle has three equal angles, it may also be called an equiangular triangle.

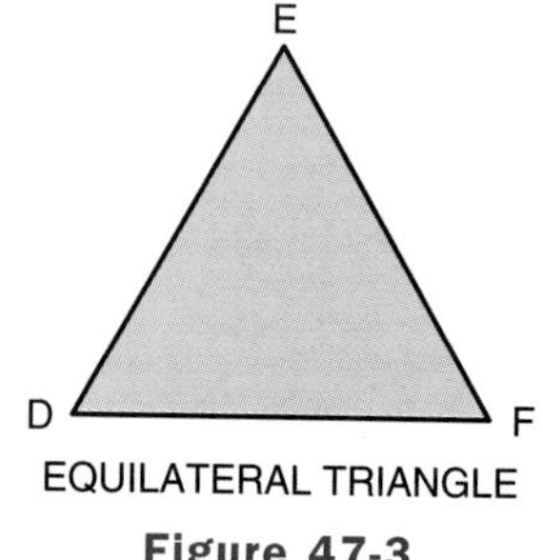

Figure 47-3

A *right triangle* has a right or 90° angle. The symbol for a right angle is a small square placed at the vertex of the angle. The side opposite the right angle is called the *hypotenuse*. The other two sides are called *legs*. Right triangle HJK is shown in Figure 47-4 with ∠H = 90° and JK is the hypotenuse.

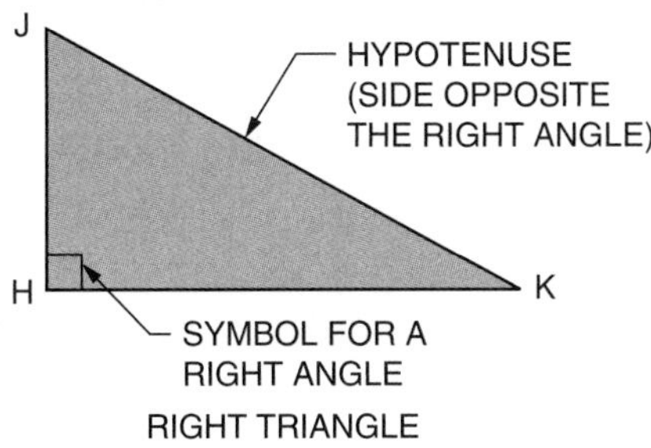

RIGHT TRIANGLE

Figure 47-4

Angles of a Triangle

➤ **Principle 6**

The sum of the angles of any triangle is equal to 180°. This principle is applied in many practical applications.

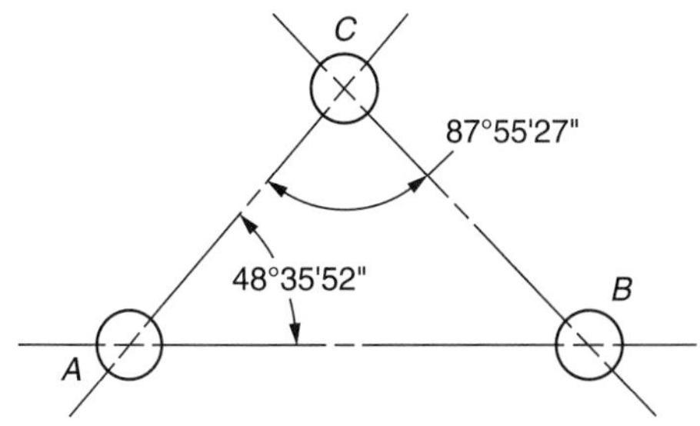

Figure 47-5

Example Angles A, B, and C in Figure 47-5 are hole centerline angles. Angle A = 48°35'52" and Angle C = 87°55'27". Determine ∠B.

∠B = 180° − (48°35'52" + 87°55'27")

180 [−] [(] 48 [° ' "] 35 [° ' "] 52 [° ' "] [+] 87 [° ' "] 55 [° ' "] 27 [° ' "] [)] [=] [SHIFT] [←] 43°28°41, 43°28'41" Ans

or 180 [−] [(] 48 [° ' "] 35 [° ' "] [▶] 52 [° ' "] [▶] [▶] [+] 87 [° ' "] 55 [° ' "] [▶] 27 [° ' "] [▶] [▶] [)] [° ' "] [◀] [ENTER] [ENTER] 43°28'41" Ans

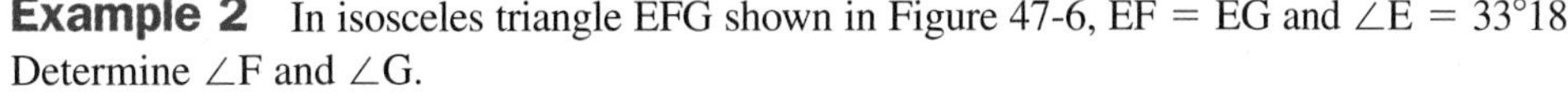

Example 2 In isosceles triangle EFG shown in Figure 47-6, EF = EG and ∠E = 33°18'. Determine ∠F and ∠G.

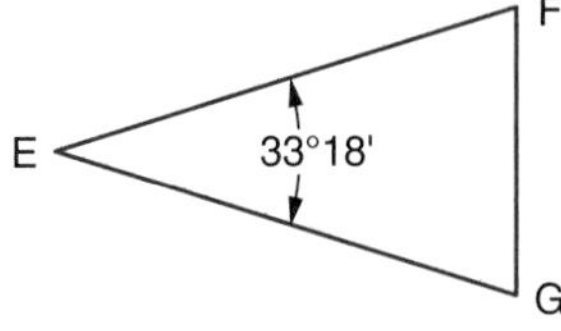

Figure 47-6

∠E + ∠F + ∠G = 180°

180° − ∠E = ∠F + ∠G

180° − 33°18' = ∠F + ∠G

146°42' = ∠F + ∠G

Since ∠F = ∠G, ∠F and ∠G each = $\frac{146°42'}{2}$ = 73°21' Ans

Example 3 In Figure 47-7, triangle HJK is equilateral. Determine ∠H, ∠J, and ∠K.

∠H + ∠J + ∠K = 180°

Since ∠H = ∠J = ∠K, each angle = $\frac{180°}{3}$ = 60° Ans

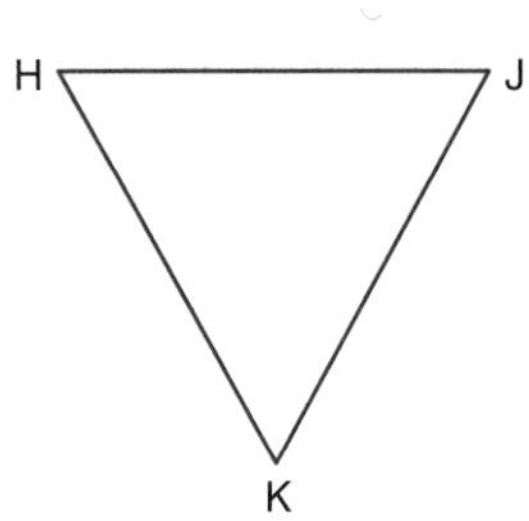

Figure 47-7

Corresponding Parts of Triangles

It is essential to develop the ability to identify corresponding angles and sides of two or more triangles. Corresponding sides and angles between triangles are not determined by the positions of the triangles. The smallest angle of a triangle lies opposite the shortest side and the largest angle of a triangle lies opposite the longest side. *Corresponding angles* between two triangles are determined by comparing the lengths of the sides that lie opposite the angles. *Corresponding sides* between two triangles are determined by comparing the sizes of the angles that lie opposite the sides.

Example 1 In triangle ABC shown in Figure 47-8, determine the longest, next longest, and shortest sides.

The longest side is CB since it lies opposite the largest angle, 107°. Ans

The next longest side is AB since it lies opposite the next largest angle, 43°. Ans

The shortest side is AC since it lies opposite the smallest angle, 30°. Ans

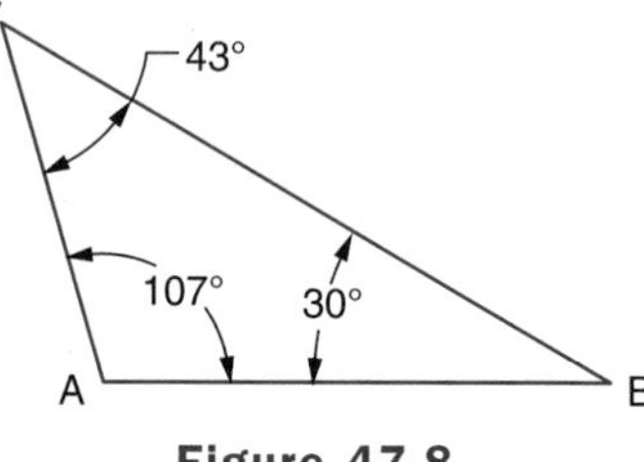

Figure 47-8

Example 2 In triangle DEF shown in Figure 47-9, determine the largest, next largest, and smallest angle. All dimensions are in inches.

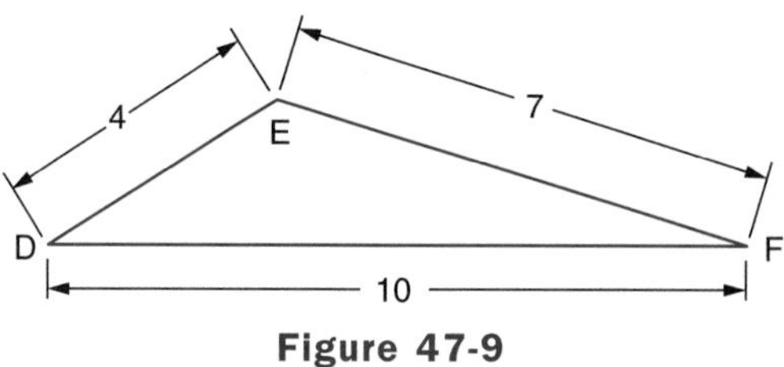

Figure 47-9

The largest angle is ∠E since it lies opposite the longest side, 10 inches. Ans

The next largest angle is ∠D since it lies opposite the next longest side, 7 inches. Ans

The smallest angle is ∠F since it lies opposite the shortest side, 4 inches. Ans

Example 3 In triangles ABC and FED shown in Figure 47-10, determine the pairs of corresponding angles between the two triangles. All dimensions are in millimeters.

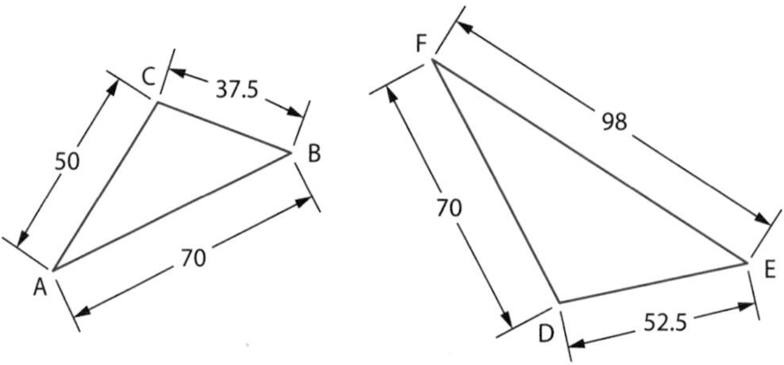

Figure 47-10

Angle C corresponds to ∠D since each angle lies opposite the longest side of each triangle.

Angle B corresponds to ∠E since each angle lies opposite the next longest side of each triangle.

Angle A corresponds to ∠F since each angle lies opposite the shortest side of each triangle.

Application

Identify each of the triangles 1 through 8 as scalene, isosceles, equilateral, or right.

1. ______________

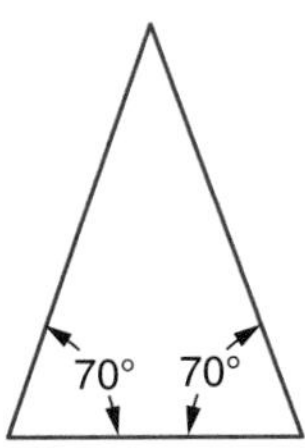

2. All dimensions are in inches. ______________

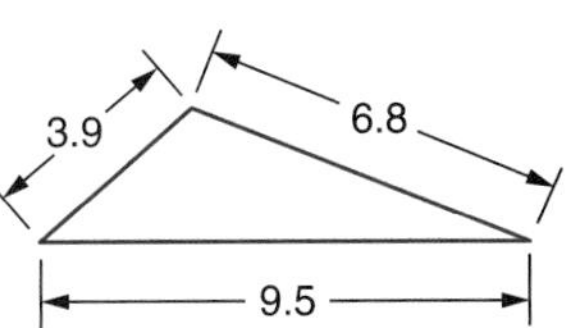

3. ______________

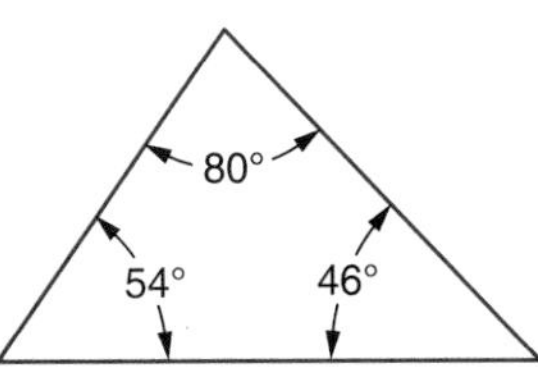

4. All dimensions are in millimeters. ______________

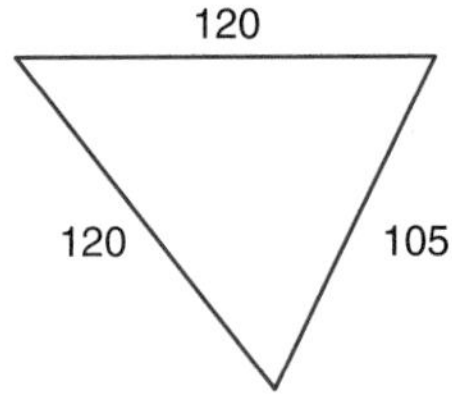

5. ______________

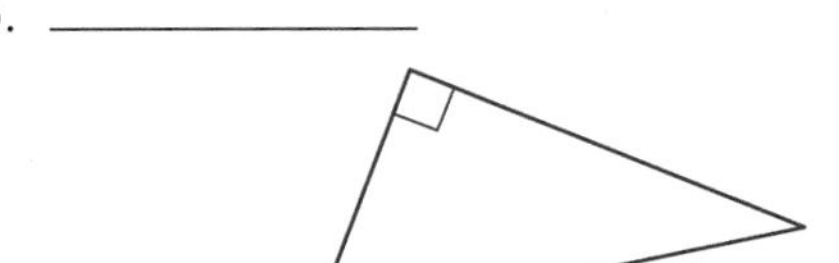

6. All dimensions are in millimeters. ______________

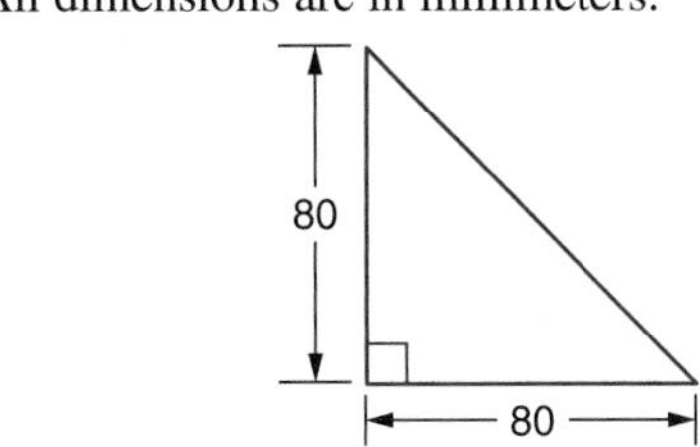

7. ______________

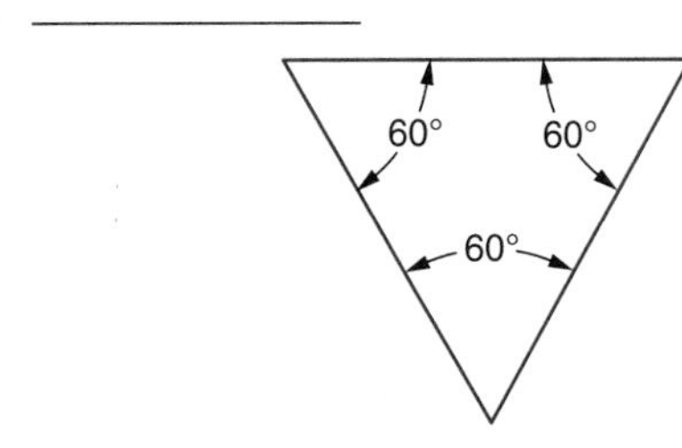

8. All dimensions are in inches. ______________

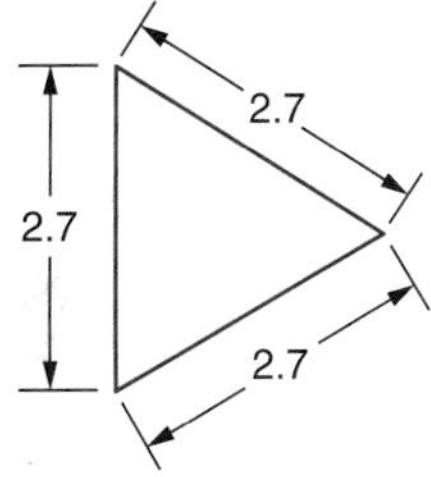

Angles of a Triangle

Solve the following problems.

9. Find the value of ∠A + ∠B + ∠C. ______________

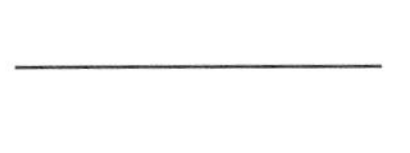

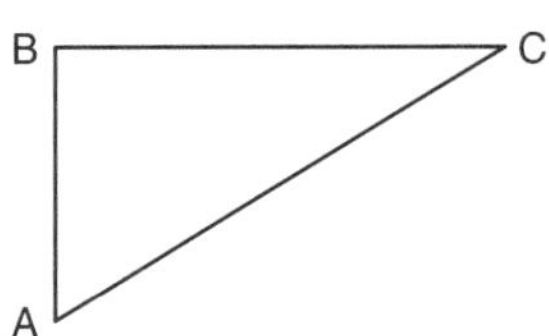

10. Find the value of the unknown angles for these given angle values.

   a. If ∠1 = 56° and ∠2 = 86°, find ∠3. ______________

   b. If ∠2 = 81° and ∠3 = 46°, find ∠1. ______________

11. Find the value of the unknown angles for these given angle values.

   a. If ∠4 = 32°43' and ∠5 = 119°17', find ∠6. ______________

   b. If ∠5 = 123°17'13" and ∠6 = 27°, find ∠4. ______________

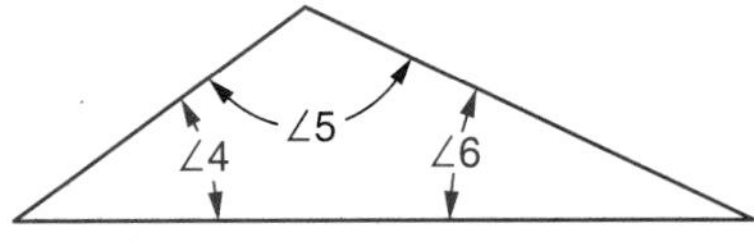

12. Find the value of the unknown angles for these given angle values.

    a. If $\angle A = 19°43'$, find $\angle B$. ____________

    b. If $\angle B = 67°58'$, find $\angle A$. ____________

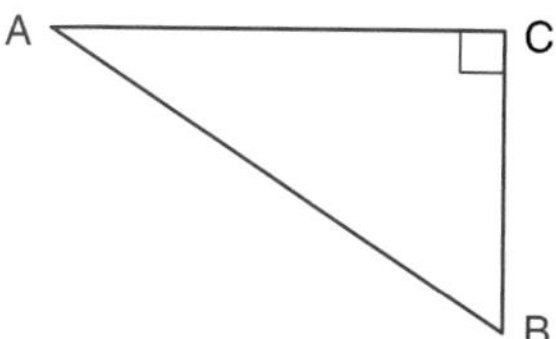

13. In triangle ABC, BC = 17.3 inches.

    a. Find AB. ____________

    b. Find AC. ____________

14. In triangle EFG, find the value of the unknown angles for these given angle values. All dimensions are in inches.

    a. If $\angle E = 81°$, find $\angle G$. ____________

    b. If $\angle G = 83°27'$, find $\angle F$. ____________

15. Find the value of the unknown angles for these given angle values. All dimensions are in millimeters.

    a. If $\angle 3 = 17°$, find $\angle 1$. ____________

    b. If $\angle 3 = 25°19'$, find $\angle 2$. ____________

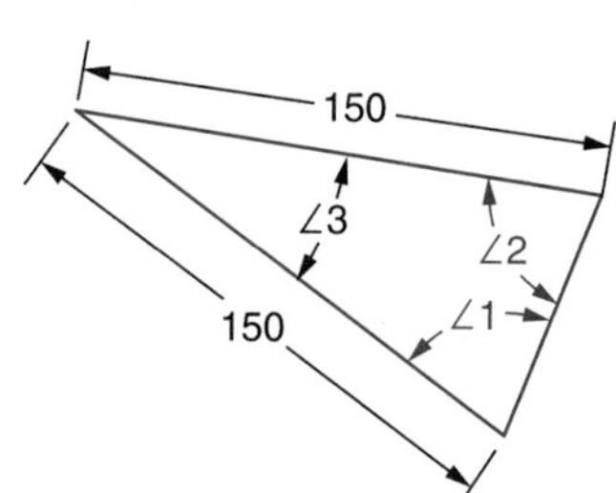

16. All dimensions are in inches.

    a. Find $\angle 3$. ____________

    b. Find $\angle 4$. ____________

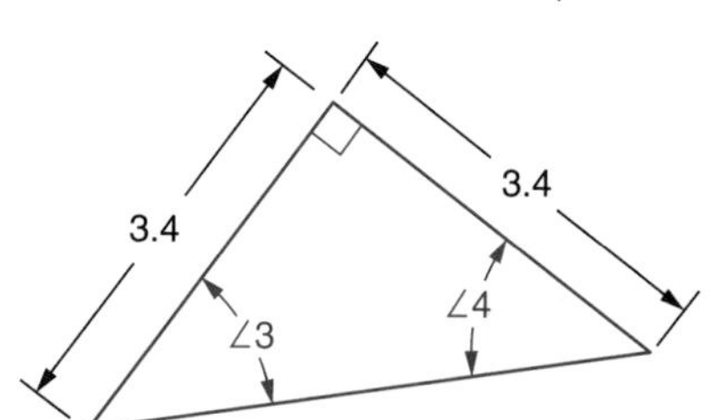

17. Find the value of the unknown angles for these given angle values.

    a. If $\angle 1 = 26°$ and $\angle 3 = 48°$, find $\angle 2$. ____________

    b. If $\angle 1 = 28°$ and $\angle 2 = 15°$, find $\angle 3$. ____________

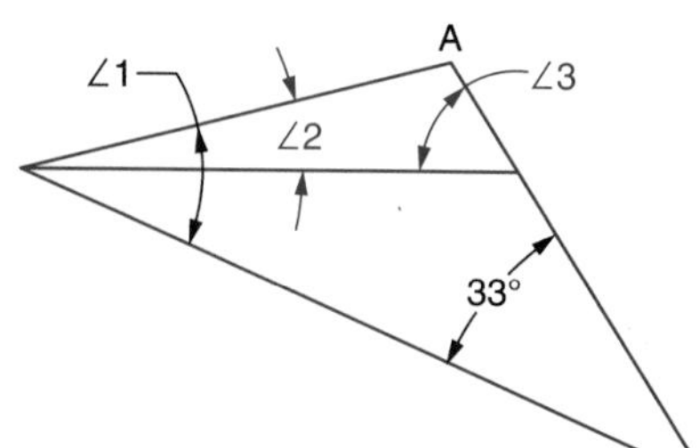

18. Hole centerlines AB ∥ CD.

    a. If $\angle 1 = 86°32'$, find $\angle 2$. ____________

    b. If $\angle 2 = 67°47'$, find $\angle 1$. ____________

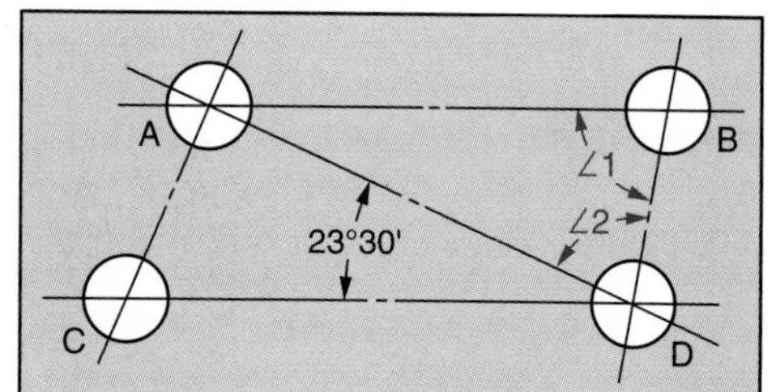

19. Find the value of the unknown angles listed.

    a. ∠3 ____________

    b. ∠4 ____________

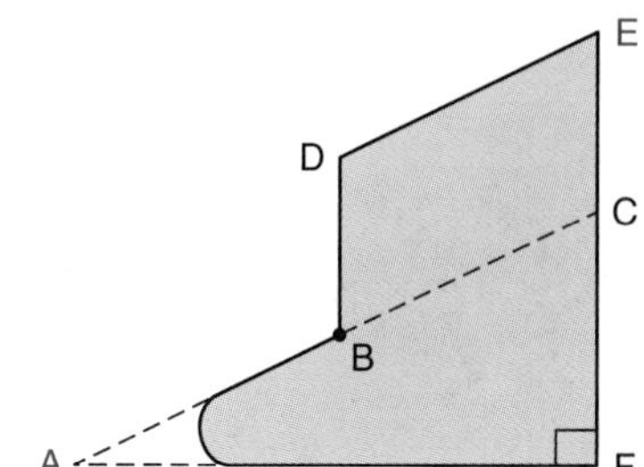

20. AB ∥ DE, BC is an extension of AB.

    a. If ∠E = 66°43', find ∠A. ____________

    b. If ∠A = 19°07', find ∠E. ____________

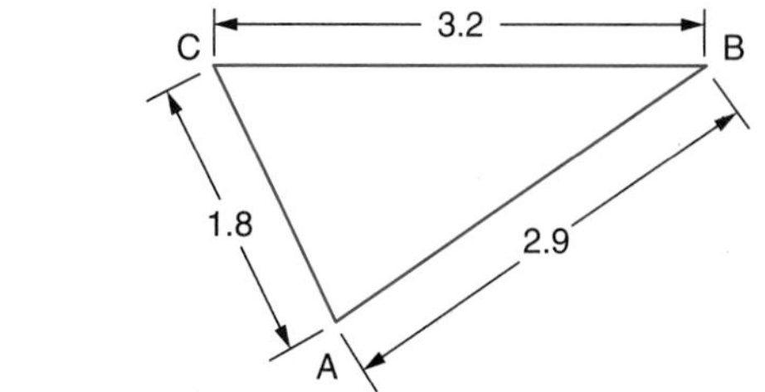

Corresponding Parts of Triangles

Determine the answers to the following problems, which are based on corresponding parts.

21. All dimensions are in inches.

    a. Find the largest angle. ____________

    b. Find the next largest angle. ____________

    c. Find the smallest angle. ____________

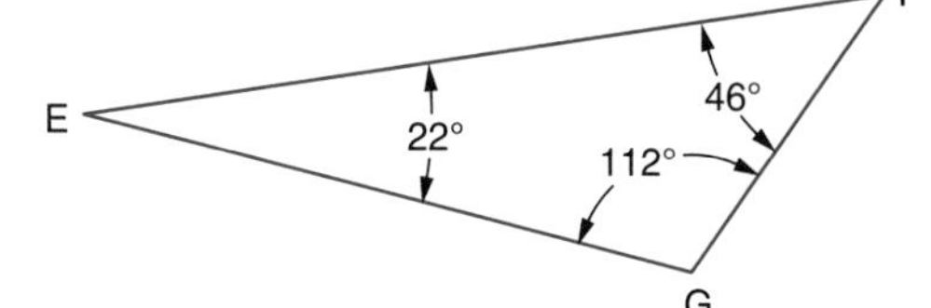

22. Refer to triangle EFG.

    a. Find the shortest side. ____________

    b. Find the next shortest side. ____________

    c. Find the longest side. ____________

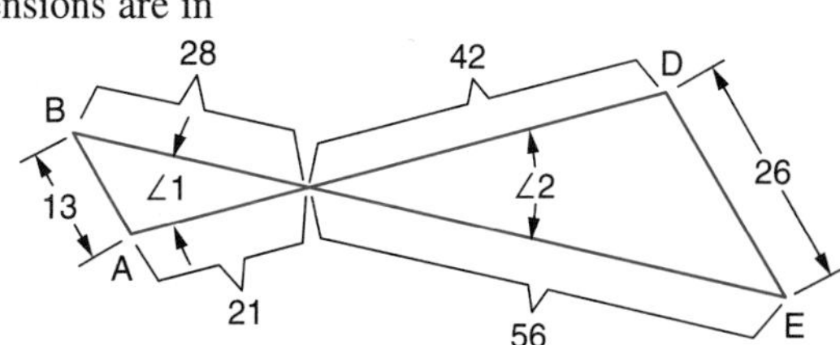

23. Identify the angle that corresponds with each angle listed. All dimensions are in millimeters.

    a. ∠A ____________ 

    b. ∠B ____________

    c. ∠1 ____________

24. Identify the angle that corresponds with each angle listed. All dimensions are in inches.

    a. ∠F ____________

    b. ∠G ____________

    c. ∠H ____________

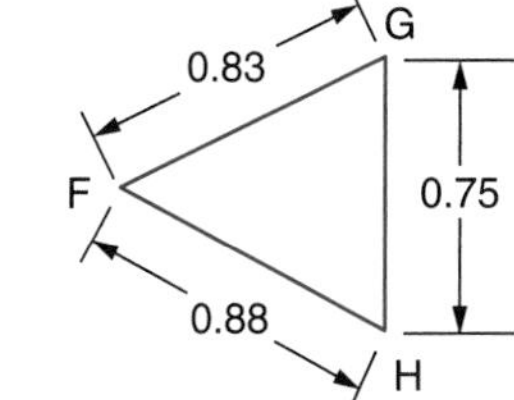

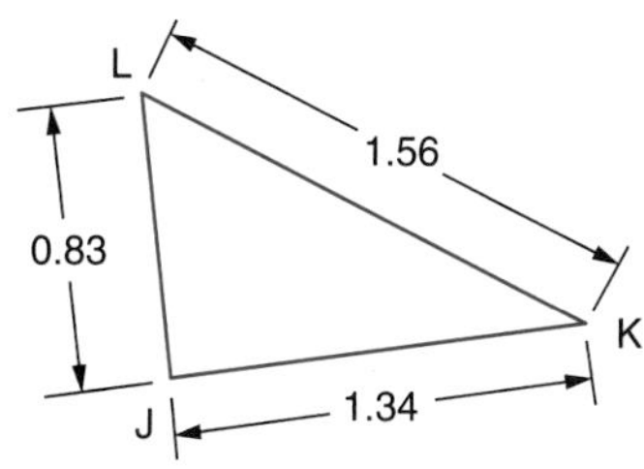

UNIT 48

Geometric Principles for Triangles and Other Common Polygons

Objectives ***After studying this unit you should be able to***

- Identify similar triangles and compute unknown angles and sides.
- Compute angles and sides of isosceles, equilateral, and right triangles.
- Determine interior angles of any polygon.

Congruent Triangles

Two triangles are *congruent* if they are identical in size and shape. If one congruent triangle is placed on top of the other, they fit together exactly. The symbol ≅ means congruent.

Corresponding Parts of Congruent Triangles are Equal.

Example In Figure 48-1, △ABC ≅ △DEF.

Figure 48-1

Corresponding parts of congruent triangles are equal.

$$\angle A = \angle D, \angle B = \angle E, \text{ and } \angle C = \angle F.$$
$$AB = DE, AC = DF, \text{ and } BC = EF.$$

In Figure 48-2, △GHI ≅ △JKL.

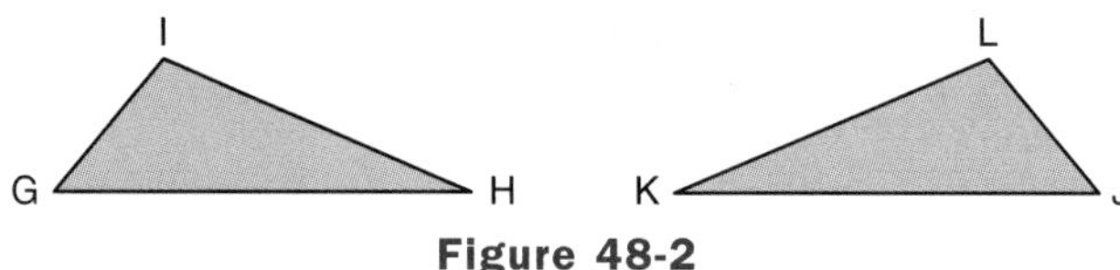

Figure 48-2

Notice that in order to get the triangles to fit together exactly, one of them will need to be "flipped over." Corresponding parts of these two congruent triangles are:

$$\angle G = \angle J, \angle H = \angle K, \text{ and } \angle I = \angle L.$$
$$GH = JK, GI = JL, \text{ and } HI = KL.$$

Similar Figures

Stated in a general way, *similar figures* are figures that are alike in shape but different in size. For example, a photograph is similar to the object that is photographed.

In machine technology, engineering drawings made to scale are similar to the objects they represent. Often, scale drawings are in the form of *similar polygons* or combinations of similar polygons. Recall that a *polygon* is a closed plane figure formed by three or more straight line segments. *Similar polygons* have the same number of sides, equal corresponding angles, and proportional corresponding sides. The symbol ~ means similar. A triangle is a three-sided polygon; it is the simplest kind of polygon.

Similar Triangles

Two triangles are similar if their corresponding angles are equal; their corresponding sides will also be proportional.

Example 1 Triangles ABC and DEF in Figure 48-3 have equal corresponding angles.

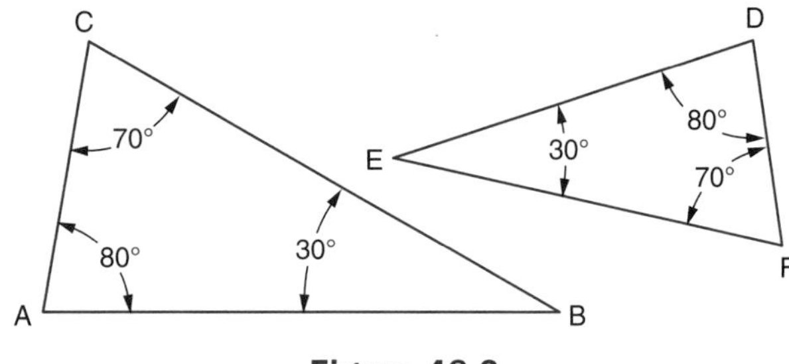

Figure 48-3

Two triangles are *similar* if their corresponding angles are equal.

$\triangle ABC \sim \triangle DEF$

Example 2 The lengths of the sides of triangles HJK and LMN in Figure 48-4 are given in inches.

$$\frac{HJ}{LM} = \frac{JK}{MN} = \frac{HK}{LN}$$

$$\frac{2}{4} = \frac{4}{8} = \frac{5}{10}$$

$$\frac{1}{2} = \frac{1}{2} = \frac{1}{2}$$

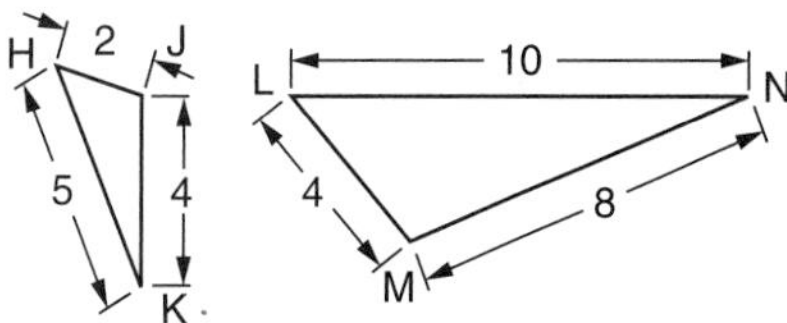

Figure 48-4

The corresponding sides are proportional.

$\triangle HJK \sim \triangle LMN$

Example 3 In Figure 48-5, $\triangle PRS \sim \triangle TWY$.

All linear dimensions are in millimeters.

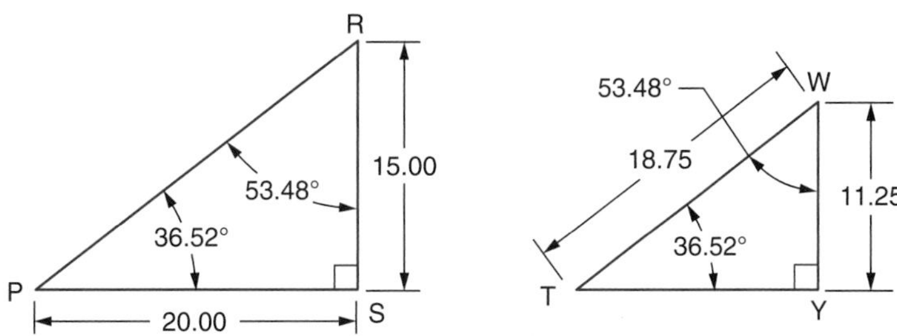

Figure 48-5

a. Determine the length of side PR.

b. Determine the length of side TY.

Set up proportions and solve for the unknown sides, PR and TY.

a. $\frac{WY}{RS} = \frac{WT}{PR}$

$\frac{11.25 \text{ mm}}{15.00 \text{ mm}} = \frac{18.75 \text{ mm}}{PR}$

$(11.25 \text{ mm})\, PR = 15.00 \text{ mm}\,(18.75 \text{ mm})$

$PR = \frac{15.00 \text{ mm}\,(18.75 \text{ mm})}{11.25 \text{ mm}}$

$PR = 25.00 \text{ mm}$ Ans

b. $\frac{WY}{RS} = \frac{TY}{PS}$

$\frac{11.25 \text{ mm}}{15.00 \text{ mm}} = \frac{TY}{20.00 \text{ mm}}$

$(15.00 \text{ mm})\, TY = 20.00 \text{ mm}\,(11.25 \text{ mm})$

$TY = \frac{20.00 \text{ mm}\,(11.25 \text{ mm})}{15.00 \text{ mm}}$

$TY = 15.00 \text{ mm}$ Ans

➤ Principle 7

Two triangles are similar if their sides are respectively parallel.

Given: AB ∥ DE, AC ∥ DF, and BC ∥ EF in Figure 48-6.

Conclusion: △ABC ~ △DEF.

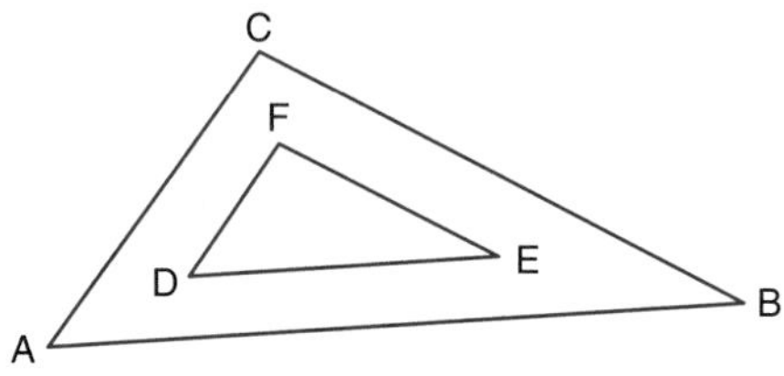

Figure 48-6

➤ **Two triangles are similar if their sides are respectively perpendicular.**

Given: HJ ⊥ LM, HK ⊥ LP, and JK ⊥ MP in Figure 48-7.

Conclusion: △HJK ~ △LMP.

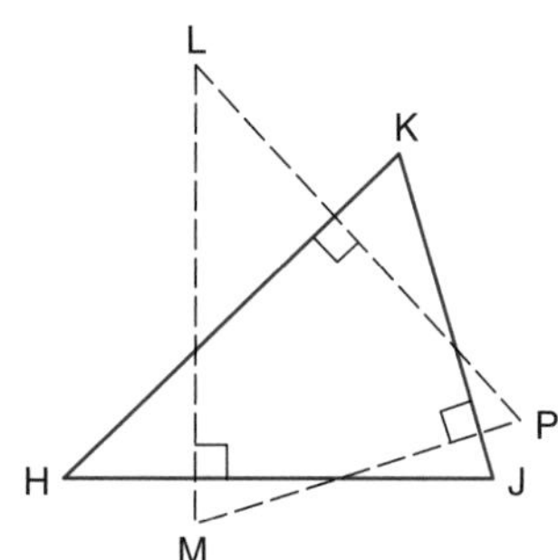

Figure 48-7

➤ **Within a triangle, if a line is drawn parallel to one side, the triangle formed is similar to the original triangle.**

Given: DE ∥ BC in Figure 48-8.

Conclusion: △ADE ~ △ABC.

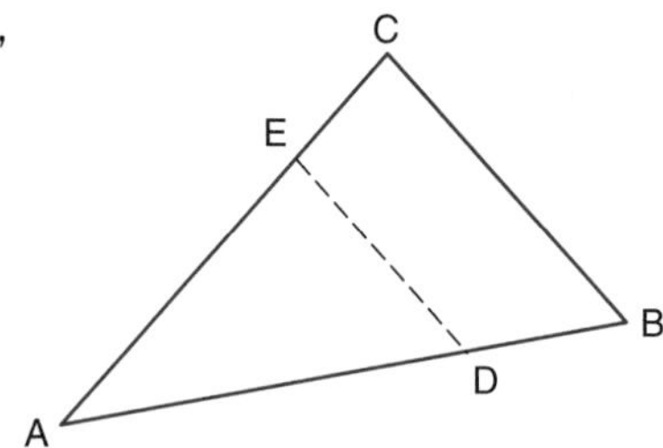

Figure 48-8

➤ **In a right triangle, if a line is drawn from the vertex of the right angle perpendicular to the opposite side, the two triangles formed and the original triangle are similar.**

Given: In Figure 48-9, we have right △HFG, FL ⊥ HG.

Conclusion: △FLH ~ △GLF ~ △GFH.

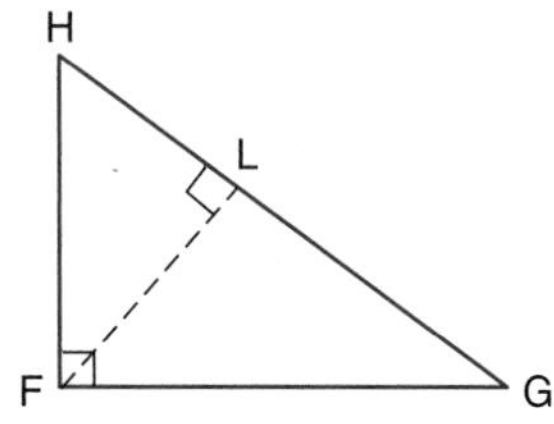

Figure 48-9

Isosceles, Equilateral, and Right Triangles

➤ **Principle 8**

In an isosceles triangle, an altitude to the base bisects the base and the vertex angle.

An altitude is a line drawn from a vertex perpendicular to the opposite side.
To *bisect* means to divide into two equal parts.

Given: Isosceles △ABC in Figure 48-10 with AC = CB and line CD the altitude to base AB.

Conclusion: AD = BD and ∠1 = ∠2.

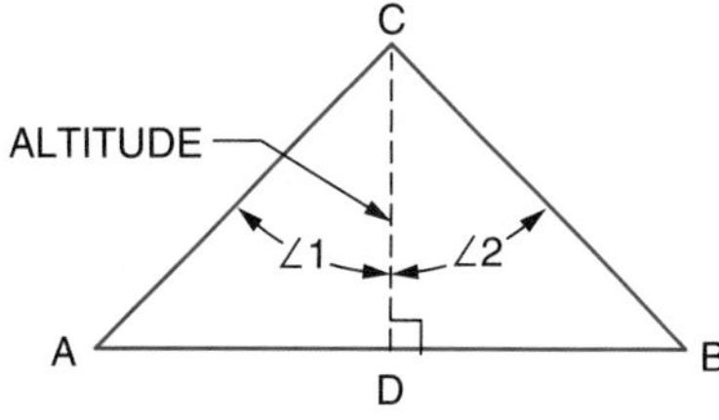

Figure 48-10

➤ **In an equilateral triangle, an altitude to any side bisects the side and the vertex angle.**

Given: Equilateral △EFG in Figure 48-11 with EH the altitude to FG.

Conclusion: FH = GH and ∠3 = ∠4.

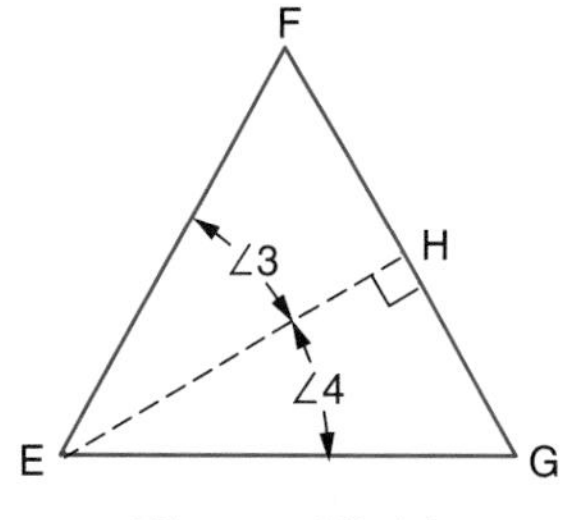

Figure 48-11

➤ **Principle 9 (Pythagorean Theorem)**

In a right triangle, the square of the hypotenuse is equal to the sum of the squares of the other two sides or legs. If two sides of a right triangle are known, the third side can be calculated.

This principle, called the *Pythagorean Theorem,* is often used for solving machine technology problems.

Example 1 In the right △ABC shown in Figure 48-12, dimensions a and b are given; the centerlines meet at right angles at C. To determine the distance between holes A and B, distance c must be computed. Dimensions are given in inches.

Side c is the hypotenuse. Substitute the given values for sides a and b and solve for distance c.

$$c^2 = a^2 + b^2$$

$$c = \sqrt{a^2 + b^2}$$

$$c = \sqrt{(6.027 \text{ in})^2 + (8.139 \text{ in})^2}$$

$$c = \sqrt{36.3247 \text{ in}^2 + 66.2433 \text{ in}^2}$$

$$c = \sqrt{102.5680 \text{ in}^2}$$

$$c = 10.128 \text{ in} \quad \text{Ans (rounded)}$$

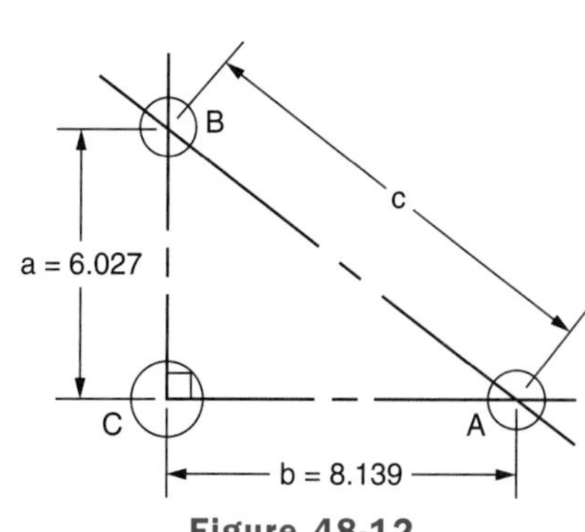

Figure 48-12

$$c = \sqrt{(6.027 \text{ in})^2 + (8.139 \text{ in})^2}$$

[√] [(] 6.027 [x²] [+] 8.139 [x²] [)] [=] 10.12758856, 10.128 in Ans (rounded)

or [2nd] [√] [(] 6.027 [x²] [+] 8.139 [x²] [)] [ENTER] 10.12758856, 10.128 in Ans (rounded)

Example 2 In the right $\triangle EFG$ shown in Figure 48-13, $f = 5.800$ inches and hypotenuse $g = 7.200$ inches. Determine side e.

Side g is the hypotenuse. $g^2 = e^2 + f^2$

Substitute the given values, $(7.200 \text{ in})^2 = e^2 + (5.800 \text{ in})^2$

$51.840 \text{ sq in} = e^2 + 33.640 \text{ sq in}$

rearrange the equation, $18.200 \text{ sq in} = e^2$

and solve for e. $\sqrt{18.200 \text{ sq in}} = e$

$e = 4.266$ in Ans (rounded)

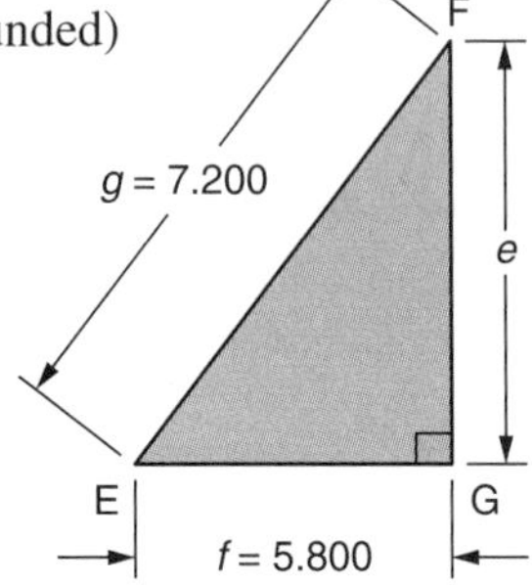

Figure 48-13

$7.200^2 = e^2 + 5.800^2$

Rearrange the equation in terms of e:

$\sqrt{7.200^2 - 5.800^2} = e$

[√] [(] 7.2 [x²] [−] 5.8 [x²] [)] [=] 4.266145802,

$e = 4.266$ in Ans (rounded)

or [2nd] [√] [(] 7.2 [x²] [−] 5.8 [x²] [)] [=] 4.266145802,

$e = 4.266$ in Ans (rounded)

Polygons Other Than Triangles

The types of polygons most common to machine trade applications in addition to triangles are the quadrilaterals of squares, rectangles, and parallelograms. *Quadrilaterals* are four-sided polygons. Regular hexagons also have wide application. A *regular polygon* is one that has equal sides and equal angles.

A *square* is a regular four-sided polygon (quadrilateral). Each angle equals 90°. In the square ABCD shown in Figure 48-14, AB = BC = CD = AD and $\angle A = \angle B = \angle C = \angle D = 90°$.

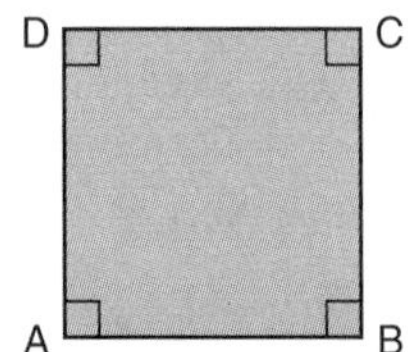

Figure 48-14

A *rectangle* is a four-sided polygon (quadrilateral) with opposite sides parallel and equal. Each angle equals 90°. In the rectangle EFGH shown in Figure 48-15, EF ∥ GH, EH ∥ FG; EF = GH, EH = FG; $\angle E = \angle F = \angle G = \angle H = 90°$.

Figure 48-15

A *parallelogram* is a four-sided polygon (quadrilateral) with opposite sides parallel and equal. Opposite angles are equal. In the parallelogram ABCD shown in Figure 48-16, AB ∥ CD, AD ∥ BC; AB = CD, AD = BC; $\angle A = \angle C, \angle B = \angle D$.

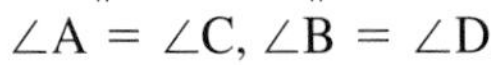

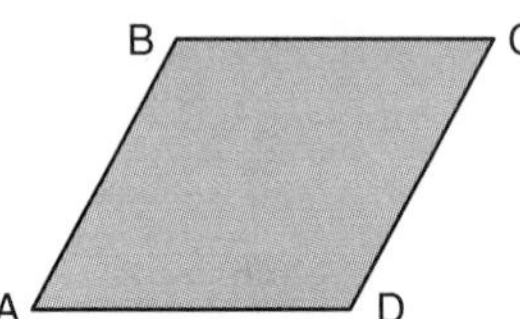

Figure 48-16

A *regular hexagon* is a six-sided figure with all sides equal and all angles equal. In the regular hexagon ABCDEF shown in Figure 48-17, AB = BC = CD = DE = EF = AF, and $\angle A = \angle B = \angle C = \angle D = \angle E = \angle F$.

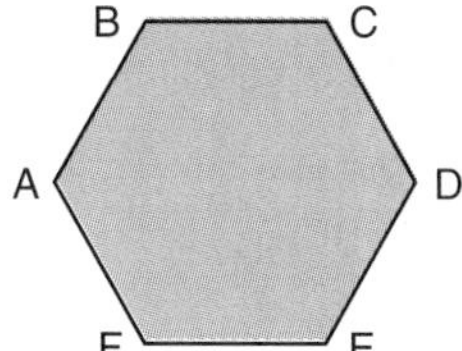

Figure 48-17

➤ Principle 10

The sum of the interior angles of a polygon of N sides is equal to (N − 2) times 180°.

Example 1 In Figure 48-18 is quadrilateral EFGH, with $\angle E = 72°$, $\angle F = 95°$, $\angle G = 108°$. Determine $\angle H$.

Since EFGH has 4 sides, N = 4.
The sum of the 4 angles = $(4 - 2)180° = 2(180°) = 360°$.
Add the 3 given angles and subtract from 360° to find $\angle H$.

$$\angle H = 360° - (\angle E + \angle F + \angle G)$$
$$\angle H = 360° - (72° + 95° + 108°)$$
$$\angle H = 360° - 275°$$
$$\angle H = 85° \quad \text{Ans}$$

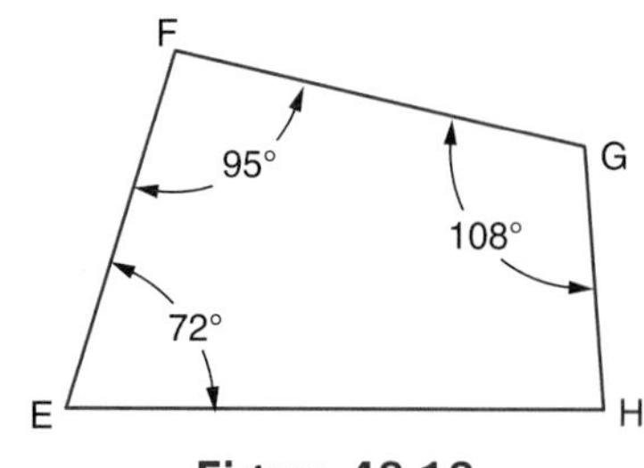

Figure 48-18

Example 2 Refer to polygon ABCDEF in Figure 48-19 and determine $\angle 1$.

Since ABCDEF has 6 sides, N = 6.
The sum of the 6 angles = $(6 - 2)180° = 4(180°) = 720°$.
Find $\angle 2$.

$$\angle 2 = 360° - 114.02° = 245.98°$$

Add the 5 known interior angles and subtract from 720° to find $\angle 1$.

$$\angle 1 = 720° - (57.65° + 245.98° + 40.18° + 77.26° + 90°)$$
$$\angle 1 = 720° - 511.07°$$
$$\angle 1 = 208.93° \quad \text{Ans}$$

720 [−] [(] 57.65 [+] 245.98 [+] 40.18 [+] 77.26 [+] 90 [)] [=] 208.93
$\angle 1 = 208.93°$ Ans

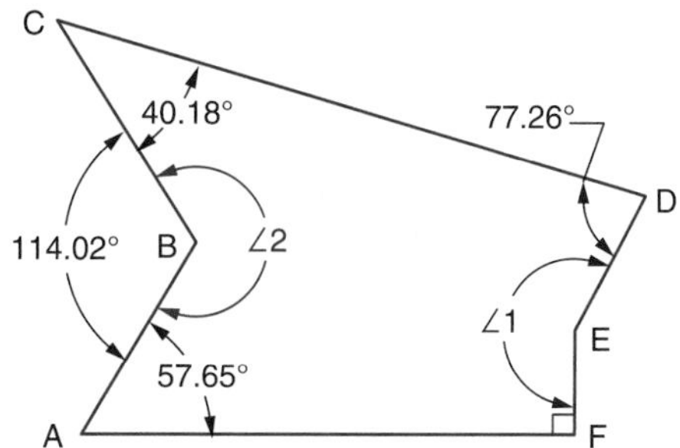

Figure 48-19

Application

Similar Triangles

1. Determine which of the following pairs of triangles (A through F) are similar. All linear dimensions are in inches.

   The similar pairs of triangles are ______________________________

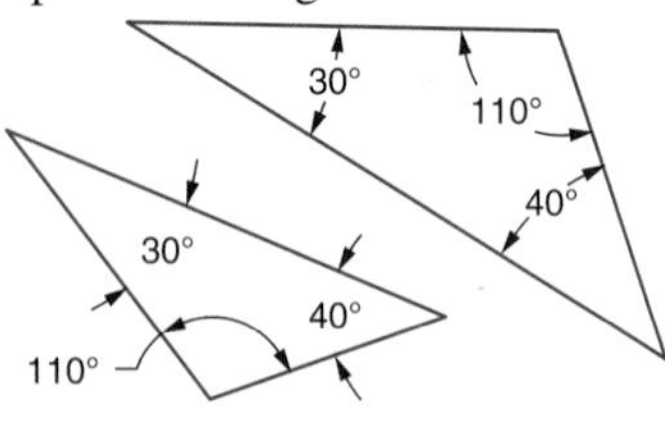

PAIR A

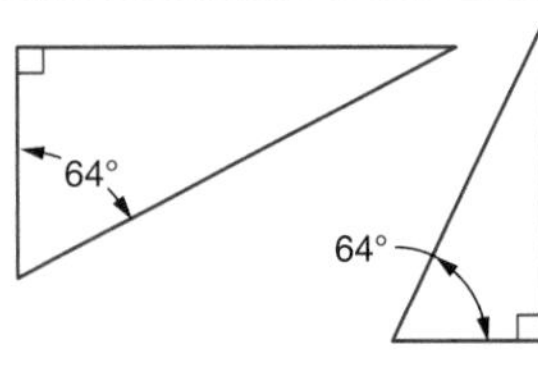

PAIR B

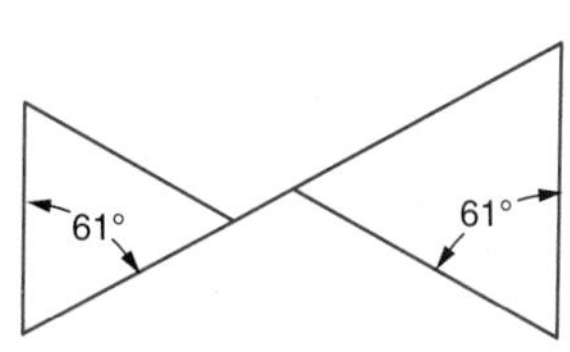

PAIR C

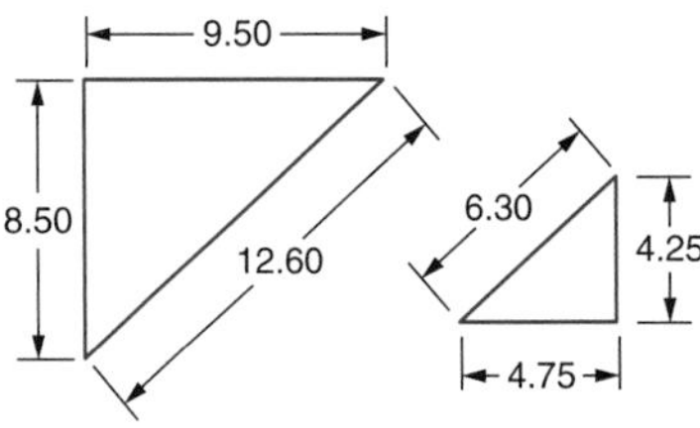

PAIR D

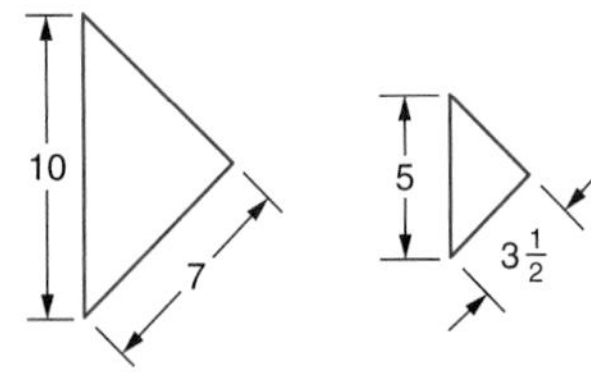

PAIR E

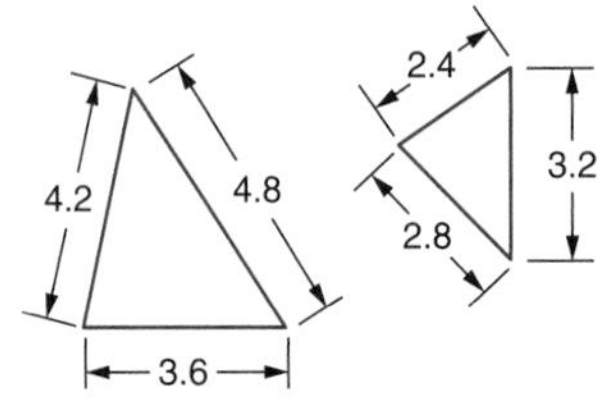

PAIR F

Solve the following problems.

2. In △ABC and △DEF, ∠A = ∠D, ∠B = ∠E, ∠C = ∠F. All dimensions are in inches.

   a. Find AC. ______________

   b. Find DE. ______________

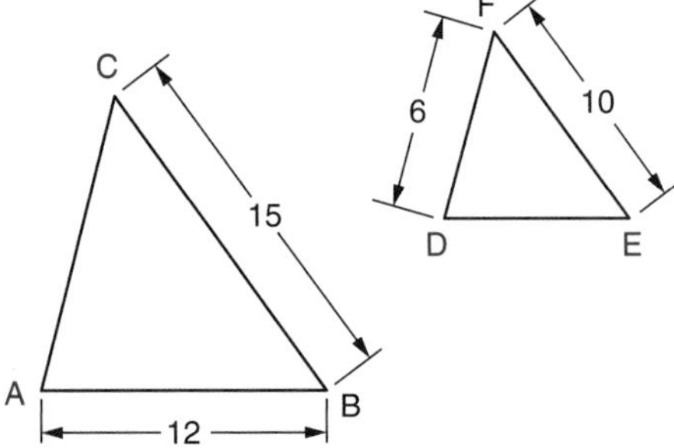

3. In the figure, ∠H = ∠P, ∠J = ∠M, ∠K = ∠L. All dimensions are in millimeters. Round the answers to 2 decimal places.

   a. Find HK. ______________

   b. Find LM. ______________

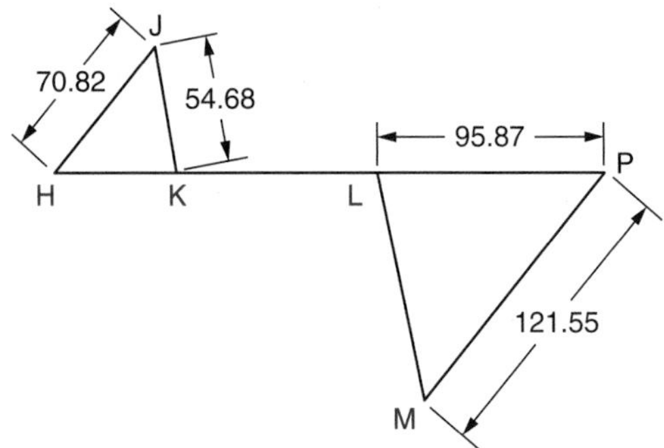

4. In △ABC and △DEF, AB ∥ DE, AC ∥ DF, BC ∥ EF.

   a. Find ∠A. ____________

   b. Find ∠F. ____________

   c. Find ∠B. ____________

   d. Find ∠E. ____________

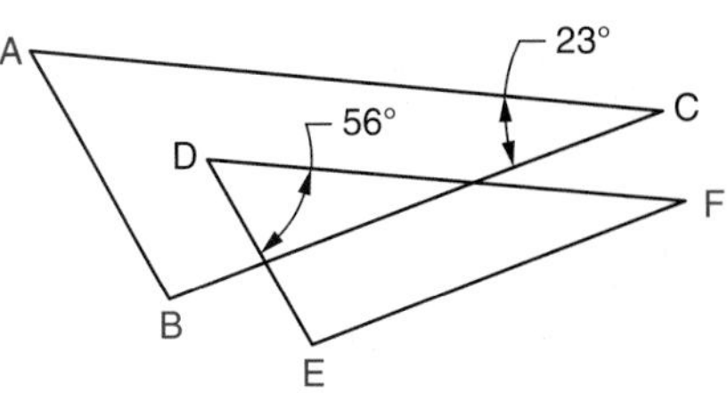

5. Use the figure to find the value of the following angles.

   a. Find ∠1. ____________

   b. Find ∠2. ____________

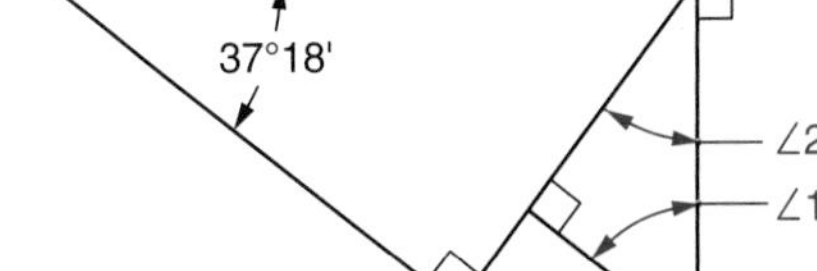

6. In △HJK, PM ∥ JK.

   a. Find ∠HPM. ____________

   b. Find ∠PMK. ____________

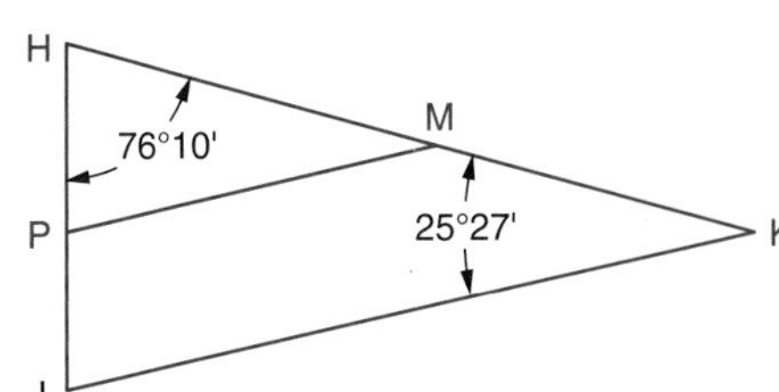

7. Refer to the figure to find these angles.

   a. ∠1 ____________

   b. ∠2 ____________

   c. ∠3 ____________

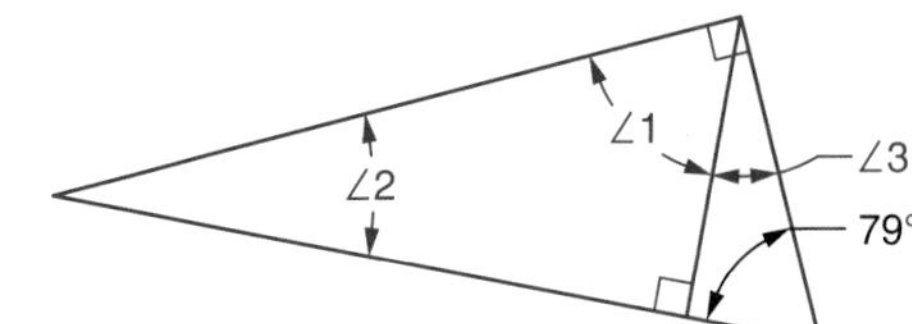

8. Refer to the figure to find these dimensions. All dimensions are in millimeters. Round the answers to 1 decimal place.

   a. Find dimension A. ____________

   b. Find dimension B. ____________

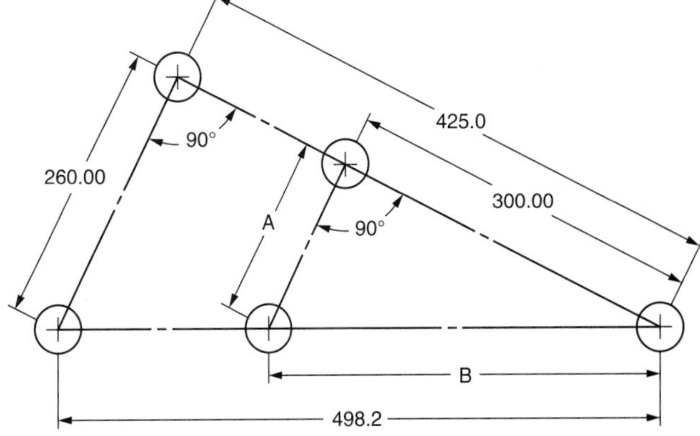

9. In this figure, AB ∥ DE and CB ∥ EF. All dimensions are in inches. Round the answers to 3 decimal places.

   a. Find x. ____________

   b. Find y. ____________

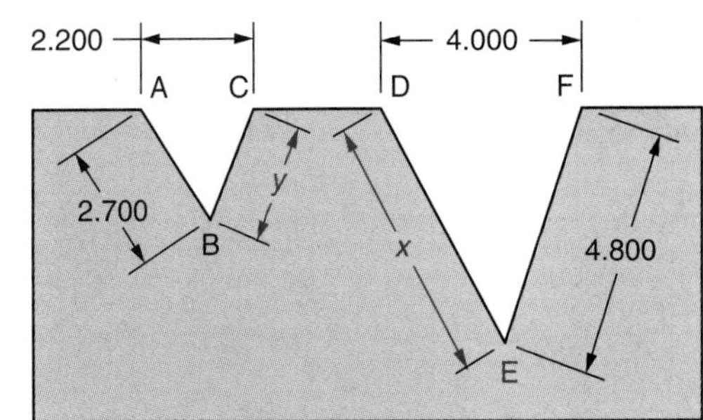

Isosceles, Equilateral, and Right Triangles

Solve the following problems.

10. All dimensions are in inches.

    a. Find x. ____________

    b. Find ∠1. ____________

12.40
12.40
∠1
38°40'
x
18.4

11. All dimensions are in millimeters.

    a. Find x. ____________

    b. Find y. ____________

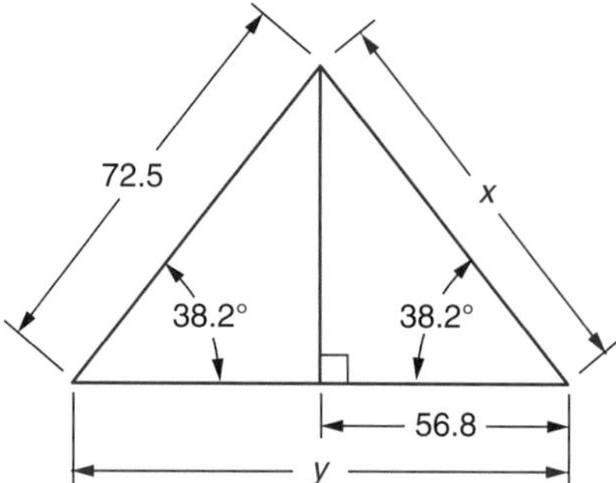

12. All dimensions are in inches.

    a. Find ∠1. ____________

    b. Find ∠2. ____________

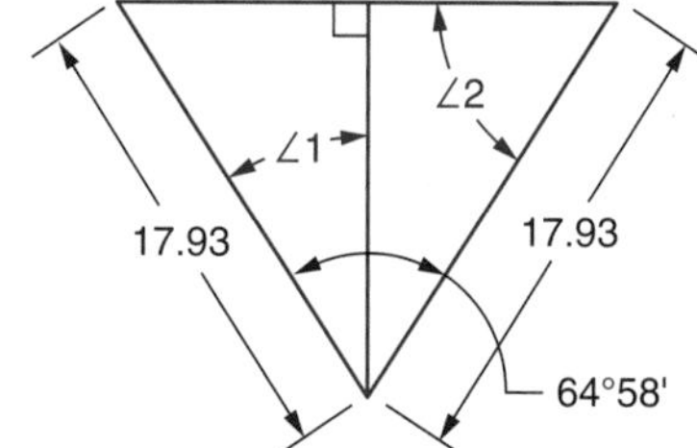

13. All dimensions are in millimeters.

    a. Find x. ____________

    b. Find y. ____________

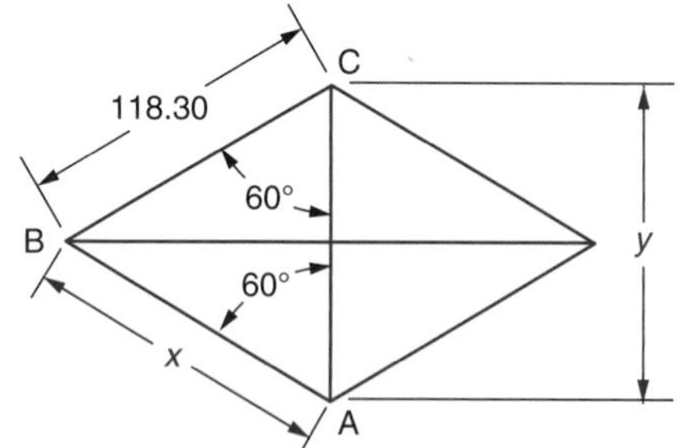

14. All dimensions are in inches.

    a. Find ∠1.

    b. Find x.

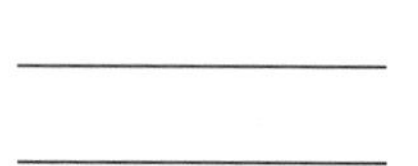

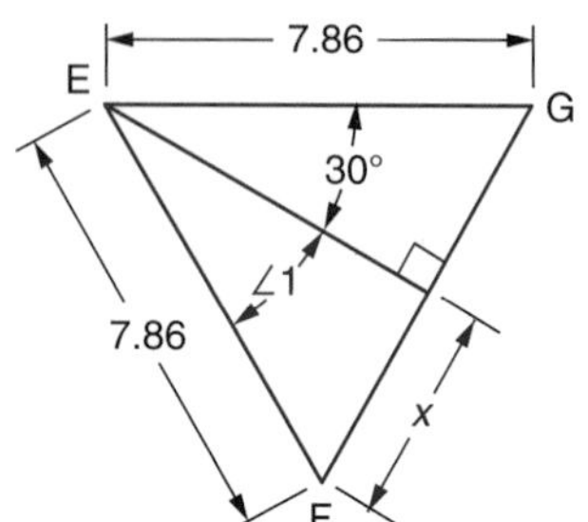

15. Refer to this figure. Using the given values, find the values of x.

   a. If d = 9" and e = 12", find x. ____________

   b. If d = 3" and e = 4", find x. ____________

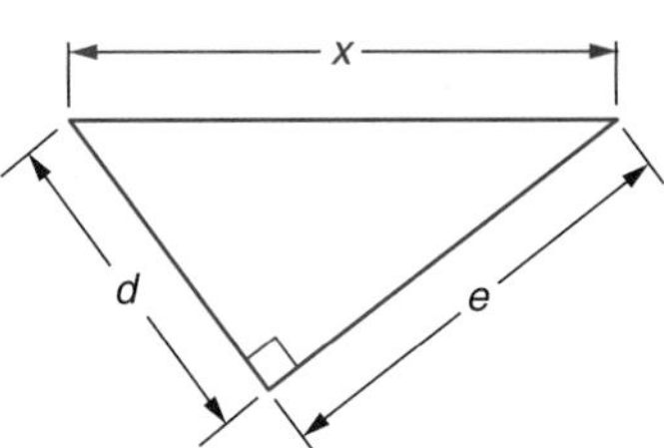

16. Using the figure and these given values, find the values of y. Round the answers to the nearest whole millimeter.

   a. If g = 108 mm and m = 123 mm, find y. ____________

   b. If g = 153.70 mm and m = 170 mm, find y. ____________

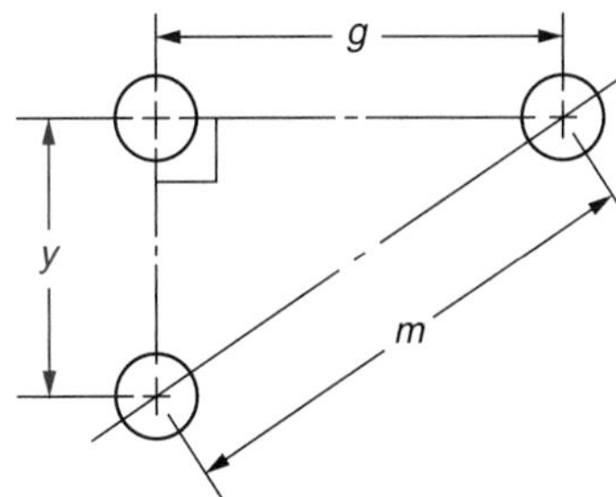

17. Using the figure and these given values, find the values of y.

   a. If Radius A = 360.00 mm and x = 480.00 mm, find y. ____________

   b. If Radius A = 216.00 mm and x = 288.00 mm, find y. ____________

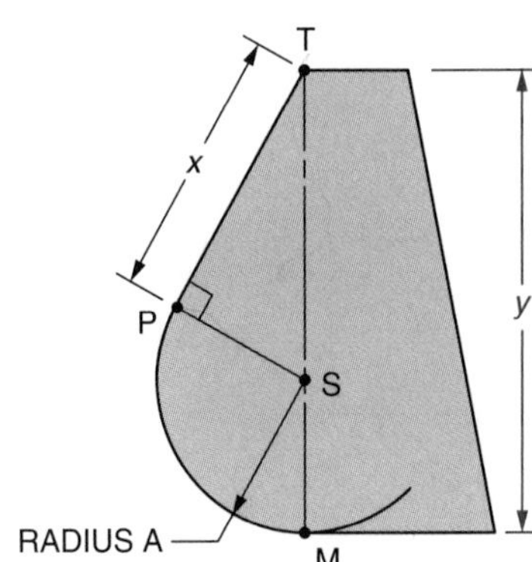

18. Three holes are drilled in the plate shown. All dimensions are in inches. Determine dimensions A and B to 3 decimal places.

   a. A = ____________

   b. B = ____________

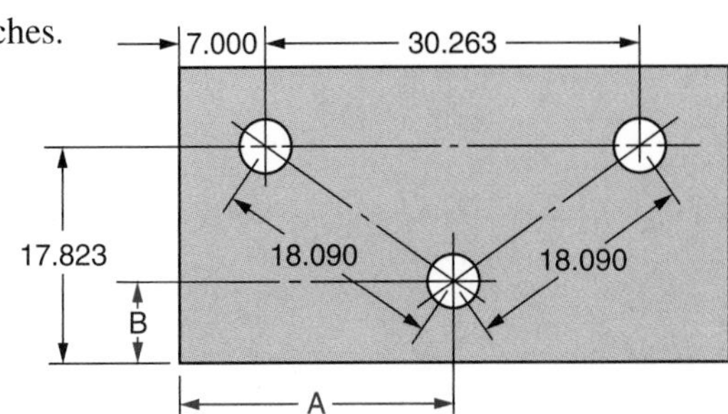

19. All dimensions are in inches. Round the answers to 3 decimal places.

   a. If y = 2.800", find x. ____________

   b. If y = 3.000", find x. ____________

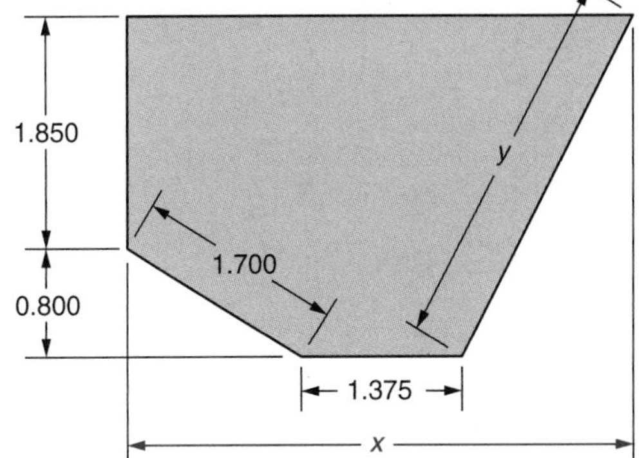

20. All dimensions are in inches. Round the answers to 3 decimal places.

    a. If $y = 2.145"$, find x. __________

    b. If $y = 2.265"$, find x. __________

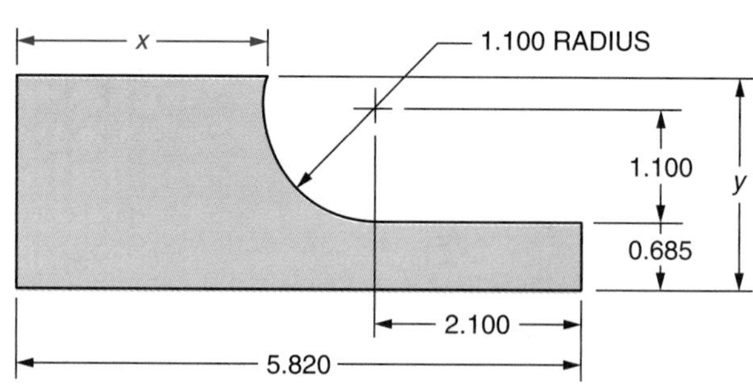

Other Polygons

Solve the following problems.

21. A template is shown. All dimensions are in millimeters. Determine length x and length y to two decimal places.

    x __________

    y __________

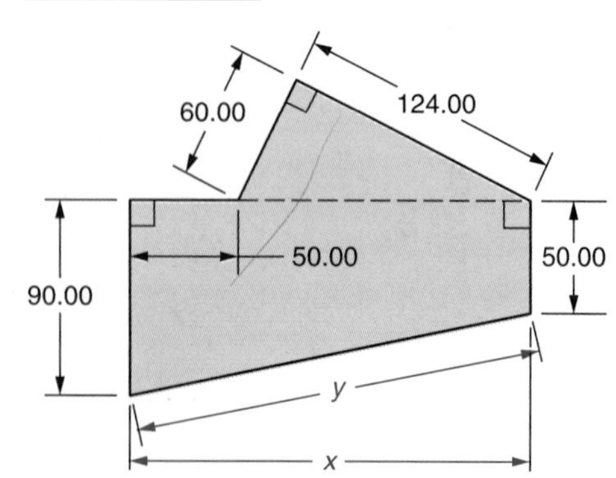

22. Refer to polygon ABCD.

    a. If ∠2 = 87.0°, find ∠1. __________

    b. If ∠1 = 114.0°, find ∠2. __________

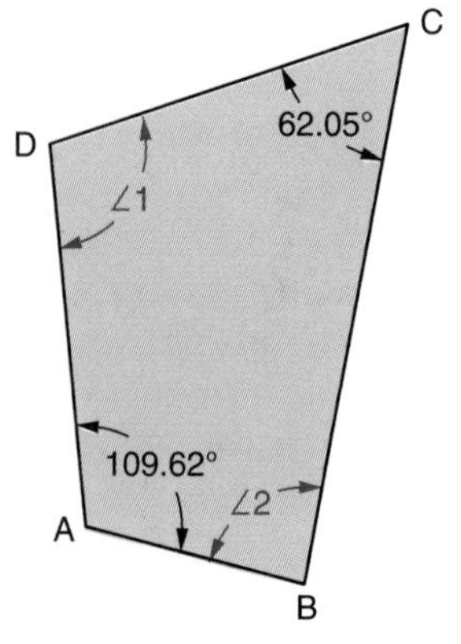

23. Use the angle values given.

    a. If ∠1 = 114°, find ∠2. __________

    b. If ∠2 = 83°, find ∠1. __________

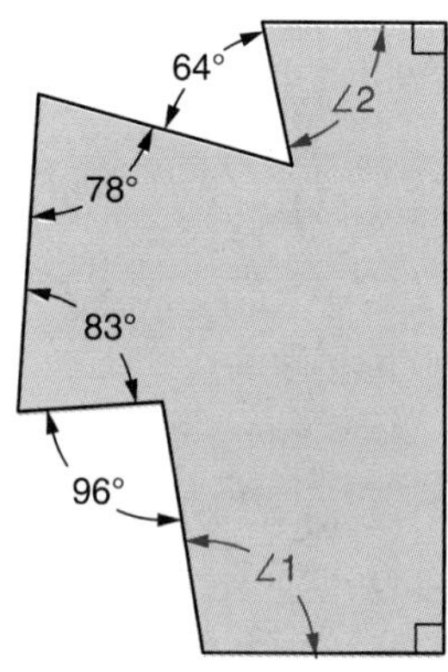

24. Use the angle values given to find ∠2.

    a. If ∠1 = 37°, find ∠2. __________

    b. If ∠1 = 29°, find ∠2. __________

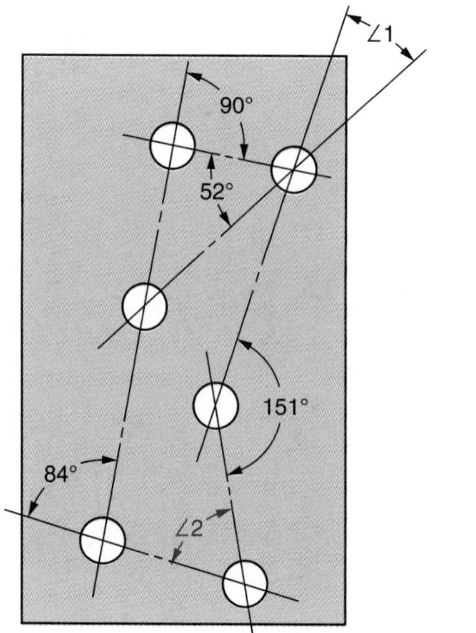

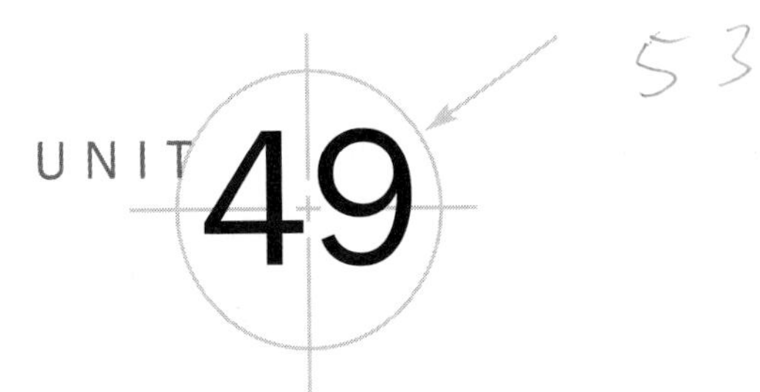

Introduction to Circles

Objectives ***After studying this unit you should be able to***

- Identify parts of a circle.
- Solve problems by using geometric principles that involve chords, arcs, central angles, perpendiculars, and tangents.

Circles are the simplest of all closed curves and their basic properties are readily understood. Holes are often laid out on bolt circles, and rotary tables move in a circular motion. Machines operate by the circular motion of gears. Parts are machined with cutting tools and/or work pieces revolving in a circular path.

Definitions

A *circle* is a closed curve on which every point is equally distant from a fixed point called the *center*.

Refer to Figure 49-1 for the following definitions:

The *circumference* is the length of the curved line that forms the circle.

A *chord* is a straight line segment that joins two points on the circle. AB is a chord.

A *diameter* is a chord that passes through the center of a circle. CD is a diameter.

A *radius* (plural radii) is a straight line segment that connects the center of the circle with a point on the circle. The radius is one-half the diameter. OE is a radius.

Figure 49-1

Refer to Figure 49-2 for the following definitions:

An *arc* is that part of a circle between any two points on the circle. The symbol ⌒ written above the letters means arc. $\overset{\frown}{AB}$ is an arc. Any two points on a circle divide the circle into two arcs. A *semicircle* is an arc that is one-half of a circle. If the arcs are not equal, the smaller is a *minor arc* and the larger is a *major arc*. In Figure 49-2, $\overset{\frown}{AB}$ is the minor arc and $\overset{\frown}{APB}$ is the major arc. Unless otherwise stated, $\overset{\frown}{AB}$ means the minor arc $\overset{\frown}{AB}$.

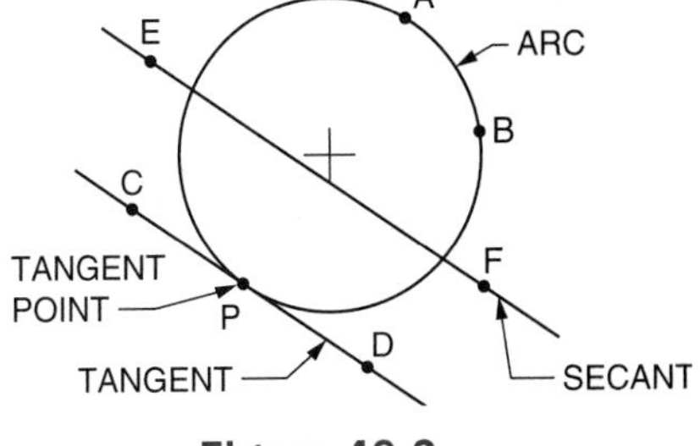

Figure 49-2

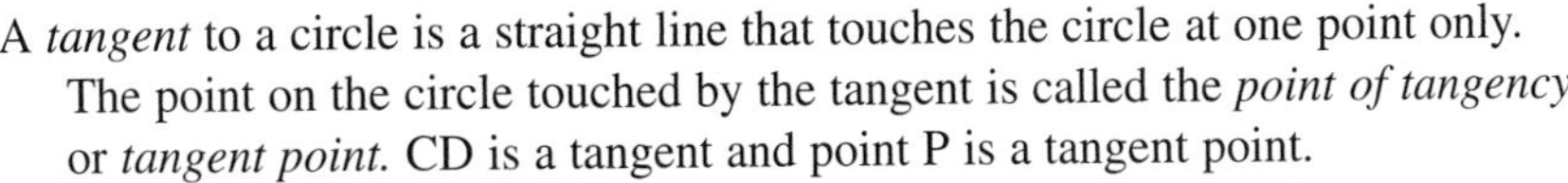
A *tangent* to a circle is a straight line that touches the circle at one point only. The point on the circle touched by the tangent is called the *point of tangency* or *tangent point.* CD is a tangent and point P is a tangent point.

A *secant* is a straight line that passes through a circle and intersects the circle at two points. EF is a secant.

Refer to Figure 49-3 for the following definitions:

A *segment* is a figure formed by an arc and the chord joining the end points of the arc. The shaded figure ABC is a segment.

A *sector* is a figure formed by two radii and the arc intercepted by the radii. The shaded figure EOF is a sector.

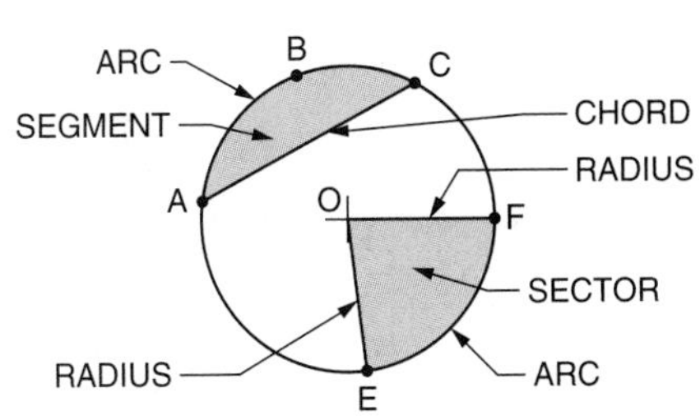

Figure 49-3

Refer to Figure 49-4 for the following definitions:

A *central angle* is an angle whose vertex is at the center of a circle and whose sides are radii. Angle MON is a central angle.

An *inscribed angle* is an angle in a circle whose vertex is on the circle and whose sides are chords. Angle SRT is an inscribed angle.

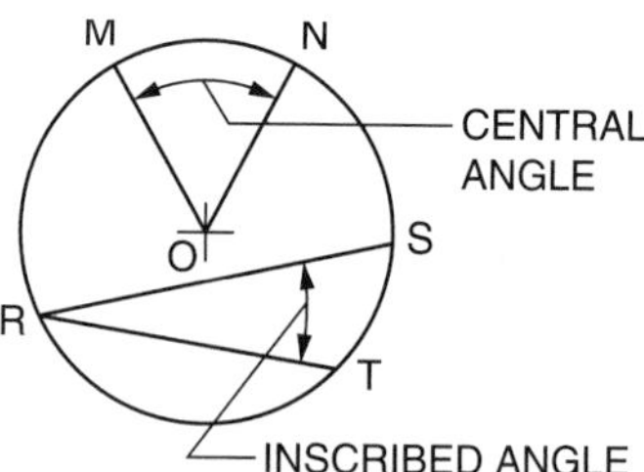

Figure 49-4

Circumference Formula

A polygon is *inscribed* in a circle when each vertex of the polygon is a point of the circle. In Figure 49-5, regular polygons are inscribed in circles. As the number of sides increases, the perimeter increases and approaches the circumference.

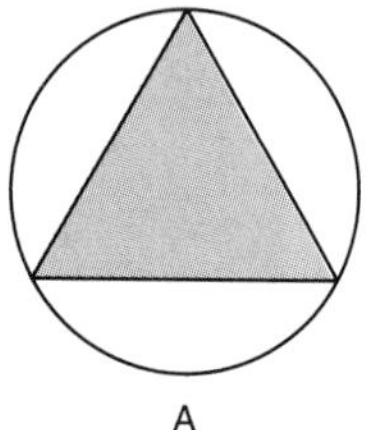

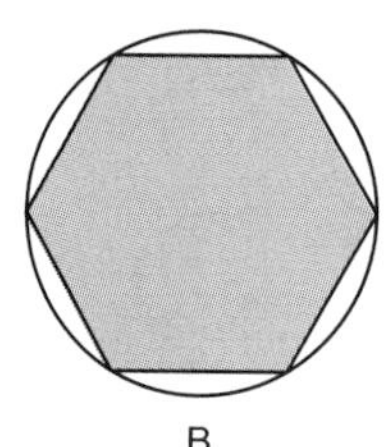

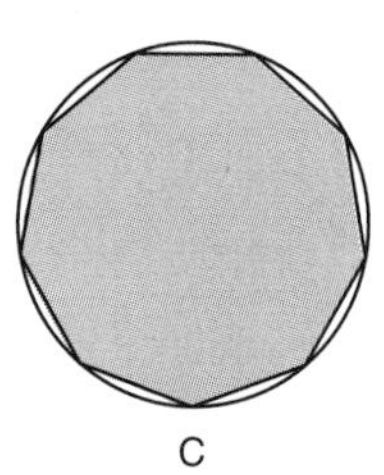

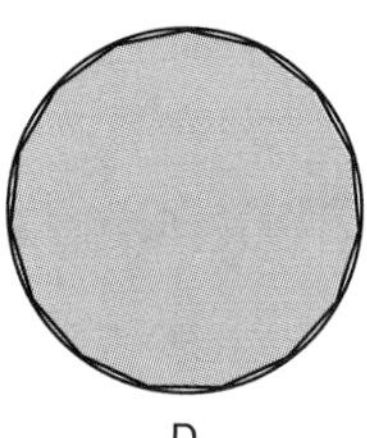

Figure 49-5

An important relationship exists between the circumference and the diameter of a circle. As the number of sides of an inscribed polygon increases, the perimeter approaches a certain number times the diameter. This number is called *pi.* The symbol for pi is π. No matter how many sides an inscribed polygon has, the value of π cannot be expressed exactly with digits. Pi is called an irrational number.

The circumference of a circle is equal to pi (π) times the diameter or two pi times the radius. Generally, for the degree of precision required in machining applications, a value of 3.1416 is used for π if a calculator is not available.

$$C = \pi d \quad \text{or} \quad C = 2\pi r$$

where C = circumference
π = pi
d = diameter
r = radius

Example 1 Compute the circumference of a circle with a 50.70 mm diameter. $C = \pi d = 3.1416(50.70 \text{ mm}) = 159.28 \text{ mm}$ Ans (rounded)

As presented on page 148, depressing the pi key ([π]) displays the value of pi to 10 digits (3.141592654) on most calculators. Recall that [π] is the alternate function on many calculators.

$C = \pi d, C = \pi\ (50.70 \text{ mm})$

[π] [×] 50.7 [=] 159.2787475

$C = 159.28 \text{ mm}$ Ans (rounded)

Example 2 Determine the radius of a circle which has a circumference of 14.860 inches.

$C = 2\pi r$

$14.860 \text{ in} = 2(3.1416)(r)$

$r = 2.365 \text{ in}$ Ans (rounded)

$$r = \frac{14.860 \text{ in}}{2\pi}$$

14.86 [÷] [(] 2 [×] [π] [)] [=] 2.365042454,

$r = 2.365 \text{ in}$ Ans (rounded)

Geometric Principles

➤ Principle 11

In the same circle or in equal circles, equal chords cut off equal arcs.

Given: In Figure 49-6, Circle A = Circle B and chords CD = EF = GH = MS.

Conclusion: $\overset{\frown}{CD} = \overset{\frown}{EF} = \overset{\frown}{GH} = \overset{\frown}{MS}$.

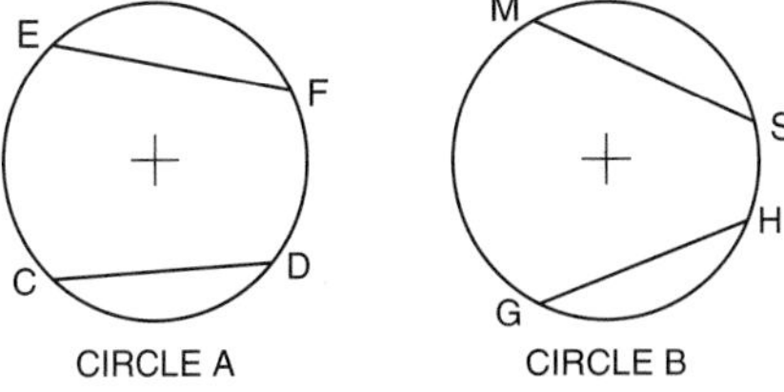

Figure 49-6

➤ Principle 12

In the same circle or in equal circles, equal central angles cut off equal arcs.

Given: In Figure 49-7, Circle D = Circle E and $\angle 1 = \angle 2 = \angle 3 = \angle 4$.

Conclusion: $\overset{\frown}{AB} = \overset{\frown}{FG} = \overset{\frown}{HK} = \overset{\frown}{MP}$

and $\overset{\frown}{AG} = \overset{\frown}{BF} = \overset{\frown}{HM} = \overset{\frown}{KP}$.

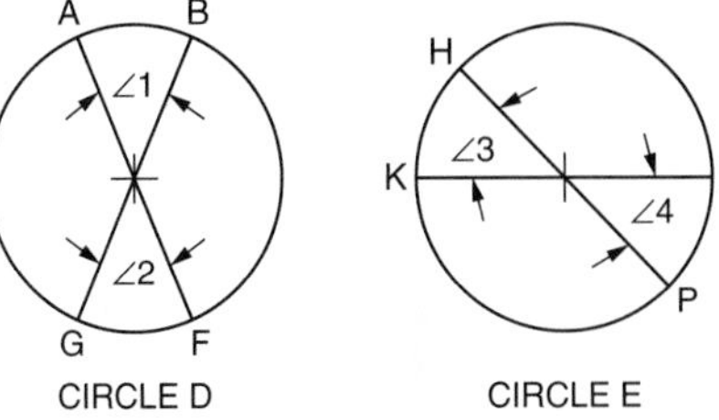

Figure 49-7

➤ Principle 13

In the same circle or in equal circles, two central angles have the same ratio as the arcs that are cut off by the angles.

Example In Figure 49-8, Circle A = Circle B. If $\angle COD = 90°$, $\angle EOF = 50°$, $\overset{\frown}{CD} = 1.400"$, and $\overset{\frown}{GH} = 2.100"$, determine (a) the length of $\overset{\frown}{EF}$ and (b) the size of $\angle GOH$.

a. Set up a proportion between $\overset{\frown}{CD}$ and $\overset{\frown}{EF}$ with their respective central angles. Solve for $\overset{\frown}{EF}$.

$$\frac{\angle COD}{\angle EOF} = \frac{\overset{\frown}{CD}}{\overset{\frown}{EF}}$$

$$\frac{90°}{50°} = \frac{1.400"}{\overset{\frown}{EF}}$$

$$90\overset{\frown}{EF} = 50(1.400")$$

$$\overset{\frown}{EF} = \frac{50(1.400")}{90}$$

$$\overset{\frown}{EF} = 0.778" \quad \text{Ans}$$

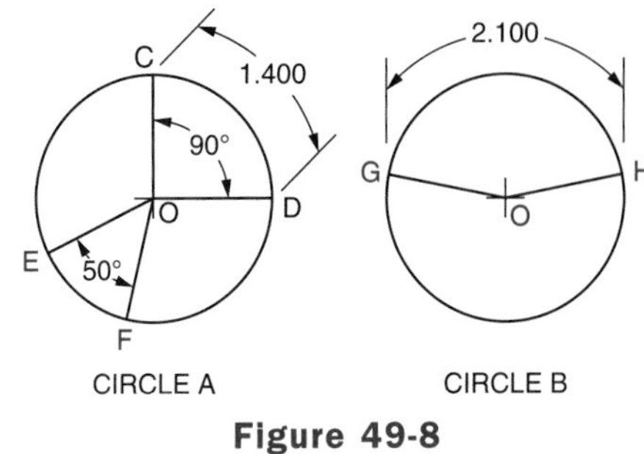

Figure 49-8

b. Set up a proportion between $\overset{\frown}{CD}$ and $\overset{\frown}{GH}$ with their central angles. Solve for $\angle GOH$.

$$\frac{\angle COD}{\angle GOH} = \frac{\overset{\frown}{CD}}{\overset{\frown}{GH}}$$

$$\frac{90°}{\angle GOH} = \frac{1.400"}{2.100"}$$

$$1.400(\angle GOH) = 90°(2.100)$$

$$\angle GOH = \frac{90°(2.100)}{1.400}$$

$$\angle GOH = 135° \quad \text{Ans}$$

The *perpendicular bisector* of a line segment AB is the line through the midpoint and perpendicular to AB. For example, in Figure 49-9, CD is the perpendicular bisector of AB.

Figure 49-9

➤ **Principle 14**

A line drawn from the center of a circle perpendicular to a chord bisects the chord and the arc cut off by the chord.

Given: In Figure 49-9, diameter DE $\perp$ chord AB.

Conclusion: AC = BC and $\overset{\frown}{AD} = \overset{\frown}{BD}$ and $\overset{\frown}{AE} = \overset{\frown}{BE}$.

➤ **The perpendicular bisector of a chord passes through the center of a circle.**

Given: In Figure 49-9, DE is the perpendicular bisector of chord AB.

Conclusion: DE passes through the center, O, of the circle.

The use of Principle 14 with the Pythagorean Theorem (Principle 9) has wide practical application in the machine trades.

Example Holes A, B, and C are to be drilled in the plate shown. The centers of holes A and C lie on a 280.00-mm diameter circle. The center of hole B lies on the intersection of chord AC and segment OB, which is perpendicular to AC. Compute working dimensions F, G, and H. All dimensions are in millimeters.

Compute dimension F: Applying Principle 14, AC is bisected by OB.

AB = BC = 250.00 mm ÷ 2 = 125.00 mm:

F = 200.00 mm − 125.00 mm = 75.00 mm Ans

Compute dimension G.

G = 200.00 mm + 125.00 mm = 325.00 mm Ans

Compute dimension H.

In right △ABO, AB = 125.00 mm, AO = 280.00 mm ÷ 2 = 140.00 mm

Figure 49-10

Compute OB by applying the Pythagorean Theorem (Principle 9).

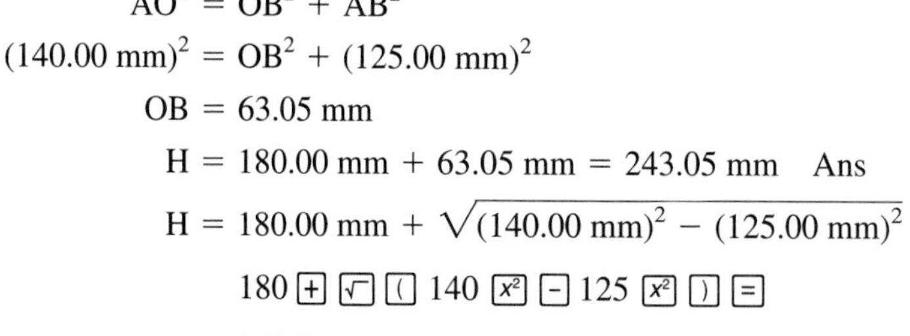

$$AO^2 = OB^2 + AB^2$$

$$(140.00 \text{ mm})^2 = OB^2 + (125.00 \text{ mm})^2$$

$$OB = 63.05 \text{ mm}$$

$$H = 180.00 \text{ mm} + 63.05 \text{ mm} = 243.05 \text{ mm} \quad \text{Ans}$$

$$H = 180.00 \text{ mm} + \sqrt{(140.00 \text{ mm})^2 - (125.00 \text{ mm})^2}$$

180 [+] [√] [(] 140 [x²] [−] 125 [x²] [)] [=]

243.0476011, 243.05 mm Ans (rounded)

➤ **Principle 15**

A line perpendicular to a radius at its extremity is tangent to the circle.
A tangent is perpendicular to a radius at its tangent point.

Example 1 Given: Line AB $\perp$ to radius CO at point C in Figure 49-11.

Conclusion: Line AB is a tangent.

Example 2 Given: In Figure 49-11, tangent DE passes through point F of radius FO.

Conclusion: Tangent DE $\perp$ radius FO.

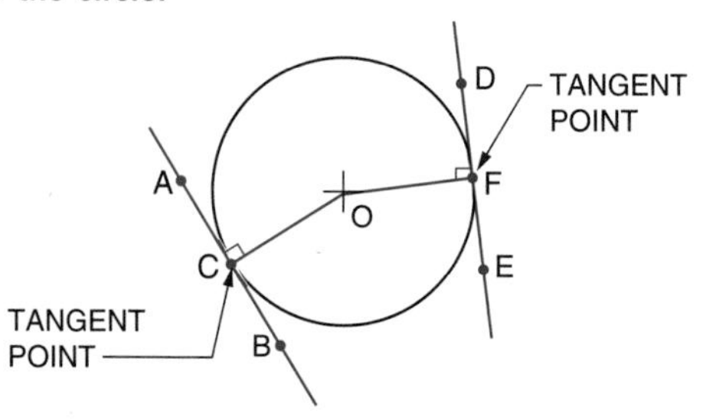

Figure 49-11

➤ **Principle 16**

Two tangents drawn to a circle from a point outside the circle are equal. The angle at the outside point is bisected by a line drawn from the point to the center of the circle.

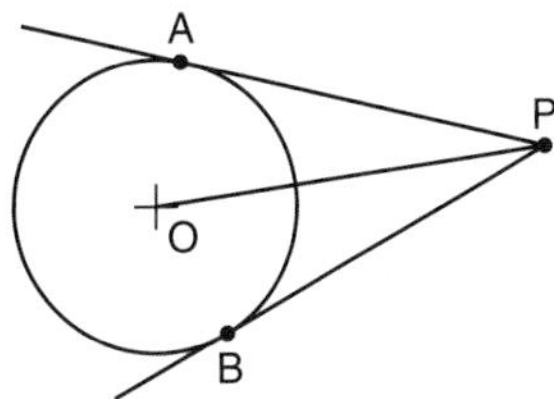

Figure 49-12

Example 1 Given: In Figure 49-12, tangents AP and BP are drawn to the circle from point P.

Conclusion: AP = BP.

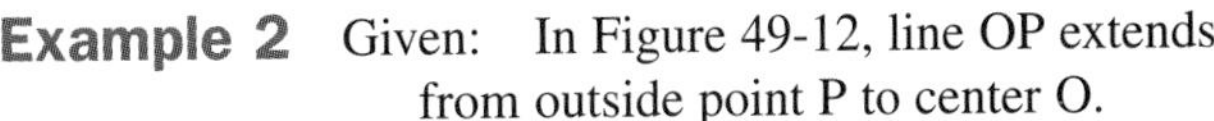

Example 2 Given: In Figure 49-12, line OP extends from outside point P to center O.

Conclusion: ∠APO = ∠BPO.

➤ **Principle 17**

If two chords intersect inside a circle, the product of the two segments of one chord is equal to the product of the two segments of the other chord.

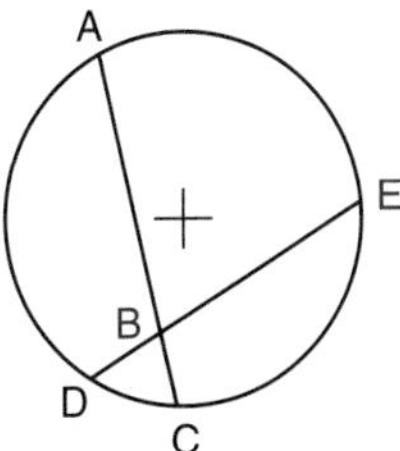

Figure 49-13

Example 1 Given: Chords AC and DE intersect at point B in Figure 49-13.

Conclusion: AB(BC) = BD(BE).

Example 2 In Figure 49-13, if AB = 7.5 inches, BC = 2.8 inches, and BD = 2.1 inches, determine the length of BE.

$$AB(BC) = BD(BE)$$
$$7.5(2.8) = 2.1(BE)$$
$$21.0 = 2.1BE$$
$$BE = 10.0 \text{ inches} \quad \text{Ans}$$

Application

Definitions

Name each of the parts of circles for the following problems.

1. a. AB ________
   b. CD ________
   c. EO ________
   d. Point O ________

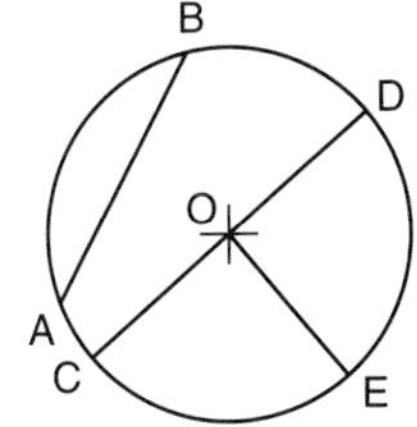

2. a. $\overset{\frown}{GF}$ ________
   b. HK ________
   c. LM ________
   d. GF ________
   e. Point P ________

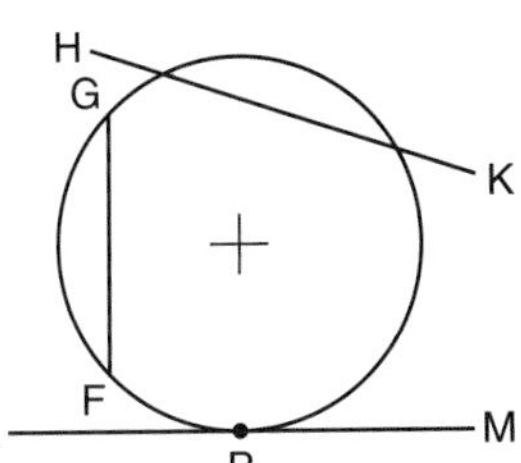

3. a. M ________
   b. P ________
   c. SO ________
   d. TO ________
   e. RW ________
   f. $\overset{\frown}{RW}$ ________

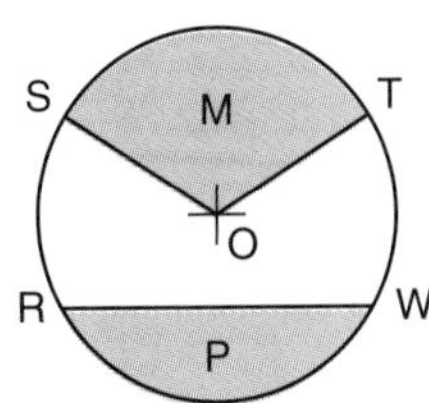

4. a. ∠1 ________
   b. ∠2 ________
   c. AO ________
   d. CD ________
   e. CE ________
   f. $\overset{\frown}{AB}$ ________

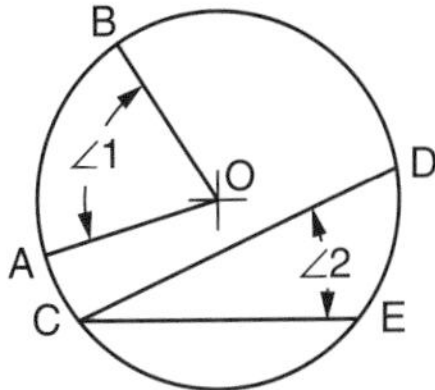

Circumference Formula

Use $C = \pi d$ or $C = 2\pi r$ where C = circumference

π = 3.1416 if not using a calculator

d = diameter

r = radius

5. Determine the unknown value for each of the following problems. Round the answers to 3 decimal places.

a. If d = 6.500", find C. ____________

b. If d = 30.000 mm, find C. ____________

c. If r = 18.600 mm, find C. ____________

d. If r = 2.930", find C. ____________

e. If C = 35.000", find d. ____________

f. If C = 218.000 mm, find d. ____________

g. If C = 327.000 mm, find r. ____________

h. If C = 7.680", find r. ____________

6. Determine the length of wire, in feet, in a coil of 60.0 turns. The average diameter of the coil is 30.0 inches. Round the answer to the nearest whole foot. ____________

7. A pipe with a wall thickness of 6.00 millimeters has an outside diameter of 79.20 millimeters. Compute the inside circumference of the pipe. Round the answer to 2 decimal places. ____________

8. The flywheel of a machine has a 0.80-meter diameter and revolves 240.0 times per minute. How many meters does a point on the outside of the flywheel rim travel in 5.0 minutes? Round the answer to the nearest whole meter. ____________

Geometric Principles

Solve the following problems. These problems are based on Principles 11–14, although a problem may require the application of two or more of any of the principles. Round the answers to 3 decimal places where necessary unless otherwise stated.

9. Determine the length of belt required to connect the two pulleys shown. All dimensions are in inches. Round the answer to 2 decimal places. ____________

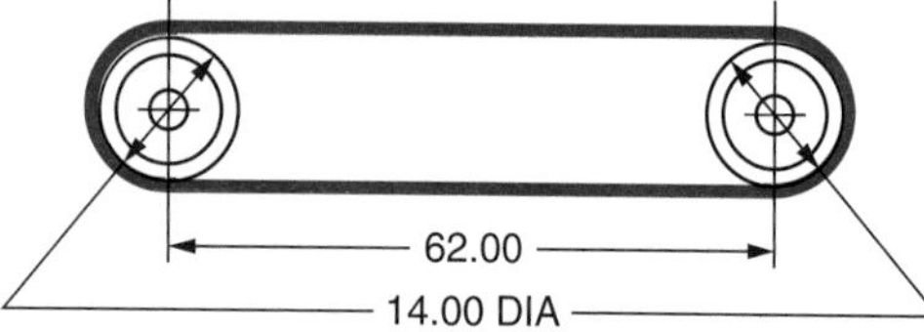

10. $\triangle ABC$ is equilateral. All dimensions are in inches.

a. Find $\widehat{AB}$. ____________

b. Find $\widehat{BC}$. ____________

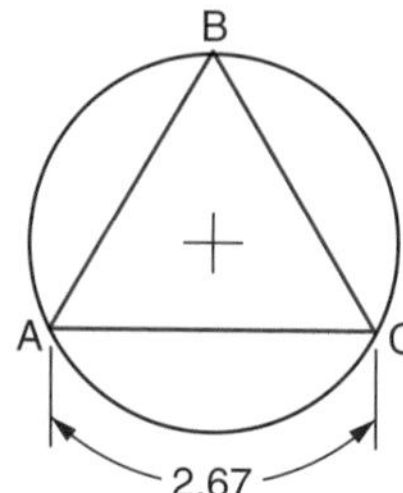

11. All dimensions are in inches.

a. Find $\widehat{AB}$. ____________

b. Find $\widehat{BC}$. ____________

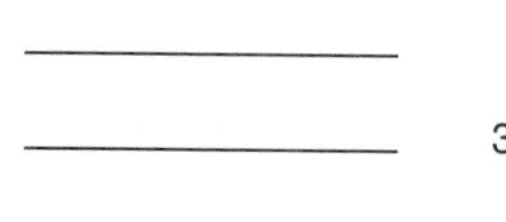

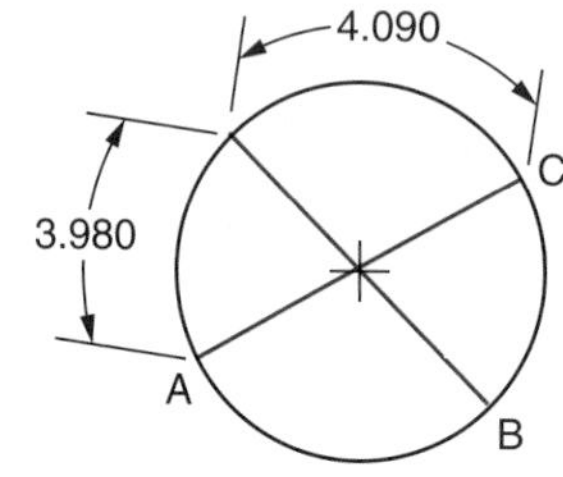

12. a. If $\overset{\frown}{EF}$ = 160 mm, find $\overset{\frown}{HP}$. __________

    b. If $\overset{\frown}{HP}$ = 284 mm, find $\overset{\frown}{EF}$. __________

    Round the answer to the nearest whole millimeter.

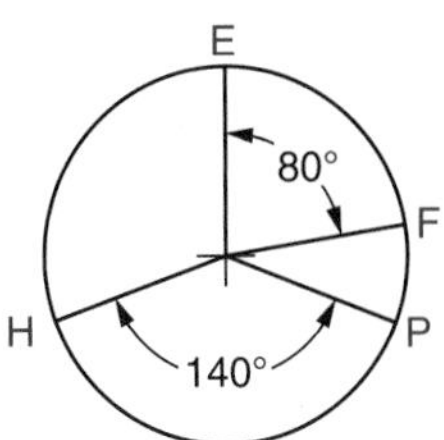

13. a. If $\overset{\frown}{SW}$ = 4.800" and $\overset{\frown}{TM}$ = 5.760", find ∠1. __________

    b. If $\overset{\frown}{TM}$ = 4.128" and $\overset{\frown}{SW}$ = 2.064", find ∠1. __________

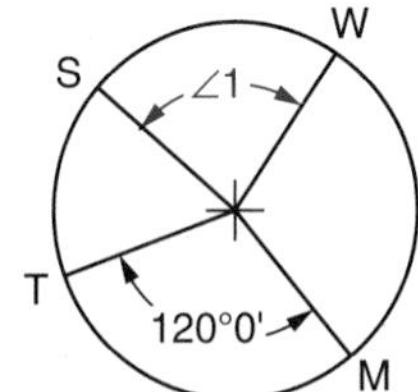

14. a. If AB = 5.378" and $\overset{\frown}{AC}$ = 3.782", find (1) DB and (2) $\overset{\frown}{ACB}$.

    (1) __________ (2) __________

    b. If DB = 3.017" and $\overset{\frown}{ACB}$ = 7.308", find (1) AB and (2) $\overset{\frown}{CB}$.

    (1) __________ (2) __________

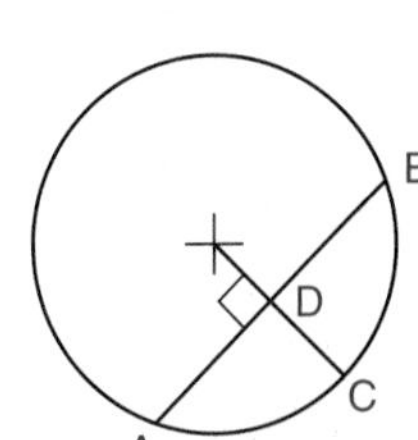

15. Find $\overset{\frown}{HK}$ when $\overset{\frown}{EF}$ = 21.23 mm. __________

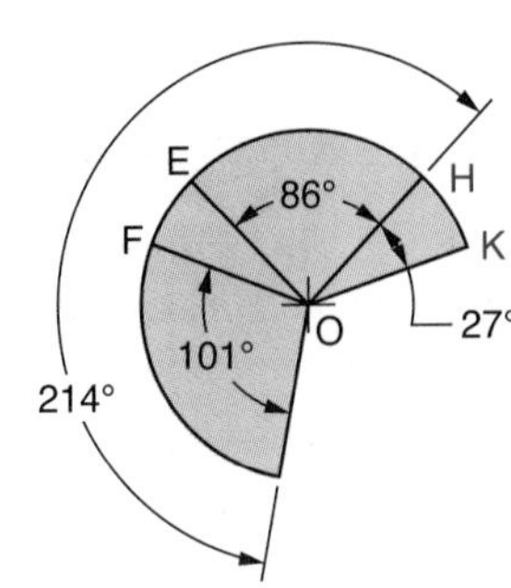

16. All dimensions are in inches.

    a. If ∠1 = 240°0', find $\overset{\frown}{ABC}$. __________

    b. If $\overset{\frown}{ABC}$ = 2.300", find ∠1. __________

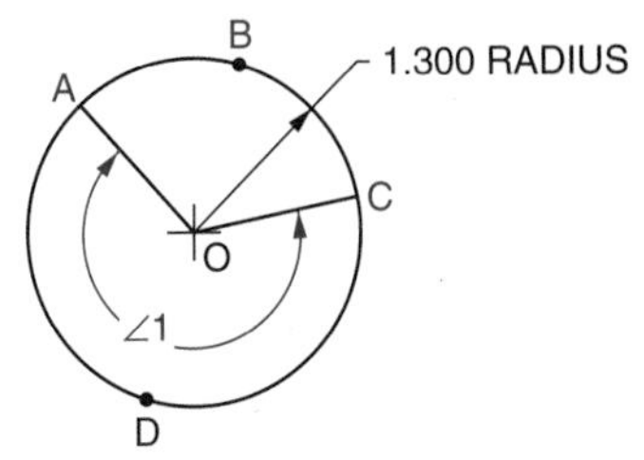

17. All dimensions are in inches.

    a. If x = 5.100", find ∠1. __________

    b. If x = 4.750", find ∠1. __________

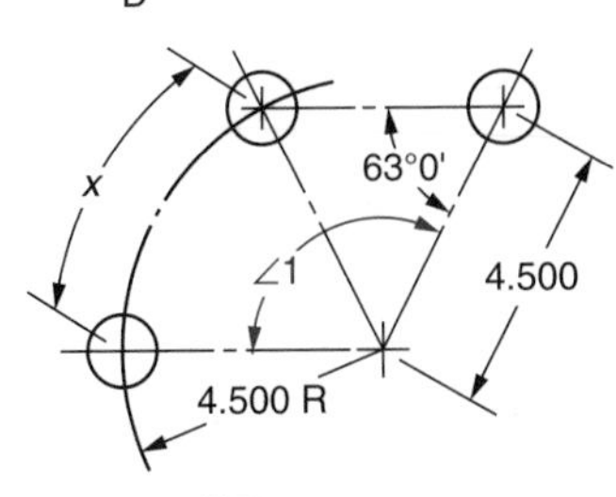

18. a. If radius x = 7.500" and y = 4.500", find PM. __________

    b. If radius x = 8.000" and y = 4.800", find PM. __________

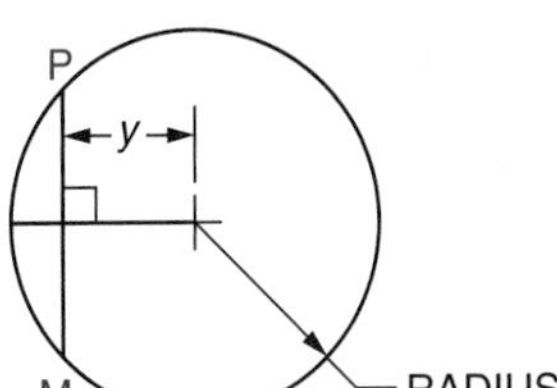

19. The circumference of this circle is 14.400".

    a. If x = 3.200", find ∠1. __________

    b. If ∠1 = 36°0', find x. __________

20. Determine the centerline distance between hole A and hole B for these values.

    a. Radius x = 8.000" and DO = 2.100". __________

    b. Radius x = 1.200" and DO = 0.700". __________

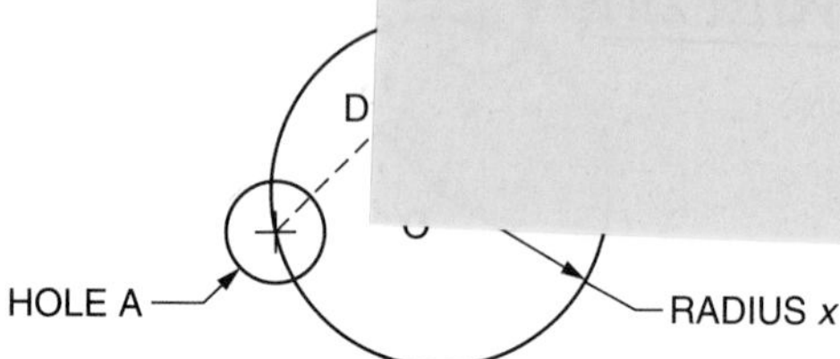

Solve the following problems. These problems are based on Principles 15–17, although a problem may require the application of two or more of any of the principles. Round the answers to 3 decimal places where necessary unless otherwise stated.

21. Point P is a tangent point and ∠1 = 107°18'.

    a. If ∠2 = 41°21', find (1) ∠E and (2) ∠F .

       (1) __________ (2) __________

    b. If ∠2 = 48°20', find (1) ∠E and (2) ∠F.

       (1) __________ (2) __________

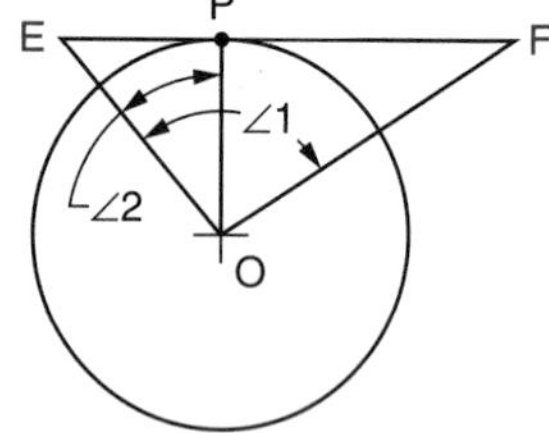

22. AB and CB are tangents.

    a. If y = 137.20 mm and ∠ABC = 67.0°, find (1) ∠1 and (2) x.

       (1) __________ (2) __________

    b. If x = 207.70 mm and ∠1 = 33.8°, find (1) ∠ABC and (2) y.

       (1) __________ (2) __________

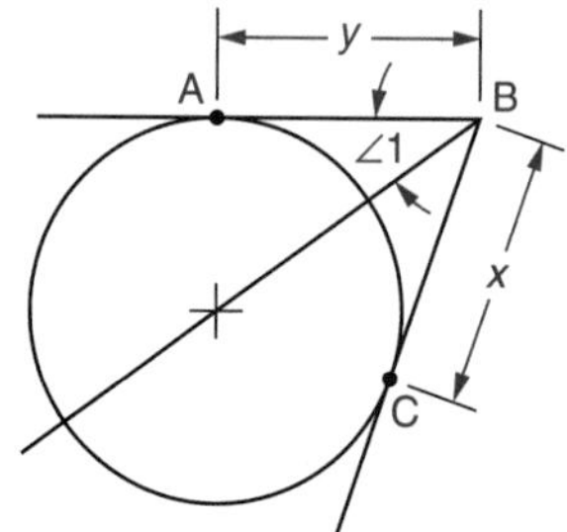

23. Point A is a tangent point of the V-groove cut and pin shown. All dimensions are in inches.

    a. If y = 1.400", find x. __________

    b. If y = 1.800", find x. __________

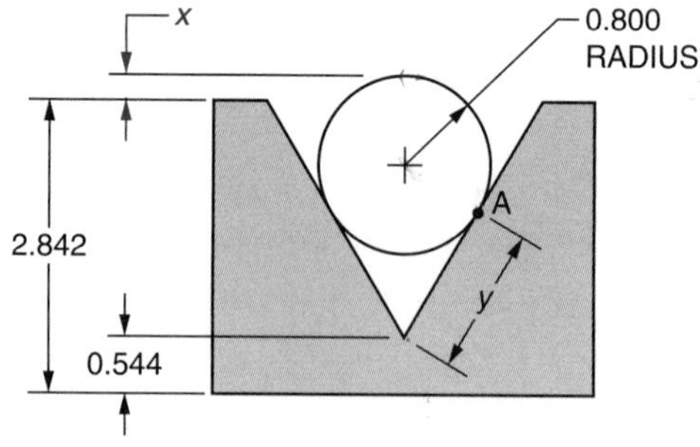

24. Points E, G, and F are tangent points.

    a. If ∠1 = 109°, find ∠2. __________

    b. If ∠1 = 118°45', find ∠2. __________

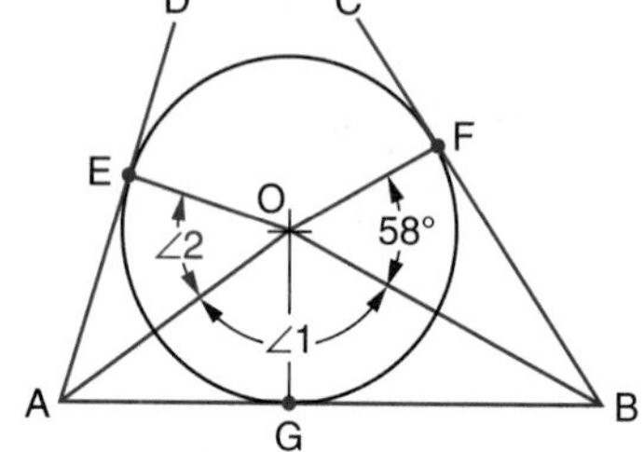

25. All dimensions are in millimeters. Round the answers to 2 decimal places.

    a. If EK = 150.00 mm, find GK. ____________

    b. If GK = 120.00 mm, find EK. ____________

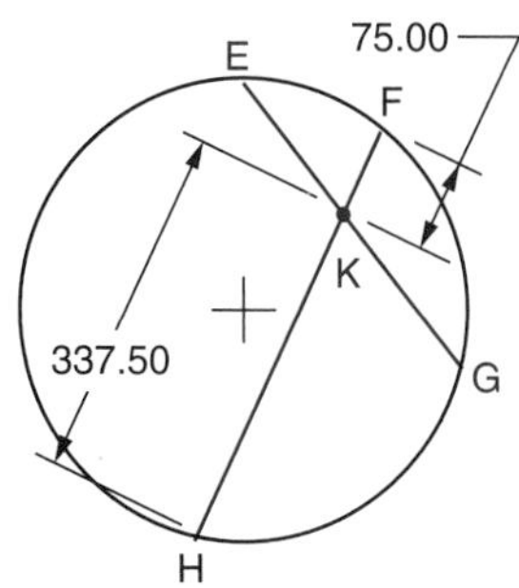

26. All dimensions are in inches.

    a. If PT = 1.800", find x. ____________

    b. If PT = 2.000", find x. ____________

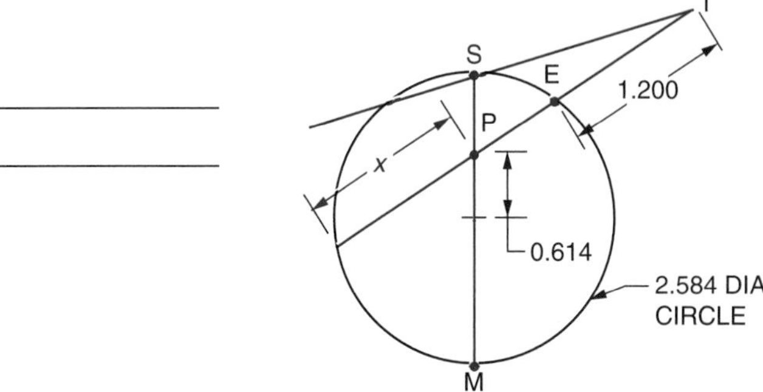

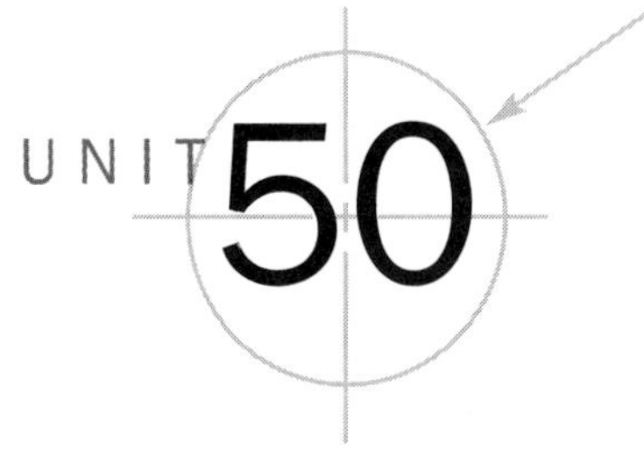

Arcs and Angles of Circles, Tangent Circles

Objectives ***After studying this unit you should be able to***

- Solve problems by using geometric principles that involve angles formed inside, on, and outside a circle.
- Solve problems by using geometric principles that involve internally and externally tangent circles.

The geometric principles of arcs and angles of circles and tangent circles have wide application in machine technology. For example, circle arc and angle principles are used in computing working dimension locations for machining holes and in determining angle rotations for positioning workpieces.

Angles Formed Inside a Circle

➤ **Principle 18**

A central angle is equal to its intercepted arc.

(An *intercepted arc* is an arc that is cut off by a central angle.)

Given: $\overset{\frown}{AB} = 78°$ in Figure 50-1.

Conclusion: $\angle AOB = 78°$.

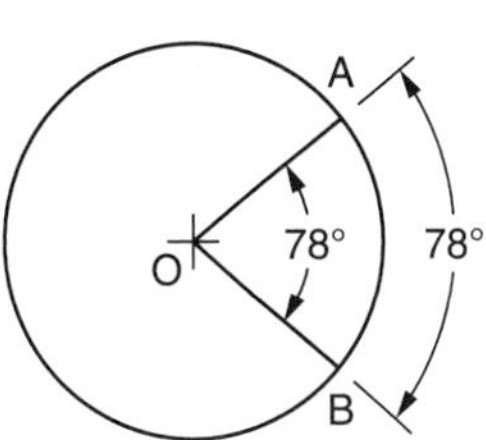

Figure 50-1

➤ **An angle formed by two chords that intersect inside a circle is equal to one-half the sum of its two intercepted arcs.**

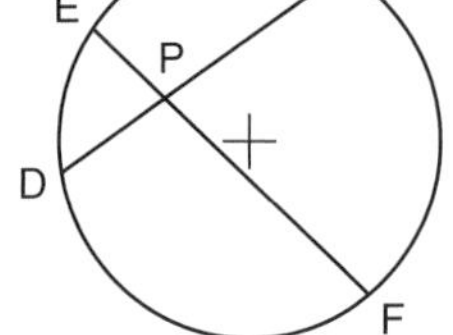

Figure 50-2

Example 1 Given: Chords CD and EF intersect at point P in Figure 50-2.

Conclusion: $\angle EPD = \frac{1}{2}(\overset{\frown}{CF} + \overset{\frown}{DE})$.

Example 2 In Figure 50-2, if $\overset{\frown}{CF} = 106°$ and $\overset{\frown}{ED} = 42°$, determine ∠EPD.

$\angle EPD = \frac{1}{2}(106° + 42°) = 74°$ Ans

Example 3 In Figure 50-2, if ∠EPD = 64°12' and $\overset{\frown}{CF} = 95°58'$, determine $\overset{\frown}{DE}$.

$$\angle EPD = \frac{1}{2}(\overset{\frown}{CF} + \overset{\frown}{DE})$$
$$64°12' = \frac{1}{2}(95°58' + \overset{\frown}{DE})$$
$$64°12' = 47°59' + \frac{1}{2}\overset{\frown}{DE}$$
$$16°13' = \frac{1}{2}\overset{\frown}{DE}$$
$$\overset{\frown}{DE} = 32°26' \quad \text{Ans}$$

$\overset{\frown}{DE} = 2 \times 64°12' - 95°58'$

2 [×] 64 [° ' "] 12 [° ' "] [−] 95 [° ' "] 58 [° ' "] [=] 32°26°0", DE = 32°26' Ans

or 2 [×] 64 [° ' "] 12 [° ' "] [▶] [−] 95 [° ' "] 58 [° ' "] [▶] [° ' "] [◀] [ENTER] [ENTER] 32°26'0" Ans

➤ **An inscribed angle is equal to one-half of its intercepted arc.**

Given: $\overset{\frown}{AC} = 105°$ in Figure 50-3.

Conclusion: $\angle ABC = \frac{1}{2}\overset{\frown}{AC} = \frac{1}{2}(105°) = 52°30'$.

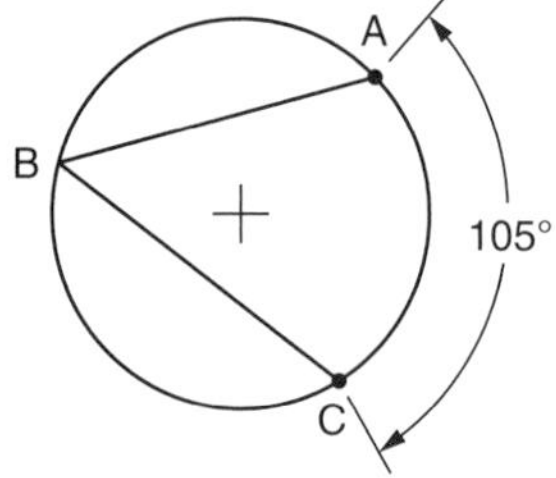

Figure 50-3

Arc Length Formula

Consider a complete circle as an arc of 360°. The ratio of the number of degrees of an arc to 360° is the fractional part of the circumference that is used to find the length of an arc. **The ratio of the length of an arc to the circumference of a circle is the same as the ratio of the number of degrees of the arc to 360°.**

$$\frac{\text{Arc Length}}{\text{Circumference}} = \frac{\text{Arc Degrees}}{360°} \quad \text{or} \quad \frac{\text{Arc Length}}{\text{Circumference}} = \frac{\text{Central Angle}}{360°}$$

If you know the radius of the circle, these two formulas can be rewritten as:

$$\text{Arc Length} = \frac{\text{Arc Degrees}}{360°}(2\pi r) \quad \text{or} \quad \text{Arc Length} = \frac{\text{Central Angle}}{360°}(2\pi r)$$

Example 1 $\overset{\frown}{AC} = 130.00°$ and the radius is 120.00 mm in Figure 50-4. Determine the arc length $\overset{\frown}{AC}$ to 2 decimal places.

$$\text{Arc Length} = \frac{\text{Arc Degrees}}{360°}(2\pi r)$$

$$\text{Arc Length} = \frac{130.00°}{360°}[2(3.1416)(120.00\text{ mm})]$$

$$\text{Arc Length} = 272.27\text{ mm} \quad \text{Ans (rounded)}$$

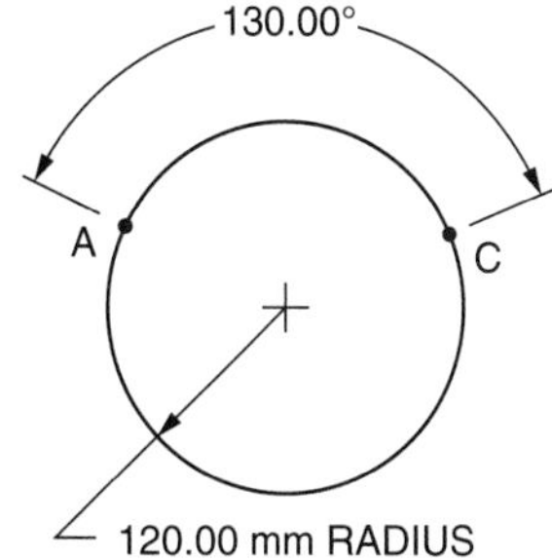

Figure 50-4

130 [÷] 360 [×] 2 [×] [π] [×] 120 [=] 272.2713633, 272.27 mm Ans (rounded)

Example 2 The arc length of $\overset{\frown}{DF}$ is 8.426" and the radius is 5.021" in Figure 50-5. Determine ∠1. All dimensions are in inches. Give the answer in degrees and minutes.

$$\frac{\text{Arc Length}}{\text{Circumference}} = \frac{\text{Central Angle}}{360°}$$

$$\frac{\text{Arc Length}}{2\pi r} = \frac{\angle 1}{360°}$$

$$\frac{8.426''}{2(3.1416)(5.021'')} = \frac{\angle 1}{360°}$$

$$\frac{(8.426'')(360°)}{2(3.1416)(5.021'')} = \angle 1$$

$$\angle 1 = 96°09' \quad \text{Ans (rounded)}$$

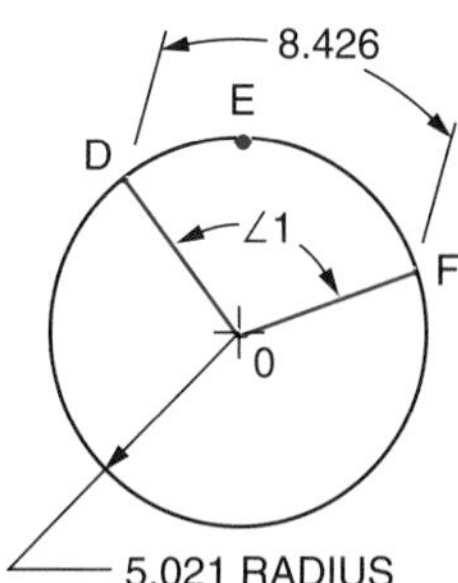

Figure 50-5

$$\frac{8.426''(360°)}{2\pi(5.021'')} = \text{Central Angle}$$

8.426 [×] 360 [÷] [(] 2 [×] [π] [×] 5.021 [)] [=] [SHIFT] [←]
96°9°3.65, 96°9' Ans (rounded)

or 8.426 [×] 360 [÷] [(] 2 [×] [π] [×] 5.021 [)] [ENTER] [° ' "] [◄] [ENTER] [ENTER] 96°9'3.648", 96°9' Ans (rounded)

Angles Formed on a Circle

➤ **Principle 19**

An angle formed by a tangent and a chord at the tangent point is equal to one-half of its intercepted arc.

Example 1 In Figure 50-6, tangent CD meets chord AB at tangent point A and $\overset{\frown}{AEB} = 110°$. Determine ∠CAB.

$$\angle CAB = \frac{1}{2}\overset{\frown}{AEB} = \frac{1}{2}(110°) = 55° \quad \text{Ans}$$

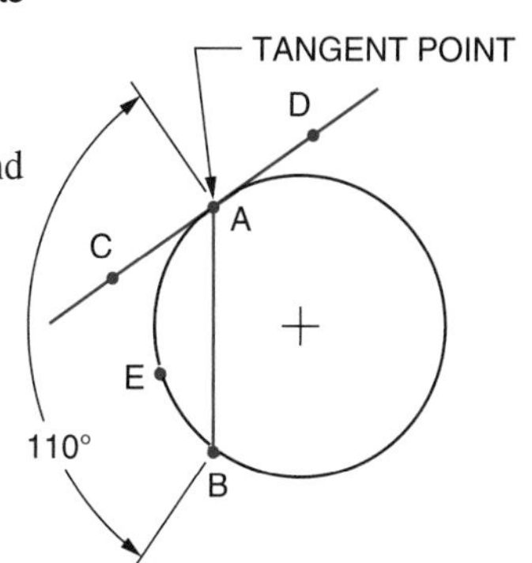

Figure 50-6

Example 2 In Figure 50-7, the centers of 3 holes lie on line ABC. Line ABC is tangent to circle O at hole-center B. The hole-center D, of a fourth hole, lies on the circle. Determine $\angle ABD$.

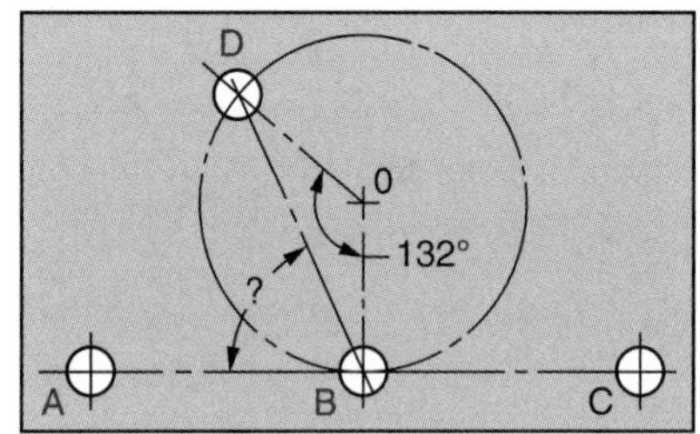

Figure 50-7

A central angle is equal to its intercepted arc (Principle 18).

$$\overset{\frown}{DB} = \angle DOB = 132°$$

Apply Principle 19.

$$\angle ABD = \frac{1}{2}\overset{\frown}{DB} = \frac{1}{2}(132°) = 66° \quad \text{Ans}$$

Angles Formed Outside a Circle

➤ **Principle 20**

An angle formed at a point outside a circle by two secants, two tangents, or a secant and a tangent is equal to one-half the difference of the intercepted arcs.

Two Secants

Example 1 Given: Secants AP and DP meet at point P and intercept $\overset{\frown}{BC}$ and $\overset{\frown}{AD}$ in Figure 50-8.

Conclusion: $\angle P = \frac{1}{2}(\overset{\frown}{AD} - \overset{\frown}{BC})$.

Example 2 In Figure 50-8, if $\overset{\frown}{AD} = 85°40'0''$ and $\overset{\frown}{BC} = 39°17'0''$, find $\angle P$.

$$\angle P = \frac{1}{2}(\overset{\frown}{AD} - \overset{\frown}{BC}) = \frac{1}{2}(85°40'0'' - 39°17'0'') = \frac{1}{2}(46°23'0'')$$
$$= 23°11'30'' \quad \text{Ans}$$

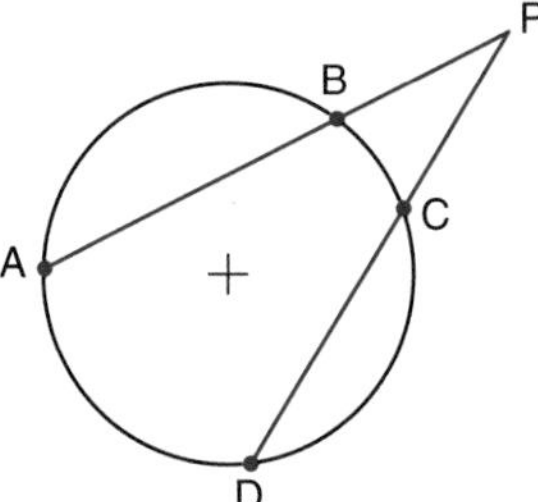

Figure 50-8

Example 3 If $\angle P = 28°$ and $\overset{\frown}{BC} = 40°$ in Figure 50-8, determine $\overset{\frown}{AD}$.

$$\angle P = \frac{1}{2}(\overset{\frown}{AD} - \overset{\frown}{BC})$$
$$28° = \frac{1}{2}(\overset{\frown}{AD} - 40°)$$
$$\overset{\frown}{AD} = 96° \quad \text{Ans}$$

Two Tangents

Example 1 Given: Tangents DP and EP meet at point P in Figure 50-9 and intercept $\overset{\frown}{DE}$ and $\overset{\frown}{DCE}$.

Conclusion: $\angle P = \frac{1}{2}(\overset{\frown}{DCE} - \overset{\frown}{DE})$.

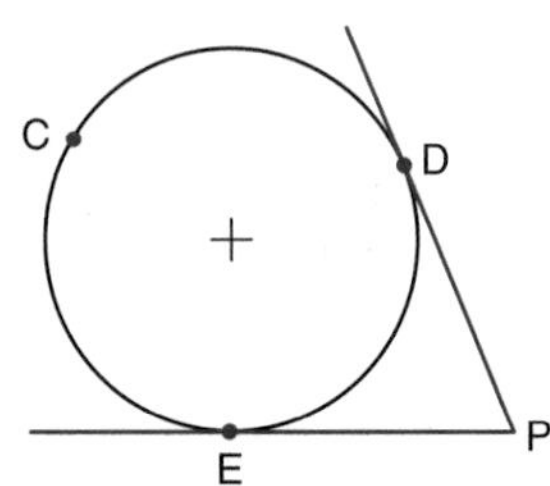

Figure 50-9

Example 2 If $\overset{\frown}{DCE} = 253°37'$ and $\overset{\frown}{DE} = 106°23'$ in Figure 50-9, determine ∠P.

$$\angle P = \frac{1}{2}(\overset{\frown}{DCE} - \overset{\frown}{DE}) = \frac{1}{2}(253°37' - 106°23') = \frac{1}{2}(147°14')$$
$$= 73°37' \quad \text{Ans}$$

A Tangent and a Secant

Example 1 Given: Tangent AP and secant CP meet at point P and intercept $\overset{\frown}{AC}$ and $\overset{\frown}{AB}$ in Figure 50-10.

Conclusion: $\angle P = \frac{1}{2}(\overset{\frown}{AC} - \overset{\frown}{AB})$.

Example 2 In Figure 50-10, if $\overset{\frown}{AC} = 126°38'$ and $\angle P = 28°50'$, determine $\overset{\frown}{AB}$.

$$\angle P = \frac{1}{2}(\overset{\frown}{AC} - \overset{\frown}{AB})$$

$$28°50' = \frac{1}{2}(126°38' - \overset{\frown}{AB})$$

$$28°50' = 63°19' - \frac{1}{2}\overset{\frown}{AB}$$

$$\frac{1}{2}\overset{\frown}{AB} = 63°19' - 28°50'$$

$$\overset{\frown}{AB} = 2(63°19' - 28°50')$$

$$\overset{\frown}{AB} = 68°58' \quad \text{Ans}$$

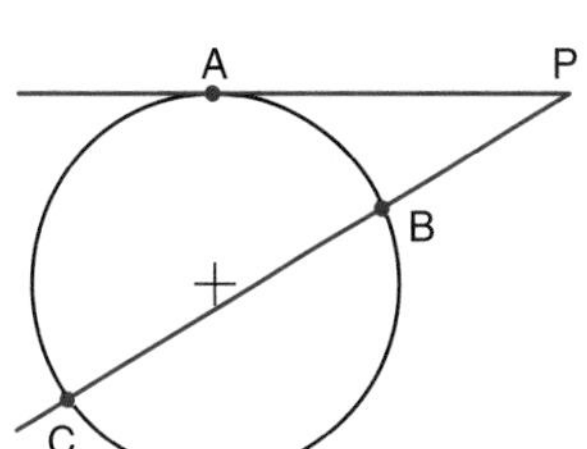

Figure 50-10

2 [×] [(] 63 [° ′ ″] 19 [° ′ ″] [−] 28 [° ′ ″] 50 [° ′ ″] [)] [=]

68°58°0, $\overset{\frown}{AB} = 68°58'$ Ans

or 2 [×] [(] 63 [° ′ ″] 19 [° ′ ″] [▶] [−] 28 [° ′ ″] 50 [° ′ ″] [▶] [)] [° ′ ″] [◀] [ENTER] [ENTER] 68°58'0",

AB = 68°58' Ans

Internally and Externally Tangent Circles

Two circles that are tangent to the same line at the same point are tangent to each other. Circles can be either internally or externally tangent.

Internally tangent—Two circles are internally tangent if both circles are on the same side of the common tangent line. (See Figure 50-11.)

Externally tangent—Two circles are externally tangent if the circles are on opposite sides of the common tangent line. (See Figure 50-12.)

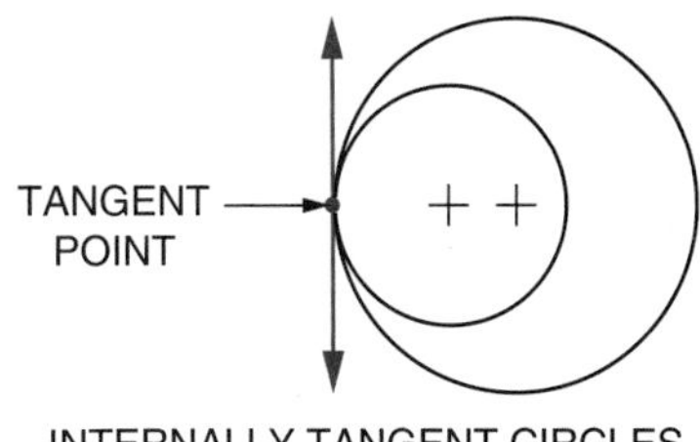

INTERNALLY TANGENT CIRCLES

Figure 50-11

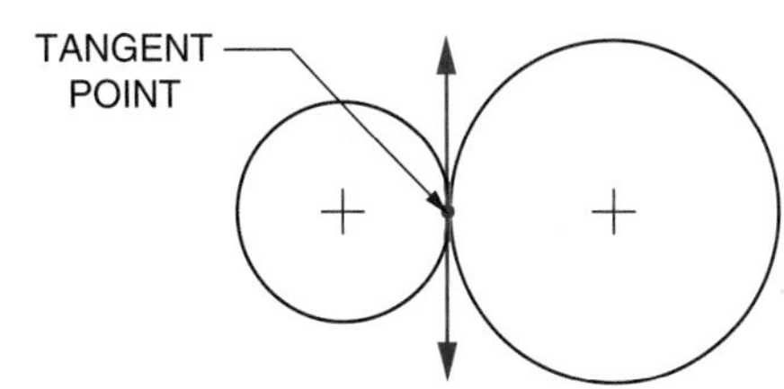

EXTERNALLY TANGENT CIRCLES

Figure 50-12

➤ **Principle 21**

If two circles are either internally or externally tangent, a line connecting the centers of the circles passes through the point of tangency and is perpendicular to the tangent line.

Internally Tangent Circles

Example Given: Circle D and Circle E in Figure 50-13 are internally tangent at point C. D is the center of Circle D and E is the center of Circle E. Line AB is tangent to both circles at point C.

Conclusion: An extension of line DE passes through tangent point C and line CDE ⊥ tangent line AB.

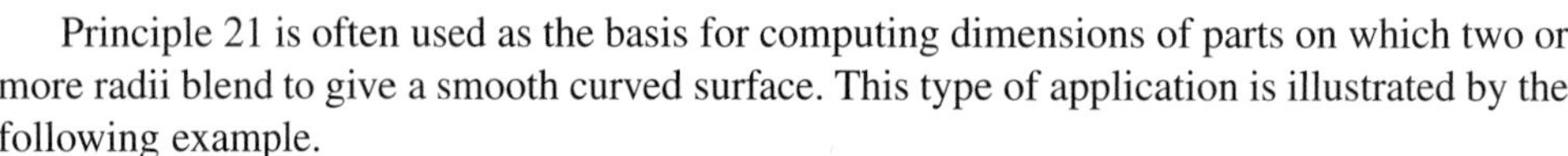

Figure 50-13

Principle 21 is often used as the basis for computing dimensions of parts on which two or more radii blend to give a smooth curved surface. This type of application is illustrated by the following example.

Example A part is to be machined as shown in Figure 50-14. The proper locations of the two radii will result in a smooth curve from point A to point B.

➤ **Note:** The curve from A to B is not an arc of one circle; it is made up of arcs from two different sized circles. In order to make the part, the location to the center of the 12.000-inch radius (dimension *x*) must be determined. Compute *x*. All dimensions are in inches.

Refer to Figure 50-15.

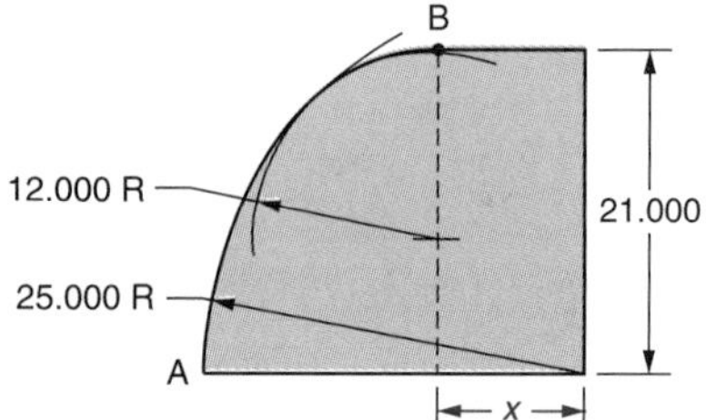

Figure 50-14

The 12.000" radius arc and the 25.000" radius arc are internally tangent. Apply Principle 21. A line connecting arc centers F and H passes through tangent point C.

Tangent point C is the endpoint of the 25.000" radius, CH = 25.000".

Tangent point C is the endpoint of the 12.000" radius, CF = 12.000".

FH = 25.000" − 12.000" = 13.000"

Since BE is vertical and AH is horizontal, ∠FEH = 90°. In right △FEH, FH = 13.000", FE = 21.000" − BF = 9.000".

Apply the Pythagorean Theorem (Principle 9) to compute EH.

$$FH^2 = EH^2 + FE^2$$

$$(13.000 \text{ in})^2 = EH^2 + (9.000 \text{ in})^2$$

$$EH = 9.381 \text{ in}$$

$$x = EH = 9.381 \text{ in} \quad \text{Ans}$$

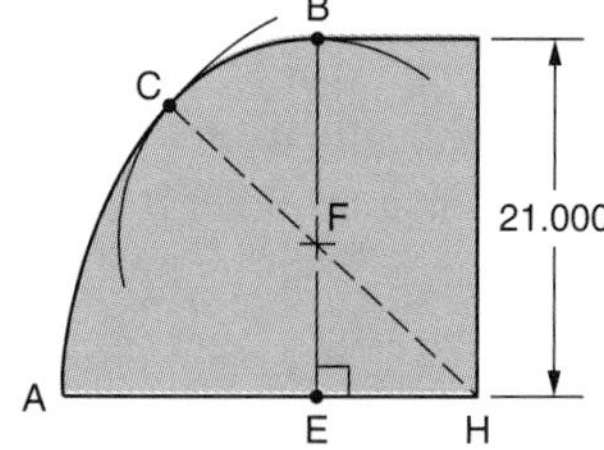

Figure 50-15

Externally Tangent Circles

Example 1 Given: Circle D and Circle E are externally tangent at point C in Figure 50-16. D is the center of Circle D and E is the center of Circle E. Line AB is tangent to both circles at point C.

Conclusion: Line DE passes through tangent point C and line DE ⊥ tangent line AB at point C.

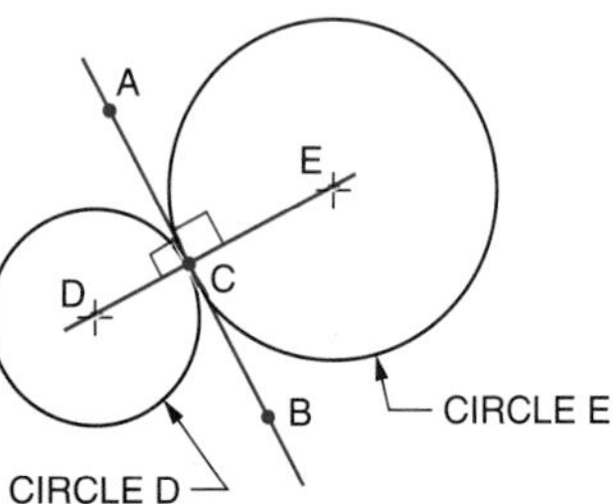

Figure 50-16

Example 2 Three holes are to be bored in a steel plate as shown in Figure 50-17. The 42.00-mm and 61.40-mm diameter holes are tangent at point D. CD is the common tangent line. Determine the distances between hole centers (AB, AC, and BC). All dimensions are in millimeters. Round the answers to 2 decimal places.

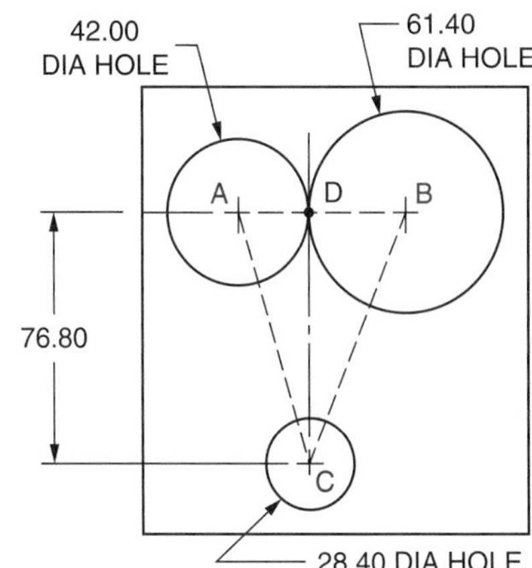

Figure 50-17

Compute AB. Apply Principle 21.
Since AB connects the centers of two tangent circles, AB passes through tangent point D.

$$AB = AD + DB = 21.00 \text{ mm} + 30.70 \text{ mm}$$
$$= 51.70 \text{ mm} \quad \text{Ans}$$

Compute AC and BC.
Since AB connects the centers of two tangent circles, AB $\perp$ tangent line DC. Triangle ADC and triangle BDC are right triangles. Apply the Pythagorean Theorem (Principle 9).

In right $\triangle$ADC, AD = 21.00 mm and DC = 76.80 mm.

$$AC^2 = AD^2 + DC^2$$
$$AC^2 = (21.00 \text{ mm})^2 + (76.80 \text{ mm})^2$$
$$AC = 79.62 \text{ mm} \quad \text{Ans}$$

$$AC = \sqrt{(21.00 \text{ mm})^2 + (76.80 \text{ mm})^2}$$

[√] [(] 21 [x^2] [+] 76.8 [x^2] [)] [=] 79.61934438

AC = 79.62 mm Ans (rounded)

In right $\triangle$BDC, DB = 30.70 mm and DC = 76.80 mm.

$$BC^2 = DB^2 + DC^2$$
$$BC^2 = (30.70 \text{ mm})^2 + (76.80 \text{ mm})^2$$
$$BC = 82.71 \text{ mm} \quad \text{Ans}$$

$$BC = \sqrt{(30.70 \text{ mm})^2 + (76.80 \text{ mm})^2}$$

[√] [(] 30.7 [x^2] [+] 76.8 [x^2] [)] [=] 82.70870571

BC = 82.71 mm Ans (rounded)

Application

Arc Length Formula

Determine the unknown value for each of the following problems. Round the answers to 3 decimal places.

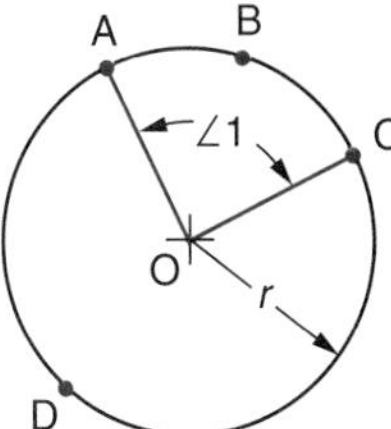

1. $\overparen{ABC} = 90°0'$ and $r = 3.500$ in. Find arc length $\overparen{ABC}$. ________
2. $\overparen{ABC} = 85.00°$ and $r = 60.000$ mm. Find arc length $\overparen{ADC}$. ________
3. Arc length $\overparen{ABC} = 510.000$ mm and $r = 120.000$ mm. Find $\angle 1$. ________
4. Arc length $\overparen{ADC} = 22.700$ in and $r = 5.200$ in. Find $\angle 1$. ________
5. Arc length $\overparen{ABC} = 18.750$ in and $\angle 1 = 72°0'$. Find r. ________
6. Arc length $\overparen{ABC} = 620.700$ mm and $\angle 1 = 69.30°$. Find r. ________

Geometric Principles

Solve the following problems. These problems are based on Principles 18 through 21, although a problem may require the application of two or more of any of the principles. Where necessary, round linear answers in inches to 3 decimal places and millimeters to 2 decimal places. Round angular answers in decimal degrees to 2 decimal places and degrees and minutes to the nearest minute.

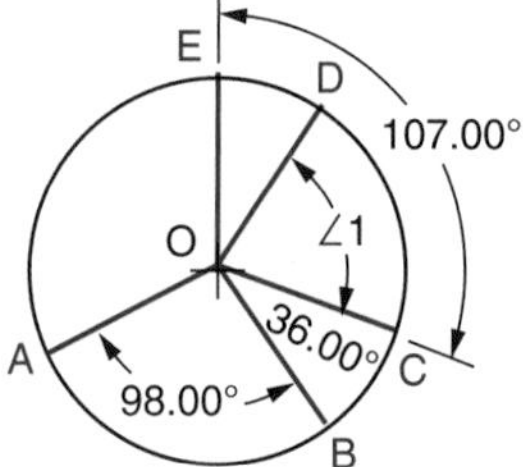

7. a. If $\angle 1 = 76.00°$, find
   (1) $\overset{\frown}{DC}$ ______
   (2) $\angle EOD$ ______
   (3) $\overset{\frown}{AC}$ ______
   b. If $\angle 1 = 63.76°$, find
   (1) $\overset{\frown}{DC}$ ______
   (2) $\angle EOD$ ______
   (3) $\overset{\frown}{BD}$ ______

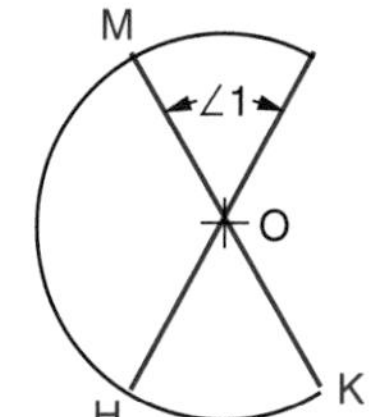

8. a. If $\angle 1 = 63°$, find
   (1) $\overset{\frown}{HK}$ ______
   (2) $\overset{\frown}{HM}$ ______
   b. If $\angle 1 = 59°47'$, find
   (1) $\overset{\frown}{HK}$ ______
   (2) $\overset{\frown}{HM}$ ______

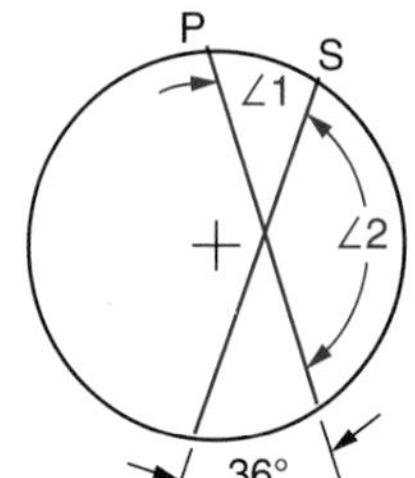

9. a. If $\overset{\frown}{PS} = 46°$, find
   (1) $\angle 1$ ______
   (2) $\angle 2$ ______
   b. If $\overset{\frown}{PS} = 39°$, find
   (1) $\angle 1$ ______
   (2) $\angle 2$ ______

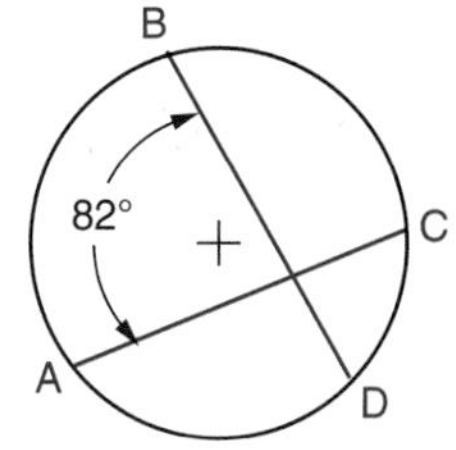

10. a. If $\overset{\frown}{DC} = 35°$, find $\overset{\frown}{AB}$. ______
    b. If $\overset{\frown}{AB} = 127°$, find $\overset{\frown}{DC}$. ______

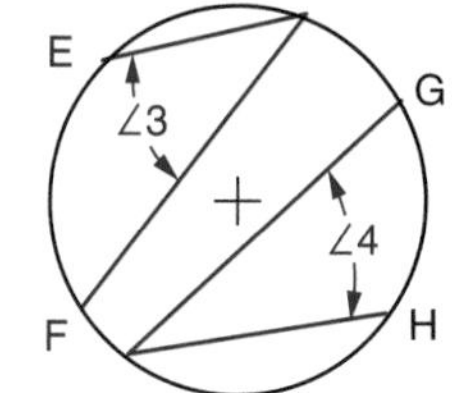

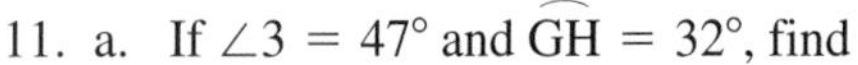

11. a. If $\angle 3 = 47°$ and $\overset{\frown}{GH} = 32°$, find
    (1) $\overset{\frown}{EF}$ ______
    (2) $\angle 4$ ______
    b. If $\angle 4 = 17°53'$ and $\overset{\frown}{EF} = 103°$, find
    (1) $\angle 3$ ______
    (2) $\overset{\frown}{GH}$ ______

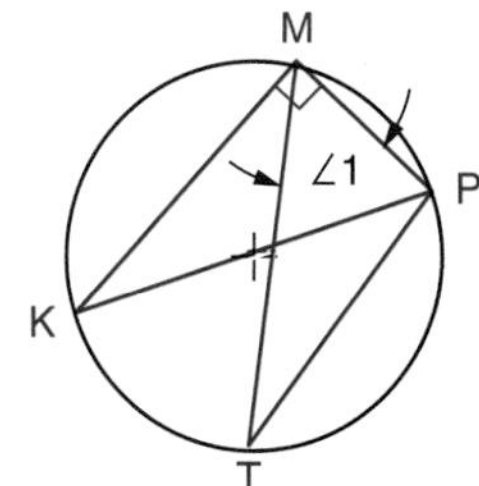

12. a. If $\angle 1 = 25°$ and $\overset{\frown}{MPT} = 95°$, find
    (1) $\overset{\frown}{KTP}$ ______
    (2) $\overset{\frown}{PT}$ ______
    (3) $\overset{\frown}{MP}$ ______

b. If $\angle 1 = 17°30'$ and $\overset{\frown}{MPT} = 103°$, find

(1) $\overset{\frown}{KPT}$ ______

(2) $\overset{\frown}{PT}$ ______

(3) $\overset{\frown}{MP}$ ______

13\. a. If $\overset{\frown}{AB} = 116°$, find

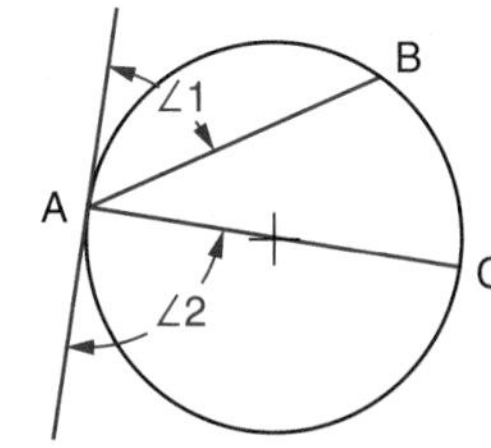

(1) $\angle 1$ ______

(2) $\angle 2$ ______

b. If $\overset{\frown}{AB} = 112°56'$, find

(1) $\angle 1$ ______

(2) $\angle 2$ ______

14\. a. If $\overset{\frown}{EF} = 84°$, find

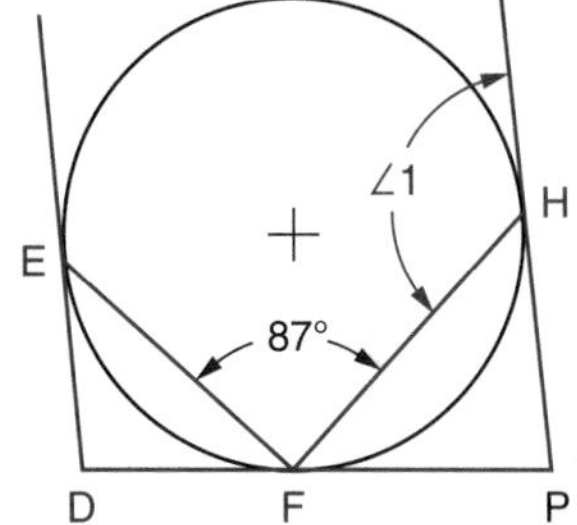

(1) $\angle EFD$ ______

(2) $\overset{\frown}{HF}$ ______

(3) $\angle 1$ ______

b. If $\overset{\frown}{EF} = 79°$, find

(1) $\angle EFD$ ______

(2) $\overset{\frown}{HF}$ ______

(3) $\angle 1$ ______

15\. a. If $\overset{\frown}{ST} = 20°18'$ and $\overset{\frown}{SM} = 38°07'$, find

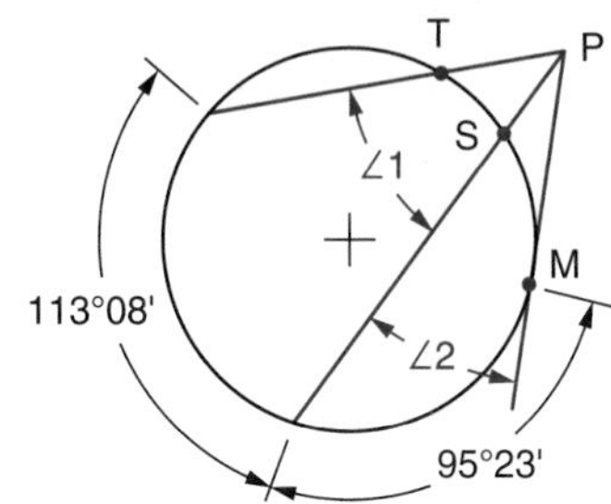

(1) $\angle 1$ ______

(2) $\angle 2$ ______

b. If $\overset{\frown}{ST} = 25°17'$ and $\overset{\frown}{SM} = 35°24'$, find

(1) $\angle 1$ ______

(2) $\angle 2$ ______

16\. a. If $\overset{\frown}{AB} = 72°20'$ and $\overset{\frown}{CD} = 50°18'$, find

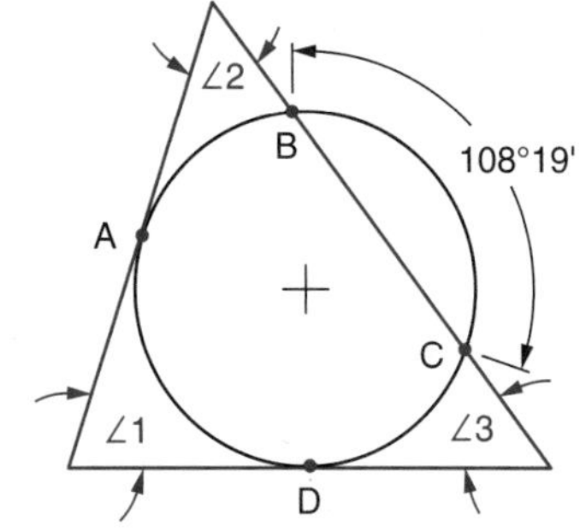

(1) $\angle 1$ ______

(2) $\angle 2$ ______

(3) $\angle 3$ ______

b. If $\overset{\frown}{CD} = 43°15'$ and $\overset{\frown}{AD} = 106°05'$, find

(1) $\angle 1$ ______

(2) $\angle 2$ ______

(3) $\angle 3$ ______

17\. a. If $\angle 1 = 24.00°$ and $\angle 2 = 60.00°$, find

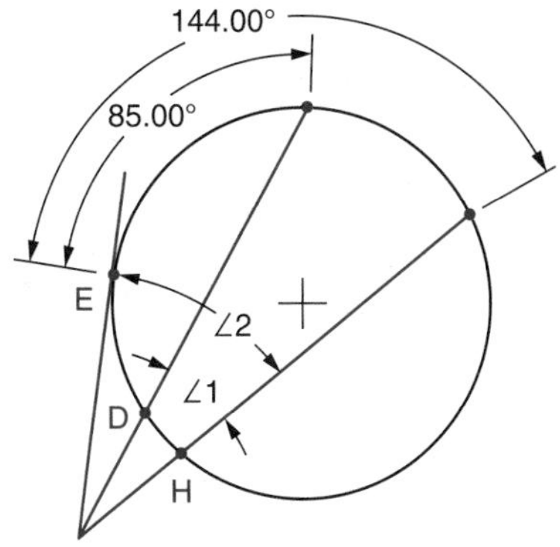

(1) $\overset{\frown}{DH}$ ______

(2) $\overset{\frown}{EH}$ ______

b. If $\angle 1 = 29.00°$ and $\angle 2 = 64.00°$, find

(1) $\overset{\frown}{DH}$ ______

(2) $\overset{\frown}{EH}$ ______

18. a. If Dia A = 3.756" and Dia B = 1.622", find x. ____________
    b. If x = 0.975" and Dia B = 1.026", find Dia A. ____________

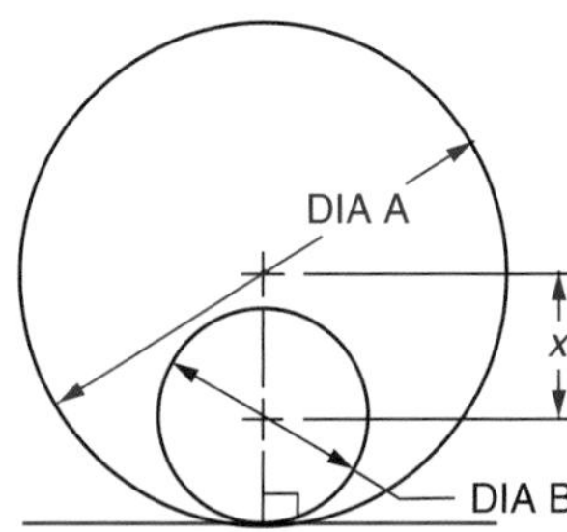

19. a. If x = 24.93 mm and y = 28.95 mm, find Dia A. ____________
    b. If x = 78.36 mm and y = 114.48 mm, find Dia A. ____________

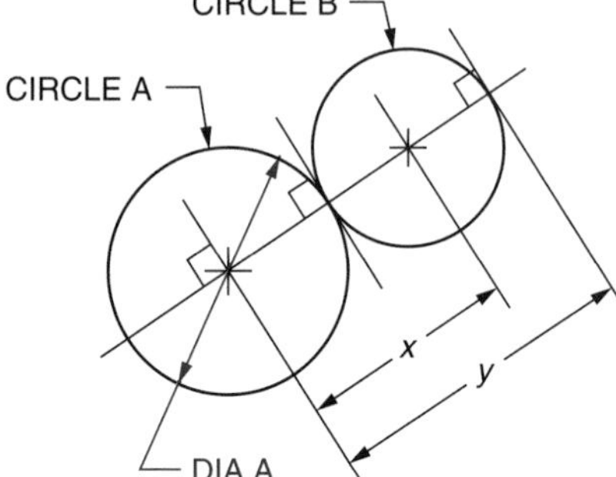

20. a. If ∠1 = 67°00' and ∠2 = 93°00', find
    (1) $\overset{\frown}{AB}$ ____________
    (2) $\overset{\frown}{DE}$ ____________
    b. If ∠1 = 75°00' and ∠2 = 85°00', find
    (1) $\overset{\frown}{AB}$ ____________
    (2) $\overset{\frown}{DE}$ ____________

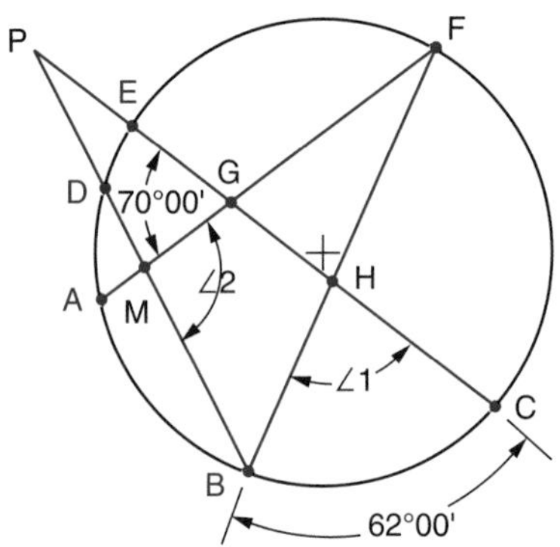

21. All dimensions are in inches.
    a. If Dia A = 1.000", find x. ____________
    b. If Dia A = 0.800", find x. ____________

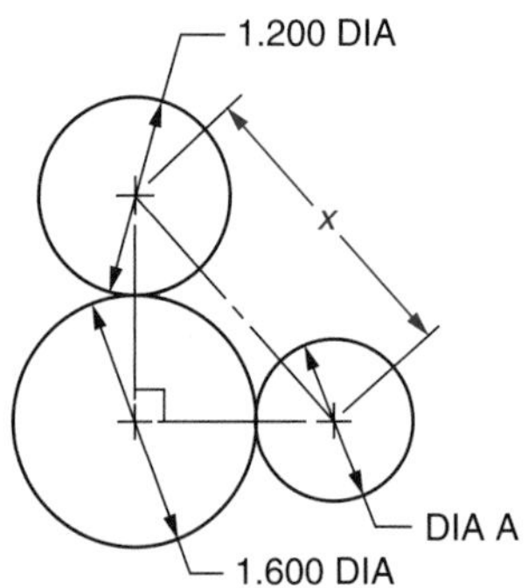

22. Determine the length of x for Gage A and Gage B. All dimensions are in inches.
    a. Gage A: y = 0.350", find x. ____________
    b. Gage B: y = 0.410", find x. ____________

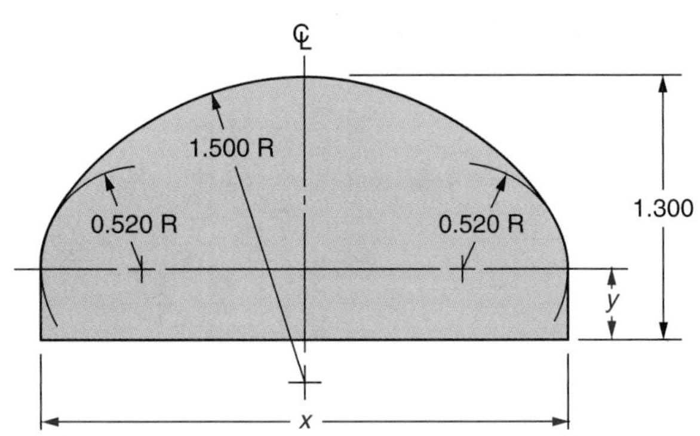

23. AC is a diameter.

    a. If $\angle 2 = 22°00'$, find $\angle 1$. ____________

    b. If $\angle 2 = 30°54'$, find $\angle 1$. ____________

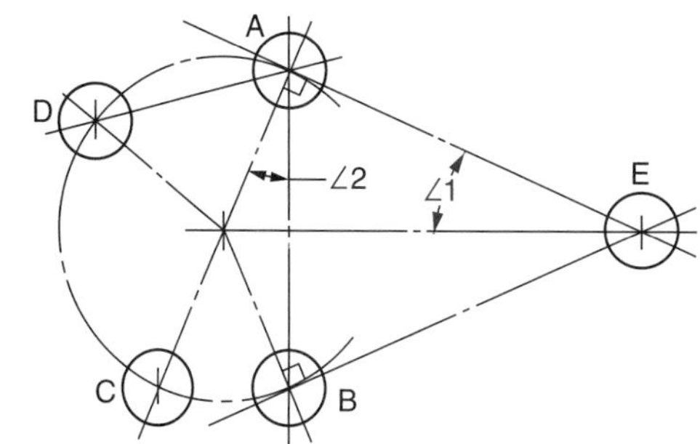

24. Three posts are mounted on the fixture shown. Each post is tangent to the arc made by the 0.650-inch radius. Determine (a) dimension A and (b) dimension B.

    ➤ **Note:** The fixture is symmetrical (identical) on each side of the horizontal centerline (℄).

    All dimensions are in inches. a. ____________

    b. ____________

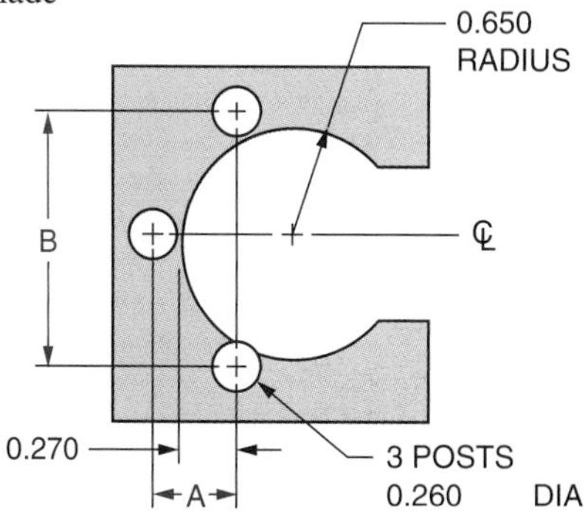

25. Points A, B, C, D, and E are tangent points.

    a. If $\overset{\frown}{AB} = 46.00°$ and $\overset{\frown}{DE} = 66.00°$, find $\angle 1$. ____________

    b. If $\overset{\frown}{AB} = 53.00°$ and $\overset{\frown}{DE} = 70.00°$, find $\angle 1$. ____________

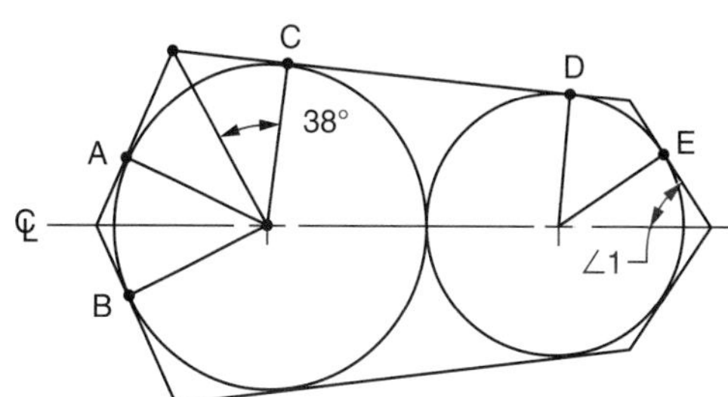

26. Three holes are to be located on the layout shown. The 72.40-mm and 30.80-mm diameter holes are tangent at point T and TA is the common tangent line between the two holes. Determine (a) dimension C and (b) dimension D.

    a. ____________

    b. ____________

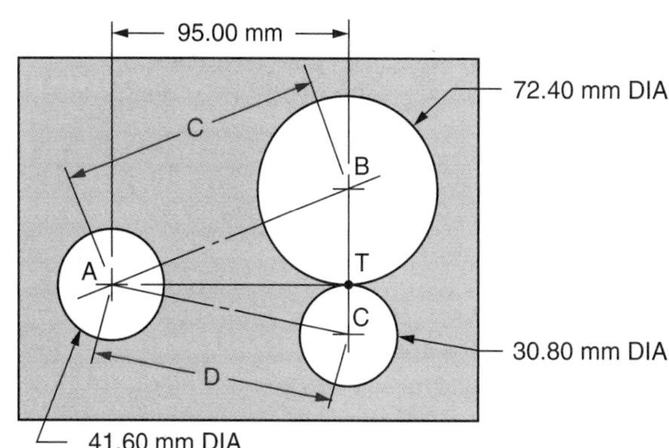

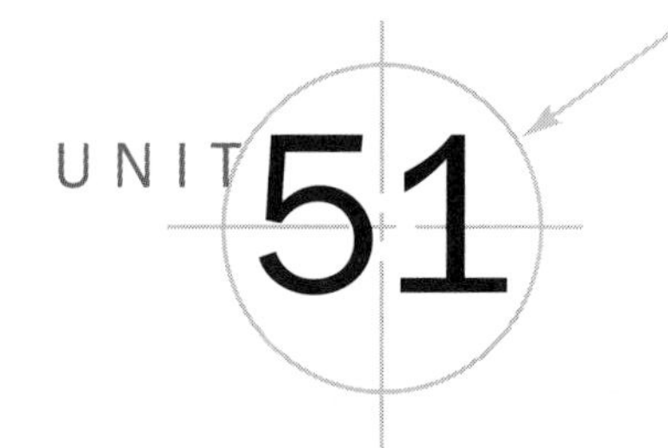

Fundamental Geometric Constructions

Objectives ***After studying this unit you should be able to***

- Make constructions that are basic to the machine trades.
- Lay out typical machine shop problems using the methods of construction.

A knowledge of basic geometric constructions done with a compass or dividers and a steel rule is required of a machinist in laying out work. The constructions are used in determining stock allowances and reference locations on castings, forgings, and sheet stock.

For certain jobs where wide dimensional tolerances are permissible, the most practical and efficient way of producing a part may be by scribing and centerpunching locations. Layout dimensions are sometimes used as a reference for machining complex parts that require a high degree of precision. Locations lightly scribed on a part are used as a precaution to ensure that the part or table movement is in the proper direction. It is particularly useful in operations that require part rotation or repositioning.

There are many geometric constructions, some of which are relatively complex. The constructions presented in this book are those that are most basic and common to a wide range of practical applications.

Common Marking Tools

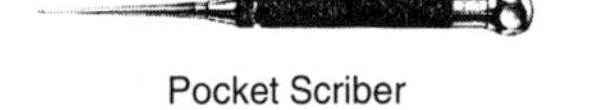

Pocket Scriber
(Courtesy of L. S. Starrett Company)

Center Punch
(Courtesy of L. S. Starrett Company)

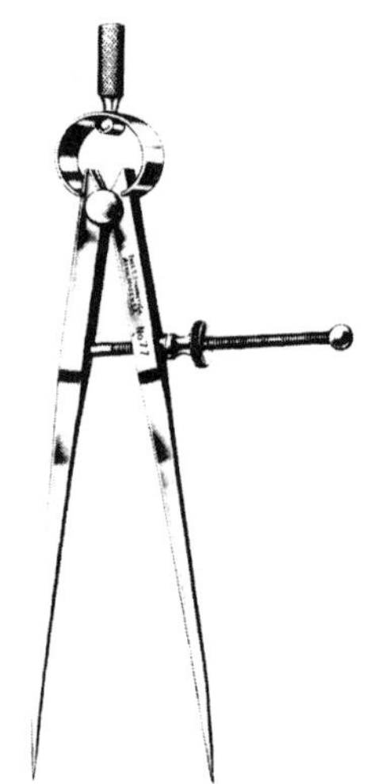

Dividers
(Courtesy of L. S. Starrett Company)

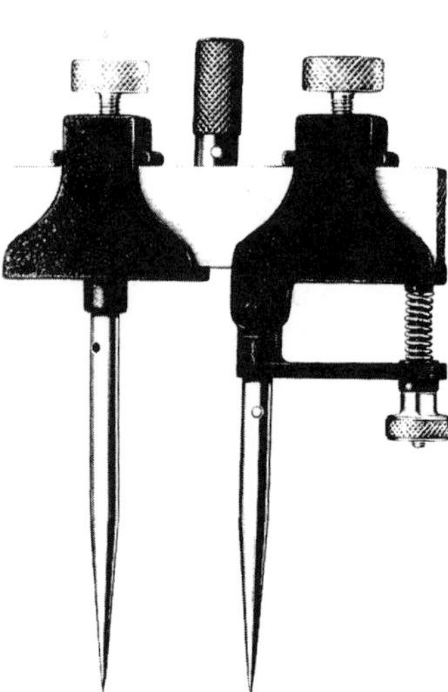

Trammels
(Courtesy of L. S. Starrett Company)

Figure 51-1

Construction 1 To Construct a Perpendicular Bisector of a Line Segment

Required: Construct a perpendicular bisector to line segment AB.

Procedure

- With endpoint A as a center and using a radius equal to more than half AB, draw arcs above and below AB as in Figure 51-2(a).
- With endpoint B as a center and with the same radius used at A, draw arcs above and below AB that intersect the first pair of arcs as shown in Figure 51-2(b).
- Draw a connecting line between the intersection of the arcs above and below AB. Line CD in Figure 51-2(c) is perpendicular to AB and point O is the midpoint of AB.

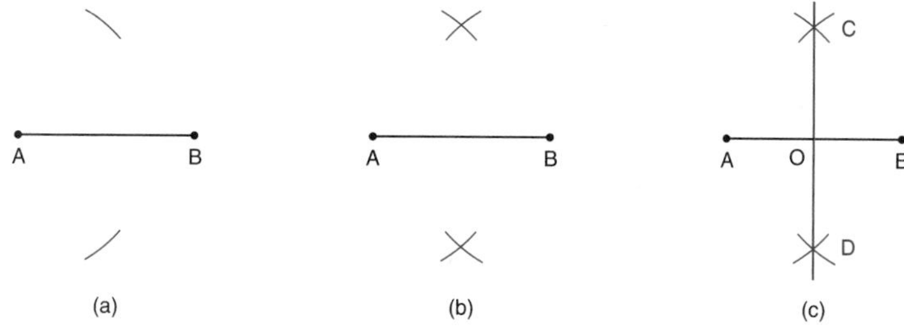

Figure 51-2

Practical Application

Locate the center of a circle.

Solution: The perpendicular bisector of a chord passes through the center of the circle. The center of a circle is located by drawing two chords and constructing a perpendicular bisector to each chord. The intersection of the two perpendicular bisectors locates the center of the circle. The construction lines are shown in Figure 51-3.

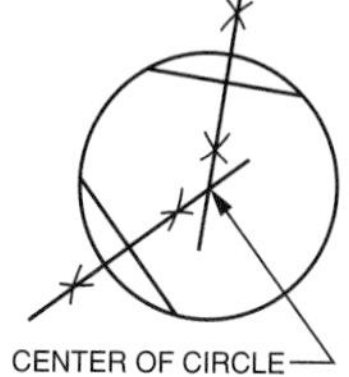

Figure 51-3

Construction 2 To Construct a Perpendicular to a Line Segment at a Given Point on the Line Segment

Required: Construct a perpendicular at point O on line segment AB.

Procedure

- With given point O as a center, and with a radius of any convenient length, draw arcs intersecting AB at points C and D as in Figure 51-4(a).
- With C as a center, and with a radius greater than OC, draw an arc. With D as a center, and with the same radius used at C, draw an arc that intersects the first arc at E as shown in Figure 51-4(b).
- Draw a line connecting point E and point O. Line EO in Figure 51-4(c) is perpendicular to line AB at point O.

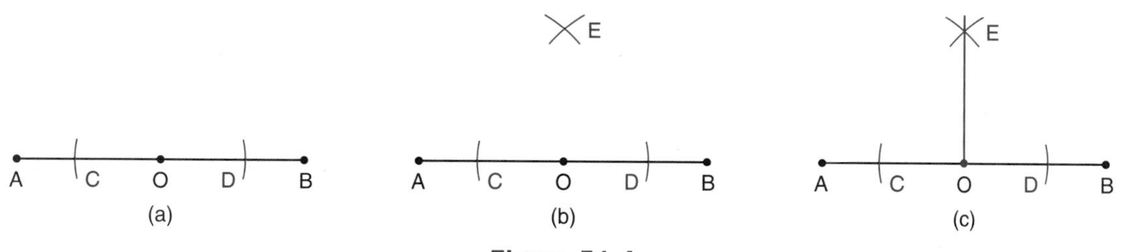

Figure 51-4

Practical Application

A triangular piece is to be scribed and cut. The piece is laid out as follows:

The 22-inch base is measured and marked off.

The $10\frac{7}{64}$-inch distance is measured and marked off at point A on the baseline.

From point A, a perpendicular to the baseline is constructed.

The construction lines are shown.

The $7\frac{1}{2}$-inch distance is measured and marked off at point B on the constructed perpendicular.

Lines are scribed connecting vertex B with the endpoints of the baseline.

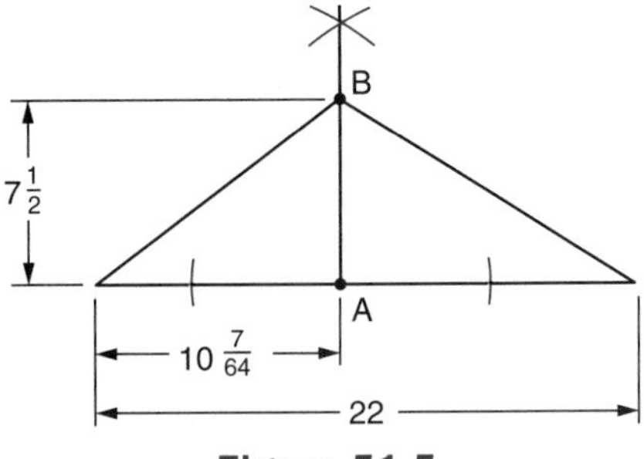

Figure 51-5

Construction 3 To Construct a Line Parallel to a Given Line at a Given Distance

Required: Construct a line parallel to line AB at a given distance of 1 inch.

Procedure

- Set the compass to the required distance (1 inch) as in Figure 51-6(b).
- With any points C and D as centers on AB, draw arcs with the given distance (1 inch) as the radius as shown in Figure 51-6(c).
- Draw a line, EF, that touches each arc at one point (the tangent point). Line EF in Figure 51-6(d) is parallel to line AB and EF is 1 inch from AB.

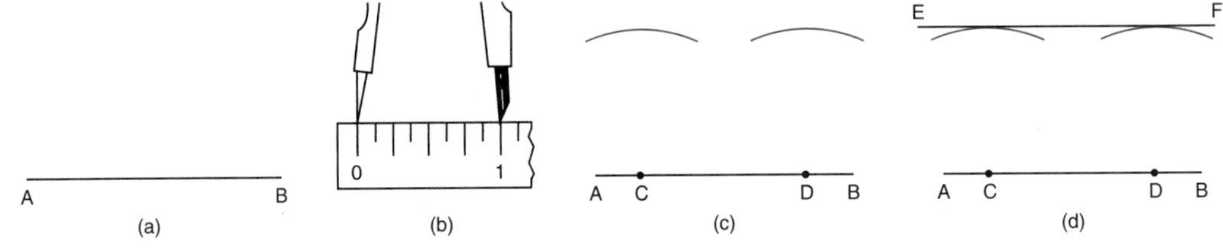

Figure 51-6

Practical Application

The cutout shown in the drawing in Figure 51-7 is laid out on a sheet as follows:

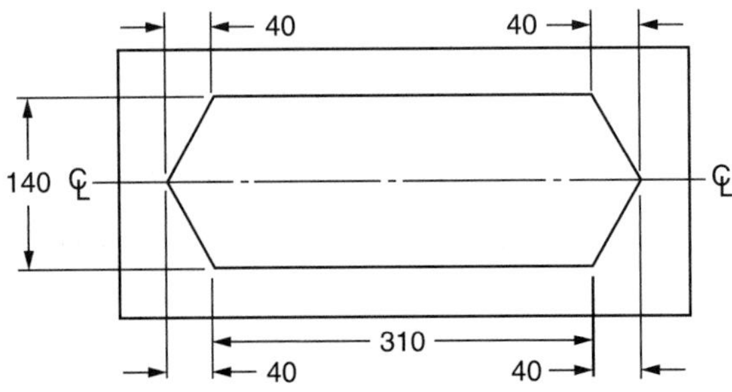

Figure 51-7

All dimensions are in millimeters.

The centerline (℄) is scribed and the 310-mm distance is marked off. Points A and B are the endpoints of the 310-mm segment.

From points A and B perpendiculars are constructed. The perpendiculars are extended more than 70 mm (140 mm ÷ 2) above and below AB (see Figure 51-8).

With points C and D as centers on AB, 70-mm radius arcs are drawn above and below AB. A line is scribed above and a line is scribed below AB touching the pairs of arcs. The lines are extended to intersect the perpendiculars constructed. The points of intersection are E, F, G, and H.

From point A and from point B on AB, 40-mm distances are marked off. Point J and point K are the endpoints.

Lines are scribed connecting J to E and G and connecting K to F and H. Scribed figure JEFKHG is the required cutout.

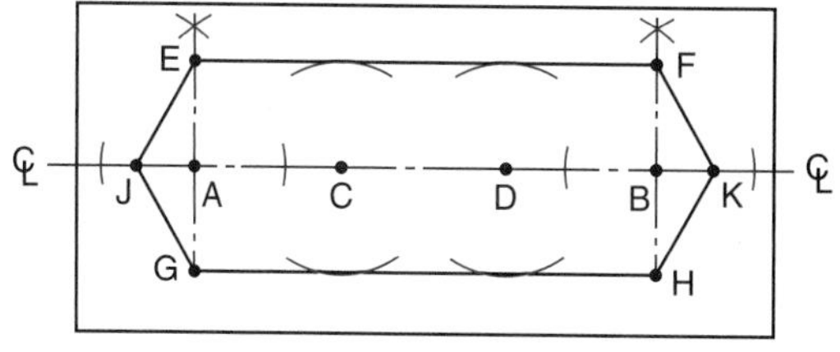

Figure 51-8

Construction 4 To Bisect a Given Angle

Required: Bisect ∠ABC.

Procedure

- With point B as the center, draw an arc intersecting sides BA and BC at points D and E as in Figure 51-9(a).
- With D as the center and with a radius equal to more than half the distance DE, draw an arc. With E as the center, and with the same radius, draw an arc. The intersection of the two arcs is point F as shown in Figure 51-9(b).
- Draw a line from point B to point F. Line BF in Figure 51-9(c) is the bisector of ∠ABC.

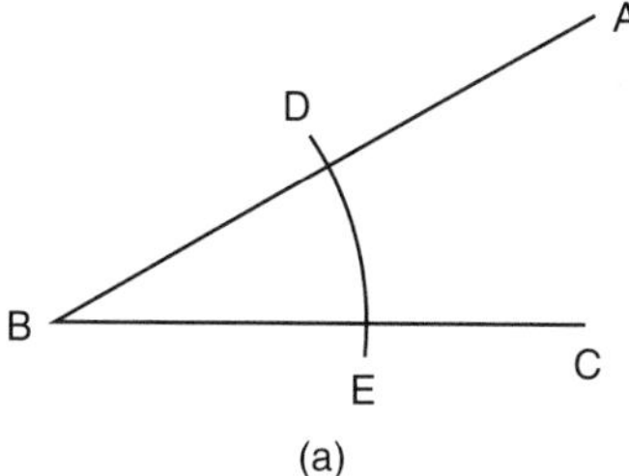

(a)

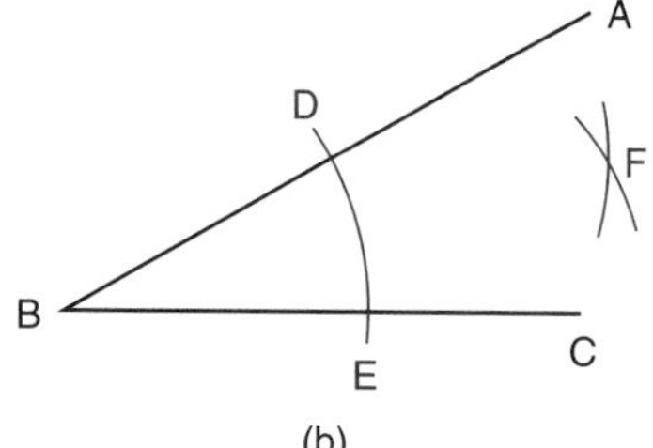

(b)

(c)

Figure 51-9

Practical Application

The centers of the three $\frac{1}{2}$-inch diameter holes in the mounting plate shown in Figure 51-10 are located and center punched. Two $\frac{1}{4}$-inch diameter holes are located by constructing the bisector of ∠ABC as shown and marking and center punching the $1\frac{3}{8}$-inch and $4\frac{1}{4}$-inch hole center locations on the bisector.

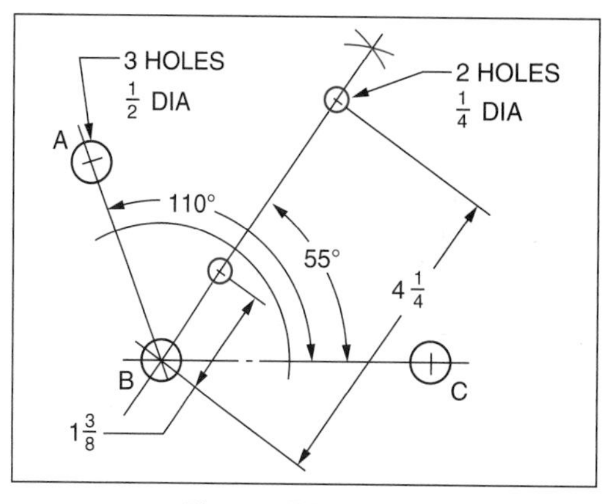

Figure 51-10

Construction 5 To Construct Tangents to a Circle from an Outside Point

Required: Construct tangents to given circle O from given outside point P.

Procedure

- Draw a line segment connecting center O and point P. Bisect OP. Point A is the midpoint of OP as in Figure 51-11(a).
- With point A as the center and AP as a radius, draw arcs intersecting circle O at points B and C as shown in Figure 51-11(b). Points B and C are tangent points.
- Connect points B and P, and C and P. Line segments BP and CP in Figure 51-11(c) are tangents.

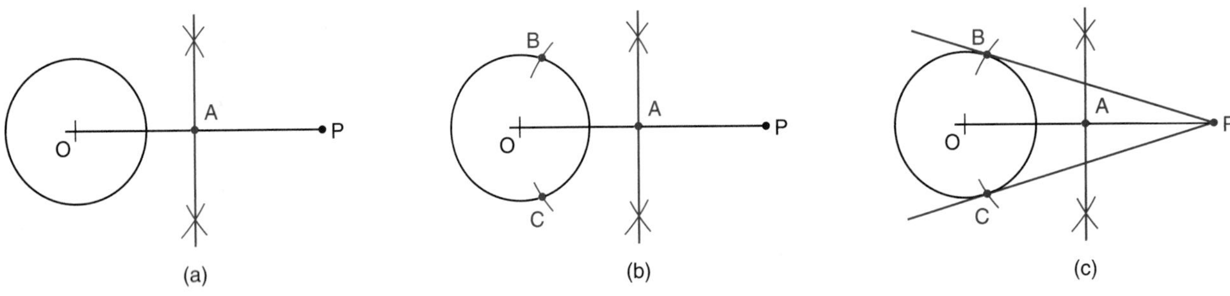

Figure 51-11

Practical Application

A piece is to be made as shown in the drawing in Figure 51-12. All dimensions are in millimeters. The piece is laid out as follows and shown in Figure 51-13.

A baseline is scribed and AB (170 mm) is marked off.

Distance OA (152 mm) is set on dividers and, with OA as the radius, an arc is scribed. Distance OB (104 mm) is set on dividers and, with OB as the radius, an arc is scribed to intersect with the OA radius arc. The intersection of the arcs locates center O of the 42-mm radius circle.

Dividers are set to the 42-mm radius dimension, and the circle is scribed from center O.

Tangents to the circle from points A and B are constructed resulting in tangent points C and D and tangent line segments AC and BD. The piece is now laid out and ready to be cut to the scribed lines.

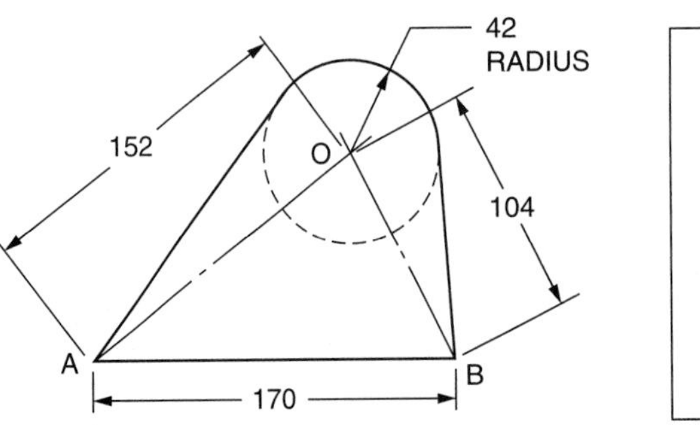

Figure 51-12

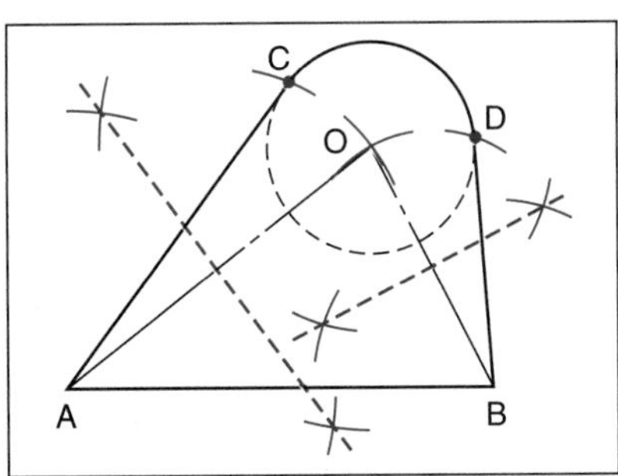

Figure 51-13

Construction 6 To Divide a Line Segment into a Given Number of Equal Parts

Required: Divide line segment AB into three equal parts.

Procedure

- From point A, draw line AC forming any convenient angle with AB as in Figure 51-14(a).
- On AC, with a compass, lay off any three equal segments, AD, DE, and EF (Figure 51-14(a)).
- Connect point F with point B as in Figure 51-14(b). With centers at points F, E, and D, draw arcs of equal radii. The arc with a center at point F intersects AC at point G and BF at point H. Set distance GH on the compass and mark off this distance on the other two arcs. The points of intersection are K and M.
- Connect points E and K, and D and M, extending the lines past AB. Line AB in Figure 51-14(c) is divided into three equal segments; AP = PS = SB.

➤ **Note:** Line segment AB can be divided into any required number of equal segments by laying off the required number of equal segments on AC and following the procedure given.

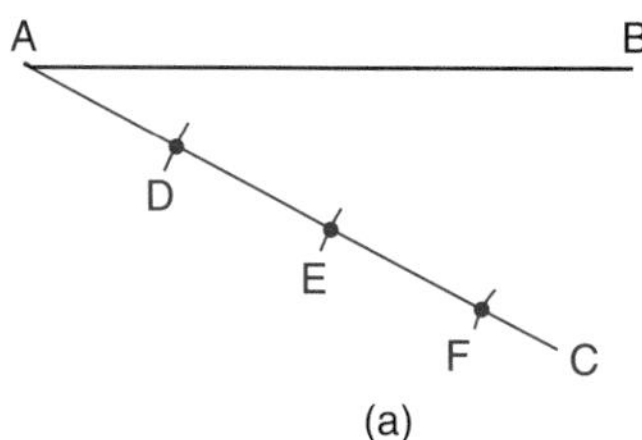

(a)

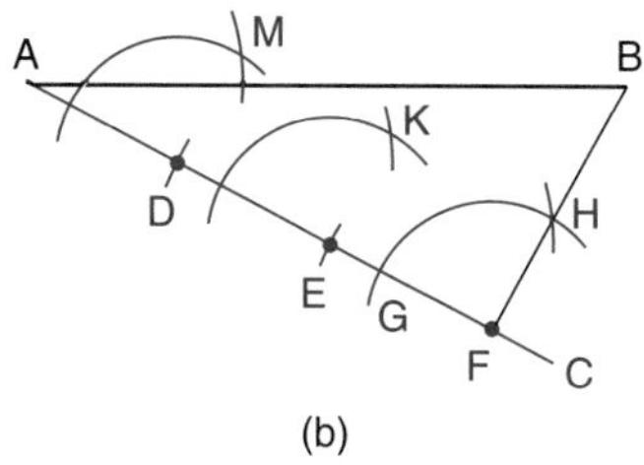

(b)

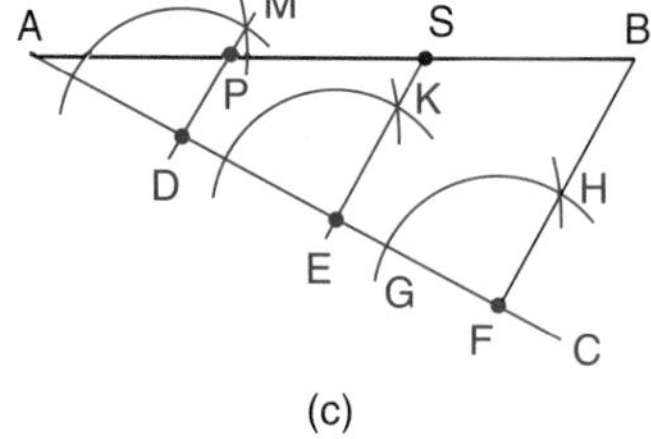

(c)

Figure 51-14

Practical Application

Six holes are to be equally spaced within a distance of $2\frac{11}{16}$ inches. Since six holes are required, there will be five equal spaces between holes. Dividing $2\frac{11}{16}$ inches by 5 results in fractional distances that are difficult to accurately measure or transfer, such as $\frac{8.6}{16}$ inch, $\frac{17.2}{32}$ inch, or $\frac{34.4}{64}$ inch. By careful construction, the hole centers are accurately located as shown in Figure 51-15.

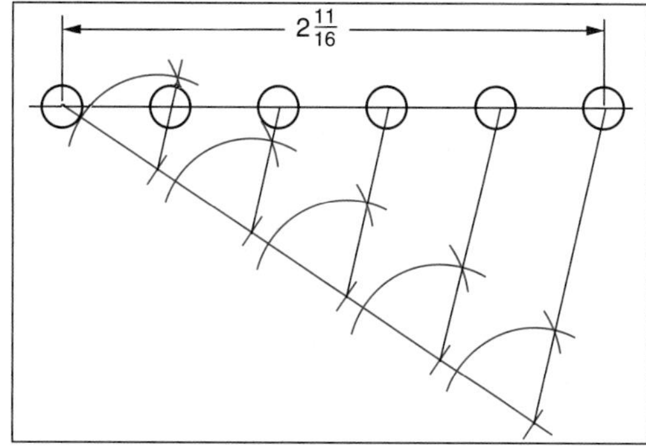

Figure 51-15

Application

Construction 1 and 2 Applications

Show construction lines and arcs for each of these problems.

1. Trace each line segment in problems a through d and construct perpendicular bisectors to each segment.

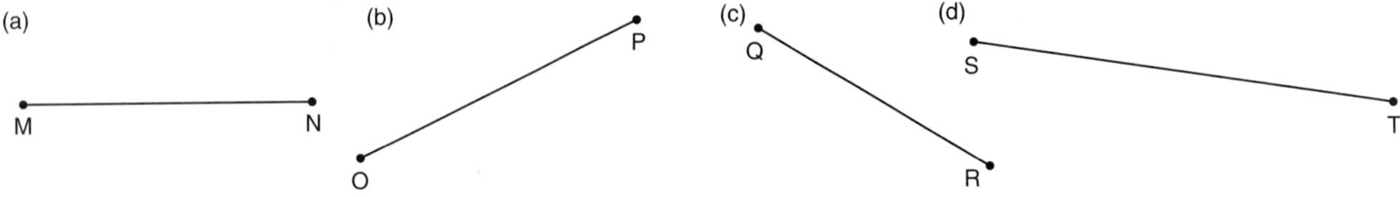

2. Trace each line in problems a through c and construct perpendiculars to each line at the given points on the lines.

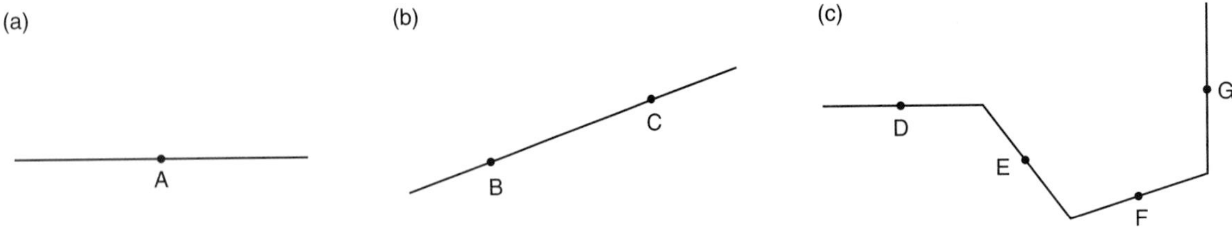

3. With a compass, draw a circle 2 inches in diameter. By construction, locate the center of the circle.
4. Lay out a figure as follows:

   a. Draw a horizontal line and mark off a distance of $2\frac{1}{2}$ inches. Label the left endpoint of the $2\frac{1}{2}$-inch line segment point A and label the right endpoint point D.

   b. From point A and above point A, construct a perpendicular to AD. Mark off a distance of $1\frac{7}{8}$ inches on the perpendicular from point A. Label the top endpoint point B.

   c. From point B and to the right of point B, construct a perpendicular to AB. Mark off a distance of $2\frac{1}{2}$ inches on the perpendicular from point B. Label the right endpoint point C.

   d. From point C and below point C, construct a perpendicular to BC. Mark off a distance of $1\frac{7}{8}$ inches on the perpendicular from point C. If your constructions are accurate the $1\frac{7}{8}$-inch distance marked off coincides with point D. What kind of a figure is formed by this construction? __________

Construction 3 and 4 Applications

Show construction lines and arcs for each of these problems.

5. Trace each of the lines in problems a through d and construct a line parallel to each line at a distance of $1\frac{1}{2}$ inches.

a.

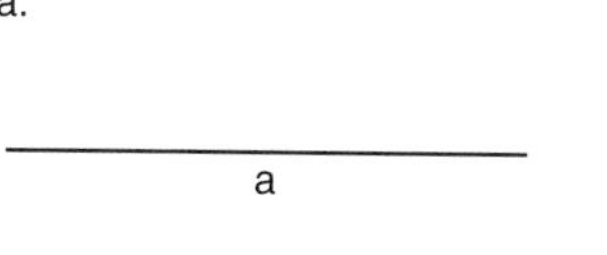

b.

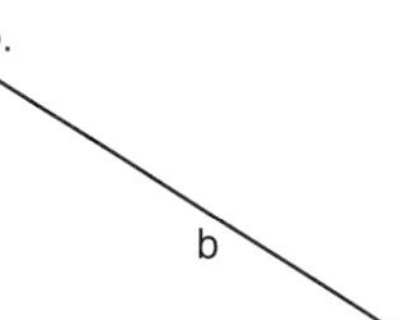

c.

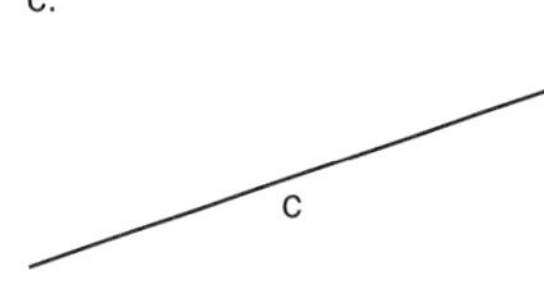

d.

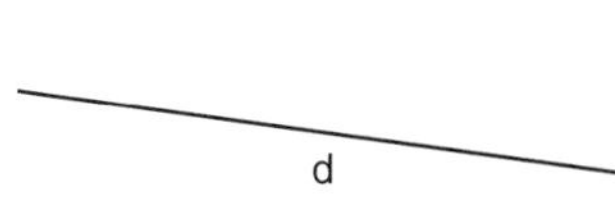

6. Trace each of the angles a through c and construct a bisector to each.

a.

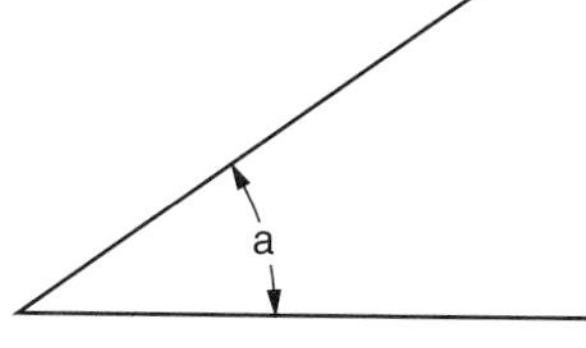

b.

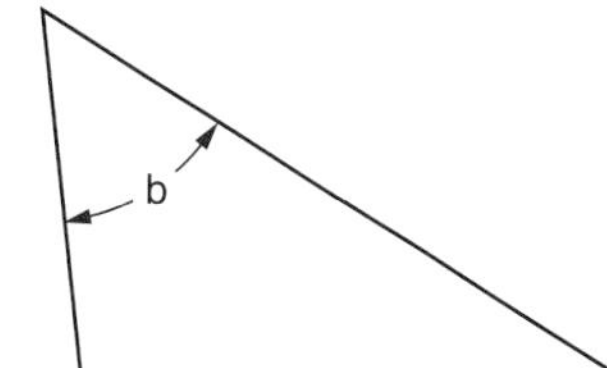

c.

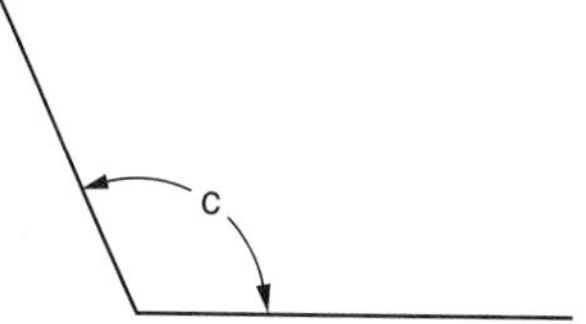

7. Lay out the following angles by construction. Check the angle with a protractor but do not lay out angles with a protractor.

a. 45° b. 22°30' c. 67°30' d. $157\frac{1°}{2}$ e. $168\frac{3°}{4}$

8. Lay out the plate shown in Figure 51-16. Make the layout full size using construction methods. Use a protractor only for checking. All dimensions are in inches.

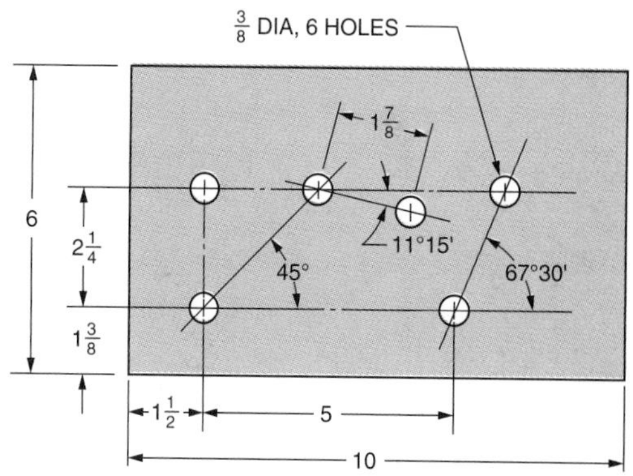

Figure 51-16

9. Lay out the gage shown in Figure 51-17. Make the layout full size using construction methods. Use a protractor only for checking. All dimensions are in inches.

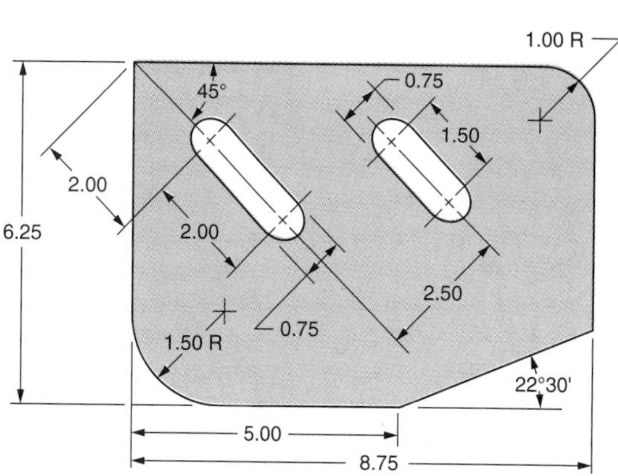

Figure 51-17

Construction 5 and 6 Applications

Show construction lines and arcs for each of these problems.

10. Trace each circle and point in problems a through c and construct tangents to the circles from the given points.

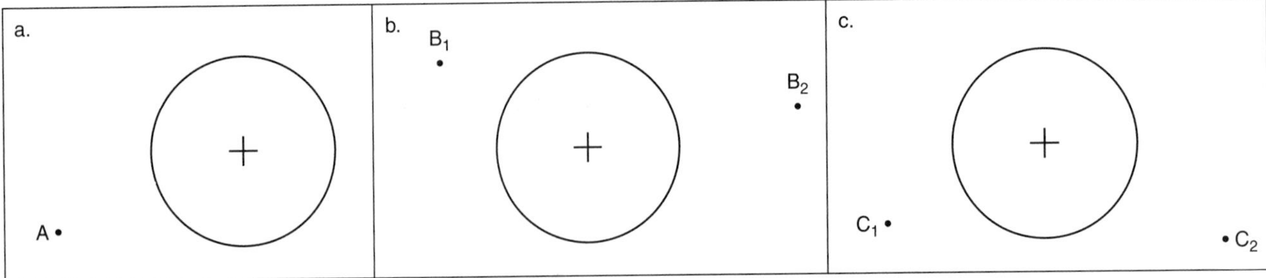

11. Trace each line segment of problems a, b, and c. Divide the given lines into the designated number of segments by means of construction.

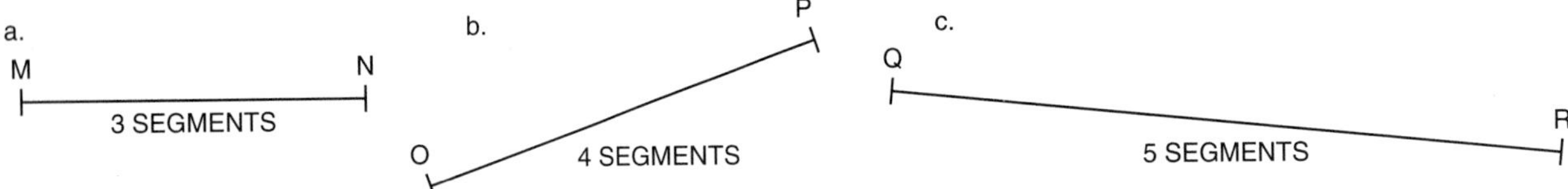

12. Lay out the template shown in Figure 51-18. Make the layout full size using construction methods. All dimensions are in inches.

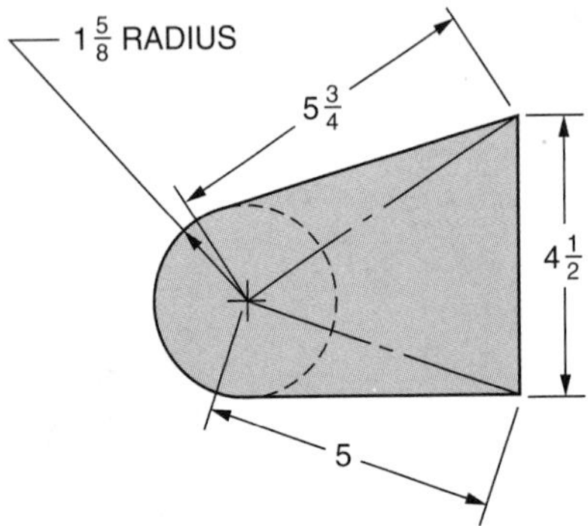

Figure 51-18

13. Lay out the cutout shown in Figure 51-19. Make the layout full size using construction methods. All dimensions are in inches.

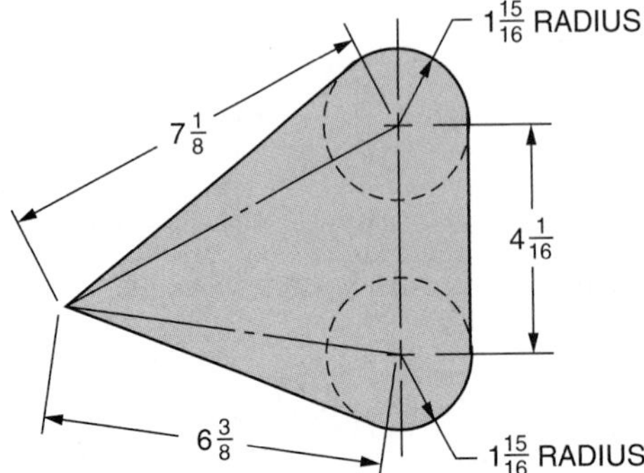

Figure 51-19

14. Trace the plate shown in Figure 51-20. Lay out three sets of holes by construction. Follow the given directions.

Directions:

- Bisect ∠A and construct 4 equally spaced $\frac{3}{16}$-inch diameter holes. Make the first hole $\frac{7}{8}$ inch from point A and the last hole $2\frac{7}{16}$ inches from point A.
- Bisect ∠B and construct 8 equally spaced $\frac{1}{4}$-inch diameter holes. Make the first hole $\frac{3}{4}$ inch from point B and the last hole $3\frac{11}{16}$ inches from the first hole.
- Bisect ∠C and construct 4 equally spaced $\frac{3}{16}$-inch diameter holes. Make the first hole $\frac{9}{16}$ inch from point C and the last hole $2\frac{11}{16}$ inches from point C.

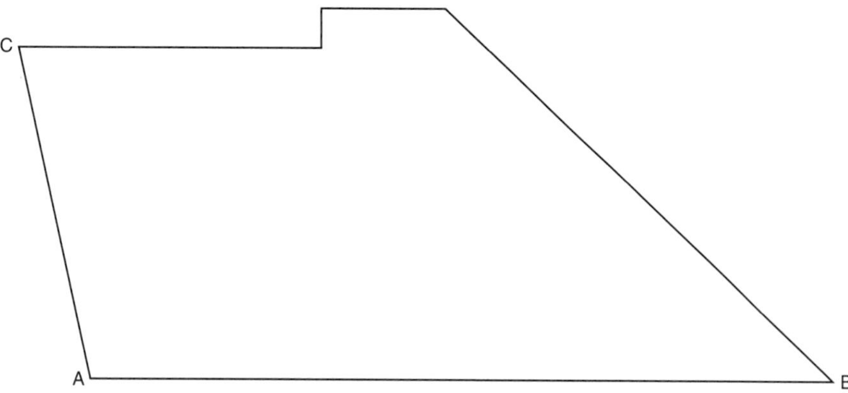

Figure 51-20

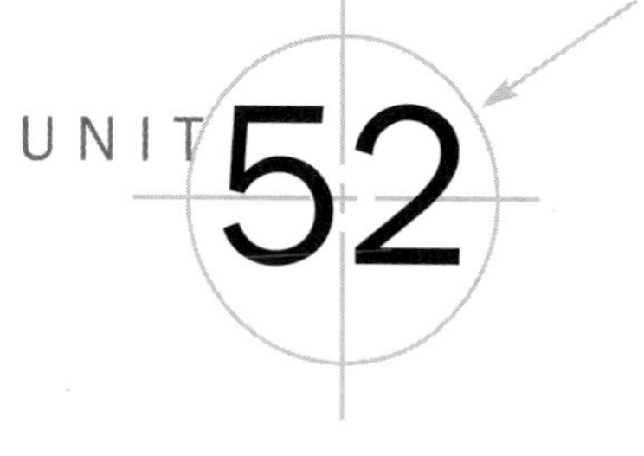

Achievement Review—Section Five

Objective ***You should be able to solve the exercises and problems in this Achievement Review by applying the principles and methods covered in Units 44–51.***

1. Add, subtract, multiply, or divide each of the following exercises as indicated.

a. 37°18' + 86°23'	________	e. 4(27°23')	________
b. 38°46' + 23°43'	________	f. 3(7°23'43")	________
c. 136°36'28" − 94°17'15"	________	g. 87° ÷ 2	________
d. 58°14' − 44°58'	________	h. 103°20' ÷ 4	________

2. Determine $\angle A$. ______________

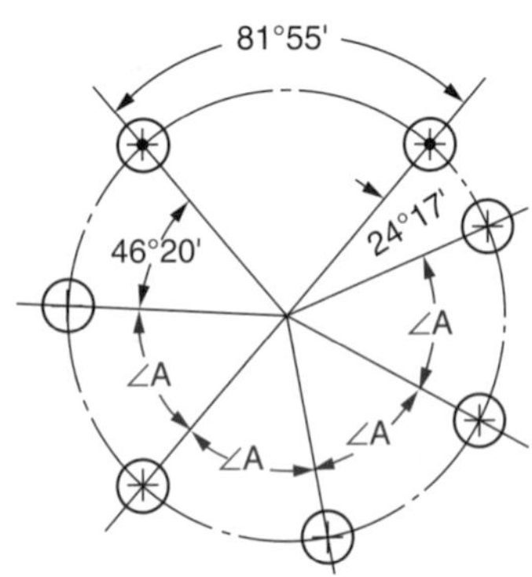

3. Given: The sum of all angles = 720°00'00"
   $\angle 3 = \angle 4 = \angle 5 = \angle 6$.
   $\angle 1 = \angle 2 = 68°42'18"$.
   Determine $\angle 3$. ______________

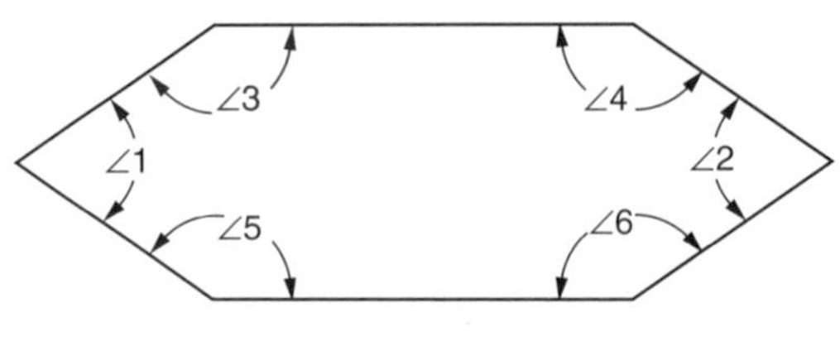

4. Express 68.85° as degrees and minutes. ______________
5. Express 64.1420° as degrees, minutes, and seconds. ______________
6. Express 37°23' as decimal degrees to 2 decimal places. ______________
7. Express 103°38'43" as decimal degrees to 4 decimal places. ______________
8. Using a simple protractor, measure each of the angles, $\angle 1$ through $\angle 7$, to the nearer degree. It may be necessary to extend sides of angles.

   $\angle 1$ = ______________

   $\angle 2$ = ______________

   $\angle 3$ = ______________

   $\angle 4$ = ______________

   $\angle 5$ = ______________

   $\angle 6$ = ______________

   $\angle 7$ = ______________

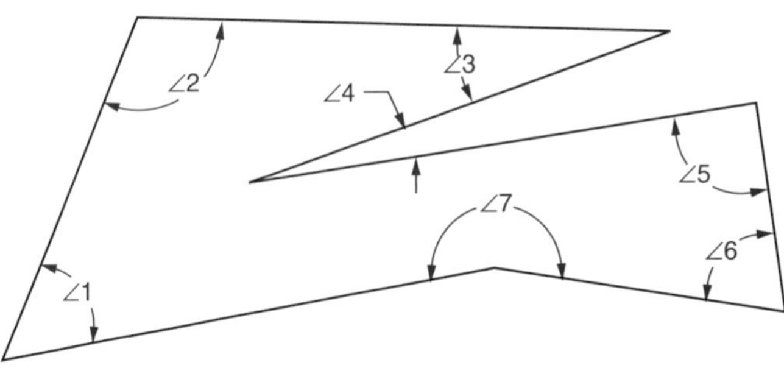

9. Write the values of the settings shown in the following vernier protractor scales.

   a. ______________ b. ______________ c. ______________

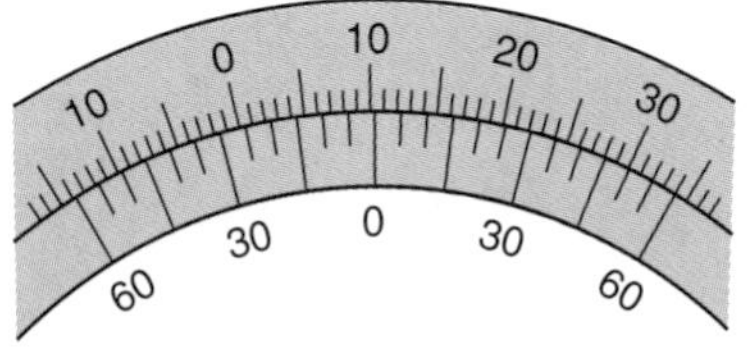

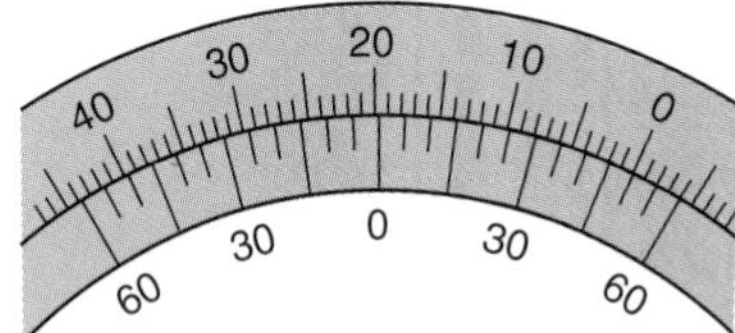

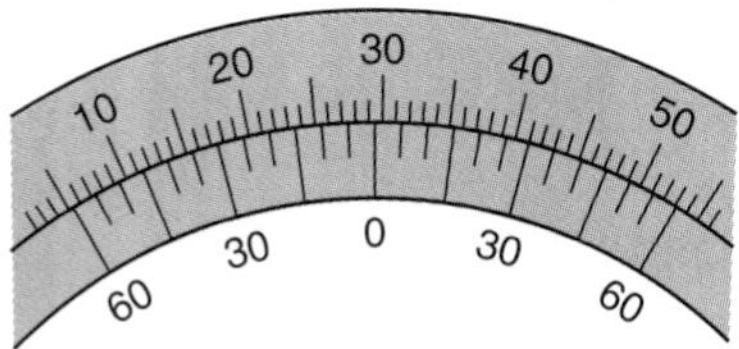

10. Write the complement of each of the following angles.

    a. 67° ______________ b. 17°41' ______________ c. 54°47'53" ______________

11. Write the supplement of each of the following angles.

    a. 41° ______________ b. 99°32' ______________ c. 103°03'27" ______________

12. Given: AB ∥ CD and EF ∥ GH. Determine the value of each angle, ∠1 through ∠10, to the nearer minute.

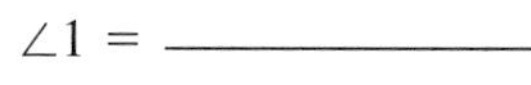

∠1 = ____________

∠2 = ____________

∠3 = ____________

∠4 = ____________

∠5 = ____________

∠6 = ____________

∠7 = ____________

∠8 = ____________

∠9 = ____________

∠10 = ____________

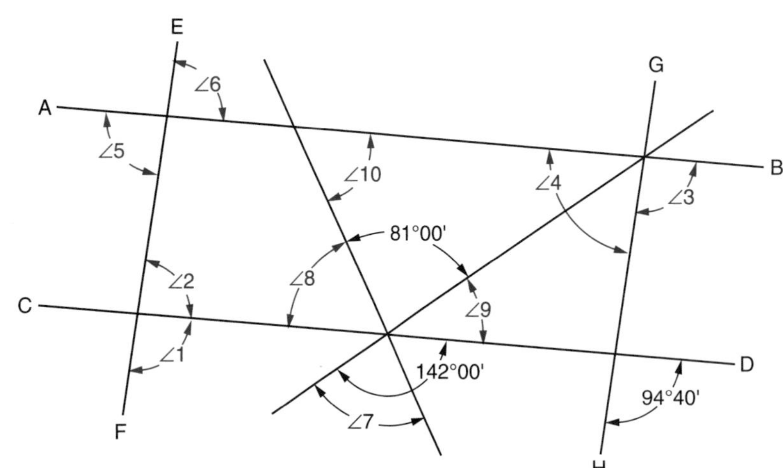

13. a. Determine:

(1) ∠1 ____________

(2) Side a ____________

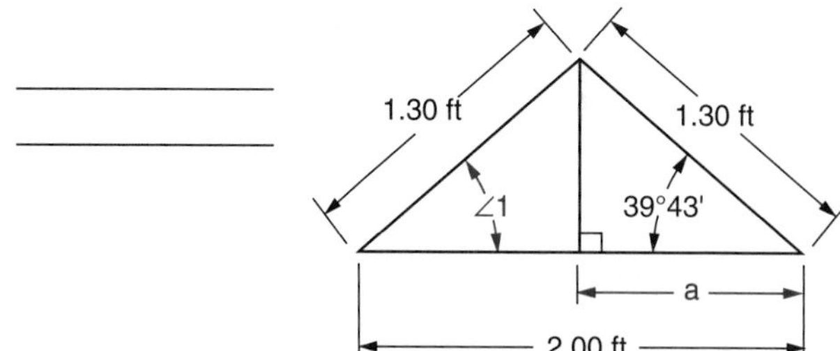

b. Determine:

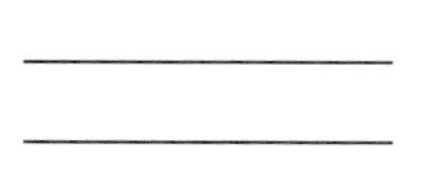

(1) ∠1 ____________

(2) Side b ____________

(3) Side c ____________

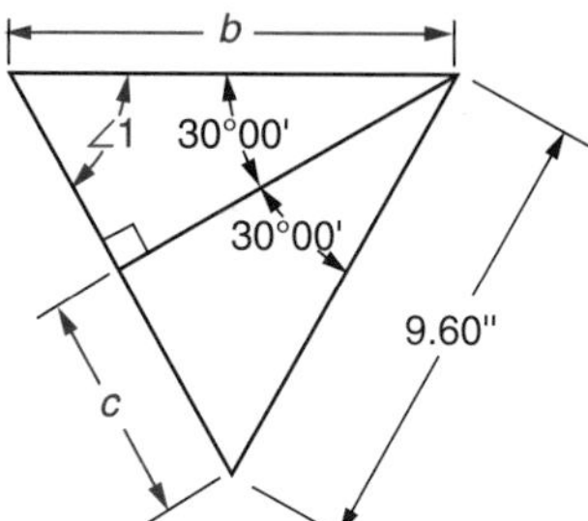

c. Determine:

(1) ∠1 ____________

(2) ∠2 ____________

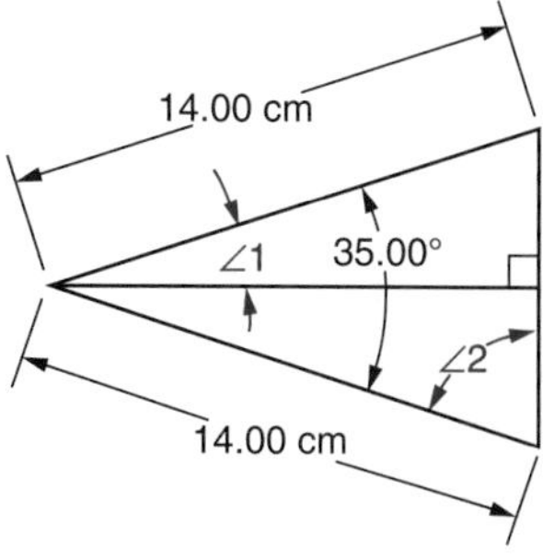

14. a. Given: $a = 8.400"$ and $b = 9.200"$. Find c. ____________

b. Given: $b = 90.00$ mm and $c = 150.00$ mm. Find a. ____________

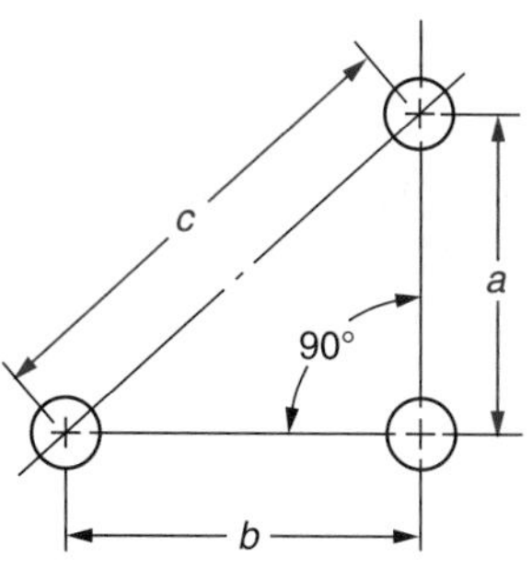

15. Compute $\angle 1$. ____________

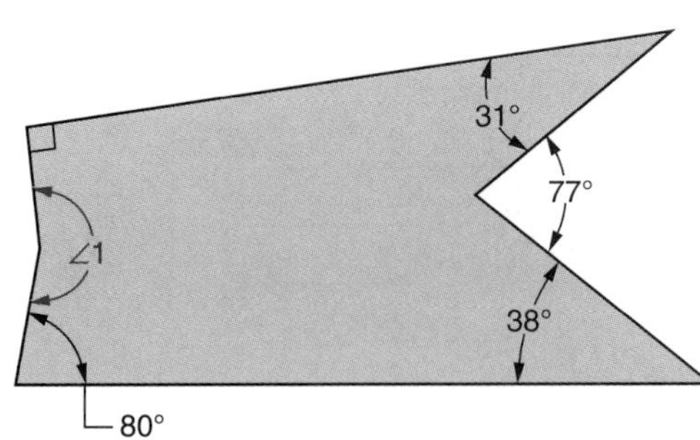

16. Determine the circumference of a circle that has a 5.360-inch radius. Round the answer to 3 decimal places. ____________

17. Determine the diameter of a circle that has a 360.00-millimeter circumference. Round the answer to 2 decimal places. ____________

18. a. Given: CD = 184 mm and $\overset{\frown}{CE}$ = 118 mm.

   Determine CF and $\overset{\frown}{CD}$.

   CF = ____________

   $\overset{\frown}{CD}$ = ____________

   b. Given: FD = 26 mm and $\overset{\frown}{CD}$ = 78 mm.

   Determine CD and $\overset{\frown}{ED}$.

   CD = ____________

   $\overset{\frown}{ED}$ = ____________

19. a. Given: EB = 5.150".

   Determine AE. ____________

   b. Given: AE = 4.200".

   Determine AB. ____________

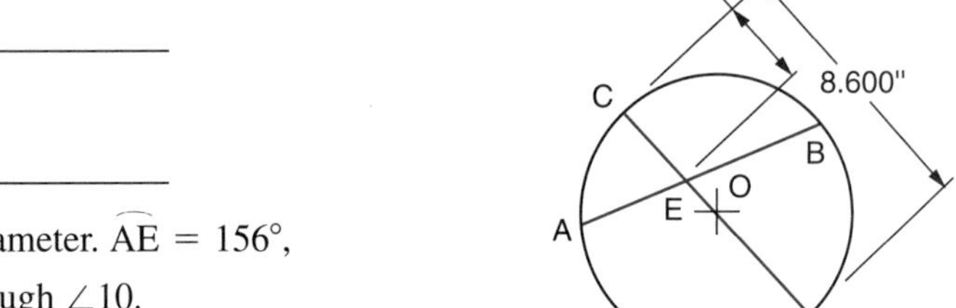

20. Given: Points A and E are tangent points. EB is a diameter. $\overset{\frown}{AE}$ = 156°, $\overset{\frown}{CE}$ = 140°, and $\overset{\frown}{ED}$ = 60°. Determine angles $\angle 1$ through $\angle 10$.

$\angle 1$ = ____________

$\angle 2$ = ____________

$\angle 3$ = ____________

$\angle 4$ = ____________

$\angle 5$ = ____________

$\angle 6$ = ____________

$\angle 7$ = ____________

$\angle 8$ = ____________

$\angle 9$ = ____________

$\angle 10$ = ____________

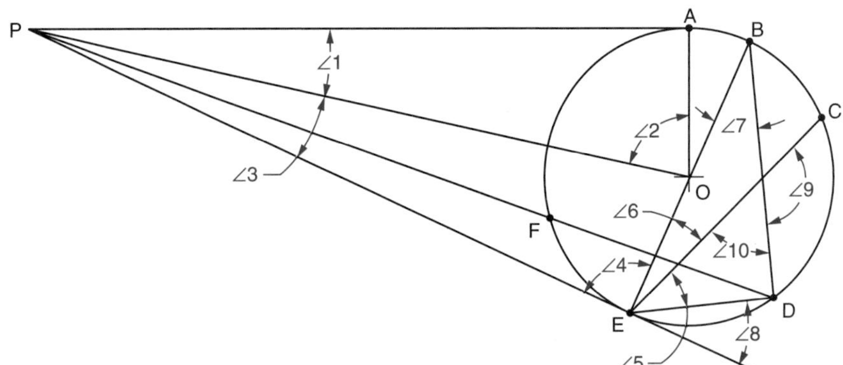

21. a. Given: $\overset{\frown}{AC}$ = 110° and r = 4.700".

   Compute arc length $\overset{\frown}{AC}$ to 3 decimal places. ____________

   b. Given: Arc length $\overset{\frown}{ABC}$ = 478.60 mm and r = 105.00 mm.

   Compute $\angle 1$ to 2 decimal places. ____________

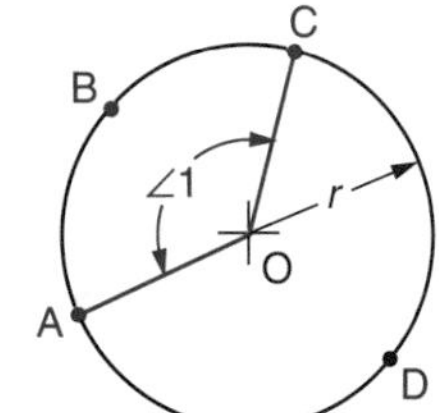

22. a. Given: Dia H = 14.520" and d = 8.300".
Compute Dia M. ____________

b. Given: Dia M = 36.900", e = 15.840", and d = 12.620".
Compute f. ____________

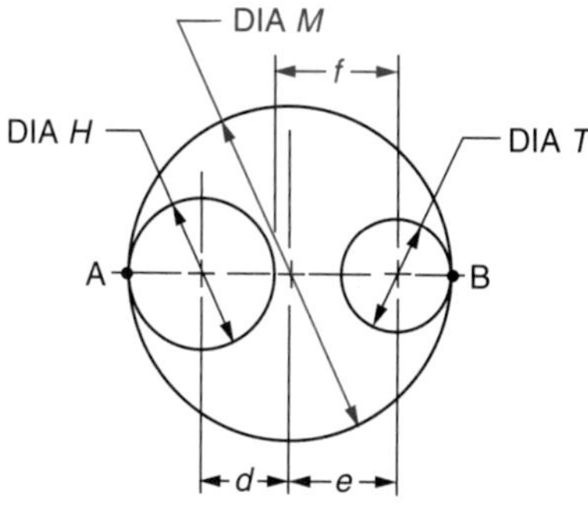

23. Given: ∠CAD = 38°, ∠BEC = 40°, $\overset{\frown}{AC}$ = 130°, and $\overset{\frown}{CE}$ = 134°. Determine angles ∠1 through ∠10.

∠1 = ____________
∠2 = ____________
∠3 = ____________
∠4 = ____________
∠5 = ____________
∠6 = ____________
∠7 = ____________
∠8 = ____________
∠9 = ____________
∠10 = ____________

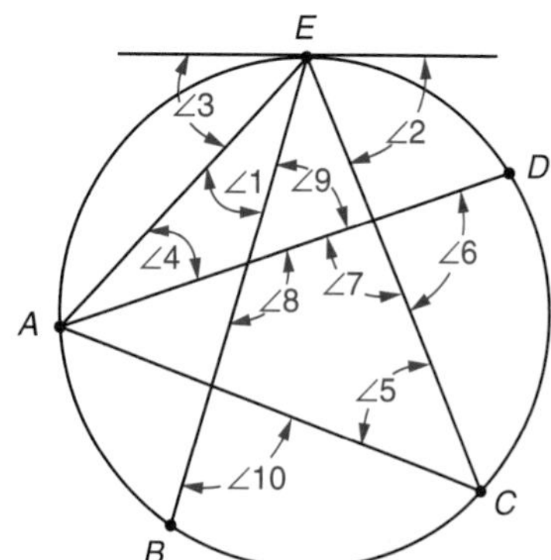

24. a. Given: x = 3.60 inches and y = 5.10 inches. Compute DIA A to 2 decimal places.
b. Given: DIA A = 8.76 inches and x = 10.52 inches. Compute t to 2 decimal places.

a. ____________
b. ____________

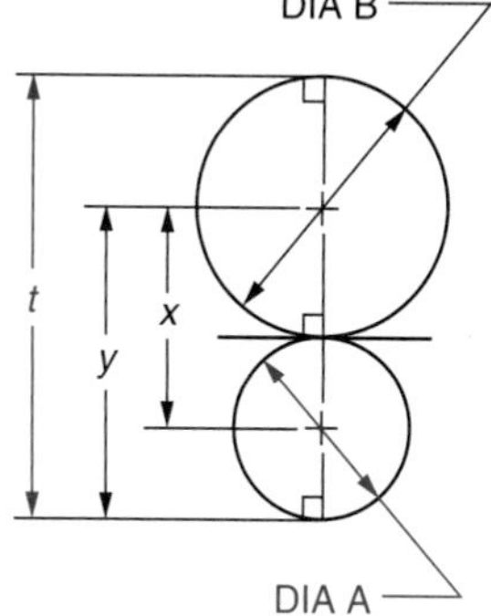

25. Points A and C are tangent points, DC is a diameter, $\overset{\frown}{AC}$ = 116°, $\overset{\frown}{EC}$ = 140°, $\overset{\frown}{EF}$ = 64°, and $\overset{\frown}{CH}$ = 42°. Determine angles ∠1 through ∠10.

∠1 = ____________
∠2 = ____________
∠3 = ____________
∠4 = ____________
∠5 = ____________
∠6 = ____________
∠7 = ____________
∠8 = ____________
∠9 = ____________
∠10 = ____________

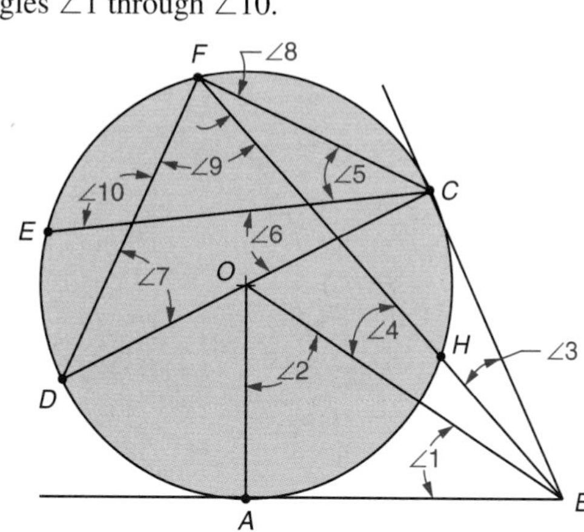

26. A flat is cut on a circular piece as shown. Determine the distance from the center of the circle to the flat, dimension *x*. ______

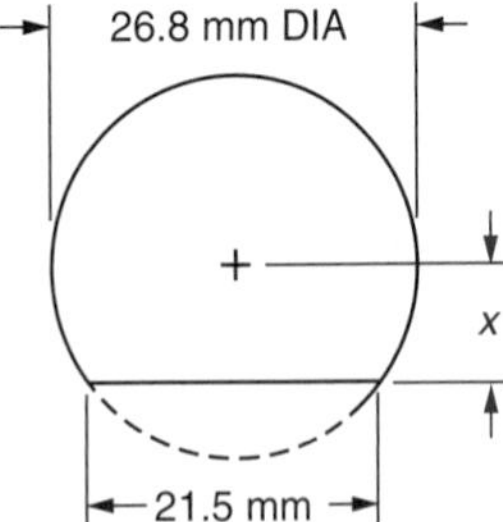

27. A spur gear is shown. Pitch circles of spur gears are the imaginary circles of meshing gears that make contact with each other. A pitch diameter is the diameter of a pitch circle. Circular pitch is the length of the arc measured on the pitch circle between the centers of two adjacent teeth. Determine the circular pitch of a spur gear that has 26 teeth and a pitch diameter of 4.1250 inches. Express the answer to 4 decimal places. ______

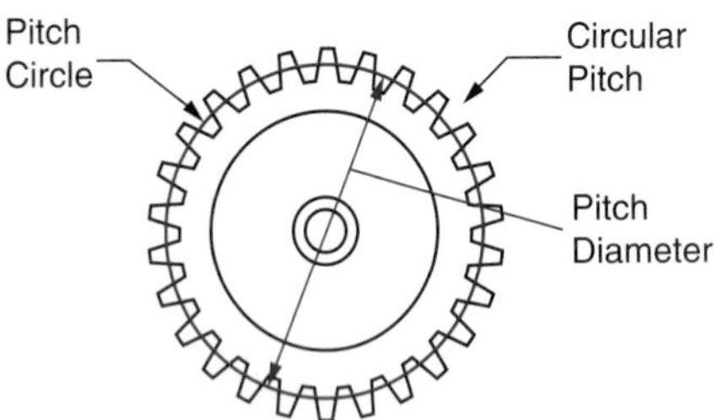

28. Determine the arc length from point *C* to point *D* on the template shown. ______

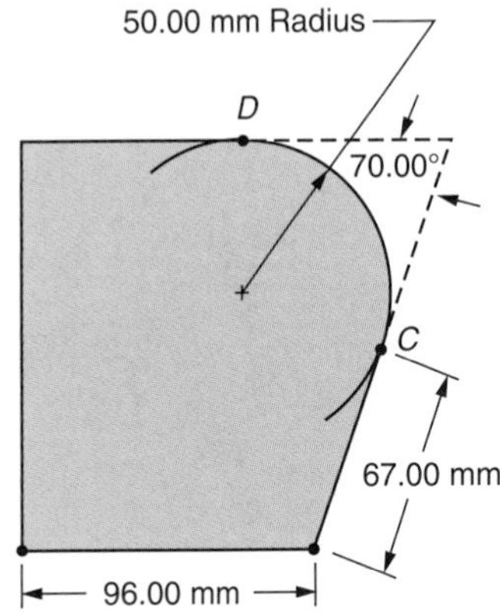

29. In the layout shown, points *E*, *F*, and *G* are tangent points. Determine lengths *OA*, *OB*, and *OC*.

OA = ______
OB = ______
OC = ______

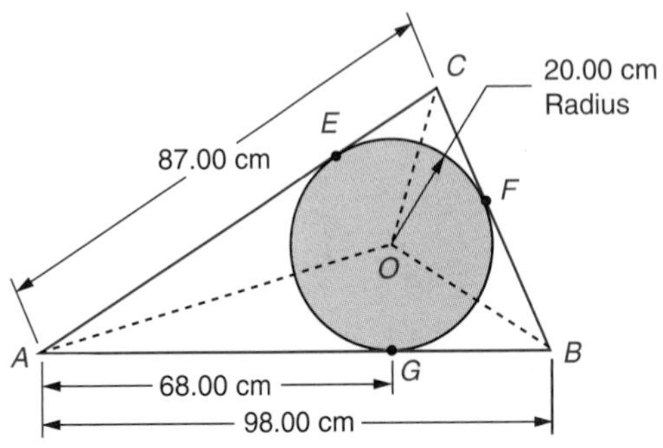

30. Determine dimension x to 3 decimal places. ____________

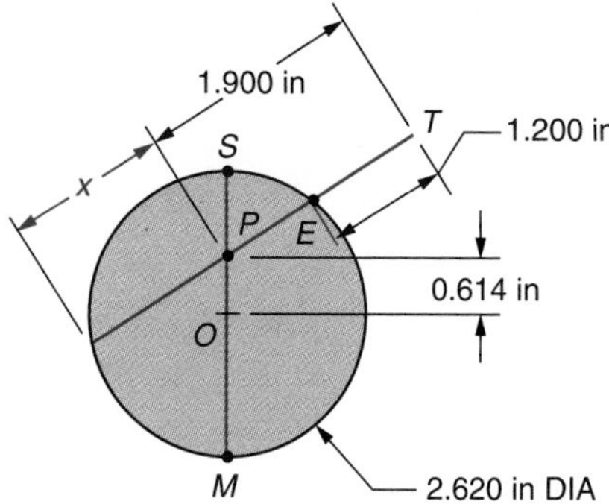

31. Refer to the drill jig shown. Determine ∠1. ____________

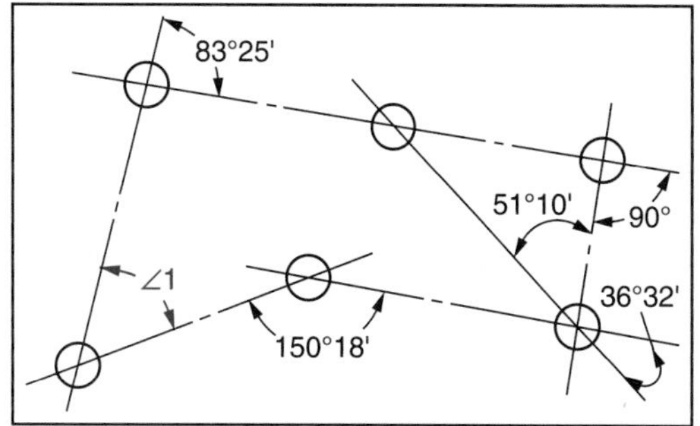

32. Circle O has a diameter AB of 14.50 inches and a chord CD of 8.00 inches. Chord CD is perpendicular to diameter AB. How far is chord CD from the center O of the circle? Express the answer to 2 decimal places. ____________

➤ **Note:** It is helpful to sketch and label this problem.

33. Refer to the figure shown. Determine dimension x. ____________

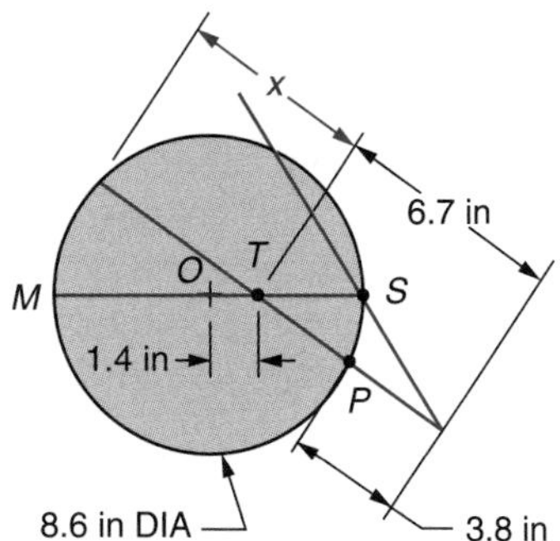

34. Lay out the template shown. Make the layout full size using construction methods. Do not use a protractor. All dimensions are in inches.

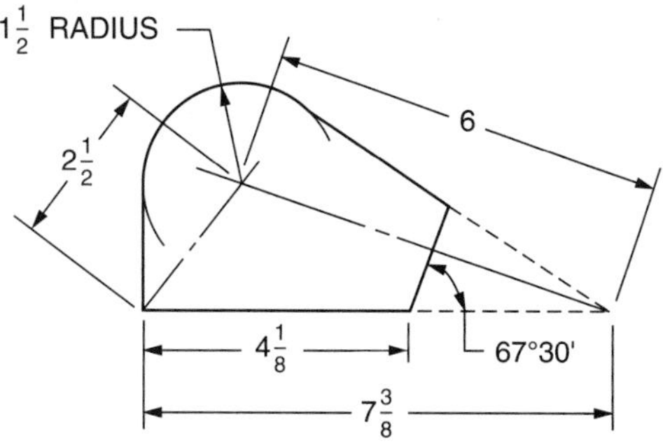

SECTION 6

Geometric Figures: Areas and Volumes

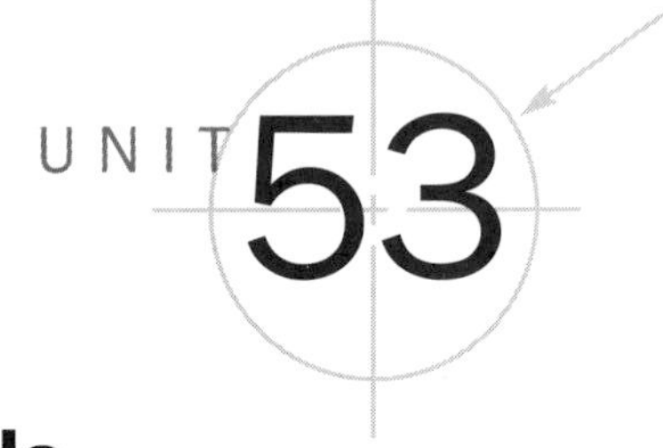

Areas of Rectangles, Parallelograms, and Trapezoids

Objectives ***After studying this unit you should be able to***

- Express given customary area measures in larger and smaller units.
- Express given metric area measures in larger and smaller units.
- Convert between customary area measures and metric area measures.
- Compute areas, lengths, and widths of rectangles.
- Compute areas, bases, and heights of parallelograms.
- Compute areas, both bases, and heights of trapezoids.
- Compute areas of more complex figures (composite figures) that consist of two or more common polygons.

As previously stated, in machine technology linear or length measure is used more often than area and volume measure. However, the ability to compute areas and volumes is required in determining job-material quantities and costs. Often, before a product is manufactured, part weights are computed. Volumes of simple geometric figures and combinations of figures (composite figures) must be calculated before weights can be determined.

Section 6 presents area and volume measure of two-dimensional and three-dimensional geometric figures and practical area and volume applications.

Customary and Metric Units of Surface Measure (Area)

A surface is measured by determining the number of surface units contained in it. A surface is two-dimensional. It has a length and width, but no thickness. Both length and width must be expressed in the same unit of measure. Area is computed as the product of two linear measures and is expressed in square units. For example, 2 inches $\times$ 4 inches $=$ 8 square inches (8 sq in or 8 in^2).

The surface enclosed by a square that is 1 inch on a side is 1 square inch (1 sq in or in^2). The surface enclosed by a square that is 1 foot on a side is 1 square foot (1 sq ft or 1 ft^2).

The reduced drawing in Figure 53-1 shows a square inch and a square foot. Observe that 1 linear foot = 12 linear inches, but 1 square foot = 12 inches × 12 inches or 144 square inches.

The table below has the common customary units of surface measure that might be used in a machine shop.

CUSTOMARY UNITS OF AREA MEASURE
1 square foot (sq ft or ft^2) = 144 square inches (sq in or in^2)
1 square yard (sq yd or yd^2) = 9 square feet (sq ft or ft^2)
1 square yard (sq yd or yd^2) = 1728 square inches (sq in or in^2)

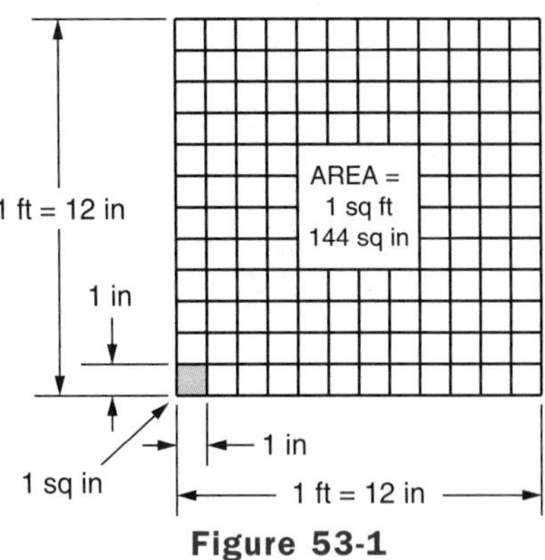

Figure 53-1

Expressing Customary Area Measure Equivalents To express a given customary unit of area as a larger customary unit of area, either divide the given area by the number of square units contained in one of the smaller units or multiply by a unit ratio.

Example Express 728 square inches as square feet.

METHOD 1

Since 144 sq in = 1 sq ft, divide 728 by 144.

728 ÷ 144 ≈ 5.06; 728 sq in ≈ 5.06 sq ft Ans

METHOD 2

The appropriate unit ratio is $\frac{1 \text{ sq ft}}{144 \text{ sq in}}$.

$$728 \cancel{\text{sq in}} \times \frac{1 \text{ sq ft}}{144 \cancel{\text{sq in}}} = \frac{728 \text{ sq ft}}{144} \approx 5.06 \text{ sq ft} \quad \text{Ans}$$

To express a given customary unit of area as a smaller customary unit of area, either multiply the given area by the number of square units contained in one of the larger units or multiply by a unit ratio.

Example Express 1.612 square yards as square inches.

Multiply 1.612 square yards by the two unit ratios $\frac{9 \text{ sq ft}}{1 \text{ sq yd}}$ and $\frac{144 \text{ sq in}}{1 \text{ sq ft}}$.

$$1.612 \cancel{\text{sq yd}} \times \frac{9 \cancel{\text{sq ft}}}{1 \cancel{\text{sq yd}}} \times \frac{144 \text{ sq in}}{1 \cancel{\text{sq ft}}} \approx 2089 \text{ sq in} \quad \text{Ans}$$

Metric Units of Surface Measure (Area)

The method of computing surface measure is the same in the metric system as it is in the customary system. The product of two linear measures produces square measure. The only difference is in use of metric rather then customary units. For example, 2 millimeters × 4 millimeters = 8 square millimeters.

Surface measure symbols are expressed as linear measure symbols with an exponent of 2. For example, 4 square meters is written as 4 m^2, 25 square millimeters is written 25 mm^2.

The basic metric unit of area is the square meter. The surface enclosed by a square that is 1 meter on a side is 1 square meter (1 m^2). The surface enclosed by a square that is 1 millimeter on a side is 1 square millimeter (1 mm^2).

One linear meter = 1000 linear millimeters, but 1 square meter (1 m^2) = 1000 millimeters × 1000 millimeters or 1 000 000 square millimeters (1 000 000 mm^2). Also, since 1 linear meter = 100 linear centimeters, 1 square meter = 100 centimeters × 100 centimeters or 10 000 square centimeters (10 000 cm^2).

To make numbers easier to read they may be divided into groups of three, separated by spaces (or thin spaces), as in 12 345, but not commas or points. This applies to digits on both sides of the decimal marker (0.901 234 56). Numbers with four digits may be written either with the space (5 678) or without it (5678).

This practice not only makes large numbers easier to read, but also allows all countries to keep their custom of using either a point or a comma as decimal marker. For example, engine size in the United States is written as 3.2 L and in Germany as 3,2 L. The space prevents possible confusion and sources of error.

The table below has the common metric units of surface measure that might be used in a machine shop.

METRIC UNITS OF AREA MEASURE
1 square millimeter (mm^2) = 0.000 001 square meter (m^2)
1 square millimeter (mm^2) = 0.01 square centimeter (cm^2)
1 square centimeter (cm^2) = 0.000 1 square meter (m^2)
1 000 000 square millimeters (mm^2) = 1 square meter (m^2)
100 square millimeters (mm^2) = 1 square centimeter (cm^2)
10 000 square centimeters (cm^2) = 1 square meter (m^2)

Expressing Metric Area Measure Equivalents To express a given metric unit of area as a larger metric unit of area, divide the given area by the number of square units contained in the smaller unit or multiply by the appropriate unit ratio.

Example Express 840.5 square centimeters as square meters.

METHOD 1

Since 10 000 cm^2 = 1 m^2, divide 840.5 by 10 000.

$840.5 \div 10\,000 = 0.08405$; $840.5 \text{ cm}^2 \approx 0.08 \text{ m}^2$ Ans

METHOD 2

The appropriate unit ratio is $\dfrac{1 \text{ m}^2}{10\,000 \text{ cm}^2}$.

$$840.5 \cancel{\text{cm}^2} \times \frac{1 \text{ m}^2}{10\,000 \cancel{\text{cm}^2}} = \frac{840.5 \text{ m}^2}{10\,000} = 0.08405 \text{ m}^2 \approx 0.08 \text{ m}^2 \quad \text{Ans}$$

To express a given metric unit of area as a smaller metric unit of area, multiply the given area by the number of square units contained in one of the larger units.

Example Express 47.6 square centimeters (cm^2) as square millimeters (mm^2). From the table, we see that 1 cm^2 = 100 mm^2.

$$47.6 \cancel{\text{cm}^2} \times \frac{100 \text{ mm}^2}{1 \cancel{\text{cm}^2}} = 47.6 \times 100 \text{ mm}^2 = 4760 \text{ mm}^2;\ 47.6 \text{ cm}^2 = 4760 \text{ mm}^2 \quad \text{Ans}$$

Conversion Between Metric and Customary Systems

In technical work it is sometimes necessary to change from one measurement system to the other. Use the following metric-customary conversions for the area of an object. Since 1 inch is defined to be 2.54 cm, the conversion between square inches and square centimeters is exact. The other conversions are approximations.

METRIC-CUSTOMARY AREA CONVERSIONS
1 square inch (sq in or in^2) = 6.4516 cm^2
1 square foot (sq ft or ft^2) ≈ 0.0929 m^2
1 square yard (sq yd or yd^2) ≈ 0.8361 m^2

Example Convert 12.75 ft^2 to square centimeters.

Since 12.75 ft^2 is to be expressed in square centimeters, multiply by the unit ratios $\frac{0.0929 \text{ m}^2}{1 \text{ ft}^2}$ and $\frac{10\,000 \text{ cm}^2}{1 \text{ m}^2}$.

$$12.75 \text{ ft}^2 = 12.75 \cancel{\text{ft}^2} \times \frac{0.0929 \cancel{\text{m}^2}}{1 \cancel{\text{ft}^2}} \times \frac{10\,000 \text{ cm}^2}{1 \cancel{\text{m}^2}} = 11\,844.75 \text{ cm}^2, \quad 11\,840 \text{ cm}^2 \quad \text{Ans}$$

Areas of Rectangles

A *rectangle* is a four-sided polygon with opposite sides equal and parallel and with each angle equal to a right angle. The *area of a rectangle* is equal to the product of the length and width.

$$A = lw \qquad \text{where } A = \text{area}$$
$$l = \text{length}$$
$$w = \text{width}$$

Example 1 A rectangle cross-section of a steel bar is 24 millimeters long and 14 millimeters wide.

Find the cross-sectional area of the bar.

$$A = lw$$
$$A = 24 \text{ mm} \times 14 \text{ mm}$$
$$A = 336 \text{ mm}^2 \quad \text{Ans}$$

Example 2 A metal stamping is shown in Figure 53-2 on the following page. All dimensions are in inches. Determine the area of the stamping.

Divide the figure into rectangles. One way of dividing the figure is shown in Figure 53-3. To find the cross-sectional area, compute the area of each rectangle and add the three areas.

- Area of rectangle ①

  $A = 10.500 \text{ in} \times 4.000 \text{ in}$

  $A = 42 \text{ sq in}$

- Area of rectangle ②

  Length = 10.500 in + 7.500 in = 18.000 in

  Width = 9.625 in − 4.000 in = 5.625 in

  $A = 18.000 \text{ in} \times 5.625 \text{ in} = 101.25 \text{ sq in}$

- Area of rectangle ③

  $A = 11.190 \text{ in} \times 4.500 \text{ in} = 50.355 \text{ sq in}$

Total cross-sectional area

42 sq in + 101.25 sq in + 50.355 sq in ≈ 193.6 sq in Ans (rounded)

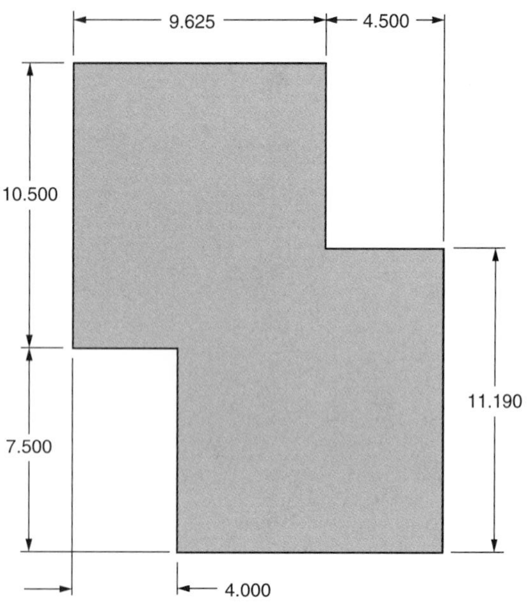

Figure 53-2

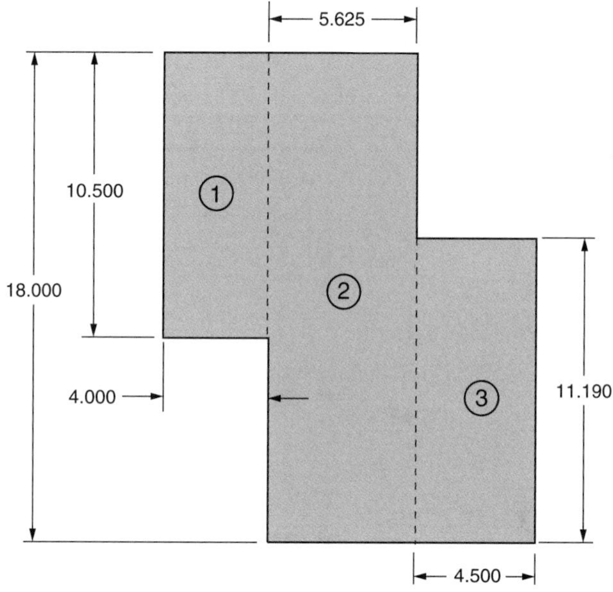

Figure 53-3

Areas of Parallelograms

A *parallelogram* is a four-sided polygon with opposite sides parallel and equal. The *area of a parallelogram* is equal to the product of the base and height. An *altitude* is a segment perpendicular to the line containing the base drawn from the side opposite the base. The *height* is the length of the altitude.

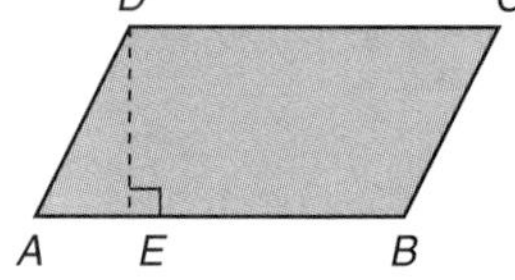

Figure 53-4

$$A = bh \qquad \text{where} \quad A = \text{area}, \quad b = \text{base}, \quad h = \text{height}$$

In Figure 53-4, *AB* is a base, and *DE* is a height of parallelogram *ABCD*.

$$\text{Area of parallelogram } ABCD = AB(DE)$$

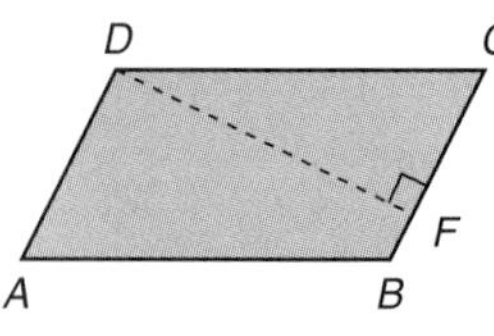

Figure 53-5

In Figure 53-5, *BC* is a base, and *DF* is a height of parallelogram *ABCD*.

$$\text{Area of parallelogram } ABCD = BC(DF)$$

Example 1 What is the area of a parallelogram with a 152.3 millimeter base and a 40.5 millimeter height?

$A = 152.3 \text{ mm} \times 40.5 \text{ mm}$

$= 6170 \text{ mm}^2$ Ans (rounded)

Example 2 A drawing of a baseplate is shown in Figure 53-6. The plate is made of number 2 gage (thickness) aluminum, which weighs 3.4 pounds per square foot. Find the weight of the plate to the nearest tenth pound. All dimensions are in inches.

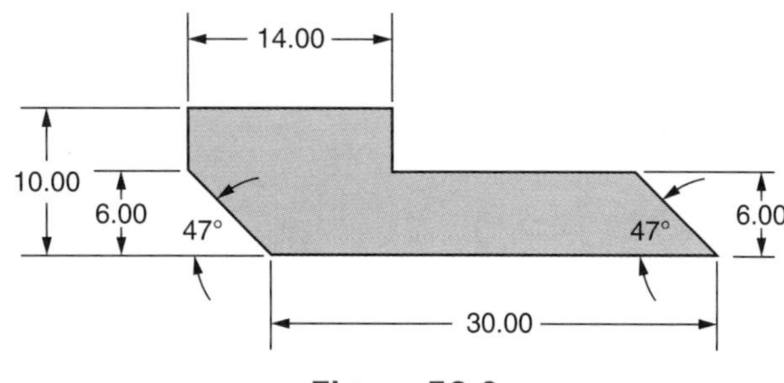

Figure 53-6

The area of the plate must be found. By studying the drawing, one method for finding the area is to divide the figure into a rectangle and a parallelogram as shown in Figure 53-7.

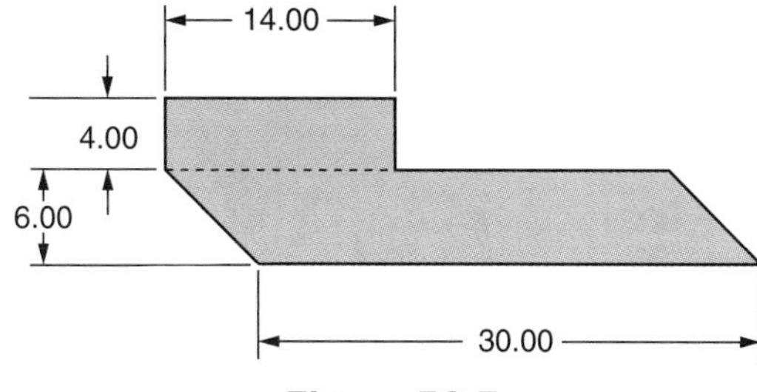

Figure 53-7

- Find the area of the rectangle.

  $A = 14.00 \text{ in} \times 4.00 \text{ in}$

  $A = 56 \text{ sq in}$

- Find the area of the parallelogram.

  $A = 30.00 \text{ in} \times 6.00 \text{ in}$

  $A = 180 \text{ sq in}$

- Find the total area of the plate.

  Total area = 56 sq in + 180 sq in = 236 sq in

Compute the weight.

- Find the area in square feet.

  Since 1 foot equals 12 inches, 1 square foot equals 12 inches squared.

  $(12 \text{ in})^2 = 144 \text{ in}^2$ or 144 sq in

  1 sq ft = 144 sq in

  236 sq in ÷ 144 sq in ≈ 1.64, 1.64 sq ft

  Weight of plate ≈ 1.64 sq ft × 3.4 lb/sq ft ≈ 5.6 lb Ans

Areas of Trapezoids

A *trapezoid* is a four-sided polygon that has only two sides parallel. The parallel sides are called *bases*. The *area of a trapezoid* is equal to one-half the product of the height and the sum of the bases.

$$A = \frac{1}{2}h(b_1 + b_2) \quad \text{where} \quad A = \text{area}, \quad h = \text{height}, \quad b_1 \text{ and } b_2 = \text{bases}$$

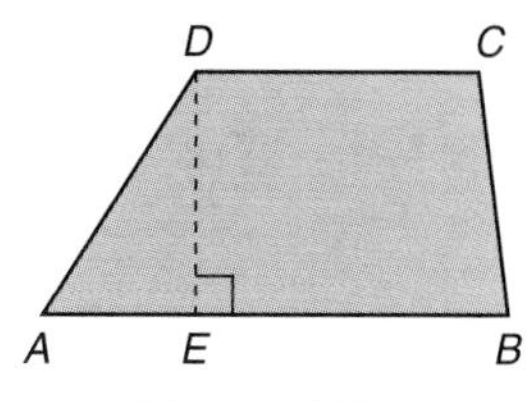

Figure 53-8

In Figure 53-8, DE is the height, and AB and DC are the bases of trapezoid $ABCD$.

$$\text{Area of trapezoid } ABCD = \frac{1}{2}DE\,(AB + DC)$$

Example 1 What is the area of a trapezoid that has bases of 7.000 inches and 3.800 inches and a height of 4.200 inches?

$A = \frac{1}{2}(4.200 \text{ in})(7.000 \text{ in} + 3.800 \text{ in})$

$A = \frac{1}{2}(4.200 \text{ in})(10.800 \text{ in})$

$A = 22.68$ sq in Ans

Example 2 The area of a trapezoid is 376.58 square centimeters. The height is 16.25 centimeters, and one base is 35.56 centimeters. Find the other base.

Substitute values in the formula for the area of a trapezoid and solve.

$376.58 \text{ cm}^2 = \frac{1}{2}(16.25 \text{ cm})(35.56 \text{ cm} + b_2)$

$376.58 \text{ cm}^2 = 8.125 \text{ cm } (35.56 \text{ cm} + b_2)$

$376.58 \text{ cm}^2 = 288.925 \text{ cm}^2 + 8.125 \text{ cm}(b_2)$

$87.655 \text{ cm}^2 = 8.125 \text{ cm}(b_2)$

$b_2 \approx 10.79 \text{ cm}$ Ans

Example 3 Find the area of the template shown in Figure 53-9. All dimensions are in inches. Express the answer to 1 decimal place.

Divide the template into simpler figures. One way is to divide the template into two rectangles and a trapezoid as shown in Figure 53-10.

- Find area ① (rectangle).

  $A = 4.20 \text{ in} \times 2.00 \text{ in}$

  $A = 8.40$ sq in

- Find area ② (trapezoid).

  Height = 8.30 in − 6.50 in = 1.80 in

  First base = 6.80 in

  Second base = 2.00 in

  $A = \frac{1}{2}(1.80 \text{ in})(6.80 \text{ in} + 2.00 \text{ in})$

  $A = \frac{1}{2}(1.80 \text{ in})(8.80 \text{ in})$

  $A = 7.92$ sq in

- Find area ③ (rectangle).

  $A = 6.50 \text{ in} \times 6.80 \text{ in}$

  $A = 44.2$ sq in

- Find the total area of template.

  Total Area = 8.40 sq in + 7.92 sq in + 44.2 sq in

  Total Area ≈ 60.5 sq in Ans (rounded)

  4.2 [×] 2 [+] .5 [×] 1.8 [×] [(] 6.8 [+] 2 [)] [+] (4.2 × 2: sq in ①; .5 × 1.8 × (6.8 + 2): sq in ②)

  6.5 [×] 6.8 [=] 60.52, 60.5 Ans (rounded) (6.5 × 6.8: sq in ③)

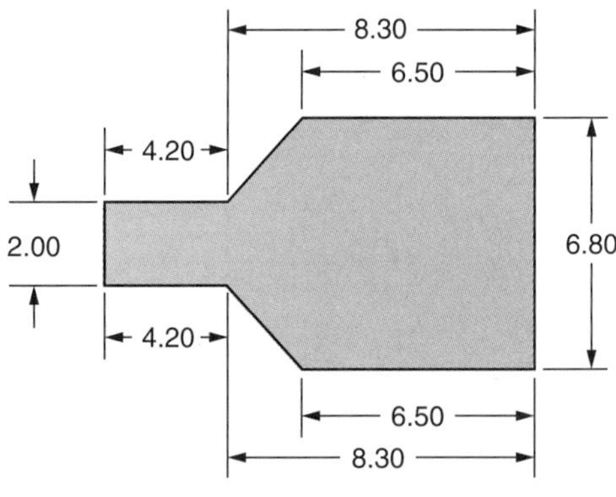

Figure 53-9

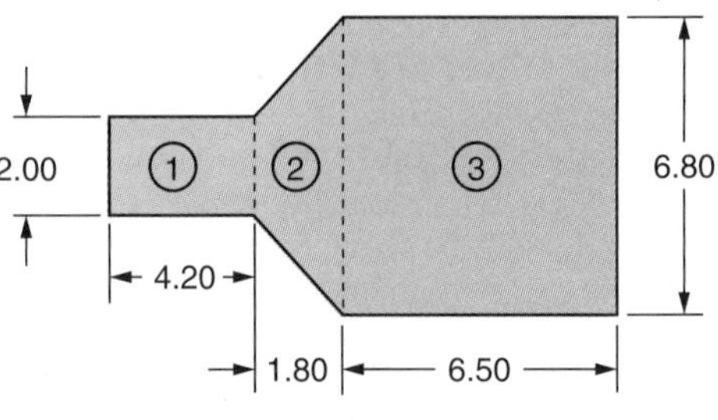

Figure 53-10

Application

Equivalent Customary Units of Area Measure

Express each area as indicated. Round each answer to the same number of significant digits as in the original quantity.

1. 196 square inches as square feet ______
2. 1085 square inches as square feet ______
3. 45.8 square feet as square yards ______
4. 2.02 square feet as square yards ______
5. 2300 square inches as square yards ______
6. 0.624 square foot as square inches ______
7. 4.30 square yards as square feet ______
8. 0.612 square yard as square inches ______

Equivalent Metric Units of Area Measure

Express each area as indicated. Round each answer to the same number of significant digits as in the original quantity.

9. 500 square millimeters as square centimeters ______
10. 2470 square millimeters as square centimeters ______
11. 38 250 square centimeters as square meters ______
12. 7520 square centimeters as square meters ______
13. 2.3 square meters as square millimeters ______
14. 5.74 square centimeters as square millimeters ______
15. 0.902 square centimeters as square millimeters ______
16. 0.0075 square meters as square millimeters ______

Conversion Between Metric and Customary Units of Area Measure

Express each area as indicated. Round each answer to the same number of significant digits as in the original quantity.

17. 18.5 square feet as square centimeters ______
18. 47.75 square inches as square millimeters ______
19. 3.9 square yards as square meters ______
20. 1.20 square feet as square millimeters ______
21. 680 square millimeters as square inches ______
22. 370.8 square inches as square centimeters ______
23. 18.75 m^2 as square feet ______
24. 18.75 m^2 as square yards ______

Areas of Rectangles

Find the unknown area, length, or width for each of the rectangles 25 through 36. Where necessary, round the answers to 1 decimal place.

	Length	Width	Area
25.	8.5 in	6.0 in	
26.	5.0 m	9.0 m	
27.	2.6 in		11.7 sq in
28.		23 mm	200.1 mm^2
29.	0.4 m		0.2 m^2
30.		0.086 in	0.136 sq in

	Length	Width	Area
31.	26.2 mm		366.8 mm^2
32.		39.8 in	31.7 sq in
33.	64.2 mm		3762 mm^2
34.	2.95 in	0.76 in	
35.	7.4 ft		6.7 sq ft
36.		125.0 ft	26,160 sq ft

Solve these problems. Round the answers to 2 decimal places unless otherwise specified.

37. A rectangular strip is 9.00 inches wide and 6.50 feet long. Find the area of the strip in square feet. ________

38. A sheet metal square contains 729 square inches. Find the length of each side of the square sheet. ________

39. The cost of a rectangular plate of aluminum 3'0" wide and 4'0" long is $45.00. Find the cost of a rectangular plate 6'0" wide and 8'0" long, using the same stock. ________

40. Find the area of the sheet metal pieces shown. ________

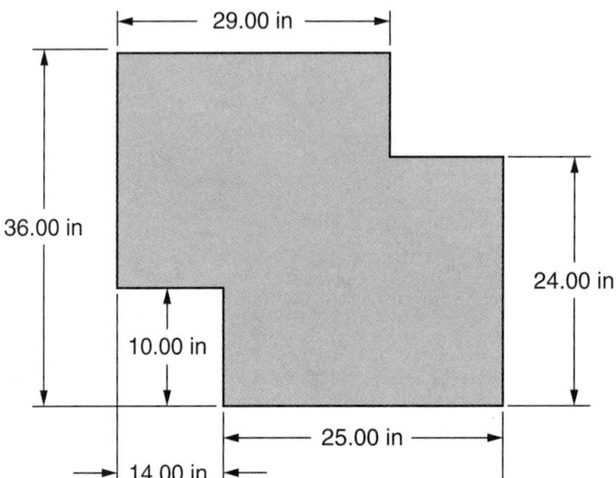

41. The support base shown has an area of 6350 square millimeters. Determine dimension x to the nearest tenth millimeter. ________

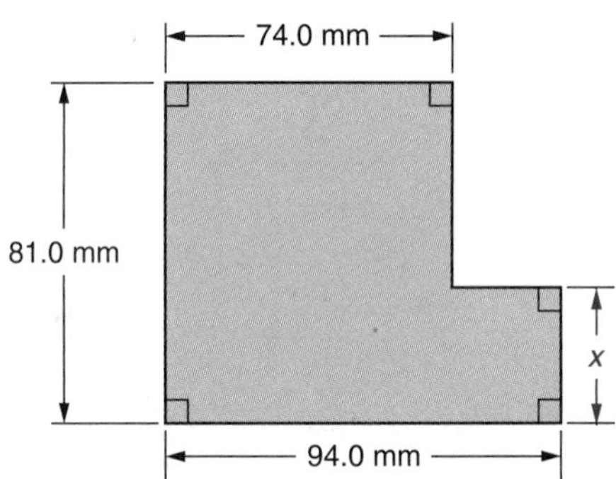

42. The rectangular cross-sectional area of a metal brace is to have an area of 2250 square millimeters. The length is to be one and one-half times the width. Compute the length and width dimensions to the nearest millimeter. ________

Areas of Parallelograms and Composite Figures

Find the unknown area, base, or height for each of the parallelograms 43 through 54. Where necessary, round the answers to one decimal place.

	Base	Height	Area
43.	20.00 mm	5.20 mm	
44.	6.00 in	9.80 in	
45.	26.0 in		486.2 sq in
46.		37.4 mm	2057 mm^2
47.	0.07 m		0.014 m^2
48.	24.0 in	4.50 in	

	Base	Height	Area
49.	18.5 in		312.5 sq in
50.		0.60 ft	5.1 sq ft
51.	56.00 mm	6.80 mm	
52.	17.00 in	18.30 in	
53.		38.0 mm	1887.2 mm^2
54.	0.38 m	0.266 m	

Solve these problems. Round the answers to two decimal places unless otherwise specified.

55. The cross section of the piece of tool steel shown is in the shape of a parallelogram. Find the cross-sectional area. All dimensions are in inches.

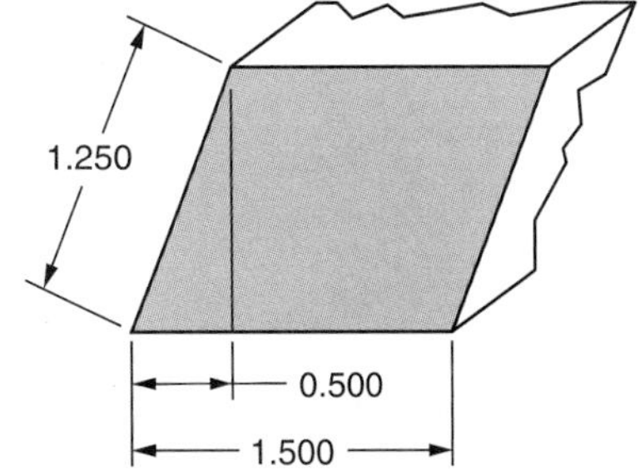

56. Two cutouts in the shape of parallelograms are stamped in a strip of metal as shown. Segment *AB* is parallel to segment *CD*, and dimension *E* equals dimension *F*. Compare the areas of the two cutouts.

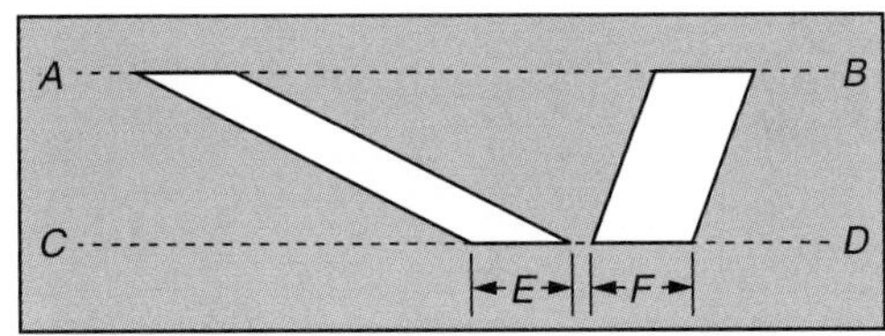

57. An oblique groove is cut in a block as shown. Before the groove was cut, the top of the block was in the shape of a rectangle. Determine the area of the top after the groove is cut.

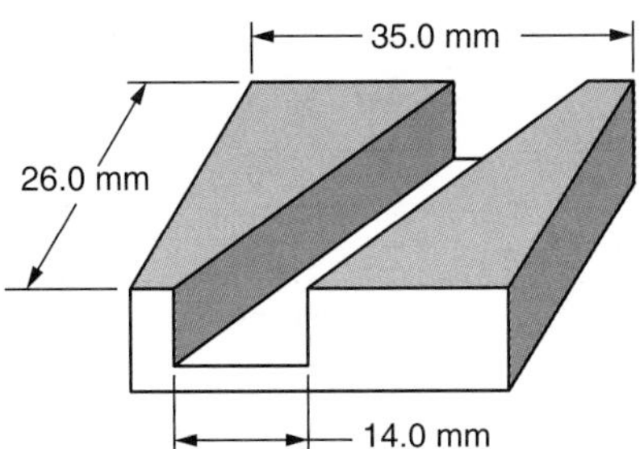

58. Find the area of the template shown. ____________

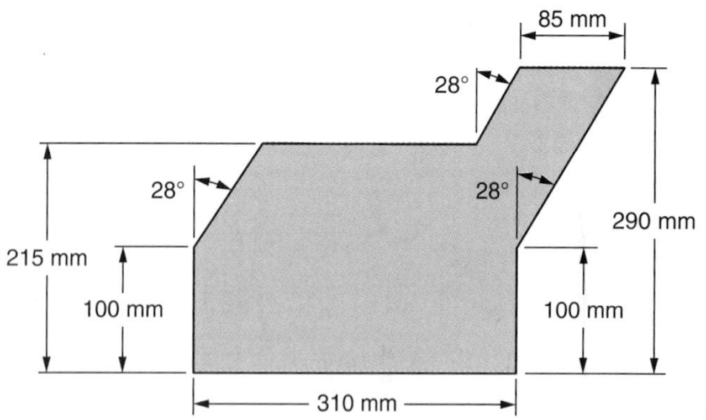

Areas of Trapezoids and Composite Figures

Find the unknown area, height, or base for each of the trapezoids 59 through 70. Where necessary, round the answers to 1 decimal place.

		Bases		
	Height (h)	b_1	b_2	Area (A)
59.	8.00 in	16.00 in	10.00 in	
60.	28.0 mm	47.0 mm	38.0 mm	
61.		8.00 ft	4.00 ft	64.0 sq ft
62.	1.2 ft		5.5 ft	7.7 sq ft
63.	0.6 m	0.8 m		0.4 m^2
64.		56.0 m	48.00 m	738.4 m^2

		Bases		
	Height (h)	b_1	b_2	Area (A)
65.	18.70 in	36.00 in	28.40 in	
66.	38.0 mm		8.7 mm	210.9 mm^2
67.	0.1 mm	1.2 m	0.6 m	
68.		66.37 in	43.86 in	2125 sq in
69.	0.3 m	0.8 m		0.2 m^2
70.	14.00 in	20.00 in	3.200 in	

71. A cross section of an aluminum bar in the shape of a trapezoid is shown. All the dimensions are in inches.

 a. Find the cross-sectional area of the bar. ____________

 b. Find the length of side AB. Round the answer to the nearest hundredth inch. ____________

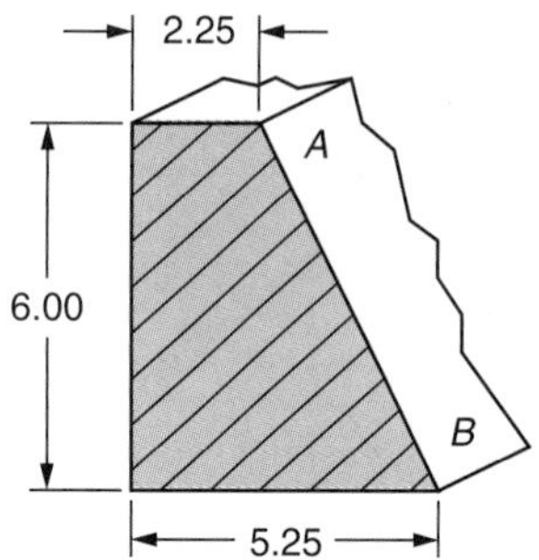

72. Determine the area of the metal plate shown. All dimensions are in inches. Round the answer to the nearest tenth square foot. ______

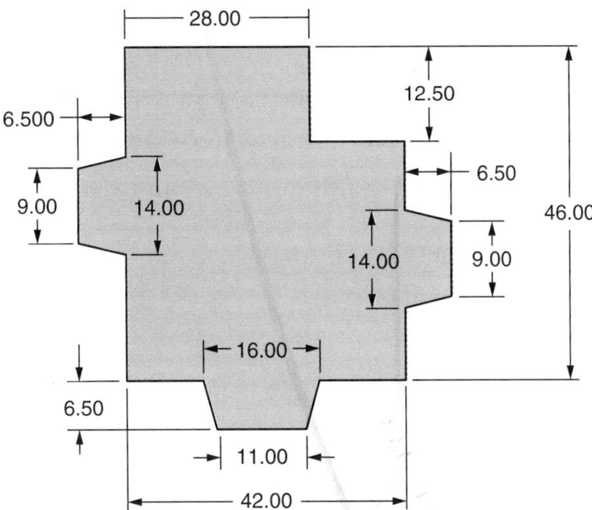

73. An industrial designer decided that the front plate of an appliance should be in the shape of an isosceles trapezoid with an area of 0.420 square meters. To give the desired appearance, the lower base dimension is to be equal to the height dimension, and the upper base dimension is to be equal to three-quarters of the lower base dimension. Compute the dimensions of the height and each base. Round answers to the nearest tenth millimeter.

74. One of the examples showed how to find the area of the metal stamping in Figure 53-2. Another method is to find the area of the large rectangular and subtract the areas of the rectangles that are removed. In Figure 53-11, the rectangles that will be removed are marked ① and ②. Use this subtraction method to determine the area of the stamping.

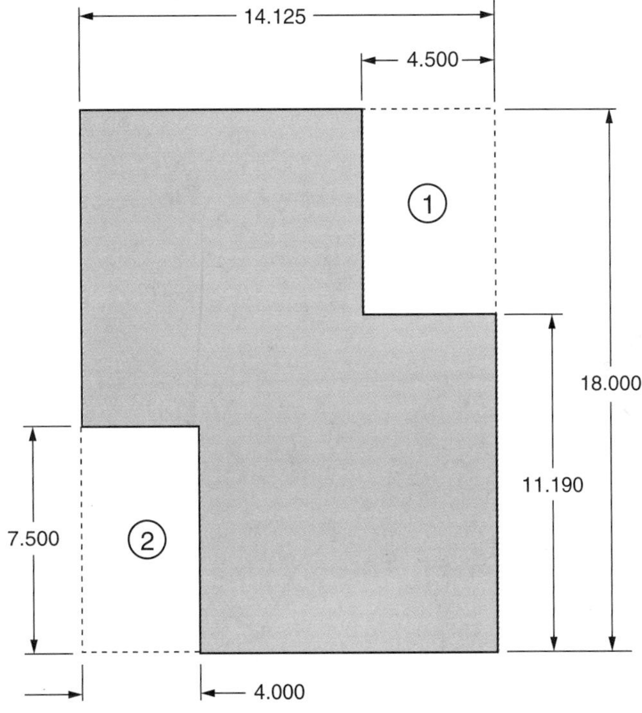

Figure 53-11

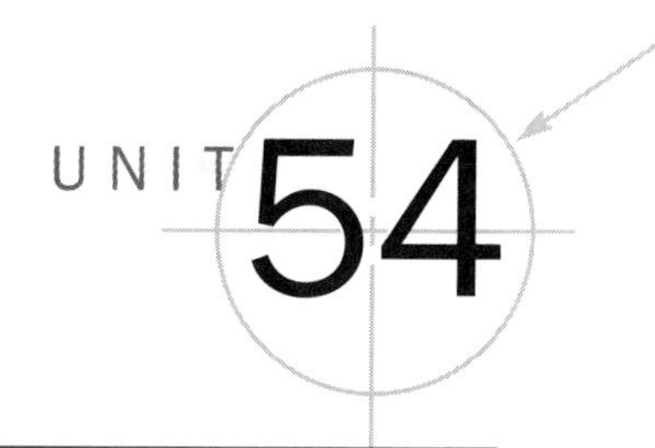

Areas of Triangles

Objectives ***After studying this unit you should be able to***

- Compute areas of triangles given the base and height.
- Compute areas of triangles given three sides.
- Compute bases and heights given triangle areas.
- Compute areas of more complex figures (composite figures) that consist of two or more common polygons.

Machined products are often in the shape of a triangle or can be divided into triangles, rectangles, parallelograms, and trapezoids.

Areas of Triangles Given the Base and Height

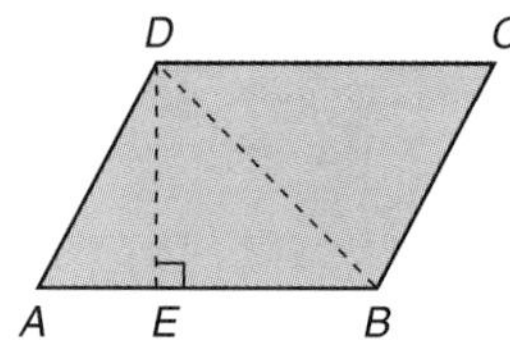

Figure 54-1

In parallelogram *ABCD* shown in Figure 54-1, segment *DE* is the altitude to the base *AB*. Diagonal *DB* divides the parallelogram into two congruent triangles.

$$\triangle ABD \cong \triangle CDB$$

Parallelogram *ABCD* and triangles *ABD* and *CDB* have equal bases and equal heights. The area of either triangle is equal to one-half the area of the parallelogram. The area of parallelogram $ABCD = AB(DE)$. Therefore, the area of $\triangle ABD$ or $\triangle CDB = \frac{1}{2}AB(DE)$. The *area of a triangle* is equal to one-half the product of the base and height.

$$A = \frac{1}{2}bh \qquad \text{where} \quad A = \text{area}, \; b = \text{base}, \; h = \text{height}$$

Example Find the area of the triangle shown in Figure 54-2.

$$A = \frac{1}{2}(22.0 \text{ mm})(19.0 \text{ mm})$$

$$A = 209 \text{ mm}^2 \quad \text{Ans}$$

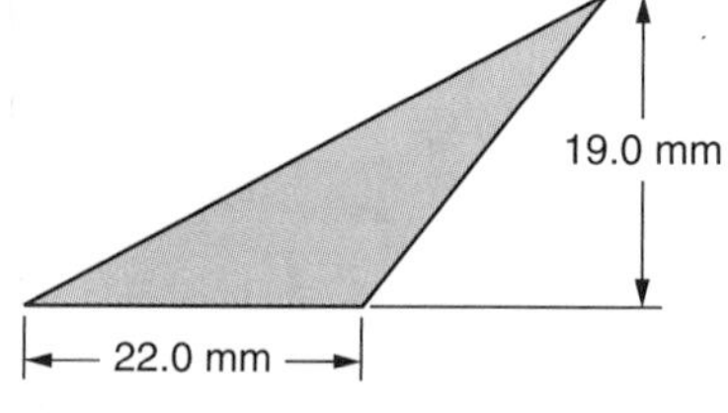

Figure 54-2

Areas of Triangles Given Three Sides

Often, three sides of a triangle are known, but a height is not known. A height can be determined by applying the Pythagorean theorem and a system of equations. However, with a calculator, it is quicker and easier to compute areas of triangles, given three sides using a formula called *Hero's* or *Heron's formula.*

Hero's (Heron's) Formula

$$A = \sqrt{s(s-a)(s-b)(s-c)} \qquad \text{where } A = \text{area}$$

$$a, b, \text{ and } c = \text{sides}$$

$$s = \frac{1}{2}(a+b+c)$$

Example Refer to the flat triangular brace shown in Figure 54-3.

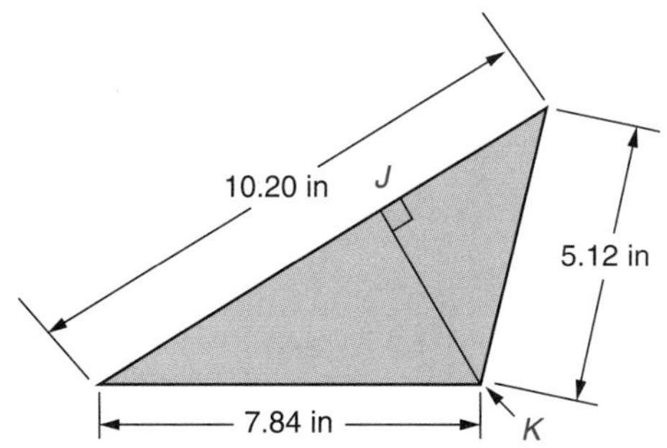

Figure 54-3

a. Find the area of the brace.

b. Find the height JK.

a. Compute the area by using Hero's formula.

$$s = \frac{1}{2}(7.84 \text{ in} + 5.12 \text{ in} + 10.20 \text{ in})$$

$$s = 11.58 \text{ in}$$

$$A = \sqrt{(11.58 \text{ in})(11.58 \text{ in} - 7.84 \text{ in})(11.58 \text{ in} - 5.12 \text{ in})(11.58 \text{ in} - 10.20 \text{ in})}$$

$$A = \sqrt{(11.58 \text{ in})(3.74 \text{ in})(6.46 \text{ in})(1.38 \text{ in})}$$

$$A \approx \sqrt{386.093 \text{ in}^4} \approx 19.65 \text{ sq in} \quad \text{Ans}$$

b. Compute height JK from the formula $A = \frac{1}{2}bh$.

$$19.65 \text{ sq in} \approx \frac{1}{2}(10.20 \text{ in})(JK)$$

$$19.65 \text{ sq in} \approx (5.10 \text{ in})(JK)$$

$$JK \approx 3.85 \text{ in} \quad \text{Ans}$$

a. Compute the area.

$s = .5$ ⊠ (7.84 ⊞ 5.12 ⊞ 10.2) $= 11.58$

√ (11.58 × (11.58 − 7.84) × (11.58 − 5.12) ×

(11.58 − 10.2)) = 19.64924569, $A \approx 19.65$ sq in Ans

b. Compute altitude JK.

$$JK = \frac{A}{\frac{1}{2}b}$$

$JK = 19.65$ ÷ (.5 × 10.2) = 3.852941176

$JK \approx 3.85$ in Ans

Application

Areas of Triangles and Composite Figures

Find the unknown area, base, or height for each of the triangles 1 through 12. Where necessary, round the answers to 1 decimal place.

	Base	Height	Area
1.	21.0 mm	17.0 mm	
2.		6.0 ft	78.2 sq ft
3.	0.2 m		0.02 m^2
4.		1.40 in	3.22 sq in
5.	0.8 m	0.4 m	
6.	18.5 in	7.25 in	

	Base	Height	Area
7.	30.5 mm		427 mm^2
8.		38.0 mm	1919 mm^2
9.	17.0 in	9.8 in	
10.		0.8 ft	16 sq ft
11.	45.41 in		249.7 sq in
12.		3.43 in	1.76 sq in

Given 3 sides, find the area of each of the triangles 13 through 20. Where necessary, round the answers to 1 decimal place unless otherwise specified.

	Side *a*	Side *b*	Side *c*
13.	4.00 in	6.00 in	8.00 in
14.	2.0 ft	5.0 ft	6.0 ft
15.	3.5 in	4.0 in	2.5 in
16.	20.0 mm	15.0 mm	25.0 mm

	Side *a*	Side *b*	Side *c*
17.	3.2 ft	3.6 ft	0.8 ft
18.	9.10 in	30.86 in	28.57 in
19.	7.20 mm	10.00 mm	9.00 mm
20.	0.5 m	1.0 m	0.8 m

13. ____________
14. ____________
15. ____________
16. ____________
17. ____________
18. ____________
19. ____________
20. ____________

Solve these problems.

21. Find the cross-sectional area of metal in the triangular tubing shown. Round the answer to the nearest whole square millimeter.

____

33.3 mm
22.4 mm
25.8 mm
38.4 mm

22. Two triangular pieces are sheared from the aluminum sheet shown. After the triangular pieces are cut, the sheet is discarded. Find the number of square feet of aluminum wasted. Round the answer to 1 decimal place.

____

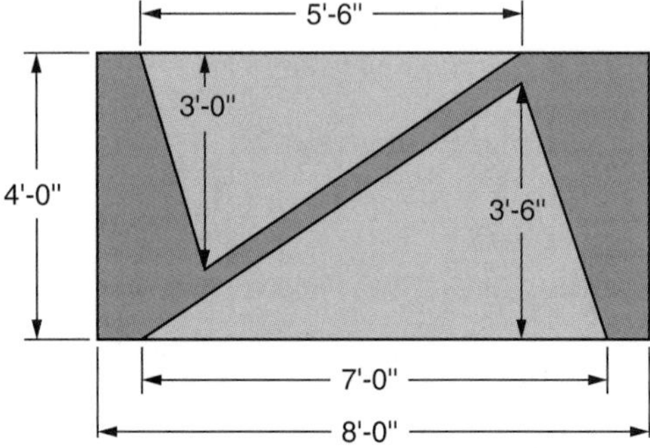

23. Determine the area of the figure shown. Round the answer to the nearest square millimeter. ______

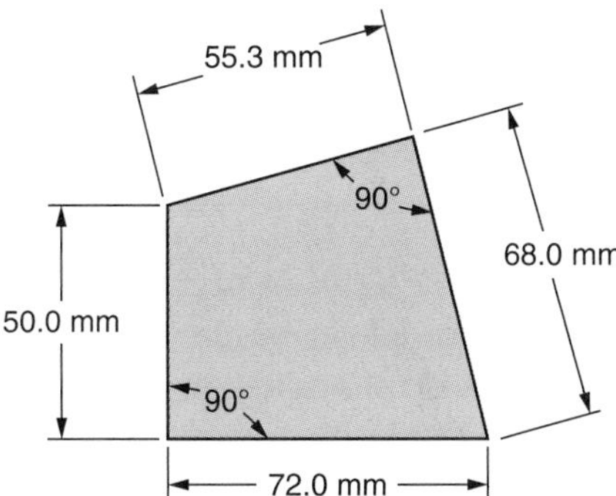

24. Find the area of the template shown. All dimensions are in inches. ______

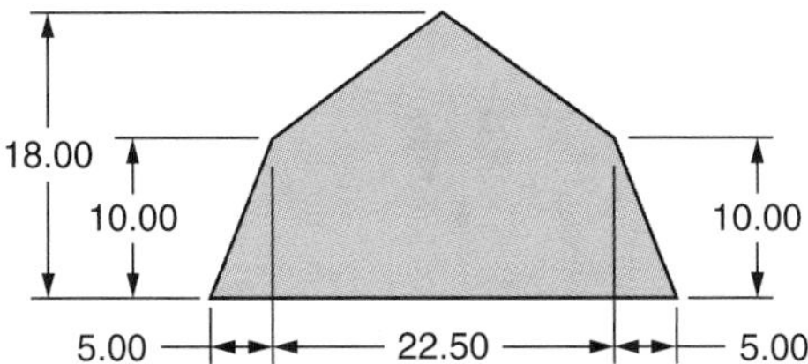

25. The area of the irregularly shaped sheet metal piece in the figure below on the left is to be determined. The longest diagonal is drawn on the figure as shown in the figure on the right. Perpendiculars are drawn to the diagonal from each of the other vertices. The perpendicular segments are measured as shown. From the measurements, the areas of each of the common polygons are computed. This is one method often used to compute areas of irregular figures. Compute the area of the sheet metal piece. ______

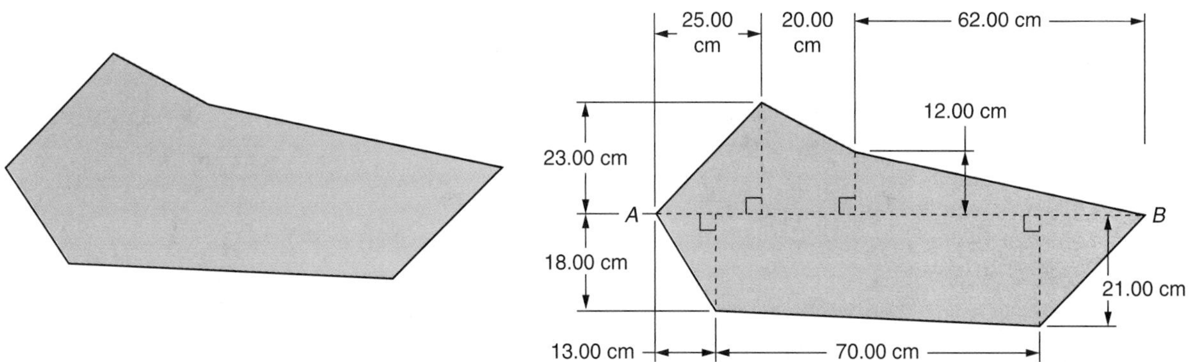

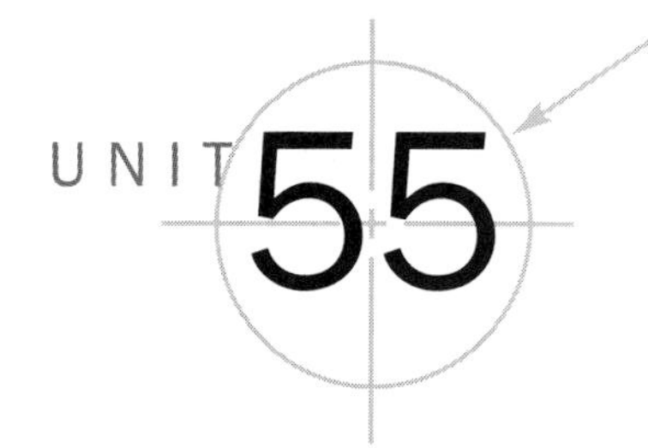

Areas of Circles, Sectors, and Segments

Objectives ***After studying this unit you should be able to***

- Compute areas, radii, and diameters of circles.
- Compute areas, radii, and central angles of sectors.
- Compute areas of segments.
- Compute areas of more complex figures that consist of two or more simple figures.

Computations of areas of circular objects are often made when planning for the production of a product. Also, many industrial material-strength calculations are based on circular cross-sectional areas of machined parts.

Areas of Circles

The *area of a circle* is equal to the product of π and the square of the radius.

$$A = \pi r^2 \qquad \text{where } A = \text{area}, \quad r = \text{radius}, \quad \pi \approx 3.1416$$

The formula for the area of a circle can be expressed in terms of the diameter. Since the radius is one-half the diameter, $\frac{d}{2}$ can be substituted in the formula for r.

$$A = \pi\left(\frac{d}{2}\right)^2 \text{ or } A = \pi\frac{d^2}{4}$$

Since $\frac{\pi}{4} \approx 0.7854$, the formula $A \approx 0.7854\, d^2$ is often used.

Example 1 Find the area of a circle that has a radius of 6.500 inches.

Substitute the values in the formula and solve.

$A = \pi r^2 \approx 3.1416(6.500 \text{ in})^2 \approx 3.1416(42.25 \text{ sq in}) \approx 132.7 \text{ sq in}$ Ans

Example 2 A circular hole is cut in a square metal plate as shown in Figure 55-1. The plate weighs 8.3 pounds per square foot. What is the weight of the plate after the hole is cut? All dimensions are in inches.

Compute the area of the square.
$A = (10.30 \text{ in})^2 = 106.09 \text{ sq in}$

Compute the area of the hole.
$A \approx 0.7854(7.00 \text{ in})^2 \approx 38.48 \text{ sq in}$

Compute the area of the plate.
$A \approx 106.09 \text{ sq in} - 38.48 \text{ sq in} \approx 67.61 \text{ sq in}$

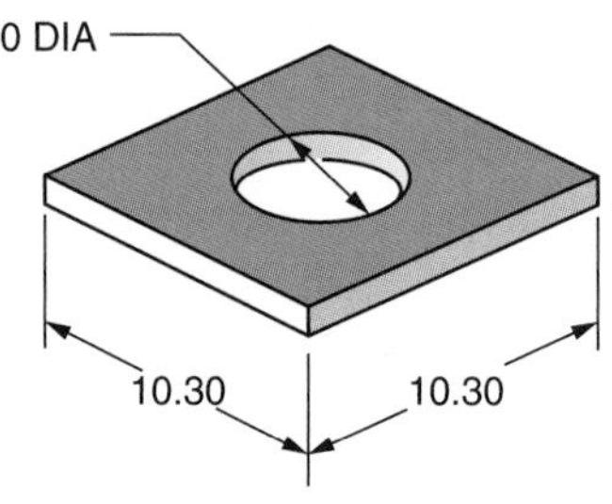

Figure 55-1

Compute the weight of the plate.
67.61 sq in ÷ 144 sq in/sq ft ≈ 0.470 sq ft

Weight ≈ 0.470 × 8.3 lb ≈ 3.9 lb Ans

(10.3 x^2 − .7854 × 7 x^2) ÷ 144 × 8.3 = 3.896700139
Weight ≈ 3.9 lb Ans

Areas of Sectors

A *sector* of a circle is a figure formed by two radii and the arc intercepted by the radii. To find the *area of sector* of a circle, first find the fractional part of a circle represented by the central angle. Then multiply the fraction by the area of the circle.

$$A = \frac{\theta}{360^\circ}(\pi r^2) \quad \text{where } A = \text{area}$$

θ = cental angle
$\pi \approx 3.1416$
r = radius

Example 1 A base plate in the shape of a sector is shown in Figure 55-2. Find the area of the plate to the nearest hundredth square foot.

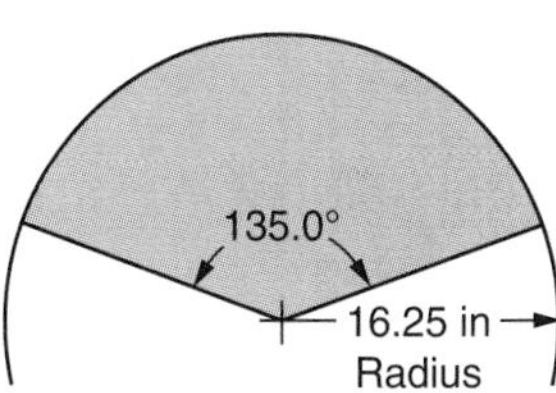

Figure 55-2

Find the area of the sector in square inches.

$$A = \frac{135.0^\circ}{360^\circ}(3.1416)(16.25 \text{ in})^2 \approx 311.09 \text{ sq in}$$

Find the area in square feet.
Since 1 sq ft = 144 sq in, 311.09 ÷ 144 ≈ 2.16 = 2.16 sq ft Ans (rounded)

135 ÷ 360 × π × 16.25 x^2 ÷ 144 =
2.160356276, 2.16 sq ft Ans (rounded)

Example 2 Pieces in the shape of sectors are to be stamped from sheet stock. Each piece is to have an area of 1080 square millimeters and a central angle of 222 degrees. Compute the length of the straight sides (radii) of a piece.

$$1080 \text{ mm}^2 = \frac{222}{360}(3.1416)r^2$$
$$1080 \text{ mm}^2 \approx 1.93732r^2$$
$$r^2 \approx 557.471 \text{ mm}^2$$
$$r \approx 23.6 \text{ mm}$$

length of straight sides ≈ 23.6 mm Ans

Areas of Segments

A *segment* of a circle is a figure formed by an arc and the chord joining the end points of the arc. In the circle shown in Figure 55-3, *the area of segment ACB* is found by subtracting the area of triangle *AOB* from the area of sector *OACB*.

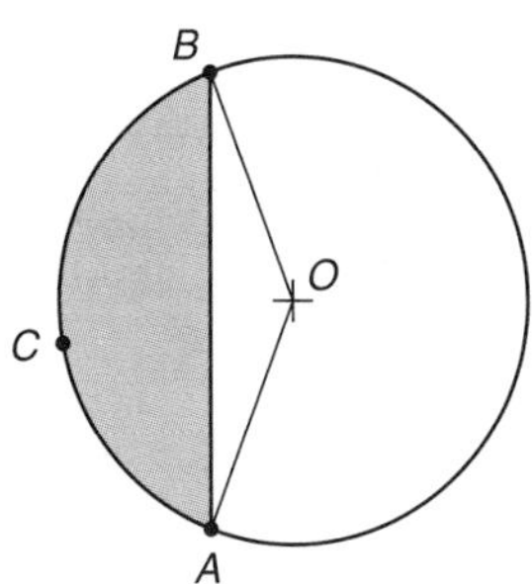

Figure 55-3

Example Segment *ACB* is cut from the circular plate shown in Figure 55-4. Find the area of the segment.

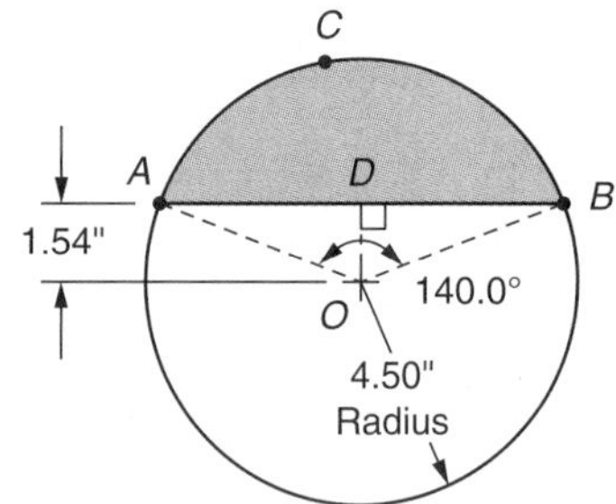

Figure 55-4

Find the area of the sector.

$$A \approx \frac{140°}{360°}(3.1416)(4.50 \text{ in})^2 \approx 24.740 \text{ sq in}$$

Find the area of isosceles triangle *AOB*.

$$c^2 = a^2 + b^2$$
$$(4.50 \text{ in})^2 = (1.54 \text{ in})^2 + AD^2$$
$$20.25 \text{ sq in} = 2.3716 \text{ sq in} + AD^2$$
$$AD^2 = 17.8784 \text{ sq in}$$
$$AD \approx 4.228 \text{ in}$$

The base *AB* of isosceles triangle *AOB* is bisected by altitude *DO*.

$AB = 2(AD) \approx 2(4.228 \text{ in}) \approx 8.456 \text{ in}$

$A \approx 0.5(8.456 \text{ in})(1.54 \text{ in}) \approx 6.511 \text{ sq in}$

Find the area of segment *ACB*.

$A \approx 24.740 \text{ sq in} - 6.511 \text{ sq in} \approx 18.2 \text{ sq in}$ Ans

Application

Areas of Circles and Composite Figures

Find the unknown area, radius, or diameter for each of the circles 1 through 12. Where necessary, round the answers to 1 decimal place.

	Radius	Diameter	Area
1.	7.000 in	—	
2.	10.80 cm	—	
3.	—	15.5 in	
4.	—	17.23 mm	
5.		—	3.4 sq ft
6.	—		218.3 cm^2

	Radius	Diameter	Area
7.		—	380.0 mm^2
8.	2.78 in	—	
9.	—	0.75 ft	
10.	—		102.6 cm^2
11.		—	36.8 sq in
12.	0.026 m	—	

13. A rectangular steel plate 15.10 inches long and 12.45 inches wide weighs 19.4 pounds. Find the weight of the plate after three 4.65-inch diameter holes are cut. Round the answer to the nearest tenth pound. ____________

14. Hydraulic pressure of 705.0 pounds per square inch is exerted on a 3.150-inch diameter piston. Find the total force exerted on the piston. Round the answer to the nearest pound. ____________

15. A circular base is shown. The base is cut from a steel plate that weighs 34 kilograms per square meter of surface area. Find the weight of the circular base. Round the answer to the nearest tenth kilogram. ______

➤ **Note:** A kilogram is a unit of weight (mass) measure in the metric system. One kilogram (kg) weighs approximately 2.2046 pounds.

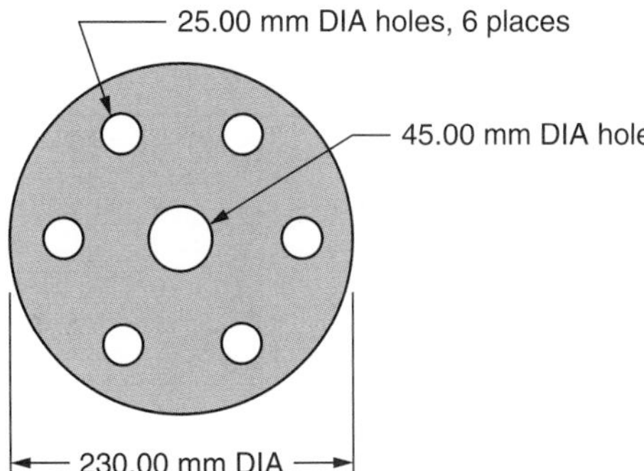

16. Find the area of the template shown. Round the answer to the nearest tenth inch. All dimensions are in inches. ______

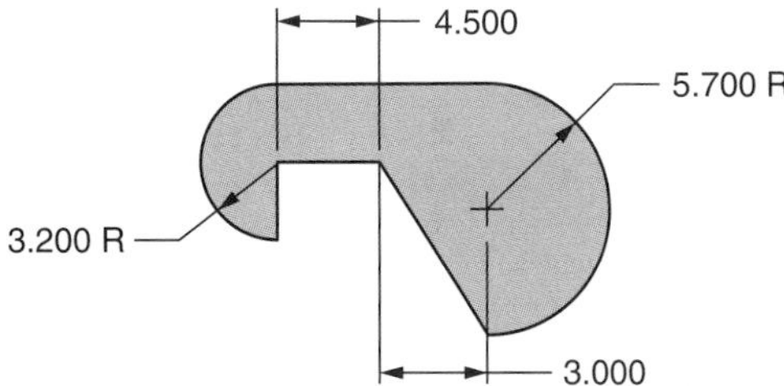

17. A force of 62,125 pounds pulls on a steel rod that has a diameter of 1800 inches. Find the force pulling on one square inch of the cross-sectional area. Round the answer to the nearest pound. ______
18. A piece shown by the shaded portion is to be cut from a square plate 128 millimeters on a side.

    a. Compute the area of the piece to be cut. Round the answer to the nearest square millimeter. ______

    b. After cutting the piece, determine the percentage of the plate that will be wasted. ______

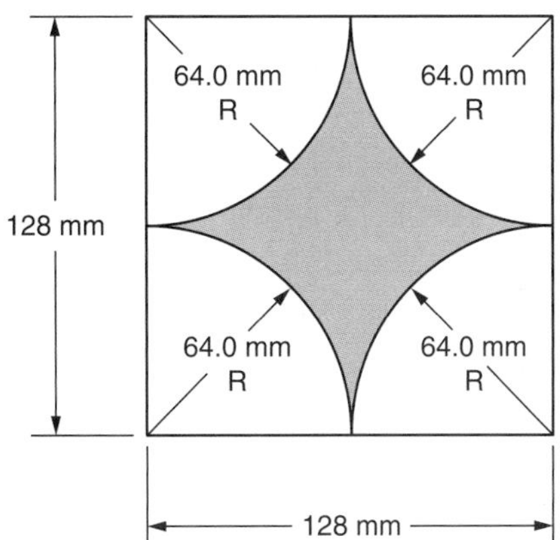

Areas of Sectors and Composite Figures

Find the unknown area, radius, or central angle for each of these sectors 19 through 30. Where necessary, round the answers to 1 decimal place.

	Radius	Central Angle	Area
19.	10.00 cm	120.0°	
20.	3.5 ft	90.0°	
21.	15.3 in	40.00°	
22.		65.0°	300.0 sq in
23.		180.5°	750.0 mm^2
24.	9.570 in		94.62 sq in

	Radius	Central Angle	Area
25.	20.25 in		1028 sq in
26.	0.2 m	220°	
27.	54.08 cm	26.30°	
28.		307.2°	79.4 sq in
29.	3.273 in		15.882 sq in
30.	150.78 mm	15.286°	

31. A cross section of a piece of round stock with a V-groove cut is shown. Find the cross-sectional area of the stock. Round the answer to the nearest square millimeter. __________

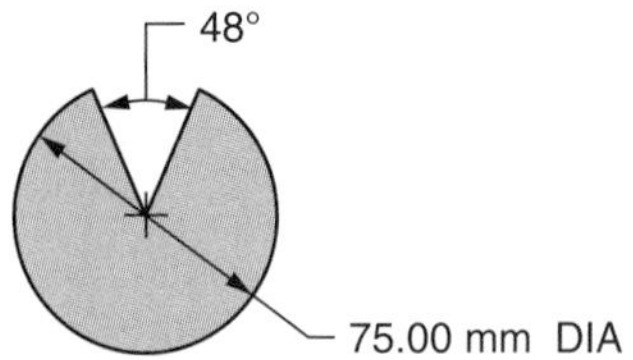

32. Three pieces, each in the shape of a sector, are cut from the rectangular sheet of steel shown. How many square meters are wasted after the 3 pieces are cut? Round the answer to the nearest tenth meter. __________

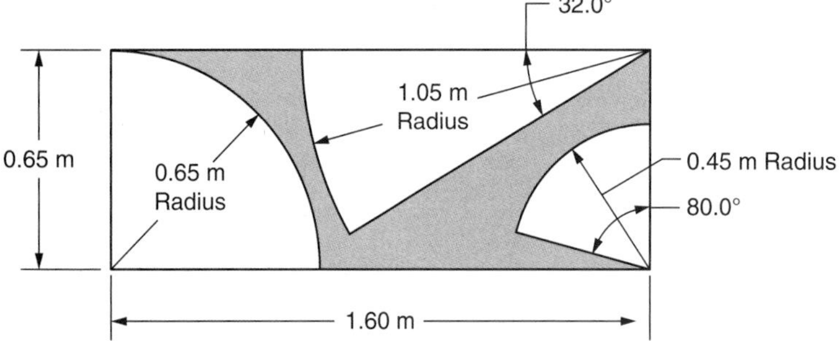

33. Determine the cross-sectional area of the machined part shown. Dimensions are in inches. Round the answer to the nearest tenth square inch. __________

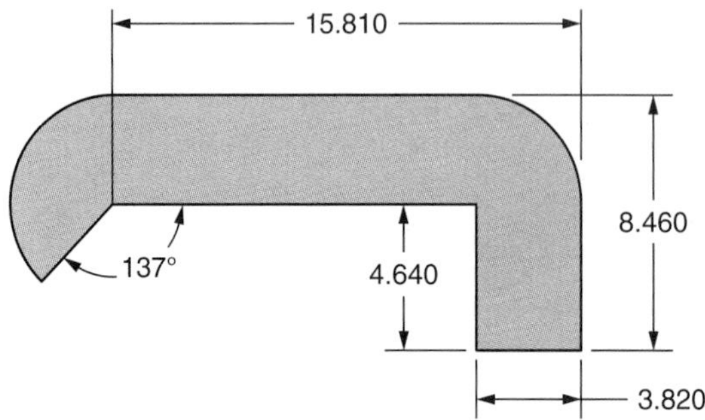

34. Find the area of the top view of the piece shown. Dimensions are in inches. Round the answer to 1 decimal place. ____________

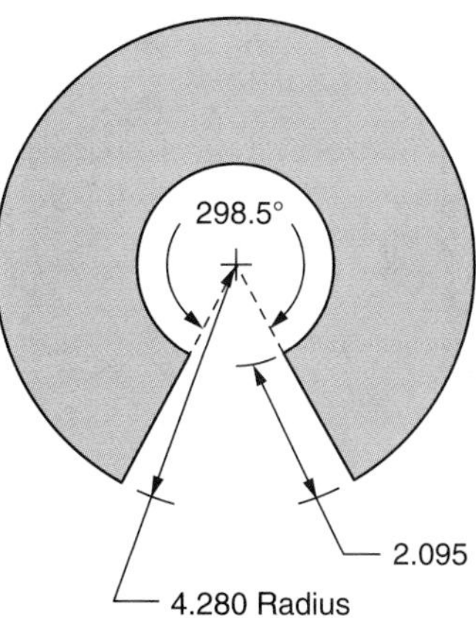

Area of Segments and Composite Figures

Find the area of each of the segments *ACB* for exercises 35 through 39.

	Area of Isosceles △*AOB*	Area of Sector *OACB*	Area of Segment *ACB*
35.	1.65 sq ft	2.48 sq ft	
36.	6.98 sq in	9.35 sq in	
37.	156 mm^2	213.5 mm^2	
38.	85.33 sq in	109.27 sq in	
39.	0.26 m^2	0.39 m^2	

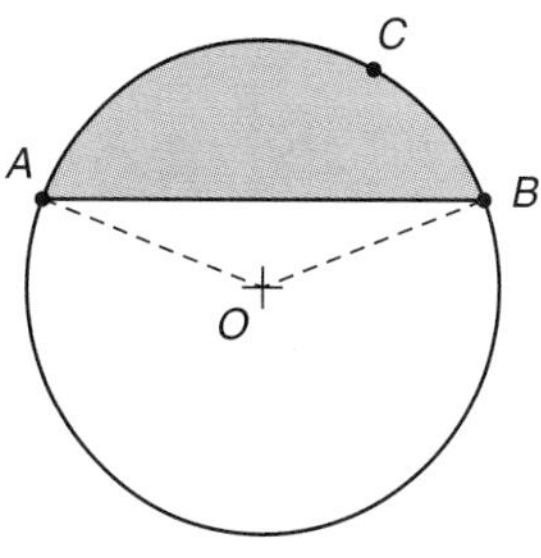

40. Find the area of the shaded segment shown. Round the answer to 1 decimal place. ____________

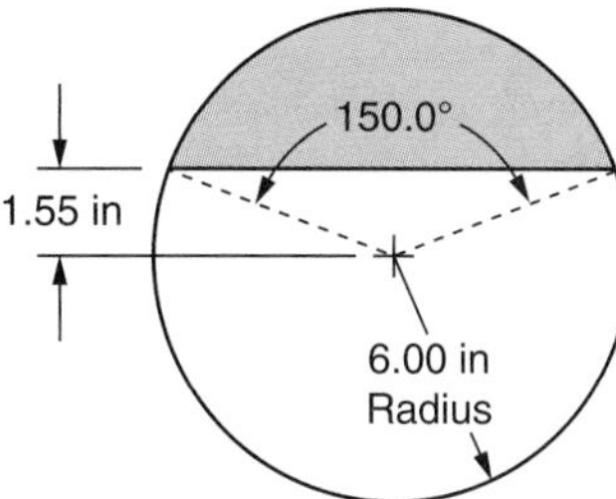

41. Compute the area of the steel insert (shaded segment) shown. Round the answer to 1 decimal place. ____________

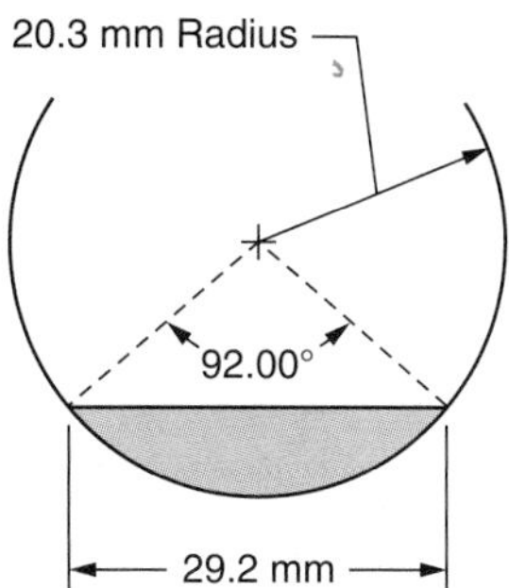

42. A pattern is shown.

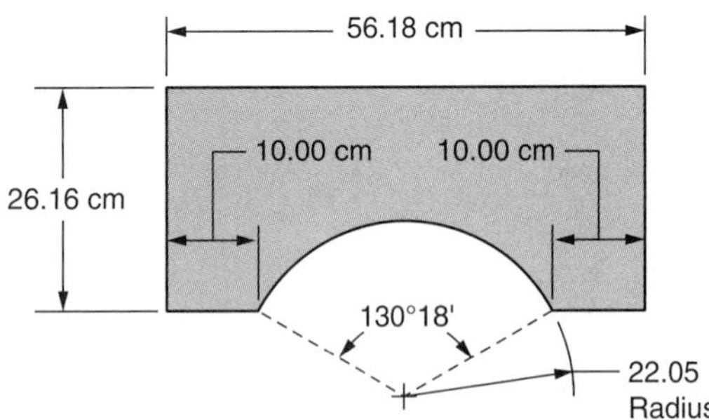

a. Find the surface area of the pattern. Round the answer to the nearest square centimeter. ______

b. The metal from which the pattern is made weighs 7.85 kilograms per square meter of area. Find the weight of the pattern. Round the answer to the nearest hundredth kilogram. ______

➤ **Note:** Since 100 cm = 1 m, $(100\text{ cm})^2 = 1\text{ m}^2$ or $10\,000\text{ cm}^2 = 1\text{ m}^2$.

43. The shaded piece shown is cut from a circular disk. Find the area of the piece. Round the answer to 1 decimal place. Dimensions are in inches. ______

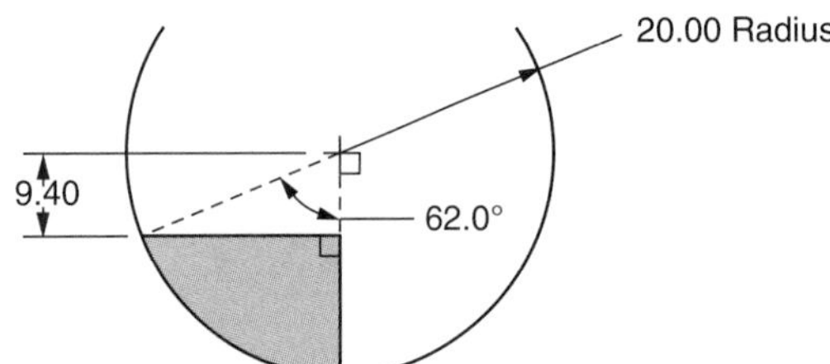

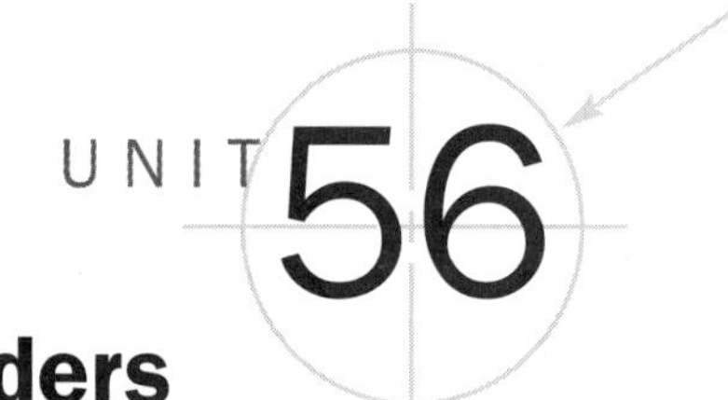

Volumes of Prisms and Cylinders

Objectives ***After studying this unit you should be able to***

- Express given customary volume measures in larger and smaller units.
- Express given metric volume measures in larger and smaller units.
- Convert between customary volume measures and metric measures.
- Compute volumes of prisms and cylinders.
- Compute heights and base areas of prisms and cylinders.
- Compute weights of prisms and cylinders.

Manufactured products are often in the shape of prisms and cylinders. Engine pistons, shafts, tubes, pipes, and food containers are a few examples of cylinders. Lengths of bar stock with triangular, square, rectangular, and hexagonal cross sections are some examples of prisms.

Many objects consist of a combination of prisms and cylinders or components of prisms and cylinders.

Customary and Metric Unit of Volume (Cubic Measure)

A solid is measured by determining the number of cubic units contained in it. A solid is three-dimensional; it has length, width, and thickness or height. Length, width, and thickness must be expressed in the same unit of measure. Volume is the product of three linear measures and is expressed in cubic units. For example, 2 inches × 3 inches × 5 inches = 30 cubic inches.

Customary Units of Volume Measure

The volume of a cube having sides 1 foot long is 1 cubic foot. The volume of a cube having sides 1 inch long is 1 cubic inch. A similar meaning is attached to the cubic yard.

A reduced illustration of a cubic foot and a cubic inch is shown in Figure 56-1. Observe that 1 linear foot equals 12 linear inches, but 1 cubic foot equals 12 inches × 12 inches × 12 inches, or 1728 cubic inches.

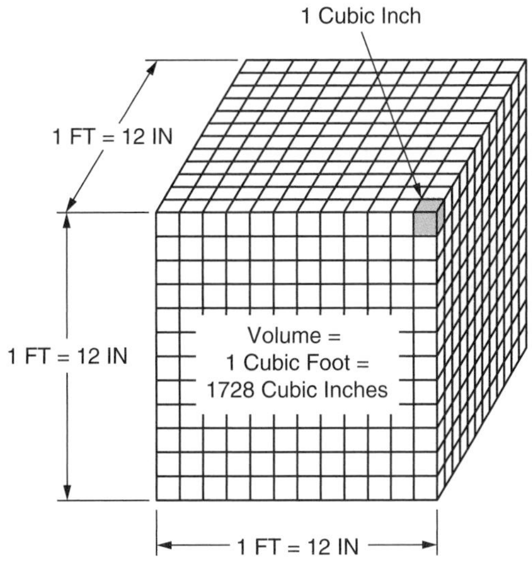

Figure 56-1

The table below has the common customary units of volume measure that might be used in a machine shop.

CUSTOMARY UNITS OF VOLUME MEASURE
1 cubic foot (cu ft or ft^3) = 1728 cubic inches (cu in or in^3)
1 cubic yard (cu yd or yd^3) = 27 cubic feet (cu ft or ft^3)

Expressing Customary Volume Equivalents To express a given customary unit of volume as a larger customary unit of volume, either divide the given area by the number of cubic units contained in one of the smaller units or multiply by a unit ratio.

Example Express 4310 cubic inches as cubic feet.

METHOD 1

Since 1728 cu in = 1 cu ft, divide 4310 by 1728.

4310 ÷ 1728 ≈ 2.494; 4310 cu in ≈ 2.49 cu ft Ans

METHOD 2

The appropriate unit ratio is $\frac{1 \text{ cu ft}}{1728 \text{ cu in}}$.

$$4310 \cancel{\text{sq in}} \times \frac{1 \text{ cu ft}}{1728 \cancel{\text{cu in}}} = \frac{4310 \text{ sq ft}}{1728} \approx 2.49 \text{ sq ft} \quad \text{Ans}$$

To express a given customary unit of volume as a smaller customary unit of volume, either multiply the given area by the number of cubic units contained in one of the larger units or multiply by a unit ratio.

Example Express 0.0197 cubic yards as cubic inches.

Multiply 0.0197 cubic yards by the two unit ratios $\frac{27 \text{ cu ft}}{1 \text{ cu yd}}$ and $\frac{1728 \text{ cu in}}{1 \text{ cu ft}}$.

$$0.0197 \cancel{\text{cu yd}} \times \frac{27 \cancel{\text{cu ft}}}{1 \cancel{\text{cu yd}}} \times \frac{1728 \text{ cu in}}{1 \cancel{\text{cu ft}}} \approx 919 \text{ cu in} \quad \text{Ans}$$

Metric Units of Volume Measure

The method of computing volume measure is the same in the metric system as in the customary system. The product of three linear measures produces cubic measure. The only difference is in the use of metric rather than customary units. For example, 2 millimeters × 3 millimeters × 5 millimeters = 30 cubic millimeters.

Volume measure symbols are expressed as linear measure symbols with an exponent of 3. For example, 6 cubic meters is written as 6 m^3, and 750 cubic millimeters is written as 750 mm^3.

The basic unit of volume is the cubic meter. The volume of a cube having sides 1 meter long is 1 cubic meter (1 m^3). The volume of a cube having sides 1 millimeter long is 1 cubic millimeter (1 mm^3).

One linear meter = 1000 linear millimeters, but 1 cubic meter (1 m^3) = 1000 millimeters × 1000 millimeters × 1000 millimeters or 1 000 000 000 cubic millimeters (1 000 000 000 mm^3) or 10^9 mm^3. Also, since one linear meter = 100 linear centimeters, 1 cubic meter = 100 centimeters × 100 centimeters × 100 centimeters or 1 000 000 cubic centimeters (1 000 000 cm^3) or 10^6 cm^3.

The table below has the common metric units of volume measure that might be used in a machine shop.

METRIC UNITS OF VOLUME MEASURE
1 cubic millimeter (mm^3) = 0.000 000 001 cubic meter (m^3)
1 cubic millimeter (mm^3) = 0.001 cubic centimeter (cm^3)
1 cubic centimeter (cm^3) = 0.000 001 cubic meter (m^3)
1 000 000 000 cubic millimeters (mm^3) = 1 cubic meter (m^3)
1000 cubic millimeters (mm^3) = 1 cubic centimeter (cm^3)
1 000 000 cubic centimeters (cm^3) = 1 cubic meter (m^3)

Expressing Metric Volume Measure Equivalents To express a given metric unit of volume as a larger metric unit of volume, divide the given volume by the number of cubic units contained in the smaller unit or multiply by the appropriate unit ratio.

Example Express 1840.5 cubic centimeters as cubic meters.

METHOD 1

Since 1 000 000 cm^3 = 1 m^3, divide 1840.5 by 1 000 000.

1840.5 ÷ 1 000 000 = 0.00184; 1840.5 cm^3 ≈ 0.00184 m^3 Ans

METHOD 2

The appropriate unit ratio is $\frac{1\text{ m}^3}{1\,000\,000\text{ cm}^3}$.

$$1840.5\ \cancel{\text{cm}^3} \times \frac{1\text{ m}^3}{1\,000\,000\ \cancel{\text{cm}^3}} = \frac{1840.5\text{ m}^3}{1\,000\,000} = 0.0018405\text{ m}^3 \approx 0.00184\text{ m}^3 \quad \text{Ans}$$

To express a given metric unit of volume as a smaller metric unit of volume, multiply the given volume by the number of cubic units contained in one of the larger units.

Example Express 81.6 cubic centimeters (cm^3) as cubic millimeters (mm^3).

From the table, we see that $1\text{ cm}^3 = 1000\text{ mm}^3$.

$$81.6\ \cancel{\text{cm}^3} \times \frac{1000\text{ mm}^3}{1\ \cancel{\text{cm}^3}} = 81.6 \times 1000\text{ mm}^3 = 81\,600\text{ mm}^3;\ 81.6\text{ cm}^3 = 81\,600\text{ mm}^3 \quad \text{Ans}$$

Conversion Between Metric and Customary Systems

In technical work it is sometimes necessary to change from one measurement system to the other. Use the following metric-customary conversions for the volume of an object.

METRIC-CUSTOMARY VOLUME CONVERSIONS
1 cubic inch (cu in or in^3) $\approx$ 16387 mm^3
1 cubic inch (cu in or in^3) $\approx$ 16.387 cm^3
1 cubic foot (cu ft or ft^3) $\approx$ 0.0283 m^3
1 cubic yard (cu yd or yd^3) $\approx$ 0.7645 m^3
1 mm^3 = 0.000061 cubic inch (cu in or in^3)
1 cm^3 = 0.061024 cubic inch (cu in or in^3)

Example Convert 612.75 cu in to cubic centimeters.

Since 612.75 cu in is to be expressed in cubic centimeters, multiply by the unit ratio $\frac{16.387\text{ cm}^3}{1\text{ cu in}}$.

$$612.75\text{ cu in} = 612.75\ \cancel{\text{cu in}} \times \frac{16.387\text{ cm}^3}{1\ \cancel{\text{cu in}}} = 10\,041.13\text{ cm}^3,\ 10\,041\text{ cm}^3 \quad \text{Ans}$$

Prisms

A *polyhedron* is a three-dimensional (solid) figure whose surfaces are polygons. In practical work, perhaps the most widely used solid is the prism. A *prism* is a polyhedron that has two identical (congruent) parallel polygon faces called *bases* and parallel lateral edges. The other sides or faces of a prism are parallelograms called *lateral faces*. A *lateral edge* is the line segment where two lateral faces meet. An *altitude* of the prism is a perpendicular segment that joins the planes of the two bases. The *height* of the prism is the length of an altitude.

Prisms are named according to the shape of their bases, such as triangular, rectangular, pentagonal, hexagonal, and octagonal. Some common prisms are shown in Figure 56-2. The parts of the prisms are identified. In a *right prism,* the lateral edges are perpendicular to the bases. Prisms A and B are examples of right prisms. In an *oblique prism,* the lateral edges are *not* perpendicular to the bases. Prisms C and D are examples of oblique prisms.

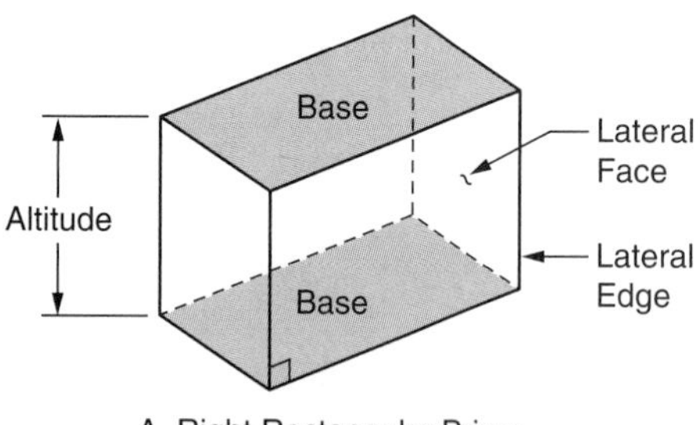

A. Right Rectangular Prism

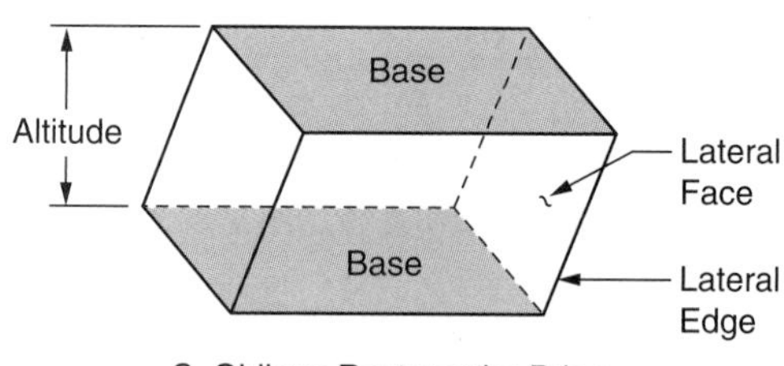

C. Oblique Rectangular Prism

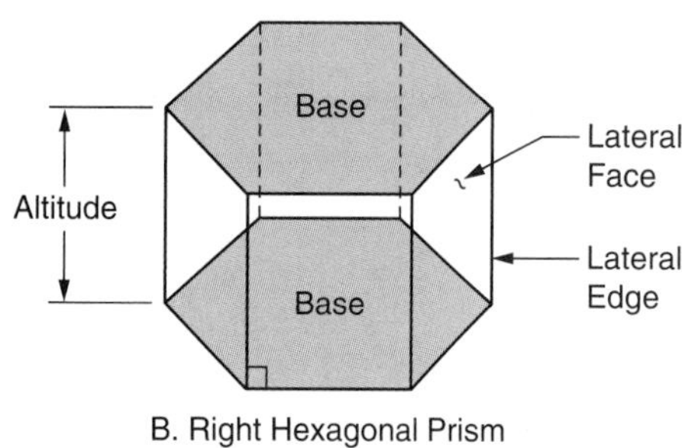

B. Right Hexagonal Prism

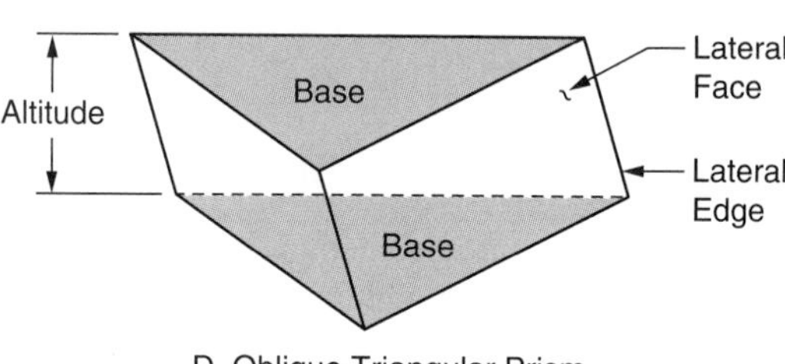

D. Oblique Triangular Prism

Figure 56-2

Volumes of Prisms

The *volume of any prism* (right or oblique) is equal to the product of the base area and height.

$$V = A_B h \qquad \text{where} \quad V = \text{volume}$$
$$A_B = \text{area of base}$$
$$h = \text{height}$$

Example 1 Compute the volume of a prism that has a base area of 34.40 square inches and a height of 16.00 inches.

$$V = 34.40 \text{ sq in} \times 16.00 \text{ in} = 550.4 \text{ cu in} \quad \text{Ans}$$

Example 2 A solid steel wedge is shown in Figure 56-3.

a. Find the volume of the wedge. Round the answer to the nearest tenth cubic centimeter.

b. The steel used for the wedge weighs 0.0080 kilogram per cubic centimeter. Find the weight of the wedge. Round the answer to the nearest tenth kilogram.

8.20 cm
15.60 cm
6.40 cm

Figure 56-3

a. Find the volume of the wedge.
Compute right triangle base area.

$$A_B = \frac{1}{2}bh$$

$$A_B = \frac{1}{2}(15.60 \text{ cm})(6.40 \text{ cm}) = 49.92 \text{ cm}^2$$

$$V = A_B h$$

$$V = 49.92 \text{ cm}^2 \times 8.20 \text{ cm} = 409.344 \text{ cm}^3,\ 409.3 \text{ cm}^3 \quad \text{Ans (rounded)}$$

b. Find the weight of the wedge.

Since 1 cm^3 weighs 0.0080 km, 409.344 cm^3 weighs 409.344×0.0080 kg

$409.344 \times 0.0080 \text{ kg} = 3.274752 \text{ kg}$

Weight = 3.3 kg Ans (rounded)

Cylinders

A *circular cylinder* is a solid that has identical (congruent) circular parallel bases. The surface between the bases is called the *lateral surface*. The *altitude* of a circular cylinder is a perpendicular segment that joins the planes of the bases. The *height* of a cylinder is the length of an altitude. The *axis* of a circular cylinder is a line that connects the centers of the bases.

In a *right circular cylinder,* the axis is perpendicular to the bases. A right circular cylinder with its parts identified is shown in Figure 56-4. Only right circular cylinders are considered in this book.

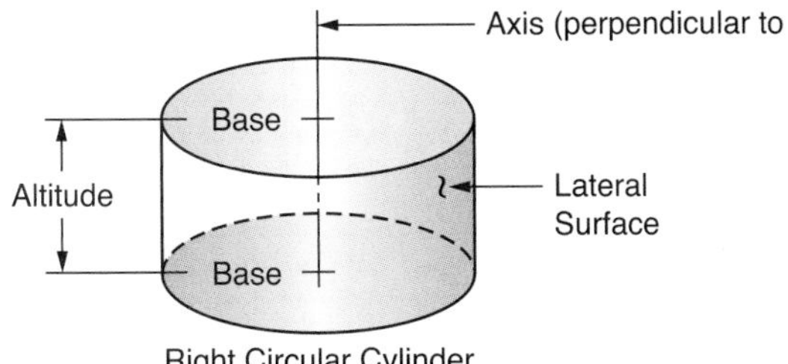

Figure 56-4

Volumes of Cylinders

As with a prism, a right circular cylinder has uniform cross-sectional area. The formula for computing volumes of right circular cylinders is the same as that of prisms. The *volume of a right circular cylinder* is equal to the product of the base area and height.

$V = A_B h$ where V = volume

A_B = area of base, $A_B = \pi r^2$ or $0.7854d^2$ (rounded)

h = height

Example 1 Find the volume of a cylinder with a base area of 30.0 square centimeters and a height of 6.0 centimeters.

$V = 30.0 \text{ cm}^2 \times 6.0 \text{ cm} = 180 \text{ cm}^3$ Ans

Example 2 A length of pipe is shown in Figure 56-5. Find the volume of metal in the pipe.

Find the area of the outside circle.

$A_B \approx 0.7854d^2$

$A_B \approx (0.7854)(4.00 \text{ in})^2 \approx 12.566371$ sq in

Find the area of the hole.

$A_B \approx (0.7854)(3.40 \text{ in})^2 \approx 9.079224$ sq in

Find the cross-sectional area.

$12.566371 \text{ sq in} - 9.079224 \text{ sq in} \approx 3.487147$ sq in

Find the volume.

$V \approx 3.487147 \text{ sq in} \times 50.0 \approx 174$ cu in Ans

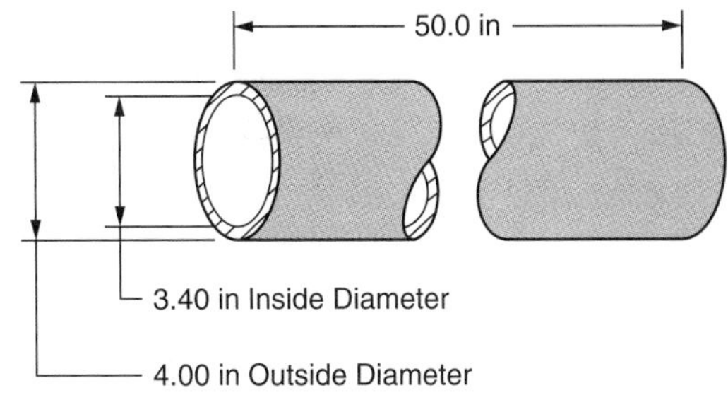

Figure 56-5

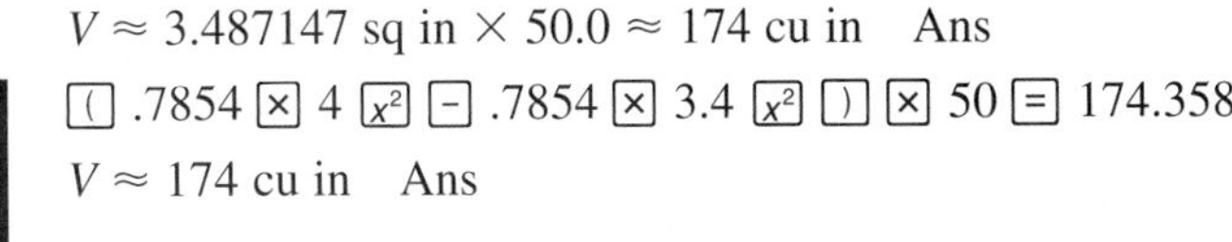

(.7854 × 4 x² − .7854 × 3.4 x²) × 50 = 174.3584,

$V \approx 174$ cu in Ans

Computing Heights and Bases of Prisms and Cylinders

The height of a prism or a right circular cylinder can be determined if the base area and volume are known. Also, the base area can be found if the height and volume are known. Substitute the known values in the volume formula and solve for the unknown value.

Example 1 Find the height of a spacer block in the shape of a right rectangular prism that has a base area of 68.40 square inches and a volume of 366.60 cubic inches.

Substitute values in the formula and solve for h.

$366.6 \text{ cu in} = 68.40 \text{ sq in} \times h$

$h \approx 5.36 \text{ in}$ Ans

Example 2 Calculate the diameter of a piston designed to have a volume of 450.0 cubic centimeters and a length (height) of 10.800 centimeters.

Find the base area.

$450.0 \text{ cm}^3 = A_B(10.800 \text{ cm})$

$A_B \approx 41.6667 \text{ cm}^2$

Find the piston diameter.

$41.6667 \text{ cm}^2 \approx 0.7854d^2$

$d^2 \approx 53.051566 \text{ cm}^2$

$d \approx 7.284 \text{ cm}$ Ans

$d =$ [√] [(] 450 [÷] 10.8 [÷] .7854 [)] [=] 7.283647688, 7.284 cm Ans (rounded)

Application

Equivalent Customary Units of Volume Measure

Express each volume as indicated. Round each answer to the same number of significant digits as in the original quantity.

1. 4320 cubic inches as cubic feet ____________
2. 850 cubic inches as cubic feet ____________
3. 117 cubic feet as cubic yards ____________
4. 182 cubic feet as cubic yards ____________
5. 12,900 cubic inches as cubic feet ____________
6. 1.650 cubic feet as cubic inches ____________
7. 0.325 cubic feet as cubic inches ____________
8. 0.1300 cubic yards as cubic inches ____________

Equivalent Metric Units of Volume Measure

Express each volume as indicated. Round each answer to the same number of significant digits as in the original quantity.

9. 2700 cubic millimeters as cubic centimeters ____________
10. 30.05 cubic millimeters as cubic centimeters ____________
11. 78 cubic centimeters as cubic millimeters ____________
12. 260.1 cubic centimeters as cubic meters ____________
13. 0.075 cubic meters as cubic centimeters ____________
14. 0.109 cubic centimeters as cubic millimeters ____________
15. 0.730 2 cubic meters as cubic millimeters ____________
16. 0.003 74 cubic meters as cubic millimeters ____________

Conversion Between Metric and Customary Units of Volume Measure

Express each volume as indicated. Round each answer to the same number of significant digits as in the original quantity.

17. 2.16 cubic inches as cubic millimeters
18. 73.06 cubic feet as cubic meters
19. 0.348 cubic feet as cubic centimeters
20. 0.004687 cubic feet as cubic millimeters
21. 273 cubic centimeters as cubic inches
22. 372.5 cubic millimeters as cubic inches
23. 137 020 cubic millimeters as cubic inches
24. 84 300 cubic centimeters as cubic feet

Volumes of Prisms

Solve these problems. Where necessary, round the answers to 2 decimal places unless otherwise specified.

25. Find the volume of a prism with a base area of 125 square inches and a height of 8 inches.
26. Compute the volume of a prism with a height of 26.500 centimeters and a base of 610.00 square centimeters.
27. One cubic inch of cast iron weighs 0.26 pound. Find the weight of a block of cast iron with a base area of 48 square inches and a height of 5.3 inches. Round the answer to the nearest pound.
28. Find the capacity in gallons of a rectangular electroplating tank with a base area of 38.50 square feet and a height of 4.25 feet. One cubic foot has a capacity of 7.5 gallons. Round the answer to the nearest gallon.
29. Find the volume of a prism that is 13.82 centimeters high and has an equilateral triangle base that has 9.08 centimeter sides. Round the answer to the nearest cubic centimeter.
30. A length of angle iron is shown.

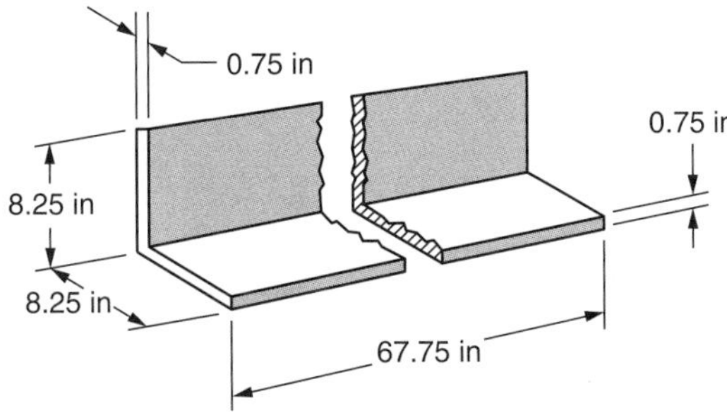

a. Find the volume of the angle iron. Round the answer to the nearest cubic inch.

b. Find the weight of the angle iron if the material weighs 490 pounds per cubic foot. Round the answer to the nearest pound.

31. A steel forging is shown.

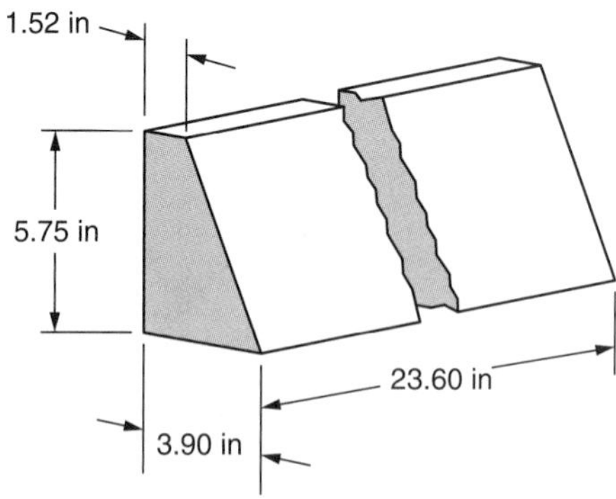

a. Find the number of cubic inches contained in the forging. Round the answer to the nearest cubic inch. ________

b. Find the weight of the forging. The steel weighs 488.5 pounds per cubic foot. Round the answer to the nearest pound. ________

32. A copper casting is in the shape of a prism with an equilateral triangle base. The length of each base side is 17.80 centimeters, and the casting height is 16.23 centimeters. Copper weighs 8.8 grams per cubic centimeter. Determine the weight of the casting in kilograms. One kilogram equals 1000 grams. Round the answer to 2 decimal places. ________

33. Determine the number of cubic centimeters of material needed for the cast plate shown. The casting is 3.20 centimeters thick. Round the answers to the nearest thousandth cubic meter. ________

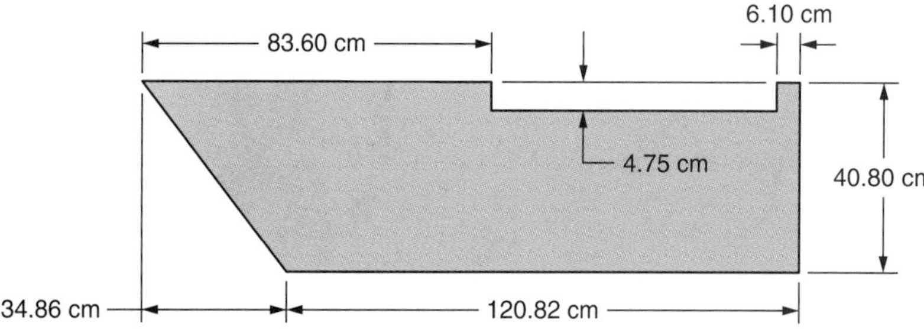

34. The steel beam shown weighs 7800 kilograms per cubic meter. Determine the weight of the beam to the nearest kilogram. ________

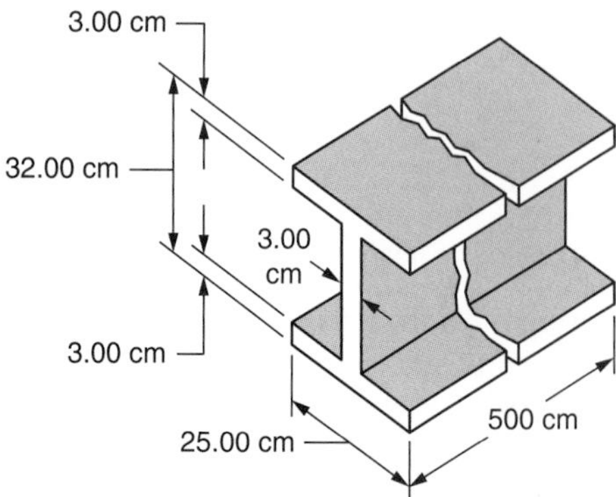

Volumes of Cylinders

Solve the following problems. Where necessary, round the answers to 2 decimal places unless otherwise specified.

35. Find the volume of a right circular cylinder with a base area of 76.00 square inches and a height of 8.600 inches. ______
36. Compute the volume of a right circular cylinder with a height of 0.40 meter and a base area of 0.30 square meter. ______
37. A cylindrical container has a base area of 154.0 square inches and a height of 16.00 inches. Find the capacity of the container in gallons. One gallon contains 231 cubic inches. ______
38. Find the volume of a steel shaft that is 59.00 centimeters long and has a diameter of 3.840 centimeters. Round the answer to 1 decimal place. ______
39. Each cylinder of a 6-cylinder engine has a 3.125-inch diameter and a piston stroke of 4.570 inches. Find the total piston displacement of the engine. Round the answer to 1 decimal place. ______
40. A 0.460-inch diameter brass rod is $8'5\frac{7''}{32}$ long. Round answers a and b to 1 decimal place.
    a. Find the volume of the rod in cubic inches. ______
    b. Compute the total weight of 40 rods. Brass weighs 0.300 pound per cubic inch. ______
41. A bronze bushing is shown.

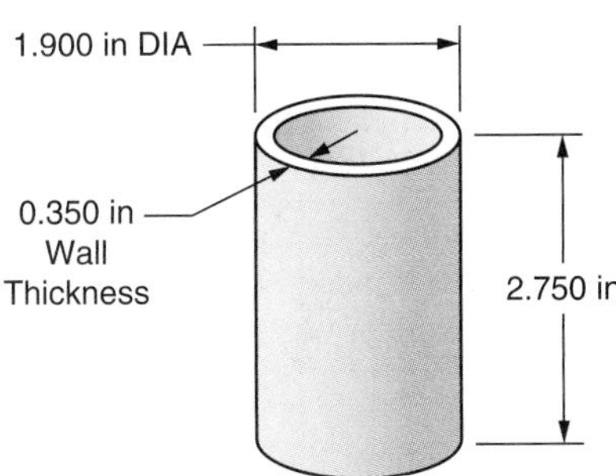

    a. Compute the volume of bronze in the bushing. ______
    b. The bushing weighs 1.50 pounds. Find the weight of 1.00 cubic inch of bronze. ______
42. Find the number of cubic centimeters of material contained in the support base plate shown. The plate is 4.73 centimeters thick. ______

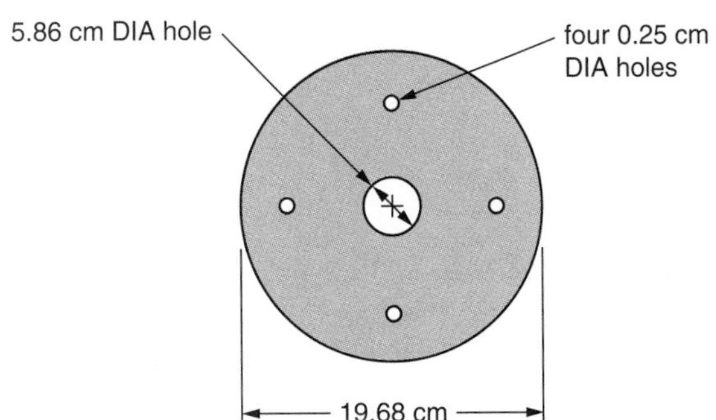

43. The brass used to make the bushing shown weighs 546.2 pounds per cubic foot. All dimensions are in inches. Find the weight of 2600 bushings. Round the answer to the nearest pound.

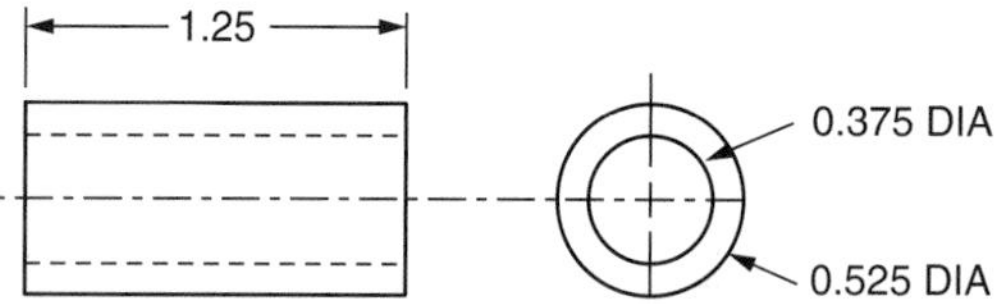

44. Determine the number of cubic millimeters of material contained in the flanged collar shown. Round the answer to the nearest cubic millimeter.

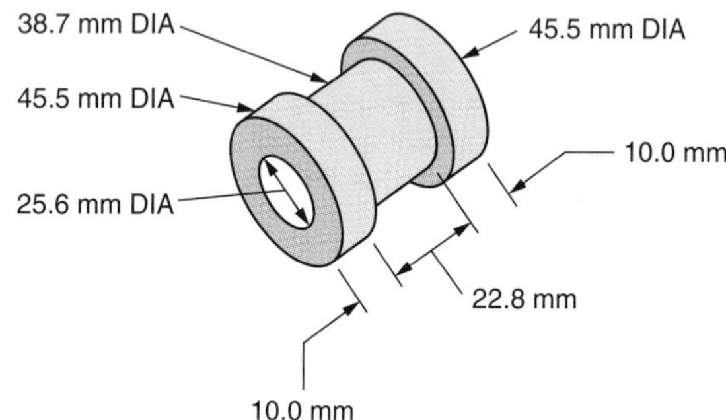

Heights and Base Areas of Prisms and Cylinders

Solve the following problems. Where necessary, round the answers to 2 decimal places unless otherwise specified.

45. Find the height of a prism that has a base area of 67.84 square inches and a volume of 512.70 cubic inches.
46. A solid right circular cylinder is 18.25 centimeters high and contains 119.62 cubic centimeters of material. Compute the cross-sectional area of the cylinder.
47. A solid steel bar 27.60 inches long has a square base. The bar has a volume of 104.00 cubic inches. Compute the length of a side of the base.
48. A cylindrical quart can has a 3.86-inch diameter. What is the height of the can? (Note: 1 gallon = 231 cubic inches; 1 gallon = 4 quarts.)
49. A cylinder container 26.20 centimeters high is designed to have a capacity of 1.76 liters. Determine the base area of the container. One liter contains 1000 cubic centimeters.
50. A triangular brass casting is shown. The block weighs 15.30 pounds. Brass weighs 526.7 pounds per cubic foot. Compute the height of the block. All dimensions are in inches.

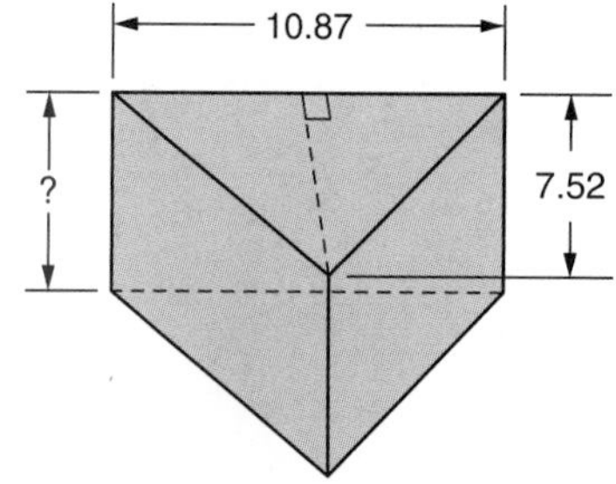

51. A rectangular aluminum plate required for a job is 4 feet 0 inches wide and 5 feet 0 inches long. The maximum allowable weight is 505.5 pounds. Aluminum weighs 168.5 pounds per cubic foot. What is the maximum thickness of the plate? Round the answer to the nearest tenth inch.

52. Determine the cross-sectional area of a regular octagon cross-section tool steel piece shown. The piece weighs 3.72 pounds. The tool steel weighs 0.284 pounds per cubic inch. ______

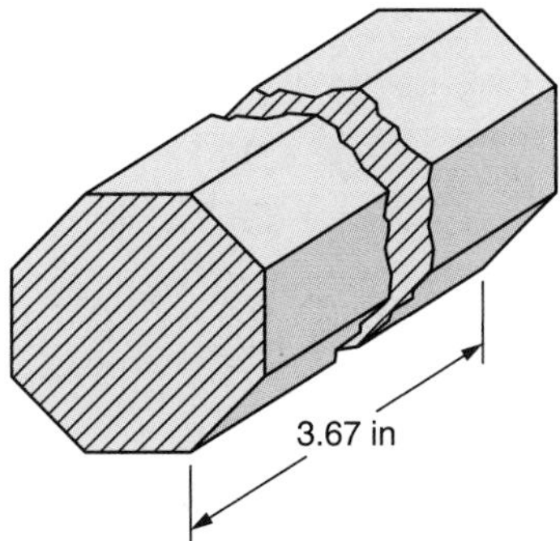

53. A circular cross-sectional piece of brass rod weighs 8.67 pounds. The piece is cut from a brass rod that weighs 3.66 pounds per foot of length. The piece has a volume of 56.67 cubic inches. Find the area of circular base. ______

54. Find the length of a piece of bar stock with a regular hexagon cross section with 0.875-inch sides. The piece has a volume of 31.2 cubic inches. The formula $A \approx 2.598s^2$ is used to determine the area of a regular hexagon where A is the area and s is the length of a side. ______

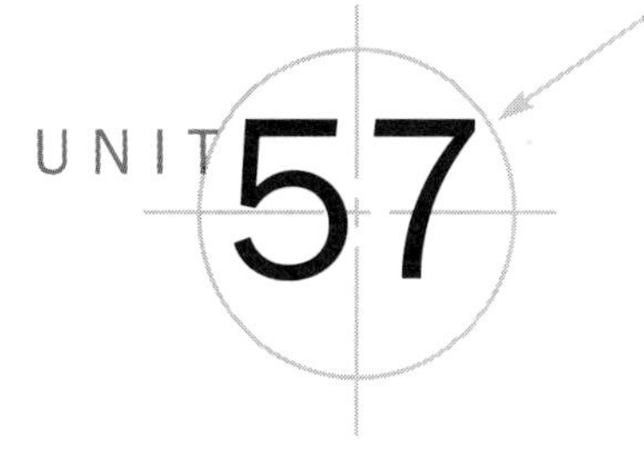

Volumes of Pyramids and Cones

Objectives ***After studying this unit you should be able to***

- Compute volumes of pyramids and cones.
- Compute heights, bases, and weights of pyramids and cones.
- Compute volumes of frustums of pyramids and cones.
- Compute heights, bases, and weights of frustums of pyramids and cones.

Tapered shafts, conical pulleys and clutches, conical compression springs, and roller bearings are a few of the many applications of portions (frustums) of cones. Storage containers are often fabricated in the shape of a frustum of a pyramid.

Pyramids

A *pyramid* is polyhedron whose base can be any polygon, and the other faces are triangles that meet at a common point called the *vertex* of the pyramid. The triangular faces that meet at the vertex are called *lateral faces*. A *lateral edge* is the line segment where two lateral faces meet. The *altitude* of a pyramid is the perpendicular segment from the vertex to the plane of the base. The *height* is the length of the altitude.

Pyramids are named according to the shape of the bases, such as triangular, quadrangular, pentagonal, hexagonal, and octagonal. In a *regular pyramid,* the base is a regular polygon, and the lateral edges are all equal in length. Only regular pyramids are considered in this book. Some common regular pyramids are shown in Figure 57-1. The parts of the pyramids are identified.

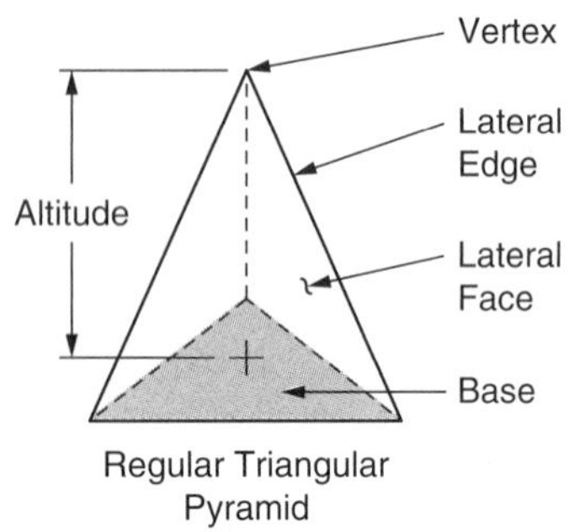

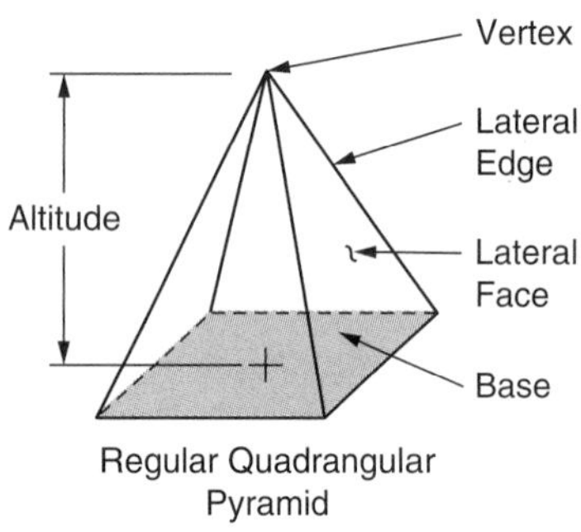

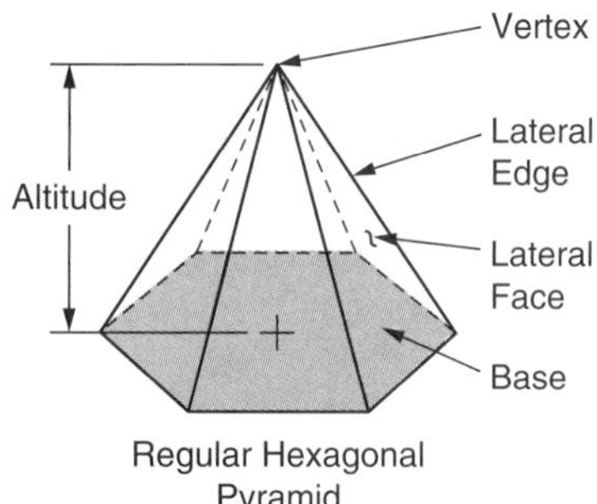

Figure 57-1

Cones

A *circular cone* is a solid figure with a circular base and a surface that tapers from the base to a point called the *vertex*. The surface lying between the base and the vertex is called the *lateral surface*. The *altitude* of a right circular cone is the perpendicular segment from the vertex to the center of the base. The *height* is the length of the altitude. The *axis* of a circular cone is a line that connects the vertex to the center of the circular base.

In a *right circular cone* the axis is perpendicular to the base. A right circular cone with the parts identified is shown in Figure 57-2. Only right circular cones are considered in this book.

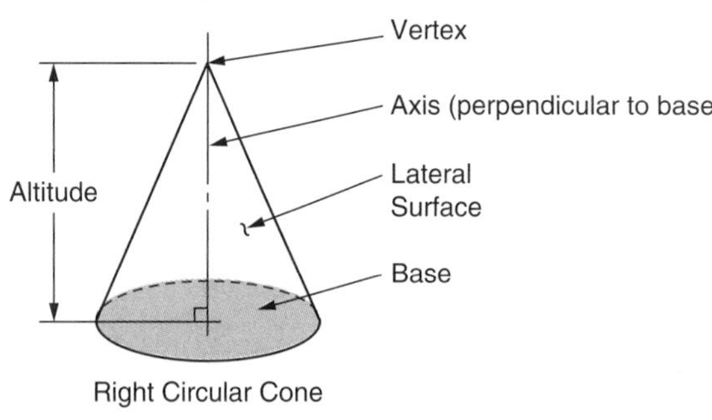

Figure 57-2

Volumes of Regular Pyramids and Right Circular Cones

Consider a prism and a pyramid that have identical base areas and altitudes. If the volumes of the prism and pyramid are measured, the volume of the pyramid will be one-third the volume of the prism. Also, if the volumes of a cylinder and a cone with identical bases and heights are measured, the volume of the cone will be one-third the volume of the cylinder. The formulas for computing volumes of prisms and right circular cylinders are the same.

Therefore, the formulas for computing volumes of regular pyramids and right circular cones are the same. The *volume of a regular pyramid or a right circular cone* equals one-third the product of the area of the base and height.

$$V = \frac{1}{3}A_B h \quad \text{where} \quad V = \text{volume}, \quad A_B = \text{area of the base}, \quad h = \text{height}$$

Example 1 Compute the volume of a pyramid that has a base of 24.0 square feet and a height of 6.0 feet.

$$V = \frac{24.0 \text{ sq ft} \times 6.0 \text{ ft}}{3} = 48 \text{ cu ft} \quad \text{Ans}$$

Example 2 A bronze casting in the shape of a right circular cone is 14.52 inches high and has a base diameter of 10.86 inches.

a. Find the volume of bronze required for the casting.

b. Find the weight of the casting to the nearest pound. Bronze weighs 547.9 pounds per cubic foot. Recall that there are 1728 cubic inches per cubic foot.

a. $V \approx \dfrac{(3.1416)(5.43 \text{ in})^2(14.52 \text{ in})}{3} \approx 448.3 \text{ cu in}$ Ans

b. 448.3 cu in ÷ 1728 cu in/cu ft ≈ 0.2594 cu ft
0.2594 cu ft × 547.9 lb/cu ft ≈ 142 lb Ans

a. V = [π] [×] 5.43 [x^2] [×] 14.52 [÷] 3 [=] 448.3269989, 448.3 cu in Ans (rounded)

b. Weight = 448.3 [÷] 1728 [×] 547.9 [=] 142.1432697, 142 lb Ans (rounded)

Computing Heights and Bases of Regular Pyramids and Right Circular Cones

As with prisms and cylinders, heights and base areas of regular pyramids and right circular cones of known volume are readily determined. Substitute known values in the volume formula and solve for the unknown value.

Example 1 The volume of a regular pyramid is 270 cubic centimeters, and the height is 18 centimeters.

Compute the base area.

Substitute the values in the formula and solve.

$$270 \text{ cm}^3 = \frac{A_B(18 \text{ cm})}{3}$$

$$A_B = 45 \text{ cm}^2 \quad \text{Ans}$$

Example 2 A disposable plastic drinking cup is designed in the shape of a right circular cone. The cup holds $\frac{1}{3}$ pint (9.63 cubic inches) of liquid when full. The rim (base) diameter is 3.60 inches. Find the cup depth (height).

Substitute the values in the formula and solve.

$$9.63 \text{ cu in} \approx \frac{3.1416(1.80 \text{ in})^2(h)}{3}$$

$$h \approx 2.84 \text{ in} \quad \text{Ans}$$

Frustums of Pyramids and Cones

When a pyramid or a cone is cut by a plane parallel to the base, the part that remains is called a *frustum*. A frustum has two bases, upper and lower.

The *larger base* is the base of the cone or pyramid. The *smaller base* is the circle or polygon formed by the parallel cutting plane. The smaller base of the pyramid has the same shape as the larger base. The two bases are similar. The *altitude* is the perpendicular segment that joins the planes of the bases. The *height* is the length of the altitude. A frustum of a pyramid and a frustum of a cone with their parts identified are shown in Figure 57-3.

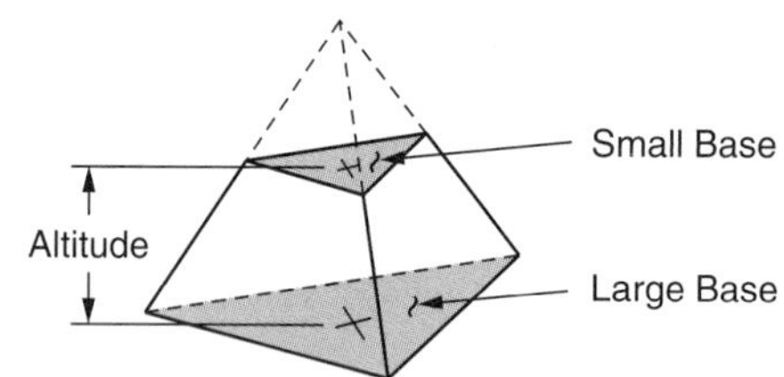

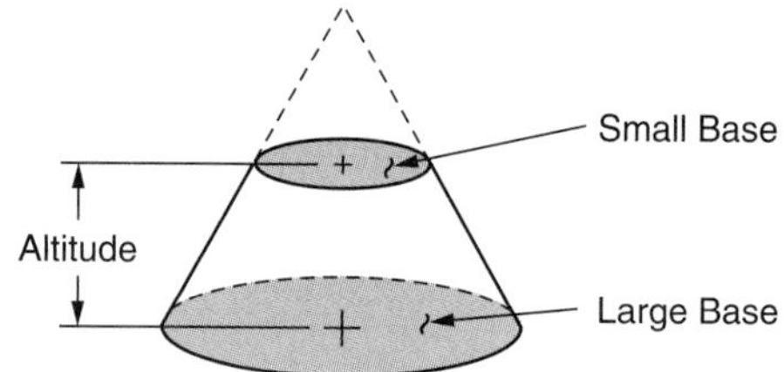

Figure 57-3

Volumes of Frustums of Regular Pyramids and Right Circular Cones

The *volume of the frustum of a pyramid or cone* is computed from the formula

$$V = \frac{1}{3}h\left(A_B + A_b + \sqrt{A_BA_b}\right)$$

where V = volume of the frustum of a pyramid or cone
h = height
A_B = area of larger base
A_b = area of smaller base

The formula for the volume of a frustum of a right circular cone is expressed in this form.

$$V = \frac{1}{3}\pi h(R^2 + r^2 + Rr)$$

where V = volume of a right circular cone
h = height
R = radius of larger base
r = radius of smaller base

Example 1 A container is designed in the shape of a frustum of a pyramid with square bases as shown in Figure 57-4.

a. Find the volume of the container in cubic feet.

b. Compute the capacity (number of gallons) of liquid that the container can hold when full.

➤ **Note:** One cubic foot contains 7.5 gallons.

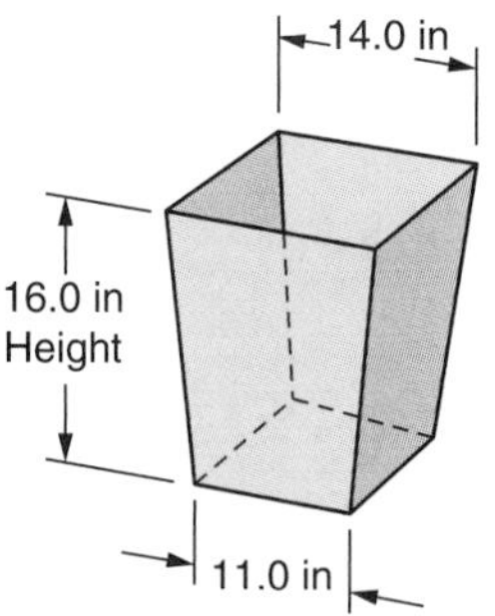

Figure 57-4

a. Find the volume.
Find the larger base area.

$A_B = (14.0 \text{ in})^2 = 196 \text{ sq in}$

Find the smaller base area.

$A_b = (11.0 \text{ in})^2 = 121 \text{ sq in}$

Find the volume.

$$V = \frac{(16.0 \text{ in})\left[196 \text{ sq in} + 121 \text{ sq in} + \sqrt{(196 \text{ sq in})(121 \text{ sq in})}\right]}{3}$$

$$V = \frac{(16.0 \text{ in})(196 \text{ sq in} + 121 \text{ sq in} + 154 \text{ sq in})}{3}$$

$V = 2512 \text{ cu in}$

Express the volume in cubic feet.

$2512 \text{ cu in} \div 1728 \text{ cu in/cu ft} \approx 1.45 \text{ cu ft}$ Ans

b. Compute the capacity of the container in gallons.

$1.45 \text{ cu ft} \times 7.5 \text{ gal/cu ft} \approx 10.9 \text{ gal}$ Ans

Example 2 A tapered shaft is shown in Figure 57-5.

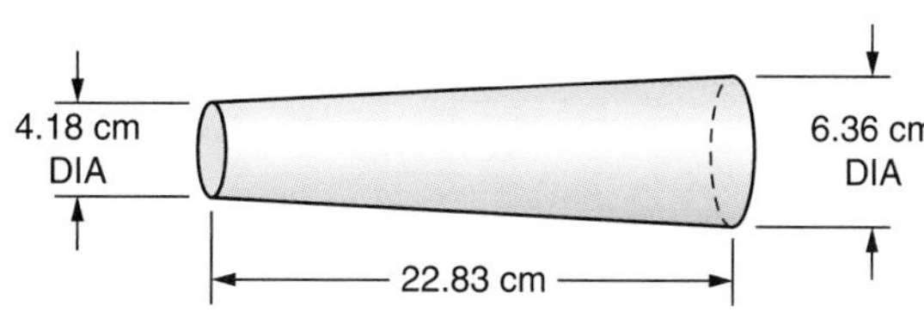

Figure 57-5

a. Find the number of cubic centimeters of steel contained in the shaft.

b. Find the weight of the shaft. The steel in the shaft weighs 0.0078 kilogram per cubic centimeter.

a. Find the volume.

$$V \approx \frac{(3.1416)(22.83 \text{ cm})[(3.18 \text{ cm})^2 + (2.09 \text{ cm})^2 + (3.18 \text{ cm})(2.09 \text{ cm})]}{3}$$

$\approx 505 \text{ cm}^3$ Ans

b. Compute the weight.

$505 \text{ cm}^3 \times 0.0078 \text{ kg/cm}^3 \approx 3.9 \text{ kg}$ Ans

a. $V =$ [π] [×] 22.83 [×] [(] 3.18 [x^2] [+] 2.09 [x^2] [+] 3.18 [×] 2.09 [)] [=]

[÷] 3 [=] 505.0870048

$V \approx 505 \text{ cm}^3$ Ans

b. Weight = 505 [×] .0078 [=] 3.939, 3.9 kg Ans (rounded)

Application

Volumes of Pyramids and Right Circular Cones

Solve these problems. Where necessary, round the answers to 2 decimal places unless otherwise specified.

1. Compute the volume of a regular pyramid with a base area of 236.90 square inches and a height of 12.84 inches. ________
2. Find the volume of a right circular cone with a base area of 38.60 square centimeters and a height of 5.000 centimeters. ________
3. Find the volume of a regular pyramid with a height of 10.80 inches and a base area of 98.00 square inches. ________
4. A container is in the shape of a right circular cone. The base area is 63.60 square inches and the height is 7.65 inches. Compute the capacity of the container in gallons. One gallon contains 231 cubic inches. ________
5. A brass casting is in the shape of a right circular cone with a base diameter of 8.26 centimeters and a height of 18.36 centimeters. Find the volume. Round the answer to the nearest cubic centimeter. ________
6. Two solid pieces of aluminum in the shape of right circular cones with different base diameters are machined. The heights of both pieces are 6 inches. The base of the smaller piece is 2 inches in diameter. The base of the larger piece is twice as large, or 4 inches in diameter. How many times heavier is the larger piece than the smaller? ________
7. A vessel is in the shape of a right circular cone. This vessel contains liquid to a depth of 12.8 centimeters, as shown. How many liters of liquid must be added in order to fill the vessel? One liter contains 1000 cubic centimeters. ________

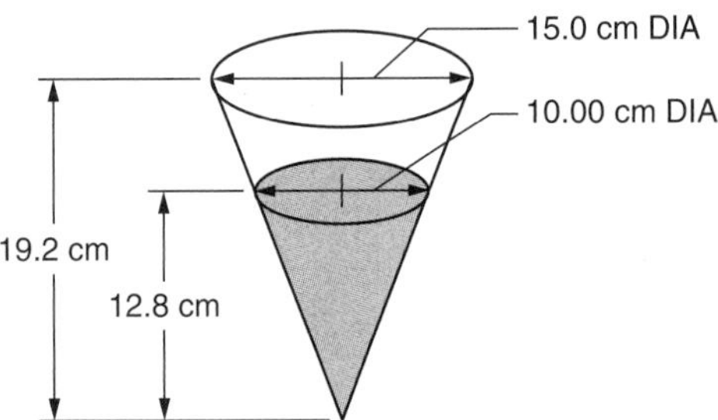

8. A solid iron casting is in the shape of a regular pyramid with a triangular base. Each of the 3 base sides is 5.70 inches long and the casting is 4.65 inches high. The casting weighs 5.82 pounds.

   a. Determine the volume of material in the casting. ________

   b. What is the weight of one cubic inch of cast iron? ________

9. The No. 3 taper standard lathe center shown has a tungsten-carbide tip with the dimensions given. How many cubic inches of carbide are there in the exposed part of the tip? ________

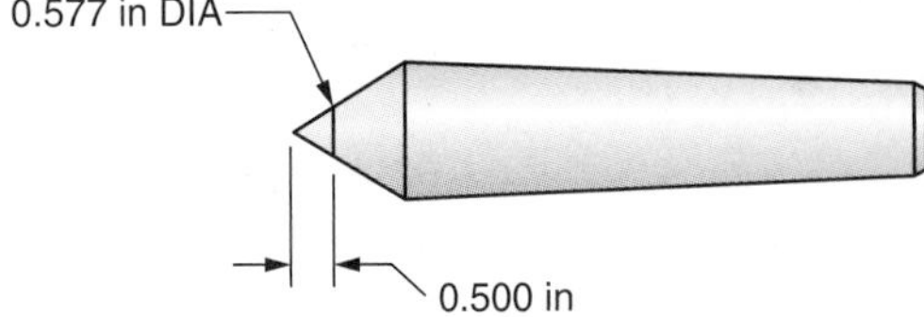

10. A piece in the shape of a pyramid with a regular octagon (8 sided) base is machined from a solid block of bronze. Each side of the octagon base is 9.36 inches long. The height of the piece is 7.08 inches. The octagon base area $\approx 4.828s^2$ where s is the length of a side of the octagon.

    a. Determine the volume of the piece. Round the answer to the nearest cubic inch. __________

    b. Determine the weight of the piece. Round the answer to the nearest pound. __________

    ➤ **Note:** One cubic foot of the bronze used weighs 547.9 pounds per cubic foot.

Heights and Bases of Regular Pyramids and Right Circular Cones

Solve these problems. Where necessary, round the answers to 2 decimal places unless otherwise specified.

11. Compute the height of a regular pyramid with a base area of 32.87 square inches and a volume of 152.08 cubic inches. __________
12. A right circular cone 12.7 centimeters high contains 198.7 cubic centimeters of material. Find the area of the base of the cone. Round the answer to the nearest square centimeter. __________
13. The base area of a thin-walled casting in the shape of a regular pyramid is 49 square inches. The casting contains 323 cubic inches of air space. How high is the casting? Round the answer to the nearest inch. __________
14. A container with a capacity of 6.00 gallons is in the shape of a right circular cone. The container is 17.80 inches high. Find the base area of the container. One gallon contains 231 cubic inches. __________
15. A regular pyramid has a volume of 1152 cubic centimeters and a height of 8.64 centimeters. The base is a square. Determine the length of a base side. __________
16. Find the base diameter of a right circular cone that has a volume of 922.4 cubic centimeters and a height of 14.85 centimeters. __________
17. A steel forging in the shape of a regular pyramid weighs 204.8 pounds. The base of the pyramid is a square with 9.71-inch sides. Compute the height of the forging. The steel in the forging weighs 490.5 pounds per cubic foot. __________
18. The volume of a regular pyramid with an equilateral triangle base is 3174 cubic centimeters. The pyramid height is 16.91 centimeters. Find the length of each side of the triangular base. __________

Volumes of Frustums of Regular Pyramids and Right Circular Cones

19. The frustum of a right circular cone has the larger base area equal to 31.76 square inches and the smaller base area equal to 14.05 square inches. The height is 16.29 inches. Compute the volume. Round the answer to the nearest cubic inch. __________
20. The container is in the shape of a frustum of a right circular cone. The smaller base area is 426 square centimeters and the larger base area is 876 square centimeters. The height is 29.5 centimeters. Compute the capacity of the container in liters. One liter contains 1000 cubic centimeters. Round the answers to the nearest tenth liter. __________
21. A steel forging is in the shape of a frustum of a regular pyramid. It has a larger base area of 58.30 square inches and a smaller base area of 40.0 square inches. The height is 5.10 inches. Find the weight of the forging. The steel weights 490.3 pounds per cubic foot. Round the answer to the nearest tenth pound. __________
22. Find the volume of the frustum of a right circular cone with a 155.68-centimeter radius base and a 126.98-centimeter radius base. The frustum height is 24.52 centimeters. Round the answer to the nearest hundredth cubic meter. __________

23. A hollow aluminum casting in the shape of a frustum of a regular pyramid with square bases is shown. All given dimensions are outside dimensions. The casting wall thickness is 0.62 centimeter. Find the number of cubic centimeters of material that can be held in the casting. Round the answer to the nearest cubic centimeter. __________

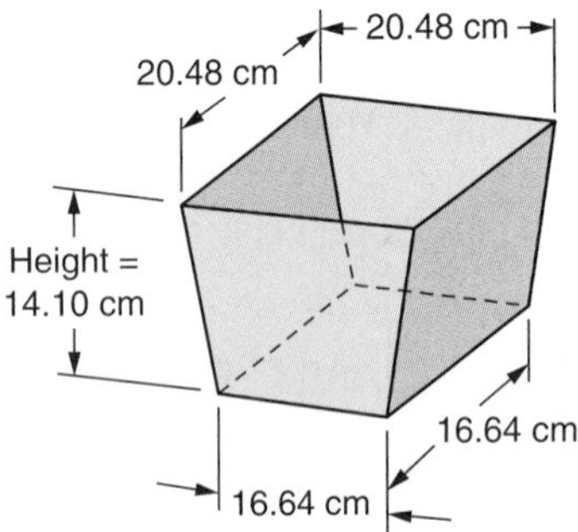

24. The side view of a tapered steel shaft is shown. The length of the shaft is reduced from 18.40 inches to 13.60 inches. How many cubic inches of stock are removed? Round the answer to 1 decimal place. __________

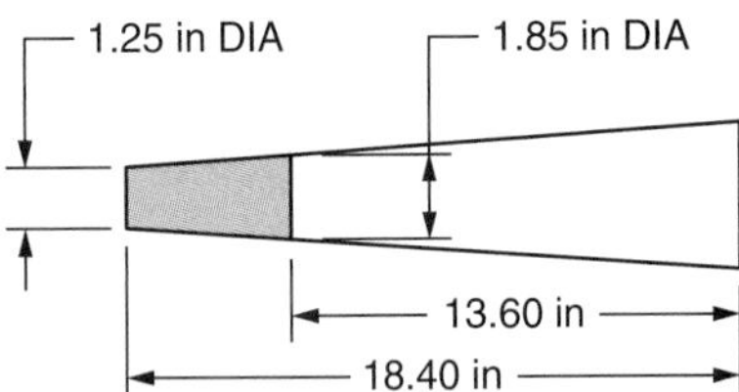

25. A zinc casting is in the shape of a frustum of a right circular cone. The larger base area is 2.80 inches in diameter and the smaller base area is 2.30 inches in diameter. The height is 3.50 inches. Round the answers for a and b to 1 decimal point.

    a. Compute the volume of the casting. __________

    b. Determine the weight of the casting. Zinc weighs 0.256 pound per cubic inch. __________

26. A piece in the shape of a frustum of a pyramid with regular octagon bases is 23.84 centimeters high. The length of each side of the larger base is 8.17 centimeters, and the length of each side of the smaller base is 6.77 centimeters. If a side (s) of a regular octagon is known, its area (A) can be computed by the formula $A \approx 4.828s^2$. Determine the volume of the piece. Round the answer to the nearest cubic centimeter. __________

27. Find the volume of a hollow machined steel piece shown. Round the answer to 1 decimal place. __________

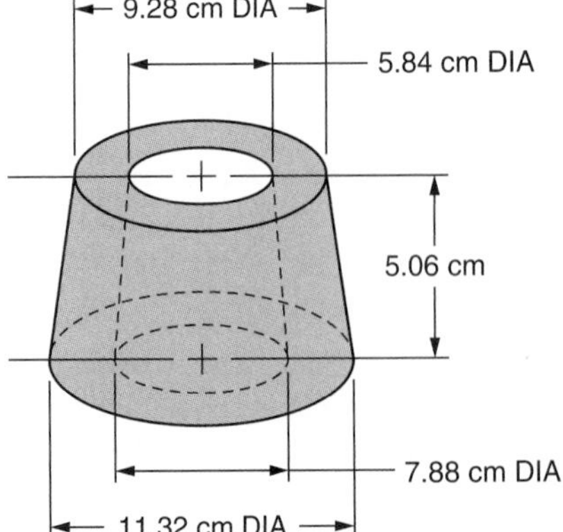

28. Parts are to be produced in the shape of a frustum of a regular pyramid with triangular bases. The larger base is to have each side 6.77 inches long and the smaller base is to have each side 4.98 inches long. Each part is to contain 240.5 cubic inches. Determine the required height. Round the answer to 2 decimal places. __________

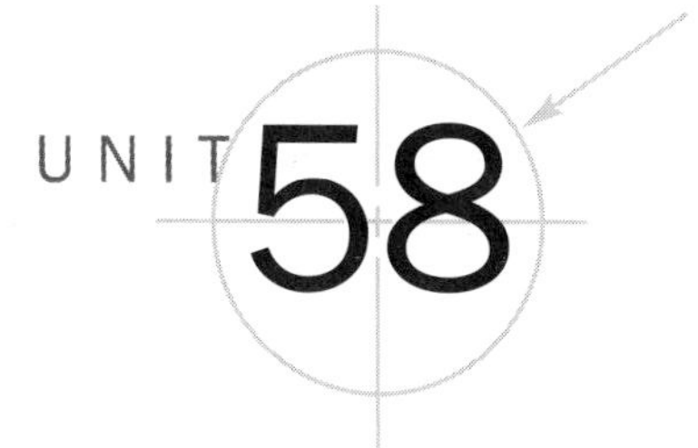

Volumes of Spheres and Composite Solid Figures

Objectives ***After studying this unit you should to able to***

- Compute volumes of spheres.
- Compute capacities and weights of spheres.
- Compute volumes of composite solids.
- Compute capacities and weights of composite solids.

Spheres

A *sphere* is a solid figure bounded by a curved surface such that every point on the surface is the same distance (equidistant) from a point called the *center*. A round ball, such as a baseball or basketball, is an example of a sphere.

The *radius* of a sphere is the length of any segment from the center to any point on the surface. A *diameter* is a segment through the center with its endpoints on the curved surface. The diameter of a sphere is twice the radius.

If a plane cuts through (intersects) a sphere and does *not* go through the center, the section is called a *small circle*. As intersecting planes move closer to the center, the circular sections get larger. A plane that cuts through (intersects) the center of a sphere is called a *great circle*. A great circle is the largest circle that can be cut by an intersecting plane. The sphere and the great circle have the same center. The circumference of the great circle is the circumference of the sphere. If a plane is passed through the center of a sphere, the sphere is cut into two equal parts. Each part is a half sphere, called a *hemisphere*. A sphere with its parts identified is shown in Figure 58-1.

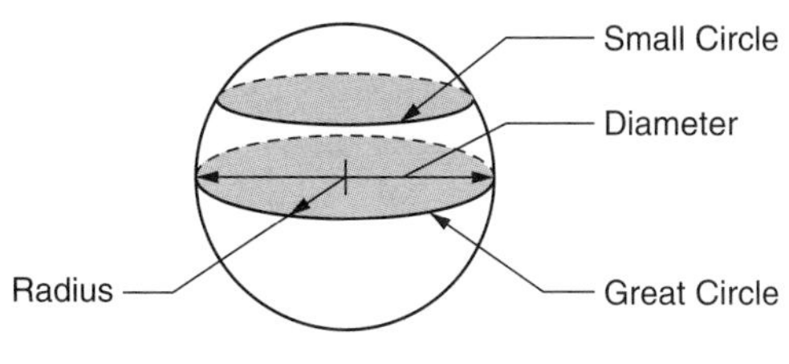

Figure 58-1

Volume of a Sphere

The volume of a sphere is found by the following formula.

$$V = \frac{4}{3}\pi r^3 \qquad \text{where} \quad V = \text{volume of the sphere}$$
$$r = \text{radius of the sphere}$$

Example A stainless steel ball bearing contains balls that are each 1.80 centimeters in diameter.

Find the volume of a ball.

Find the weight of a ball to the nearest gram. Stainless steel weighs 7.88 grams per cubic centimeter.

a. Find the volume.

$$V = \frac{4(3.1416)(0.900 \text{ cm})^3}{3} \approx 3.05 \text{ cm}^3 \quad \text{Ans}$$

b. Find the weight.

$3.05 \text{ cm}^3 \times 7.88 \text{ g/cm}^3 \approx 24 \text{ g}$ Ans

4 [×] [π] [×] .9 [y^x] 3 [÷] 3 [=] 3.053628059 [×] 7.88 [=] 24.06258911

(3.053628059 ↑ Volume; 24.06258911 ↑ Weight)

a. $V \approx 3.05 \text{ cm}^3$ Ans

b. Weight ≈ 24 g Ans

Volumes of Composite Solid Figures

A shaft or a container may be a combination of a cylinder and the frustum of a cone. A round-head rivet is a combination of a cylinder and a hemisphere. Objects of this kind are called *composite solid figures or composite space figures.*

To compute *volumes of composite solid figures,* it is necessary to determine the volume of each simple solid figure separately. The individual volumes are then added or subtracted.

Example The side view of a flanged shaft is shown. Find the volume of metal in the shaft in Figure 58-2.

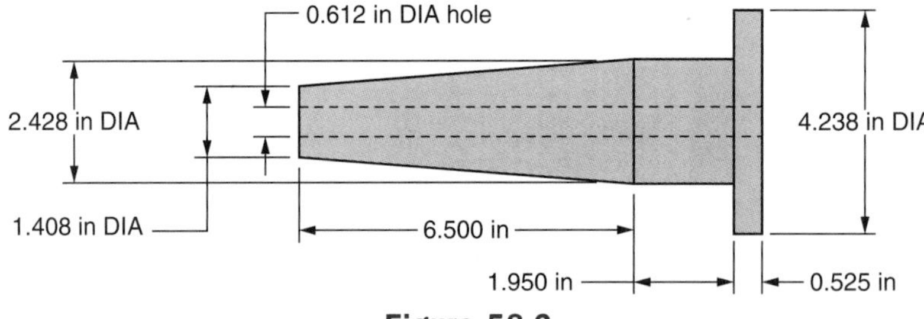

Figure 58-2

- Find the volume of the 6.500-inch long frustum of a cone.

$$V = \frac{1}{3}\pi h(R^2 + r^2 + Rr)$$

$$R = 0.5(2.428 \text{ in}) = 1.214 \text{ in}$$

$$r = 0.5(1.408 \text{ in}) = 0.704 \text{ in}$$

$$V_1 \approx \frac{(3.1416)(6.500 \text{ in})[(1.214 \text{ in})^2 + (0.704 \text{ in})^2 + (1.214 \text{ in})(0.704 \text{ in})]}{3}$$

$$V_1 \approx \frac{(3.1416)(6.500 \text{ in})(1.474 \text{ sq in} + 0.4956 \text{ sq in} + 0.8547 \text{ sq in})}{3}$$

$$V_1 \approx 19.223 \text{ cu in}$$

- Find the volume of the 2.428-inch diameter cylinder.

  $V = A_b h$

  $r = 0.5(2.428 \text{ in}) = 1.214 \text{ in}$

  $V_2 \approx [3.1416(1.214 \text{ in})^2](1.950 \text{ in}) \approx 9.029 \text{ cu in}$

- Find the volume of the 4.238-inch diameter cylinder.

  $V = A_b h$

  $V_3 = [3.1416(2.119 \text{ in})^2](0.525 \text{ in}) \approx 7.406 \text{ cu in}$

- Find the volume of the 0.612-inch diameter through hole.

  $V = A_b h$

  $r = 0.5(0.612 \text{ in}) = 0.306 \text{ in}$

  $h = 6.500 \text{ in} + 1.950 \text{ in} + 0.525 \text{ in} = 8.975 \text{ in}$

  $V_4 = [3.1416(0.306 \text{ in})^2](8.975 \text{ in}) \approx 2.640 \text{ cu in}$

- Find the volume of the metal.

  $V_T \approx 19.223 \text{ cu in} + 9.029 \text{ cu in} + 7.406 \text{ cu in} - 2.640 \text{ cu in} \approx 33.018 \text{ cu in}$

  $V_T \approx 33.0 \text{ cu in}$ Ans

$V_1 \approx$ [π] [×] 6.5 [×] [(] 1.214 [x^2] [+] .704 [x^2] [+] 1.214 [×] .704 [)] [=]
[÷] [3] [=] 19.22282111

$V_2 \approx$ [π] [×] 1.214 [x^2] [×] 1.95 [=] 9.028630039

$V_3 \approx$ [π] [×] 2.119 [x^2] [×] .525 [=] 7.405784826

$V_4 \approx$ [π] [×] .306 [x^2] [×] [(] 6.5 [+] 1.95 [+] .525 [)] [=] 2.640141373

Volume of metal: $V_T \approx$ 19.223 [+] 9.0286 [+] 7.4058 [−] 2.6401 [=] 33.0173

$V_T \approx 33.0 \text{ cu in}$ Ans

Application

Volumes of Spheres

Compute the volume of each sphere 1 through 8. Round the answer to 2 decimal places.

1. 2.15-centimeter radius ________
2. 0.28-meter diameter ________
3. 7.60-inch diameter ________
4. 1.2-foot radius ________
5. 4.78-inch radius ________
6. 0.075-meter diameter ________
7. 16.2-centimeter diameter ________
8. 1-foot, 3-inch diameter ________

Solve these problems.

9. A thrust bearing contains 18 steel balls. The steel used weighs 0.283 pound per cubic inch. The diameter of each ball is 0.240 inch. Compute the total weight of balls in the bearing. Round the answer to 3 decimal places. ________
10. A vat in the shape of a hemisphere with an 18.00-inch inside diameter contains liquid. What is the capacity of the vat in gallons? There are 231 cubic inches per gallon. Round the answer to 2 decimal places. ________
11. Spheres are formed from molten bronze. The diameter of the mold in which the spheres are formed is 6.26 centimeters. When the bronze spheres solidify (turn solid) they shrink by 6% of the molten-state volume. Compute the volume of the sphere after the bronze solidifies. Round the answer to 1 decimal place. ________

12. A company produces spherical copper parts with 1.46-inch diameters. Each part weighs 0.523 pound. A larger spherical copper part 2.27 inches in diameter is planned. Determine the weight of the larger part. Round the answer to 2 decimal places. ____________

13. Pieces in the shape of spheres are to be made of lead castings. Each piece weighs 27.50 pounds. Find the required diameter of a casting. Lead weighs 707.7 pounds per cubic foot. Round the answer to the nearest hundredth inch. ____________

14. A plastic products manufacturer produces plastic container covers in the shape of hemispheres that are 0.05 centimeter thick. The outside diameter of a cover is 38.60 centimeters. The material expense for the covers is based on a cost of $0.0014 per cubic centimeter. What is the material cost for a production run of 55,000 covers? Round the answer to the nearest dollar. ____________

Volumes of Composite Solid Figures

Solve these problems.

15. Find the weight of the steel baseplate shown. Steel weighs 490 pounds per cubic foot. Round the answer to the nearest pound. ____________

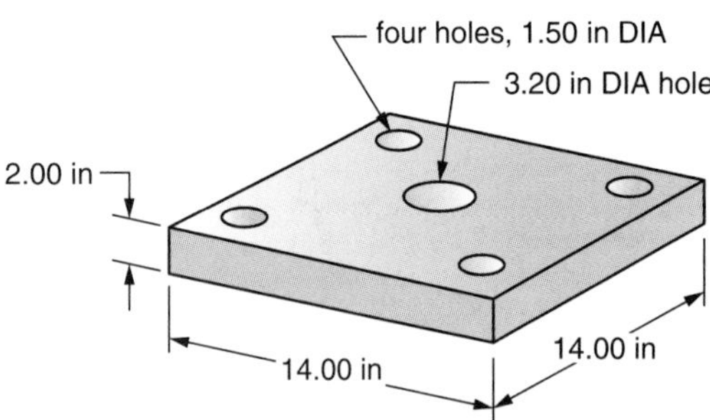

16. Compute the capacity, in liters, of the container shown. Round the answer to the nearest tenth liter. One liter contains 1000 cubic centimeters. ____________

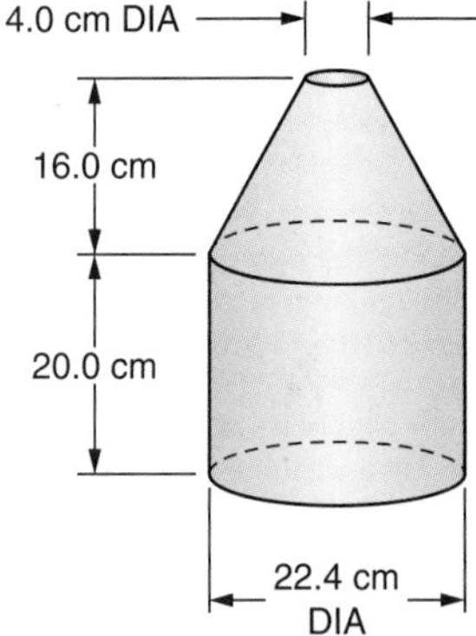

17. A seamless brass tube and brass flange assembly is shown. The tube is pressed fit into the full 1.250-inch plate thickness. The brass used weighs 0.305 pound per cubic inch. Find the total weight of the assembly. Round the answer to the nearest tenth pound. ____________

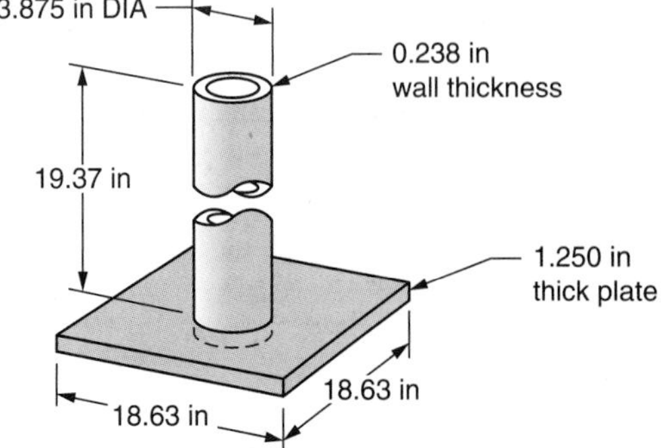

18. Find the number of cubic centimeters of material contained in the jig bushing shown. Round the answer to the nearest tenth cubic centimeter. ____________

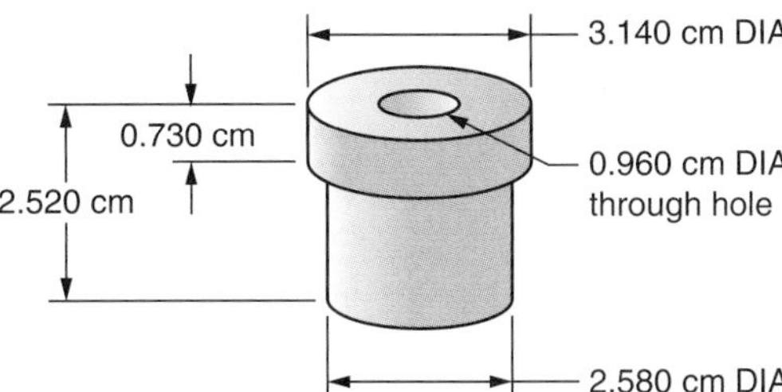

19. Find the weight of the cast-iron angle iron shown. Cast iron weighs 0.284 pounds per cubic inch. All dimensions are in inches. Round the answer to the nearest tenth pound. ____________

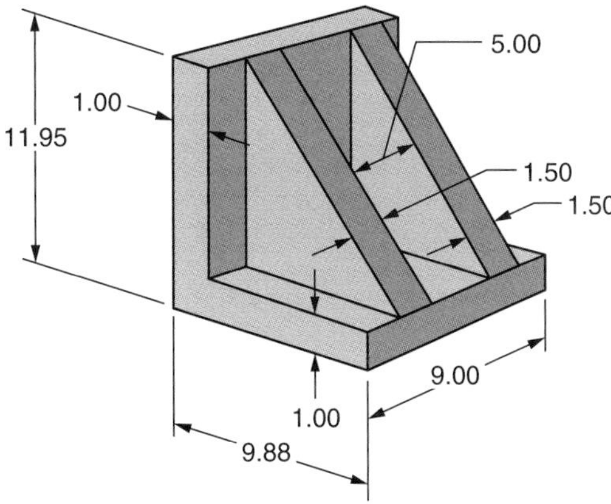

20. Compute the volume of the machined piece shown. All dimensions are in inches. Round the answer to the nearest tenth cubic inch. ____________

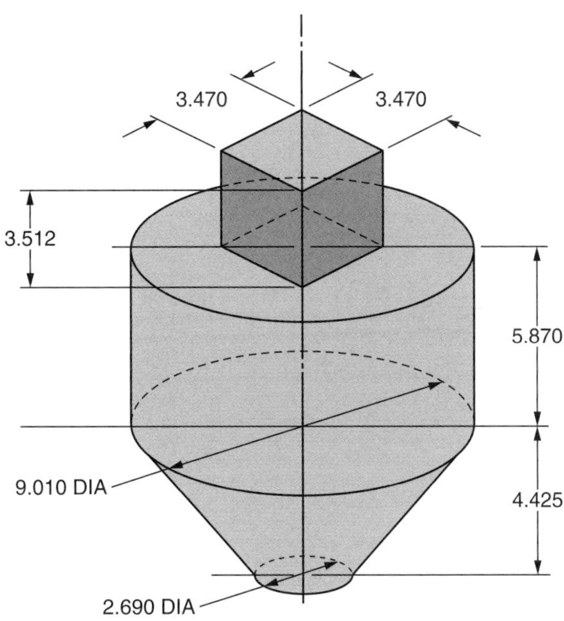

21. Compute the weight of the steel bearing washer shown. Consider the 4 ribs as triangular prisms, disregarding the slight curvatures connecting to the surface of the cylindrical shapes. The steel used weighs 0.283 pound per cubic inch. All dimensions are in inches. Round the answer to the nearest hundredth pound. __________

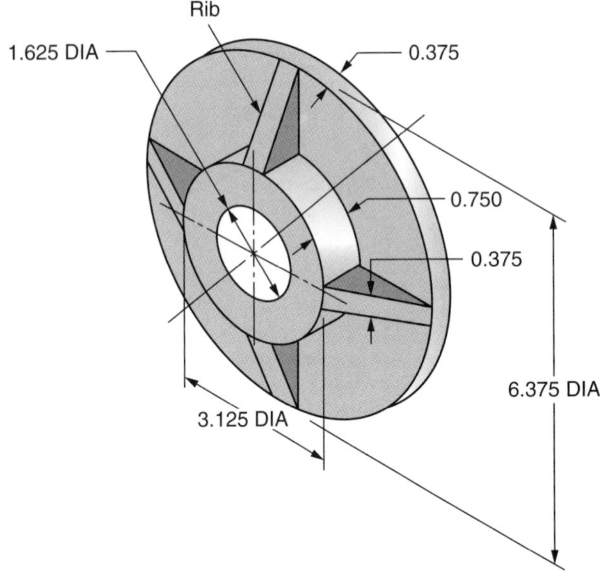

22. Compute the number of cubic centimeters of material in the locating saddle shown. Round the answer to the nearest cubic centimeter. __________

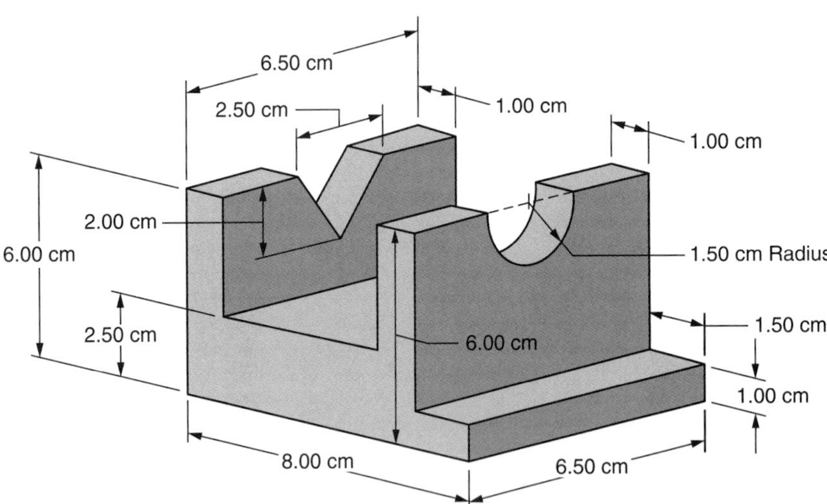

Achievement Review—Section Six

Objective ***You should be able to solve problems in this Achievement Review by applying the principles and methods covered in Units 53–58.***

Express each area in problems 1 through 6 as indicated. Round each answer to the same number of significant digits as in the original quantity.

1. 356.4 square inches as square feet ________
2. 1.030 square feet as square inches ________
3. 1.04 square centimeters as square millimeters ________
4. 125 000 square centimeters as square meters ________
5. 508 square millimeters as square inches ________
6. 5.92 square inches as square centimeters ________

The following problems, 7 through 19, involve rectangles, parallelograms, trapezoids, and triangles. Given certain values, find the unknown value for each. Where necessary, round the answers to 1 decimal place.

7. Rectangle: length = 17.500 in, width = 9.000 in, area = ________
8. Rectangle: width = 0.20 m, area = 0.12 m^2, length = ________
9. Rectangle: length = 43.70 mm, area = 851.6 mm^2 , width = ________
10. Parallelogram: base = 17.71 cm, height = 12.07 cm, area = ________
11. Parallelogram: base = 3.56 in, area = 13.83 sq in, height = ________
12. Parallelogram: height = 0.25 ft, area = 0.22 sq ft, base = ________
13. Trapezoid: height = 14.75 cm, base (b_1) = 23.06 cm, base (b_2) = 17.66 cm, area = ________
14. Trapezoid: height = 0.42 m, base (b_1) = 2.98 m, area = 0.87 m^2, base (b_2) = ________
15. Trapezoid: base (b_1) = 24.82 in, base (b_2) = 21.77 in, area = 202.7 sq in, height = ________
16. Triangle: base = 7.60 cm, height = 5.50 cm, area = ________
17. Triangle: height = 4.36 in, area = 32.7 sq in, base = ________
18. Triangle: side a = 12.62 cm, side b = 8.04 cm, side c = 16.56 cm, area = ________
19. Pieces in the shape of parallelograms are stamped from rectangular strips of stock as shown. If 24 pieces are stamped from a strip, how many square inches of strip are wasted? Round the answer to 1 decimal place. ________

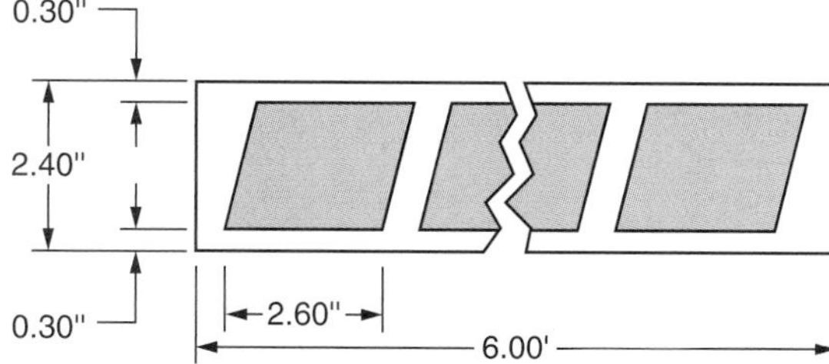

20. The cross section of a dovetail slide is shown. Before the dovetail was cut, the cross section was rectangular in shape. Find the cross-sectional area of the dovetail slide. Round the answer to 1 decimal place. ______

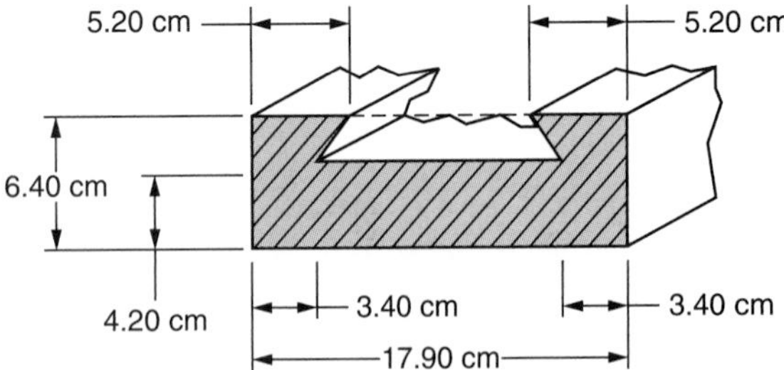

21. Compute the area of the pattern shown. Round the answer to the nearest square centimeter. ______

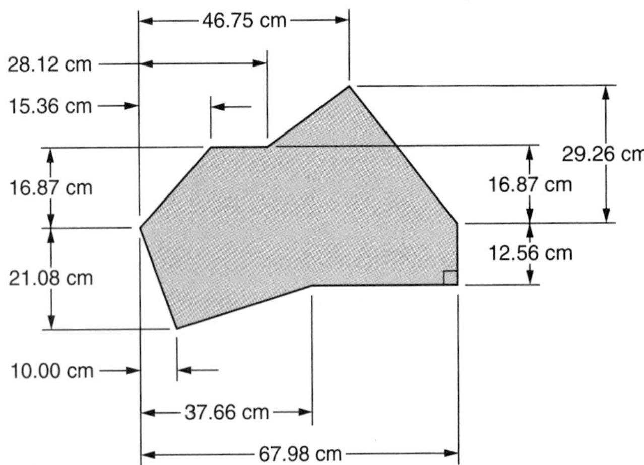

The following problems, 22 through 27, involve circles and sectors. Where necessary, round the answer to 1 decimal place.

22. Circle: radius = 14.86 in, area = ______
23. Circle: diameter = 28.60 cm, area = ______
24. Circle: area = 0.62 sq ft, radius = ______
25. Sector: radius = 5.50 in, central angle = 120.0°, area = ______
26. Sector: central angle = 230°25', area = 54.36 sq in, radius = ______
27. Sector: radius = 0.200 m, area = 0.0300 m^2, central angle = ______

Find the area of each of the segments *ACB* for problems 28 through 30. Refer to the figure shown. Round the answer to 1 decimal place.

	Area of Isosceles △*AOB*	Area of Sector *OACB*	Area of Segment *ACB*
28.	58.0 cm^2	72.3 cm^2	
29.	135.3 sq in	207.8 sq in	
30.	587.0 sq in	719.5 sq in	

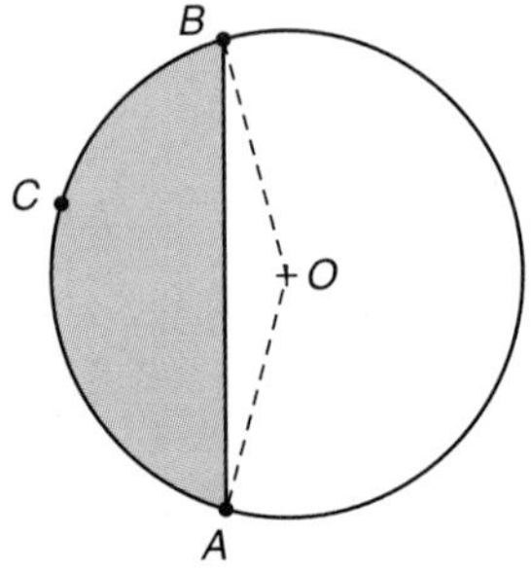

31. Find the area of the shaded segment shown. Round the answer to 2 decimal places. ________

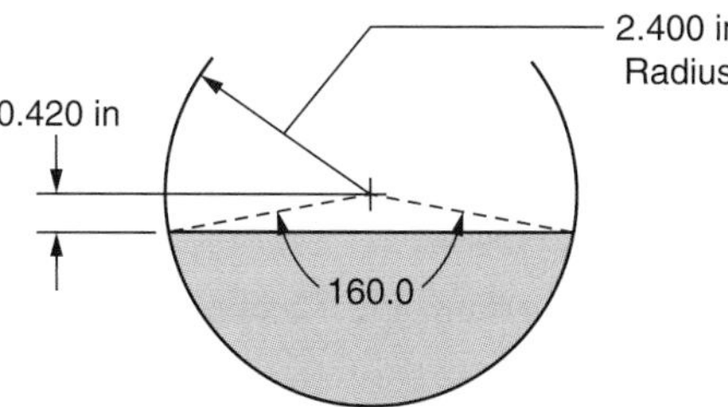

32. A circular plate with a 15.00-inch radius is cut from a rectangular aluminum piece that is 3'0.0" wide and 4'0.0" long. Before the circular plate was cut, the rectangular piece weighed 30.00 pounds. Find the weight of the circular plate. Round the answer to the nearest tenth pound. ________
33. Compute the cross-sectional area of the grooved block shown. Round the answer to 1 decimal place. ________

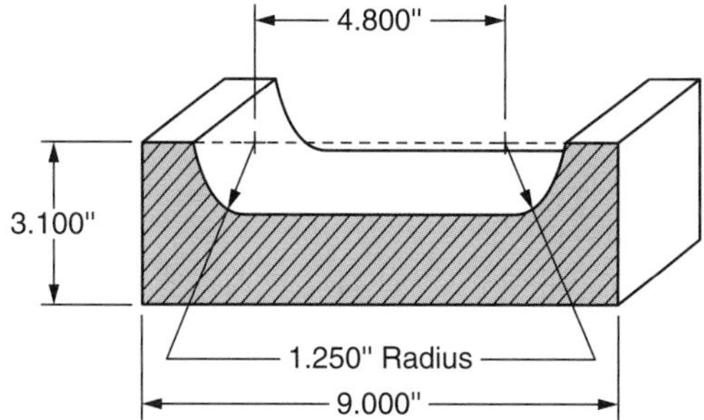

34. Find the area of the pattern shown. Round the answer to the nearest square centimeter. ________

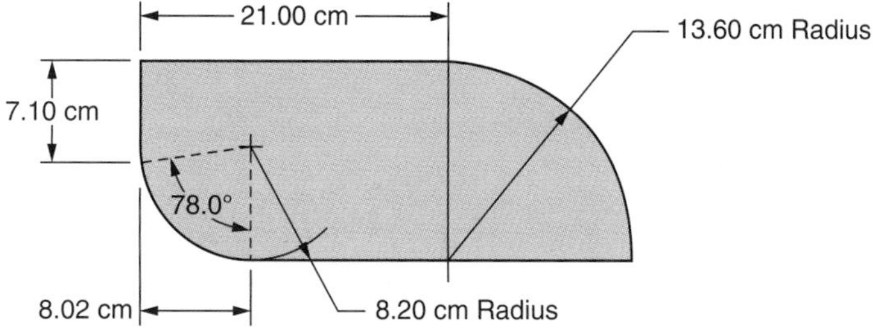

Express each volume in problems 35 through 40 as indicated. Round each answer to the same number of significant digits as in the original quantity.

35. 1.250 cubic feet as cubic inches ________
36. 518.4 cubic inches as cubic feet ________
37. 7.49 cubic centimeters as cubic millimeters ________
38. 29 040 cubic centimeters as cubic meters ________
39. 4.28 cubic inches as cubic centimeters ________
40. 29.8 cubic millimeters as cubic inches ________

Solve these prism and cylinder problems. Where necessary, round the answers to 2 decimal places unless otherwise specified.

41. Compute the volume of a prism with a base area of 220.0 square centimeters and a height of 7.600 centimeters. ________
42. Find the volume of a right circular cylinder that has a height of 4.600 inches and a base area of 53.00 square inches. ________

43. Compute the height of a prism with a base area of 2.7 square feet and a volume of 4.86 cubic feet. __________

44. A solid right cylinder 9.55 centimeters high contains 1910 cubic centimeters of material. Compute the cross-sectional area of the cylinder. __________

45. Each side of a square steel plate is 8.70 inches long. The plate is 2.85 inches thick. Compute the number of cubic inches of steel contained in the plate. __________

46. A circular aluminum casting 2.63 inches thick has a 22.08-inch diameter. The aluminum used for the casting weighs 168.5 pounds per cubic foot. Find the weight of the casting. Round the answer to the nearest pound. __________

47. Find the capacity, in gallons, of a cylindrical container with a base diameter of 6.75 inches and a height of 8.12 inches. One cubic foot contains 7.5 gallons. __________

48. A cylindrical 1.000-liter vessel has a base diameter of 5.00 centimeters. How high is the vessel? __________

49. A steel forging is shown. Compute the number of cubic centimeters of steel contained in the forging. __________

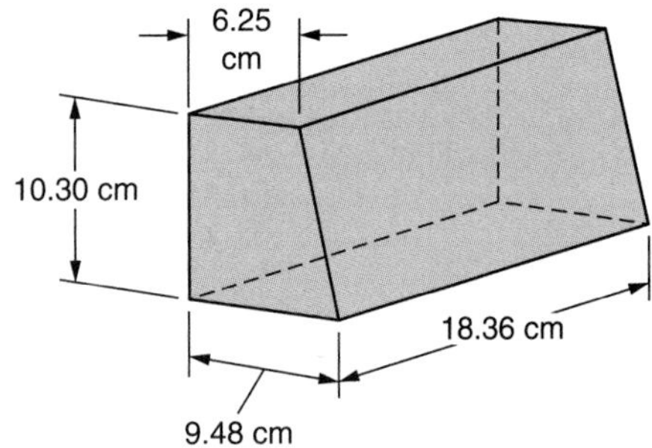

50. A length of brass pipe is shown.

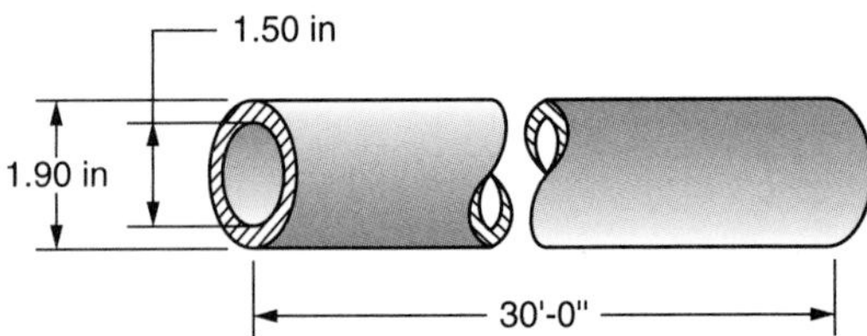

a. Find the number of cubic inches of brass contained in the pipe. Round the answer to the nearest cubic inch. __________

b. Brass weighs 526 pounds per cubic foot. What is the weight of the pipe? Round the answer to the nearest pound. __________

Solve these pyramid and cone problems. Where necessary, round the answer to 2 decimal places unless otherwise specified.

51. Find the volume of a right circular cone with a base area of 0.800 square inch and a height of 1.240 inches. __________

52. Compute the volume to the nearest hundredth cubic meter of a regular pyramid that has a height of 13.70 centimeters and a base area of 2010.3 square centimeters. __________

53. A regular pyramid with a base area of 54.6 square inches contains 210.5 cubic inches of material. Find the height of the pyramid. __________

54. Compute the base area of a right circular cone that is 15.8 centimeters high and has a volume of 1070 cubic centimeters. __________

55. The frustum of a right circular cone has a larger base area of 40.0 square centimeters and a smaller base area of 19.0 square centimeters. The height is 22.0 centimeters. Find the volume. Round the answer to the nearest cubic centimeter. __________

56. The frustum of a regular pyramid with square bases has a larger base perimeter of 26.10 inches and a smaller base area of 18.60 square inches. The height is 5.63 inches. Find the volume. ________

57. A solid brass casting in the shape of a right circular cone has a base diameter of 4.36 inches and a height of 3.94 inches. Find the weight of the casting. Brass weighs 0.302 pound per cubic inch. ________

58. A vessel in the shape of a right circular cone has a capacity of 0.690 liter. The base diameter is 12.30 centimeters. What is the height of the vessel? One liter contains 1000 cubic centimeters. ________

59. The stainless steel front of a nozzle assembly is shown. Determine the weight of the piece. The stainless steel used weighs 7.8 grams per cubic centimeter. Round the answer to the nearest tenth gram. ________

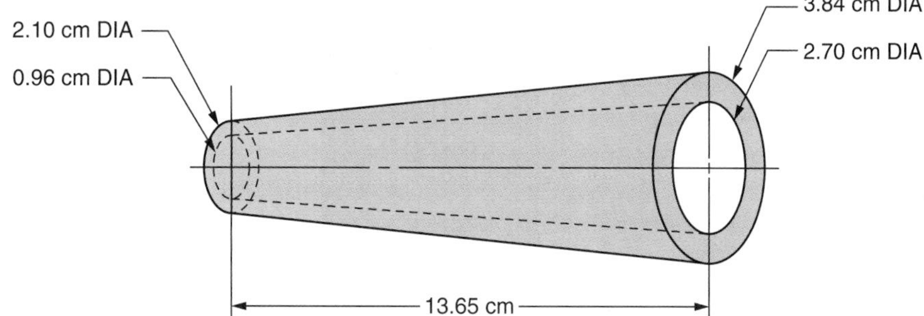

60. A cast brass container in the shape of a frustum of a regular pyramid with square bases is shown. The casting wall thickness is 1.40 centimeters. All dimensions shown are outside dimensions. Compute the capacity of the container in liters. One liter contains 1000 cubic centimeters. Round the answer to the nearest tenth liter. ________

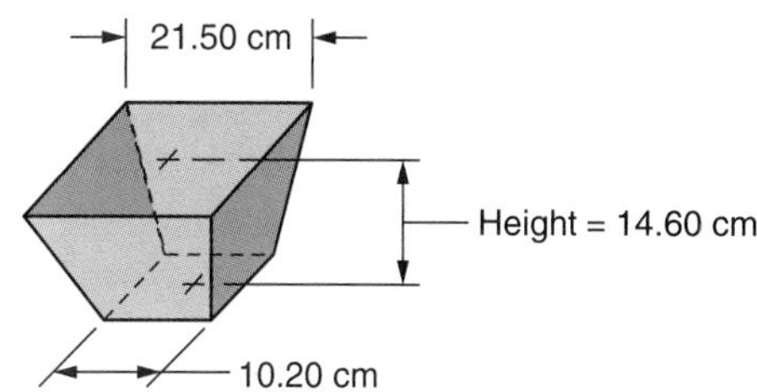

For each sphere, problems 61 through 65, determine the volume or diameter.

61. Diameter = 3.87 inches; Volume = Round the answer to the nearest tenth cubic inch. ________

62. Diameter = 9.050 centimeters; Volume = Round the answer to the nearest cubic centimeter. ________

63. Diameter = 0.308 foot; Volume = Round the answer to the nearest hundredth cubic foot. ________

64. Volume = 42.98 cubic inches; Diameter = Round the answer to the nearest hundredth inch. ________

65. Volume = 775 cubic centimeters; Diameter = Round the answer to the nearest tenth centimeter. ________

66. The side view of a steel roundhead (hemisphere) rivet is shown. What is the weight of the rivet? Steel weighs 0.283 pound per cubic inch. Round the answer to 3 decimal places. ________

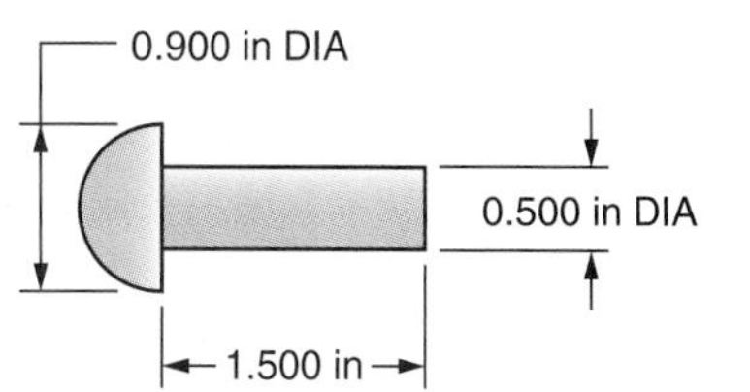

67. The material cost of a solid bronze sphere with a diameter of 3.80 centimeters is \$1.05. Compute the material cost of a solid bronze sphere with a 5.70-centimeter diameter. Round the answer to the nearest cent. ________

68. A hollow glass sphere has an outside circumference (great circle) of 23.80 centimeters. The wall thickness of the sphere is 0.5 centimeter. Glass weighs 1.50 grams per cubic centimeter. Round the answers for a and b to 1 decimal place.
    a. Find the number of cubic centimeters of glass contained in the sphere. ________
    b. Compute the weight of the sphere. ________

69. A solid aluminum support base is shown. The top and the bottom sections are in the shape of prisms with square bases. The middle section is in the shape of a frustum of a regular pyramid with square bases. Compute the volume in cubic feet to the nearest hundredth cubic foot. ________

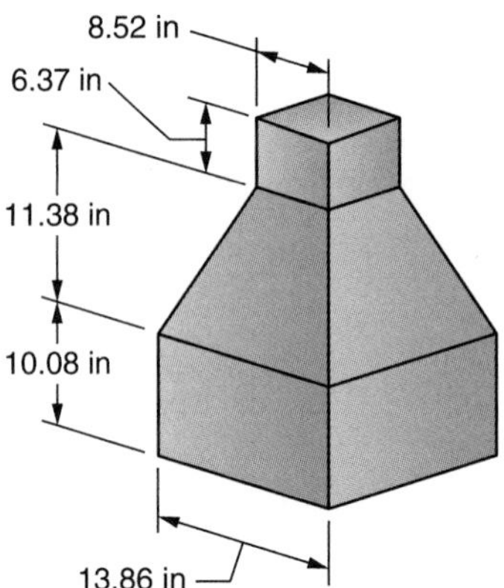

70. Compute the number of cubic centimeters of iron required for the cast-iron plate shown. The plate is 3.50 centimeters thick. Round the answer to the nearest cubic centimeter. ________

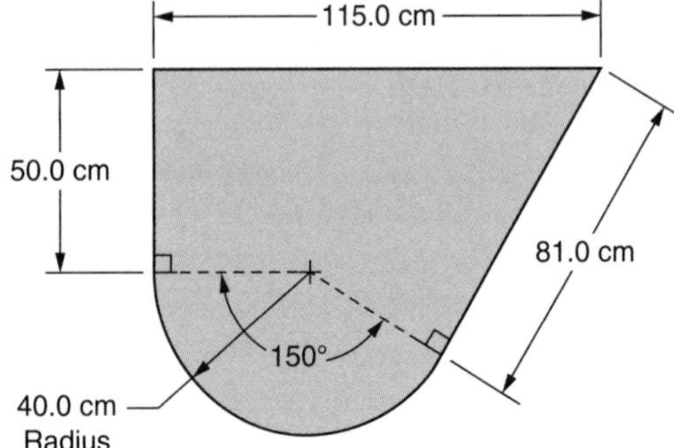

Trigonometry

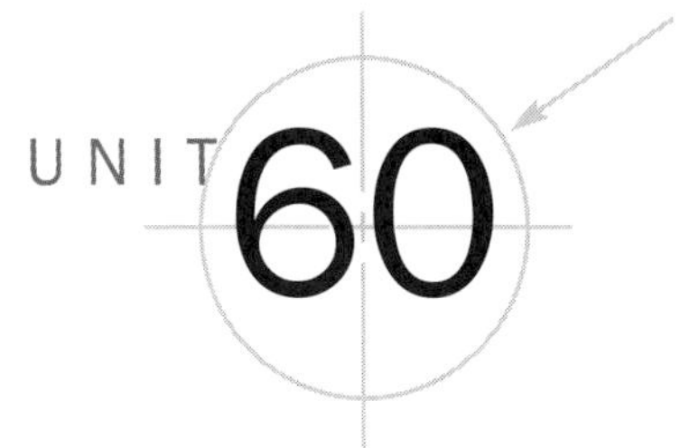

Introduction to Trigonometric Functions

Objectives ***After studying this unit you should be able to***

- Identify the sides of a right triangle with reference to any angle.
- State the ratios of the six trigonometric functions in relation to given triangles.
- Find functions of angles given in decimal degrees and degrees, minutes, and seconds.
- Find angles in decimal degrees and degrees, minutes, and seconds of given functions.

Trigonometry is the branch of mathematics that is used to compute unknown angles and sides of triangles. The word *trigonometry* is derived from the Greek words for triangle and measurement. Trigonometry is used in the design of products. It is also used in the planning. setting up, and processing of manufactured products.

The machines that produce the products could not be made without the use of trigonometry. It is important that machinists, tool and die makers, drafters, designers, and related occupations apply trigonometric principles.

Practical machine shop problems are often solved by using a combination of elements of algebra, geometry, and trigonometry. Therefore, it is essential to develop the ability to analyze a problem in order to relate and determine the mathematical principles that are involved in its solution. Then the problem must be worked in clear, orderly steps, based on mathematical facts.

When solving a problem, it is important to understand the trigonometric operations involved rather than to mechanically "plug in" values. Attempting to solve trigonometry problems without understanding the principles involved will prove to be unsuccessful, particularly in practical shop applications such as those found later in the text.

Ratio of Right Triangle Sides

In a right triangle, the ratio of two sides of the triangle determines the sizes of the angles, and the angles determine the ratio of two sides. Refer to the triangles shown in Figure 60-1. The size of angle A is determined by the ratio of side a to side b. When side $a = 1$ inch and side $b = 2$ inches, the ratio of a to b is 1:2 or 1/2 as in Figure 60-1(a). If side a is increased to 2 inches and side b remains 2 inches, the ratio of a to b is 1:1 or 1/1 as in Figure 60-1(b). Observe the increase in angle A as the ratio changed from 1/2 to 1/1 (see Figure 60-1(c)).

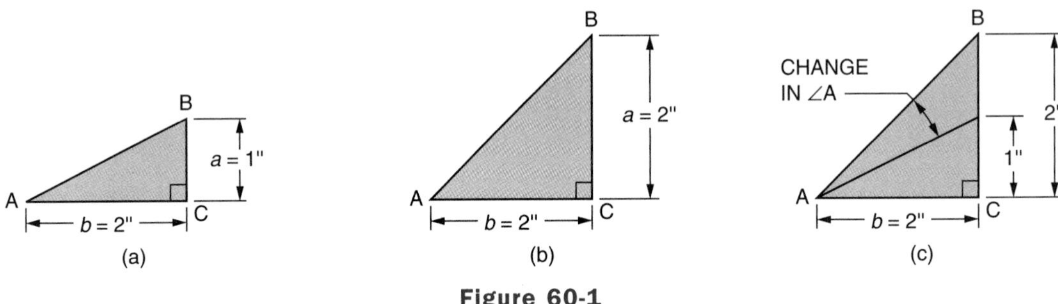

Figure 60-1

Identifying Right Triangle Sides by Name

The sides of a right triangle are named opposite side, adjacent side, and hypotenuse. The *hypotenuse* (hyp) is always the side opposite the right angle. It is always the longest side of a right triangle. The positions of the opposite and adjacent sides depend on the reference angle. The *opposite side* (opp side) is opposite the reference angle and the *adjacent side* (adj side) is next to the reference angle.

In Figure 60-2, ∠A is shown as the reference angle, side b is the adjacent side, and side a is the opposite side. In Figure 60-3, ∠B is the reference angle, side b is the opposite side, and side a is the adjacent side. It is important to be able to identify the opposite and adjacent sides of right triangles in reference to any angle regardless of the positions of the triangles.

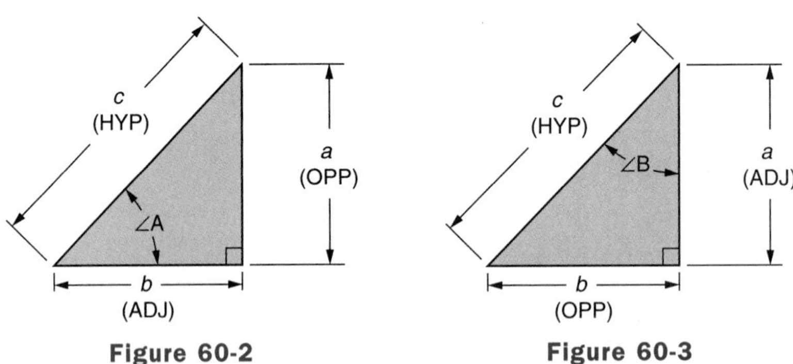

Figure 60-2 **Figure 60-3**

Trigonometric Functions: Ratio Method

There are two methods of defining trigonometric functions: the unity or unit circle method and the ratio method. Only the ratio method is presented in this book.

Since a triangle has three sides and a ratio is the comparison of any two sides, there are six different ratios. The names of the ratios are the sine, cosine, tangent, cotangent, secant, and cosecant.

The six trigonometric functions are defined in this table in relation to the triangle in Figure 60-4. The reference angle is A, the adjacent side is *b*, the opposite side is *a*, and the hypotenuse is *c*.

Function	Symbol	Definition of Function
sine of Angle A	sin A	$\sin A = \frac{\text{opp side}}{\text{hyp}} = \frac{a}{c}$
cosine of Angle A	cos A	$\cos A = \frac{\text{adj side}}{\text{hyp}} = \frac{b}{c}$
tangent of Angle A	tan A	$\tan A = \frac{\text{opp side}}{\text{adj side}} = \frac{a}{b}$
cotangent of Angle A	cot A	$\cot A = \frac{\text{adj side}}{\text{opp side}} = \frac{b}{a}$
secant of Angle A	sec A	$\sec A = \frac{\text{hyp}}{\text{adj side}} = \frac{c}{b}$
cosecant of Angle A	csc A	$\csc A = \frac{\text{hyp}}{\text{opp side}} = \frac{c}{a}$

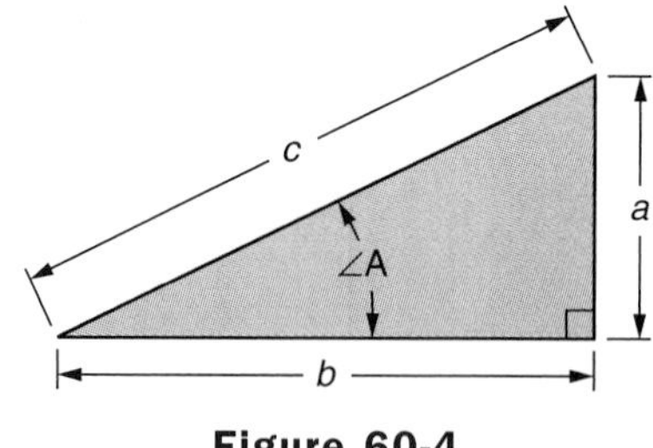

Figure 60-4

To properly use trigonometric functions, it is essential to know that the function of an angle depends upon the ratio of the sides and **not** the size of the triangle. The value of functions of similar triangles are the same regardless of the sizes of the triangles since the sides of similar triangles are proportional. For example, in the similar triangles shown in Figure 60-5, the function values of angle A are the same for the three triangles. The equality of the tangent function is shown. Each of the other five functions have equal values for the three similar triangles.

$$\text{In } \triangle ABC, \tan \angle A = \frac{0.500}{1.000} = 0.500$$

$$\text{In } \triangle ADE, \tan \angle A = \frac{0.800}{1.600} = 0.500$$

$$\text{In } \triangle AFG, \tan \angle A = \frac{1.200}{2.400} = 0.500$$

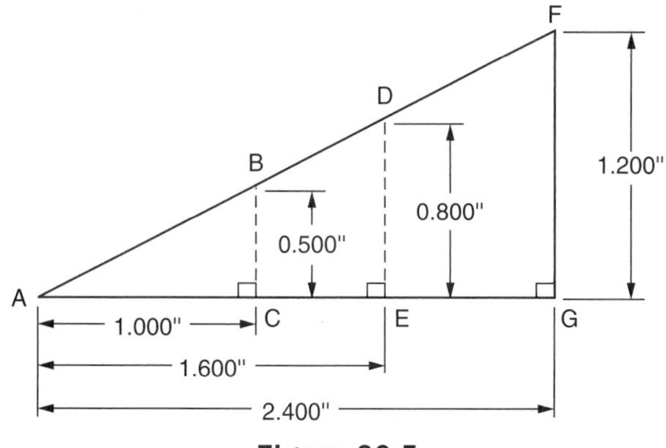

Figure 60-5

Customary and Metric Units of Angular Measure

As discussed in Unit 44, angular measure in the customary system is generally expressed in degrees and minutes or in degrees, minutes, and seconds for very precise measurements. In the metric system, the decimal degree is the preferred unit of measure. Unless otherwise specified, degrees and minutes or degrees, minutes, and seconds are to be used when solving customary system units of measure problems. Decimal degrees are to be used when solving metric system units of measure problems.

Determining Functions of Given Angles and Determining Angles of Given Functions

Calculator Applications

Determining functions of given angles or angles of given functions is readily accomplished using a calculator. As previously stated, calculator procedures vary among different makes of calculators. Also, different models of the same make calculator vary in some procedures. Generally, where procedures differ, there are basically two different procedures. Where relevant, both procedures are shown. However, because of the many makes and models of calculators, some procedures on your calculator may differ from the procedures shown. If so, it is essential that you refer to your user's guide or owner's manual.

The trigonometric keys, [sin], [cos], and [tan], calculate the sine, cosine, and tangent of the angle in the display. An angle can be measured in degrees, radians, or gradients. **When calculating functions of angles measured in degrees, be certain that the calculator is in the degree mode.** A calculator is in the degree mode when the abbreviation DEG or D appears in the display when the calculator is turned on.

Procedure for Determining the Sine, Cosine, and Tangent Functions

The appropriate function key, [sin], [cos], or [tan], is pressed first, and then the value of the angle is entered.

Examples Round each answer to 5 decimal places.

1. Determine the sine of 43°.

   [sin] 43 [=] 0.6819983601, 0.68200 Ans

2. Determine the cosine of 6.034°.

   [cos] 6.034 [=] 0.9944596918, 0.99446 Ans

3. Determine the tangent of 51.9162°.

   [tan] 51.9162 [=] 1.276090171, 1.27609 Ans

4. Determine the sine of 61°49'.

   [sin] 61 [° ' "] 49 [° ' "] [=] 0.8814408742, 0.88144 Ans

   or [sin] 61 [° ' "] [ENTER] 49 [° ' "] [▶] [ENTER] [ENTER]

   0.8814408742, 0.88144 Ans

5. Determine the tangent of 32°7'23".

   [tan] 32 [° ' "] 7 [° ' "] 23 [° ' "] [=]

   0.6278596985, 0.62786 Ans

   or [tan] 32 [° ' "] [ENTER] 7 [° ' "] [▶] [ENTER] 23 [° ' "] [▶] [▶] [ENTER] [ENTER]

   0.6278596985, 0.62786 Ans

Procedure for Determining the Cosecant, Secant, and Cotangent Functions

The cosecant, secant, and cotangent functions are reciprocal functions. The cosecant is the reciprocal of the sine.

$$\csc \angle A = \frac{1}{\sin \angle A}$$

The secant is the reciprocal of the cosine.

$$\sec \angle A = \frac{1}{\cos \angle A}$$

The cotangent is the reciprocal of the tangent.

$$\cot \angle A = \frac{1}{\tan \angle A}$$

Cosecants, secants, and cotangents are computed with the reciprocal key, [x^{-1}].

The appropriate function key, [sin], [cos], [tan], is pressed first. The value of the angle is entered next, then press [=], press [x^{-1}], press [=].

Examples Round each answer to 5 decimal places.

1. Determine the cosecant of 57.16°.

   [sin] 57.16 [=] [x^{-1}] [=] 1.190209506, 1.19021 Ans

2. Determine the secant of 13.795°.

   [cos] 13.795 [=] [x^{-1}] [=] 1.029701649, 1.02970 Ans

3. Determine the cotangent of 78.63°.

   [tan] 78.63 [=] [x^{-1}] [=] 0.2010905402, 0.20109 Ans

4. Determine the cosecant of 24°51'.

   [sin] 24 [° ' "] 51 [° ' "] [=] [x^{-1}] [=] 2.379569353, 2.37957 Ans

   or [sin] 24 [° ' "] [ENTER] 51 [° ' "] [▶] [ENTER] [ENTER] [x^{-1}] [ENTER]

   2.379569353, 2.37957 Ans

5. Determine the secant of 43°36'25".

   [cos] 43 [° ' "] 36 [° ' "] 25 [° ' "] [=] [x^{-1}] [=]

   1.381047089, 1.38105 Ans

   or [cos] 43 [° ' "] [ENTER] 36 [° ' "] [▶] [ENTER] 25 [° ' "] [▶] [▶] [ENTER] [ENTER]

   [x^{-1}] [ENTER] 1.381047089, 1.38105 Ans

Angles of Given Functions

Determining the angle of a given function is the inverse of determining the function of a given angle. When a certain function value is known, the angle can be found easily.

The term *arc* is often used as a prefix to any of the names of the trigonometric functions, such as arcsine, arctangent, etc. Such expressions are called inverse functions and they mean angles. For example, if sin 30°15' = 0.503774, then 30°15' = arcsin 0.503774 or 30°15' is the angle whose sine is 0.503774.

Arcsin is often written as $\sin^{-1}$, arccos is written as $\cos^{-1}$, and arctan is written as $\tan^{-1}$.

Procedure for Determining Angles of Given Functions

The procedure for determining angles of given functions varies somewhat with the make and model of calculator. With most calculators, the inverse functions are shown as second functions [$\sin^{-1}$], [$\cos^{-1}$], and [$\tan^{-1}$] above the function keys [sin], [cos], and [tan].

With some calculators, the function value is entered before the function key is pressed. With other calculators, the function key is pressed before the function value is entered.

The following examples show the procedure for determining angles of given functions. All examples show the procedures where [$\sin^{-1}$], [$\cos^{-1}$], and [$\tan^{-1}$] are the second functions. Remember, for certain calculators it is necessary to substitute [2nd] or [INV] in place of [SHIFT].

Examples

1. Find the angle whose tangent is 1.902. Round the answer to 2 decimal places.

   [SHIFT] [$\tan^{-1}$] 1.902 [=] 62.2662961, 62.27° Ans

2. Find the angle whose sine is 0.21256. Round the answer to 2 decimal places.

   [SHIFT] [$\sin^{-1}$] .21256 [=] 12.27241712, 12.27° Ans

3. Find the angle whose cosine is 0.732976. Give the answer in degrees, minutes, and seconds.

   [SHIFT] [$\cos^{-1}$] .732976 [=] [SHIFT] [←] 42°51'48.72", 42°51'49" Ans

   or [2nd] [$\cos^{-1}$] .732976 [ENTER] [° ' "] ◀ [ENTER] [ENTER] 42°51'49" Ans

Angles for the reciprocal functions—cosecant, secant, and cotangent are calculated using the reciprocal key, [1/x] or [x^{-1}].

Examples

1. Find the angle whose secant is 1.2263. Round the answer to 2 decimal places.

   [SHIFT] [$\cos^{-1}$] 1.2263 [x^{-1}] [=] 35.36701576, 35.37° Ans

2. Find the angle whose cotangent is 0.4166. Give the answer in degrees and minutes.

   [SHIFT] [$\tan^{-1}$] .4166 [x^{-1}] [=] [SHIFT] [←] 67°23'0.2", 67°23' Ans

   or [2nd] [$\tan^{-1}$] .4166 [x^{-1}] [° ' "] ◀ [ENTER] [ENTER] 67°23'0.2", 67°23' Ans

Application

Identifying Right Triangle Sides by Name

With reference to ∠1, name each of the sides of the following triangles as opposite, adjacent, or hypotenuse.

1. Name sides r, x, and y.

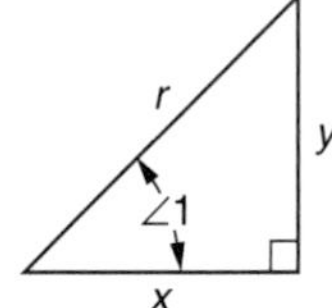

r = ______________
x = ______________
y = ______________

2. Name sides r, x, and y.

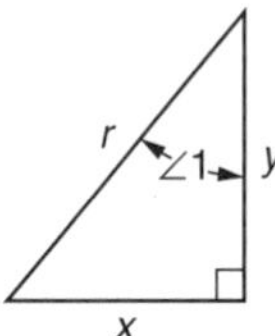

r = ______________
x = ______________
y = ______________

3. Name sides a, b, and c.

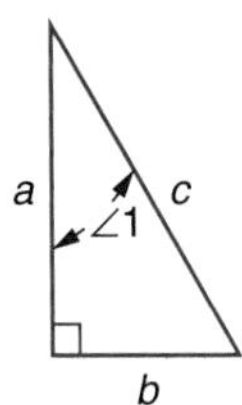

a = ______
b = ______
c = ______

4. Name sides a, b, and c.

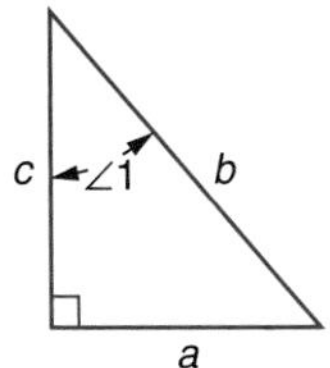

a = ______
b = ______
c = ______

5. Name sides a, b, and c.

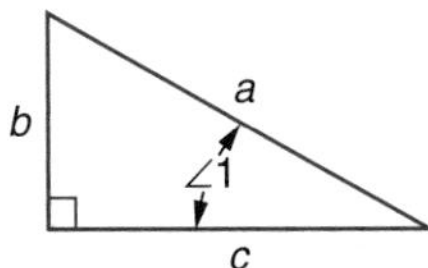

a = ______
b = ______
c = ______

6. Name sides d, m, and p.

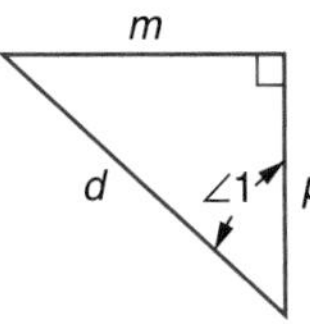

d = ______
m = ______
p = ______

7. Name sides d, m, and p.

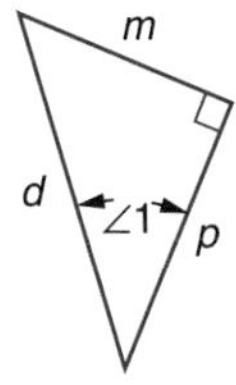

d = ______
m = ______
p = ______

8. Name sides e, f, and g.

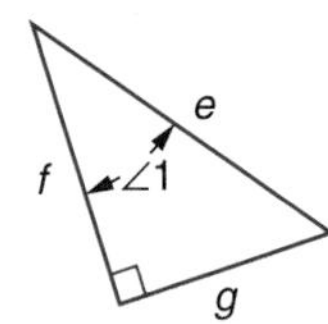

e = ______
f = ______
g = ______

9. Name sides h, k, and l.

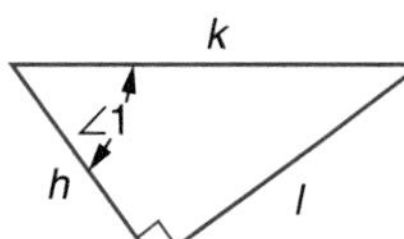

h = ______
k = ______
l = ______

10. Name sides h, k, and l.

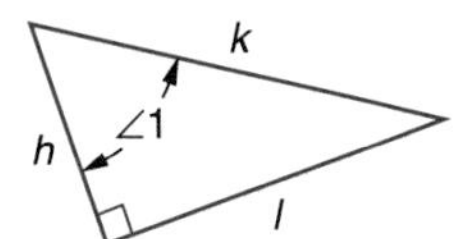

h = ______
k = ______
l = ______

11. Name sides m, p, and s.

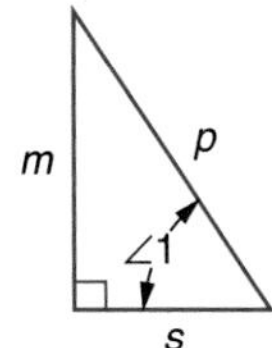

m = ______
p = ______
s = ______

12. Name sides m, p, and s.

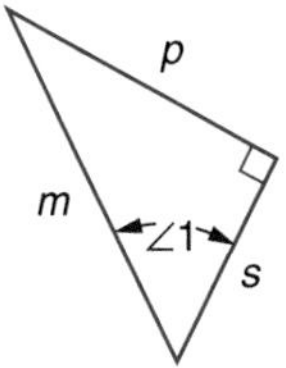

m = ______
p = ______
s = ______

13. Name sides m, r, and t.

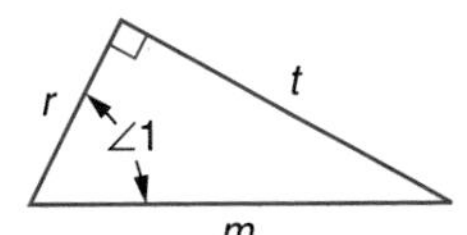

m = ______
r = ______
t = ______

14. Name sides m, r, and t.

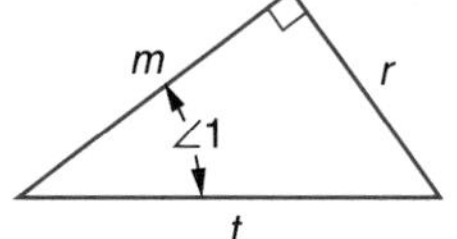

m = ______
r = ______
t = ______

15. Name sides *f*, *g*, and *h*.

$f =$ ______________

$g =$ ______________

$h =$ ______________

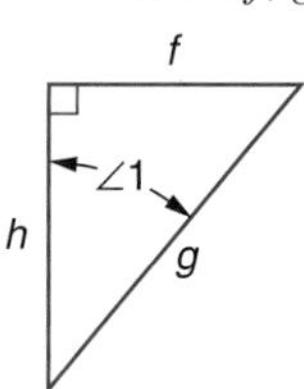

16. Name sides *f*, *g*, and *h*.

$f =$ ______________

$g =$ ______________

$h =$ ______________

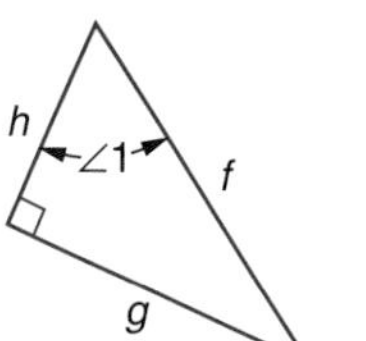

Trigonometric Functions

The sides of each of the following triangles are labeled with different letters. State the ratio of each of the six functions in relation to $\angle 1$ for each of the triangles. For example, for the triangle in exercise number 17, $\sin \angle 1 = \frac{y}{r}$, $\cos \angle 1 = \frac{x}{r}$, $\tan \angle 1 = \frac{y}{x}$, $\cot \angle 1 = \frac{x}{y}$, $\sec \angle 1 = \frac{r}{x}$, and $\csc \angle 1 = \frac{r}{y}$.

17. ______________ ______________

______________ ______________

______________ ______________

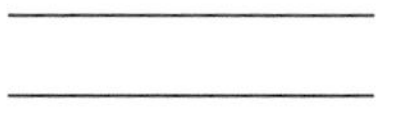

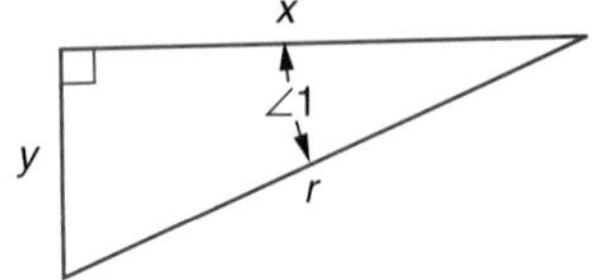

18. ______________ ______________

______________ ______________

______________ ______________

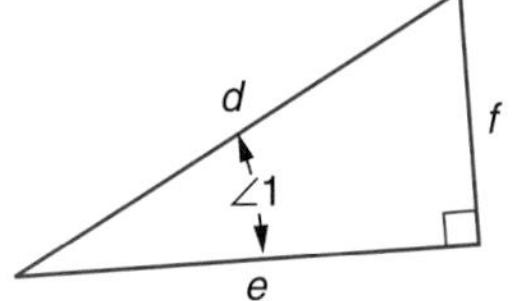

19. ______________ ______________

______________ ______________

______________ ______________

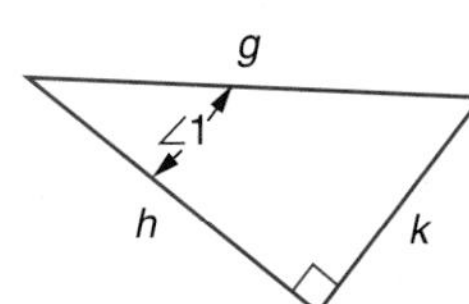

20. ______________ ______________

______________ ______________

______________ ______________

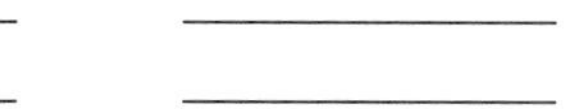

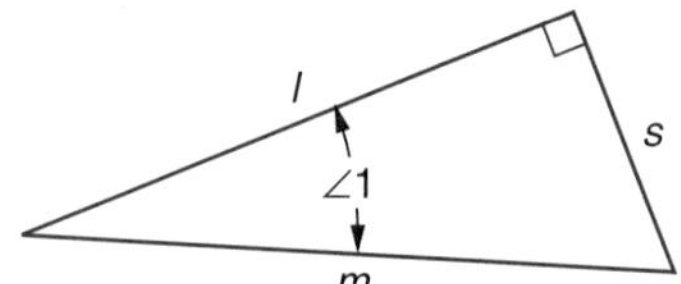

21. ______________ ______________

______________ ______________

______________ ______________

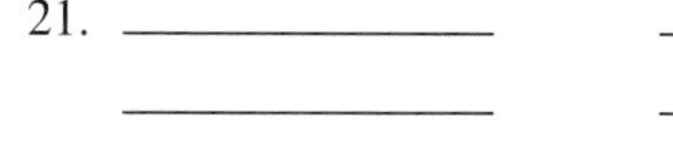

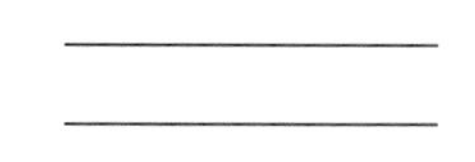

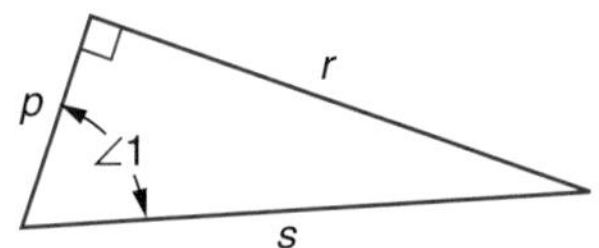

22. ______________ ______________

______________ ______________

______________ ______________

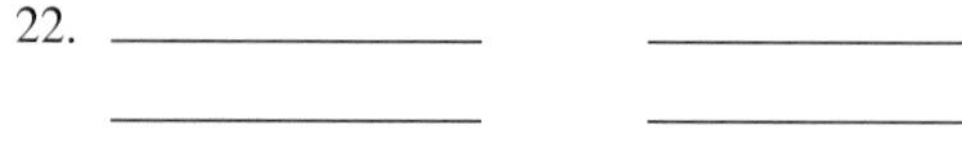

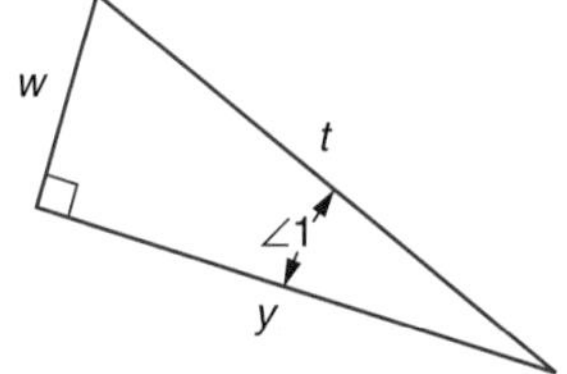

23. Three groups of triangles are given below. Each group consists of four triangles. Within each group, name the triangles—a, b, c, or d—in which angles A are equal.

Group 1 ____________

a. A, 2", 1" b. A, 6", 3" c. A, 4", 3" d. A, 1", 0.5"

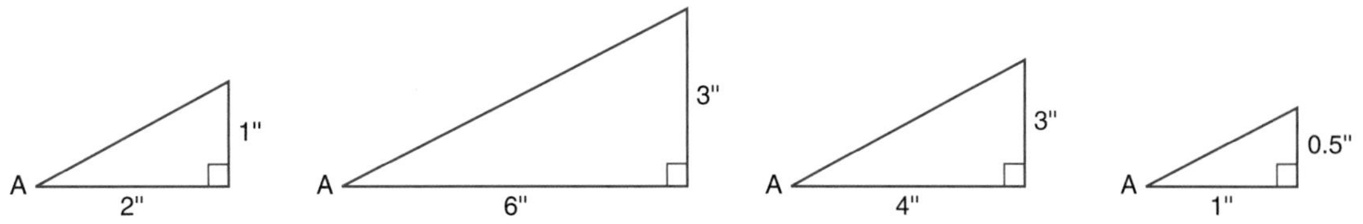

Group 2 ____________

a. A, 30 mm, 40 mm b. A, 60 mm, 80 mm c. A, 15 mm, 20 mm d. A, 15 mm, 20 mm

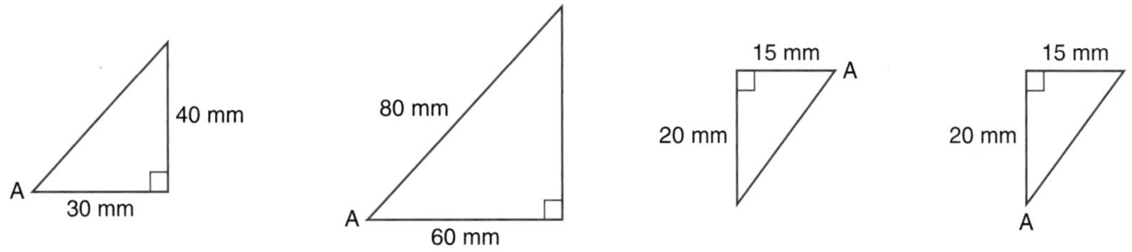

Group 3 ____________

a. A, 4", 10" b. A, 10", 4" c. A, 5", 2" d. A, 12.5", 5"

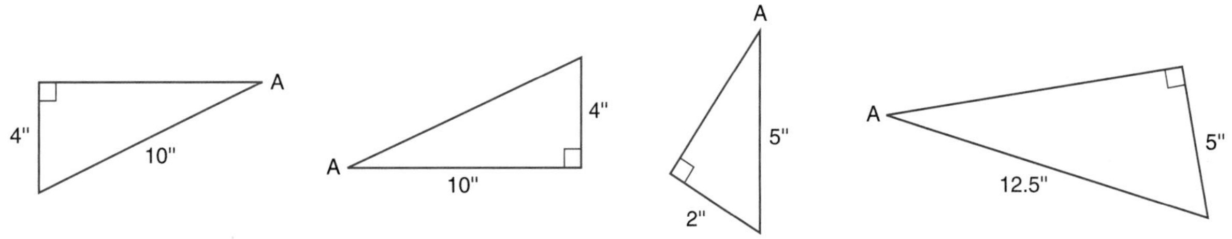

Determine the sine, cosine, or tangent functions of the following angles. Round the answers to 5 decimal places.

24. sin 36° ____________
25. cos 53° ____________
26. tan 47° ____________
27. cos 18° ____________
28. sin 79° ____________
29. cos 4° ____________
30. tan 65.18° ____________
31. sin 27.06° ____________
32. tan 12.92° ____________
33. cos 4.63° ____________
34. tan 73.86° ____________
35. sin 50.05° ____________
36. cos 16.77° ____________
37. sin 0.86° ____________
38. tan 59.89° ____________
39. cos 60.605° ____________
40. cos 77.144° ____________
41. tan 10°18' ____________
42. sin 26°29' ____________
43. sin 6°53' ____________
44. cos 19°42' ____________
45. sin 71°59' ____________
46. tan 42°36' ____________
47. sin 20°28' ____________
48. cos 6°16' ____________
49. tan 37°26'12" ____________
50. tan 9°4'50" ____________
51. cos 86°30'38" ____________
52. sin 53°46'19" ____________
53. tan 70°51'44" ____________

Determine the cosecant, secant, or cotangent functions of the following angles. Round the answers to 5 decimal places.

54. csc 27° ________
55. sec 56° ________
56. cot 19° ________
57. cot 36.97° ________
58. sec 77.08° ________
59. csc 6.904° ________
60. cot 31.081° ________
61. sec 20°16' ________
62. csc 46°27' ________
63. csc 76°0'15" ________
64. cot 2°58'59" ________

Determine the value of angle A in decimal degrees for each of the given functions. Round the answers to the nearest hundredth of a degree.

65. sin A = 0.83692 ________
66. cos A = 0.23695 ________
67. tan A = 0.59334 ________
68. cos A = 0.97370 ________
69. tan A = 3.96324 ________
70. sin A = 0.77376 ________
71. sin A = 0.02539 ________
72. tan A = 1.56334 ________
73. tan A = 0.09632 ________
74. cos A = 0.20893 ________
75. cos A = 0.87736 ________
76. sin A = 0.10532 ________
77. cos A = 0.38591 ________
78. tan A = 0.67871 ________
79. sin A = 0.63634 ________
80. cos A = 0.05332 ________
81. sec A = 1.58732 ________
82. csc A = 2.08363 ________
83. cot A = 0.89538 ________
84. cot A = 6.06790 ________
85. csc A = 5.93632 ________
86. sec A = 1.02353 ________

Determine the value of angle A in degrees and minutes for each of the given functions. Round the answers to the nearest minute.

87. cos A = 0.23076 ________
88. tan A = 0.56731 ________
89. sin A = 0.92125 ________
90. tan A = 4.09652 ________
91. cos A = 0.03976 ________
92. sin A = 0.09741 ________
93. sin A = 0.73204 ________
94. tan A = 0.95300 ________
95. cos A = 0.00495 ________
96. cos A = 0.89994 ________
97. sin A = 0.30536 ________
98. tan A = 7.60385 ________
99. cos A = 0.69304 ________
100. tan A = 3.03030 ________
101. sin A = 0.70705 ________
102. cos A = 0.90501 ________
103. csc A = 1.38630 ________
104. sec A = 5.05377 ________
105. cot A = 0.27982 ________
106. csc A = 2.02103 ________
107. sec A = 9.90778 ________
108. cot A = 8.03012 ________

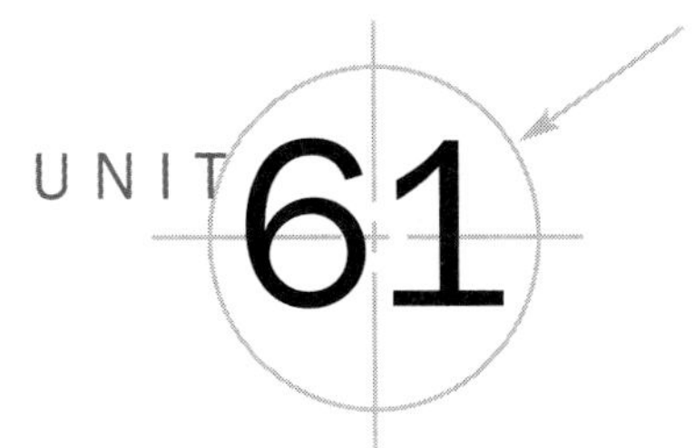

Analysis of Trigonometric Functions

Objectives ***After studying this unit you should be able to***

- Determine the variations of functions as angles change.
- Compute cofunctions of complementary angles.

Variation of Functions

As the size of an angle increases, the sine, tangent, and secant functions increase while the cofunctions (cosine, cotangent, cosecant) decrease. As the reference angles approach 0° or 90°, the function variation can be shown. These examples illustrate variations of an increasing function and a decreasing function for a reference angle that is increasing in size.

➤ **Note:** Use Figure 61-1 for Examples 1 and 2.

OP_1 and OP_2 are radii of the arc of a circle and $\angle 1$ is smaller than $\angle 2$.

$$OP_1 = OP_2 = r$$

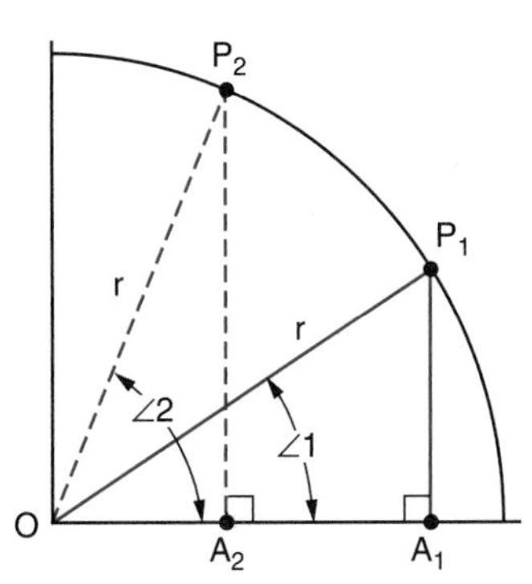

Figure 61-1

Example 1 Variation of an increasing function; the sine function.

$$\text{The sine of an angle} = \frac{\text{opposite side}}{\text{hypotenuse}}$$

$$\sin \angle 1 = \frac{A_1P_1}{r}$$

$$\sin \angle 2 = \frac{A_2P_2}{r}$$

A_2P_2 is greater than A_1P_1; therefore, $\sin \angle 2$ is greater than $\sin \angle 1$. Observe that if $\angle 1$ decreases to 0°, side $A_1P_1 = 0$.

$$\sin 0° = \frac{0}{r} = 0$$

If $\angle 2$ increases to 90°, size $A_2P_2 = r$.

$$\sin 90° = \frac{r}{r} = 1$$

Conclusion: As an angle increases from 0° to 90°, the sine of the angle increases from 0 to 1.

Example 2 Variation of a decreasing function; the cosine function.

$$\text{The cosine of an angle} = \frac{\text{adjacent side}}{\text{hypotenuse}}$$

$$\cos \angle 1 = \frac{OA_1}{r}$$

$$\cos \angle 2 = \frac{OA_2}{r}$$

OA_2 is less than OA_1; therefore, cos ∠2 is less than cos ∠1. Observe that if ∠1 decreases to 0°, side $OA_1 = r$.

$$\cos 0° = \frac{r}{r} = 1$$

If ∠2 increases to 90°, size $OA_2 = 0$.

$$\cos 90° = \frac{0}{r} = 0$$

Conclusion: As an angle increases from 0° to 90°, the cosine of the angle decreases from 1 to 0.

It is helpful to sketch figures for all functions in order to further develop an understanding of the relationship of angles and their functions. Particular attention should be given to functions of angles close to 0° and 90°.

A summary of the variations taken from the table of trigonometric functions is shown for an angle increasing from 0° to 90°.

As an angle increases from 0° to 90°	
sin increases from 0 to 1	cos decreases from 1 to 0
tan increases from 0 to ∞	cot decreases from ∞ to 0
sec increases from 1 to ∞	csc decreases from ∞ to 1

The cotangent of 0°, cosecant of 0°, tangent of 90°, and secant of 90° involve division by zero; since division by zero is not possible, these values are undefined. Although they are undefined, the values are often written as ∞.

➤ **Note:** These undefined values are displayed as "Error" on a calculator.

The symbol ∞ means infinity. Infinity is the quality of existing beyond or being greater than any countable value. It cannot be used for computations at this level of mathematics.

Rather than attempt to treat ∞ as a value, think of the tangent and secant functions not at an angle of 90°, but at angles very close to 90°. Observe that as an angle approaches 90°, the tangent and secant functions get very large. Think of the cotangent and cosecant functions not at an angle of 0°, but as very small angles close to 0°. Observe that as an angle approaches 0°, the cotangent and cosecant functions get very small.

Functions of Complementary Angles

Two angles are complementary when their sum is 90°. For example, 20° is the complement of 70° and 70° is the complement of 20°. In the triangle shown in Figure 61-2, ∠A is the complement of ∠B and ∠B is the complement of ∠A. The six functions of the angle and the cofunctions of the complementary angle are shown.

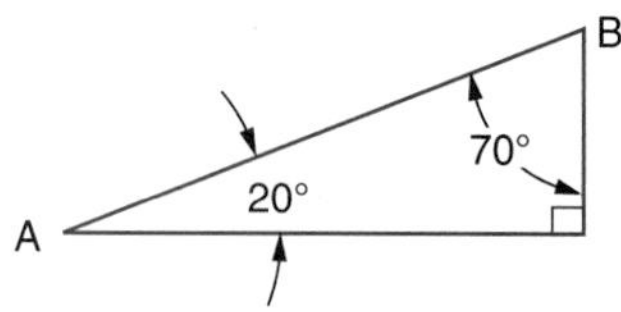

Figure 61-2

sin 20° = cos 70° ≈ 0.34202	cos 20° = sin 70° ≈ 0.93969
tan 20° = cot 70° ≈ 0.36397	cot 20° = tan 70° ≈ 2.7475
sec 20° = csc 70° ≈ 1.0642	csc 20° = sec 70° ≈ 2.9238

A function of an angle is equal to the cofunction of the complement of the angle.

The complement of an angle equals 90° minus the angle. The relationships of the six functions of angles and the cofunctions of the complementary angles are shown.

sin A = cos (90° − A)	cos A = sin (90° − A)
tan A = cot (90° − A)	cot A = tan (90° − A)
sec A = csc (90° − A)	csc A = sec (90° − A)

Examples For each function of an angle, write the cofunction of the complement of the angle.

1. sin 30° = cos (90° − 30°) = cos 60° Ans
2. cot 10° = tan (90° − 10°) = tan 80° Ans
3. tan 72.53° = cot (90° − 72.53°) = cot 17.47° Ans
4. sec 40°20' = csc (90° − 40°20') = csc (89°60' − 40°20') = csc 49°40' Ans
5. cos 90° = sin (90° − 90°) = sin 0° Ans

Application

Variation of Functions

Refer to Figure 61-3 in answering exercises 1 through 7. It may be helpful to sketch figures.

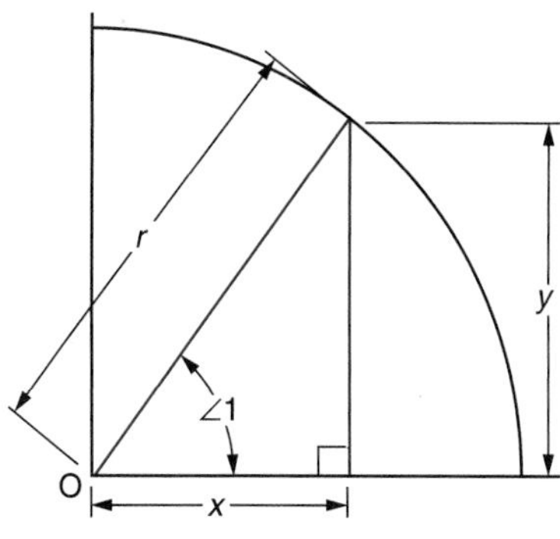

Figure 61-3

1. When ∠1 is almost 90°:
   a. how does side y compare to side r? ______
   b. how does side x compare to side r? ______
   c. how does side x compare to side y? ______
2. When ∠1 is 90°:
   a. what is the value of side x? ______
   b. how does side y compare to side r? ______
3. When ∠1 is slightly greater than 0°:
   a. how does side y compare to side r? ______
   b. how does side x compare to side r? ______
   c. how does side x compare to side y? ______

4. When $\angle 1$ is 0°:

   a. what is the value of side y? __________

   b. how does side x compare to side r? __________

5. When side x = side y:

   a. what is the value of $\angle 1$? __________

   b. what is the value of the tangent function? __________

   c. what is the value of the cotangent function? __________

6. When side x = side r:

   a. what is the value of the sine function? __________

   b. what is the value of the secant function? __________

   c. what is the value of the cosine function? __________

   d. what is the value of the tangent function? __________

7. When side y = side r:

   a. what is the value of the sine function? __________

   b. what is the value of the cotangent function? __________

   c. what is the value of the cosine function? __________

   d. what is the value of the cosecant function? __________

For each exercise, functions of two angles are given. Which of the functions of the two angles is greater? Do **not** use a calculator.

8. sin 38°; sin 43° __________
9. tan 17°; tan 18° __________
10. cos 78°; cos 85° __________
11. cot 40°; cot 36° __________
12. sec 5°; sec 8° __________
13. csc 22°; csc 25° __________
14. tan 21°40'; tan 12°50' __________
15. cos 81°19'; cos 81°20' __________
16. sin 0.42°; sin 0.37° __________
17. csc 40.50°; csc 40.45° __________
18. cot 27°23'; cot 87°0' __________
19. sec 55°; sec 54°50' __________

Functions of Complementary Angles

For each function of an angle, write the cofunction of the complement of the angle.

20. tan 23° __________
21. sin 49° __________
22. cos 26° __________
23. sec 82° __________
24. cot 35° __________
25. csc 51° __________
26. cos 90° __________
27. sin 0° __________
28. tan 57.5° __________
29. cos 12.2° __________
30. cot 7°10' __________
31. cos 36°06' __________
32. sin 0°38' __________
33. sin 5.89° __________
34. cos 3.76° __________
35. cot 0° __________
36. tan 90° __________
37. sec 43°19' __________
38. cos 0.01° __________
39. sin 89°59' __________

For each exercise, functions and cofunctions of two angles are given. Which of the functions or cofunctions of the two angles is greater? Do **not** use a calculator.

40. cos 48°; sin 18° __________
41. cos 55°; sin 40° __________
42. tan 30°; cot 65° __________
43. tan 30°; cot 45° __________
44. sec 42°; csc 58° __________
45. sec 43°; csc 58° __________
46. sin 14°; cos 78° __________
47. sin 12°; cos 75° __________
48. cot 89°10'; tan 1°20' __________
49. cot 87°50'; tan 2°40' __________
50. sin 0.2°; cos 89.9° __________
51. sin 0.2°; cos 89.0° __________

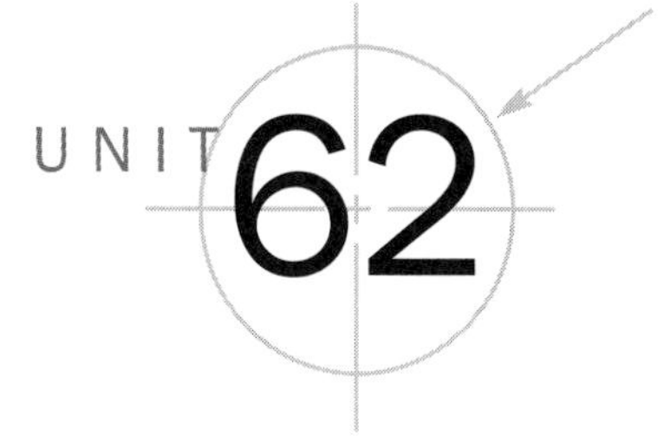

Basic Calculations of Angles and Sides of Right Triangles

Objectives *After studying this unit you should be able to*

- Compute an unknown angle of a right triangle when two sides are known.
- Compute an unknown side of a right triangle when an angle and a side are known.

Determining an Unknown Angle When Two Sides of a Right Triangle Are Known

In order to solve for an unknown angle of a right triangle where neither acute angle is known, at least two sides must be known. An understanding of the procedures required for solving for unknown angles is essential to manufacturing and related occupations.

Procedure for Determining an Unknown Angle When Two Sides Are Given

- In relation to the desired angle, identify two given sides as adjacent, opposite, or hypotenuse.
- Determine the functions that are ratios of the sides identified in relation to the desired angle.

➤ **Note:** Two of the six trigonometric functions are ratios of the two known sides. Either of the two functions can be used. Both produce the same value for the unknown.

- Choose one of the two functions; substitute the given sides in the ratio.
- Determine the angle that corresponds to the quotient of the ratio.

When sides are given in inches (customary units), compute the angle to the nearer minute. When sides are given in millimeters (metric units), compute the angle to the nearer hundredth degree unless otherwise specified.

➤ **Note:** With the calculator examples given in this unit and in the trigonometry units that follow, generally two basic procedures are shown for each example. If your calculator functions differ from these procedures, you may find it necessary to refer to your User's Guide.

Example 1 Determine $\angle A$ of the right triangle shown in Figure 62-1 to the nearer minute.

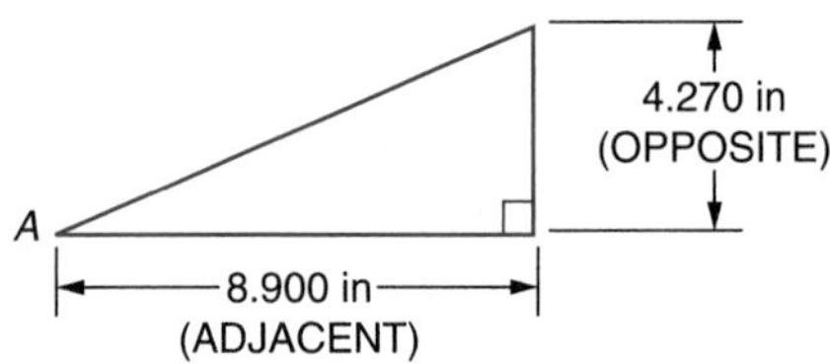

Figure 62-1

Solution In relation to $\angle$A, the 8.900-inch side is the adjacent side, and the 4.270-inch side is the opposite side.

Determine the two functions whose ratios consist of the adjacent and opposite sides. Then, tan $\angle A$ = opposite side/adjacent side, and cot $\angle A$ = adjacent side/opposite side. Either the tangent or cotangent function can be used.

Choosing the tangent function: $\tan \angle A = \dfrac{4.270 \text{ in}}{8.900 \text{ in}}$.

Determine the angle whose tangent function is the quotient of $\dfrac{4.270}{8.900}$.

$\angle A$ = [SHIFT] [tan⁻¹] [(] 4.27 [÷] 8.9 [)] [=] [SHIFT] [←] 25°37'49.95",
25°38' Ans

or $\angle A$ = [2nd] [tan⁻¹] [(] 4.27 [÷] 8.9 [)] [° ' "] [◄] [ENTER] [ENTER] 25°37'49.9",
25°38' Ans

Example 2 Determine $\angle$B of the right triangle shown in Figure 62-2 to the nearest hundredth degree.

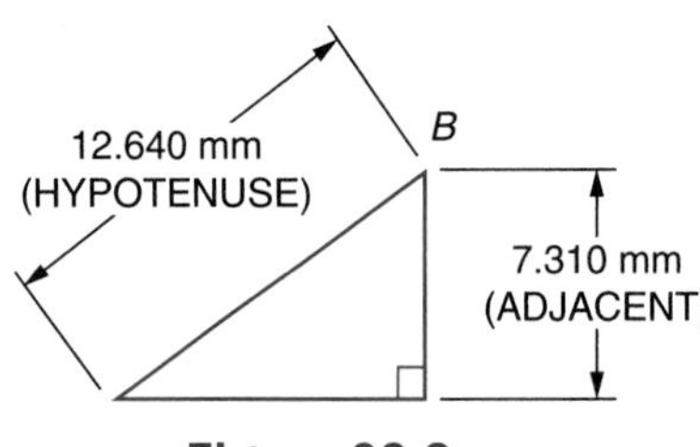

Figure 62-2

Solution In relation to $\angle B$, the 12.640-millimeter side is the hypotenuse, and the 7.310-millimeter side is the adjacent side.

Determine the two functions whose ratios consist of the adjacent side and the hypotenuse. Then, cos $\angle B$ = adjacent side/hypotenuse; and sec $\angle B$ = hypotenuse/adjacent side. Either the cosine or secant function can be used.

Choosing the cosine function: $\cos \angle B = \dfrac{7.310 \text{ mm}}{12.640 \text{ mm}}$.

Determine the angle whose cosine function is the quotient of $\dfrac{7.310}{12.640}$.

$\angle B$ = [SHIFT] [cos⁻¹] [(] 7.31 [÷] 12.64 [)] [=] 54.66733748,
54.67° Ans

or $\angle B$ = [2nd] [cos⁻¹] [(] 7.31 [÷] 12.64 [)] [ENTER] 54.66733748,
54.67° Ans

Example 3 Determine ∠1 and ∠2 of the triangle shown in Figure 62.3 to the nearest minute.

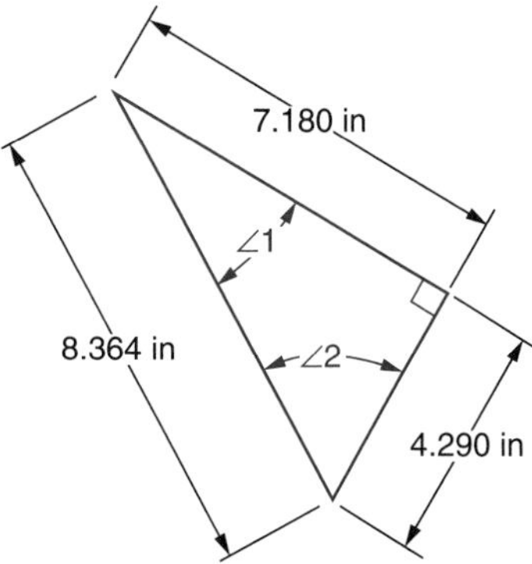

Figure 62-3

Solution Compute either ∠1 or ∠2. Choose any two of the three given sides for a ratio. In relation to ∠1, the 4.290-inch side is the opposite side, and the 8.364-inch side is the hypotenuse.

Determine the two functions whose ratios consist of the opposite side and the hypotenuse. Then, sin ∠1 = opposite side/hypotenuse, and csc ∠1 = hypotenuse/opposite. Either the sine or cosecant can be used.

Choosing the sine function: $\sin \angle 1 = \dfrac{4.290 \text{ in}}{8.364 \text{ in}}$.

Determine the angle whose sine function is the quotient of $\dfrac{4.290}{8.364}$.

∠1 = [SHIFT] [sin⁻¹] [(] 4.29 [÷] 8.364 [)] [=] [SHIFT] [←]
30°51'28.89", 30°51' Ans

or ∠1 = [2nd] [sin⁻¹] [(] 4.29 [÷] 8.364 [)] [° ' "] ◀ [ENTER] [ENTER]
30°51'28.89", 30°51' Ans

Since ∠1 + ∠2 = 90°, ∠2 = 90° − 30°51', ∠2 = 59°9' Ans

Determining an Unknown Side When an Acute Angle and One Side of a Right Triangle Are Known

In order to solve for an unknown side of a right triangle, at least an acute angle and one side must be known.

Procedure for Determining an Unknown Side When an Acute Angle and One Side of a Right Triangle Are Known

- Identify the given side and the unknown side as adjacent, opposite, or hypotenuse in relation to the given angle.
- Determine the trigonometric functions that are ratios of the sides identified in relation to the given angle.

➤ **Note:** Two of the six functions will be found as ratios of the two identified sides. Either of the two functions can be used. Both produce the same value for the unknown. If the unknown side is made the numerator of the ratio, the problem is solved by multiplication. If the unknown side is made the denominator of the ratio, the problem is solved by division.

- Choose one of the two functions and substitute the given side and given angle.
- Solve as a proportion for the unknown side.

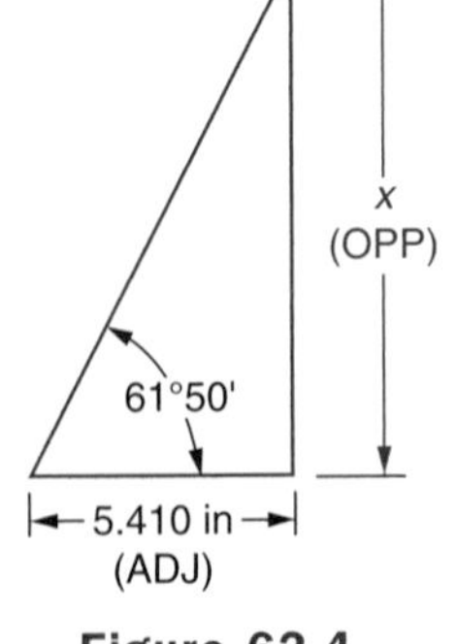

Figure 62-4

Example 1 Determine side x of the right triangle shown in Figure 62-4. Round the answer to 3 decimal places.

Solution In relation to the 61°50' angle, the 5.410-inch side is the adjacent side and side x is the opposite side.

Determine the two functions whose ratios consist of the adjacent and opposite sides. Tan 61°50' = opposite side/adjacent side, and cot 61°50' = adjacent side/opposite side. Either the tangent or cotangent function can be used.

Choosing the tangent function: $\tan 61°50' = \dfrac{x}{5.410 \text{ in}}$.

Solve as a proportion.

$$\frac{\tan 61°50'}{1} = \frac{x}{5.410 \text{ in}}$$

$$x = \tan 61°50'(5.410 \text{ in})$$

$x =$ [tan] 61 [° ' ''] 50 [° ' ''] [×] 5.41 [=] 10.10371739, 10.104 in Ans

or $x =$ [tan] 61 [° ' ''] [ENTER] 50 [° ' ''] [◀] [ENTER] [ENTER] [×] 5.41 [ENTER] 10.10371739, 10.104 in Ans

Example 2 Determine side r of the right triangle shown in Figure 62-5. Round the answer to 3 decimal places.

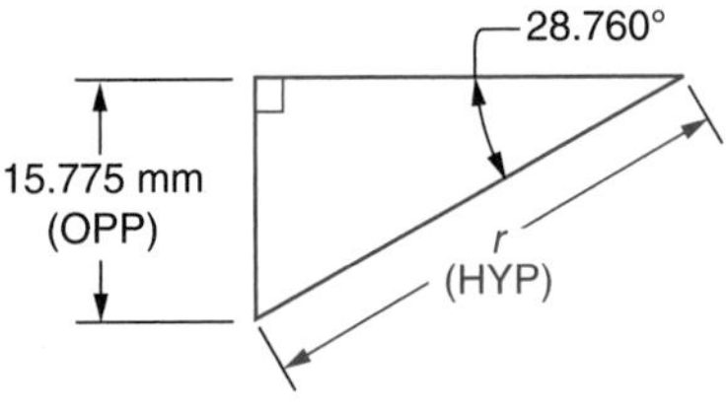

Figure 62-5

Solution In relation to the 28.760° angle, the 15.775-millimeter side is the opposite side and side r is the hypotenuse.

Determine the two functions whose ratios consist of the opposite side and the hypotenuse. Sin 28.760° = opposite side/hypotenuse, and csc 28.760° = hypotenuse/opposite side. Either the sine or cosecant function can be used.

Choosing the sine function: $\sin 28.760° = \dfrac{12.775 \text{ mm}}{r}$.

Solve as a proportion.

$$\frac{\sin 28.760°}{1} = \frac{15.775 \text{ mm}}{r}$$

$$r = \frac{15.775 \text{ mm}}{\sin 28.760°}$$

$r =$ 15.775 [÷] [sin] 28.76 [=] 32.78659364, 32.787 mm Ans

Example 3 Determine side x, side y, and $\angle 1$ of the right triangle shown in Figure 62-6. Round the answers to 3 decimal places.

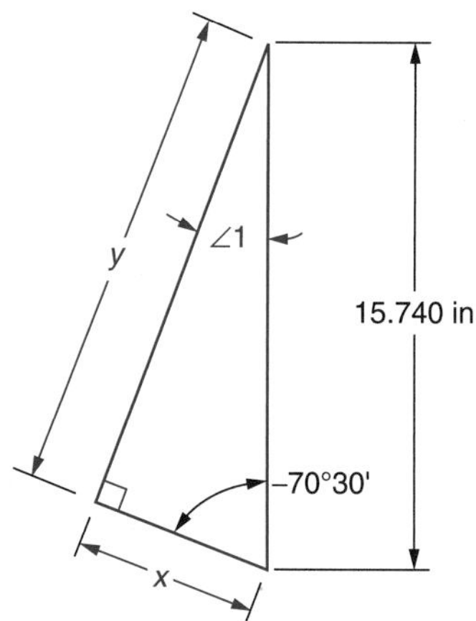

Figure 62-6

Solution Compute either side x or side y. Choosing side x, in relation to the 70°30' angle, side x is the adjacent side. The 15.740-inch side is the hypotenuse.

Determine the two functions whose ratios consist of the adjacent side and the hypotenuse. Either the cosine or secant function can be used.

Choosing the cosine function: $\cos 70°30' = \dfrac{x}{15.740 \text{ in}}$.

Solve as a proportion.

$$\frac{\cos 70°30'}{1} = \frac{x}{15.740 \text{ in}}$$

$$x = \cos 70°30°'(15.740 \text{ in})$$

x = [cos] 70 [° ' "] 30 [° ' "] [×] 15.74 [=] 5.254119964, 5.254 in Ans

or x = [cos] 70 [° ' "] [ENTER] 30 [° ' "] [▶] [ENTER] [ENTER] [×] 15.74 [ENTER] 5.254119964, 5.254 in Ans

Solve for side y by using either a trigonometric function or the Pythagorean Theorem. If the Pythagorean Theorem is used to determine y, then $y^2 = (15.740)^2 - (5.254)^2$ and $y = \sqrt{(15.740)^2 - (5.254)^2}$. In cases like this, it is generally more convenient to solve for the side by using a trigonometric function. In relation to the 70°30' angle, side y is the opposite side. The 15.740-inch side is the hypotenuse.

Determine the two functions whose ratios consist of the opposite side and the hypotenuse. Either the sine or cosecant function can be used.

Choosing the sine function: $\sin 70°30' = \dfrac{y}{15.740 \text{ in}}$.

➤ **Note:** Since side x has been calculated, it can be used with the 70°30' angle to determine side y. However, it is better to use the given 15.740-inch hypotenuse rather than the calculated side x. Whenever possible, use given values rather than calculated values when solving problems. The calculated values could have been incorrectly computed or improperly rounded off resulting in an incorrect answer.

Solve as a proportion.

$$\frac{\sin 70°30'}{1} = \frac{y}{15.740 \text{ in}}$$

$y =$ [sin] 70 [° ' "] 30 [° ' "] [×] 15.74 [=] 14.83717707, 14.837 in Ans

or $y =$ [sin] 70 [° ' "] [ENTER] 30 [° ' "] [▶] [ENTER] [ENTER] [×] 15.74 [ENTER] 14.83717707, 14.837 in Ans

Determine ∠1: ∠1 = 90° − 70°30' = 19°30' Ans

Application

Determining an Unknown Angle When Two Sides of a Right Triangle Are Known

Solve the following problems. Compute angles to the nearer minute in triangles with customary unit sides. Compute angles to the nearer hundredth degree in triangles with metric unit sides.

1. Determine ∠A. ______________

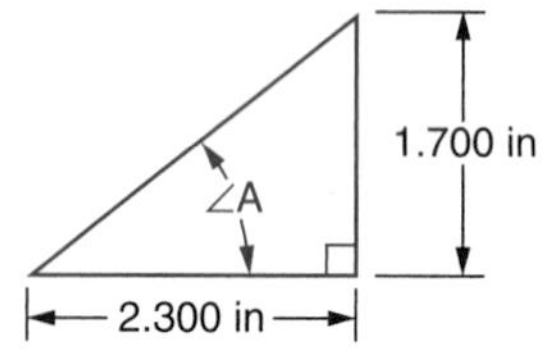

2. Determine ∠B. ______________

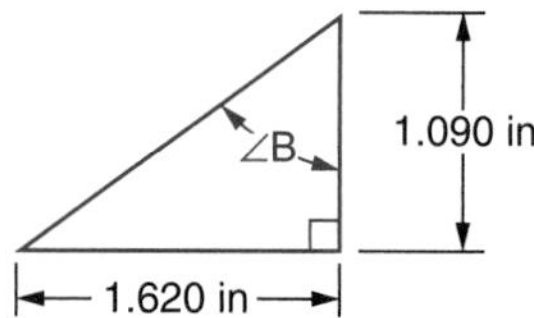

3. Determine ∠1. ______________

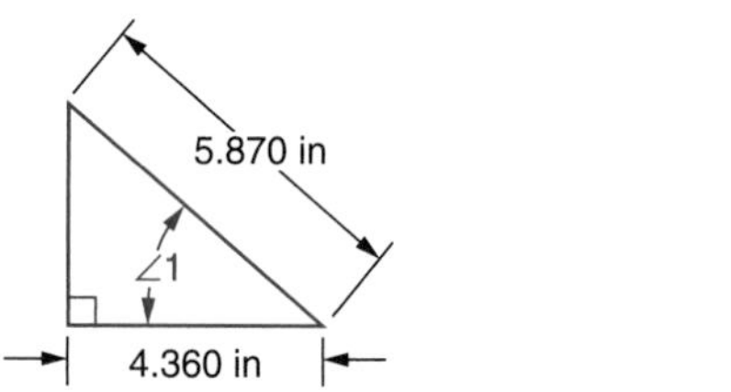

4. Determine ∠x. ______________

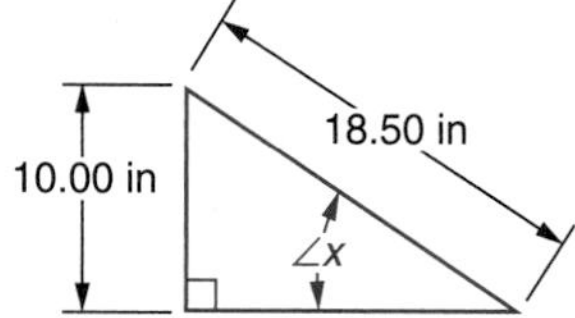

5. Determine ∠1. ______________

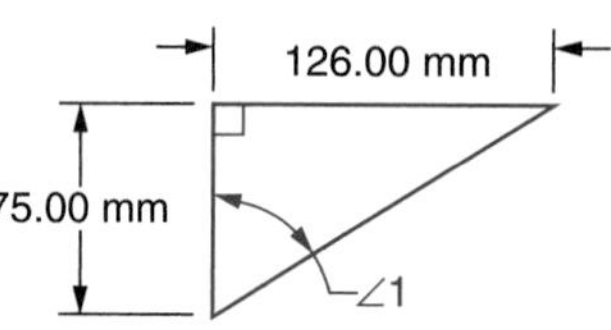

6. Determine ∠A. ______________

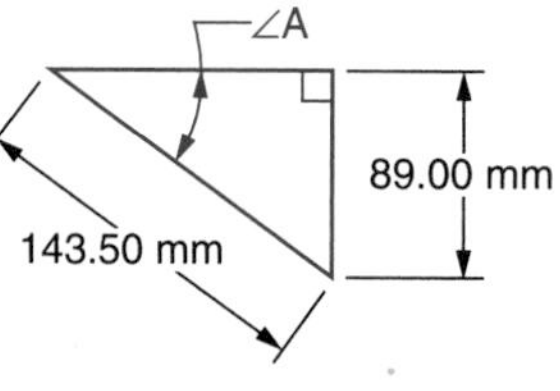

7. Determine ∠y. ______________

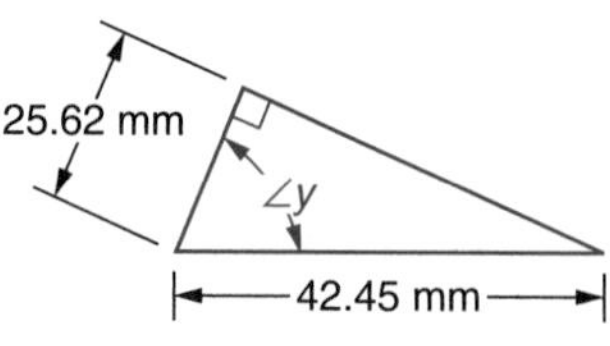

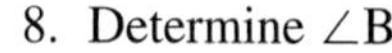

8. Determine ∠B. ______________

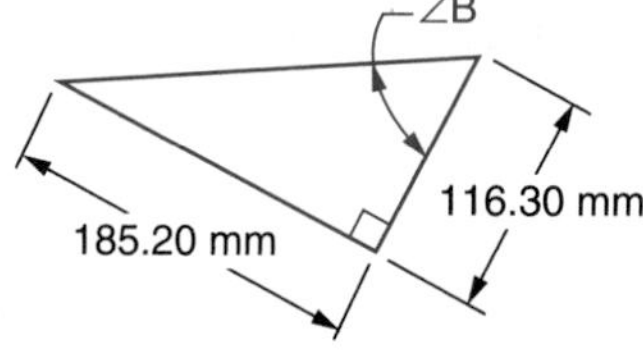

9. a. Determine $\angle 1$. ______________
   b. Determine $\angle 2$. ______________

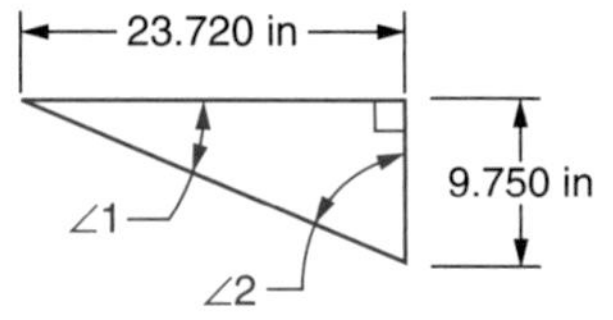

10. a. Determine $\angle A$. ______________
    b. Determine $\angle B$. ______________

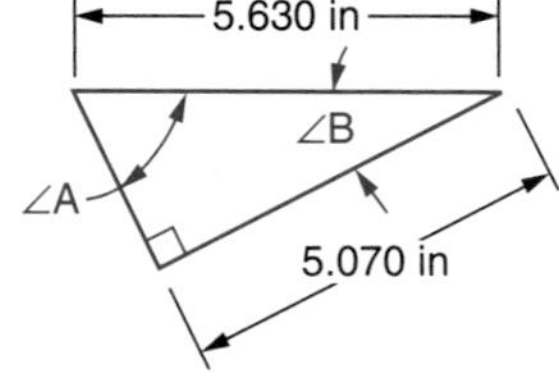

11. a. Determine $\angle x$. ______________
    b. Determine $\angle y$. ______________

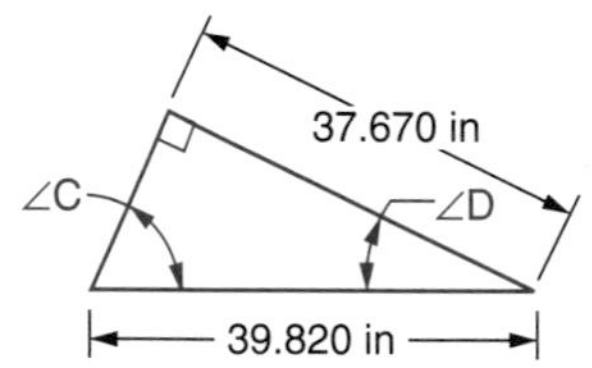

12. a. Determine $\angle C$. ______________
    b. Determine $\angle D$. ______________

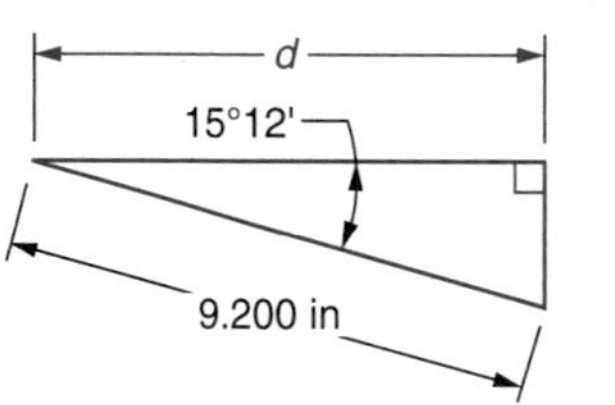

Determining an Unknown Side When an Acute Angle and One Side of a Right Triangle Are Known

Solve the following problems. Compute the sides to 3 decimal places in triangles dimensioned in customary units. Compute the sides to 2 decimal places in triangles dimensioned in metric units.

13. Determine side b. ______________

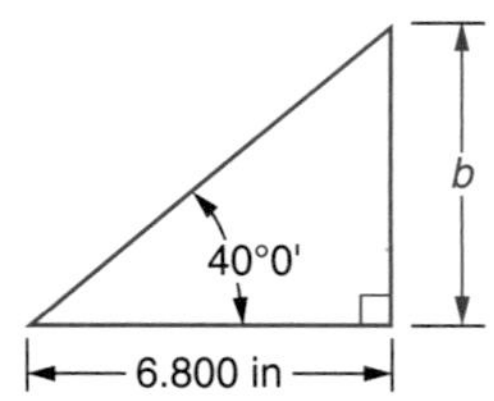

14. Determine side c. ______________

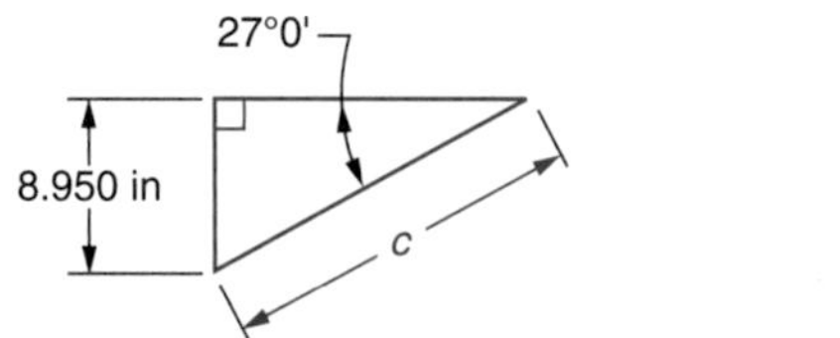

15. Determine side x. ______________

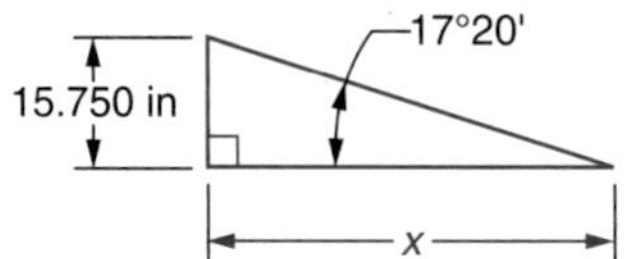

16. Determine side d. ______________

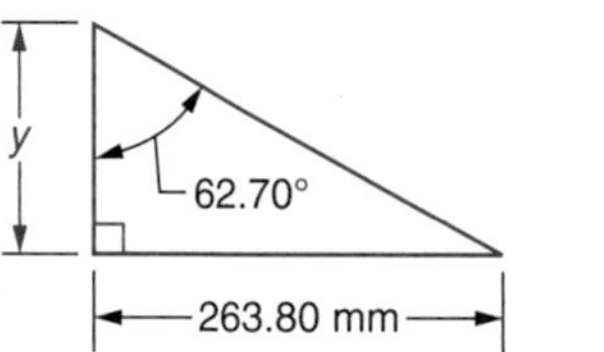

17. Determine side y. ______________

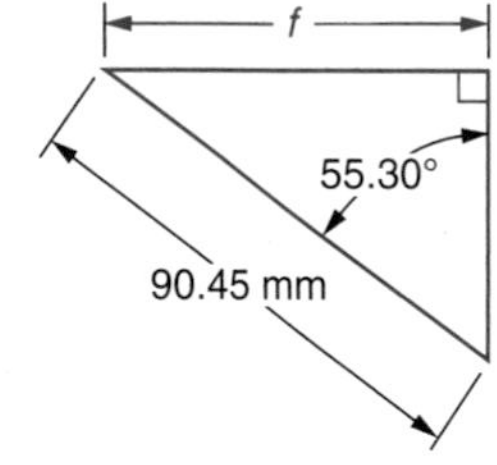

18. Determine side f. ______________

19. Determine side p. ____________

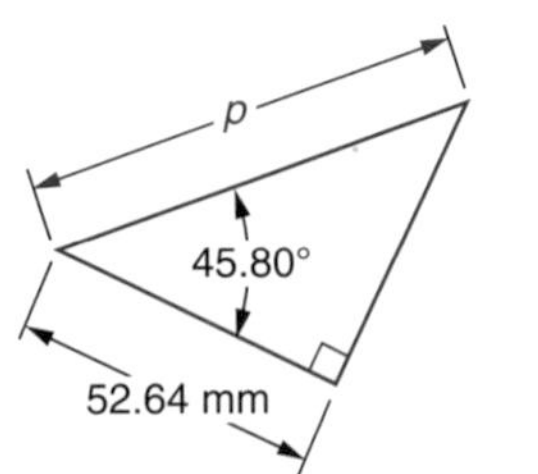

20. Determine side y. ____________

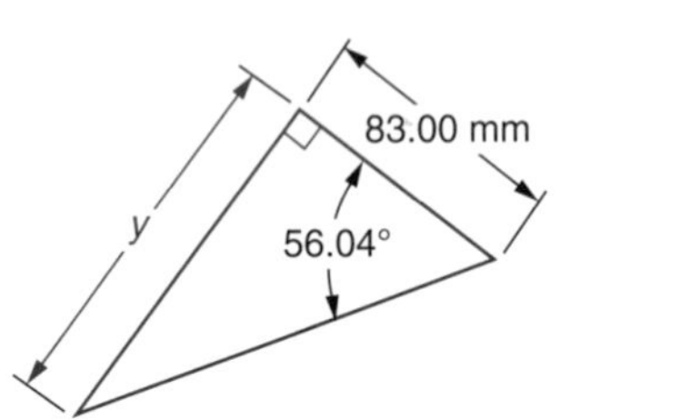

21. a. Determine side d. ____________

    b. Determine side e. ____________

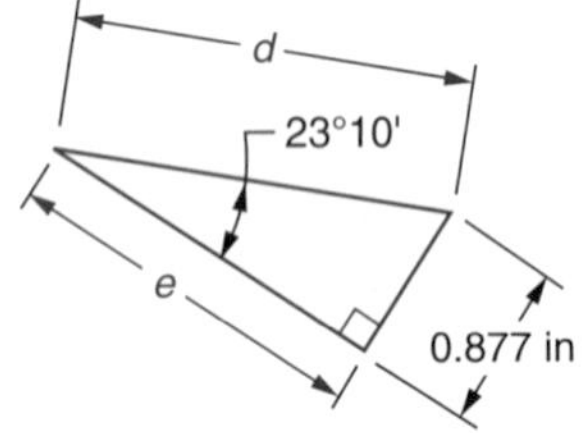

22. a. Determine side s. ____________

    b. Determine side t. ____________

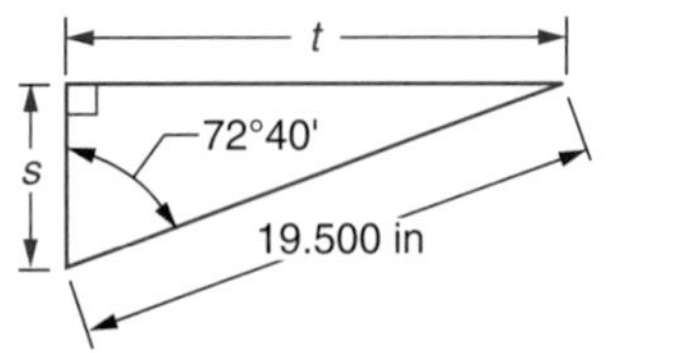

23. a. Determine side x. ____________

    b. Determine side y. ____________

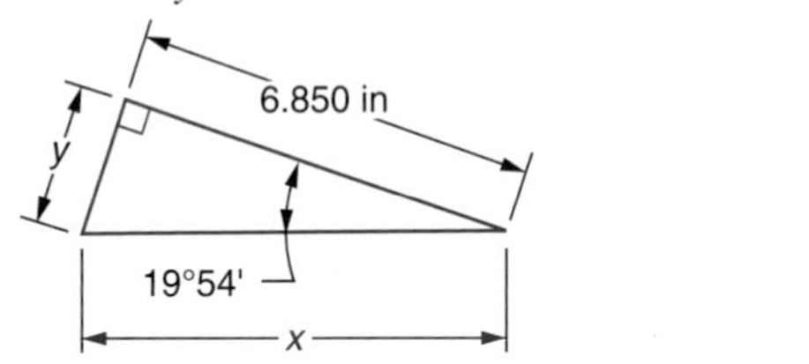

24. a. Determine side p. ____________

    b. Determine side n. ____________

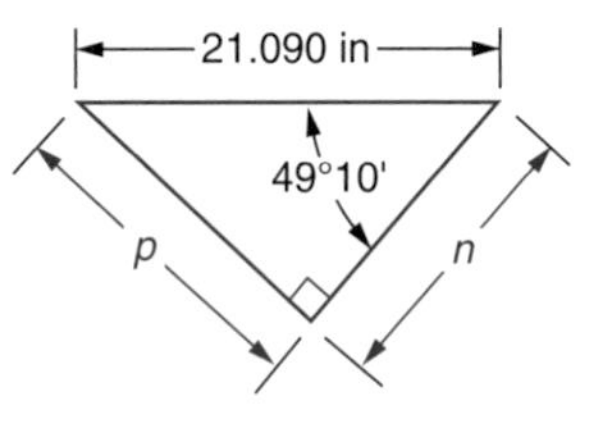

Determining Unknown Sides and Angles

Solve the following problems. For triangles dimensioned in customary units, compute the sides to 3 decimal places and the angles to the nearer minute. For triangles dimensioned in metric units, compute the sides to 2 decimal places and the angles to the nearer hundredth degree.

25. a. Determine ∠B. ____________

    b. Determine side x. ____________

    c. Determine side y. ____________

26. a. Determine ∠1. ____________

    b. Determine ∠2. ____________

    c. Determine side a. ____________

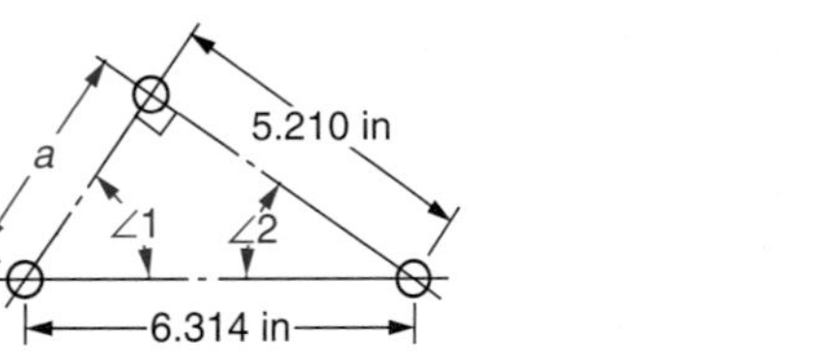

27. a. Determine side a. ____________

    b. Determine side b. ____________

    c. Determine ∠2. ____________

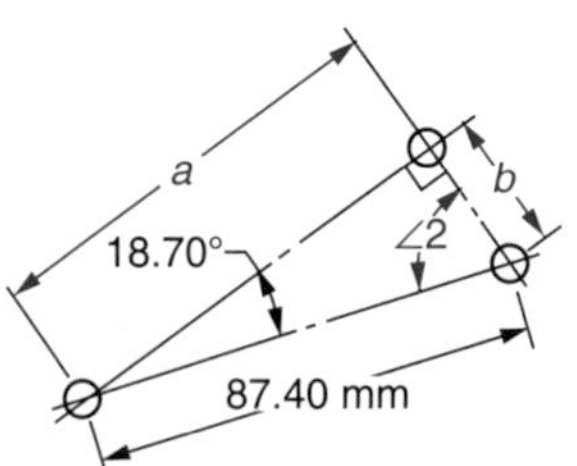

28. a. Determine ∠A. ____________

    b. Determine ∠B. ____________

    c. Determine side r. ____________

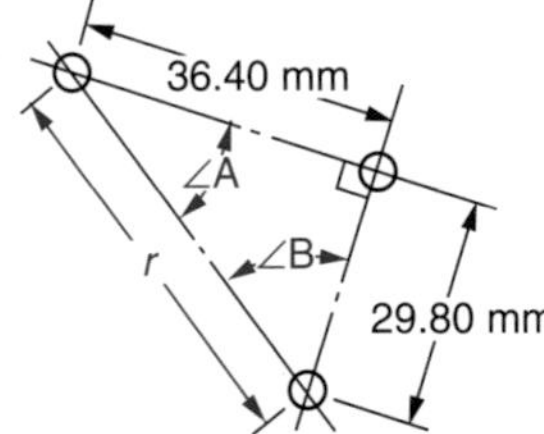

29. a. Determine ∠B. __________

b. Determine side *b*. __________

c. Determine side *c*. __________

30. a. Determine ∠D. __________

b. Determine ∠E. __________

c. Determine side *m*. __________

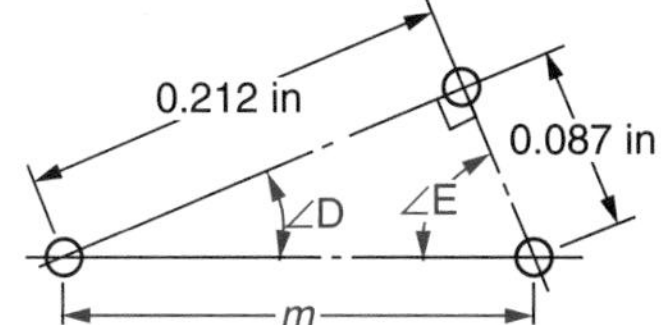

31. a. Determine ∠1. __________

b. Determine side *g*. __________

c. Determine side *h*. __________

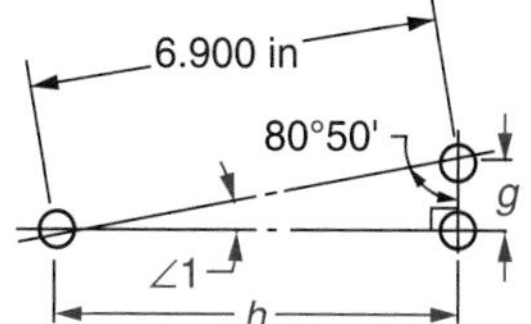

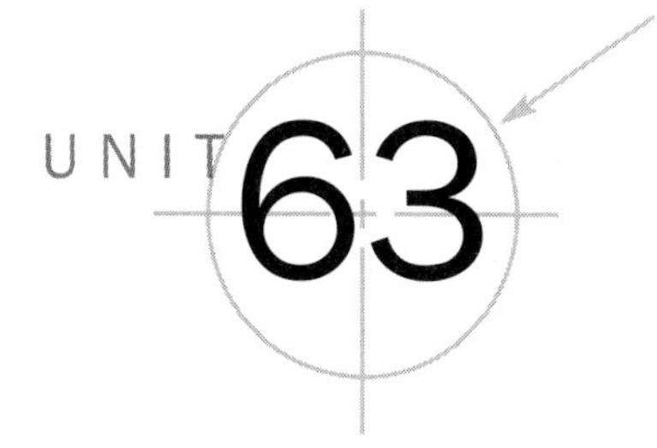

Simple Practical Machine Applications

Objective ***After studying this unit you should be able to***

- Solve simple machine technology problems that require the projection of auxiliary lines and the use of geometric principles and trigonometric functions.

Method of Solution

In the previous unit, you solved for unknown angles and sides of right triangles. Emphasis was placed on developing an understanding of and the ability to apply proper procedures in solving for angles and sides. No attempt was made to show the many practical applications of right-angle trigonometry.

The examples discussed in this unit are simple practical shop applications of right-angle trigonometry, although they may not be given directly in the form of right triangles. To solve most of the examples, it is necessary to project auxiliary lines to produce a right triangle. The unknown, or a dimension required to compute the unknown, is part of the triangle. The auxiliary lines may be projected between given points, or from given points. The lines may be projected parallel or perpendicular to centerlines, tangents, or other reference lines.

It is important to study carefully the procedures and the use of auxiliary lines as they are applied to the examples that follow. The same basic method is used in solving many similar machine shop problems. A knowledge of both geometric principles and trigonometric functions and the ability to relate and apply them to specific situations are required in solving many machine technology problems.

Sine Bar and Sine Plate

Sine bars and sine plates are used to measure angles that have been cut in parts and to position parts that are to be cut at specified angles. One end of the sine bar or plate is raised with gage blocks in order to set a desired angle. The most common sizes of bars and plates are 5 inches and 10 inches between rolls. In setting angles, the sine bar or the top plate of the sine plate is the hypotenuse of a right triangle, and the gage blocks are the opposite side in reference to the desired angle.

Example 1 Determine the gage block height x that is required to set an angle of 24°21' with a 5-inch sine bar as shown in Figure 63-1.

$$\sin 24°21' = \frac{\text{gage block height } x}{\text{sine bar length}}$$

$$\frac{\sin 24°21'}{1} = \frac{x}{5 \text{ in}}$$

$x = (\sin 24°21')5$

$x =$ [sin] 24 [° ' "] 21 [° ' "] [=] [×] 5 [=] 2.061547756, 2.0615 in Ans

or $x =$ [sin] 24 [° ' "] [ENTER] 21 [° ' "] [▶] [ENTER] [ENTER] [×] 5 [ENTER] 2.061547756, 2.0615 in Ans

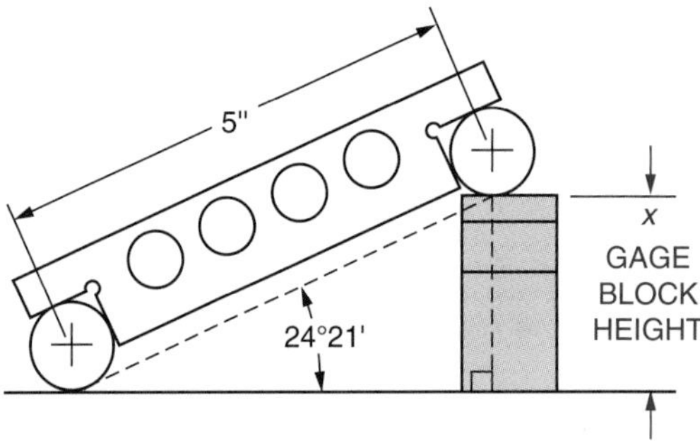

Figure 63-1

Example 2 Determine the angle set on a 10-inch sine plate using a gage block height of 3.0625 inches as shown in Figure 63-2.

$$\sin \angle x = \frac{3.0625 \text{ in}}{10 \text{ in}}$$

Determine the angle whose sine function is the quotient of $\frac{3.0625}{10}$.

$\angle x =$ [SHIFT] [$\sin^{-1}$] [(] 3.0625 [÷] 10 [)] [=] [SHIFT] [←] 17°50'0.18", 17°50' Ans

or $\angle x =$ [2nd] [$\sin^{-1}$] [(] 3.0625 [÷] 10 [)] [° ' "] [◀] [ENTER] [ENTER] 17°50'0.18", 17°50' Ans

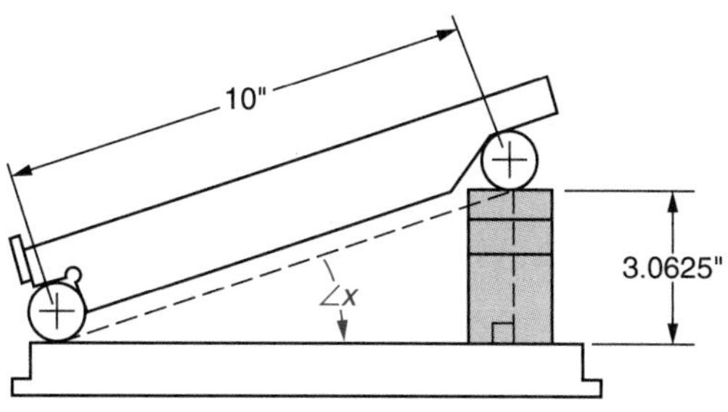

Figure 63-2

Tapers and Bevels

Example 1 Determine the included taper angle of the shaft shown in Figure 63-3. All dimensions are in inches.

The problem must be solved by using a figure in the form of a right triangle. Therefore, project line AB from point A parallel to the centerline. Right △ABC is formed in which ∠BAC is one-half the included taper angle. Side AB = 10.500".

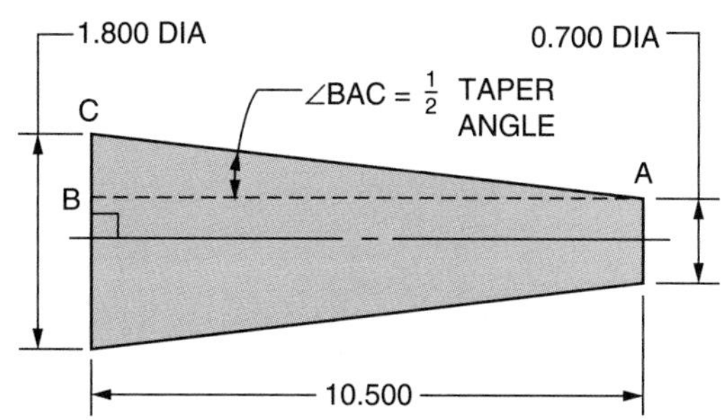

Figure 63-3

$$\text{Side BC} = \frac{1.800" - 0.700"}{2} = 0.550"$$

Using sides AB and BC, solve for ∠BAC. $\tan \angle BAC = \frac{BC}{AB} = \frac{0.550"}{10.500"}$

Determine the angle whose tangent function is the quotient of $\frac{0.550}{10.500}$.

$\angle x =$ SHIFT tan⁻¹ (.55 ÷ 10.5) =
SHIFT ← 2°59'55", 3°0'

or $\angle x =$ 2nd tan⁻¹ (.55 ÷ 10.5) °' " ENTER ENTER
2°59'55", 3°0'

The included taper angle = 2(3°0') = 6°0' Ans

Example 2 Determine diameter x of the part shown in Figure 63-4. All dimensions are in millimeters.

Project line DE from point D parallel to the centerline, in order to form right △DEF.

Side DE = 21.80 mm − 7.50 mm = 14.30 mm

∠EDF = 32.50°

Using side DE and ∠EDF, solve for side EF.

$$\tan \angle EDF = \frac{EF}{DE}$$

$$\tan 32.50° = \frac{EF}{14.30 \text{ mm}}$$

EF = tan 32.50°(14.30 mm)

EF = [tan] 32.5 [×] 14.3 [=] 9.11010473, 9.11 mm

or EF = [tan] 32.5 [ENTER] [×] 14.3 [ENTER] 9.11010473, 9.11 mm

DIA $x \approx$ 26.25 mm − 2(9.11 mm) ≈ 8.03 mm Ans

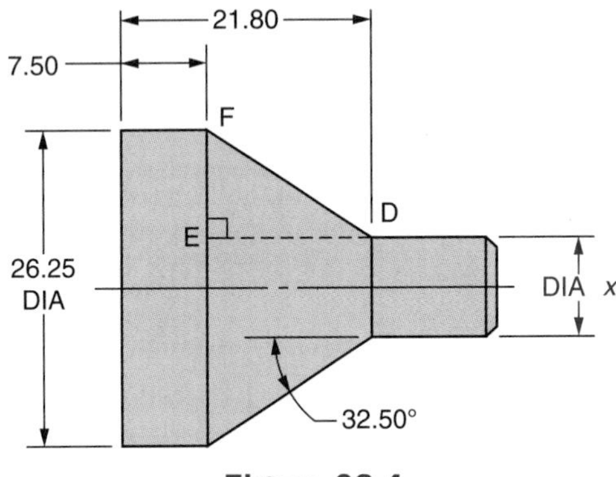

Figure 63-4

Isosceles Triangle Applications: Distance Between Holes and V-Slots

The solutions to many practical trigonometry problems are based on recognizing figures as isosceles triangles. In an isosceles triangle, an altitude to the base bisects the base and the vertex angle.

Example 1 In Figure 63-5, five holes are equally spaced on a 5.200-inch diameter circle. Determine the straight line distance between two consecutive holes.

Project radii from center O to hole centers A and B.

Project a line from A to B.

$$\angle AOB = \frac{360°}{5} = 72°$$

Since OA = OB, △AOB is isosceles. Project line OC to AB from center O.
Line OC bisects ∠AOB and side AB.

In right △AOC, $\angle AOC = \frac{72°}{2} = 36°$

$$AO = \frac{5.200 \text{ in}}{2} = 2.600 \text{ in}$$

Solve for side AC.

$$\sin \angle AOC = \frac{AC}{AO}$$

$$\sin 36° = \frac{AC}{2.600 \text{ in}}$$

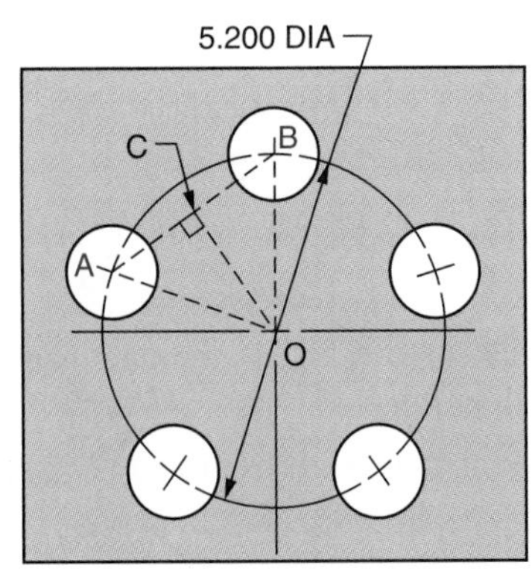

Figure 63-5

AC = sin 36°(2.600 in)

AC = [sin] 36 × 2.6 [=] 1.528241656, 1.528 in

or AC = [sin] 36 [ENTER] [×] 2.6 [ENTER] 1.528241656, 1.528 in

AB ≈ 2(1.528 in) ≈ 3.056 in Ans

Example 2 Determine the depth of cut x required to machine the V-slot shown in Figure 63-6. All dimensions are in inches.

Connect a line between points R and T. Side RS = TS; therefore, △RST is isosceles.

Project line SM from point S to RT. Side RT and ∠RST are bisected. In right △RMS,

$$\angle RSM = \frac{62°46'}{2} = 31°23'$$

$$RM = \frac{3.856 \text{ in}}{2} = 1.928 \text{ in}$$

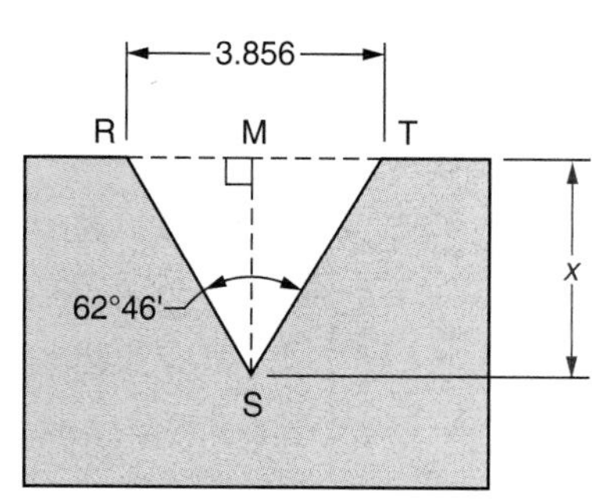

Figure 63-6

Solve for depth of cut MS.

$$\tan \angle RSM = \frac{RM}{MS}$$

$$\tan 31°23' = \frac{1.928 \text{ in}}{MS}$$

MS = 1.928 in ÷ tan 31°23'

MS = 1.928 [÷] [tan] 31 [° ' "] 23 [° ' "] = 3.160638084, 3.161 in

or MS = 1.928 [÷] [tan] 31 [° ' "] 23 [° ' "] [▶] [ENTER] [ENTER] 3.160638084, 3.161 in

x ≈ MS = 3.161 in Ans

Tangents to Circles Applications: V-Blocks, Thread Wire Checking Dimensions, Dovetails, and Angle Cuts

A tangent is perpendicular to a radius of a circle at its tangent point. Solutions to many applied trigonometry problems are based on this principle.

Example 1 A 75.00-millimeter diameter pin is used to inspect the groove machined in the block shown in Figure 63-7. Determine dimension x. The sides of the groove are equal. All dimensions are in millimeters.

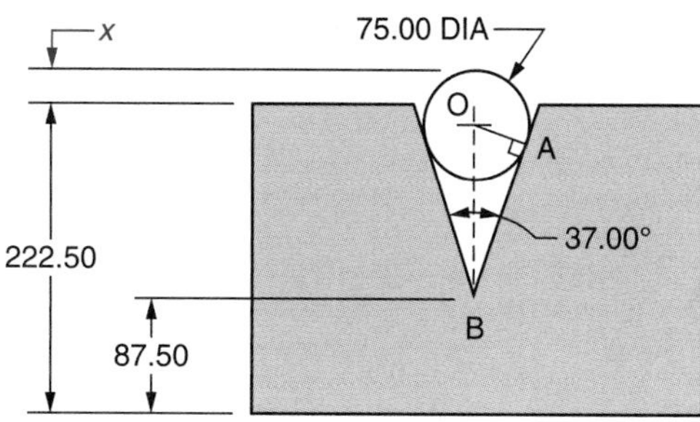

Figure 63-7

Project a line from center O to point B. Project radius AO from center O to tangent point A. Since a radius is ⊥ to a tangent line at the point of tangency, △AOB is a right triangle. In

right △AOB, $OA = \frac{75.00 \text{ mm}}{2} = 37.50 \text{ mm}$

Since the angle formed by two tangents to a circle from an outside point is bisected by a line from the point to the center of the circle, $\angle ABO = \frac{37.00°}{2} = 18.50°$

Solve for side OB.

$$\sin \angle ABO = \frac{OA}{OB}$$

$$\sin 18.50° = \frac{37.50 \text{ mm}}{OB}$$

OB = 37.50 mm ÷ sin 18.50°
OB = 37.5 [÷] [sin] 18.5 [=] 118.1829489, 118.18 mm

Find the height from the base of the block to the top of the pin.

87.50 mm + OB + radius of pin =

87.50 mm + 118.18 mm + 37.50 mm = 243.18 mm

x ≈ 243.18 mm − 222.50 mm = 20.68 mm Ans

Example 2 An internal dovetail is shown in Figure 63-8. Two pins or balls are used to check the dovetail for both location and angular accuracy. Calculate check dimension *x*. All dimensions are in inches.

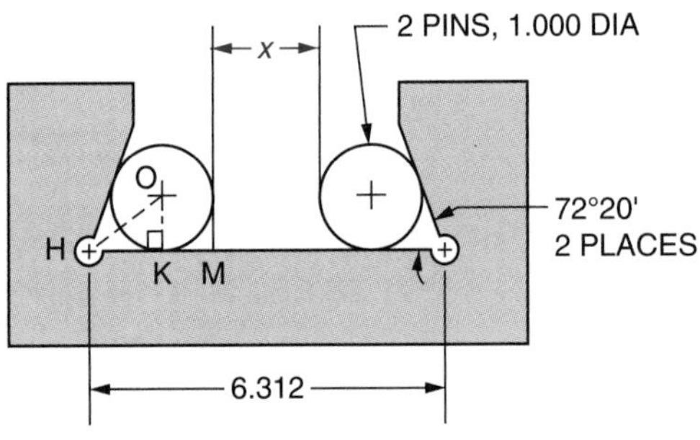

Figure 63-8

Project line HO from point H to the pin center O; HO bisects the 72°20' angle. Project a radius from point O to the point of tangency K; ∠HKO is a right angle since a radius is perpendicular to a tangent at the point of tangency.

In right △HOK, $\angle KHO = \frac{72°20'}{2} = 36°10'$

$$KO = \frac{1.000 \text{ in}}{2} = 0.500 \text{ in}$$

Solve for side HK. $\tan \angle KHO = \frac{KO}{HK}$

$$\tan 36°10' = \frac{0.500}{HK}$$

HK = 0.500 ÷ tan 36°10'

HK = 0.500 [÷] [tan] 36 [° ′ ″] 10 [° ′ ″] [=] 0.6839979623, 0.684 in

or HK = 0.500 [÷] [tan] 36 [° ′ ″] 10 [° ′ ″] [▸] [ENTER] [ENTER] 0.683997962, 0.684 in

HK ≈ 0.684 in

KM = pin radius = 0.500 in

HM = HK + KM ≈ 0.684 in + 0.500 in = 1.184 in

x ≈ 6.312 in − 2(HM) = 6.312 in − 2(1.184 in)

x ≈ 6.312 [−] 2 [×] 1.184 [=] 3.944 in Ans

Application

Sine Bars and Sine Plates

1. Determine the height of gage blocks required to set the following angles on a 10" sine plate.

a. 35° ______	d. 9°44' ______	g. 0°20' ______
b. 13°10' ______	e. 28°32' ______	h. 2°26' ______
c. 36°50' ______	f. 44°20' ______	i. 19°51' ______

2. Determine the height of gage blocks required to set the following angles with a 5" sine bar.

a. 40°40' ______	d. 0°30' ______	g. 39°12' ______
b. 7° ______	e. 21°57' ______	h. 44°50' ______
c. 12°10' ______	f. 13°18' ______	i. 8°17' ______

Tapers and Bevels

Solve the following problems. For customary unit dimensioned problems, calculate angles to the nearer minute and lengths to the nearer thousandths inch. For metric unit dimensioned problems, calculate angles to the nearer hundredth degree and lengths to the nearer hundredth millimeter.

3. Find the included taper $\angle x$.
   All dimensions are in inches. ______

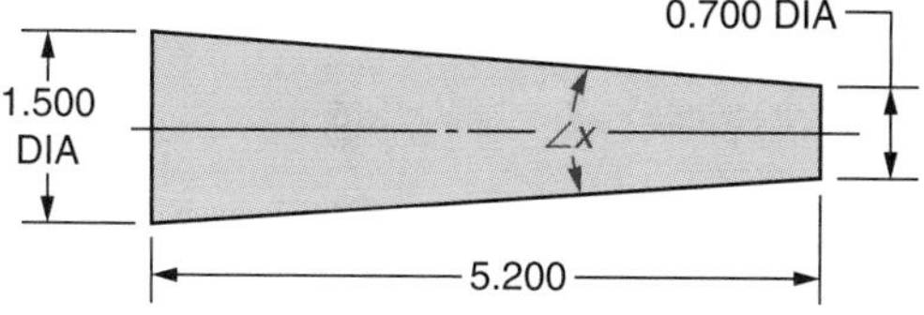

4. Find length x.
   All dimensions are in inches. ______

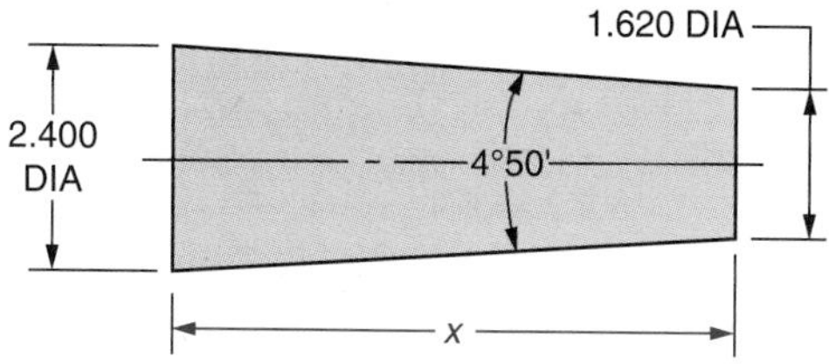

5. Find diameter *y*.
   All dimensions are in millimeters. ______________

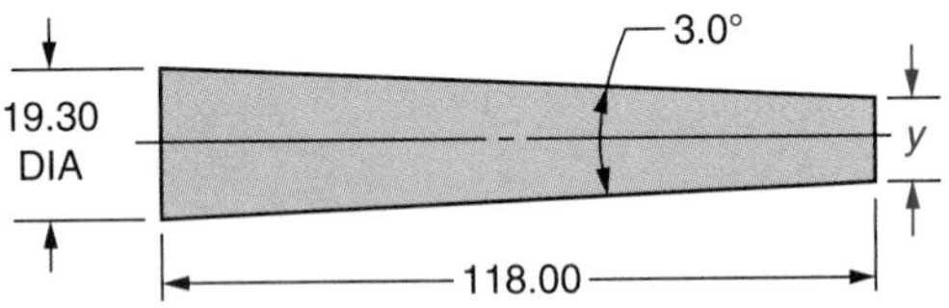

7. Find ∠*x*
   All dimensions are in inches. ______________

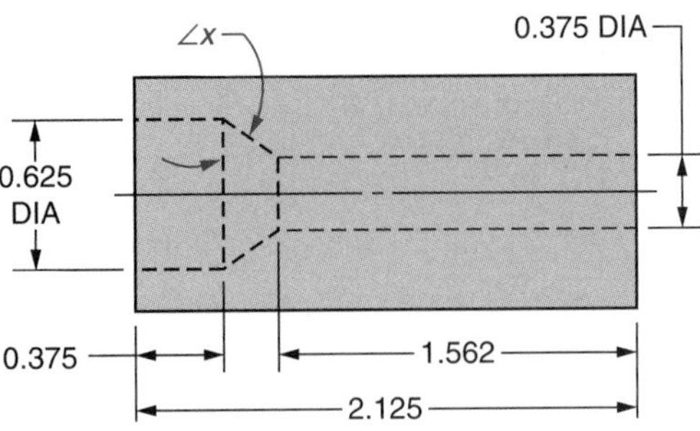

6. Find diameter *x*.
   All dimensions are in inches. ______________

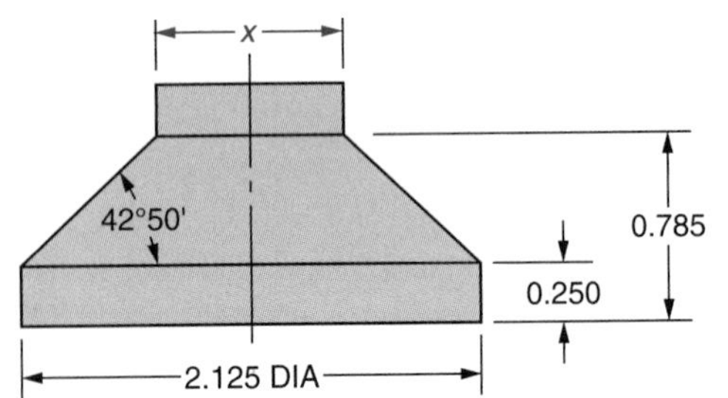

8. Find dimension *y*.
   All dimensions are in inches. ______________

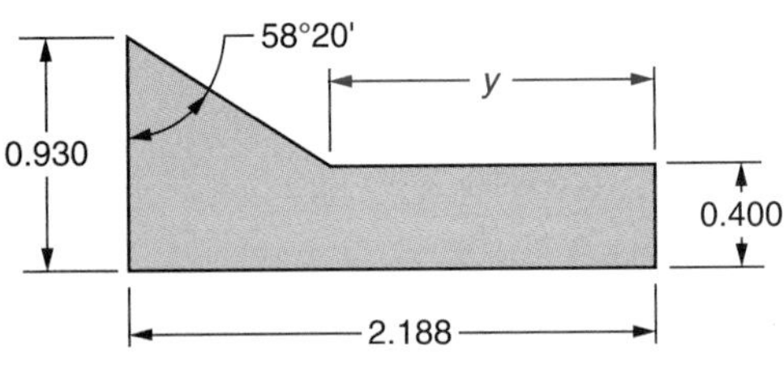

Distance Between Holes and V-Slots

9. Find center distance *y*.
   All dimensions are in millimeters. ______________

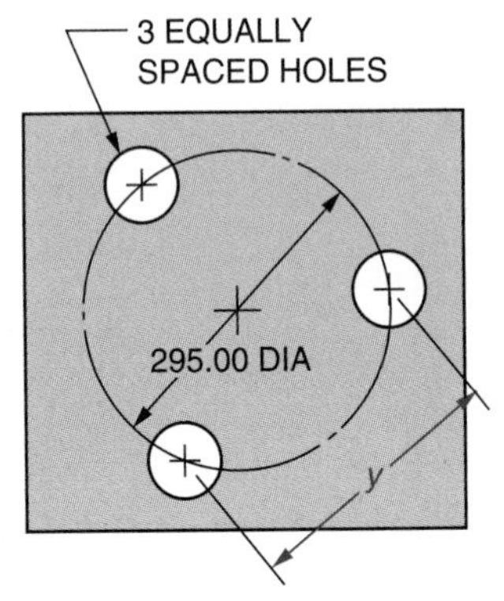

10. Find inside caliper dimension *x*.
    All dimensions are in inches. ______________

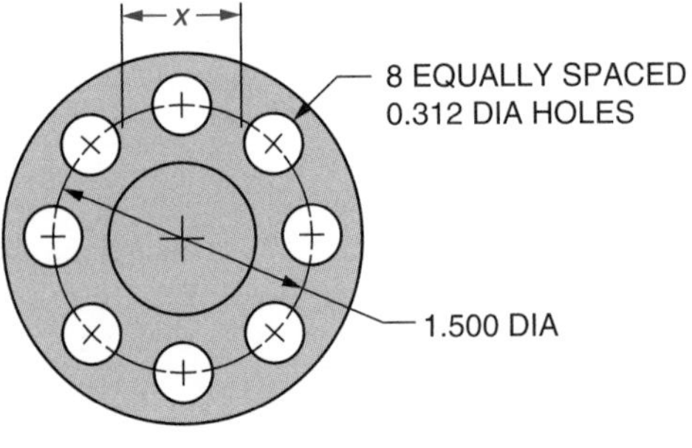

11. Find radius *r*.

    All dimensions are in millimeters. ______________

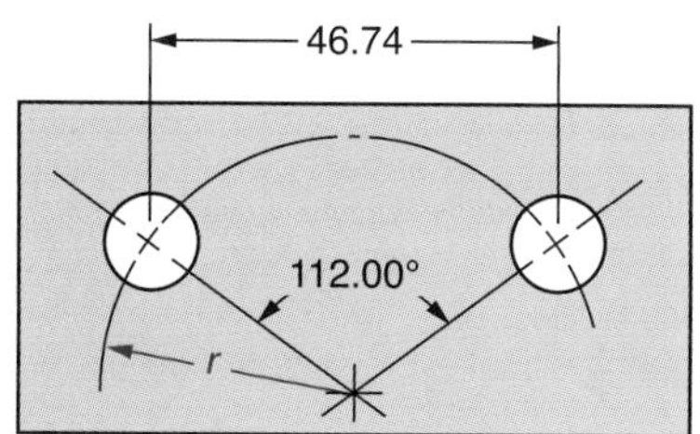

12. Find arc dimension *x*.

    All dimensions are in inches. ______________

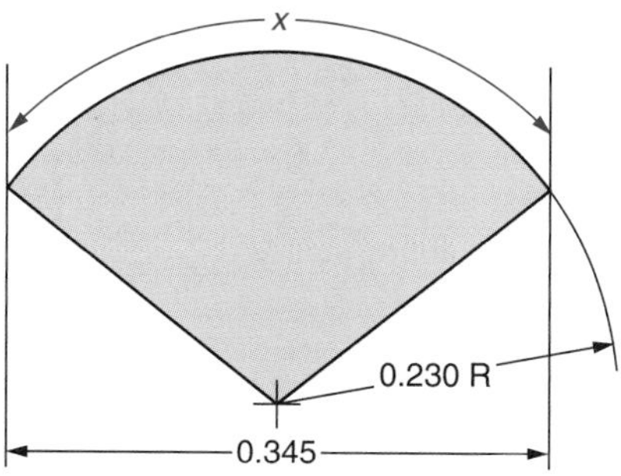

13. Find the depth of cut *x*.

    All dimensions are in inches. ______________

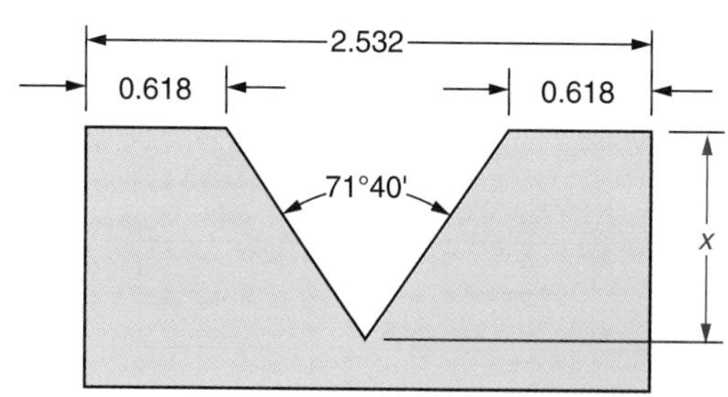

V-Blocks, Thread Wire Checking Dimensions, Dovetails, and Angle Cuts

14. Find ∠*x*.

    All dimensions are in inches. ______________

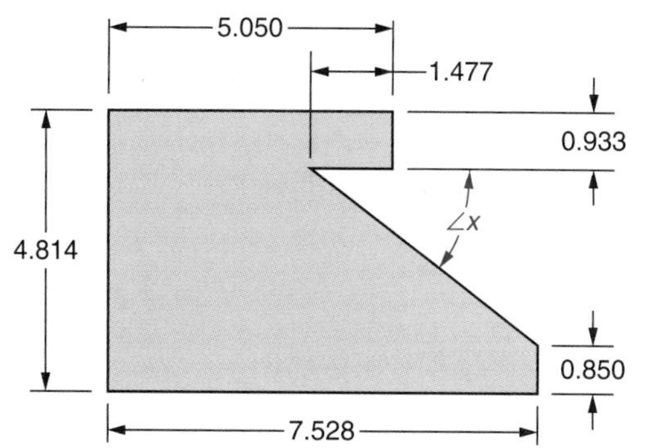

15. Find gage dimension *y*.

    All dimensions are in millimeters. ______________

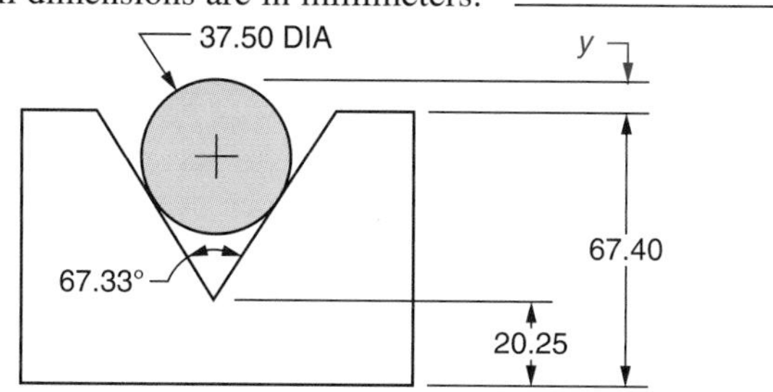

16. Find ∠*y*.

    All dimensions are in inches. ______________

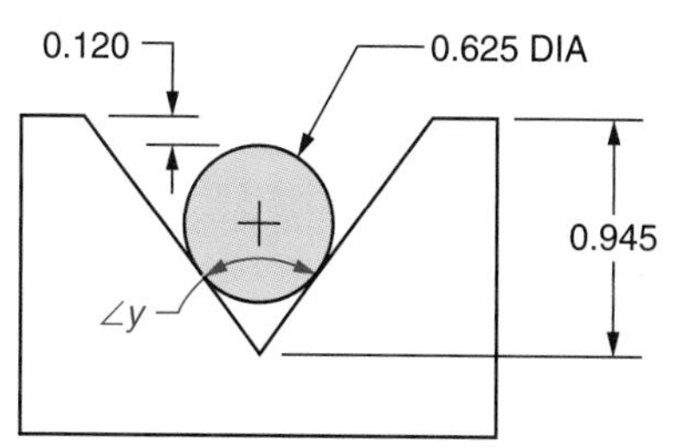

17. Find gage dimension *x*.

    All dimensions are in inches. ______________

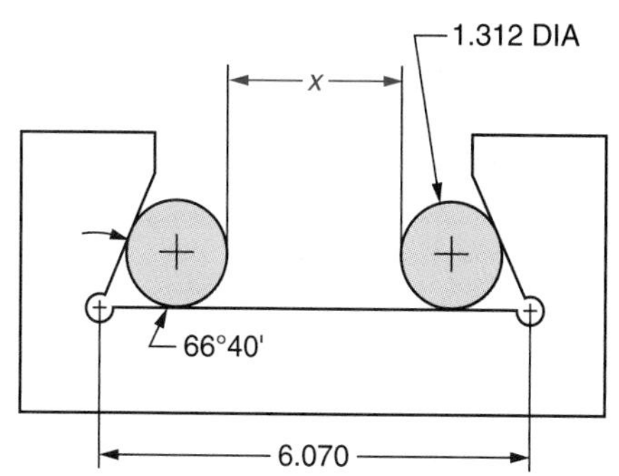

18. Find $\angle x$.
All dimensions are in millimeters. ______

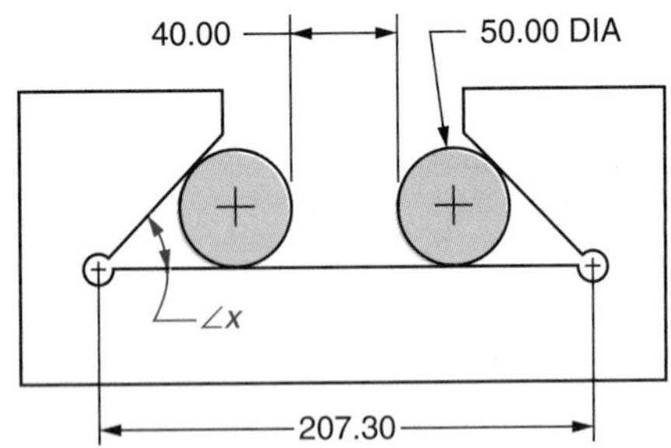

19. Find dimension y.
All dimensions are in inches. ______

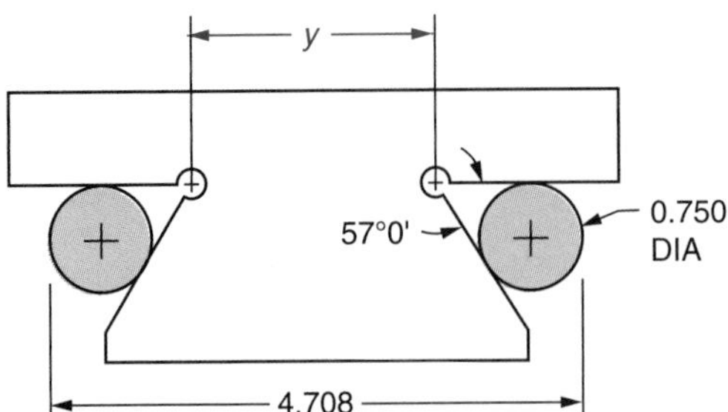

Miscellaneous Applications

20. Find $\angle y$.
All dimensions are in inches. ______

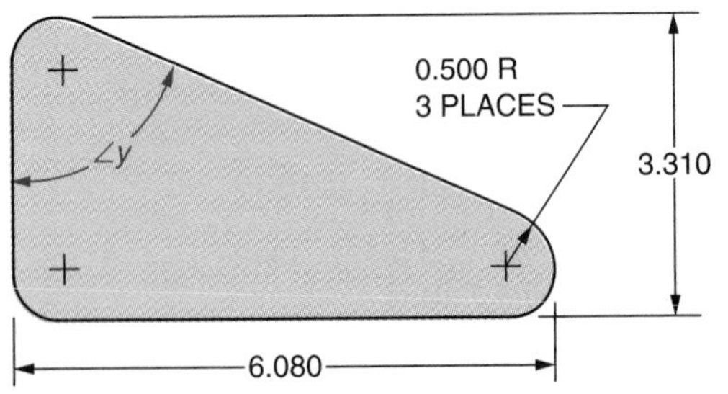

22. Find $\angle x$.
All dimensions are in inches. ______

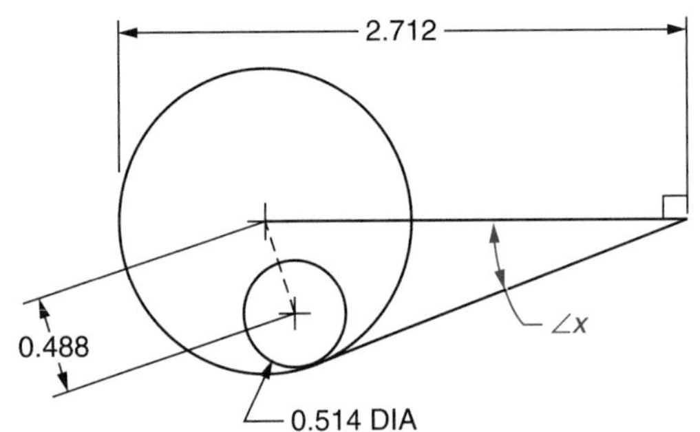

21. Find dimension x.
All dimensions are in millimeters. ______

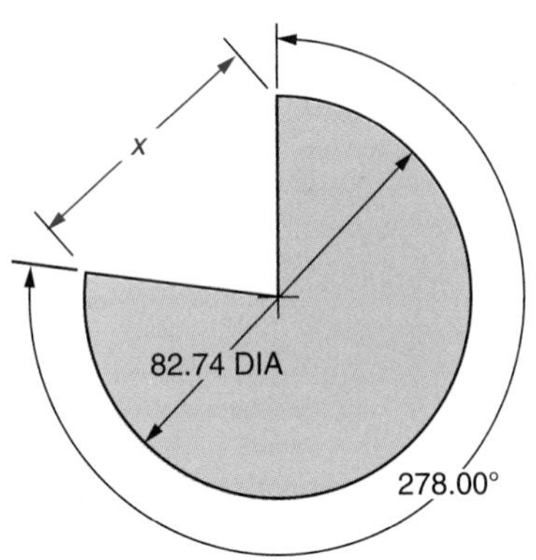

23. Find distance y.
All dimensions are in millimeters. ______

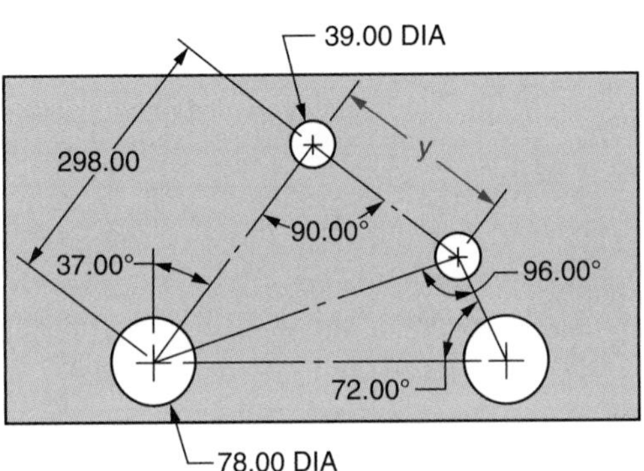

24. Find dimension y.

    All dimensions are in inches. ______________

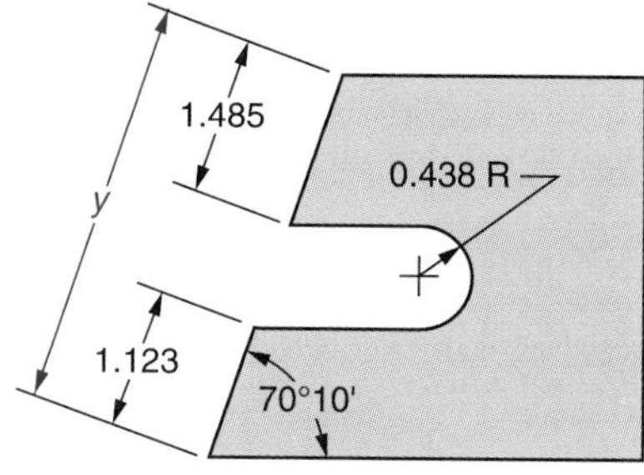

25. The length L of the point drill with included angle A can be calculated using the formula $L = k\,Ø$ where $Ø$ is the diameter of the drill and $k = \frac{1}{2}\tan\left(90° - \frac{A}{2}\right)$.

    Determine k for each of the following angles. Round your answer to three decimal places.

	A	*K*
a.	60°	
b.	82°	
c.	90°	
d.	118°	
e.	135°	

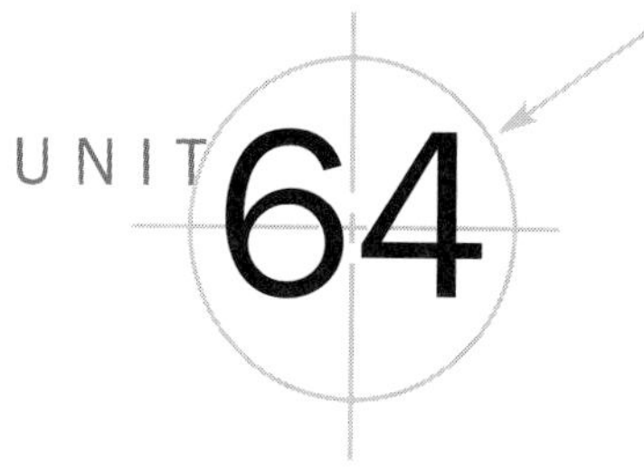

Complex Practical Machine Applications

Objective ***After studying this unit you should be able to***

- Solve complex applied machine technology problems that require forming two or more right triangles by the projection of auxiliary lines.

The problems in this unit are more challenging than those in the last unit and are typical of those found in actual practice when working directly from engineering drawings. The solutions of these problems require the projection of auxiliary lines to form two or more right triangles.

Study the procedures, which are given in detail for solving the examples. There is a common tendency to begin writing computations before analyzing the problem. This tendency must be avoided. As problems become more complex, a greater proportion of time and effort is required in the analyses. The written computations must be developed in clear and orderly steps.

Method of Solution

Analyze the problem before writing computations.

- Relate given dimensions to the unknown and determine whether other dimensions in addition to the given dimensions are required in the solution.

- Determine the auxiliary lines that are required to form right triangles that containing dimensions needed for the solution.
- Determine whether sufficient dimensions are known to obtain required values within the right triangles. If enough information is not available for solving a triangle, continue the analysis until enough information is obtained.
- Check each step in the analysis to verify that there are no gaps or false assumptions.

Write the computations.

Example 1 Determine length x of the part shown in Figure 64-1. All dimensions are in inches.

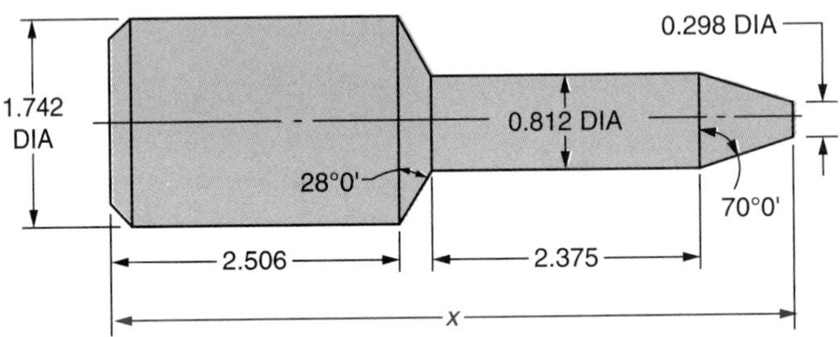

Figure 64-1

Analyze the problem:

Project auxiliary lines to form right △ABF and right △CDE in Figure 64-2. If distances AB and CD can be determined, length x can be computed.

$$x = 2.506 \text{ in} + AB + 2.375 \text{ in} + CD$$

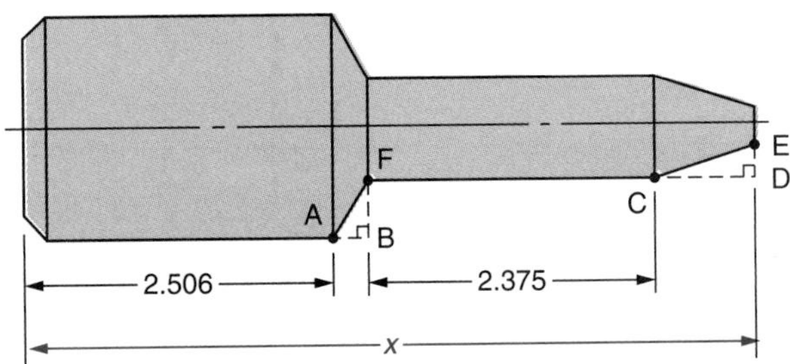

Figure 64-2

Determine whether enough information is given to solve for AB. In right △ABF:
∠FAB = 90° − 28° = 62° (complementary angles)

$$BF = \frac{1.742 \text{ in} - 0.812 \text{ in}}{2} = 0.465 \text{ in}$$

There is enough information to determine AB. Determine whether enough information is given to solve for CD. In right △CDE:
∠ECD = 90° − 70° = 20° (complementary angles)

$$DE = \frac{0.812 \text{ in} - 0.298 \text{ in}}{2} = 0.257 \text{ in}$$

There is enough information to determine CD.

Computations:

Solve for AB.

$$\tan \angle FAB = \frac{BF}{AB}$$

$$\tan 62°0' = \frac{0.465 \text{ in}}{AB}$$

$$AB = 0.465 \text{ in} \div \tan 62°0'$$

.465 [÷] [tan] 62 [=] 0.247244886

AB ≈ 0.2472 in

Solve for CD.

$$\tan CD = \frac{DE}{CD}$$

$$\tan 20°0' = \frac{0.257 \text{ in}}{CD}$$

$$CD = 0.257 \text{ in} \div 20°0'$$

or .257 [÷] [tan] 20 [=] 0.706101697

CD ≈ 0.7061 in

Solve for *x*.

$$x \approx 2.506 \text{ in} + AB + 2.375 \text{ in} + CD$$

$$x \approx 2.506 \text{ in} + 0.2472 \text{ in} + 2.375 \text{ in} + 0.7061 \text{ in}$$

$$x \approx 5.834 \text{ in} \quad \text{Ans}$$

Example 2 Determine ∠*x* of the plate shown in Figure 64-3. All dimensions are in millimeters.

➤ Note: **Generally, when solving problems that involve an arc that is tangent to one or more lines, it is necessary to project the radius of the arc to the tangent point and to project a line from the vertex of the unknown angle to the center of the arc.**

Analyze the problem:

Label the drawing as in Figure 64-4.

Project auxiliary lines between the points A and O, from point O to the tangent point B, and from point O to point C. Right △ACO and right △ABO are formed. If ∠1 and ∠2 can be computed, ∠*x* can be determined. ∠*x* = 90° − (∠1 + ∠2)

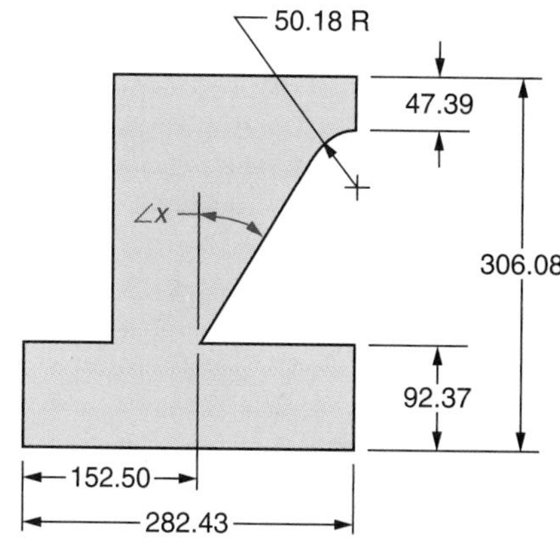

Figure 64-3

Determine whether enough information is given to solve for ∠1.

In right △ACO:

$$AC = 282.43 \text{ mm} - 152.50 \text{ mm} = 129.93 \text{ mm}$$

$$CO = 306.08 \text{ mm} - (92.37 \text{ mm} + 50.18 \text{ mm} + 47.39 \text{ mm})$$
$$= 116.14 \text{ mm}$$

There is enough information to determine ∠1

Determine whether enough information is given to solve for ∠2.

In right △ABO: BO = 50.18 mm
AO can be determined after solving for ∠1 or by using the Pythagorean theorem.
There is enough information to determine ∠2.

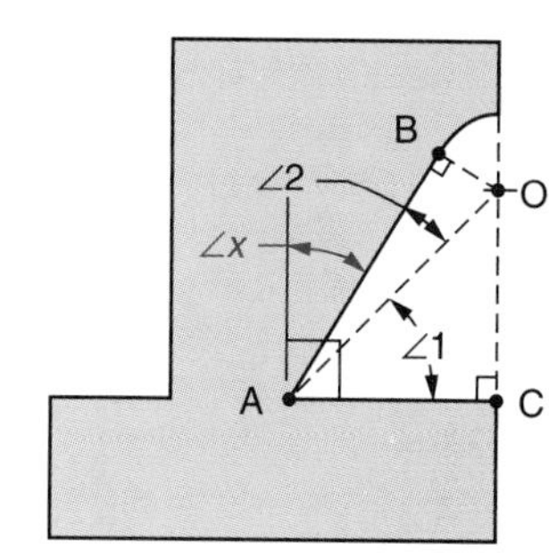

Figure 64-4

Computations:

Solve for ∠1. $\tan \angle 1 = \frac{CO}{AC} = \frac{116.14 \text{ mm}}{129.93 \text{ mm}}$

∠1 = SHIFT tan⁻¹ (116.14 ÷ 129.93) = 41.79244435

$\angle 1 = 41.79244°$

Solve for AO. $\sin \angle 1 = \frac{CO}{AO}$

$$\sin 41.79244° \approx \frac{116.14 \text{ mm}}{AO}$$

$$AO \approx \frac{116.14 \text{ mm}}{\sin 41.79244°}$$

AO ≈ 116.14 ÷ sin 41.79244 = 174.2707941

$AO \approx 174.2708 \text{ mm}$

Solve for ∠2.

$$\sin \angle 2 = \frac{BO}{AO} = \frac{50.18 \text{ mm}}{174.2708 \text{ mm}}$$

∠2 = SHIFT sin⁻¹ (50.18 ÷ 174.2708) = 16.73482688

$\angle 2 = 16.73483°$

Solve for ∠*x*.

$$\angle x = 90° - (\angle 1 + \angle 2)$$

$$\angle x \approx 90° - (41.79244° + 16.73483°) = 31.47273°$$

$\angle x \approx 31.47°$ Ans

Example 3 The front view of a piece with a V-groove is shown in Figure 64-5. A 1.250-inch diameter pin is used to check the cut for depth and angular accuracy. Compute check dimension *x*. All dimensions are in inches.

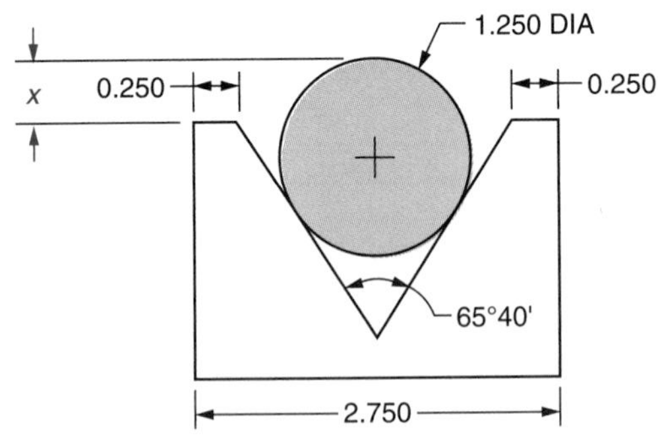

Figure 64-5

Analyze the problem:

Dimension *x* is determined by the pin size, the points of tangency where the pin touches the groove, the angle of the V-groove, and the depth of the groove. Therefore, these dimensions and locations must be part of the calculations.

In Figure 64-6 project auxiliary lines from point A through the center of the pin O, from point O to the tangent point P, and from point B horizontally to intersect vertical line AD at point C.

Right △APO and right △ACB are formed. If AO and AC can be determined, check dimension x can be computed.

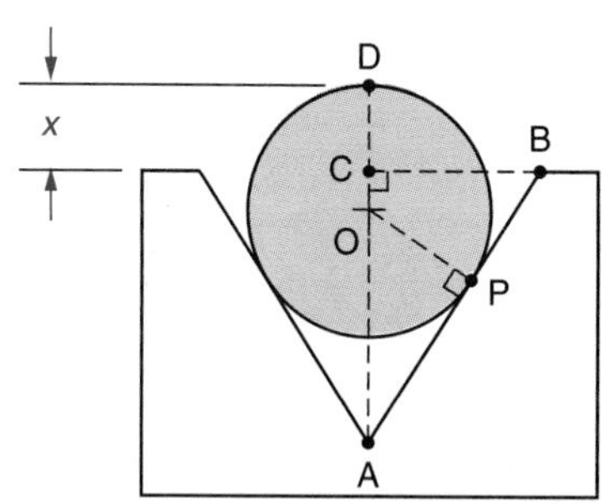

Figure 64-6

$$DO = \text{radius of pin} = 0.625 \text{ in}$$
$$x = (AO + DO) - AC$$

Determine whether enough information is given to solve for AO.

In right △APO:

$$PO = \frac{1.250 \text{ in}}{2} = 0.625 \text{ in}$$
$$\angle OAP = \frac{65°40'}{2} = 32°50'$$

There is enough information to determine AO.

Determine whether enough information is given to solve for AC.

In right △ACB:

$$BC = \frac{2.750 \text{ in}}{2} - 0.250 \text{ in} = 1.125 \text{ in}$$
$$\angle BAC = 32°50'$$

There is enough information to determine AC.

Computations:

Solve for AO.

$$\sin \angle OAP = \frac{PO}{AO}$$
$$\sin 32°50' = \frac{0.625 \text{ in}}{AO}$$
$$AO = \frac{0.625 \text{ in}}{\sin 32°50'}$$

AO ≈ .625 [÷] [sin] 32 [° ' ''] 50 [° ' ''] [SHIFT] [←] [=] 1.152717256

or AO ≈ .625 [÷] [sin] 32 [° ' ''] 50 [° ' ''] [▶] [ENTER] [ENTER] 1.152717256

AO ≈ 1.1527 in

Solve for AC.

$$\tan \angle BAC = \frac{BC}{AC}$$
$$\tan 32°50' = \frac{1.125 \text{ in}}{AC}$$
$$AC = \frac{1.125 \text{ in}}{\tan 32°50'}$$

AC = 1.125 [÷] [tan] 32 [° ' ''] 50 [° ' ''] [=] 1.743429928

or AC ≈ 1.125 [÷] [tan] 32 [° ' ''] 50 [° ' ''] [▶] [ENTER] [ENTER] 1.743429928

AC ≈ 1.7434 in

Solve for check dimension x.

$$x = (AO + DO) - AC$$
$$x = (1.1527 \text{ in} + 0.625 \text{ in}) - 1.7434 \text{ in} = 0.034 \text{ in} \quad \text{Ans}$$

Example 4 Determine $\angle x$ in the series of holes shown in the plate in Figure 64-7. All dimensions are in inches.

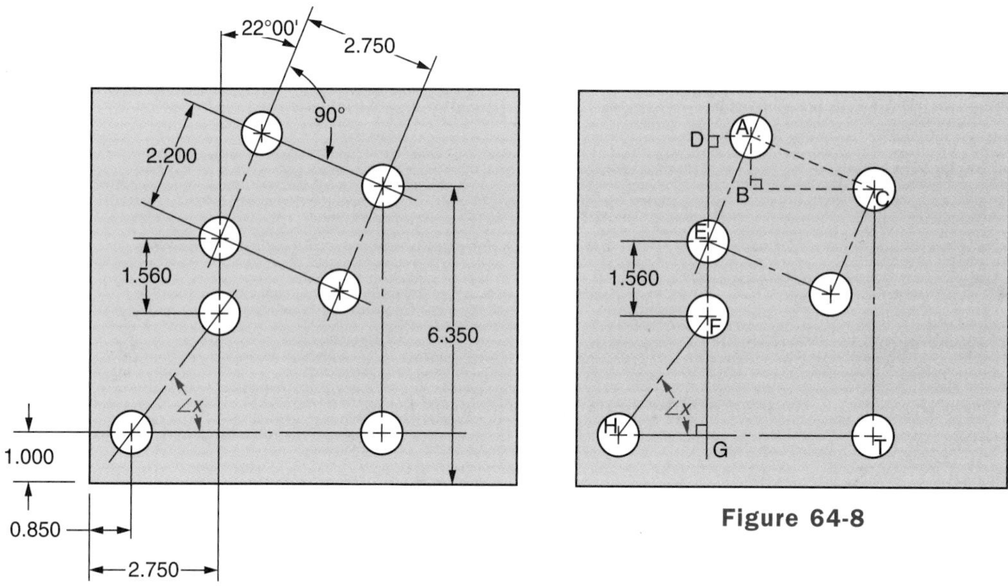

Figure 64-7

Figure 64-8

In Figure 64-8 project auxiliary lines AD, AB, BC. Right $\triangle$ABC, right $\triangle$ADE, and right $\triangle$FGH are formed. If HG and FG can be determined, $\angle x$ can be computed.

$$\text{HG} = 2.750 \text{ in} - 0.850 \text{ in} = 1.900 \text{ in}$$
$$\text{FG} = (\text{TC} + \text{AB}) - (\text{DE} + 1.560 \text{ in})$$
$$\text{FG} = [(6.350 \text{ in} - 1.000 \text{ in}) + \text{AB}] - (\text{DE} + 1.560 \text{ in})$$

Solve for AB.

In right $\triangle$ABC:

$$\text{AC} = 2.750 \text{ in}$$
$$\angle \text{ACB} = 22°00' \quad \text{(Two angles whose corresponding sides are perpendicular are equal.)}$$
$$\sin 22°00' = \frac{\text{AB}}{2.750 \text{ in}}$$

AB = sin 22°00' (2.750 in)

AB ≈ [sin] 22 [×] 2.75 [=] 1.030168132

or AB ≈ [sin] 22 [ENTER] [×] 2.75 [ENTER] 1.030168132

AB = 1.0302 in

Solve for DE.

In right $\triangle$ADE:

$$\angle \text{DEA} = 22°00'$$
$$\text{AE} = 2.200 \text{ in}$$
$$\cos 22°00' = \frac{\text{DE}}{2.200 \text{ in}}$$

DE = cos 22°0' (2.200 in)

DE ≈ [cos] 22 [×] 2.2 [=] 2.03980448

DE = 2.0398 in

or DE ≈ [cos] 22 [ENTER] [×] 2.2 [ENTER] 2.03980448

Solve for FG.

$$FG = [(6.350 \text{ in} - 1.000 \text{ in}) + AB] - (DE + 1.560 \text{ in})$$
$$FG = (5.350 \text{ in} + 1.0302 \text{ in}) - (2.0398 \text{ in} + 1.560 \text{ in}) = 2.7804 \text{ in}$$

Solve for $\angle x$.

$$\tan \angle x = \frac{FG}{HG} = \frac{2.7804 \text{ in}}{1.900 \text{ in}}$$

$\angle x$ = [SHIFT] [tan⁻¹] [(] 2.7804 [÷] 1.9 [)] [=] [SHIFT] [←] 55°39'11"

or $\angle x$ = [2nd] [tan⁻¹] [(] 2.7804 [÷] 1.9 [)] [° ' "] ◀ [ENTER] [ENTER] 55°39'11"

$\angle x \approx 55°39'$ Ans

Example 5 Determine dimension x of the template shown in Figure 64-9. All dimensions are in inches.

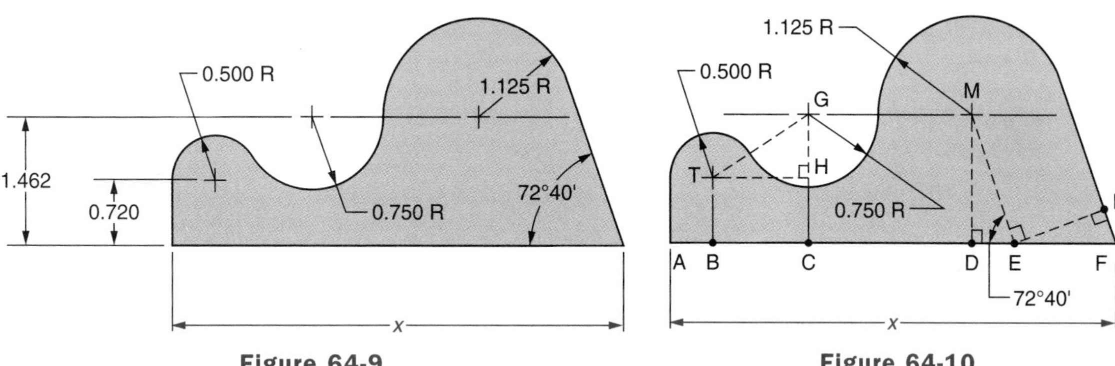

Figure 64-9 **Figure 64-10**

In Figure 64-10 project auxiliary lines to form right $\triangle$GHT, right $\triangle$DEM, and right $\triangle$EFP.

$$x = AB + BC + CD + DE + EF$$
$$AB = 0.500 \text{ in}$$
$$CD = GM = 0.750 \text{ in} + 1.125 \text{ in} = 1.875 \text{ in}$$

(A line connecting the centers of two externally tangent circles passes through the point of tangency.)

If BC, DE, and EF can be determined, x can be computed.
Solve for BC. (BC = TH)

In right $\triangle$GHT: $GH = 1.462 \text{ in} - 0.720 \text{ in} = 0.742 \text{ in}$

$GT = 0.500 \text{ in} + 0.750 \text{ in} = 1.250 \text{ in}$

(GT passes through the point of tangency.)

$$\sin \angle GTH = \frac{GH}{GT} = \frac{0.742 \text{ in}}{1.250 \text{ in}}$$

$\angle$GTH ≈ [SHIFT] [sin⁻¹] [(] .742 [÷] 1.25 [)] [=] 36.4128935

or $\angle$GTH ≈ [2nd] [sin⁻¹] [(] .742 [÷] 1.25 [)] [ENTER] 36.4128935

$\angle GTH \approx 36.41289°$

$$\tan 36.41289° = \frac{GH}{TH}$$

$$\tan 36.41289° = \frac{0.742 \text{ in}}{TH}, \quad TH = \frac{0.742 \text{ in}}{\tan 36.41289°}$$

TH ≈ .742 [÷] tan 36.41289 [=] 1.005950425

TH ≈ 1.0060 in

BC ≈ 1.0060 in

Solve for DE.

In right △DEM: ∠DEM = 72°40'
DM = 1.462 in

$$\tan \angle DEM = \frac{DM}{DE}$$

$$\tan 72°40' = \frac{1.462 \text{ in}}{DE}$$

$$DE = \frac{1.462 \text{ in}}{\tan 72°40'}$$

DE ≈ 1.462 [÷] [tan] 72 [° ' "] 40 [° ' "] [=] 0.456295531

or DE ≈ 1.462 [÷] [tan] 72 [° ' "] 40 [° ' "] ▶ [ENTER] [ENTER] 0.456295531

DE ≈ 0.4563 in

Solve for EF.

In right △EFP: ∠F = 72°40'
EP = 1.125 in
(1.125 radius is ⊥ to tangent line at the point of tangency.)

$$\sin \angle F = \frac{EP}{EF}$$

$$\sin \angle 72°40' = \frac{1.125 \text{ in}}{EF}$$

$$EF = \frac{1.125 \text{ in}}{\sin 72°40'}$$

EF ≈ 1.125 [÷] [sin] 72 [° ' "] 40 [° ' "] [=] 1.178519354

or EF ≈ 1.125 [÷] [sin] 72 [° ' "] 40 [° ' "] ▶ [ENTER] [ENTER] 1.178519354

EF = 1.1785 in

Solve for x.

$$x = AB + BC + CD + DE + EF$$

$$x \approx 0.500 \text{ in} + 1.0060 \text{ in} + 1.875 \text{ in} + 0.4563 \text{ in} + 1.1785 \text{ in}$$

$$\approx 5.016 \text{ in} \quad \text{Ans}$$

Application

Complex Practical Machine Applications

Solve the following problems. For customary unit dimensioned problems, calculate angles to the nearer minute and lengths to the nearer thousandth inch. For metric unit dimensioned problems, calculate angles to the nearer hundredth degree and lengths to the nearer hundredth millimeter.

1. Find length x.
   All dimensions are in inches. __________

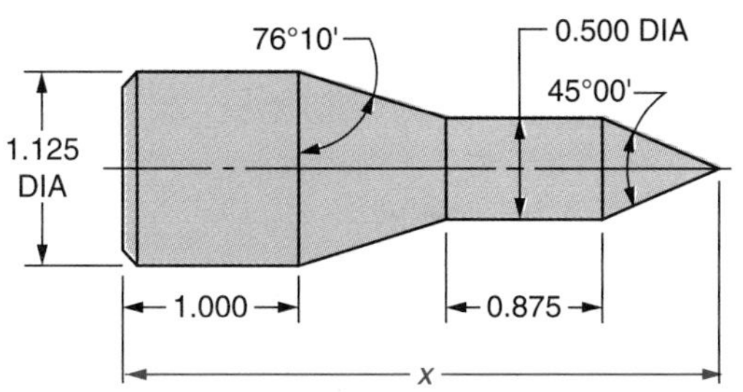

2. Find $\angle x$.
   All dimensions are in millimeters. __________

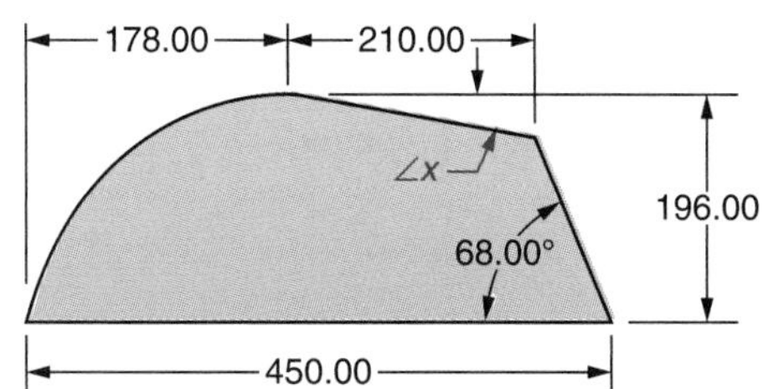

3. Find $\angle x$.
   All dimensions are in inches. __________

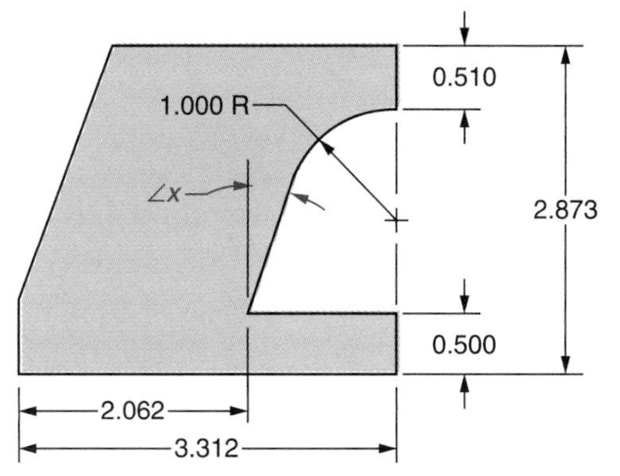

4. Find $\angle y$.
   All dimensions are in millimeters. __________

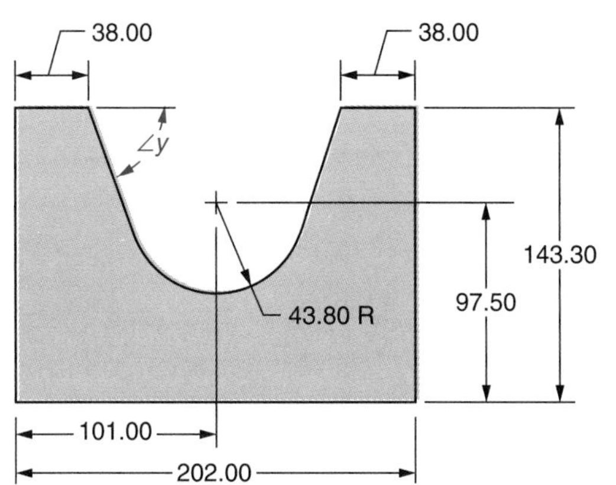

5. Find gage dimension y.
   All dimensions are in inches. __________

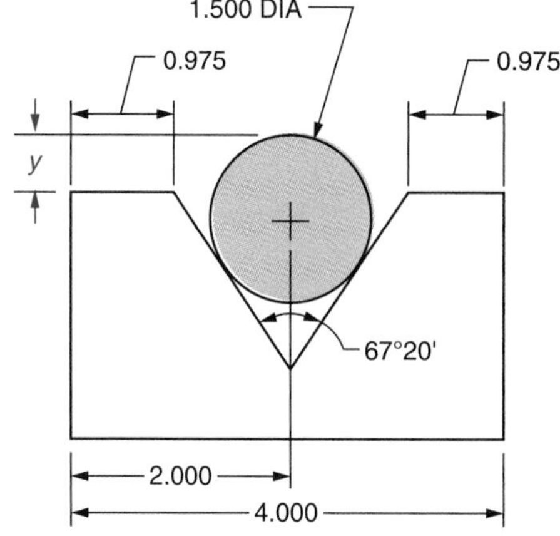

6. Find dimension x.
   All dimensions are in inches. __________

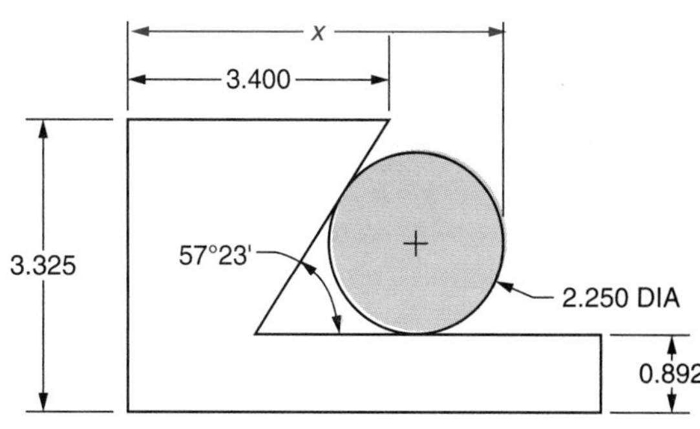

7. Find $\angle x$.
   All dimensions are in inches. __________

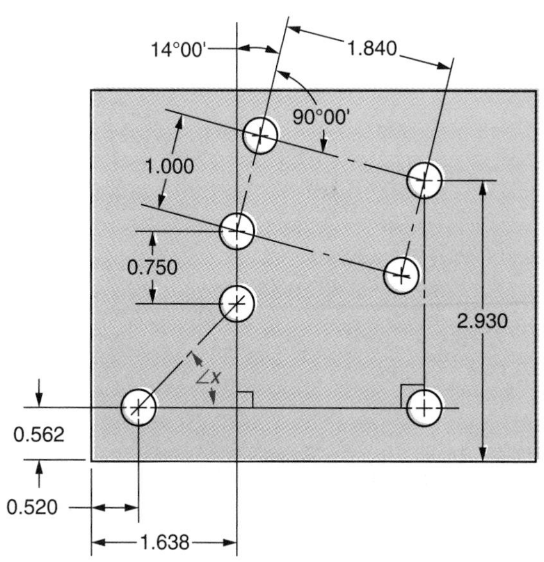

8. Find $\angle y$.
All dimensions are in inches. __________

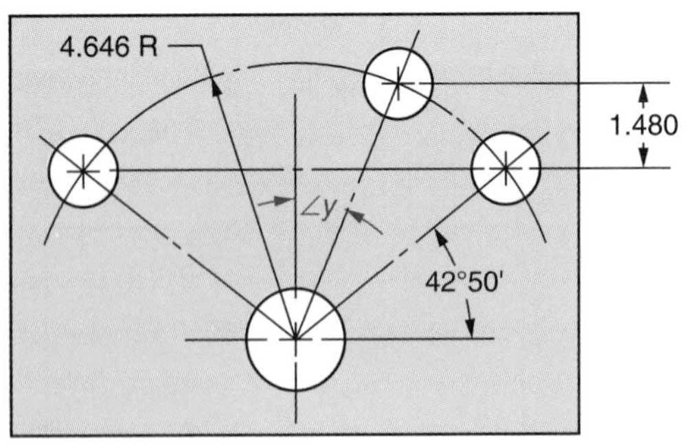

9. Find length x.
All dimensions are in millimeters. __________

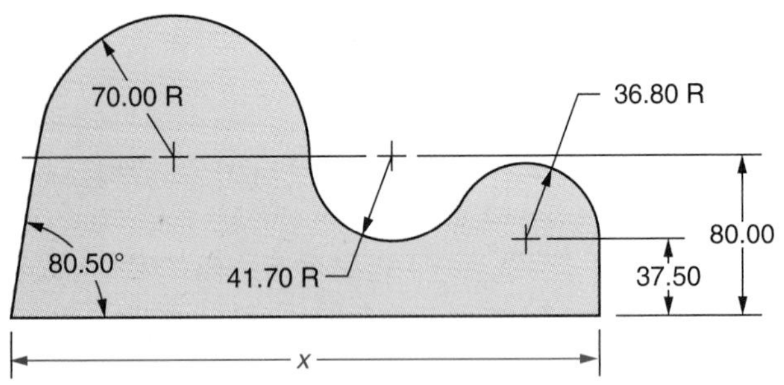

10. Find $\angle y$.
All dimensions are in inches. __________

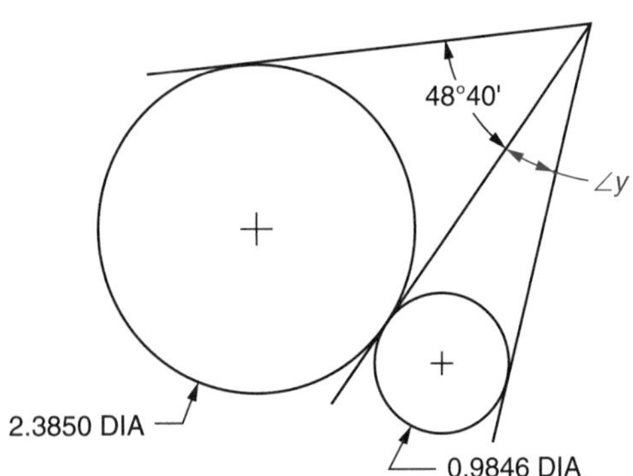

11. Find dimension x.
All dimensions are in inches. __________

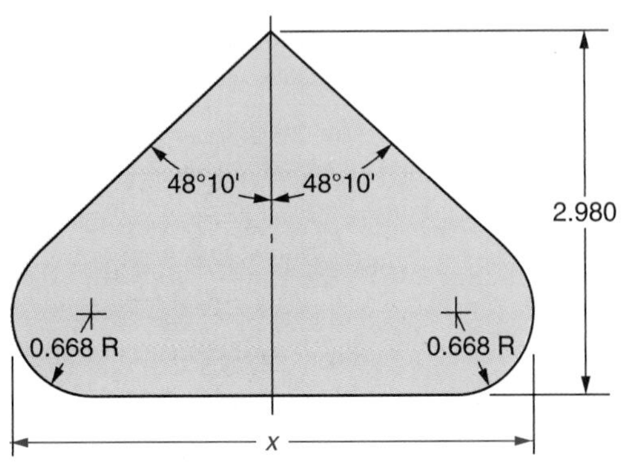

12. Find dimension y.
All dimensions are in inches. __________

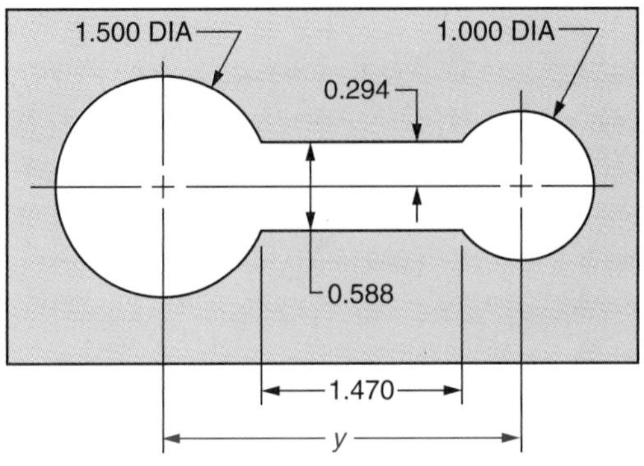

13. Find dimension y.
All dimensions are in inches. __________

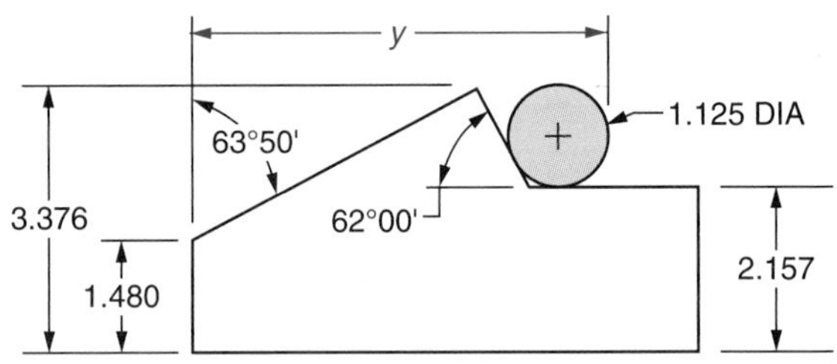

14. Find $\angle x$.
All dimensions are in millimeters. __________

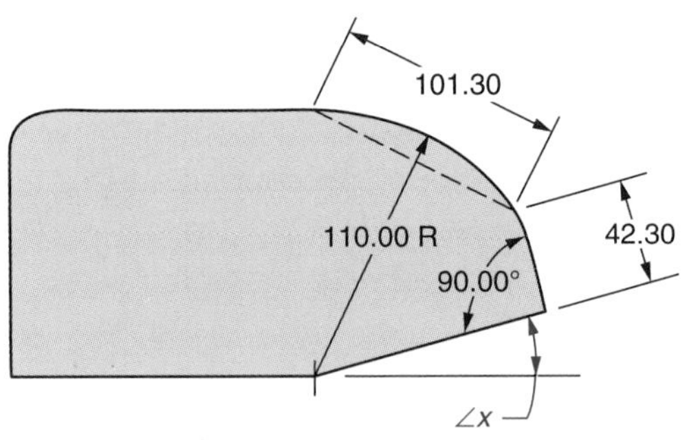

15. Find $\angle x$.
All dimensions are in inches. __________

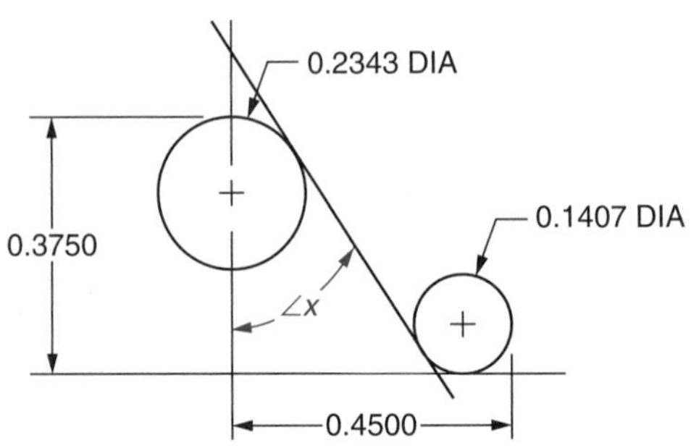

16. Find dimension *y*.
All dimensions are in inches. ______________

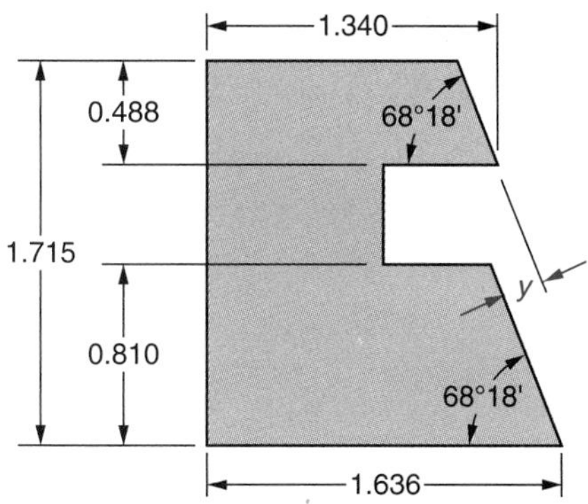

17. Find dimension *x*.
All dimensions are in inches. ______________

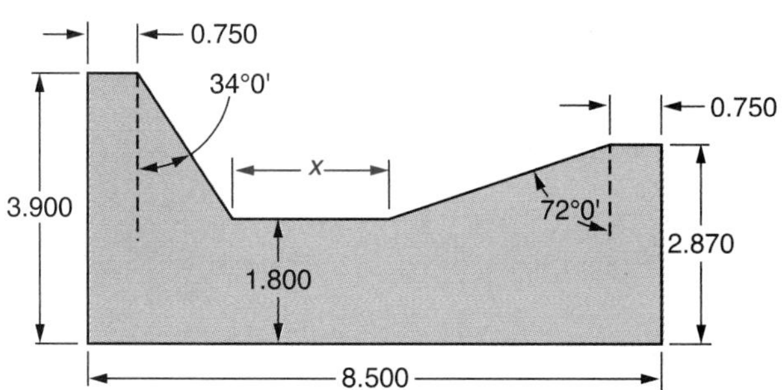

18. Find ∠*y*.
All dimensions are in millimeters. ______________

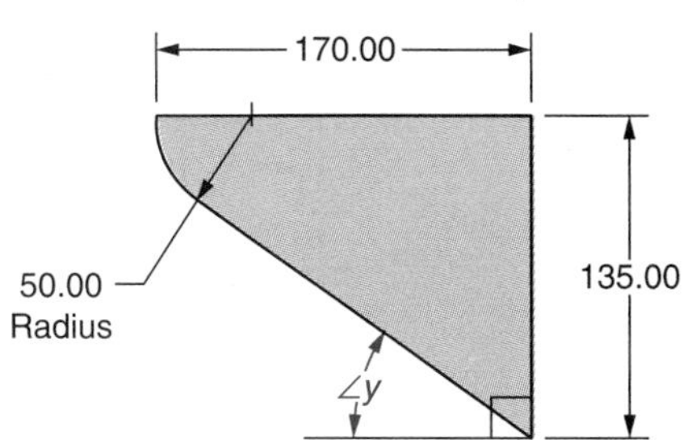

19. Find dimension *y*.
All dimensions are in inches. ______________

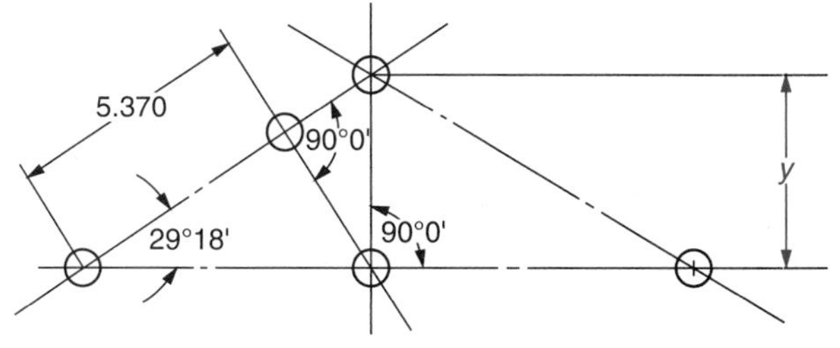

20. Find dimension *x*.
All dimensions are in inches. ______________

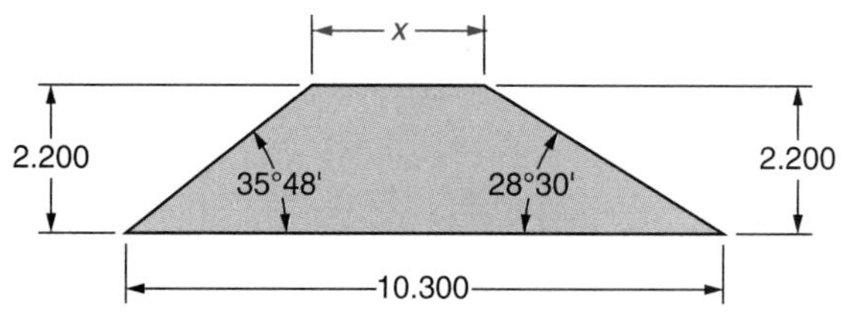

21. Find ∠*y*.
All dimensions are in inches. ______________

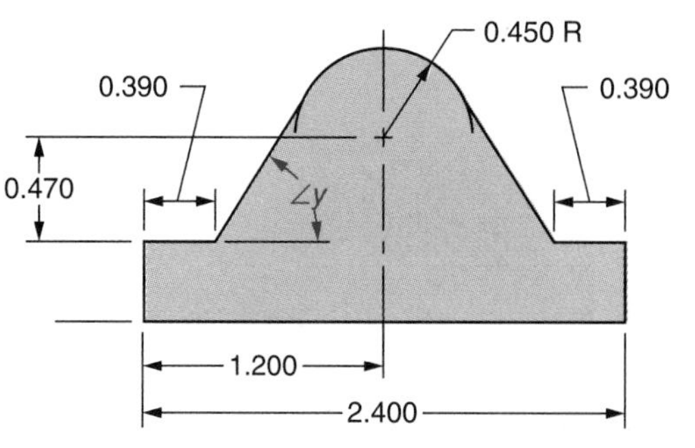

22. Find dimension *x*.
All dimensions are in millimeters. ______________

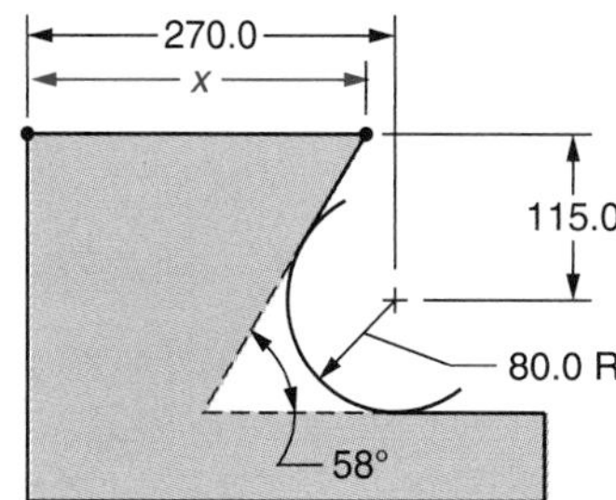

23. Find ∠*y*.
All dimensions are in millimeters. ______________

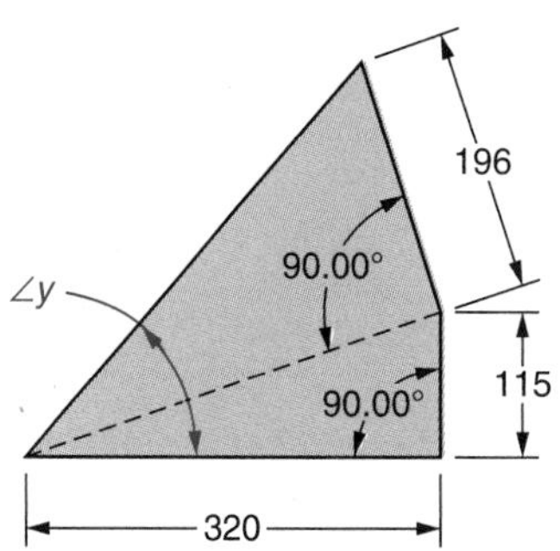

24. Find $\angle y$.
All dimensions are in inches. ______________

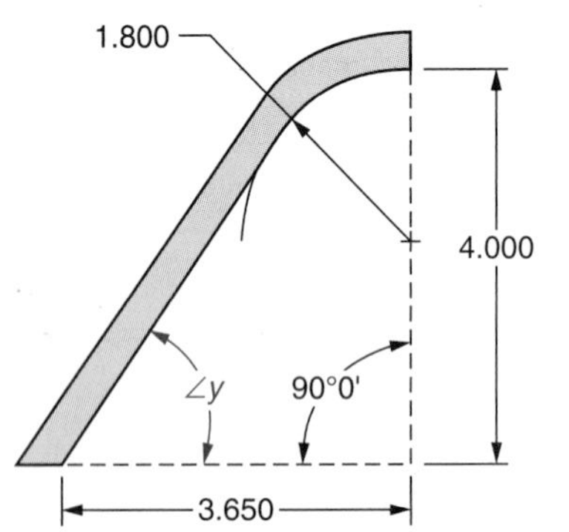

25. Find dimension y.
All dimensions are in inches. ______________

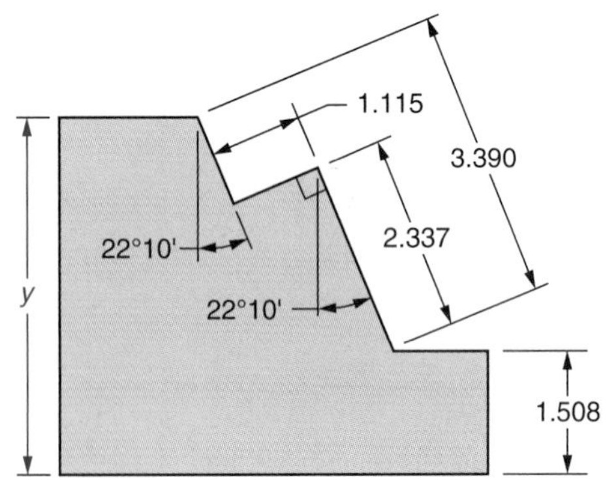

26. Find dimension x.
All dimensions are in inches. ______________

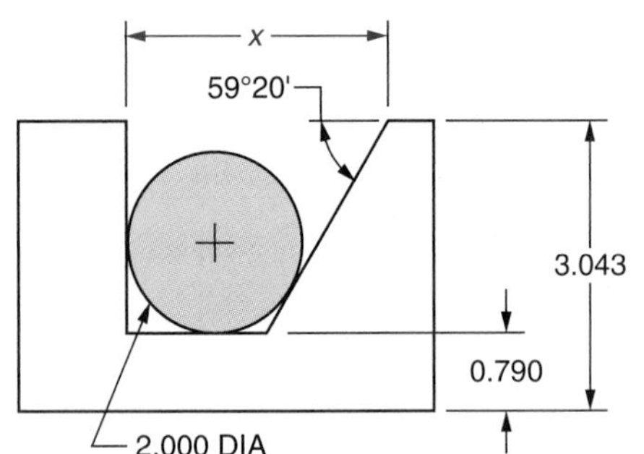

27. Find dimension x.
All dimensions are in inches. ______________

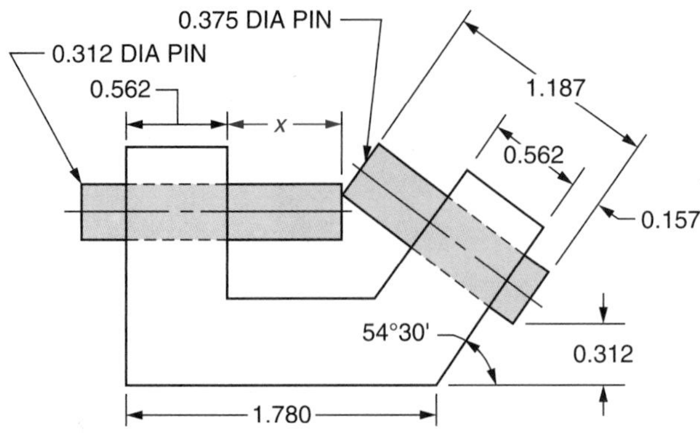

28. Find $\angle x$.
All dimensions are in millimeters. ______________

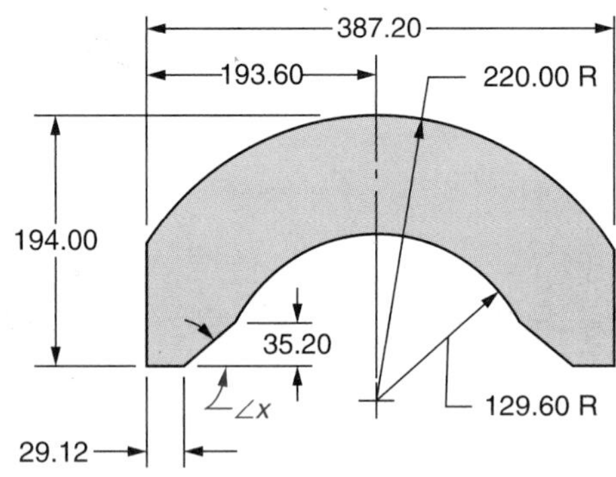

29. Find $\angle x$.
All dimensions are in inches. ______________

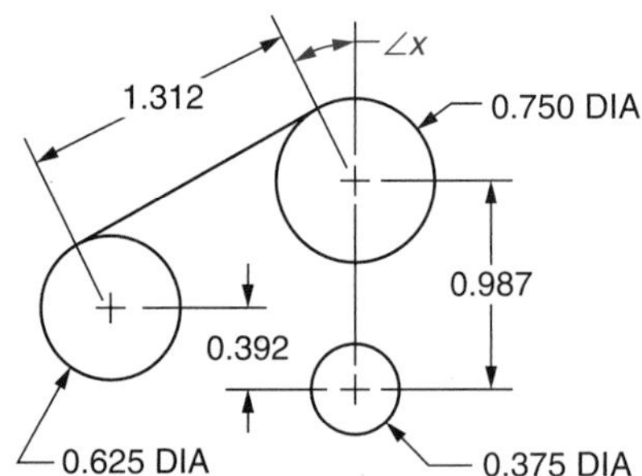

30. Find check dimension y.
All dimensions are in inches. ______________

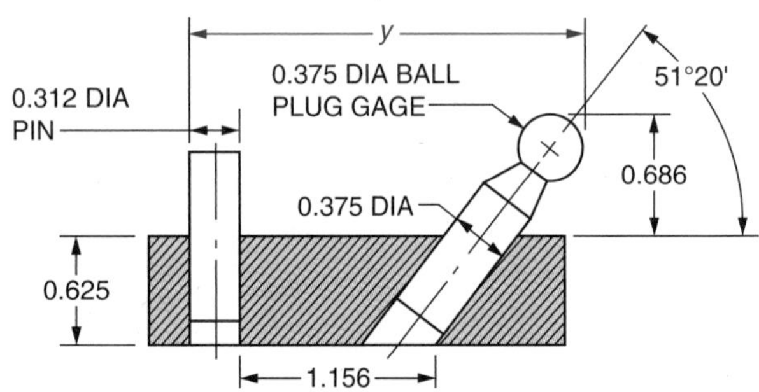

31. Find dimension *y*.
All dimensions are in inches. __________

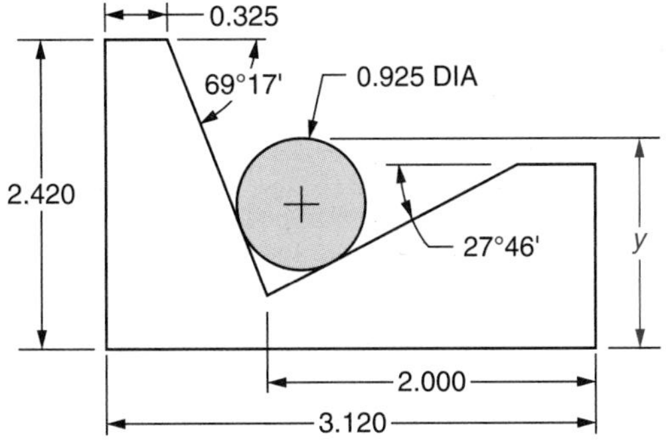

32. Find ∠*x*.
All dimensions are in millimeters. __________

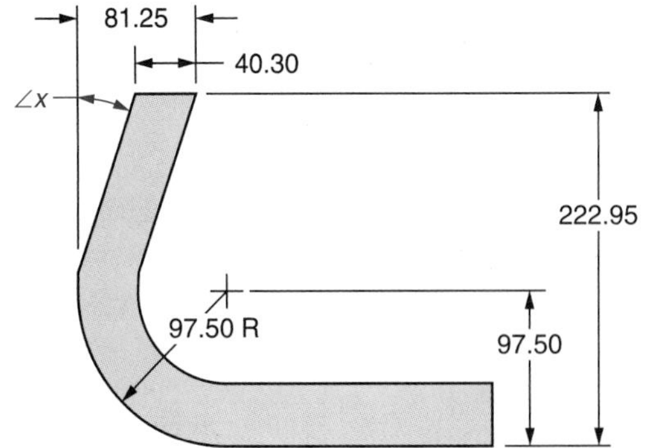

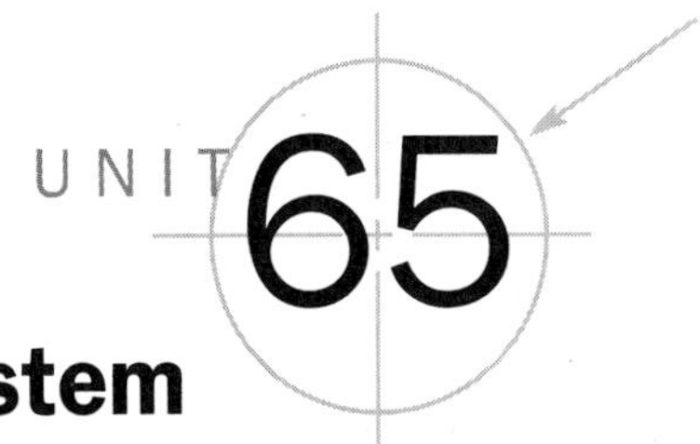

The Cartesian Coordinate System

Objective ***After studying this unit you should be able to***

- Compute functions of angles greater than 90°.

Cartesian (Rectangular) Coordinate System

It is sometimes necessary to determine functions of angles greater than 90°. In a triangle that is not a right triangle, one of the angles can be greater than 90°. Computations using functions of angles greater than 90° are often required in order to solve oblique triangle problems.

Functions of any angles are easily described in reference to the Cartesian coordinate system. A fixed point (O) called the *origin* is located at the intersection of a vertical and horizontal axes. The horizontal axis is the *x*-axis and the vertical axis is the *y*-axis. The *x* and *y* axes divide a plane into four parts that are called *quadrants.* Quadrant I is the upper right section. In a counterclockwise direction from Quadrant I are Quadrants II, III, and IV as shown in Figure 65-1.

All points located to the right of the *y*-axis have positive (+) *x* values; all points to the left of the *y*-axis have negative (−) *x* values. All points above the *x*-axis have positive (+) *y* values; all points below the *x*-axis have negative (−) *y* values. The *x* value is called the *abscissa* and the *y* value is called the *ordinate*.

The *x* and *y* values for each quadrant are listed in the table.

Quadrant I	Quadrant II	Quadrant III	Quadrant IV
$+x$	$-x$	$-x$	$+x$
$+y$	$+y$	$-y$	$-y$

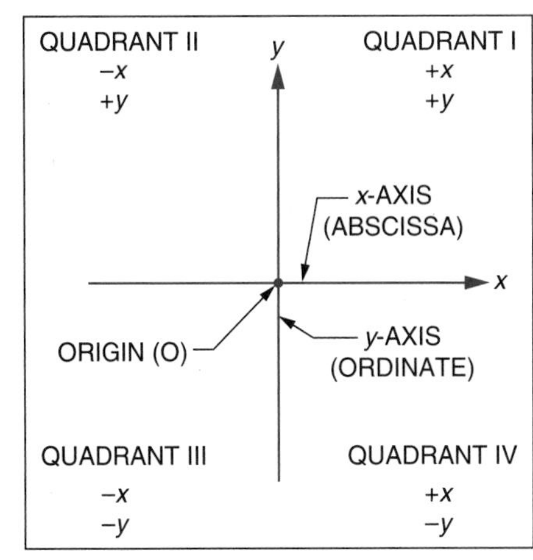

Figure 65-1

Determining Functions of Angles in Any Quadrant

As a ray is rotated through any of the four quadrants, functions of an angle are determined as follows:

- The ray is rotated in a counterclockwise direction with its vertex at the origin (O). Zero degrees is on the x-axis in Quadrant I.
- From a point on the rotated ray, a line segment is projected perpendicular to the x-axis. A right triangle is formed of which the rotated side (ray) is the hypotenuse, the projected vertical line segment is the opposite side, and the side on the x-axis is the adjacent side. The *reference angle* is the acute angle of the triangle that has the vertex at the origin (O).
- The sign of the functions of a reference angle is determined by noting the signs (+ or −) of the opposite and adjacent sides of the right triangle. The hypotenuse (r) is always positive in all four quadrants.

These examples illustrate the method of determining functions of angles greater than 90° in the various quadrants.

Example 1 Determine the sine and cosine functions of 115°.

With the endpoint of the ray (r) at the origin (O), the ray is rotated 115° in a counterclockwise direction as in Figure 65-2.

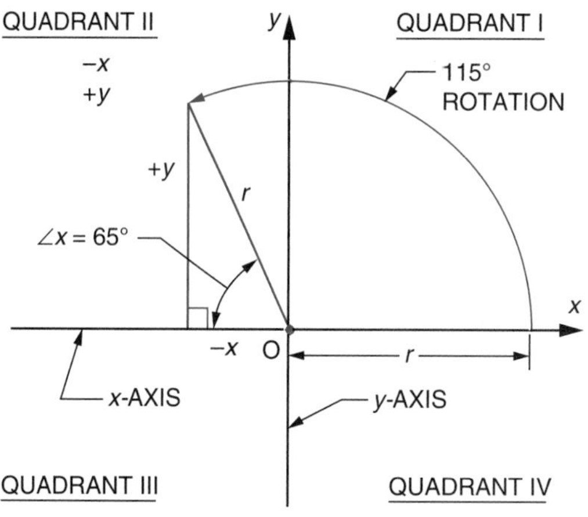

Figure 65-2

From a point on r, side y is projected perpendicular to the x-axis. In the right triangle formed, in relation to the reference angle ($\angle x$), r is the hypotenuse, y is the opposite side, and x is the adjacent side.

$$\angle x = 180° - 115° = 65°$$

Sin $\angle x = \dfrac{\text{opposite side}}{\text{hypotenuse}}$. In Quadrant II, y is positive and r is always positive. Therefore,

$\sin \angle x = \dfrac{+y}{+r}$. In Quadrant II, the sine is a positive (+) function.

$$\sin 115° = \sin (180° - 115°) = \sin 65°$$

With a calculator, functions of angles greater than 90° are computed using the same procedure as used in computing functions of acute angles.

$$\sin 115° \approx \boxed{\text{sin}}\ 115\ \boxed{=}\ 0.906307787 \quad \text{Ans}$$

$\text{Cos } \angle x = \dfrac{\text{adjacent side}}{\text{hypotenuse}}$. Side x is negative (−); therefore, $\cos \angle x = \dfrac{-x}{+r}$. Since the quotient of a negative value divided by a positive value is negative, in Quadrant II, the cosine is a negative (−) function.

$$\cos 115° = -\cos(180° - 115°) = -\cos 65°$$

$\cos 115° \approx$ [cos] 115 [=] −0.422618262 Ans

➤ **Note:** A negative function of an angle does not mean that the angle is negative; it is a negative function of a positive angle. For example, −cos 65° does not mean cos (−65°).

Example 2 Determine the tangent and secant functions of 218°.

Rotate r 218° in a counterclockwise direction.

Project side $y \perp$ to the x-axis as in Figure 65-3.

Reference $\angle x = 218° - 180° = 38°$

$$\tan \angle x = \frac{-y}{-x} = +\text{function}$$

$\tan 218° = \tan 38°$

$\tan 218° \approx$ [tan] 218 [=] 0.781285626 Ans

$$\sec \angle x = \frac{+r}{-x} = -\text{function}$$

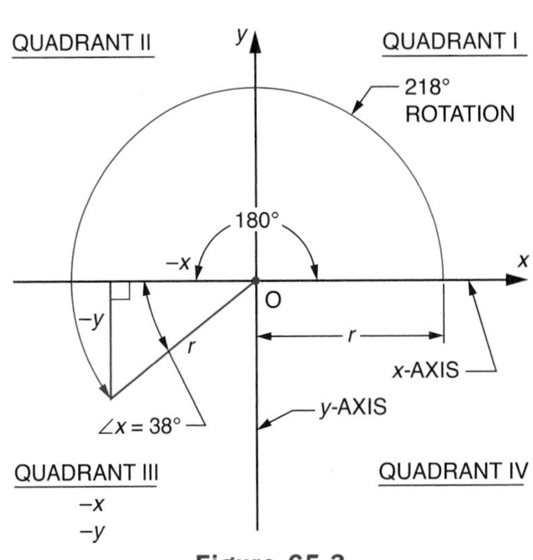

Figure 65-3

$\sec 218° = -\sec 38°$

$\sec 218° \approx$ [cos] 218 [=] [x^{-1}] [=] −1.269018215 Ans

Example 3 Determine the cotangent and cosecant functions of 310°.

Rotate 310° counterclockwise.

Project side $y \perp$ to the x-axis as shown in Figure 65-4.

Reference $\angle x = 360° - 310° = 50°$

$$\cot \angle x = \frac{+x}{-y} = -\text{function}$$

$\cot 310° = -\cot 50°$

$\cot 310° \approx$ [tan] 310 [=] [x^{-1}] [=] −0.839099631 Ans

$$\csc \angle x = \frac{+r}{-y} = -\text{function}$$

$\csc 310° = -\csc 50°$

$\csc 310° \approx$ [sin] 310 [=] [x^{-1}] [=] −1.305407289 Ans

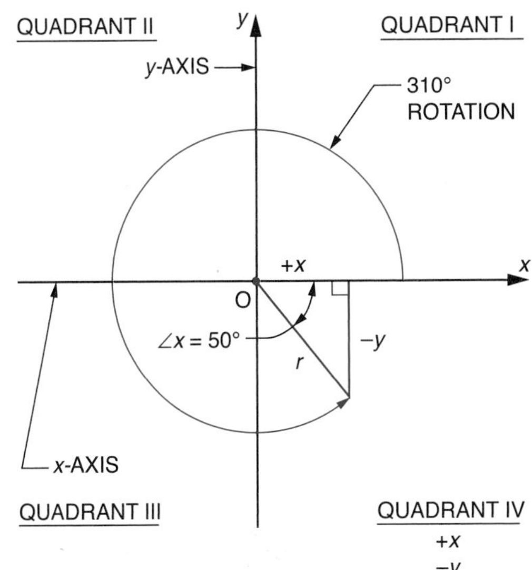

Figure 65-4

Application

Determining Functions of Angles in Any Quadrant

For each angle, sketch a right triangle. Label the sides of the triangles + or −. Determine the reference angles and functions of the angles. Determine the sine, cosine, tangent, cotangent, secant, and cosecant functions for each of these angles. Round the answers to 4 decimal places.

1. 120°
2. 207°
3. 260°
4. 172°
5. 300°
6. 350°
7. 208°50'
8. 96°42'
9. 146°10'
10. 199.40°
11. 313.17°
12. 179.90°

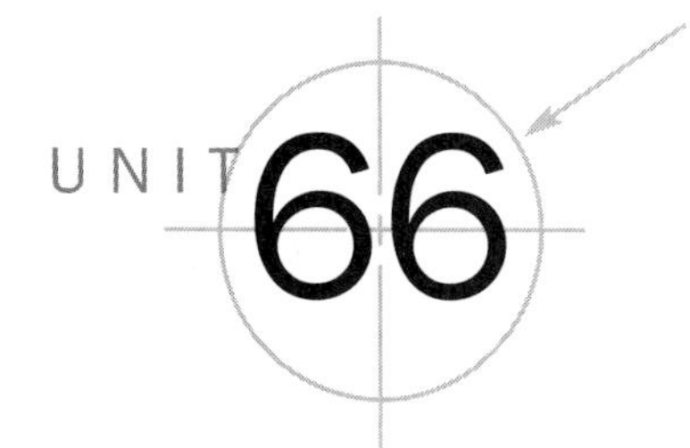

Oblique Triangles: Law of Sines and Law of Cosines

Objectives *After studying this unit you should be able to*

- Solve simple oblique triangles using the Law of Sines and the Law of Cosines.
- Solve practical shop problems by applying the Law of Sines and the Law of Cosines.

Oblique Triangles

An *oblique triangle* is one that does not contain a right angle. An oblique triangle may be either acute or obtuse. In an acute triangle, each of the three angles is acute or less than 90°. In an obtuse triangle, one of the angles is obtuse or greater than 90°. The machinist must often solve practical machine shop problems that involve oblique triangles. These problems can be reduced to a series of right triangles, but the process can be cumbersome and time consuming. Two formulas, the Law of Sines and the Law of Cosines, can be used to simplify such computations. In order to use either formula, three parts of an oblique triangle must be known; at least one part must be a side.

Law of Sines

The Law of Sines states that in any triangle, the sides are proportional to the sines of the opposite angles.

In reference to the triangle shown in Figure 66-1, the formula is stated:

$$\frac{a}{\sin A} = \frac{b}{\sin B} = \frac{c}{\sin C}$$

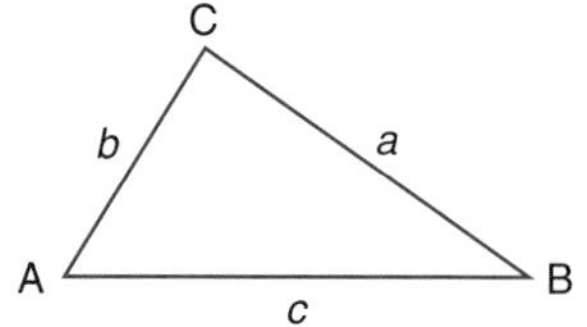

Figure 66-1

The Law of Sines is used to solve the following two kinds of problems:

- Problems where any two angles and any one side of an oblique triangle are known.
- Problems where any two sides and an angle opposite one of the given sides of an oblique triangle are known.

Solving Oblique Triangle Problems Given Two Angles and a Side, Using the Law of Sines

Example 1 Given two angles and a side, determine side x of the oblique triangle shown in Figure 66-2. All dimensions are in inches.

Since side x is opposite the 36° angle and the 3.500 inch side is opposite the 58° angle, the proportion is set up as:

$$\frac{x}{\sin 36^\circ} = \frac{3.500 \text{ in}}{\sin 58^\circ}$$

$$x = \frac{\sin 36^\circ(3.500 \text{ in})}{\sin 58^\circ}$$

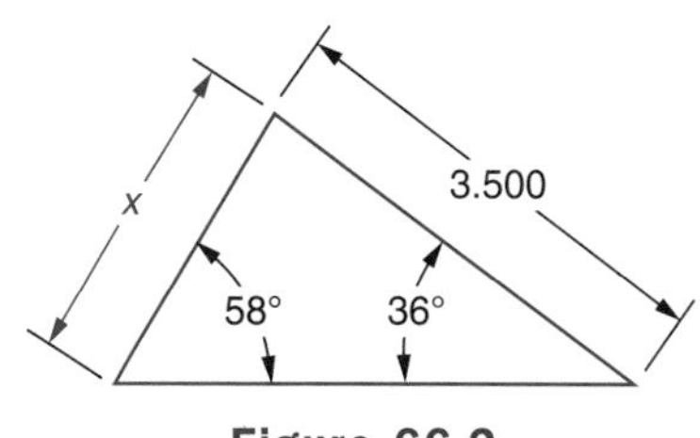

Figure 66-2

$x \approx$ [sin] 36 [×] 3.5 [÷] [sin] 58 [=] 2.425862864

or $x =$ [sin] 36 [ENTER] [×] 3.5 [÷] [sin] 58 [ENTER] 2.425862864

$x \approx 2.426$ in Ans

Example 2 Given two angles and a side of the oblique triangle shown in Figure 66-3. All dimensions are in millimeters.

a. Determine ∠A.

b. Determine side a.

c. Determine side b.

a. Determine ∠A.

$\angle A = 180° - (37.3° + 24.5°)$

$\angle A =$ 180 [−] [(] 37.3 [+] 24.5 [)] [=] 118.2

$\angle A = 118.2°$ Ans

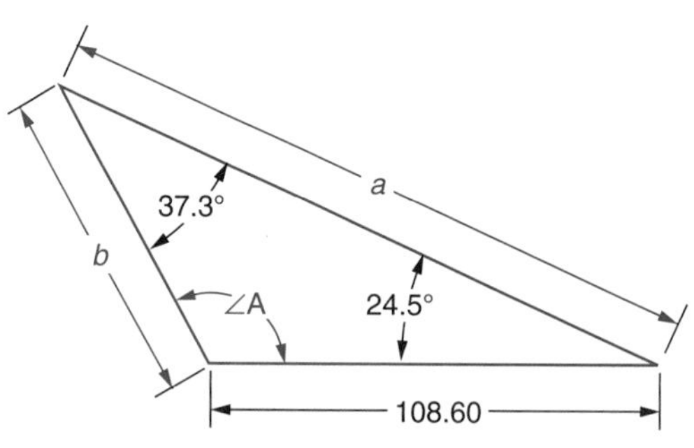

Figure 66-3

b. Determine side a. Set up a proportion and solve for side a.

$$\frac{a}{\sin 118.2°} = \frac{108.60 \text{ mm}}{\sin 37.3°}$$

$$a = \frac{\sin 118.2°(108.60 \text{ mm})}{\sin 37.3°}$$

$a \approx$ [sin] 118.2 [×] 108.6 [÷] [sin] 37.3 [=] 157.9395824

or $a \approx$ [sin] 118.2 [ENTER] [×] 108.6 [÷] [sin] 37.3 [ENTER] 157.9395824

$a \approx 157.94$ mm Ans

c. Determine side b. Set up a proportion and solve for side b.

$$\frac{b}{\sin 24.5°} = \frac{108.60 \text{ mm}}{\sin 37.3°}$$

$$b = \frac{\sin 24.5°(108.60 \text{ mm})}{\sin 37.3°}$$

$b \approx$ [sin] 24.5 [×] 108.6 [÷] [sin] 37.3 [=] 74.31773633

or $b \approx$ [sin] 24.5 [ENTER] [×] 108.6 [÷] [sin] 37.3 [ENTER] 74.31773633

$b \approx 74.32$ mm Ans

Solving Oblique Triangle Problems Given Two Sides and an Angle Opposite One of the Given Sides, Using the Law of Sines

A special condition exists when solving certain problems in which two sides and an angle opposite one of the sides is given. If triangle data are given in word form or if a triangle is inaccurately sketched, there may be two solutions to a problem.

It is possible to have two different triangles with the same two sides and the same angle opposite one of the given sides. A situation of this kind is called an *ambiguous case*. The following example illustrates the ambiguous case or a problem with two solutions.

Example (*The Ambiguous Case or 2 solutions*) A triangle has a 1.5-inch side, a 2.5-inch side, and an angle of 32°, which is opposite the 1.5-inch side.

Using the given data, Figure 66-4 is accurately drawn. Observe that two different triangles are constructed using identical given data. Both $\triangle BCA$ and $\triangle DCA$ have a 1.5-inch side, a 2.5-inch side, and a 32° angle opposite the 1.5-inch side. The two different triangles are shown in Figures 66-5(a) and 66-5(b).

The only conditions under which a problem can have two solutions is when the given angle is acute and the given side opposite the given angle is smaller than the other given side. For example, in the problem illustrated the 32° angle is acute, and the 1.5-inch side opposite the 32° angle is smaller than the 2.5-inch side.

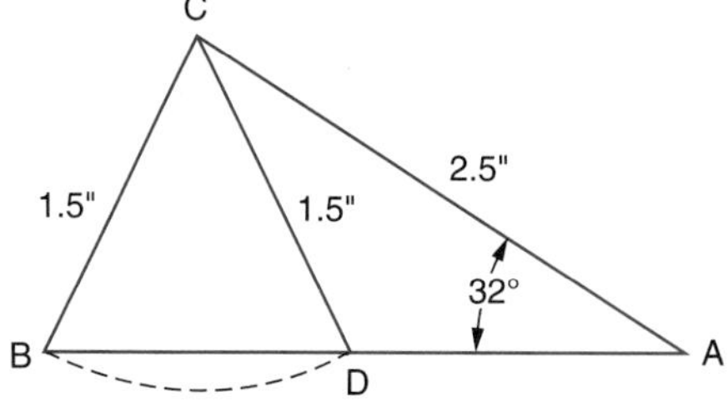

Figure 66-4

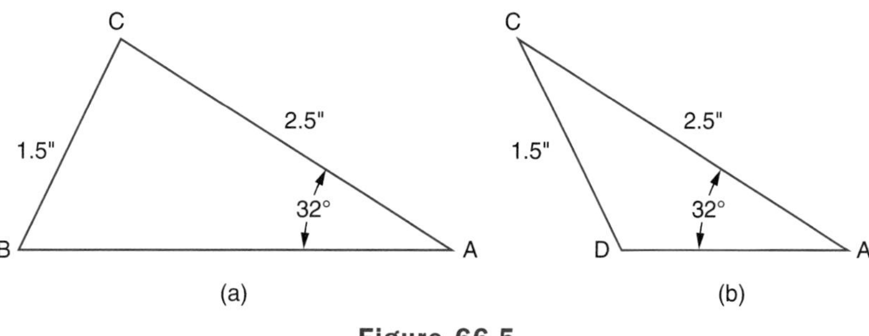

Figure 66-5

In most problems you do not get involved with two solutions. Even under the conditions in which there can be two solutions, if the problem is shown in picture form as an accurately drawn triangle, it can readily be observed that there is only one solution.

Example 1 Given two sides and an opposite angle of the oblique triangle shown in Figure 66-6. All dimensions are in inches.

a. Determine $\angle x$.

b. Determine side y.

The 6.000-inch side opposite the 63°50' angle is larger than the 4.500-inch side; therefore, there is only one solution.

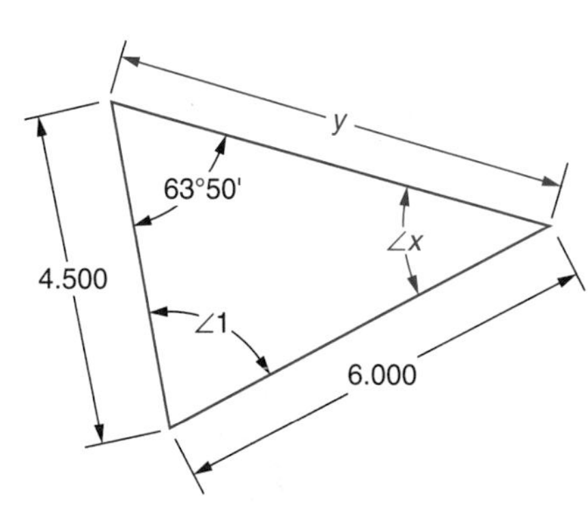

Figure 66-6

a. Determine $\angle x$.

$$\frac{4.500 \text{ in}}{\sin \angle x} = \frac{6.000 \text{ in}}{\sin 63°50'}$$

$$\sin \angle x = \frac{\sin 63°50'(4.500 \text{ in})}{6.000 \text{ in}}$$

$\sin \angle x \approx$ [sin] 63 [° ′ ″] 50 [° ′ ″] [×] 4.5 [÷] 6 [=] 0.673136307

$\angle x \approx$ [SHIFT] [sin⁻¹] .673136307 [=] [SHIFT] [←] 42°18'34"

or $\sin \angle x \approx$ [sin] 63 [° ′ ″] 50 [° ′ ″] [▶] [ENTER] [ENTER] [×] 4.5 [÷] 6 [ENTER] .0673136307

$\angle x \approx$ [2nd] [sin⁻¹] .673136307 [° ′ ″] [◀] [ENTER] [ENTER] 42°18'34.5"

$\angle x \approx 42°19'$ Ans

b. Determine side y.

In order to find the length of side y, we need to know the size of $\angle 1$, the angle opposite y.

$$\angle 1 = 180° - (63°50' + \angle x) = 180° - 106°9' = 73°51'$$

$$\frac{6.000 \text{ in}}{\sin 63°50'} = \frac{y}{\sin 73°51'}$$

$$y = \frac{\sin 73°51'(6.000 \text{ in})}{\sin 63°50'}$$

$y \approx$ [sin] 73 [° ′ ″] 51 [° ′ ″] [×] 6 [÷] [sin] 63 [° ′ ″] 50 [° ′ ″] [=] 6.421307977

or $y \approx$ [sin] 73 [° ′ ″] 51 [° ′ ″] [▶] [ENTER] [ENTER] [×] 6 [÷] [sin] 63 [° ′ ″] 50 [° ′ ″] [▶] [ENTER] [ENTER] 6.421307977

$y \approx 6.421$ in Ans

Example 2 Given two sides and an opposite angle, determine $\angle x$ of the oblique triangle shown in Figure 66-7. All dimensions are in millimeters. The figure is drawn accurately to scale.

$$\frac{140.00 \text{ mm}}{\sin 28.17°} = \frac{275.00 \text{ mm}}{\sin \angle x}$$

$$\sin \angle x = \frac{\sin 28.17°(275.00 \text{ mm})}{140.00 \text{ mm}}$$

$\sin \angle x \approx$ [sin] 28.17 [×] 275 [÷] 140 [=] 0.927318171

$\angle x \approx$ [SHIFT] [sin⁻¹] .927318171 [=] 68.02055272

or $\sin \angle x \approx$ [sin] 28.17 [ENTER] [×] 275 [÷] 140 [ENTER] .927318171

$\angle x \approx$ [2nd] [sin⁻¹] .927318171 [ENTER] 68.02055272

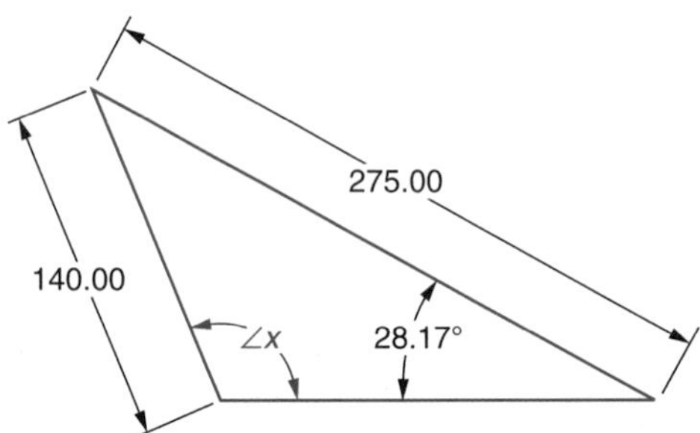

Figure 66-7

The angle that corresponds to the sine function 0.927318171 is 68.02°. Because $\angle x$ is greater than 90°, $\angle x$ = the supplement of 68.02°.

$$\angle x \approx 180° - 68.02° \approx 111.98° \quad \text{Ans}$$

Law of Cosines (Given Two Sides and the Included Angle)

In any triangle, the square of any side is equal to the sum of the squares of the other two sides minus twice the product of these two sides multiplied by the cosine of their included angle.

In reference to the triangle shown in Figure 66-8, the formula is stated:

$$a^2 = b^2 + c^2 - 2bc(\cos \text{A})$$
$$b^2 = a^2 + c^2 - 2ac(\cos \text{B})$$
$$c^2 = a^2 + b^2 - 2ab(\cos \text{C})$$

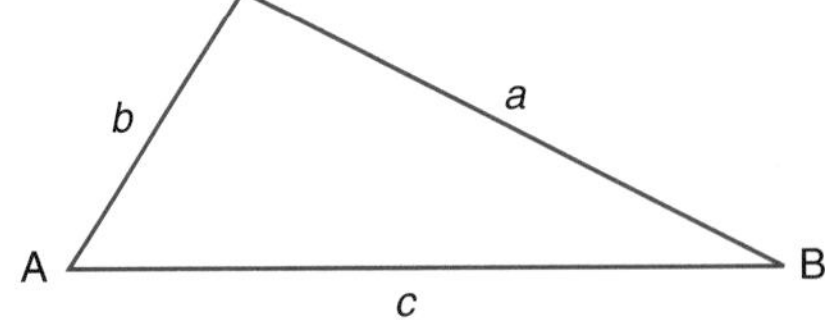

Figure 66-8

The Law of Cosines, as stated in the formulas, is used to solve the following kind of oblique triangle problems.

- Problems where two sides and the included angle of an oblique triangle are known.

➤ **Note:** An angle of an oblique triangle may be greater than 90°. Therefore, you must often determine the cosine of an angle greater than 90° and less than 180°. These angles lie in Quadrant II of the Cartesian coordinate system. Recall that the cosine of an angle between 90° and 180° equals the negative (−) cosine of the supplement of the angle. For example, the cosine of 118°10' = −cos (180° − 118°10') = −cos 61°50'.

Solving Oblique Triangle Problems Given Two Sides and the Included Angle, Using the Law of Cosines

Example 1 Given two sides and the included angle, determine side x of the oblique triangle shown in Figure 66-9. All dimensions are in millimeters. Observe that 36.83° is included between the 62.00 mm and 56.00 mm sides.

Substitute the values in their appropriate places in the formula and solve for x.

$$x^2 = (56.00 \text{ mm})^2 + (62.00 \text{ mm})^2 - 2(56.00 \text{ mm})(62.00 \text{ mm})(\cos 36.83°)$$

$$x = \sqrt{(56.00 \text{ mm})^2 + (62.00 \text{ mm})^2 - 2(56.00 \text{ mm})(62.00 \text{ mm})(\cos 36.83°)}$$

$x \approx$ [√] [(] 56 [x²] [+] 62 [x²] [−] 2 [×] 56 [×] 62 [×]
cos 36.83 [)] [=] 37.70809057

$x \approx 37.71$ mm Ans

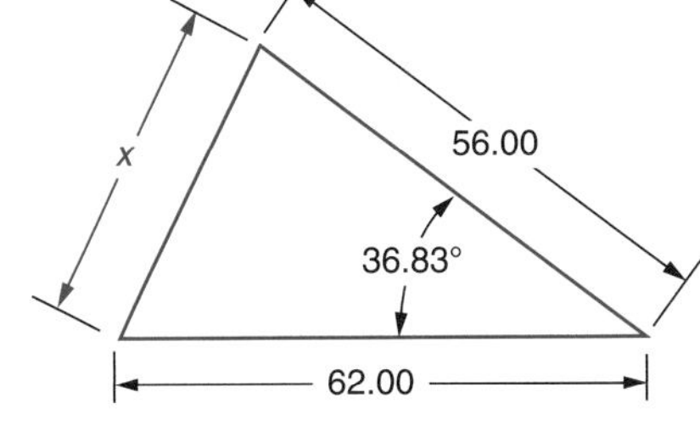

Figure 66-9

Example 2 Given two sides and the included angle of the oblique triangle shown in Figure 66-10. All dimensions are in inches.

a. Determine side a.

b. Determine $\angle B$.

c. Determine $\angle C$.

a. Solve for a, using the Law of Cosines.

$$a^2 = (3.912 \text{ in})^2 + (4.206 \text{ in})^2 - 2(3.912 \text{ in})(4.206 \text{ in})(\cos 127°26')$$

$$a = \sqrt{2(3.912 \text{ in})^2 + (4.206 \text{ in})^2 - 2(3.912 \text{ in})(4.206 \text{ in})(\cos 127°26')}$$

$a \approx$ [√] [(] 3.912 [x²] [+] 4.206 [x²] [−] 2 [×] 3.912 [×] 4.206 [×]
[cos] 127 [° ' "] 26 [° ' "] [)] [=] 7.27988697

or $a \approx$ [√] [(] 3.912 [x²] [+] 4.206 [x²] [−] 2 [×] 3.912 [×] 4.206 [×]
[cos] 127 [° ' "] 26 [° ' "] ▶ [ENTER] [)] [ENTER] 7.27988697

$a \approx 7.280$ in Ans

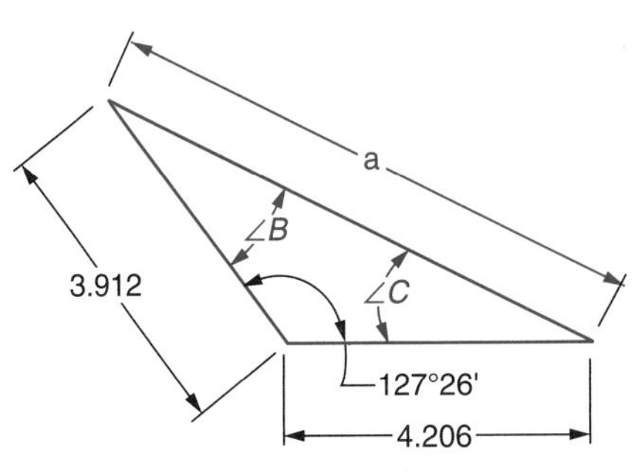

Figure 66-10

b. Solve for $\angle B$, using the Law of Sines.

$$\frac{4.206 \text{ in}}{\sin \angle B} = \frac{7.280 \text{ in}}{\sin 127°26'}$$

$$\sin \angle \text{B} = \frac{4.206 \text{ in} (\sin 127°26')}{7.280 \text{ in}}$$

$\sin \angle B \approx$ 4.206 [×] [sin] 127 [° ′ ″] 26 [° ′ ″] [÷] 7.280 [=] 0.4587666359

$\angle B \approx$ [SHIFT] [sin⁻¹] .458766636 [=] [SHIFT] [←] 27°18'27.18"

or $\sin \angle B \approx$ 4.206 [×] [sin] 127 [° ′ ″] 26 [° ′ ″] [▶] [ENTER] [ENTER] [÷] 7.28 [ENTER] 0.458766635

$\angle B =$ [2nd] [sin⁻¹] .458766636 [° ′ ″] [◀] [ENTER] [ENTER] 27°18'27.2"

$\angle B \approx 27°18'$ Ans

c. Solve for $\angle C$.

$$\angle C = 180° - (127°26' + 27°18') = 25°16' \quad \text{Ans}$$

Law of Cosines (Given Three Sides)

In any triangle, the cosine of an angle is equal to the sum of the squares of the two adjacent sides minus the square of the opposite side, divided by twice the product of the two adjacent sides.

In reference to the triangle in Figure 66-11:

$$\cos A = \frac{b^2 + c^2 - a^2}{2bc}$$

$$\cos B = \frac{a^2 + c^2 - b^2}{2ac}$$

$$\cos C = \frac{a^2 + b^2 - c^2}{2ab}$$

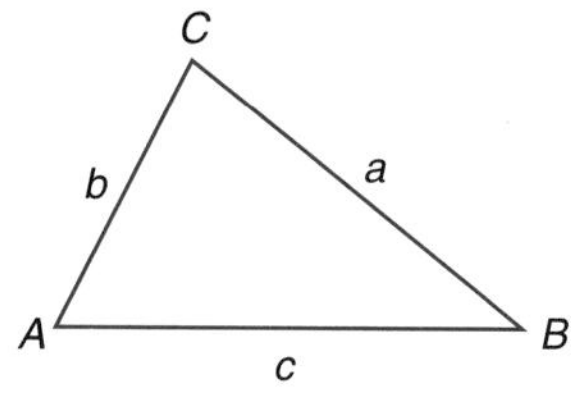

Figure 66-11

➤ **Note:** These formulas, which are stated in terms of the cosines of angles, are rearrangements of the formulas on page 463, which are stated in terms of the squares of the sides.

The Law of Cosines, as stated in the formulas, is used to solve the following kind of oblique triangle problems.

- Problems where three sides of an oblique triangle are known.

➤ **Note:** When an unknown angle is determined, its cosine function may be negative. A negative cosine function means that the angle being computed is greater than 90°. The angle lies in Quadrant II of the Cartesian coordinate system. Recall that the cosine of an angle between 90° and 180° equals the negative cosine of the supplement of the angle. For example, the cosine of 147°40' = −cos (180° − 147°40') = −cos 32°20'.

Solving Oblique Triangle Problems Given Three Sides, Using the Law of Cosines

Examples

1. Given three sides, determine $\angle A$ of the oblique triangle in Figure 66-12. All dimensions are in inches.

$$\cos \angle A = \frac{(6.400 \text{ in})^2 + (7.800 \text{ in})^2 - (4.700 \text{ in})^2}{2(6.400 \text{ in})(7.800 \text{ in})}$$

$\cos \angle A \approx$ 6.4 [x^2] [+] 7.8 [x^2] [−] 4.7 [x^2] [=] [÷] [(] 2 [×] 6.4 [×] 7.8 [)] [=] 0.798377404

$\angle A \approx$ [SHIFT] [$\cos^{-1}$] .798377404 [=] [SHIFT] [←] 37°1'28.44"

or $\angle A \approx$ [2nd] [$\cos^{-1}$] .798377404 [° ' "] [◄] [ENTER] [ENTER] 37°01'28.44"

$\angle A \approx$ 37°01' Ans

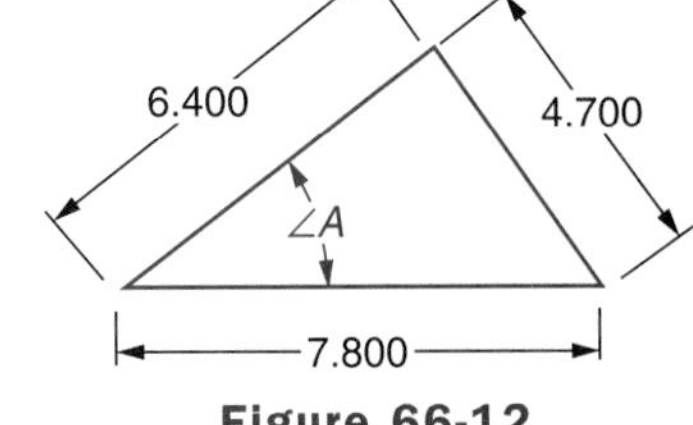

Figure 66-12

2. Given three sides, determine $\angle P$ of the oblique triangle shown in Figure 66-13. All dimensions are in millimeters.

$$\cos \angle P = \frac{(8.323 \text{ mm})^2 + (9.745 \text{ mm})^2 - (15.118 \text{ mm})^2}{2(8.323 \text{ mm})(9.745 \text{ mm})}$$

$\cos \angle P \approx$ 8.323 [x^2] [+] 9.745 [x^2] [−] 15.118 [x^2] [=] [÷] [(] 2 [×] 8.323 [×] 9.745 [)] [=] −0.396488999

$\angle P \approx$ [SHIFT] [$\cos^{-1}$] [−] .396488999 [=] 113.3588715

or $\angle P \approx$ [2nd] [$\cos^{-1}$] [(−)] .396488999 [ENTER] 113.3588715

$\angle P \approx$ 113.36° Ans

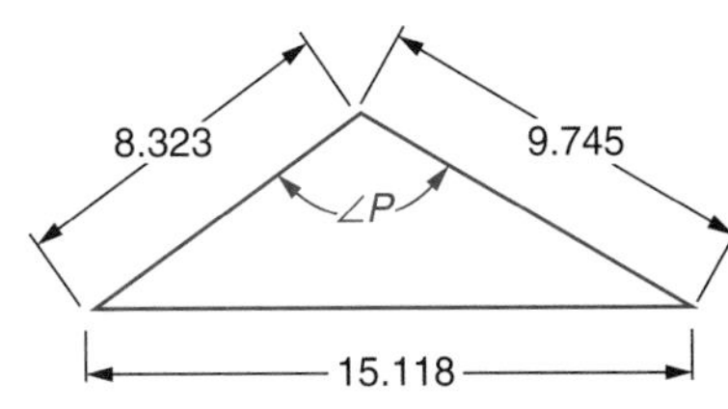

Figure 66-13

Application

For customary unit dimensioned problems, calculate angles to the nearer minute and lengths to the nearer thousandth inch. For metric unit dimensioned problems, calculate angles to the nearer hundredth degree and lengths to the nearer hundredth millimeter.

Law of Sines

Solve the following problems using the Law of Sines.

1. Find side x.

   All dimensions are in inches. ____________

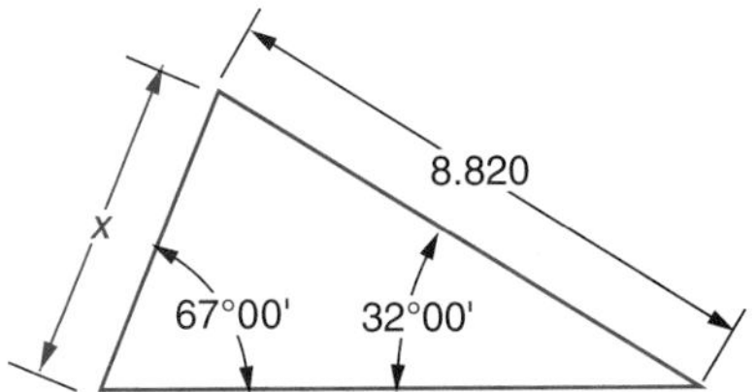

2. Find side x.

   All dimensions are in inches. ____________

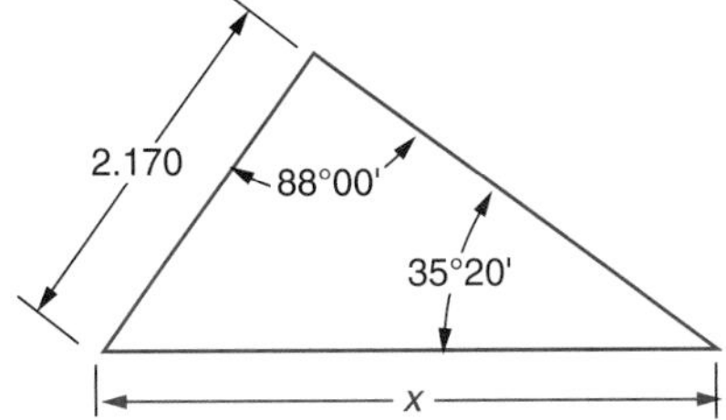

3. Find side x.

   All dimensions are in inches. __________

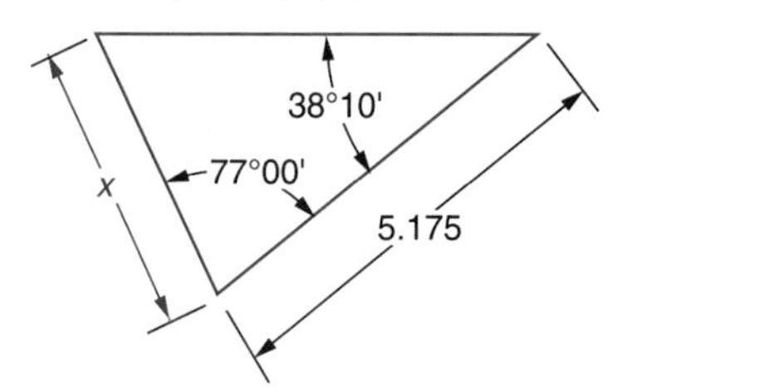

4. Find side x.

   All dimensions are in millimeters. __________

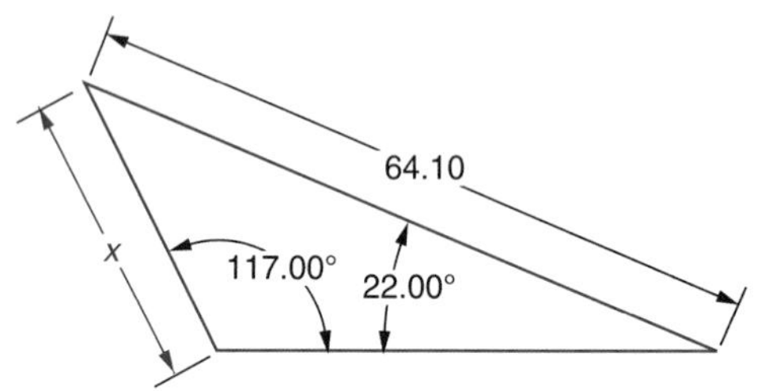

5. Find $\angle x$.

   All dimensions are in inches. __________

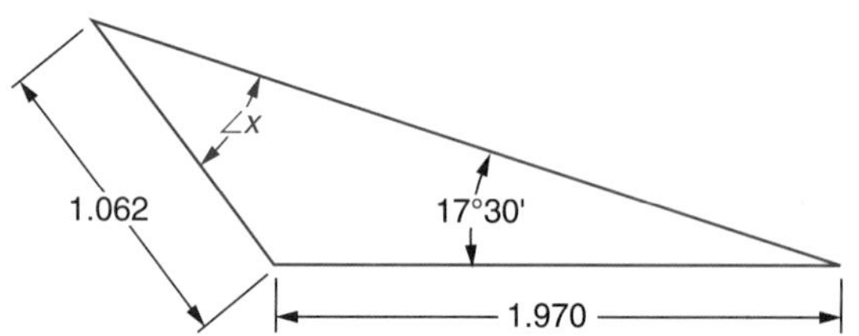

6. Find $\angle x$.

   All dimensions are in millimeters. __________

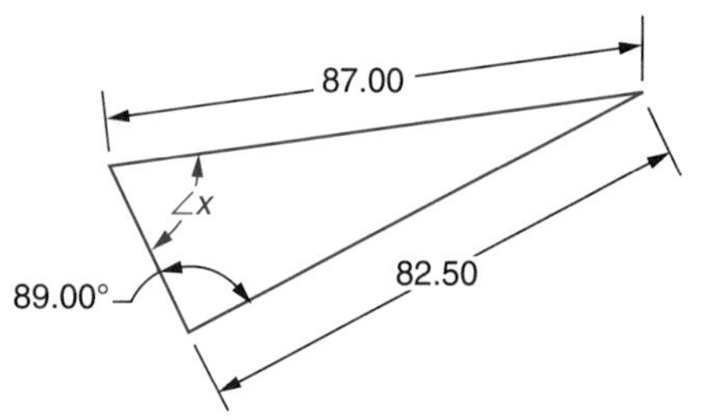

7. Find $\angle x$.

   All dimensions are in inches. __________

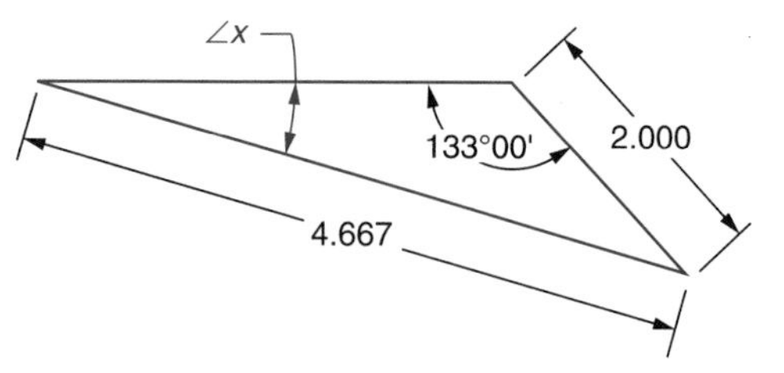

8. Find $\angle x$.

   All dimensions are in inches. __________

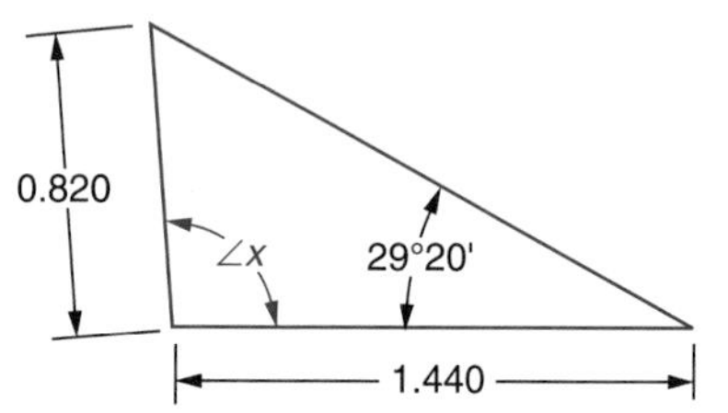

9. Find side x.

   All dimensions are in millimeters. __________

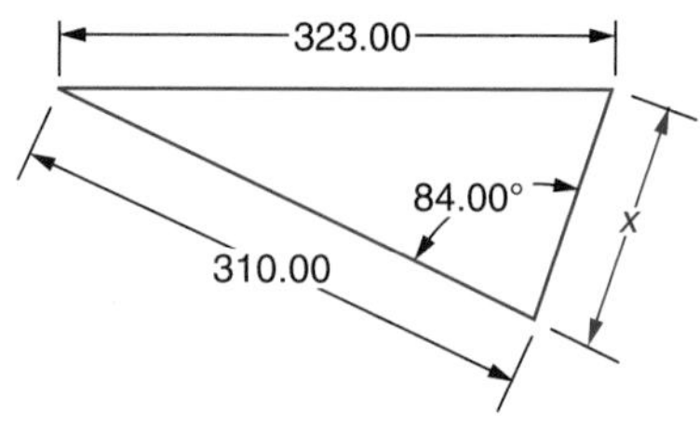

Identifying Problems with One or Two Solutions

Two sides and an angle opposite one of the sides of triangles are given in the following problems. Identify each problem as to whether it has one or two solutions. Do not solve the problems for angles and sides.

10. A 4" side, a 5" side, a 37° angle opposite the 4" side. __________
11. A 95.00-mm side, a 98.00-mm side, a 75° angle opposite the 95.00-mm side. __________
12. A 21-mm side, a 29-mm side, a 41° angle opposite the 29-mm side. __________
13. A 0.943" side, a 0.612" side, and a 62°15' angle opposite the 0.612" side. __________
14. A 2.10-ft side, a 3.05-ft side, a 29°30' angle opposite the 3.05-ft side. __________
15. A 16.35-mm side, a 23.86-mm side, a 115° angle opposite the 23.86-mm side. __________
16. An 87.60-mm side, a 124.80-mm side, a 12.90° angle opposite the 87.60-mm side. __________
17. A 34.090" side, a 35.120" side, a 46°18' angle opposite the 34.090" side. __________

Law of Cosines

Solve the following problems using the Law of Cosines.

18. Find side *x*.

    All dimensions are in inches. ____________

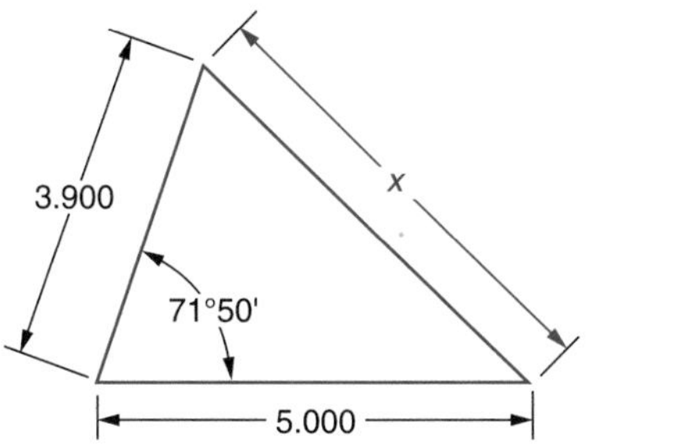

19. Find side *x*.

    All dimensions are in millimeters. ____________

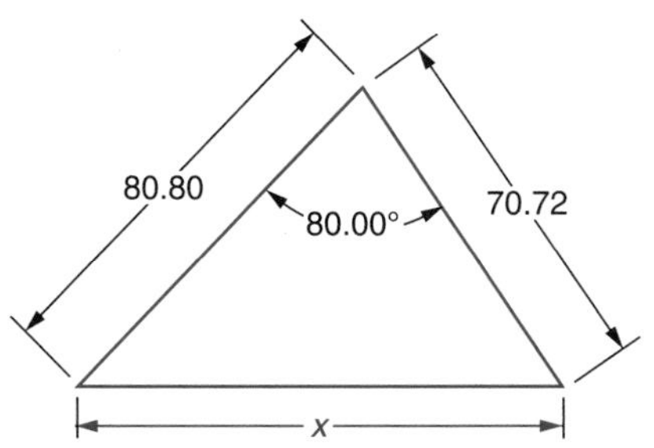

20. Find side *x*.

    All dimensions are in inches. ____________

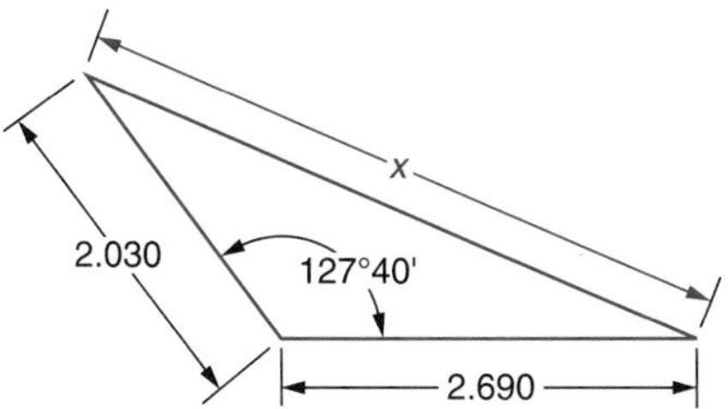

21. Find ∠*x*.

    All dimensions are in millimeters. ____________

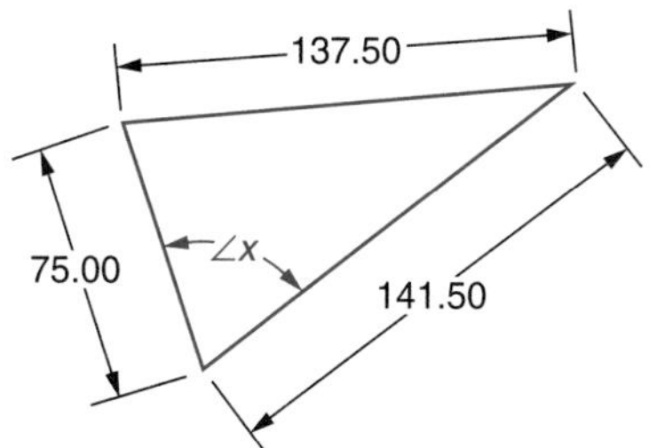

22. Find ∠*x*.

    All dimensions are in inches. ____________

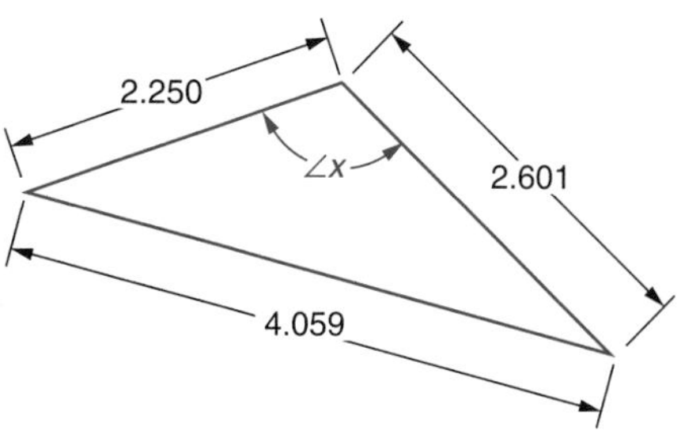

23. Find ∠*x*.

    All dimensions are in inches. ____________

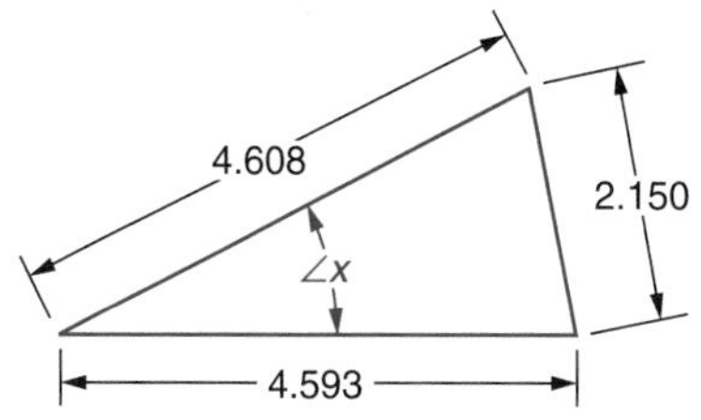

24. Find ∠*x*.

    All dimensions are in inches. ____________

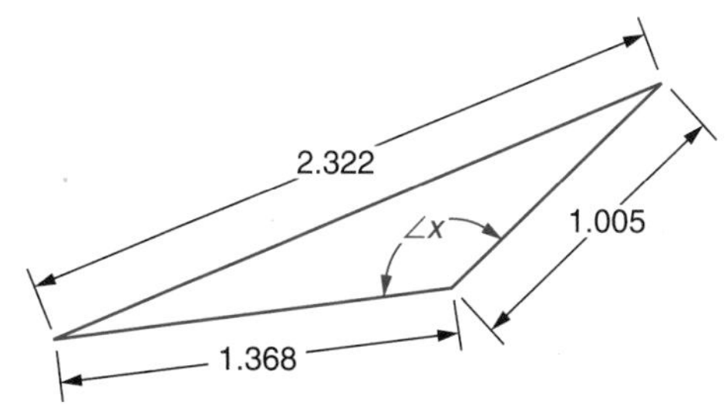

25. Find side *x*.

    All dimensions are in millimeters. ____________

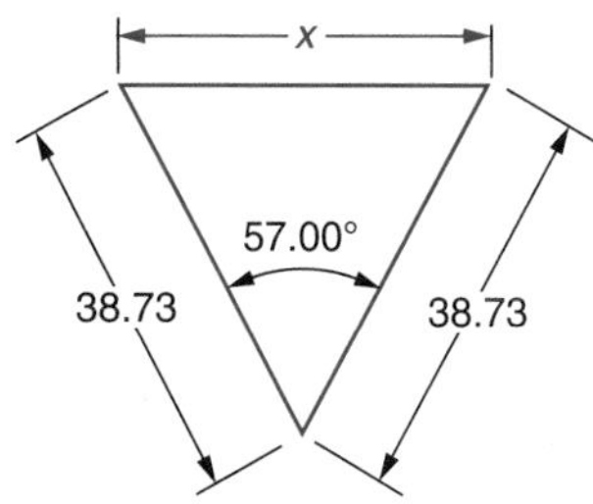

26. Find side *x*.

    All dimensions are in inches. ____________

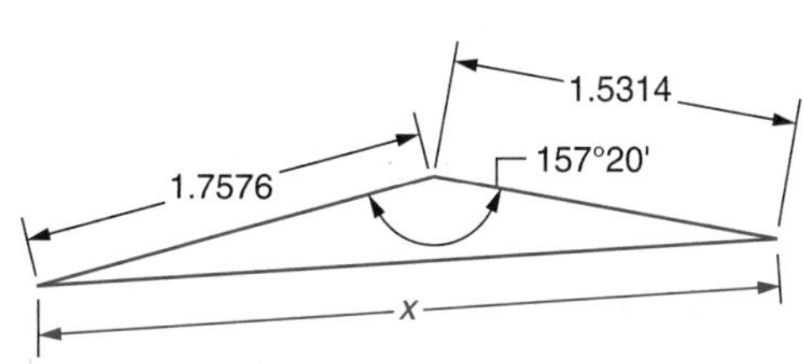

Combination of the Law of Cosines and the Law of Sines

Solve the following problems using a combination of the Law of Cosines and the Law of Sines.

27. All dimensions are in inches.

   a. Find side *x*. ____________

   b. Find ∠*y*. ____________

28. All dimensions are in inches.

   a. Find side *x*. ____________

   b. Find ∠*y*. ____________

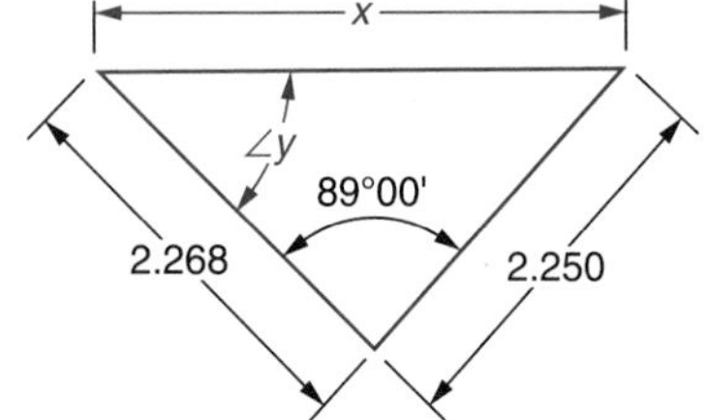

29. All dimensions are in millimeters.

   a. Find side *x*. ____________

   b. Find ∠*y*. ____________

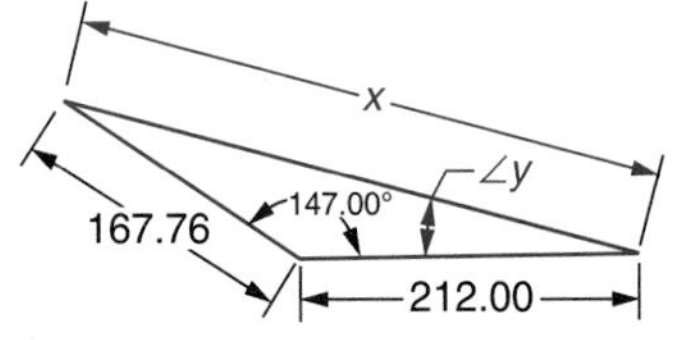

30. All dimensions are in inches.

   a. Find ∠*x*. ____________

   b. Find ∠*y*. ____________

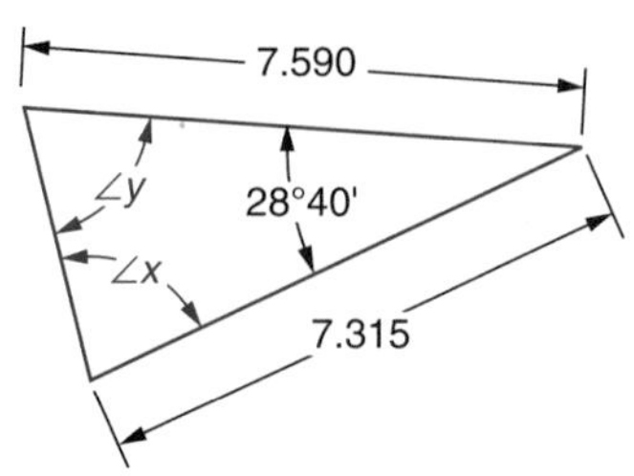

31. All dimensions are in inches.

   a. Find ∠*x*. ____________

   b. Find ∠*y*. ____________

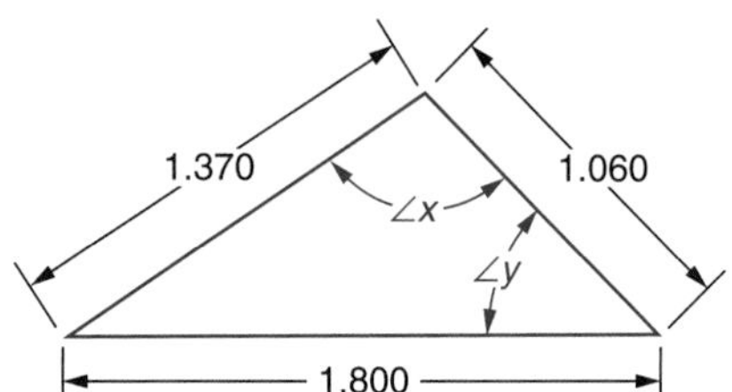

32. All dimensions are in millimeters.

   a. Find ∠*x*. ____________

   b. Find ∠*y*. ____________

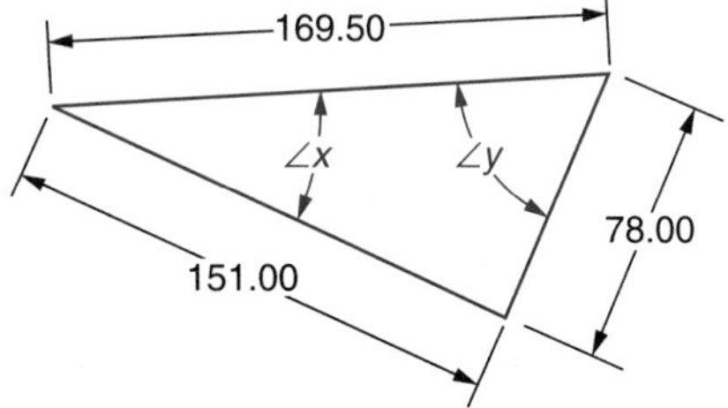

Practical Machine Shop Problems

Solve the following machine shop problems.

33. Find ∠x.

    All dimensions are in millimeters. ____________

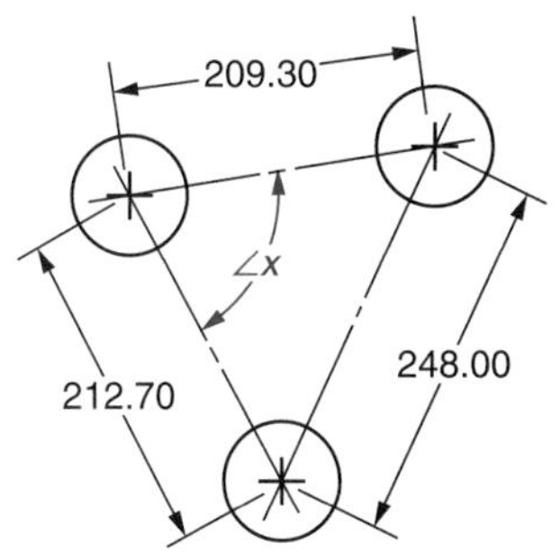

36. Find distance x.

    All dimensions are in inches. ____________

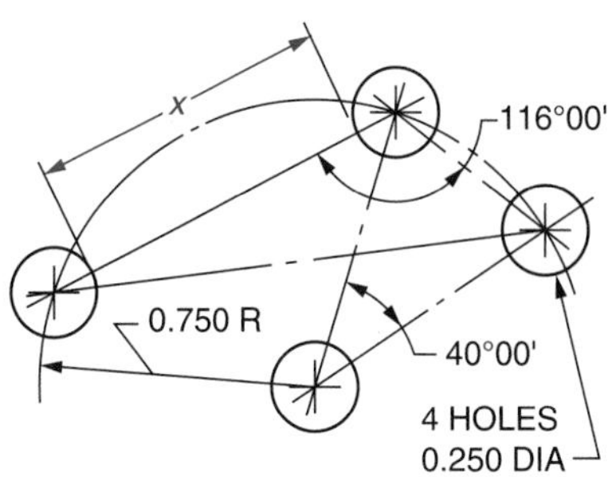

34. Find distance y.

    All dimensions are in inches. ____________

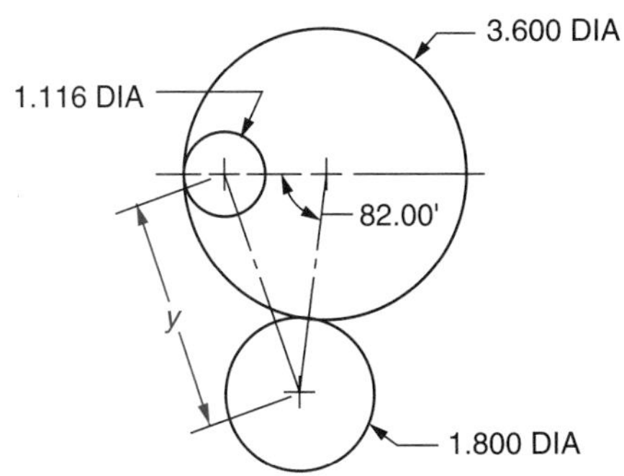

37. Find ∠x.

    All dimensions are in inches. ____________

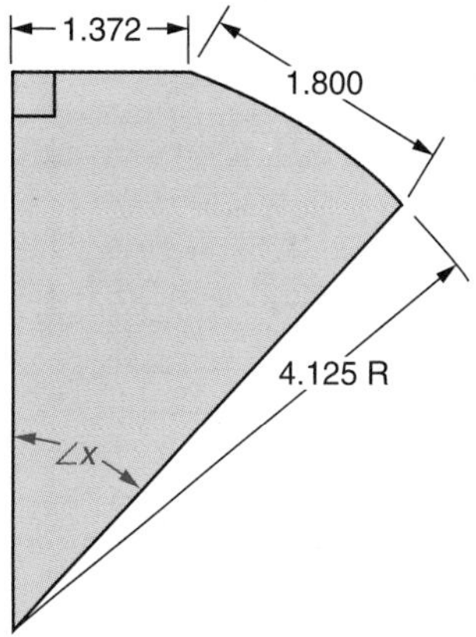

35. Three circles are to be bored in the plate shown. The 4.000-inch diameter and 5.500-inch diameter circles are each tangent to the 7.500-inch diameter circle. Determine the distance from the center of the 4.000-inch diameter circle to the center of the 5.500-inch diameter circle. ____________

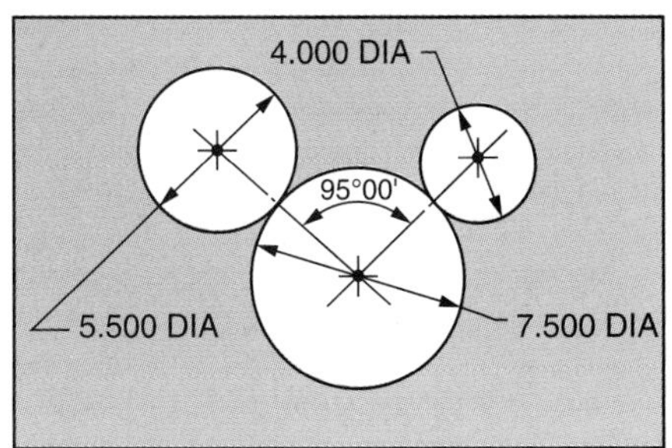

38. Find dimension y.

    All dimensions are in millimeters. ____________

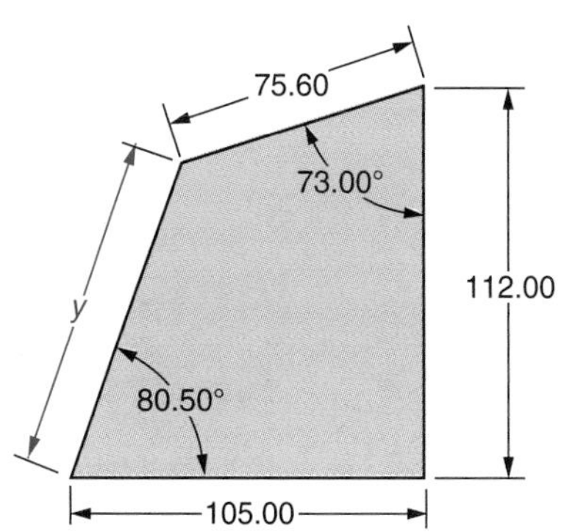

39. Find dimension *y*.
All dimensions are in inches. ______

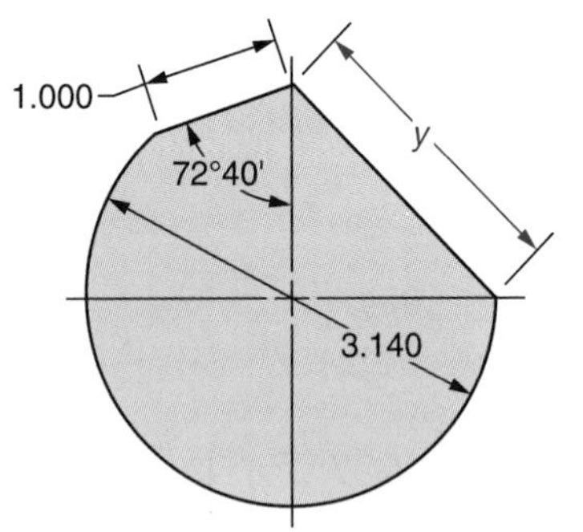

41. Find ∠*x*.
All dimensions are in inches. ______

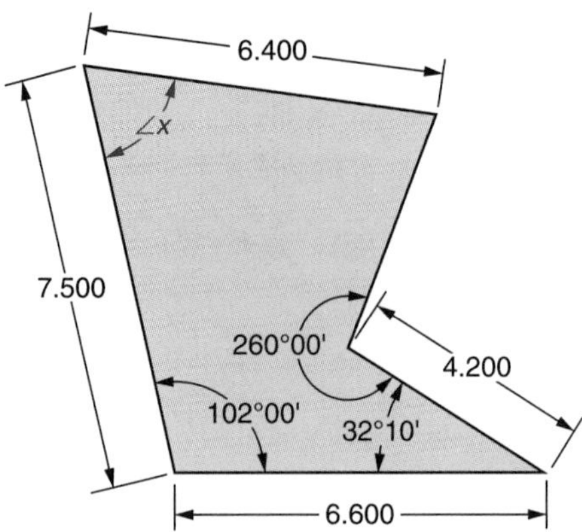

40. Find dimension *y*.
All dimensions are in inches. ______

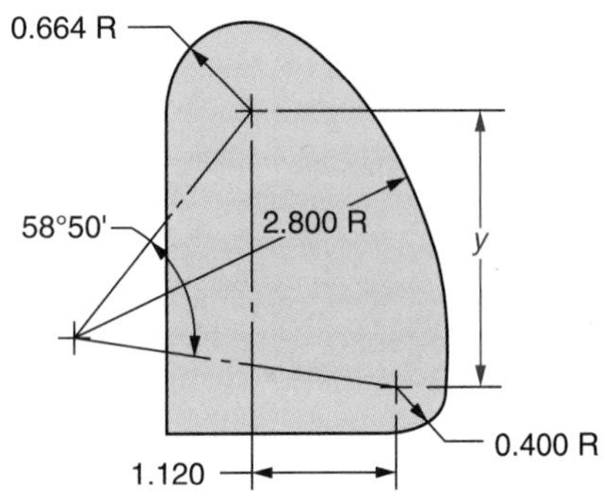

42. Find ∠*y*.
All dimensions are in millimeters. ______

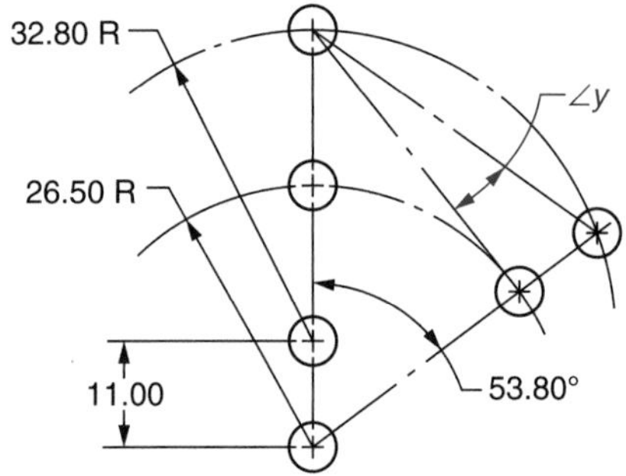

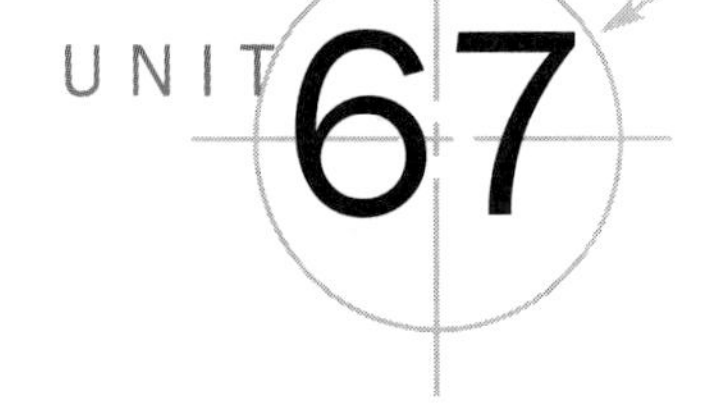

Achievement Review— Section Seven

Objective *You should be able to solve the exercises and problems in this Achievement Review by applying the principles and methods covered in Units 60–66.*

With reference to $\angle 1$, name the sides of each of the following triangles as opposite, adjacent, or hypotenuse.

1. ______________ 2. ______________ 3. ______________ 4. ______________

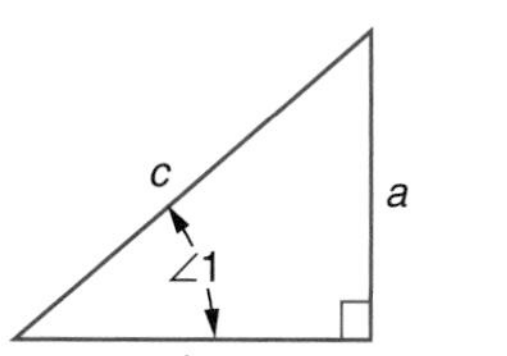

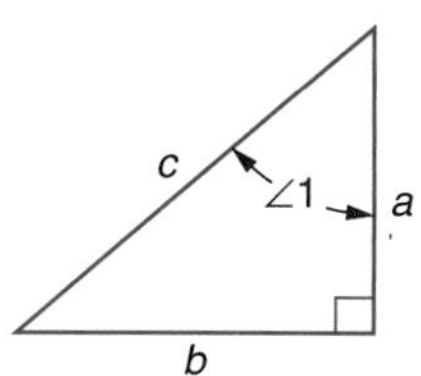

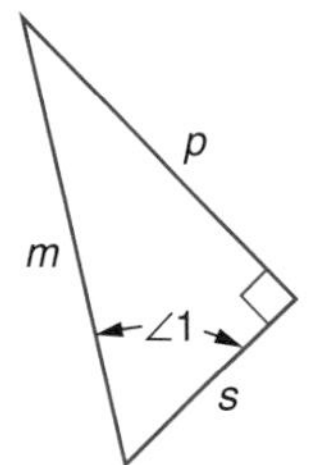

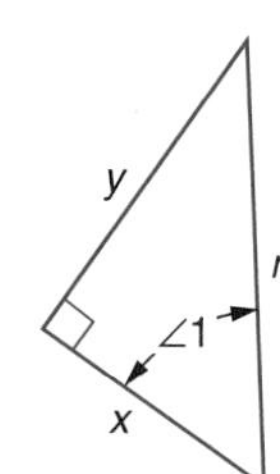

Determine the functions of the following angles. Round the answers to 4 decimal places.

5. sin 22° ______________
6. cot 46°20' ______________
7. tan 37°50' ______________
8. tan 0°21' ______________
9. cos 63°18' ______________
10. tan 74°24' ______________
11. sin 7.43° ______________
12. csc 57.82° ______________

Determine the values of $\angle$A in degrees and minutes that correspond to the following functions.

13. cos A = 0.69675 ______________
14. tan A = 0.50587 ______________
15. sin A = 0.98531 ______________
16. cot A = 1.1340 ______________
17. sec A = 1.5753 ______________
18. cos A = 0.15902 ______________

Determine the values of $\angle$A in decimal-degrees to 2 decimal places that correspond to the following functions.

19. sin A = 0.72847 ______________
20. tan A = 1.3925 ______________
21. cos A = 0.34038 ______________

For each of the following functions of angles, write the cofunction of the complement of the angle.

22. sin 36° ______________
23. tan 48°19' ______________
24. cos 16°53' ______________
25. cot 80.47° ______________

Solve the following problems. Compute angles to the nearer minute in triangles with customary unit sides. Compute angles to the nearer hundredth degree in triangles with metric unit sides. Compute sides to 3 decimal places.

26. Determine $\angle$A.
All dimensions are in inches. ______________

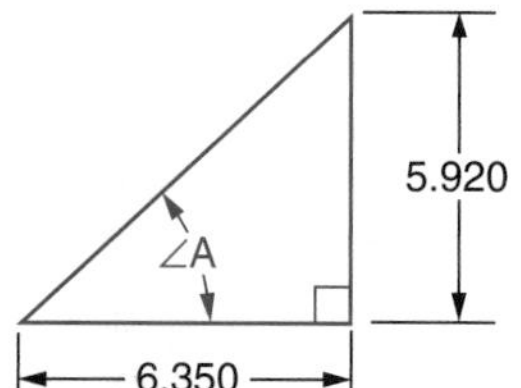

27. Determine side *a*.
All dimensions are in inches. ______________

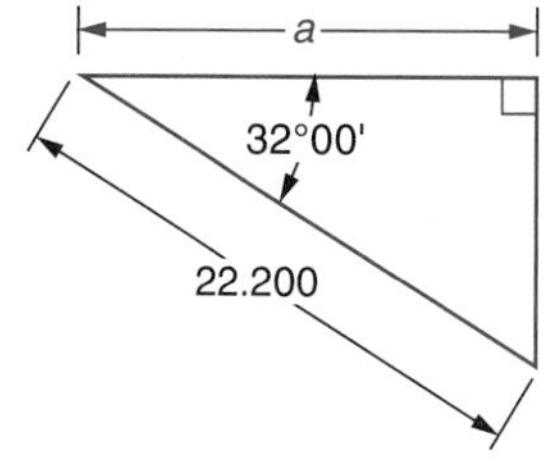

28. Determine ∠D.

All dimensions are in millimeters. __________

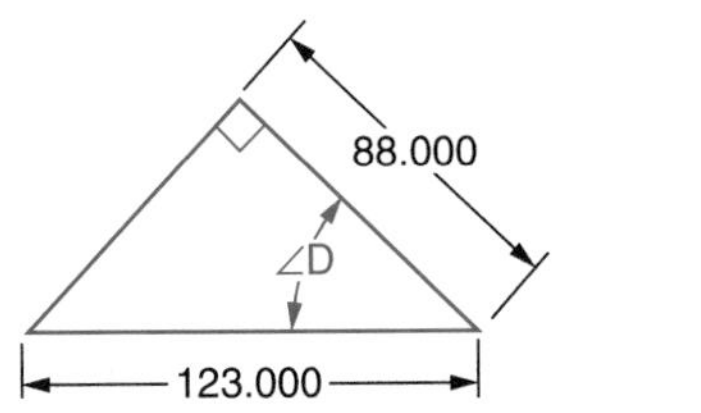

29. Determine ∠1.

All dimensions are in millimeters. __________

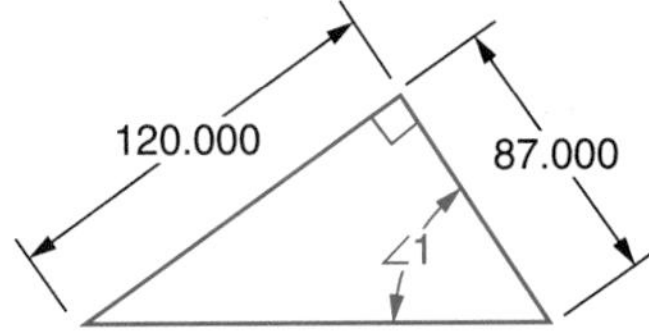

30. All dimensions are in millimeters.

a. Determine side *g*. __________

b. Determine side *h*. __________

c. Determine ∠H. __________

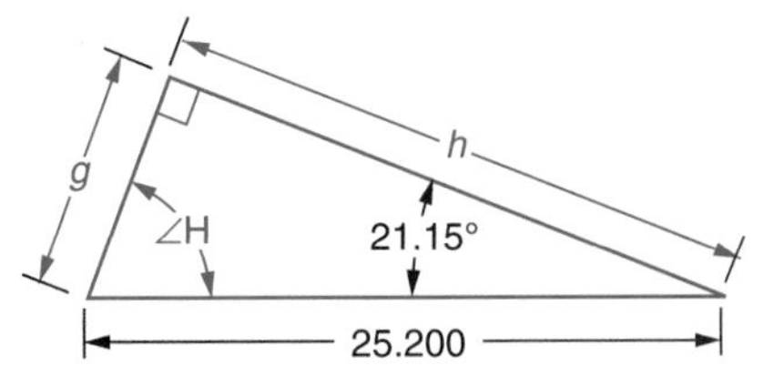

31. All dimensions are in inches.

a. Determine ∠A. __________

b. Determine ∠B. __________

c. Determine side *c*. __________

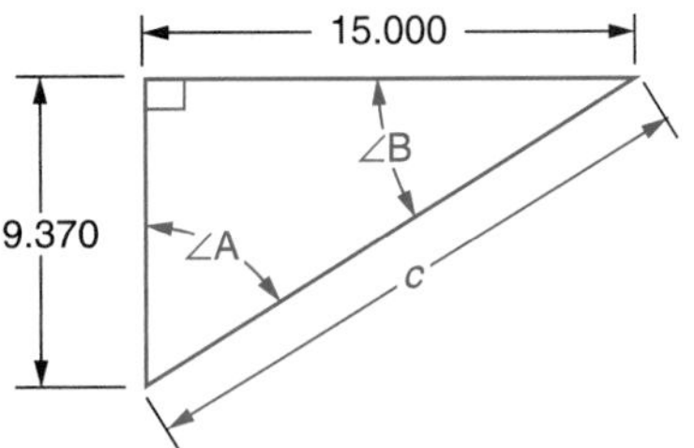

Solve the following applied right triangle problems. Compute linear values to 3 decimal places, customary unit angles to the nearer minute, and metric angles to the nearer hundredth degree.

32. All dimensions are in inches.

a. Determine dimension *c*. __________

b. Determine dimension *d*. __________

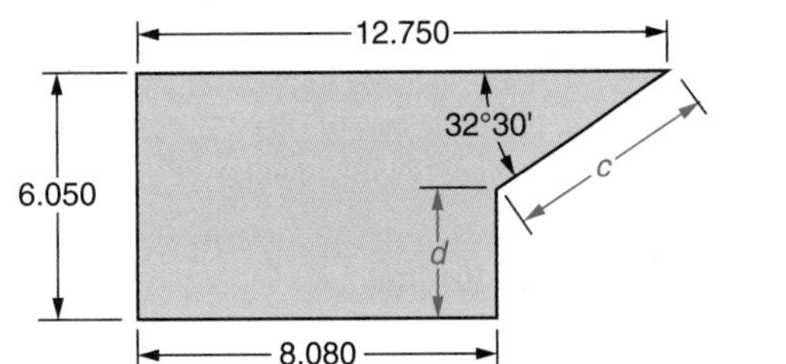

33. Determine ∠T.

All dimensions are in millimeters. __________

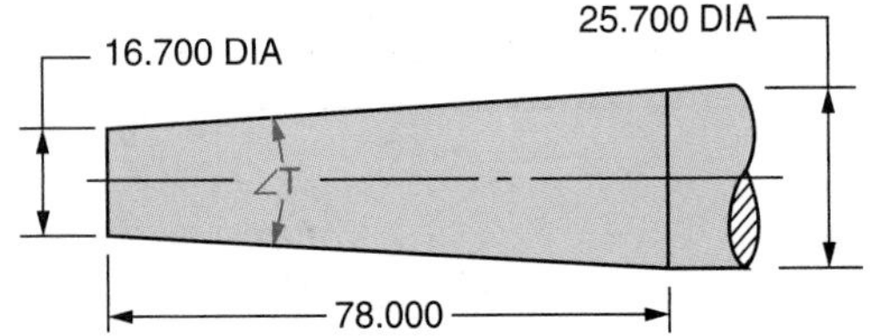

34. Determine dimension *x*.

All dimensions are in millimeters. __________

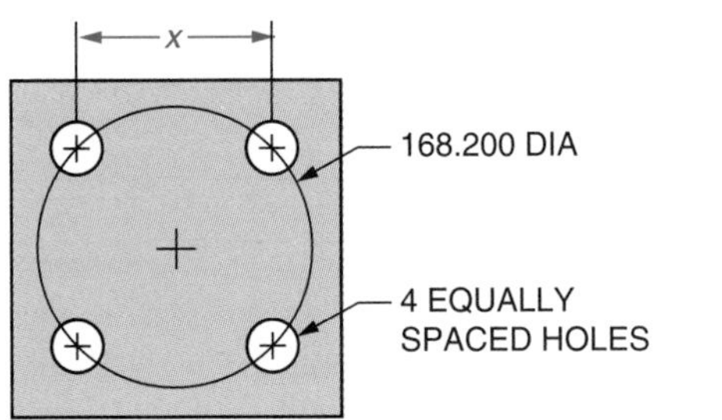

35. Determine dimension *d*.

All dimensions are in inches. __________

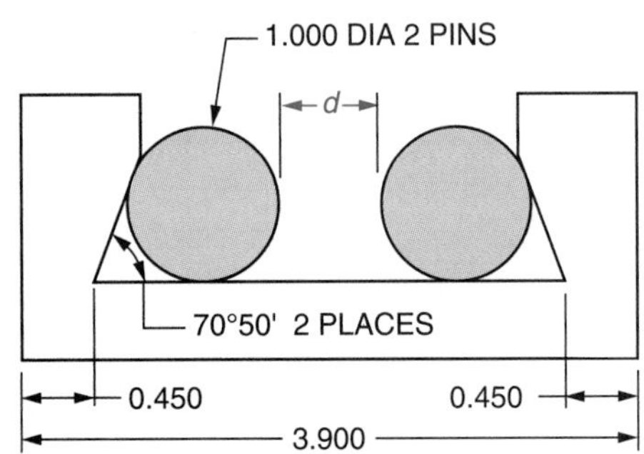

36. All dimensions are in millimeters.

    a. Determine ∠A. __________

    b. Determine distance *x*. __________

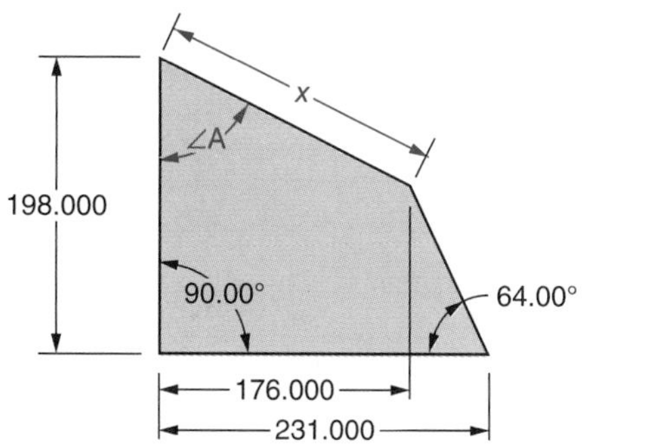

38. Determine ∠*x*.

    All dimensions are in inches. __________

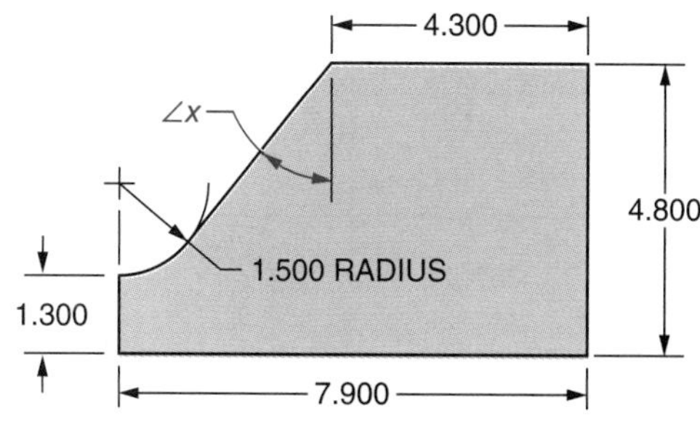

37. Determine check dimension *y*.

    All dimensions are in inches. __________

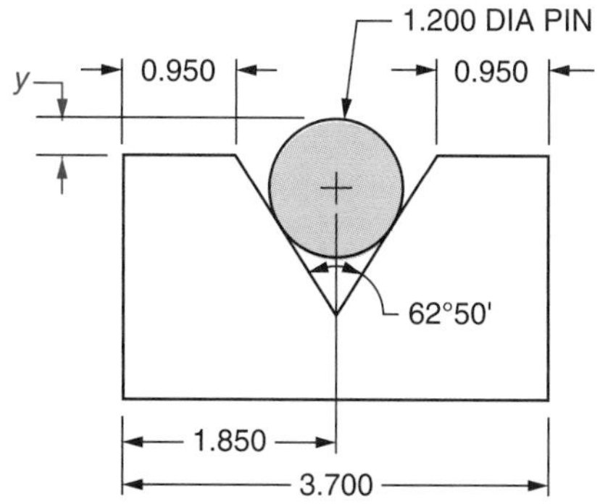

39. Determine ∠*y*.

    All dimensions are in millimeters. __________

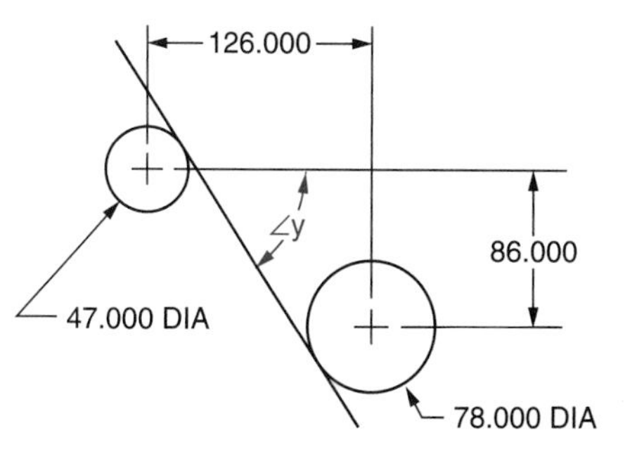

Determine the sine, cosine, tangent, cotangent, secant, and cosecant of each of the following angles.

40. 115° __________

41. 223° __________

42. 310°30' __________

Solve the following problems using the Law of Sines and/or the Law of Cosines. Compute side lengths to 3 decimal places, customary unit angles to the nearer minute, and metric unit angles to the nearer hundredth degree.

43. Determine side *a*.

    All dimensions are in millimeters. __________

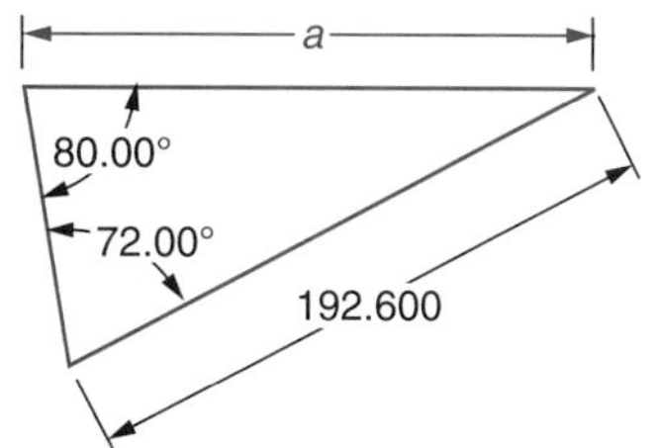

44. Determine ∠*D*.

    All dimensions are in inches. __________

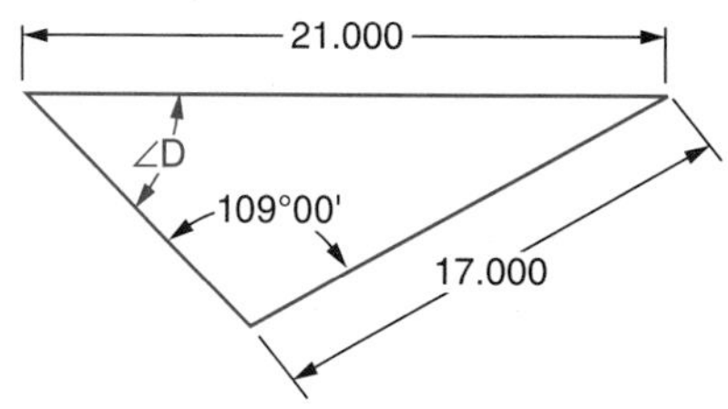

45. All dimensions are in inches.

a. Determine ∠A. ____________

b. Determine side *a*. ____________

c. Determine side *b*. ____________

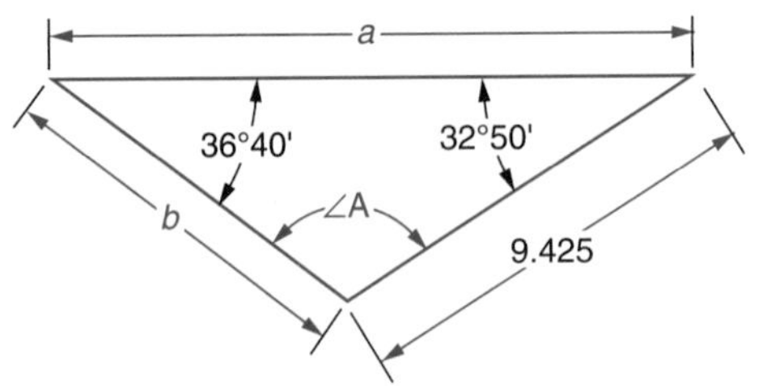

46. Determine side *d*.
All dimensions are in millimeters. ____________

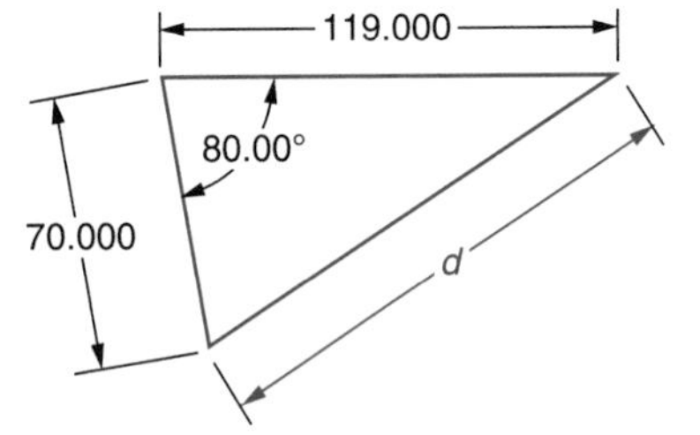

47. Determine ∠E.
All dimensions are in millimeters. ____________

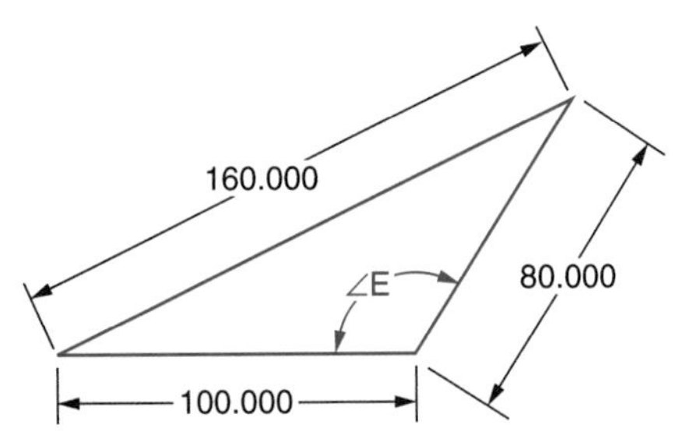

48. All dimensions are in inches.

a. Determine side *m*. ____________

b. Determine ∠N. ____________

c. Determine ∠P. ____________

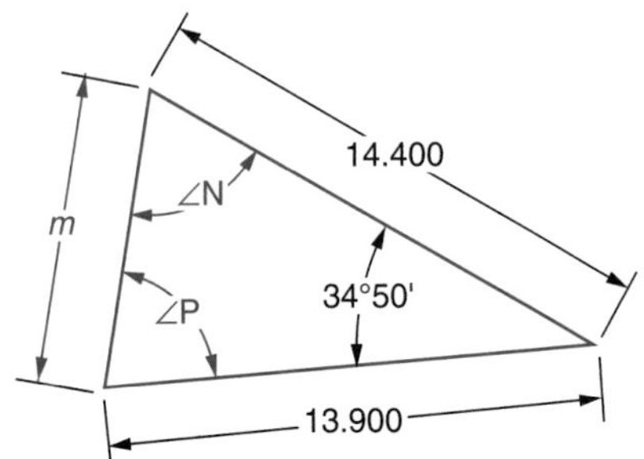

49. Determine dimension *d*.
All dimensions are in inches. ____________

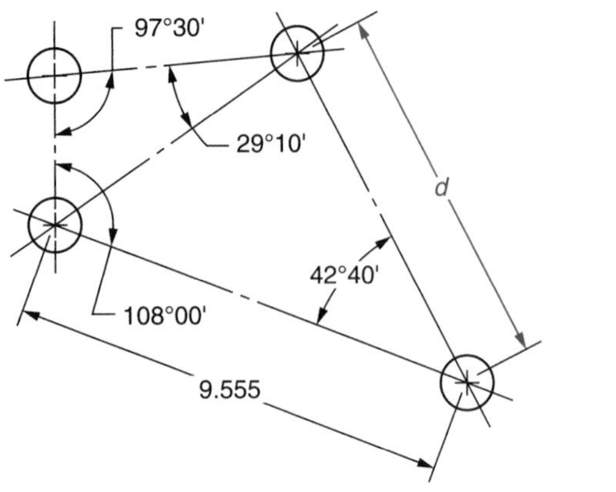

50. Determine ∠*x*.
All dimensions are in millimeters. ____________

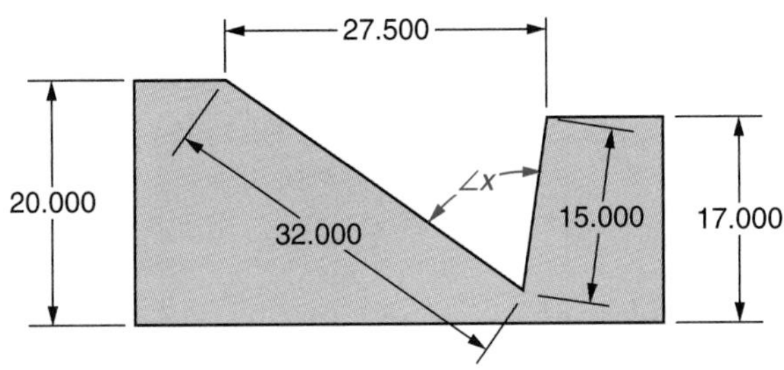

51. Determine ∠*x*.
All dimensions are in inches. ____________

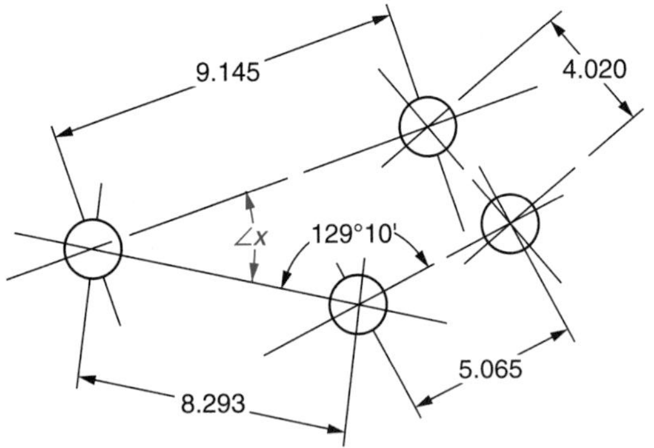

52. Determine $\angle A$.
All dimensions are in millimeters. ______________

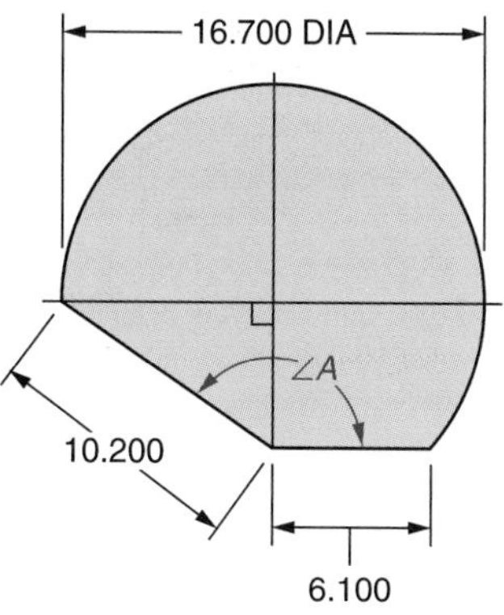

53. Determine $\angle x$.
All dimensions are in inches. ______________

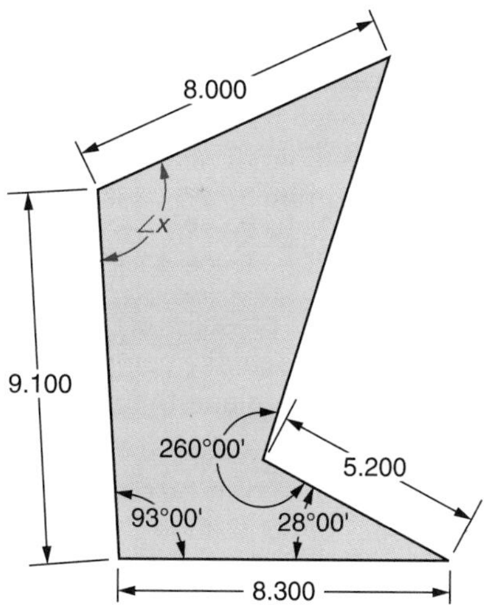

54. A piece of stock is to be machined as shown. Determine dimension b.
All dimensions are in inches. ______________

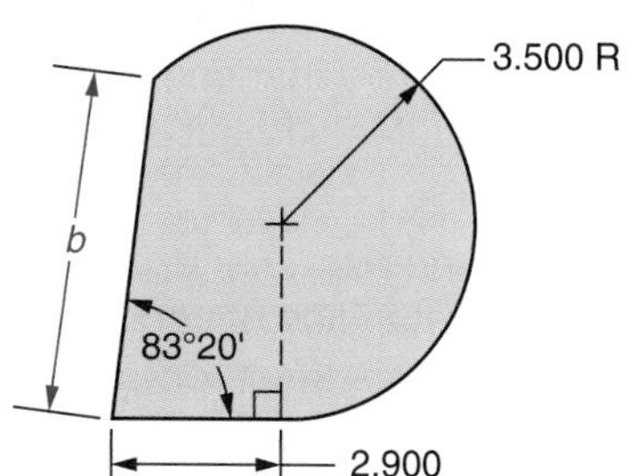

55. Determine $\angle B$
All dimensions are in inches. ______________

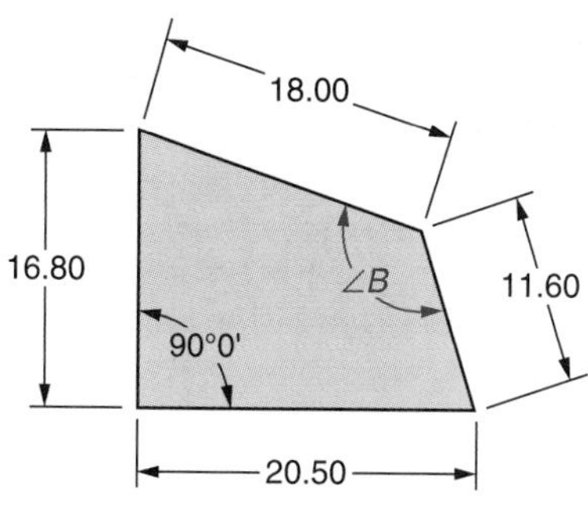

56. Determine dimension a.
All dimensions are in inches. ______________

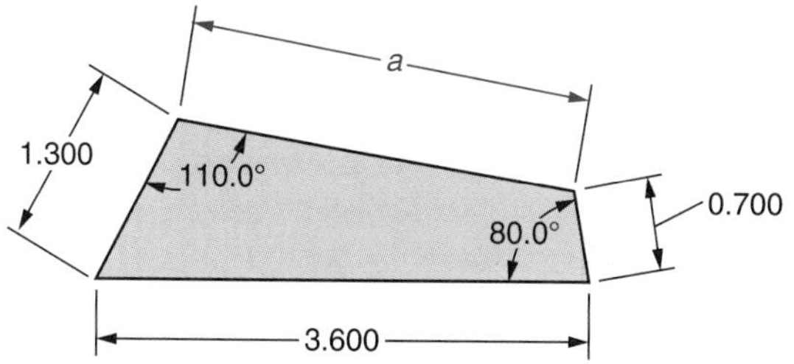

Compound Angles

UNIT 68

Introduction to Compound Angles

Objectives *After studying this unit you should be able to*

- Compute true lengths of diagonals of rectangular solids.
- Compute true angles of diagonals of rectangular solids.

In the machine trades, the application of principles of solid or three-dimensional trigonometry is commonly called *compound angles*. Generally compound angle problems require the computation of an unknown angle in a plane that is the resultant of two or more known angles lying in different planes.

Applications of compound angles are frequently required in machining fixture parts, die sections, and cutting tools. An understanding of compound angle procedures is necessary in setting up parts for drilling or boring compound-angular holes.

Often, compound angle problems are encountered when machining parts as shown on engineering drawings. Usually, the top, front, and right side views of orthographic projections are shown. Wherever applicable, compound angle examples and problems in this text are given in relation to these views.

Formulas for specific compound angle applications can be found in certain trade handbooks. These formulas are useful provided the particular compound angle applications are properly visualized and identified. There are variations in compound angle situations. Merely plugging in values in given formulas without fully visualizing the components of a problem can result in costly errors.

Certain basic compound angle situations are presented in this text. A comprehensive study of compound angles is not intended. An understanding of applications is emphasized. Visualization of a problem with its components is stressed.

Pictorial views of compound angle situations with their components located and identified in rectangular solids or pyramids are shown. The student should make sketches in pictorial form to develop understanding. Formulas should be used in the solution of a problem only after the problem has been clearly visualized.

Diagonal of a Rectangular Solid

A pictorial view of a rectangular solid with diagonal AB is shown in Figure 68-1. A rectangular solid has six rectangular faces.

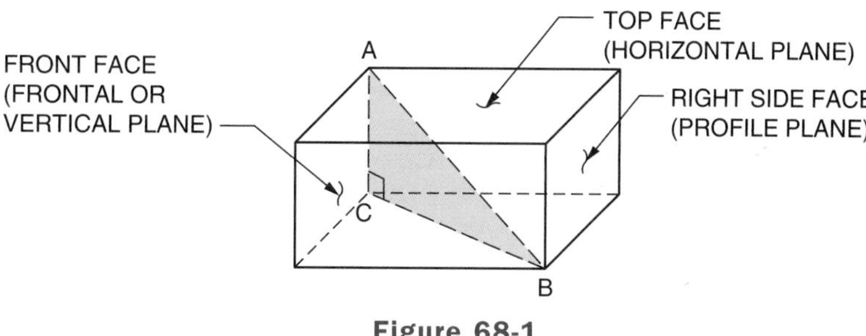

Figure 68-1

The top face (horizontal plane), front face (frontal or vertical plane), and right side face (profile plane) are identified in Figure 68-2. These faces correspond to the top, front, and right side views as they appear on an engineering drawing.

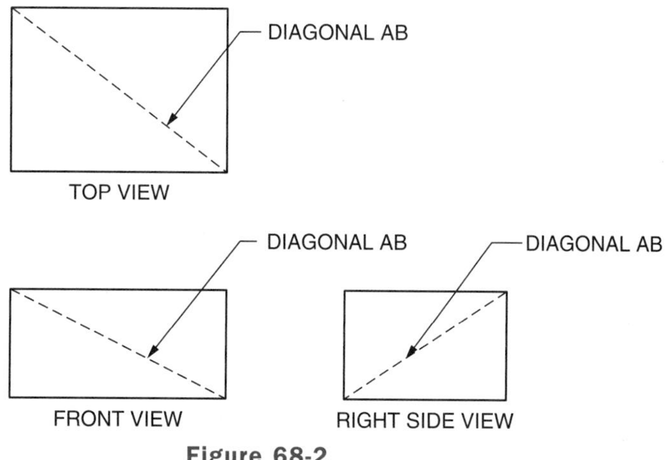

Figure 68-2

Observe that although AB appears as a diagonal in each of the three views, it does not appear in its actual (true) length in any of the views. Neither does the true angle made by AB with either a vertical or horizontal plane appear in any of the three views. The true length of a line is shown in Figure 68-3 when the line is contained in a plane that is viewed perpendicular to the line of sight.

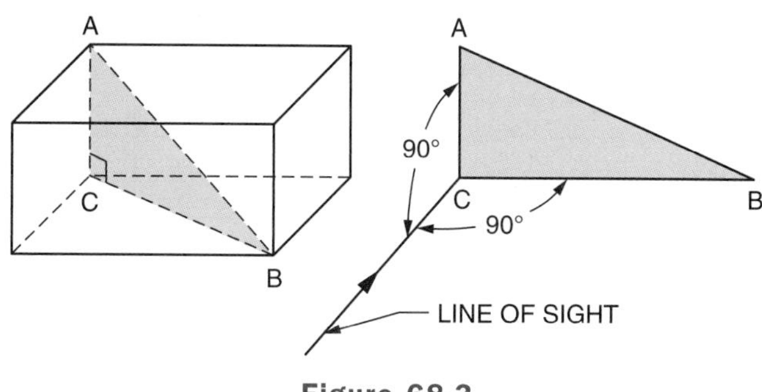

Figure 68-3

Computing True Lengths and True Angles

Example Compute true length AB and true ∠CAB shown in Figure 68-4. All dimensions are in inches.

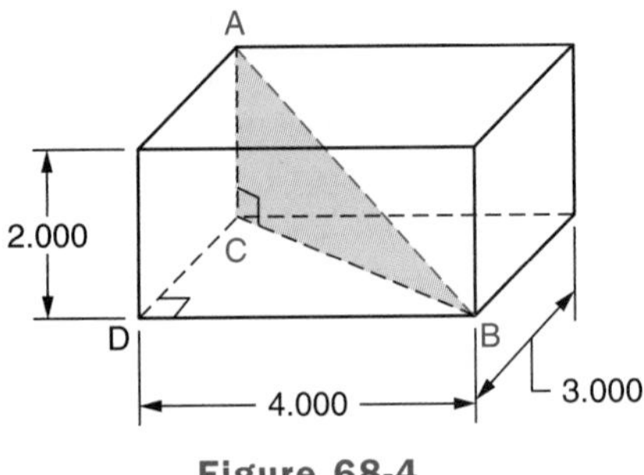

Figure 68-4

In right △CDB:
Compute CB.
Apply the Pythagorean theorem:

$$CB^2 = DB^2 + DC^2$$
$$CB^2 = (4.000\text{ in})^2 + (3.000\text{ in})^2$$
$$CB^2 = 16.000\text{ in}^2 + 9.000\text{ in}^2$$
$$CB^2 = 25.000\text{ in}^2$$
$$CB = 5.000\text{ in}$$

In right △ACB:
Compute AB.

$$AB^2 = AC^2 + CB^2$$
$$AB^2 = (2.000\text{ in})^2 + (5.000\text{ in})^2$$
$$AB^2 = 4.000\text{ in}^2 + 25.000\text{ in}^2$$
$$AB^2 = 29.000\text{ in}^2$$
$$AB = 5.385\text{ in} \quad \text{Ans}$$

Compute ∠CAB.

$$\tan \angle CAB = \frac{CB}{AC} = \frac{5.000\text{ in}}{2.000\text{ in}} = 2.5000$$

∠CAB ≈ [SHIFT] [tan⁻¹] 2.5 [=] [SHIFT] [←] 68°11'55"

or ∠CAB ≈ [2nd] [tan⁻¹] 2.5 [° ' "] ◀ [ENTER] [ENTER] 68°11'55"

∠CAB ≈ 68°12' Ans

Application

In each of the following problems a diagonal is shown within a rectangular solid.

a. Compute the true length of diagonal AB

b. Compute ∠CAB

Use this figure for #1 and #2.

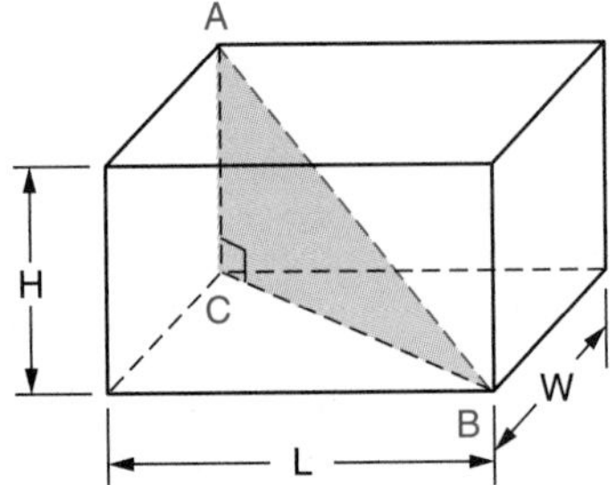

1. Given: H = 1.500 in
   L = 2.700 in
   W = 2.000 in

   a. ______________ b. ______________

2. Given: H = 50.00 mm
   L = 100.00 mm
   W = 80.00 mm

   a. ______________ b. ______________

3. Given: H = 4.340 in
L = 4.900 in
W = 4.200 in

a. ____________ b. ____________

4. Given: H = 75.00 mm
L = 90.00 mm
W = 70.00 mm

a. ____________ b. ____________

Use this figure for #3 and #4.

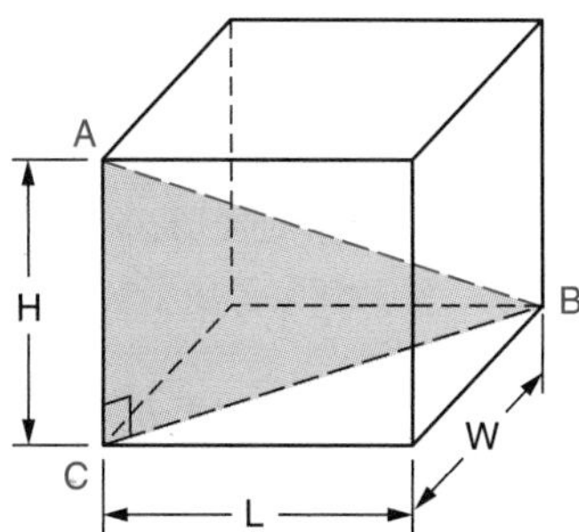

5. Given: H = 0.800 in
L = 1.400 in
W = 1.000 in

a. ____________ b. ____________

6. Given: H = 18.00 mm
L = 32.40 mm
W = 25.20 mm

a. ____________ b. ____________

Use this figure for #5 and #6.

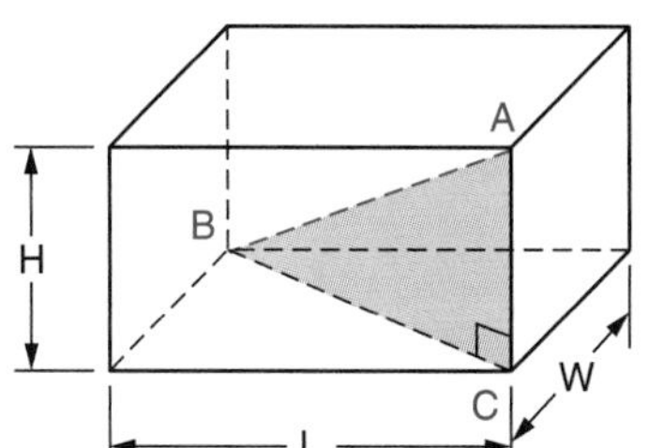

Drilling and Boring Compound-Angular Holes: Computing Angles of Rotation and Tilt Using Given Lengths

Objectives *After studying this unit you should be able to*

- Compute the angles of rotation and angles of tilt of hole axes in given rectangular solids.
- Sketch, dimension, and label compound-angular components within rectangular solids and compute angles of rotation and angles of tilt.

Computing Angles of Rotation and Angles of Tilt for Drilling and Boring Compound-Angular Holes

A part is usually positioned on an angle plate when drilling or boring compound-angular holes. In order to position a part, the angle of rotation and the angle of tilt must be computed.

The *angle of rotation,* ∠R, is the angle that the piece is rotated so the hole axis is in a plane perpendicular to the pivot axis of the angle plate to which the piece is mounted.

The *angle of tilt,* ∠T, is the angle that the angle plate is raised to put the axis of the hole in a vertical position.

The following example shows the procedure, using given length dimensions, for finding the angle of rotation and the angle of tilt.

Example Three views of a compound-angular hole are shown in Figure 69-1. All dimensions are in inches.

a. Determine the angle of rotation.

b. Determine the angle of tilt.

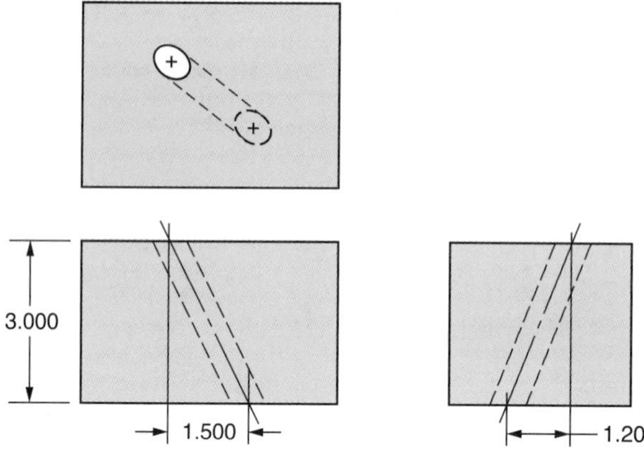

Figure 69-1

Sketch, dimension, and label a rectangular solid showing a right triangle within the solid that contains the hole axis as a side and the true angle. This is shown in Figure 69-2.

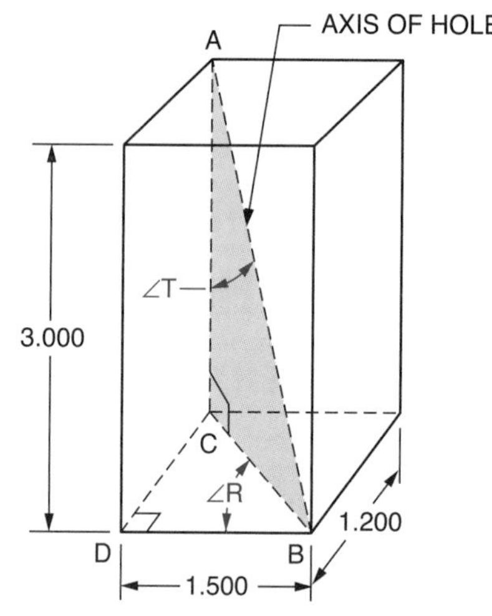

Figure 69-2

a. Compute the angle of rotation, $\angle R$.
In right $\triangle BDC$:

$$\tan \angle R = \frac{DC}{DB} = \frac{1.200 \text{ in}}{1.500 \text{ in}} = 0.80000$$

$\angle R \approx$ [SHIFT] [tan⁻¹] .8 [=] [SHIFT] [←] 38°39'35"

or $\angle R \approx$ [2nd] [tan⁻¹] .8 [ENTER] [° ' "] [◄] [ENTER] [ENTER] 38°39'35"

Angle of Rotation ($\angle R$) $\approx$ 38°40' Ans

b. Compute angle of tilt, $\angle T$.
In right $\triangle BDC$:

By the Pythagorean theorem, $CB^2 = CD^2 + DB^2$

$= (1.200 \text{ in})^2 + (1.500 \text{ in})^2$

$= 3.690$ in

$CB = \sqrt{3.690 \text{ sq in}} \approx 1.920937271$ in

$CB \approx$ [2nd] [√] 3.690 [)] [ENTER] 1.920937271

$CB \approx 1.92094$ in

In right △ACB:

$$\tan \angle T = \frac{CB}{AC}$$

$$\tan \angle T = \frac{1.92094 \text{ in}}{3.000 \text{ in}} = 0.64031$$

$\angle T \approx$ [SHIFT] [tan⁻¹] .64031 [=] [SHIFT] [←] 32°37'55"

or $\angle T \approx$ [2nd] [tan⁻¹] .64031 [° ' "] [◄] [ENTER] [ENTER] 32°37'54.6"

Angle of Title (∠T) = 32°38' Ans

Procedure for Positioning the Part on an Angle Plate for Drilling

- Rotate the part to the angle of rotation, ∠R as shown in Figure 69-3(a). Care must be taken as to whether the part is rotated to the computed ∠R or the complement of ∠R. Rotate the part 38°40' as in Figure 69-3(b).

➤ **Note:** The position of right △ACB is shown with hidden lines.

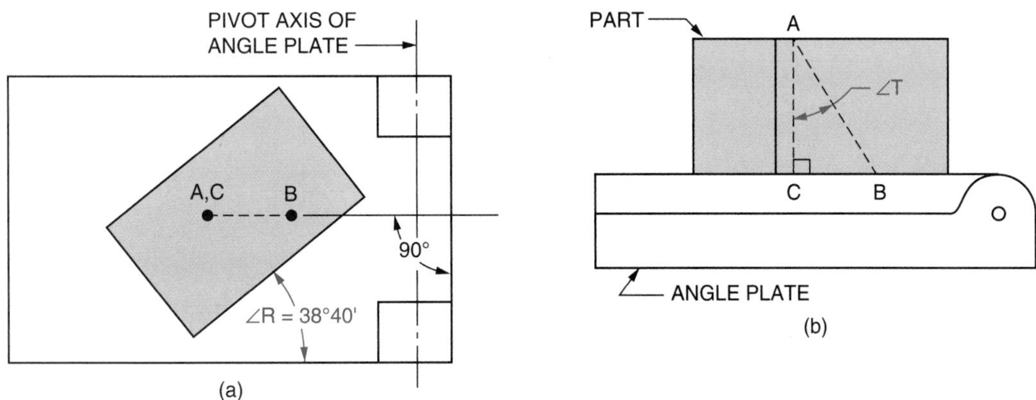

Figure 69-3

- Raise the angle plate to tilt angle, ∠T. Tilt to 32°38' as shown in Figure 69-4(a). Care must be taken as to whether the part is tilted to the computed ∠T or the complement of ∠T. Observe in Figure 69-4(b) that the position of hole axis AB is vertical.

With the part set to the angle of rotation and to the angle of tilt it is positioned to drill the hole on vertical axis AB.

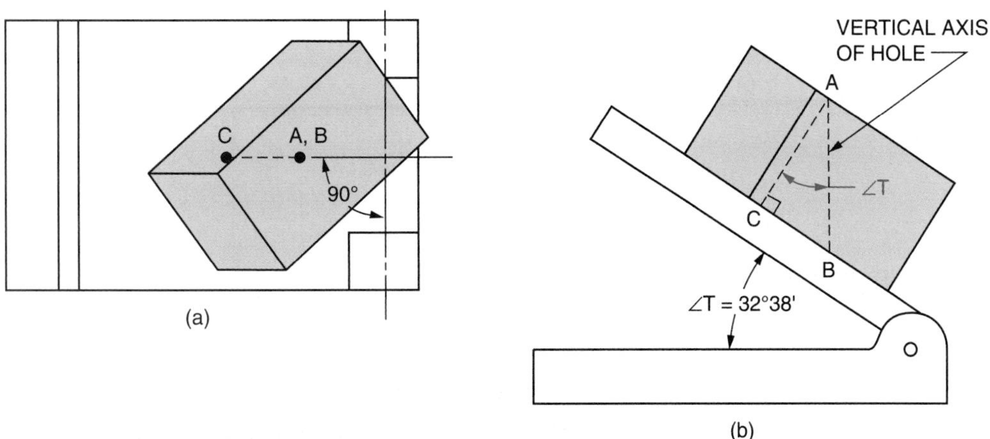

Figure 69-4

Application

In each of the following problems, the axis of a hole is shown in a rectangular solid. In order to position the hole axis for drilling, the angle of rotation and the angle of tilt must be determined. Compute angles to the nearer minute in triangles with customary unit sides. Compute angles to the nearer hundredth degree in triangles with metric unit sides.

a. Compute the angle of rotation, ∠R.

b. Compute the angle of tilt, ∠T.

1. Given: H = 2.600 in
   L = 2.400 in
   W = 1.900 in

   a. ______________ b. ______________

2. Given: H = 55.00 mm
   L = 48.00 mm
   W = 30.00 mm

   a. ______________ b. ______________

Use this figure for #1 and #2.

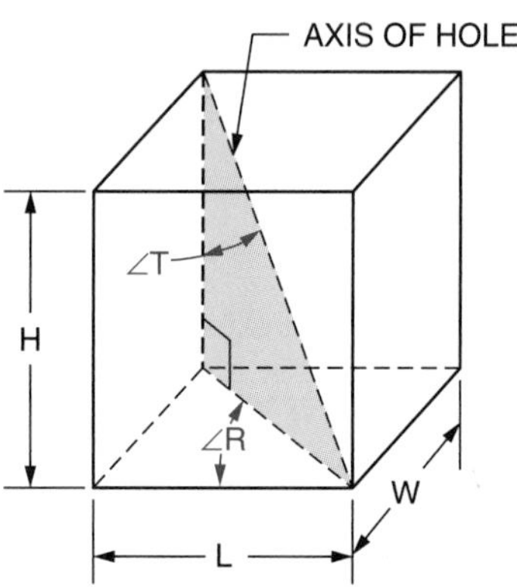

3. Given: H = 4.750 in
   L = 4.000 in
   W = 3.750 in

   a. ______________ b. ______________

4. Given: H = 42.00 mm
   L = 37.00 mm
   W = 32.00 mm

   a. ______________ b. ______________

Use this figure for #3 and #4.

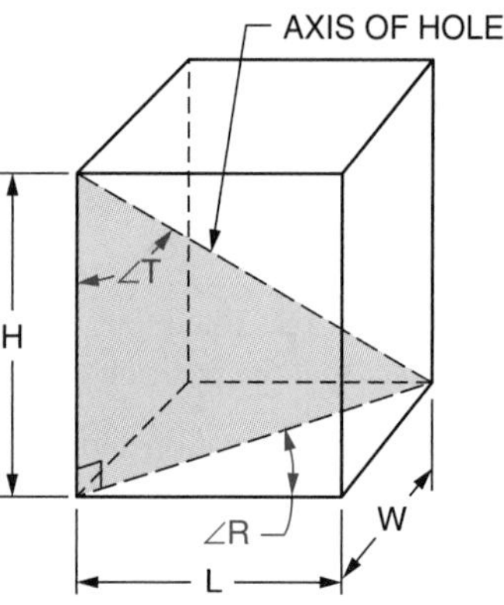

5. Given: H = 0.970 in
   L = 0.860 in
   W = 0.750 in

   a. ______________ b. ______________

6. Given: H = 22.00 mm
   L = 18.00 mm
   W = 15.00 mm

   a. ______________ b. ______________

Use this figure for #5 and #6.

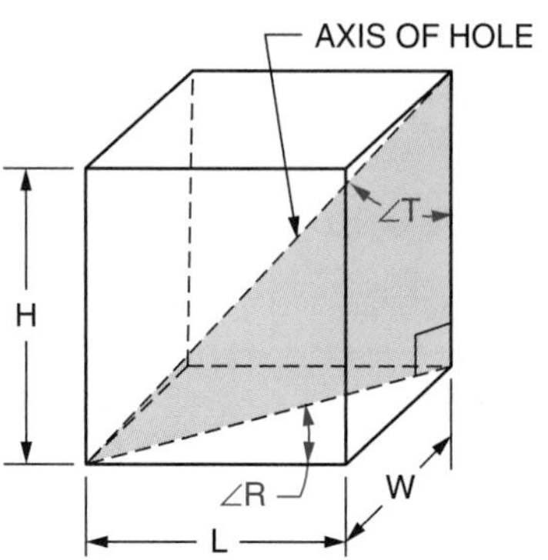

In each of the following problems, the top, front, and right side views of a compound-angular hole are shown. For each problem do the following:

a. Sketch, dimension, and label a rectangular solid. Within the solid, show the right triangle that contains the hole axis as a side and the angle of tilt. Show the position of the angle of rotation.

b. Compute the angle of rotation, ∠R.

c. Compute the angle of tilt, ∠T.

Compute angles to the nearer minute in triangles with customary unit sides. Compute angles to the nearer hundredth degree in triangles with metric unit sides.

7. All dimensions are in inches. a. *(sketch)*

b. ______________ c. ______________

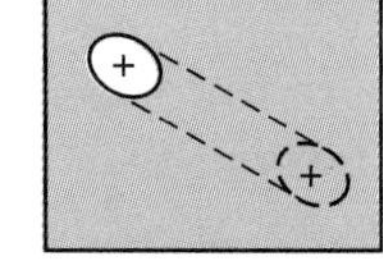

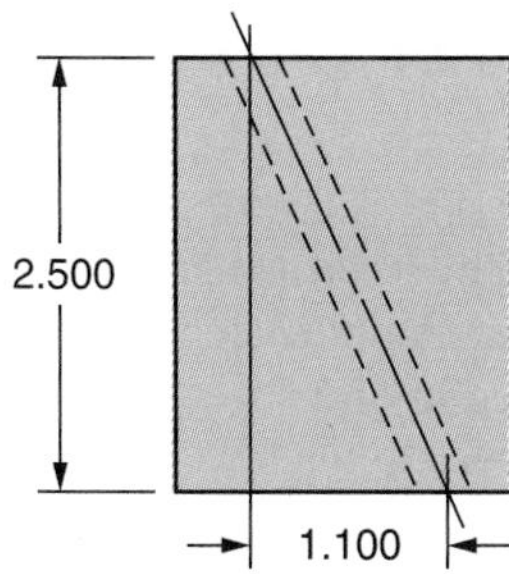

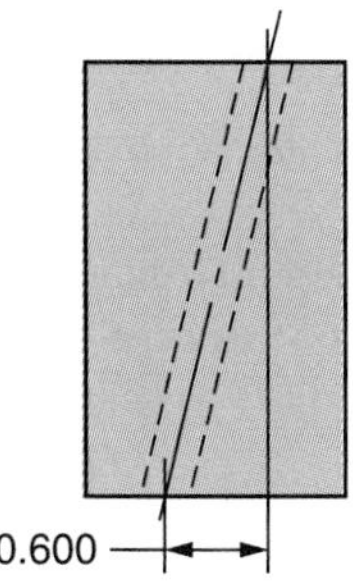

9. All dimensions are in inches. a. *(sketch)*

b. ______________ c. ______________

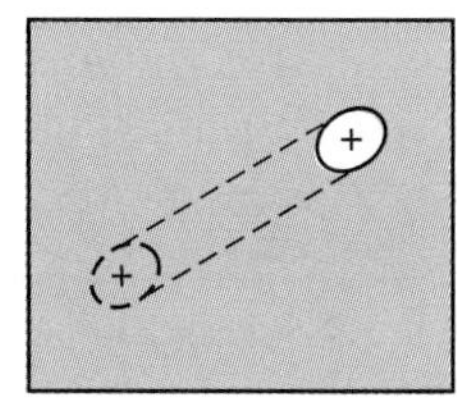

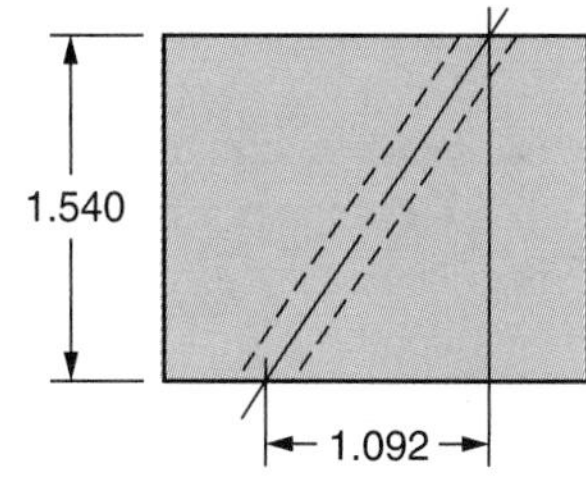

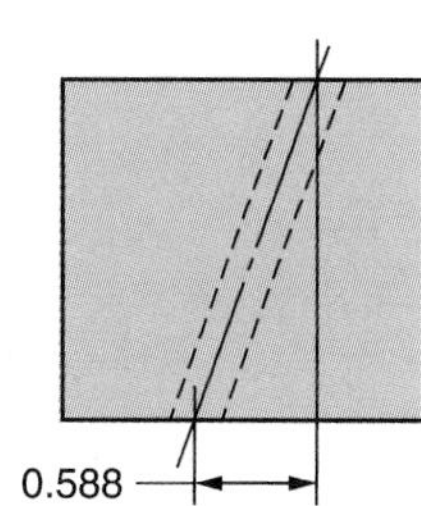

8. All dimensions are in millimeters. a. *(sketch)*

b. ______________ c. ______________

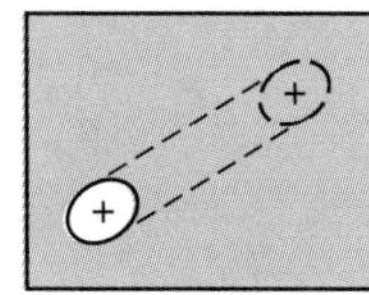

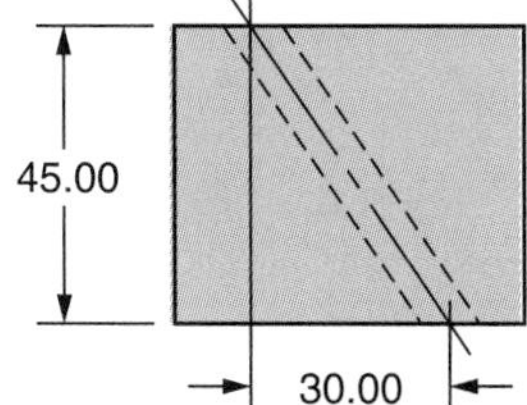

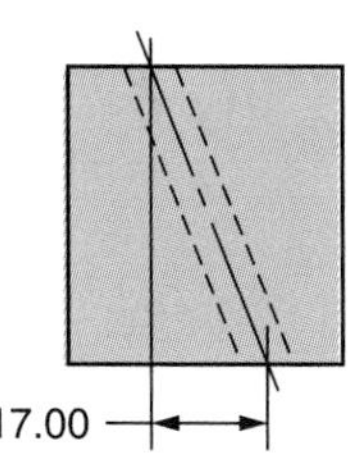

10. All dimensions are in millimeters. a. *(sketch)*

b. ______________ c. ______________

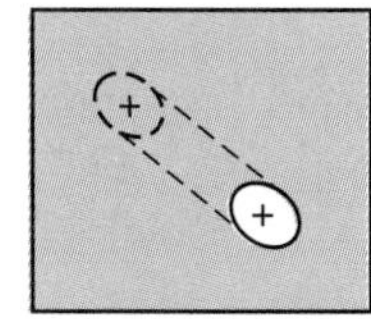

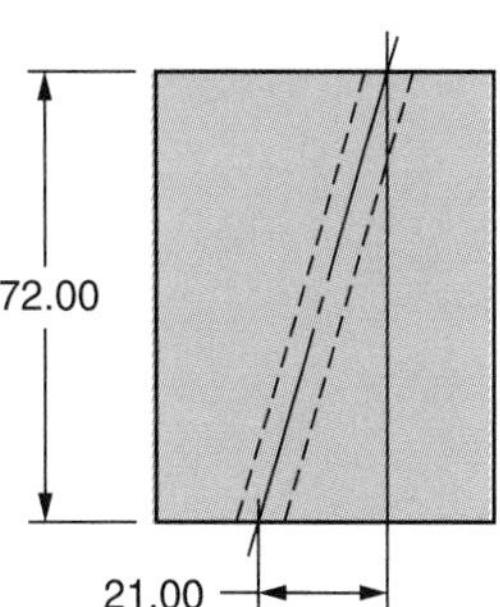

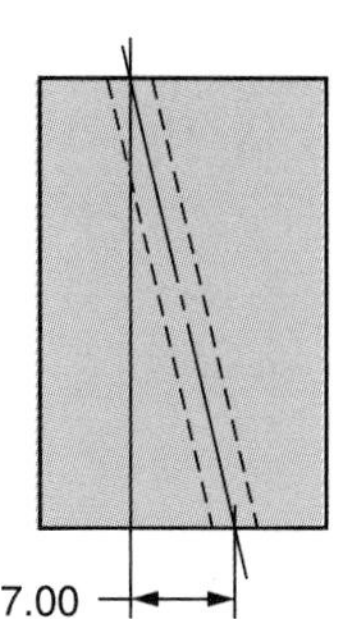

UNIT 70

Drilling and Boring Compound-Angular Holes: Computing Angles of Rotation and Tilt Using Given Angles

Objectives *After studying this unit you should be able to*

- Compute angles of rotation and angles of tilt of hole axes in rectangular solids. No length dimensions are known.
- Sketch, dimension, and label compound-angular components within rectangular solids and compute angles of rotation and angles of tilt. No length dimensions are known.
- Compute angles of rotation and angles of tilt by use of formulas.
- Compute front view and side view angles by use of formulas.

Computing Angles of Rotation and Angles of Tilt When No Length Dimensions Are Known

In certain compound angle problems no length dimensions are known; instead, angles in two different planes are known. In problems of this type where no length dimensions are known, it is necessary to assign a value of 1 (unity) to one of the sides in order to compute with trigonometric functions.

The side that is assigned a value of 1 must be a side common to two of the formed right triangles. One of the right triangles must have a known angle. The other right triangle must have either a known angle or an angle that is to be computed, ∠R or ∠T.

Example Three views of a compound-angular hole are shown in Figure 70-1. Hole angles are given in the front and right side views. No length dimensions are given.

a. Determine the angle of rotation.

b. Determine the angle of tilt.

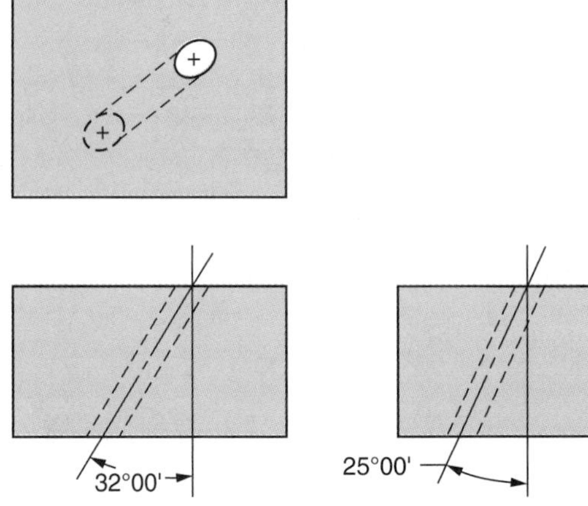

Figure 70-1

Sketch a rectangular solid in Figure 70-2. Project auxiliary lines that form right triangles containing the given angles, the axis of the hole, and the angles to be computed, ∠R and ∠T.

BC is a side of right △BCD, which contains the given 25°00' angle. BC is also a side of right △BCE, which contains ∠R that is to be computed. Make BC = 1.

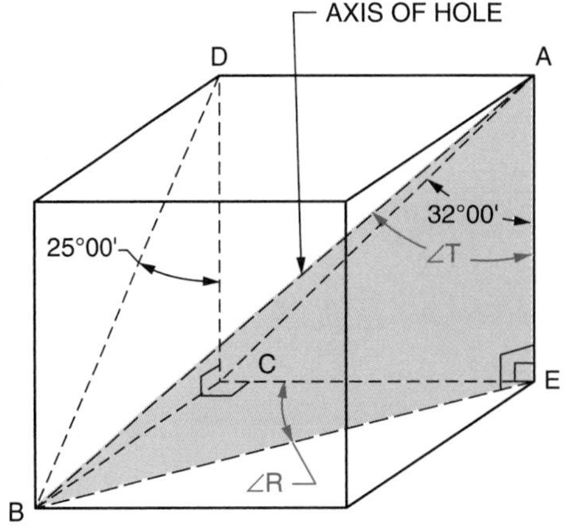

Figure 70-2

a. Compute angle of rotation, ∠R.

In right △BCD: BC = 1, ∠D = 25°00'

$$\tan 25°00' = \frac{BC}{DC}$$

$$\tan 25°00' = \frac{1}{DC}, \; DC = \frac{1}{\tan 25°00'}$$

DC ≈ 1 [÷] [tan] 25 [=] 2.144506921

DC ≈ 2.1445

In right △ACE: AE = DC = 2.1445, ∠A = 32°00'

$$\tan 32°00' = \frac{CE}{AE}$$

$$\tan 32°00' \approx \frac{CE}{2.1445}, \; CE \approx \tan 32°00' \, (2.1445)$$

CE ≈ [tan] 32 [×] 2.1445 [=] 1.340032325

or CE ≈ [tan] 32 [ENTER] [×] 2.1445 [ENTER] 1.340032325

CE ≈ 1.3400

In right △BCE: BC = 1, CE = 1.3400

$$\tan \angle R = \frac{BC}{CE} \approx \frac{1}{1.3400} \approx 0.74627$$

∠R ≈ [SHIFT] [tan⁻¹] .74627 [=] [SHIFT] [←] 36°43'58"

or ∠R ≈ [2nd] [tan⁻¹] .74627 [° ' "] [◀] [ENTER] [ENTER] 36°43'58"

∠R ≈ 36°44' Ans

b. Compute angle of tilt, ∠T.

In right △BCE: BC = 1, CE = 1.3400

$$BE = \sqrt{1^2 + 1.3400^2}$$

BE ≈ 1.6720

In right △AEB: AE = 2.1445, BE = 1.6720

$$\tan \angle T = \frac{BE}{AE}$$

$$\tan \angle T = \frac{1.6720}{2.1445} = 0.77967$$

∠T ≈ [SHIFT] [tan⁻¹] .77967 [=] [SHIFT] [←] 37°56'33"

or ∠T ≈ [2nd] [tan⁻¹] .77967 [° ' "] [◀] [ENTER] [ENTER] 37°56'33"

∠T = 37°57' Ans

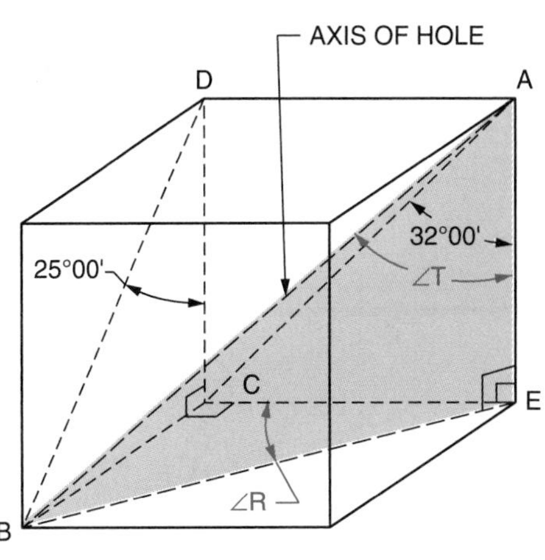

Figure 70-3

The part is rotated 36°44' (∠R) on the angle plate and the angle plate is raised 37°57' (∠T).

Formulas for Computing Angles of Rotation and Angles of Tilt Used in Drilling

Formulas for determining angles of rotation and angles of tilt have been computed. These formulas reduce the amount of computation required in solving compound angle problems. The formulas should **not** be used unless a problem is completely visualized and the method of solution shown in the previous example is fully understood.

To use formulas for the angles of rotation and angles of tilt, ∠A and ∠B shown in Figure 70-4 must be identified.

∠A is the given angle in the front view (frontal plane) in relation to the vertical.

∠B is the given angle in the side view (profile plane) in relation to the vertical.

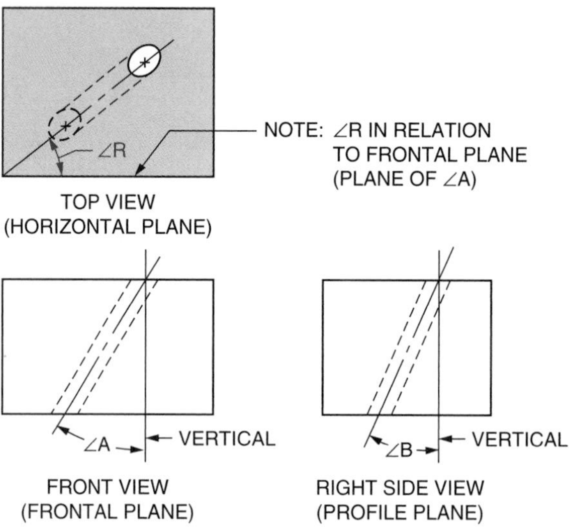

Figure 70-4

Formula for the Angle of Rotation in Relation to the Frontal Plane (Plane of ∠A) Used in Drilling

$$\tan \angle R = \frac{\tan \angle B}{\tan \angle A}$$

Formula for the Angle of Tilt Used in Drilling

$$\tan \angle T = \sqrt{\tan^2 \angle A + \tan^2 \angle B}$$

The notation $\tan^2 \angle A$ means $(\tan \angle A)^2$.

In using the formula given for the angle of rotation, ∠R must be determined in relation to the frontal plane (plane of ∠A). If ∠R is to be determined in relation to the profile plane (plane of ∠B), the complement of the computed formula ∠R must be used.

Example 1 Three views of a compound-angular hole are shown in Figure 70-5. (This is the same compound-angular hole used in the previous example.) The angle in the front view in relation to the vertical is 32°00' (∠A = 32°00'). The angle in the right side view in relation to the vertical is 25°00' (∠B = 25°00').

a. Compute ∠R.

b. Compute ∠T.

a. $\tan \angle R = \frac{\tan \angle B}{\tan \angle A}$

$\tan \angle R = \frac{\tan 25°00'}{\tan 32°00'}$

$\tan \angle R \approx$ [tan] 25 [÷] [tan] 32 [=] 0.746248246
or $\tan \angle R \approx$ [tan] 25 [ENTER] [÷] [tan] 32 [ENTER] 0.746248247
$\tan \angle R \approx 0.746248$
$\angle R \approx$ [SHIFT] [tan⁻¹] .746248 [=] [SHIFT] [←] 36°43'55"
or $\angle R \approx$ [2nd] [tan⁻¹] .746248 [° ' "] [◄] [ENTER] [ENTER] 36°43'55"
$\angle R \approx 36°44'$ Ans

b. $\tan \angle T = \sqrt{\tan^2 \angle A + \tan^2 \angle B}$

$\tan \angle T = \sqrt{\tan^2 32°00' + \tan^2 25°00'}$

$\tan \angle T \approx$ [√] [(] [(] [tan] 32 [)] [x²] [+] [(] [tan] 25 [)] [x²] [)] [=] 0.779682332

$\angle T \approx$ [SHIFT] [tan⁻¹] .779682332 [=] [SHIFT] [←] 37°56'34"
or $\angle T \approx$ [2nd] [tan⁻¹] .779682332 [° ' "] [◄] [ENTER] [ENTER] 37°56'34"
$\angle T \approx 37°57'$ Ans

Figure 70-5

Observe that the values of $\angle R = 36°44'$ and $\angle T = 37°57'$ are the same as those computed in the previous example.

Occasionally a problem requires computing a front view angle when the side view angle and the angle of tilt or rotation are known. Also, it may be required to compute a side view angle in a problem when the front view angle and angle of tilt or rotation are known. The formulas for angle of rotation and tilt are used.

Example 2 Given: $\angle B = 20°00'$, and $\angle R = 24°00'$

Compute: $\angle A$ and $\angle T$

$\tan \angle R = \dfrac{\tan \angle B}{\tan \angle A}$

$\tan 24°00' = \dfrac{\tan 20°00'}{\tan \angle A}$

$0.44523 = \dfrac{0.36397}{\tan \angle A}$

$\tan \angle A = \dfrac{0.36397}{0.44523}$

$\tan \angle A = 0.81749$

$\angle A = 39°16'$ Ans

$\tan \angle T = \sqrt{\tan^2 \angle A + \tan^2 \angle B}$

$\tan \angle T = \sqrt{\tan^2 39°16' + \tan^2 20°}$

$\tan \angle T = \sqrt{(0.81752)^2 + (0.36397)^2}$

$\tan \angle T = \sqrt{0.6683390 + 0.1324742}$

$\tan \angle T = \sqrt{0.8008132}$

$\tan \angle T = 0.89488$

$\angle T = 41°49'$ Ans

Example 3 Given: $\angle A = 40.00°$ and $\angle T = 42.50°$

Compute: $\angle B$ and $\angle R$

$\tan \angle T = \sqrt{\tan^2 \angle A + \tan^2 \angle B}$

$\tan 42.50° = \sqrt{\tan^2 40.00° + \tan^2 \angle B}$

$0.91633 = \sqrt{(0.83910)^2 + \tan^2 \angle B}$

$0.91633^2 = 0.83910^2 + \tan^2 \angle B$

$\tan^2 \angle B = 0.13557$

$\tan \angle B = 0.36820$

$\angle B = 20.21°$ Ans

$\tan \angle R = \dfrac{\tan \angle B}{\tan \angle A} = \dfrac{0.36820}{0.83910} = 0.43880$

$\angle R = 23.69°$ Ans

Application

For problems 1 through 16, compute angles to nearer minute or hundredth degree.

Computing Angles of Rotation and Tilt Without Using Drilling Formulas

In each of the following problems, 1 through 6, the axis of a hole is shown in a rectangular solid. In order to position the hole axis for drilling, the angle of rotation and the angle of tilt must be determined. Do **not** use drilling formulas in solving these problems.

a. Compute the angle of rotation, ∠R.

b. Compute the angle of tilt, ∠T.

Use this figure for #1 and #2.

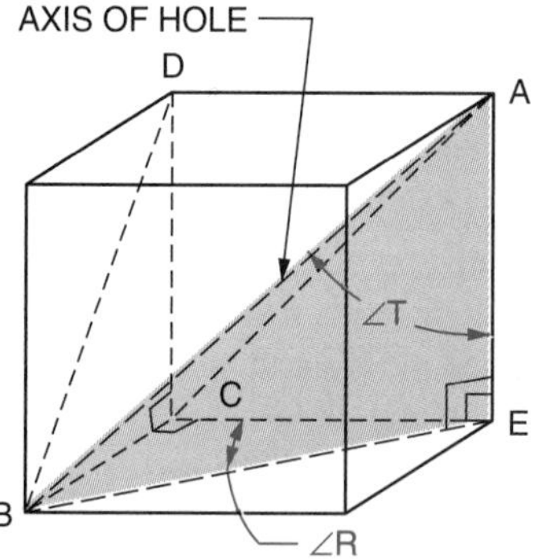

1. Given: ∠BDC = 35°00'
   ∠CAE = 42°00'

   a. ______ b. ______

2. Given: ∠BDC = 27°00'
   ∠CAE = 33°50'

   a. ______ b. ______

Use this figure for #3 and #4.

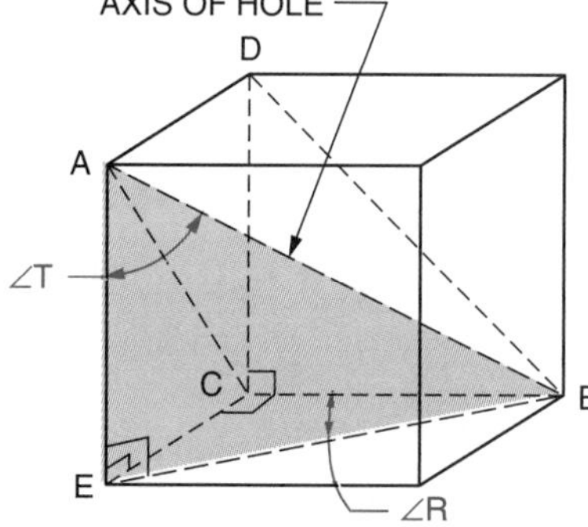

3. Given: ∠EAC = 18.25°
   ∠CDB = 31.00°

   a. ______ b. ______

4. Given: ∠EAC = 21°50'
   ∠CDB = 33°00'

   a. ______ b. ______

Use this figure for #5 and #6.

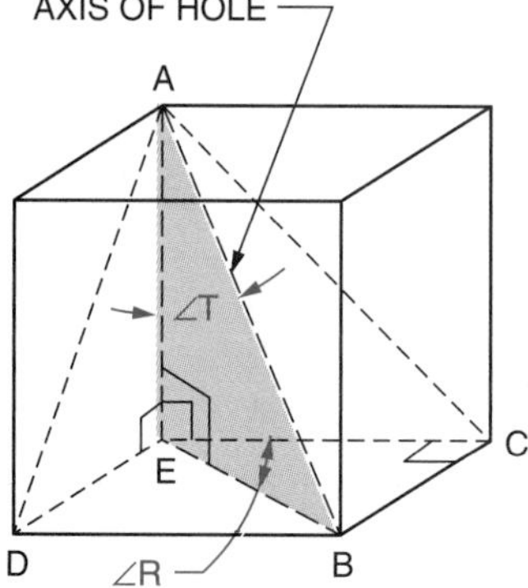

5. Given: ∠DAE = 30°00'
   ∠CAE = 42°10'

   a. ______ b. ______

6. Given: ∠DAE = 27.40°
   ∠CAE = 41.00°

   a. ______ b. ______

In each of the following problems, 7 through 10, the top, front, and right side views of a compound-angular hole are shown. Do **not** use drilling formulas in solving these problems. For each problem:

a. Sketch and label a rectangular solid. Within the solid, show the right triangle that contains the hole axis as a side and the angle of tilt. Show the position of the angle of rotation. Show the right triangles that contain the given angles.

b. Compute the angle of rotation, ∠R.

c. Compute the angle of tilt, ∠T.

7. a. *(sketch)*

b. ______________ c. ______________

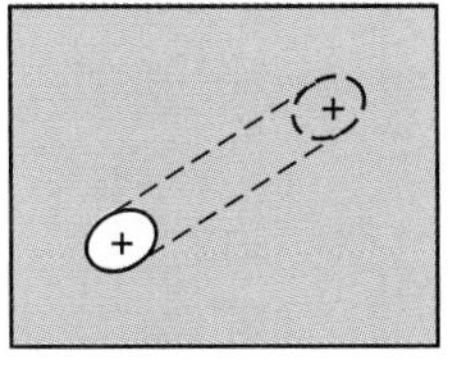

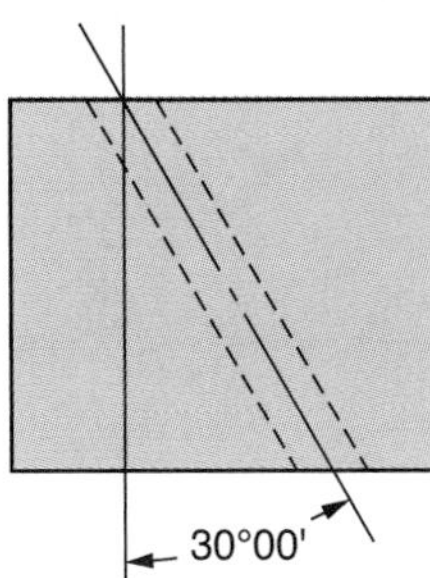

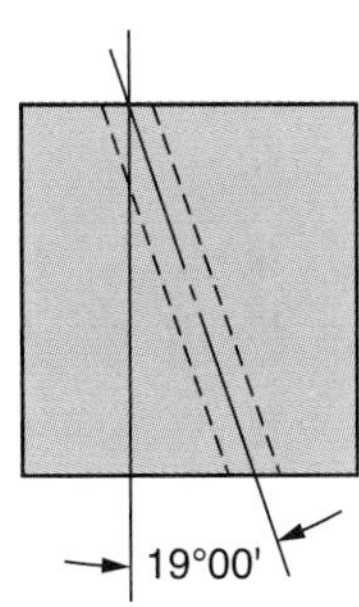

9. a. *(sketch)*

b. ______________ c. ______________

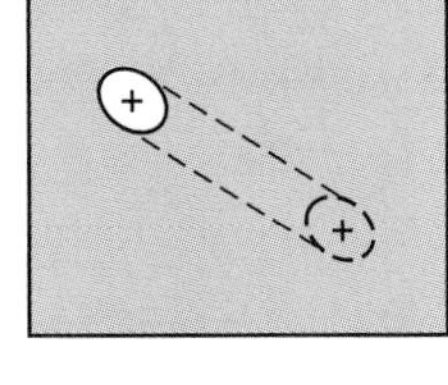

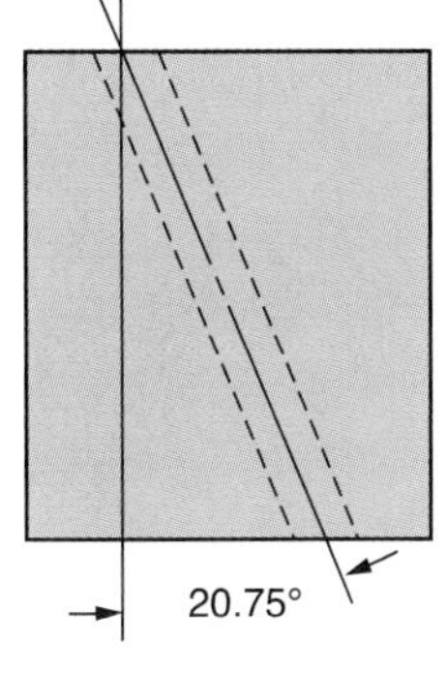

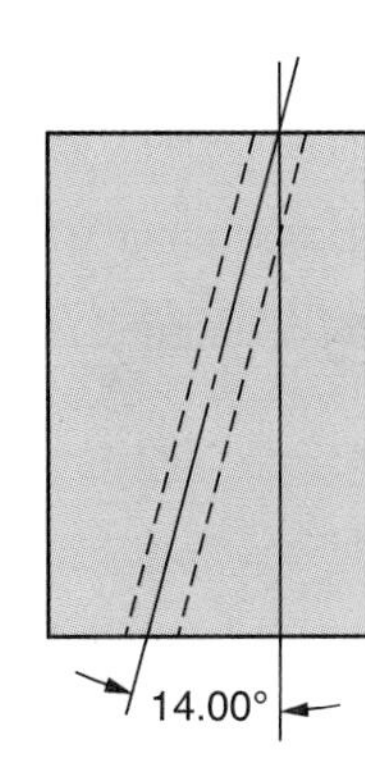

8. a. *(sketch)*

b. ______________ c. ______________

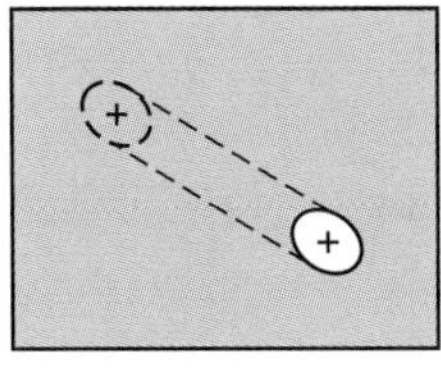

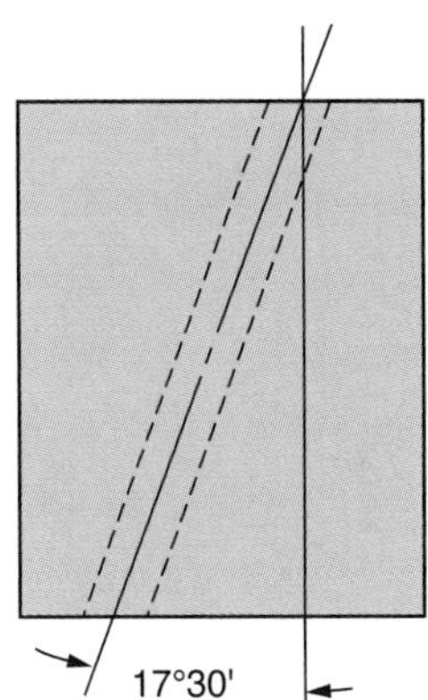

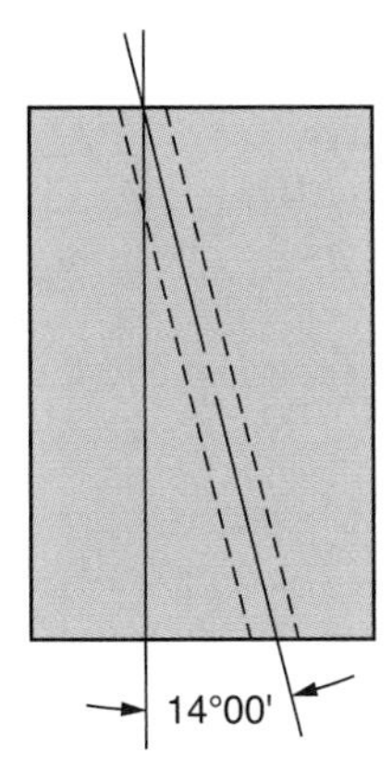

10. a. *(sketch)*

b. ______________ c. ______________

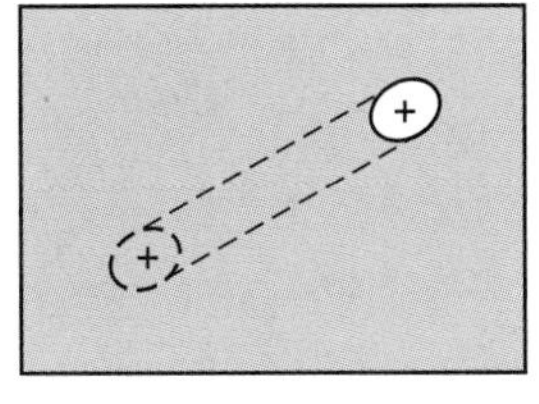

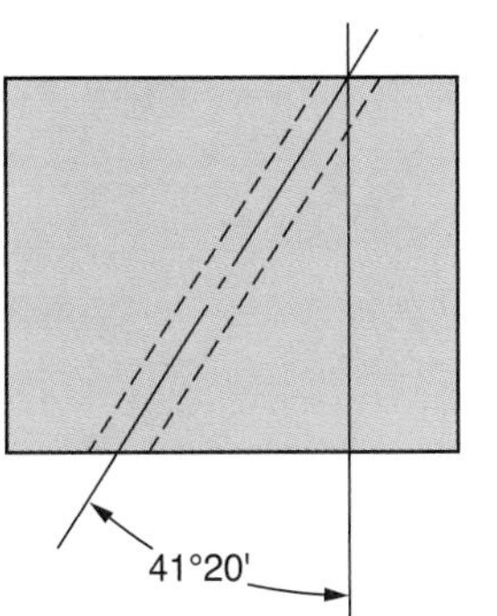

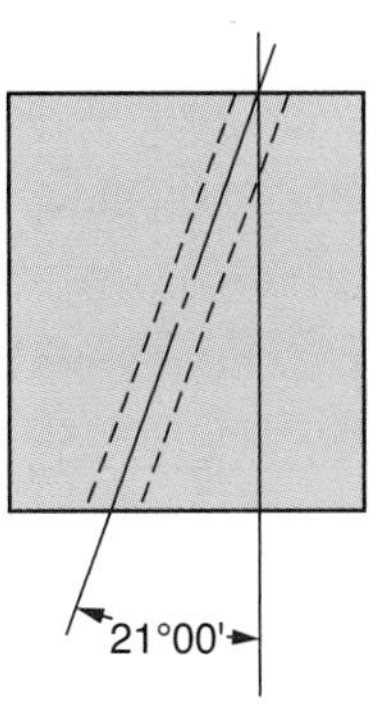

Computing Angles Using Drilling Formulas

In each of the following problems, the top, front, and right side views of a compound-angular hole are shown. Compute the required angles using these formulas.

$$\tan \angle R = \frac{\tan \angle B}{\tan \angle A}$$

$$\tan \angle T = \sqrt{\tan^2 \angle A + \tan^2 \angle B}$$

Use this figure for #11, #12, and #13.

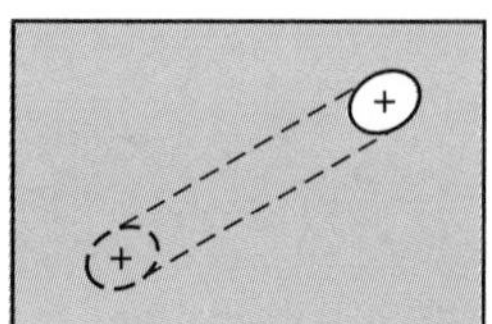

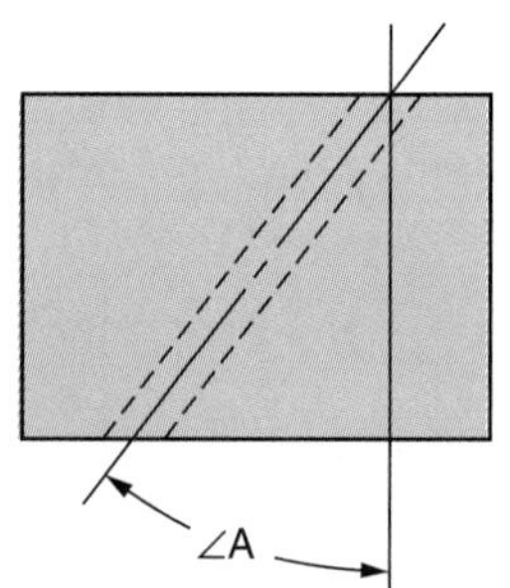

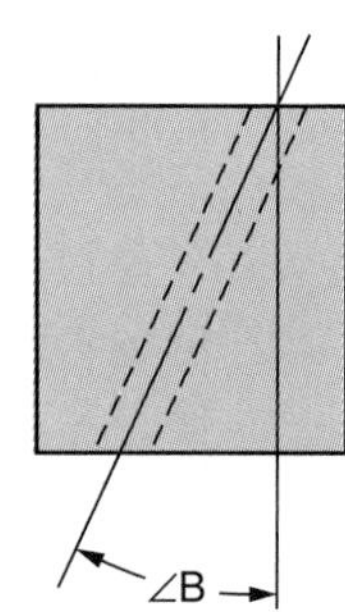

11. Given: $\angle A = 41°00'$
    $\angle B = 18°00'$
    a. Compute ∠R. ______________
    b. Compute ∠T. ______________
12. Given: $\angle B = 23°00'$
    $\angle R = 33°10'$
    a. Compute ∠A. ______________
    b. Compute ∠T. ______________
13. Given: $\angle A = 38.00°$
    $\angle T = 41.30°$
    a. Compute ∠B. ______________
    b. Compute ∠R. ______________

Use this figure for #14, #15, and #16.

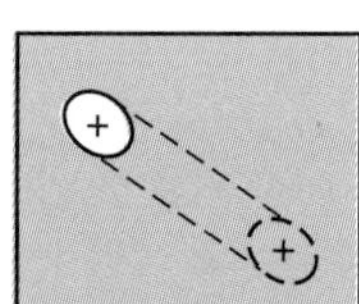

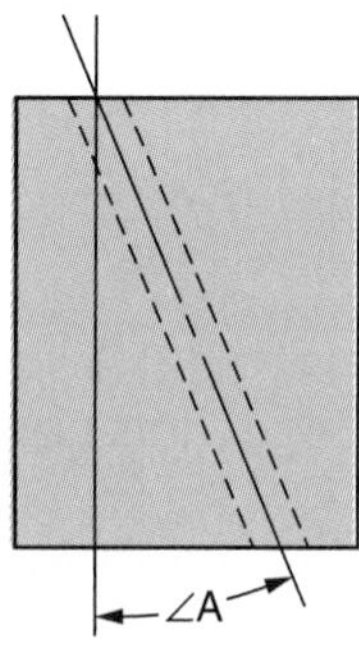

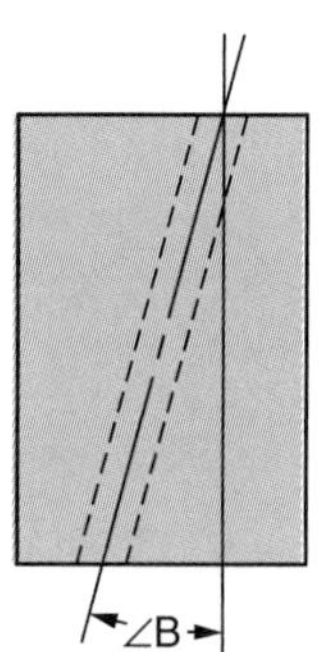

14. Given: $\angle A = 25°00'$
    $\angle B = 19°10'$
    a. Compute ∠R. ______________
    b. Compute ∠T. ______________
15. Given: $\angle B = 19.00°$
    $\angle R = 32.10°$
    a. Compute ∠A. ______________
    b. Compute ∠T. ______________
16. Given: $\angle A = 23°20'$
    $\angle T = 29°30'$
    a. Compute ∠B. ______________
    b. Compute ∠R. ______________

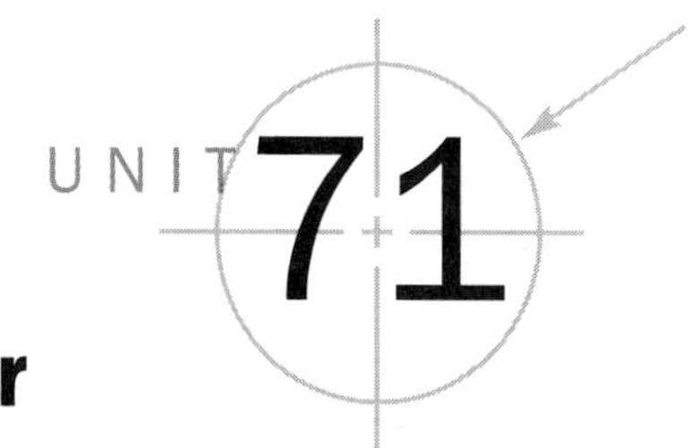

Machining Compound-Angular Surfaces: Computing Angles of Rotation and Tilt

Objectives *After studying this unit you should be able to*

- Compute angles of rotation and angles of tilt in angle plate positioning for machining compound-angular surfaces as given in rectangular solids.
- Sketch, dimension, and label compound-angular surface components within rectangular solids and compute angles of rotation and angles of tilt.
- Compute angles of rotation and angles of tilt by use of formulas.
- Compute front view and side view surface angles by use of formulas.

Machining Compound-Angular Surfaces

When the surface of a part appears as a diagonal in each of two conventional views, such as the front and right side views, setting up the part for machining involves compound angles. When just the surface (plane) must be considered in a compound angle problem, a single rotation and a single tilt are required.

The setting up of a part in a compound angle problem in which a surface (plane) and a line on the surface must both be considered is more complex. Setups of this type require single rotation and double tilt, or single tilt and double rotation.

The presentation of compound-angular surfaces in this text is limited to problems that require only single rotation and single tilt. An understanding of the procedures shown will enable you to set up most compound-angle surface-cutting problems encountered. The procedures are also the basis for the solution of more complex compound angle problems that require double tilt or double rotation.

Example Figure 71-1 shows three views of a rectangular solid block in which a compound-angular surface is to be machined.

a. Determine the angle of rotation, ∠R (see Figure 71-2).

b. Determine the angle of tilt, ∠T.

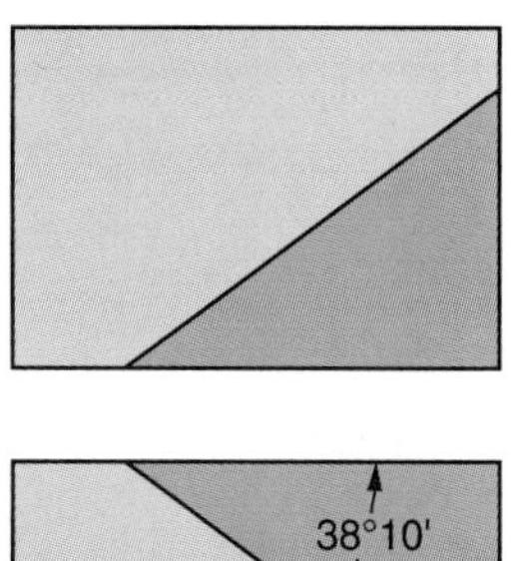

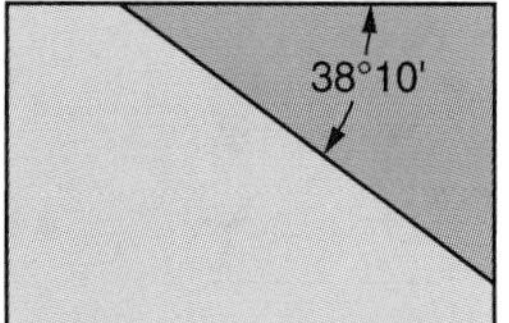

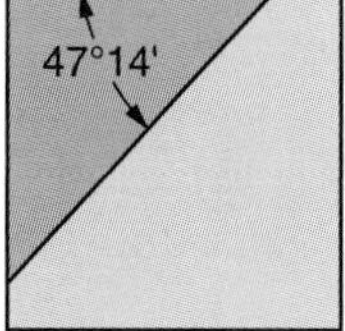

Figure 71-1

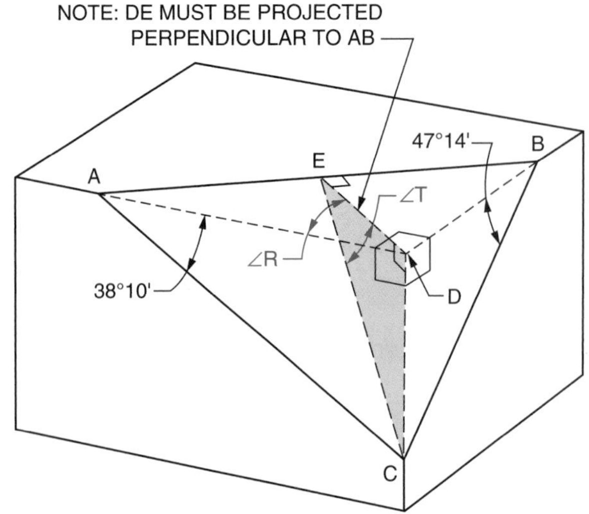

Figure 71-2

Sketch the rectangular solid and the pyramid ABCD formed by the surface ABC to be cut and the extended sides of the block. Project auxiliary lines, which form right triangles containing the given angles and the angles to be computed, ∠R and ∠T.

In cutout (pyramid) ABCD, *DE must be projected perpendicular to AB*. Right △CDE contains the angle of tilt, ∠T. The angle of rotation, ∠R, is contained in right △AED. Observe that right △AED is contained in the horizontal plane and ∠R is given in reference to line AD, which lies in both the horizontal and frontal planes.

Since no length dimensions are given, assign a value of 1 (unity) to a side that is common to two or more sides of the formed *right* triangles. One or more of the triangles must have a known angle. The other right triangle or triangles must have a known angle or an angle to be computed, ∠R or ∠T. Side DC is contained in the following three right triangles.

Right △ADC, which contains the given angle 38°10'.
Right △BDC, which contains the given angle 47°14'.
Right △CDE, which contains ∠T.

Make DC = 1.

a. In right △BDC, DC = 1, ∠B = 47°14'. Compute DB.

$$\tan 47°14' = \frac{DC}{DB}$$

$$\tan 47°14' = \frac{1}{DB}$$

$$DB = \frac{1}{\tan 47°14'}$$

DB ≈ 1 [÷] [tan] 47 [° ' "] 14 [° ' "] [=] 0.924930088

or DB ≈ 1 [÷] [tan] 47 [° ' "] 14 [° ' "] [ENTER] 0.924930088

DB ≈ 0.92493

In right △ADC, DC = 1, ∠A = 38°10'. Compute AD.

$$\tan 38°10' = \frac{DC}{AD}$$

$$\tan 38°10' = \frac{1}{AD}$$

$$DB = \frac{1}{\tan 38°10'}$$

AD ≈ 1 [÷] [tan] 38 [° ' "] 10 [° ' "] [=] 1.272295718

or AD ≈ 1 [÷] [tan] 38 [° ' "] 10 [° ' "] [ENTER] 1.272295718

AD ≈ 1.2723

In right △ADB, DB = 0.92493, AD ≈ 1.2723. Compute ∠A.

$$\tan \angle A = \frac{DB}{AD} \approx \frac{0.92493}{1.2723} \approx 0.72697$$

∠A ≈ [SHIFT] [$\tan^{-1}$] .72697 [=] [SHIFT] [←] 36°00'58"

or ∠A ≈ [2nd] [$\tan^{-1}$] .72697 [° ' "] [◀] [ENTER] [ENTER] 36°0'58"

∠A = 36°01'

In right △AED compute the angle of rotation, ∠R.

∠R and ∠A are complementary.

∠R ≈ 90° − 36°01' = 53°59' Ans

b. In right △AED, ∠DAE = ∠DAB = 36°01', AD = 1.2723. Compute DE.

$$\sin 36°01' = \frac{DE}{AD}$$

$$\sin 36°01' = \frac{DE}{1.2723}$$

$$DE = \sin 36°01'(1.2723)$$

DE ≈ [sin] 36 [° ' "] 1 [° ' "] [×] 1.2723 [=] 0.748138559

or DE ≈ [sin] 36 [° ' "] [ENTER] 1 [° ' "] [▶] [ENTER] [ENTER] [×] 1.2723 [ENTER] 0.748138559

DE ≈ 0.74814

In right △CDE, CD = 1, DE ≈ 0.74814. Compute the angle of tilt, ∠T.

$$\tan \angle T = \frac{DC}{DE}$$

$$\tan \angle T = \frac{1}{DE} = \frac{1}{0.74814} = 1.33665$$

∠T ≈ [SHIFT] [$\tan^{-1}$] 1.33665 [=] [SHIFT] [←] 53°11'54"

or ∠T ≈ [2nd] [$\tan^{-1}$] 1.33665 [° ' "] [◀] [ENTER] [ENTER] 53°11'54"

∠T ≈ 53°12' Ans

Procedure for Positioning the Part on an Angle Plate for Machining

- Rotate the part to the angle of rotation, ∠R, as shown in Figure 71-3. Care must be taken as to whether the part is rotated to the computed ∠R or the complement of ∠R. Rotate the part 53°59'.

➤ **Note:** The position of right △CDE is shown with hidden lines in Figure 71-4.

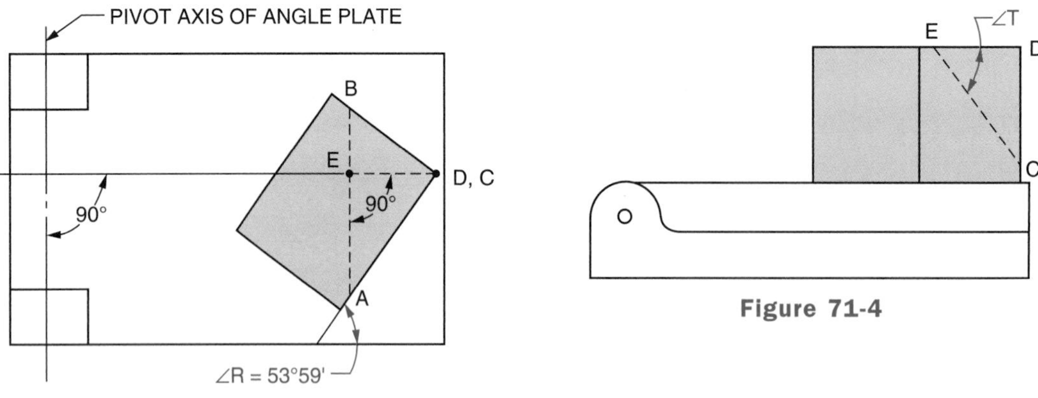

Figure 71-3

Figure 71-4

- Raise the angle plate to the tilt angle, ∠T. Tilt to 53°12' as shown in Figure 71-6. Care must be taken as to whether the part is tilted to the computed ∠T or the complement of ∠T. Observe that the position of the plane AEBC to be cut is horizontal.

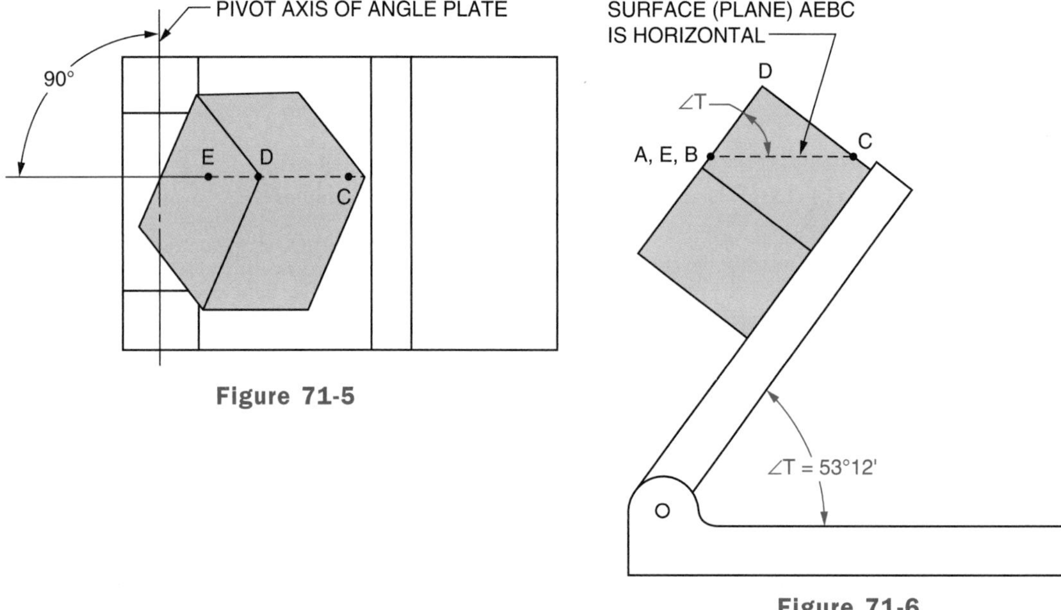

Figure 71-5

Figure 71-6

With the part set to the angle of rotation and to the angle of tilt, it is positioned to machine the surface on the horizontal plane.

Formulas for Computing Angles of Rotation and Angles of Tilt Used in Machining

As with formulas for drilling compound-angular holes, formulas for machining compound-angular surfaces should not be used until the problem is completely visualized and the method of solution shown in the previous example is fully understood.

To use formulas for the angles of rotation and angles of tilt, ∠A and ∠B shown in Figure 71-7 must be identified.

∠A is the given angle in the front view (frontal plane) in relation to the horizontal plane.

∠B is the given angle in the side view (profile plane) in relation to the horizontal plane.

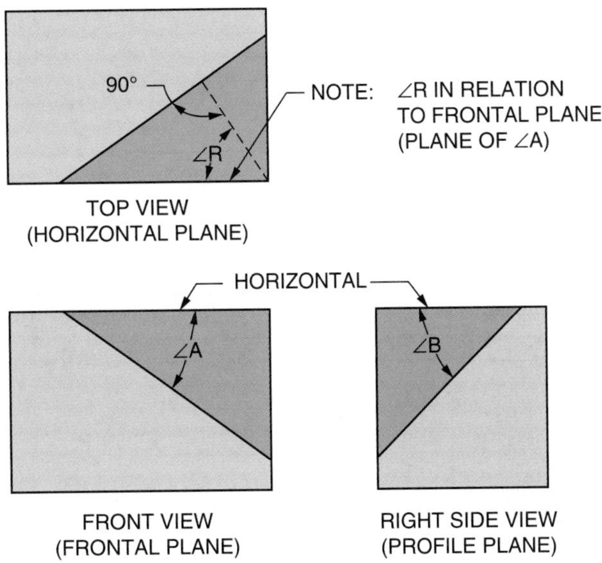

Figure 71-7

Formula for the Angle of Rotation in Relation to the Frontal Plane (Plane of ∠A) Used in Machining

$$\tan \angle R = \frac{\tan \angle B}{\tan \angle A}$$

Formula for the Angle of Tilt Used in Machining

$$\tan \angle T = \frac{\tan \angle A}{\cos \angle R}$$

In using the formula given for the angle of rotation, ∠R must be determined in relation to the frontal plane (plane of ∠A). If ∠R is to be determined in relation to the profile plane (plane of ∠B), the complement of the computed formula ∠R must be used.

Example 1 Three views of a compound-angular surface are shown in Figure 71-8. (This is the same compound-angular surface used in the previous example.) The angle in the front view in relation to the horizontal is 38°10' (∠A = 38°10'). The angle in the right side view in relation to the horizontal is 47°14' (∠B = 47°14').

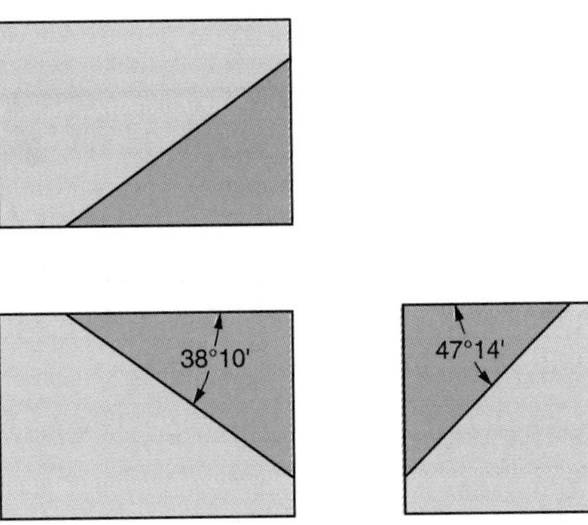

Figure 71-8

a. Compute ∠R.

b. Compute ∠T.

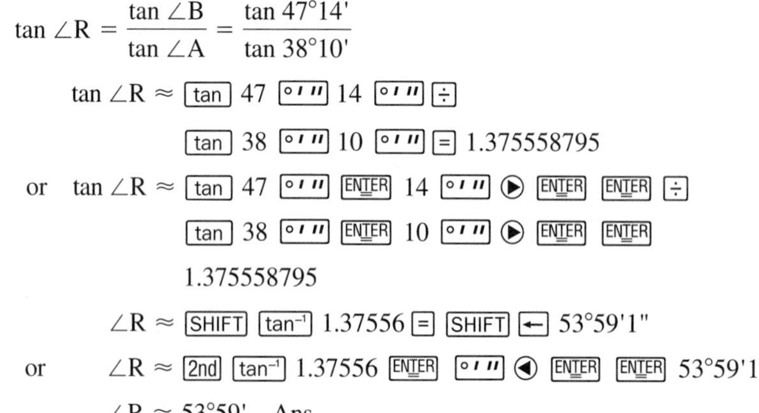

a. $\tan \angle R = \dfrac{\tan \angle B}{\tan \angle A} = \dfrac{\tan 47°14'}{\tan 38°10'}$

tan ∠R ≈ [tan] 47 [° ' "] 14 [° ' "] [÷] [tan] 38 [° ' "] 10 [° ' "] [=] 1.375558795

or tan ∠R ≈ [tan] 47 [° ' "] [ENTER] 14 [° ' "] [▶] [ENTER] [ENTER] [÷] [tan] 38 [° ' "] [ENTER] 10 [° ' "] [▶] [ENTER] [ENTER] 1.375558795

∠R ≈ [SHIFT] [tan⁻¹] 1.37556 [=] [SHIFT] [←] 53°59'1"

or ∠R ≈ [2nd] [tan⁻¹] 1.37556 [ENTER] [° ' "] [◀] [ENTER] [ENTER] 53°59'1"

∠R ≈ 53°59' Ans

b. $\tan \angle T = \dfrac{\tan \angle A}{\cos \angle R} = \dfrac{\tan 38°10'}{\cos 53°59'}$

tan ∠T ≈ [tan] 38 [° ' "] 10 [° ' "] [÷] [cos] 53 [° ' "] 59 [° ' "] [=] 1.336655293

or tan ∠T ≈ [tan] 38 [° ' "] [ENTER] 10 [° ' "] [▶] [ENTER] [ENTER] [÷] [cos] 53 [° ' "] [ENTER] 59 [° ' "] [▶] [ENTER] [ENTER] 1.336655293

∠T ≈ [SHIFT] [tan⁻¹] 1.33666 [EXE] [=] [←] 53°11'55"

or ∠T ≈ [2nd] [tan⁻¹] 1.33666 [ENTER] [° ' "] [◀] [ENTER] [ENTER] 53°11'55"

∠T ≈ 53°12' Ans

Observe that the values of ∠R = 53°59' and ∠T = 53°12' are the same as those computed in the previous example.

The same formulas for angles of rotation and tilt are used to compute an unknown front view angle when a side view angle and an angle of rotation or tilt are known. An unknown side view angle may be computed if the front view angle and angle of rotation or tilt are known.

Example 2 Given: ∠B = 18.15°, and ∠R = 27.45°.

Compute: ∠A and ∠T.

$$\tan \angle R = \frac{\tan \angle B}{\tan \angle A}$$

$$\tan 27.45° = \frac{\tan 18.15°}{\tan \angle A}$$

$$0.51946 = \frac{0.32782}{\tan \angle A}$$

$$\tan \angle A = \frac{0.32782}{0.51946}$$

$$\tan \angle A = 0.63108$$

$$\angle A = 32.25° \quad \text{Ans}$$

$$\tan \angle T = \frac{\tan \angle A}{\cos \angle R} = \frac{0.63108}{0.88741} = 0.71115$$

$$\angle T = 35.42° \quad \text{Ans}$$

Application

For problems 1 through 14, compute angles to the nearer minute or hundredth degree.

Computing Angles of Rotation and Tilt Without Using Machining Formulas

Three views of a rectangular solid block are shown in which a compound-angular surface is to be machined. A pictorial view of the block with auxiliary lines required for computations is also shown. Do **not** use machining formulas in solving these problems. For each of the following problems, 1 through 4:

a. Determine the angle of rotation, ∠R.

b. Determine the angle of tilt, ∠T.

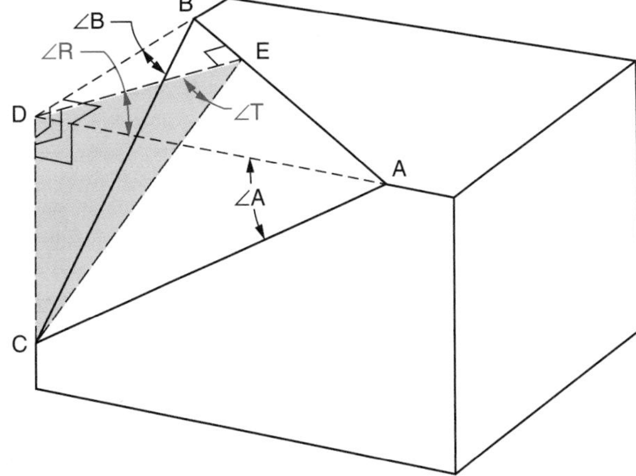

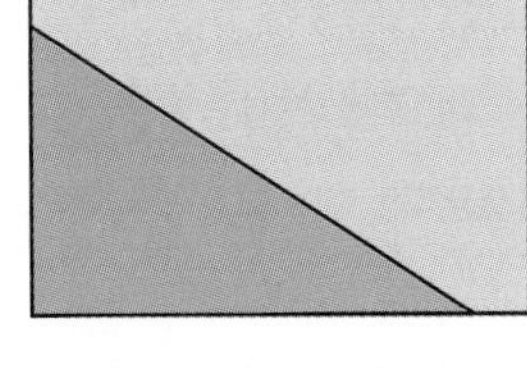

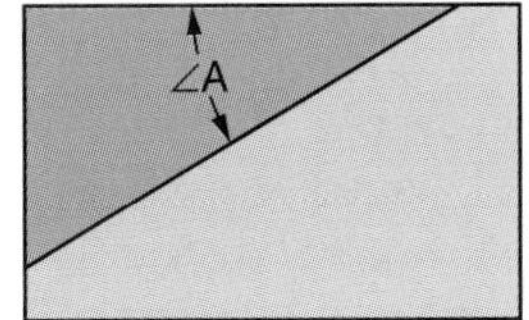

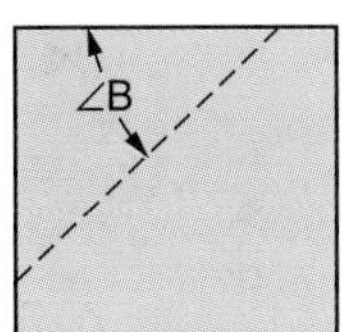

1. Given: ∠A = 32°00'
   ∠B = 44°00'

   a. ______________ b. ______________

2. Given: ∠A = 27°00'
   ∠B = 39°00'

   a. ______________ b. ______________

3. Given: ∠A = 18°10'
   ∠B = 27°50'

   a. ______________ b. ______________

4. Given: ∠A = 23.20°
   ∠B = 37.10°

   a. ______________ b. ______________

In each of the following problems, 5 through 8, the top, front, and right side views of a compound-angular surface are shown. Do **not** use machining formulas in solving these problems. For each problem:

a. Sketch and label a rectangular solid and the pyramid formed by the surface to be cut and the extended sides of the block. Show the right triangles that contain ∠T and the right triangles that contain the given angles. Identify ∠T, ∠R, and the given angles.

b. Compute the angle of rotation, ∠R.

c. Compute the angle of tilt, ∠T.

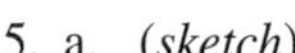

5. a. (*sketch*)

b. ______________ c. ______________

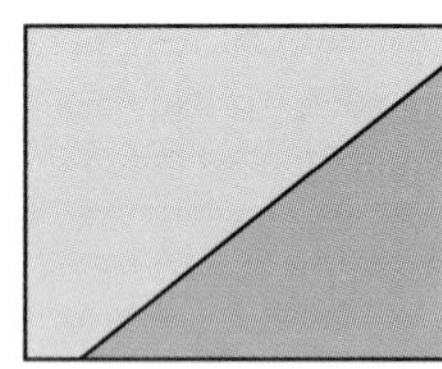

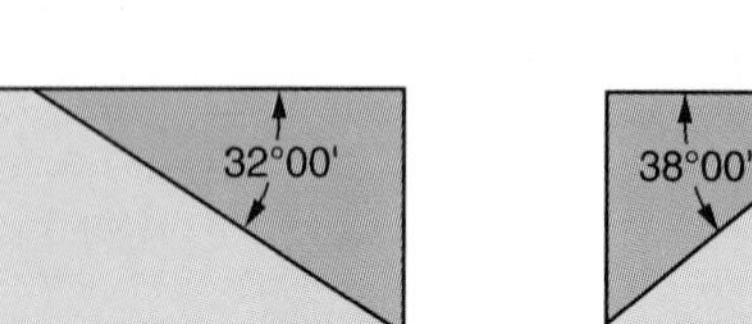

7. a. (*sketch*)

b. ______________ c. ______________

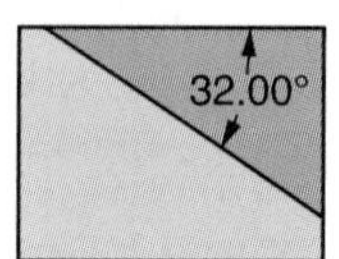

6. a. (*sketch*)

b. ______________ c. ______________

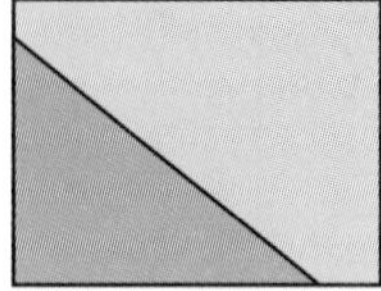

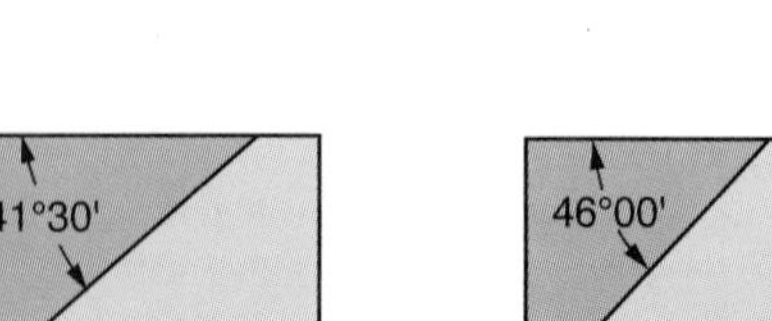

8. a. (*sketch*)

b. ______________ c. ______________

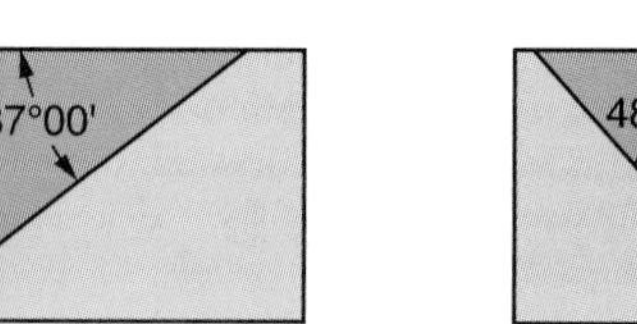

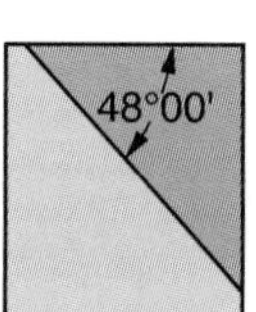

Computing Angles Using Machining Formulas

In each of the following problems, the top, front, and right side views of a compound-angular surface are shown. Compute the required angles using these formulas.

$$\tan \angle R = \frac{\tan \angle B}{\tan \angle A}$$

$$\tan \angle T = \frac{\tan \angle A}{\cos \angle R}$$

9. Given: ∠A = 39°00'
   ∠B = 44°00'

   a. Compute ∠R. ______________
   b. Compute ∠T. ______________

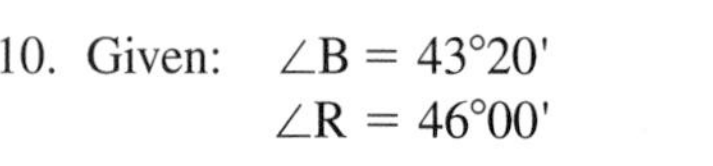

10. Given: ∠B = 43°20'
    ∠R = 46°00'

    a. Compute ∠A. ______________
    b. Compute ∠T. ______________

11. Given: ∠A = 41.20°
    ∠T = 52.00°

    a. Compute ∠R. ______________
    b. Compute ∠B. ______________

Use this figure for #9, #10, and #11.

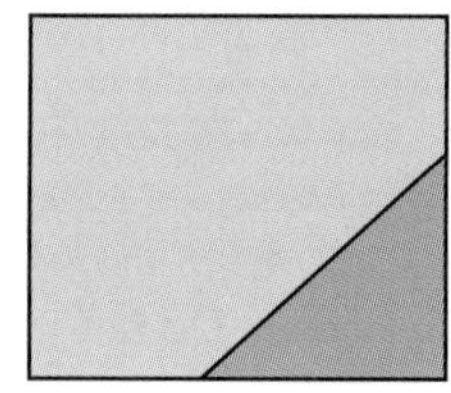

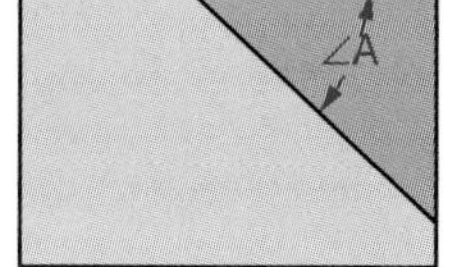

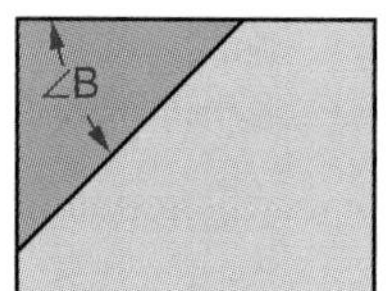

Use this figure for #12, #13, and #14.

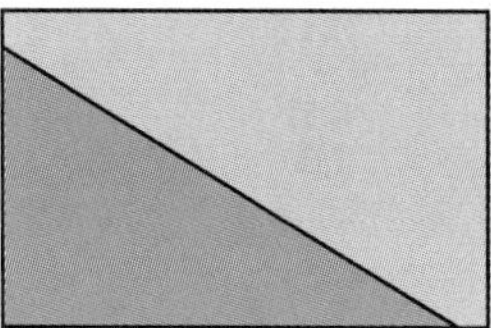

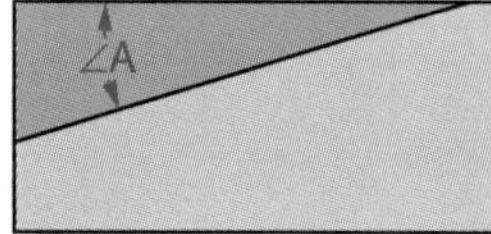

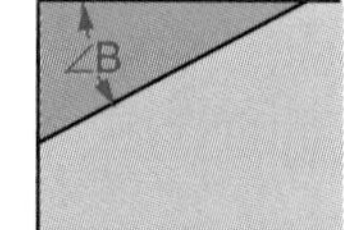

12. Given: ∠A = 19°00'
    ∠B = 23°10'
    a. Compute ∠R. ______________
    b. Compute ∠T. ______________
13. Given: ∠B = 18°00'
    ∠R = 22°00'
    a. Compute ∠A. ______________
    b. Compute ∠T. ______________
14. Given: ∠A = 15.60°
    ∠T = 26.50°
    a. Compute ∠R. ______________
    b. Compute ∠B. ______________

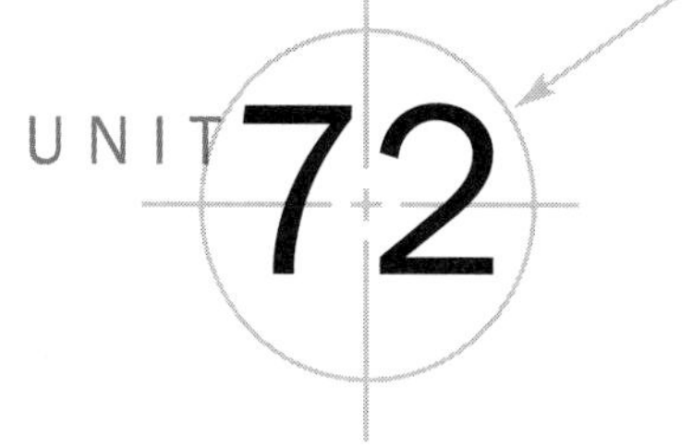

Computing Angles Made by the Intersection of Two Angular Surfaces

Objectives *After studying this unit you should be able to*

- Compute the true angles of compound-angular edges made by the intersection of two angular surfaces as given in rectangular solids.
- Sketch and label compound-angular surface edge components within rectangular solids and compute true angles.
- Compute true angles, front view angles, and side view angles by the use of formulas.

Computing Angles Made by the Intersection of Two Angular Surfaces

For design or inspection purposes, it may be required to compute angles that are made by the intersection of two cut surfaces in reference to the horizontal plane.

Example Three views of a part are shown in Figure 72-1. A pictorial view of the angular portion of the part with auxiliary lines required for computations is also shown. The surfaces are to be machined in reference to the horizontal plane at angles of 32°00' and 40°00' as shown in the front and right side views.

a. Compute ∠R.

b. Compute ∠C.

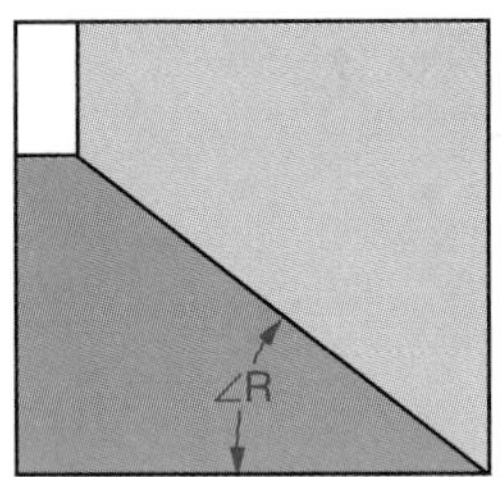

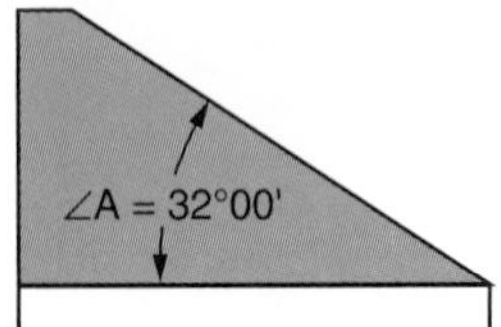

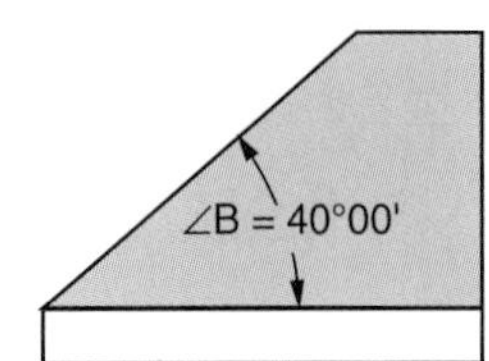

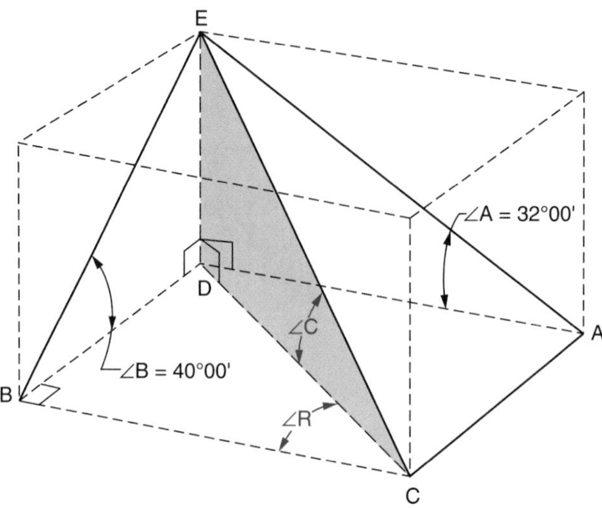

Figure 72-1

Since DE is a side of right △BDE, right △ADE, and right △CDE, make DE = 1.

a. In right △BDE, DE = 1, ∠B = 40°00'.
Compute BD:

$$\tan 40°00' = \frac{DE}{BD}$$
$$\tan 40°00' = \frac{1}{BD}$$
$$BD = \frac{1}{\tan 40°00'}$$

BD = 1 [÷] [tan] 40 [=] 1.191753593
BD ≈ 1.19175

In right △ADE, DE = 1, ∠A = 32°.
Compute DA:

$$\tan 32°00' = \frac{DE}{DA}$$
$$\tan 32°00' = \frac{1}{DA}$$
$$DA = \frac{1}{\tan 32°00'}$$

DA = 1 [÷] [tan] 32 [=] 1.600334529
DA ≈ 1.60033

In right △CBD, BD ≈ 1.19175, BC = DA ≈ 1.60033.
Compute ∠R:

$$\tan \angle R = \frac{BD}{BC}$$
$$\tan \angle R \approx \frac{1.19175}{1.60033}$$
$$\tan \angle R \approx 0.744690158$$

∠R ≈ [SHIFT] [tan⁻¹] .74469 [=] [SHIFT] [←] 36°40'29"

or ∠R ≈ [2nd] [tan⁻¹] .74469 [ENTER] [° ' "] [◀] [ENTER] [ENTER] 36°40'29"

∠R ≈ 36°40' Ans

b. In right $\triangle CBD$, $\angle R \approx 36°40'$, $BD \approx 1.1918$.

Compute DC:

$$\sin 36°40' = \frac{BD}{DC}$$

$$\sin 36°40' \approx \frac{1.1918}{DC}$$

$$DC \approx \frac{1.1918}{\sin 36°40'}$$

DC ≈ 1.1918 [÷] [sin] 36 [° ' "] 40 [° ' "] [=] 1.995784732

or DC ≈ 1.1918 [÷] [sin] 36 [° ' "] [ENTER] 40 [° ' "] [▶] [ENTER] [ENTER] 1.995784732

$DC \approx 1.9958$

In right $\triangle CDE$, $DE = 1$, $DC \approx 1.9958$.

Compute $\angle C$:

$$\tan \angle C = \frac{DE}{DC}$$

$$\tan \angle C = \frac{1}{1.9958} = 0.50105$$

∠C ≈ [SHIFT] [$\tan^{-1}$] 0.50105 [x^{-1}] [=] [SHIFT] [←] 26°36'47"

or ∠C ≈ [2nd] [$\tan^{-1}$] 0.50105 [x^{-1}] [ENTER] [° ' "] [◀] [ENTER] [ENTER] 26°36'47"

$\angle C \approx 26°37'$ Ans

Formulas for Computing Angles of Intersecting Angular Surfaces

Apply the formulas for intersecting angular surfaces only after a problem has been completely visualized and the previous method of solution is fully understood.

To use formulas for intersecting angular surfaces, $\angle A$ and $\angle B$ must be identified.

$\angle A$ is the given angle in the front view (frontal plane) in relation to the horizontal plane.

$\angle B$ is the given angle in the side view (profile plane) in relation to the horizontal plane.

Formula for ∠R in Relation to the Frontal Plane (Plane of ∠A) Used for Intersecting Angular Surfaces

$$\tan \angle R = \frac{\tan \angle A}{\tan \angle B}$$

Formula for ∠C Used for Intersecting Angular Surfaces

$$\cot \angle C = \sqrt{\cot^2 \angle A + \cot^2 \angle B}$$

In using the formula given for the angle of rotation, $\angle R$ must be determined in relation to the frontal plane (plane of $\angle A$). If $\angle R$ is to be determined in relation to the profile plane (plane of $\angle B$), the complement of the computed formula $\angle R$ must be used.

Example 1 Three conventional views and a pictorial view of the intersection of two angular surfaces are shown in Figure 72-2. (These are the same intersecting angular surfaces used in the previous example.)

a. Compute $\angle R$.

b. Compute $\angle C$.

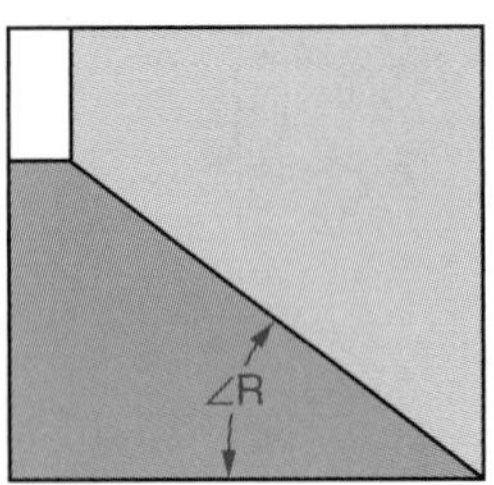

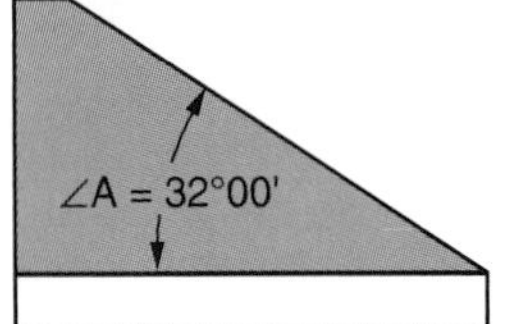

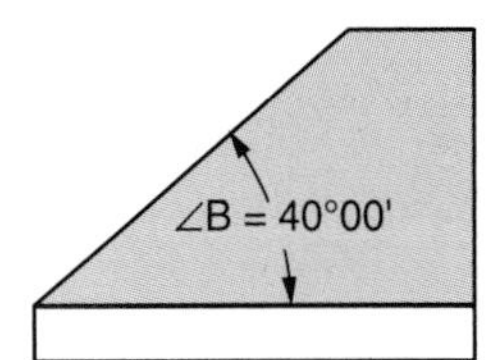

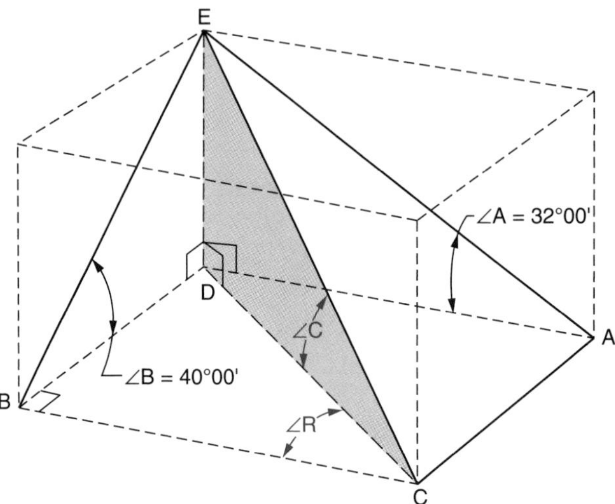

Figure 72-2

a. $\tan \angle R = \dfrac{\tan \angle A}{\tan \angle B}$

$\tan \angle R = \dfrac{\tan 32°00'}{\tan 40°00'}$

$\tan \angle R =$ [tan] 32 [)] [÷] [tan] 40 [)] [=] 0.744690295

$\angle R \approx$ [SHIFT] [tan⁻¹] .744690295 [=] [SHIFT] [←] 36°40'29"

or $\angle R \approx$ [2nd] [tan⁻¹] .744690295 [ENTER] [° ' "] ◀ [ENTER] [ENTER] 36°40'29"

$\angle R \approx 36°40'$ Ans

b. $\cot \angle C = \sqrt{\cot^2 \angle A + \cot^2 \angle B}$

$\cot \angle C = \sqrt{\cot^2 32°00' + \cot^2 40°00'}$

$\cot \angle C \approx$ [√] [(] [(] [tan] 32 [)] [x^{-1}] [x^2] [+] [(] [tan] 40 [)] [x^{-1}] [x^2] [)] [=] 1.995331359

$\angle C \approx$ [SHIFT] [tan⁻¹] 1.995331359 [x^{-1}] [=] [SHIFT] [←] 26°37'7"

or $\angle C \approx$ [2nd] [tan⁻¹] 1.995331359 [x^{-1}] [ENTER] [° ' "] ◀ [ENTER] [ENTER] 26°37'7" Ans

Observe that the values of $\angle R = 36°40'$ and $\angle C = 26°37'$ are the same as those computed in the previous example.

The same formulas for $\angle R$ and $\angle C$ may be used to compute an unknown front view angle, $\angle A$, when a side view angle, $\angle B$, and $\angle R$ or $\angle C$ are known. An unknown side view angle, $\angle B$, may be computed if the front view angle, $\angle A$, and $\angle R$ or $\angle C$ are known.

Example 2 Given: $\angle B = 35.50°$ and $\angle R = 28.30°$.

Compute: $\angle A$.

$$\tan \angle R = \frac{\tan \angle A}{\tan \angle B}$$
$$\tan 28.30° = \frac{\tan \angle A}{\tan 35.50°}$$
$$\tan \angle A = (\tan 28.30°)(\tan 35.50°)$$
$$\tan \angle A = (0.53844)(0.71329)$$
$$\tan \angle A = 0.38406$$
$$\angle A = 21.01° \quad \text{Ans}$$

Example 3 Given: $\angle A = 23°10'$ and $\angle C = 17°40'$.

Compute: $\angle B$.

$$\cot \angle C = \sqrt{\cot^2 \angle A + \cot^2 \angle B}$$
$$\cot 17°40' = \sqrt{\cot^2 23°10' + \cot^2 \angle B}$$
$$3.1397 = \sqrt{2.3369^2 + \cot^2 \angle B}$$
$$3.1397^2 = 2.3369^2 + \cot^2 \angle B$$
$$\cot \angle B = \sqrt{4.3966}$$
$$\cot \angle B = 2.0968$$
$$\angle B = 25°30' \quad \text{Ans}$$

Application

For problems 1 through 4, compute angles to nearer minute or hundredth degree.

Computing Angles Without Using Formulas for Intersecting Angular Surfaces

Three views of a part are shown in which surfaces are to be machined in reference to the horizontal plane at $\angle A$ and $\angle B$ as shown in the front and right side views. A pictorial view of the angular portion of the part with auxiliary lines required for computations is also shown. Do **not** use intersecting angular surface formulas in solving these problems. For each of the following problems, 1 through 4:

a. Compute $\angle R$.

b. Compute $\angle C$.

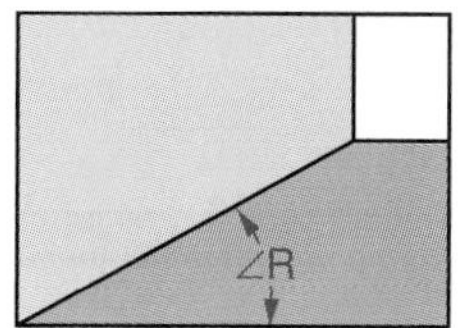

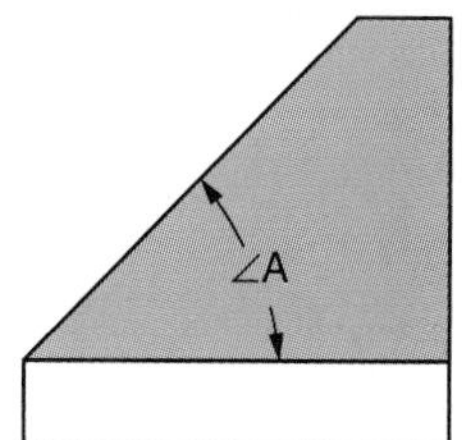

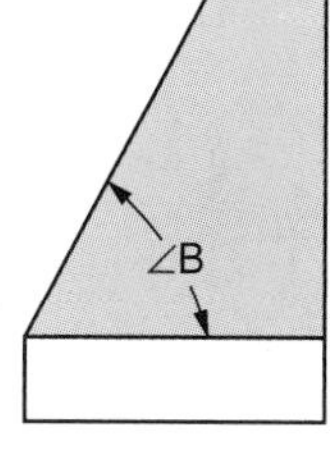

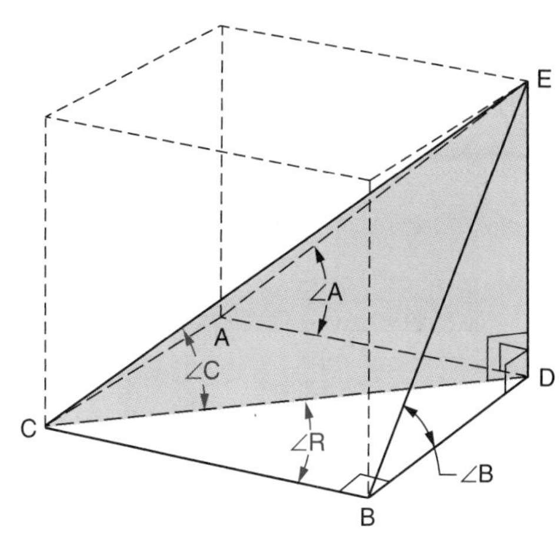

1. Given: $\angle A = 42°00'$
   $\angle B = 55°00'$

   a. ______________ b. ______________

2. Given: $\angle A = 40°00'$
   $\angle B = 48°00'$

   a. ______________ b. ______________

3. Given: $\angle A = 50°10'$
   $\angle B = 61°40'$

   a. ______________ b. ______________

4. Given: $\angle A = 43°35'$
   $\angle B = 52°70'$

   a. ______________ b. ______________

In each of problems 5 through 8, three views of a part are shown. Two surfaces are to be machined in reference to the horizontal plane at the angles shown in the front and right side views. Do **not** use intersecting angular surface formulas in solving these problems. For each problem:

a. Sketch and label a rectangular solid and the pyramid formed by the angular surface edges. Show the right triangle that contains $\angle C$ and the right triangles that contain the given angles and $\angle R$. Identify $\angle C$, $\angle R$, and the given angles.

b. Compute $\angle R$.

c. Compute $\angle C$.

5. a. (*sketch*)

   b. ______________ c. ______________

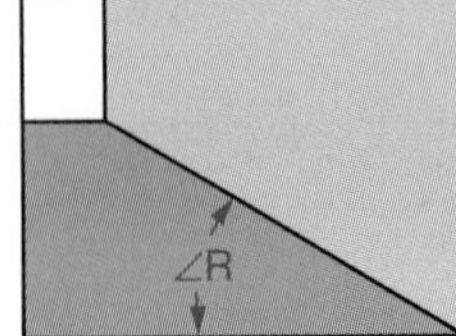

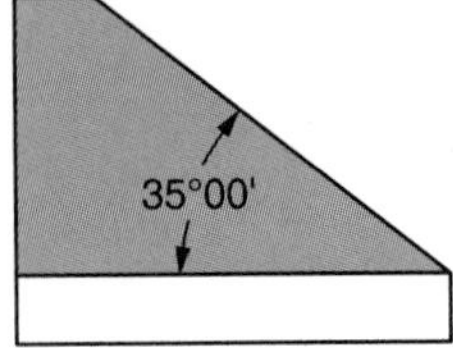

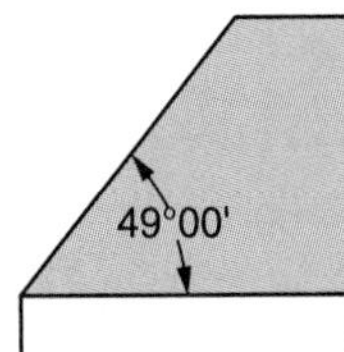

6. a. (*sketch*)

   b. ______________ c. ______________

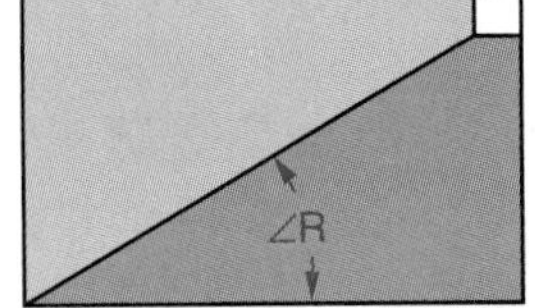

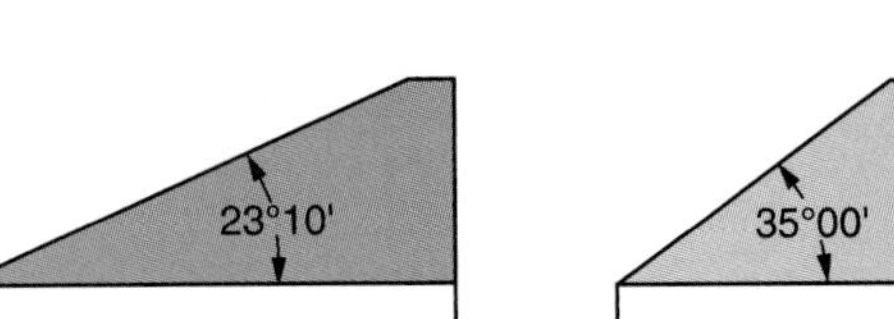

7. a. (*sketch*)

   b. ______________ c. ______________

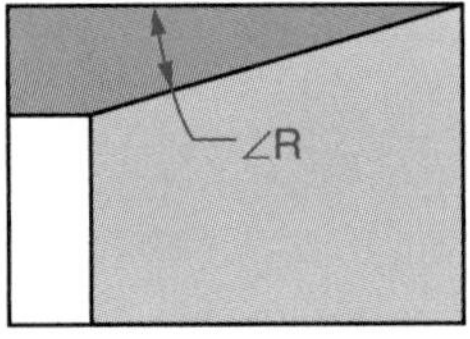

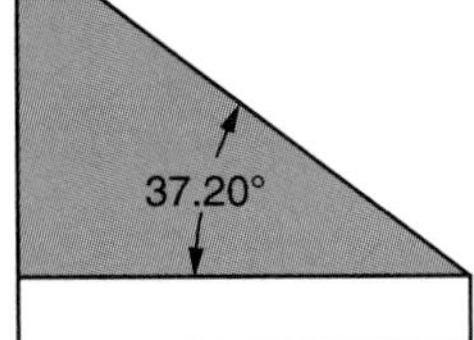

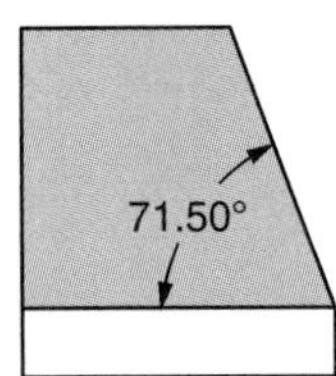

8. a. (*sketch*)

   b. ______________ c. ______________

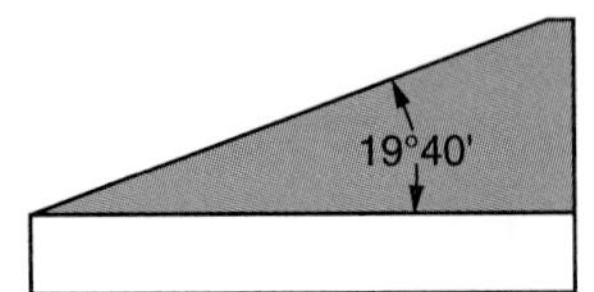

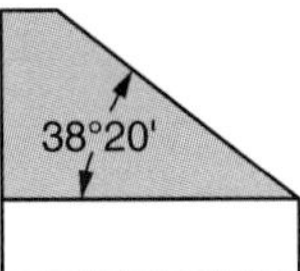

Computing Angles Using Formulas for Intersecting Angular Surfaces

In each of the following problems, three views of a part are shown. The angular surfaces are to be machined in reference to the horizontal plane at ∠A and ∠B as shown in the front and right side views. Compute the required angles using these formulas.

$$\tan \angle R = \frac{\tan \angle A}{\tan \angle B}$$

$$\cot \angle C = \sqrt{\cot^2 \angle A + \cot^2 \angle B}$$

Use this figure for #9, #10, and #11.

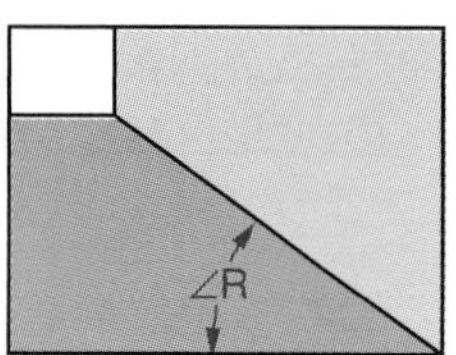

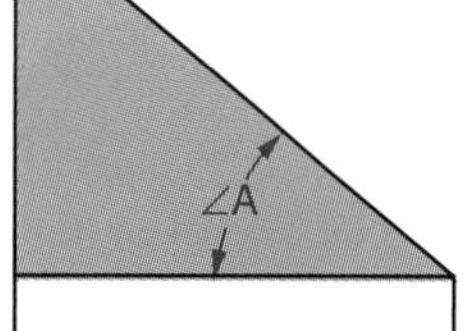

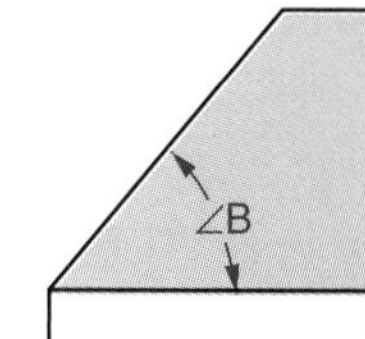

9. Given: ∠A = 36°00'
   ∠B = 43°50'
   a. Compute ∠R. ______________
   b. Compute ∠C. ______________
10. Given: ∠B = 48°10'
    ∠R = 40°00'
    a. Compute ∠A. ______________
    b. Compute ∠C. ______________
11. Given: ∠A = 31.60°
    ∠C = 28.00°
    a. Compute ∠B. ______________
    b. Compute ∠R. ______________

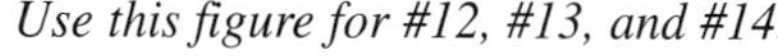
Use this figure for #12, #13, and #14.

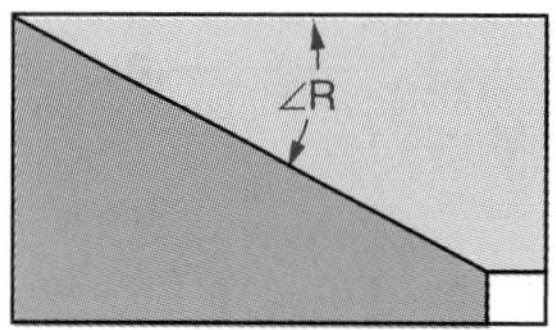

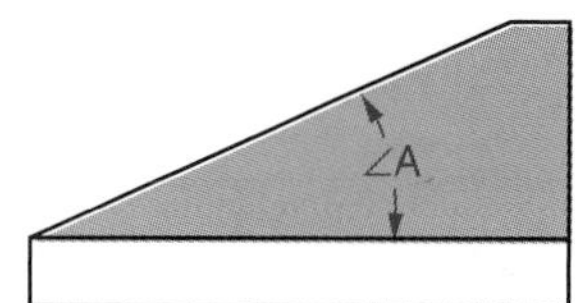

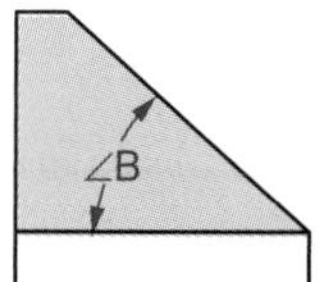

12. Given: ∠A = 17°40'
    ∠B = 25°00'
    a. Compute ∠R. ______________
    b. Compute ∠C. ______________
13. Given: ∠B = 31°00'
    ∠R = 27°50'
    a. Compute ∠A. ______________
    b. Compute ∠C. ______________
14. Given: ∠A = 16.40°
    ∠C = 14.00°
    a. Compute ∠B. ______________
    b. Compute ∠R. ______________

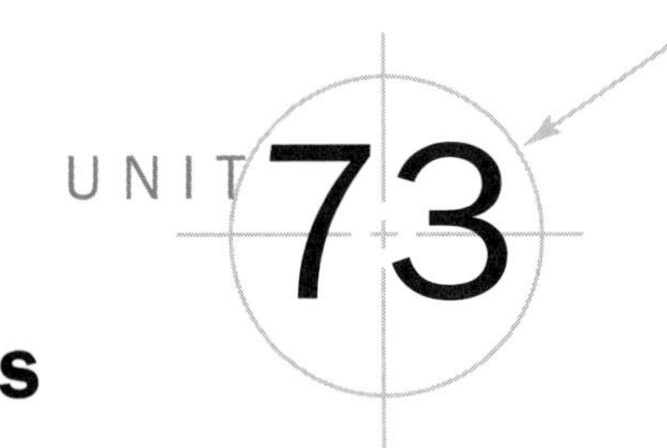

Computing Compound Angles on Cutting and Forming Tools

Objectives *After studying this unit you should be able to*

- Compute compound angles required for cutting and forming tools as given in rectangular solids.
- Sketch and label tool angular-surface-edge components within rectangular solids and compute true angles.
- Compute true angles by use of formulas.

Computing True Angles for Cutting and Forming Tools

The following examples show methods of computing compound angles that are often required in making die sections, cutting tools, and forming tools.

Example Three conventional views and a pictorial view of the angular portion of a tool are shown in Figure 73-1. Compute ∠C. The triangles in the conventional views have been labeled to help you identify them in the pictorial view.

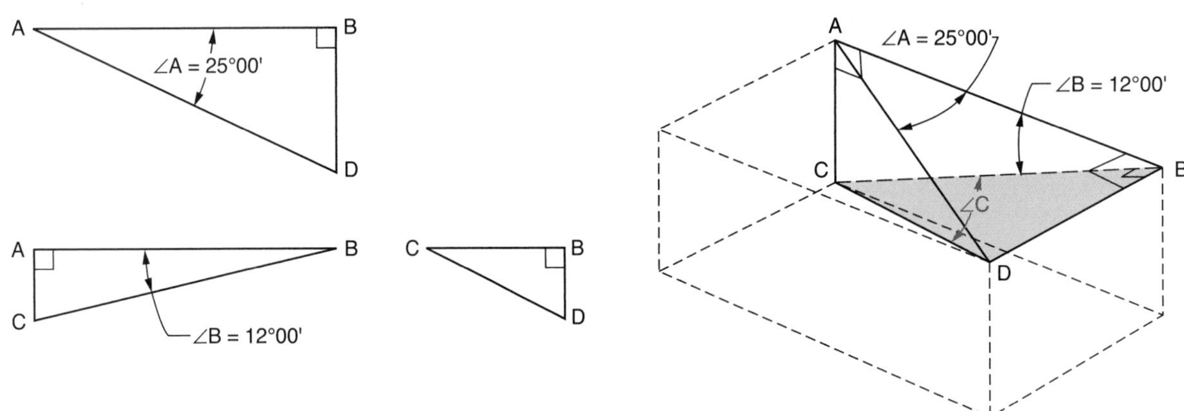

Figure 73-1

Since AB is a side of both right △ABD and right △CAB, which contain given angles of 25°00' and 12°00', make AB = 1.

In right △ABD, AB = 1, ∠A = 25°00'. Compute DB:

$$\tan 25°00' = \frac{DB}{AB}$$

$$\tan 25°00' = \frac{DB}{1}$$

DB ≈ [tan] 25 [=] 0.466307658

DB ≈ 0.46631

In right △CAB, AB = 1, ∠B = 12°00'. Compute CB:

$$\cos 12°00' = \frac{AB}{CB}$$

$$\cos 12°00' = \frac{1}{CB}$$

$$CB = \frac{1}{\cos 12°00'}$$

CB ≈ 1 [÷] [cos] 12 [=] 1.022340595

CB ≈ 1.0223

In right △CBD, DB ≈ 0.46631, CB ≈ 1.0223. Compute ∠C:

$$\tan \angle C = \frac{DB}{CB} \approx \frac{0.46631}{1.0223} \approx 0.45614$$

∠C ≈ [SHIFT] [tan⁻¹] .45614 [=] [SHIFT] [←] 24°31'11"

or ∠C ≈ [2nd] [tan⁻¹] .45614 [° ' "] ◀ [ENTER] [ENTER] 24°31'11"

∠C ≈ 24°31' Ans

Formula for Computing ∠C Used for Cutting and Forming Tools

Apply the formula for finding ∠C used for cutting and forming tools only after a problem has been completely visualized and the previous solution is fully understood.

To use the formula for cutting and forming tools, ∠A and ∠B must be identified.

∠A is the given angle in the top view (horizontal plane) in relation to the frontal plane.

∠B is the given angle in the front view (frontal plane) in relation to the horizontal plane.

In the front view, a right angle is made with either the left or right edge and the horizontal plane.

➤ **Formula for ∠C Used for Cutting and Forming Tools**

$$\tan \angle C = (\tan \angle A)(\cos \angle B)$$

Example 1 Three conventional views and a pictorial view of an angular portion of a tool are shown in Figure 73-2. Compute ∠C. (This is the same tool used in the previous example.)

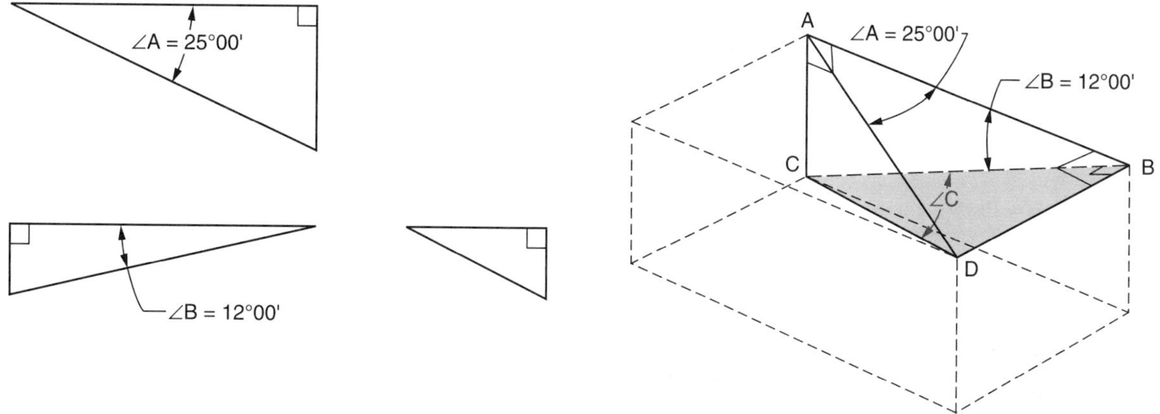

Figure 73-2

$\tan \angle C = (\tan \angle A)(\cos \angle B)$

$\tan \angle C = (\tan 25°00')(\cos 12°00')$

$\tan \angle C \approx$ [tan] 25 [×] [cos] 12 [=] 0.456117717

or $\tan \angle C \approx$ [(] [tan] 25 [)] [×] [cos] 12 [ENTER] 0.456117717

$\angle C \approx$ [SHIFT] [tan⁻¹] .456117717 [=] [SHIFT] [←] 24°31'7"

or $\angle C \approx$ [2nd] [tan⁻¹] .456117717 [° ' "] [◄] [ENTER] [ENTER] 24°31'7"

$\angle C \approx 24°31'$ Ans

Observe that the value of $\angle C = 24°31'$ is the same value as computed in the previous example.

The same formula may be used to compute an unknown top view angle, $\angle A$, when the front view angle, $\angle B$, and $\angle C$ are known. The front view angle, $\angle B$, may be computed when the top view angle, $\angle A$, and $\angle C$ are known.

Example 2 Given: $\angle B = 14.00°$, $\angle C = 28.75°$.
Compute: $\angle A$.

$$\tan \angle C = (\tan \angle A)(\cos \angle B)$$
$$\tan 28.75° = (\tan \angle A)(\cos 14.00°)$$
$$0.54862 = (\tan \angle A)(0.97030)$$
$$\tan \angle A = 0.56541$$
$$\angle A = 29.48° \quad \text{Ans}$$

Computing True Angles in Front-Clearance-Angle Applications

The following type of compound angle problem is often found in cutting tool situations where a front clearance angle is required, such as in thread cutting.

Example Three conventional views and a pictorial view of the angular portion of a cutting tool with a front clearance angle of 10°00' are shown in Figure 73-3. Compute $\angle C$.

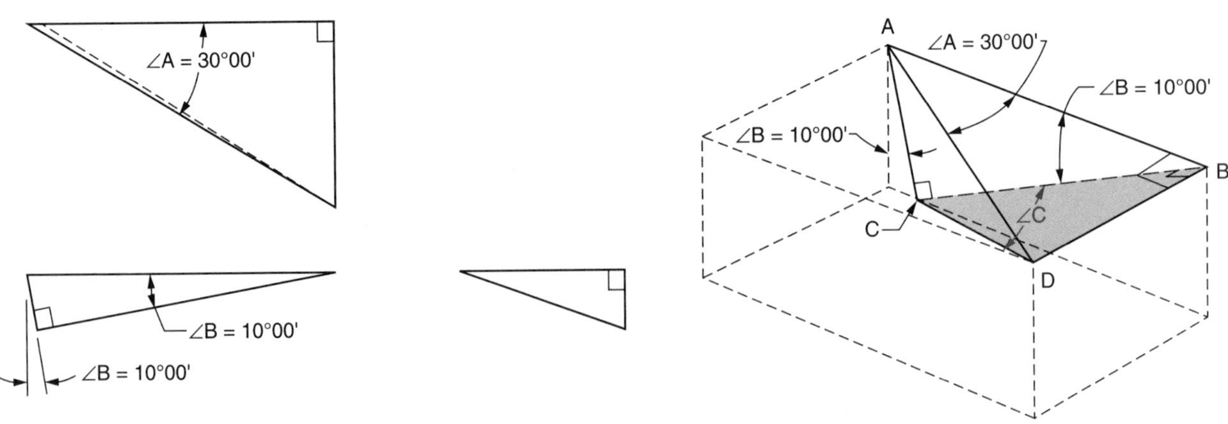

Figure 73-3

Since AB is a side of both right △ACB and right △ABD that contain given angles of 10°00' and 30°00', make AB = 1.

In right △ABD, AB = 1, ∠A = 30°00'. Compute DB:

$$\tan 30°00' = \frac{DB}{AB}$$

$$\tan 30°00' = \frac{DB}{1}$$

DB ≈ [tan] 30 [=] 0.577350269

DB ≈ 0.57735

In right △ACB, AB = 1, ∠B = 10°00'. Compute CB:

$$\cos 10°00' = \frac{CB}{AB}$$

$$\cos 10°00' = \frac{CB}{1}$$

CB ≈ [cos] 10 [=] 0.984807753

CB ≈ 0.98481

In right △CBD, DB = 0.57735, CB = 0.98481. Compute ∠C:

$$\tan \angle C = \frac{DB}{CB} \approx \frac{0.57735}{0.98481}$$

tan ∠C ≈ .57735 [÷] .98481 [=] 0.586255217

∠C ≈ [SHIFT] [tan⁻¹] .586255217 [=] [SHIFT] [←] 30°22'52"

or ∠C ≈ [2nd] [tan⁻¹] .586255217 [° ' "] [◄] [ENTER] [ENTER] 30°22'52"

∠C ≈ 30°23' Ans

Formula for Computing ∠C in Front-Clearance-Angle Applications

Apply the formula for finding ∠C in front-clearance-angle applications only after a problem has been completely visualized and the previous method of solution is fully understood.

To use the formula, ∠A and ∠B must be identified.

∠A is the given angle in the top view (horizontal plane) in relation to the frontal plane.

∠B is the given angle in the front view (frontal plane) in relation to the horizontal plane. ∠B is also the front clearance angle made with the vertical.

➤ **Formula for ∠C Used for Front-Clearance-Angle Applications**

$$\tan \angle C = \frac{\tan \angle A}{\cos \angle B}$$

Example Three conventional views and a pictorial view of the angular portion of a cutting tool with a front clearance angle of 10°00' are shown in Figure 73-4 on the next page. Compute ∠C. (This is the same front-clearance-angle application used on the previous example.)

$$\tan \angle C = \frac{\tan \angle A}{\cos \angle B} = \frac{\tan 30°00'}{\cos 10°00'}$$

tan ∠C ≈ [tan] 30 [÷] [cos] 10 [=] 0.58625683

or tan ∠C ≈ [(] [tan] 30 [)] [÷] [cos] 10 [ENTER] 0.58625683

∠C ≈ [SHIFT] [tan⁻¹] .58625683 [=] [SHIFT] [←] 30°22'53"

or ∠C ≈ [2nd] [tan⁻¹] .58625683 [° ' "] [◄] [ENTER] [ENTER] 30°22'53"

∠C ≈ 30°23' Ans

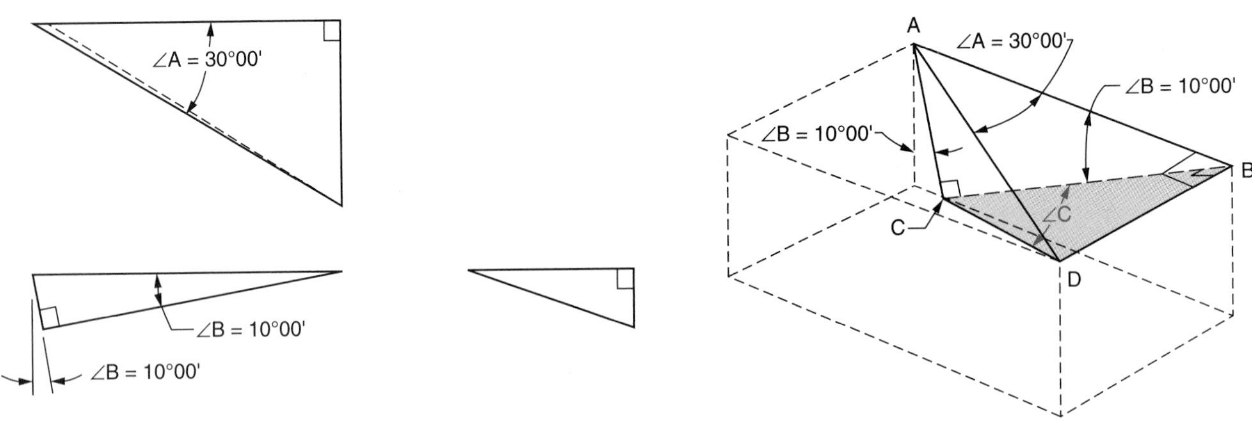

Figure 73-4

Observe that the value of ∠C = 30°23' is the same value as computed in the previous example.

Application

For problems 1 through 24, compute angles to the nearer minute or hundredth decimal degree.

Computing Angles Without Using Formulas for Cutting and Forming Tools

Three views of the angular portion of a tool are shown. A pictorial view with auxiliary lines forming the right triangles that are required for computations is also shown. Do **not** use cutting and forming tool formulas in solving these problems. For each of problems 1 through 4, compute ∠C.

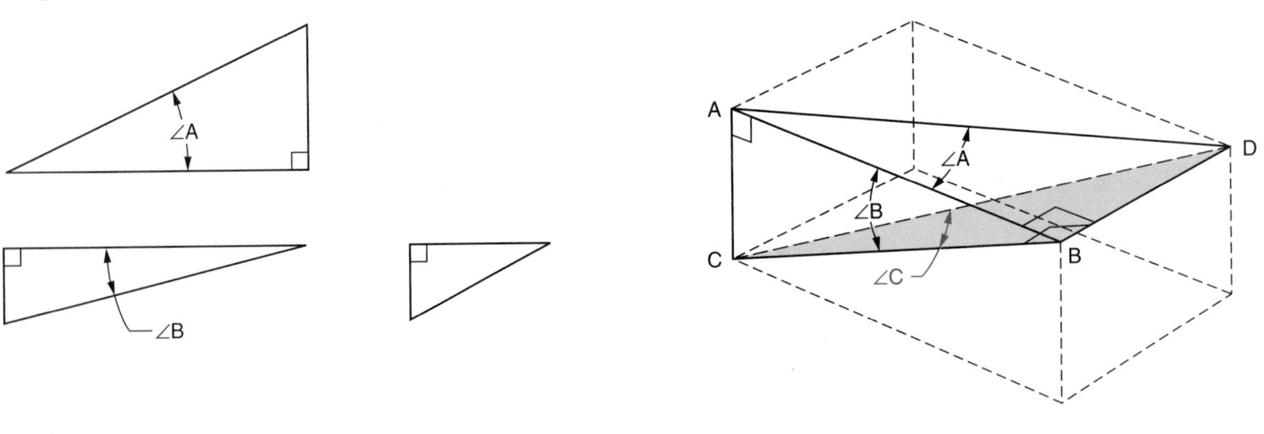

1. Given: ∠A = 30°00'
   ∠B = 15°00' __________

2. Given: ∠A = 26°00'
   ∠B = 12°00' __________

3. Given: ∠A = 27°00'
   ∠B = 11°00' __________

4. Given: ∠A = 30°00'
   ∠B = 24°00' __________

In each of the following problems, 5 through 8, three views of the angular portion of a tool are shown. Do **not** use cutting and forming tool formulas in solving these problems. For each problem:

a. Sketch and label a rectangular solid and the pyramid formed by the angular surface edges. Show the right triangles that contain ∠A, ∠B, and ∠C. Identify the angles.

b. Compute ∠C.

Use this figure for #5 and #6.

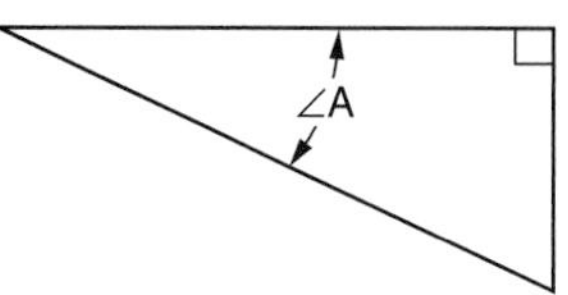

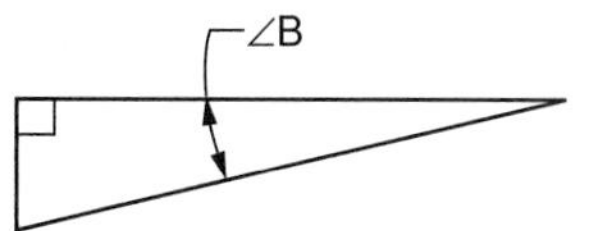

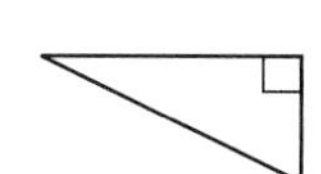

5. Given: ∠A = 24°00'
   ∠B = 15°00'

   a. (*sketch*) b. ______________

6. Given: ∠A = 20°00'
   ∠B = 8°00'

   a. (*sketch*) b. ______________

Use this figure for #7 and #8.

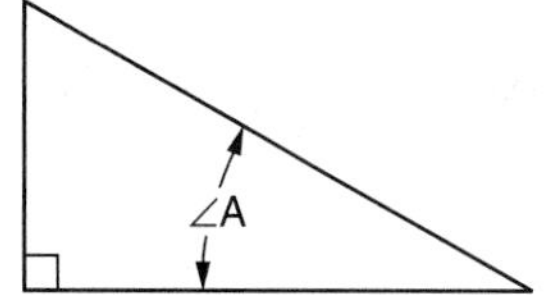

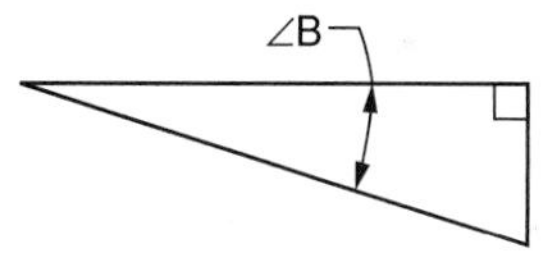

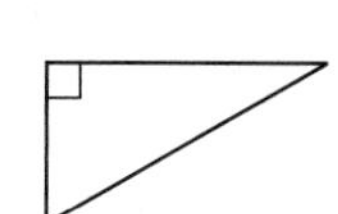

7. Given: ∠A = 30°00'
   ∠B = 12°00'

   a. (*sketch*) b. ______________

8. Given: ∠A = 23°00'
   ∠B = 10°00'

   a. (*sketch*) b. ______________

Computing Angles Using Formulas for Cutting and Forming Tools

For problems 9 through 12, compute the required angle using this formula.

tan ∠C = (tan ∠A)(cos ∠B)

Use this figure for #9 through #12.

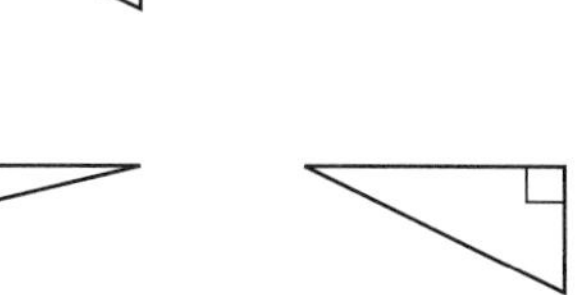

9. Given: ∠A = 33°00'
   ∠B = 14°00'

   Compute: ∠C. ______________

10. Given: ∠B = 10°00'
    ∠C = 28°30'

    Compute: ∠A. ______________

11. Given: ∠A = 26°00'
    ∠B = 14°00'

    Compute: ∠C. ______________

12. Given: ∠A = 28.00°
    ∠C = 27.20°

    Compute: ∠B. ______________

Computing Angles Without Using Front-Clearance-Application Formulas

Three views of the angular portion of a tool with front clearance are shown. A pictorial view with auxiliary lines forming the right triangles that are required for computations is also shown. Do **not** use front-clearance-application formulas for solving these problems. Compute ∠C for each of the following problems, 13 through 16.

13. Given: ∠A = 30°00'
∠B = 10°00' ____________

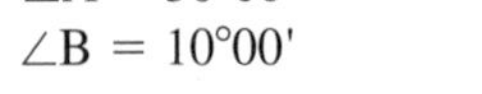

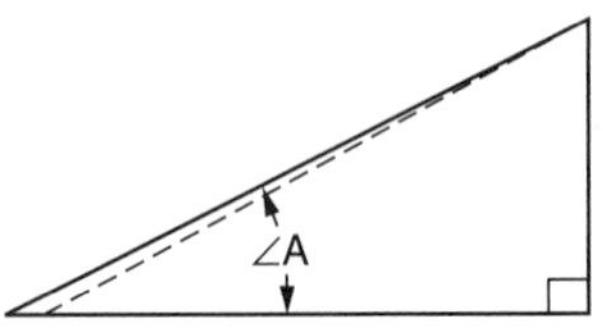

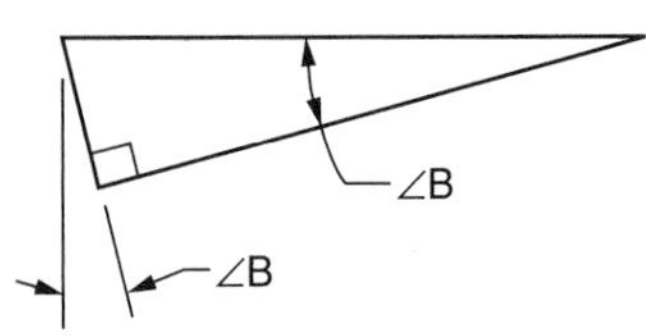

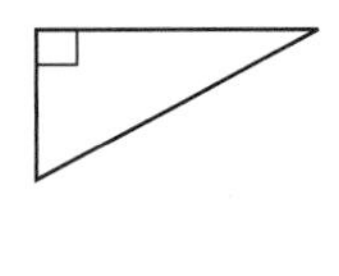

15. Given: ∠A = 33°00'
∠B = 11°00' ____________

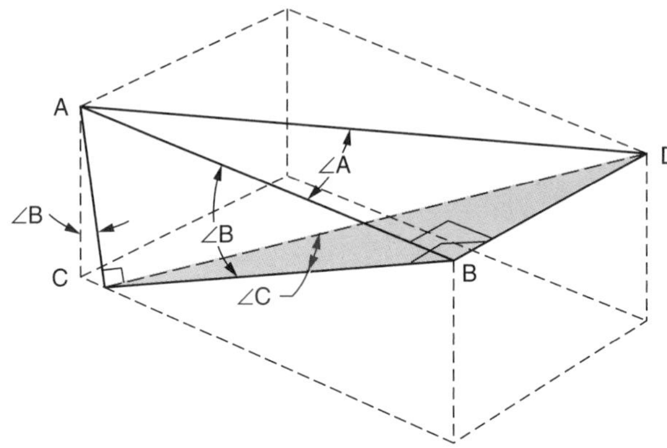

14. Given: ∠A = 28°00'
∠B = 15°00' ____________

16. Given: ∠A = 25°00'
∠B = 14°00' ____________

In each of the following problems 17 through 20, three views of the angular portion of a tool with front clearance are shown. Do **not** use front-clearance-application formulas for solving these problems. For each problem:

a. Sketch and label a rectangular solid and the pyramid formed by the angular surface edges. Show the right triangles that contain ∠A, ∠B, and ∠C. Identify the angles.

b. Compute ∠C.

17. Given: ∠A = 30°00'
∠B = 15°00'

a. (*sketch*) b. ____________

18. Given: ∠A = 38°00'
∠B = 12°00'

a. (*sketch*) b. ____________

Use this figure for #17 and #18.

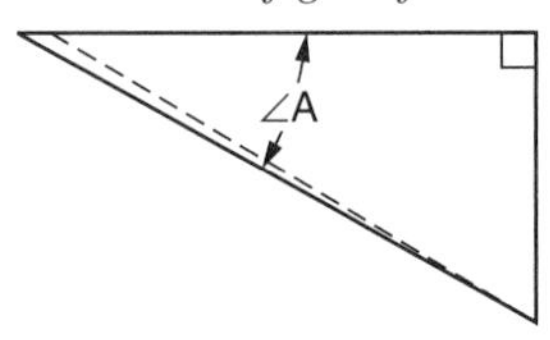

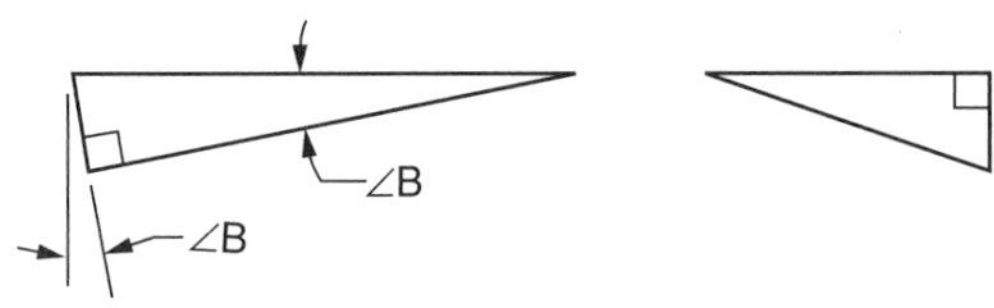

19. Given: ∠A = 32°00'
∠B = 15°00'

a. (*sketch*) b. ____________

20. Given: ∠A = 25°00'
∠B = 12°00'

a. (*sketch*) b. ____________

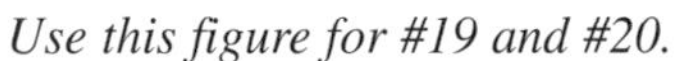

Use this figure for #19 and #20.

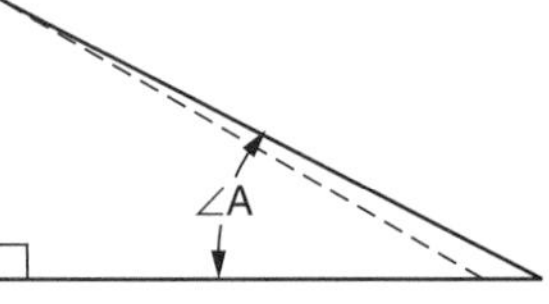

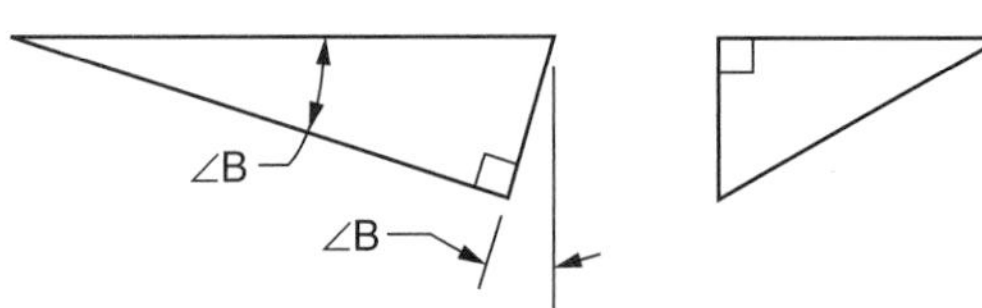

Computing Angles Using Front-Clearance-Application Formulas

For problems 21 through 24, compute the required angle using this formula.

$$\tan \angle C = \frac{\tan \angle A}{\cos \angle B}$$

Use this figure for #21 through #24.

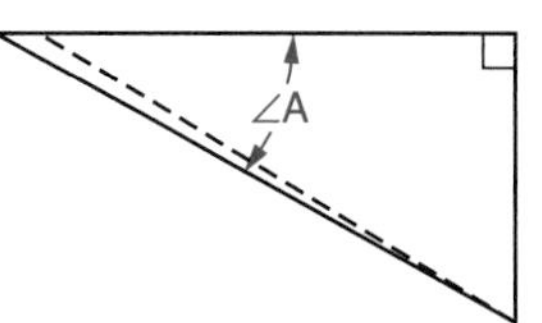

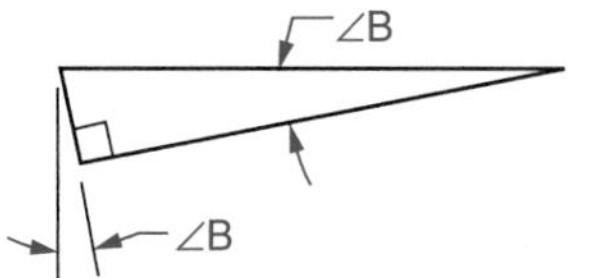

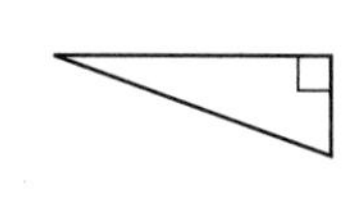

21. Given: ∠A = 25°00'
    ∠B = 9°00'
    Compute: ∠C. ______________

22. Given: ∠B = 15°00'
    ∠C = 30°40'
    Compute: ∠A. ______________

23. Given: ∠A = 34°00'
    ∠B = 8°00'
    Compute: ∠C. ______________

24. Given: ∠A = 28.00°
    ∠C = 28.60°
    Compute: ∠B. ______________

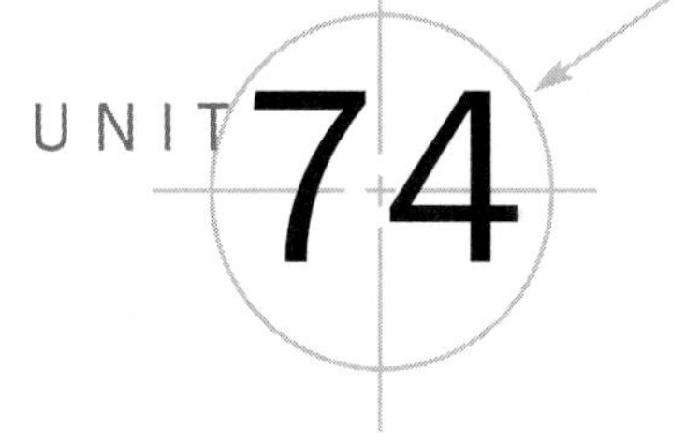

Achievement Review—Section Eight

Objective ***You should be able to solve the problems in this Achievement Review by applying the principles and methods covered in Units 68–73.***

For problems 1 through 6, compute angles to the nearer minute or hundredth degree.

1. Three views of a compound-angular hole are shown. All dimensions are in inches.
   a. Compute the angle of rotation, ∠R. ______________
   b. Compute the angle of tilt, ∠T. ______________

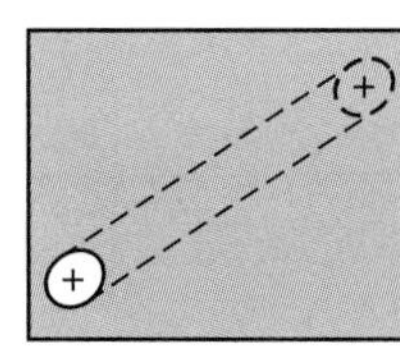

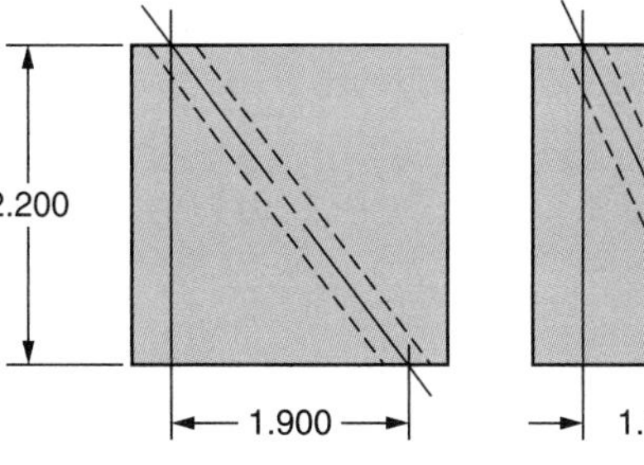

2. Three views of a compound-angular hole are shown.

   a. Compute the angle of rotation, ∠R. ______________

   b. Compute the angle of tilt, ∠T. ______________

37.30°

22.00°

3. Three views of a rectangular solid block are shown in which a compound-angular surface is to be machined.

   a. Compute the angle of rotation, ∠R. ______________

   b. Compute the angle of tilt, ∠T. ______________

44.50°

34.00°

4. Three views of a part are shown. Two surfaces are to be machined in reference to the horizontal plane at the angles shown in the front and right side views.

   a. Compute ∠R. ______________

   b. Compute ∠C. ______________

∠R

21°40'

35°00'

5. Three views of the angular portion of a tool are shown.

   Compute ∠C. ______________

28°40'

9°20'

6. Three views of the angular portion of a tool with front clearance are shown.

   Compute ∠C. ______________

27°00'

15°00'

15°00'

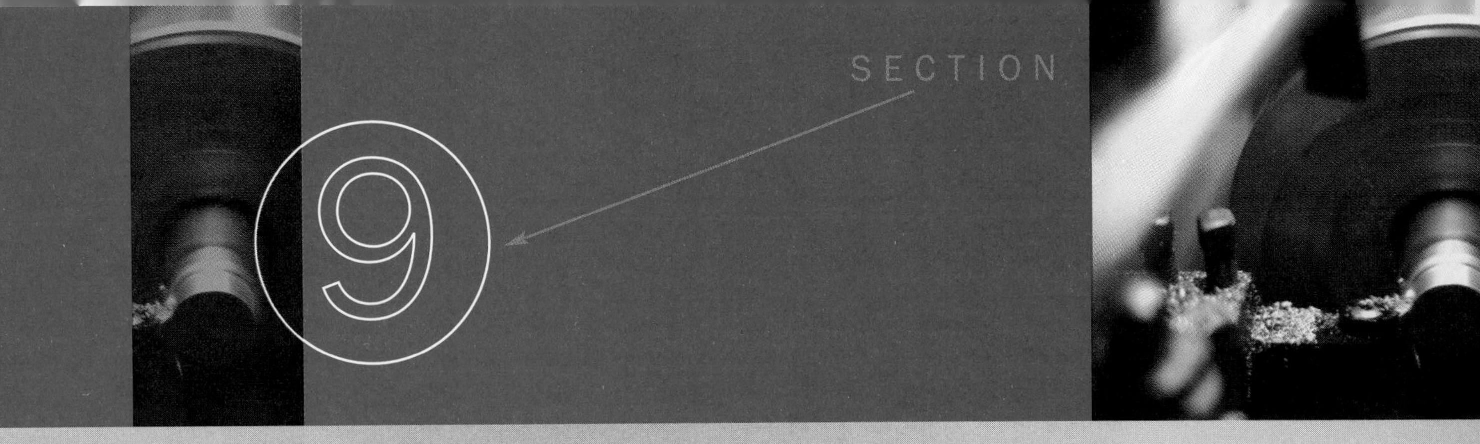

Computer Numerical Control (CNC)

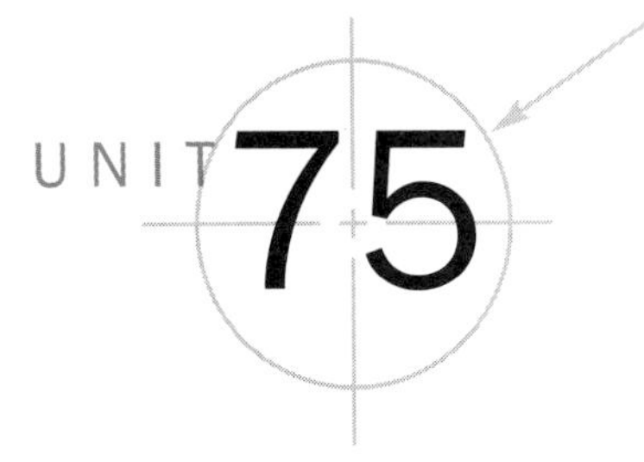

Introduction to Computer Numerical Control (CNC)

Objectives *After studying this unit you should be able to*

- Locate points in a two-axis Cartesian coordinate system.
- Plot points in a two-axis Cartesian coordinate system.
- Sketch point locations in a three-axis Cartesian coordinate system.

Numerical control is the operating of a machine using numerical commands. Computer numerical control machines (CNC) are widely used in the manufacture of machined parts. CNC machines have largely replaced manually operated machines and earlier numerical control (NC) machines. Although a machinist does not usually write a program of operations, some basics of numerical control program tool locating should be understood as it allows for basic troubleshooting.

CNC machines are designed for a wide range of applications. Before CNC machines, NC machine programs were usually coded on punched paper tape. The program had to be loaded into the machine control each time the program was run. Now, paper tape and magnetic tape are seldom used. Usually with CNC, the program is edited and stored in CNC control memory. Regardless of the method of instructing the CNC, the programs can be read from the control memory.

The most common type of CNC machines are machining centers and turning centers. A machining center is a large CNC milling machine with either a vertical or a horizontal spindle. It is capable of performing multiple operations with automatic tool changers. The machines usually have from three to five axes. A turning center is a large CNC lathe capable of performing multiple operations with automatic tool changers. They have from two to four axes. Some other types of CNC machines are grinding, flame cutting, inspection, and electrical discharge machines.

Programming

A program is a complete set of instructions for tool motion and for preparatory functions such as feed rate, type of operation, and mode of operation. Auxiliary operations such as tool changes, spindle control, and coolant control are also programmed. Before the program is written, the programmer selects the machine or machines to be used and determines the operations that will be done on a machine. The programmer determines how the part is going to be held, the tooling required, and the order of operations including calculating feeds and speeds. The programmer then writes the programs with or without the assistance of the computer. CNC machines are either manually programmed or computer-assisted programmed. In manual programming, the programmer makes the mathematical calculations required in writing the program. The CNC machine computer does not perform calculations or coding with manual programming.

With computer-assisted programming, a computer performs program calculations. Computer-assisted programming uses either language-based systems or graphic-based systems. With language-based systems, geometry and tool paths are described using a specific descriptive language. Graphic-based systems are menu-driven with the part and tool path drawn on the computer screen. Selecting from a menu, part geometry and tool paths are described. Cutting locations and offsets are calculated by the computer.

Graphic-based systems are easier to operate and are more economical than language-based systems. Programs are usually executed on personal computer-based systems. Because of relative ease of operations and cost, graphic-based systems are widely in use. Language-based systems are seldom used.

Location of Points: Two-Axis Cartesian Coordinate System

Programming is based on locating points within the Cartesian coordinate system, which is discussed in unit 65. In a plane, a point can be located from a fixed point by two dimensions. For example, a point can be located by stating that it is three units up and five units to the right of a fixed point. In machine technology applications, generally the units are either inches or millimeters. The Cartesian coordinate system gives point locations by using positive and negative values rather than locations stated as being up or down and left or right from a fixed point.

Figure 75-1 shows a two-axis Cartesian coordinate system with an x-axis and a y-axis. On a milling machine the x and y axes are perpendicular to the spindle. A point is located in reference to the origin by giving the point an x and y value. The x value is always given first. The x and y values are called the *coordinates* of the point. The following examples locate points in the Cartesian coordinate system shown.

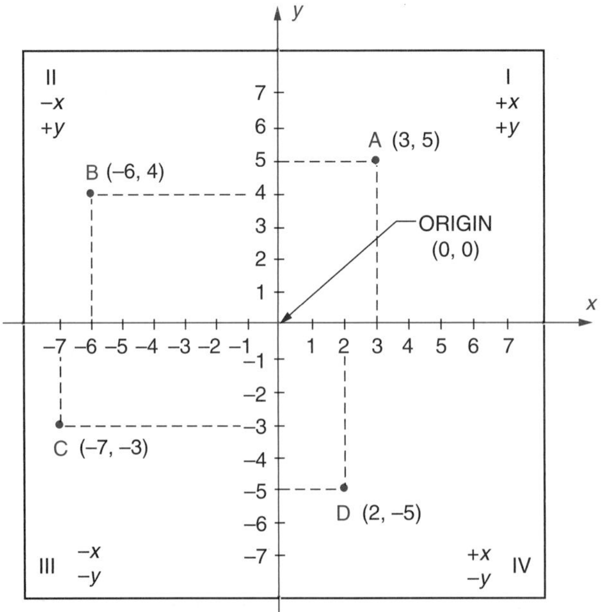

Two-Axis Cartesian Coordinate system

Figure 75-1

Example 1 Locate point A, which has coordinates of (3, 5).

The *x* value is +3 units and the *y* value is +5 units. Therefore, point A is located in Quadrant I.

Example 2 Locate point B, which has coordinates of (−6, 4).

The *x* value is −6 units and the *y* value is +4 units. Therefore, point B is located in Quadrant II.

Example 3 Locate point C, which has coordinates of (−7, −3).

The *x* value is −7 units and the *y* value is −3 units. Therefore, point C is located in Quadrant III.

Example 4 Locate point D, which has coordinates of (2, −5).

The *x* value is +2 units and the *y* value is −5 units. Therefore, point D is located in Quadrant IV.

Location of Points: Three-Axis Cartesian Coordinate System

Figure 75-2 shows a three-axis Cartesian coordinate system in which a point is considered to be in space. A point is located from a fixed point by three dimensions, *x, y,* and *z*. The *x*-axis and *y*-axis are identical to that of the two-axis coordinate system. The *z*-axis is perpendicular to the *x* and *y* axes. Most systems consider the *z* value as a positive value if it is in an upward direction from the origin. On a milling machine, the *z*-axis is parallel to the spindle; a *z*-location determines the depth of cut. On a lathe, two axes are used, the *x*-axis and *z*-axis. The *x*-axis is perpendicular to the spindle and determines part diameters. As with milling machines, the *z*-axis is parallel to the spindle and determines part lengths.

In the three-axis coordinate system, the *x* value is given first, the *y* value second, and the *z* value third. The *x, y,* and *z* values are the coordinates of the point. Figure 75-3 shows two points, point A (7, 5, 3) and point B (−6.8, −3.5, −2.2) in a three-axis coordinate system. This figure illustrates the *x, y,* and *z* locations on a vertical spindle milling machine. In machine technology applications, the *x, y,* and *z* units are usually either inches or millimeters.

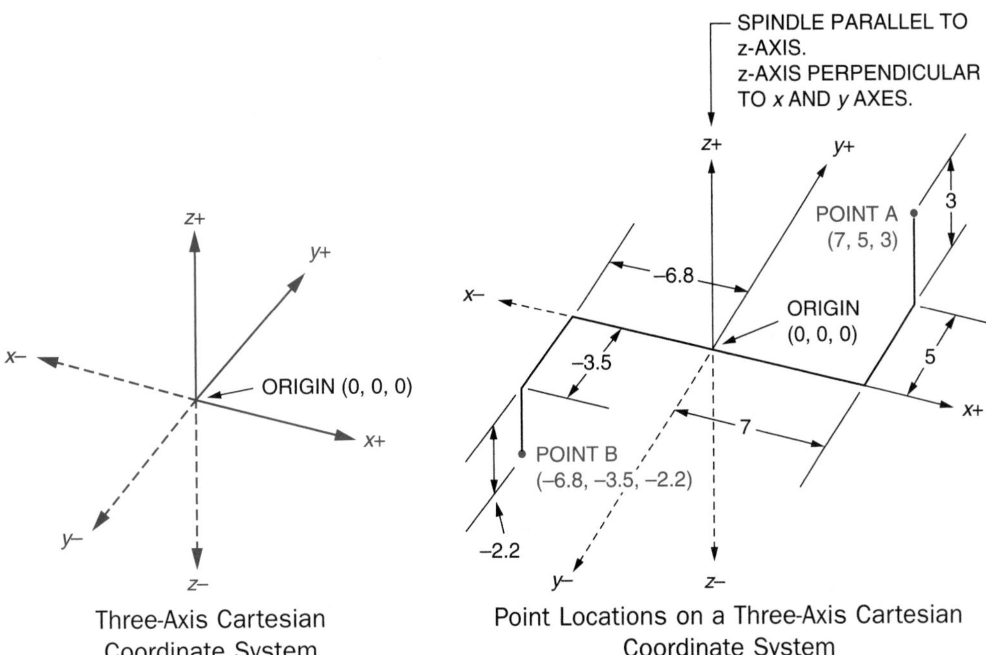

Three-Axis Cartesian Coordinate System
Figure 75-2

Point Locations on a Three-Axis Cartesian Coordinate System
Figure 75-3

Application

Plotting Points: Two-Axis Cartesian Coordinate System

1. Using graph paper, plot the following coordinates.

A = (−2, 5)	D = (0, 3)	G = (6, −8)	J = (0, 0)
B = (2, 8)	E = (−4, 0)	H = (−7, 5)	K = (−3, −4)
C = (−7, −2)	F = (−2, −2)	I = (−1, 0)	L = (9, −3)

2. Graph the following points: (−5, −5), (−3, −3), (0, 0), (2, 2), (4, 4), (7, 7). Connect these points.
   a. What kind of geometric figure is formed? ____________
   b. What is the value of the angle formed in reference to the *x*-axis? ____________
3. Graph the following points. Connect these points in the order that they are given. What kind of a geometric figure is formed? ____________

Point 1: (−9, −7)	Point 5: (4.5, 1)	Point 9: (2, 5)	Point 13: (−5, 1)
Point 2: (−6, −5.3)	Point 6: (7, 2.5)	Point 10: (0, 6)	Point 14: (−6.5, −2)
Point 3: (−3, −3.5)	Point 7: (6, 3)	Point 11: (−2, 7)	Point 15: (−8, −5)
Point 4: (1, −1)	Point 8: (4, 4)	Point 12: (−3, 5)	Point 16: (−9, −7)

Coordinates of Points: Two-Axis Cartesian Coordinate System

4. Refer to the points plotted on the Cartesian coordinate plane in Figure 75-4. Give coordinates for the following points.

A ____________
B ____________
C ____________
D ____________
E ____________
F ____________
G ____________
H ____________
I ____________
J ____________
K ____________
L ____________
M ____________
N ____________
O ____________
P ____________
Q ____________
R ____________

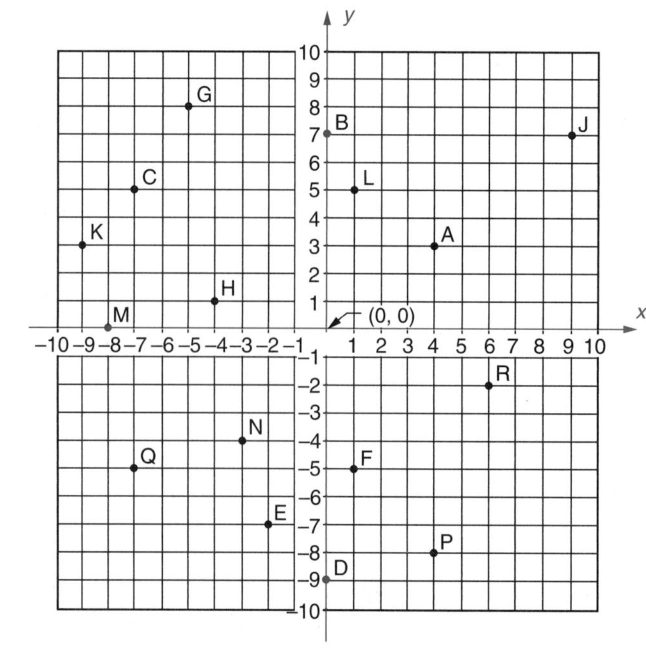

Figure 75-4

5. Refer to the points plotted on the Cartesian coordinate plane in Figure 75-5 Give coordinates for the following points.

A ______________
B ______________
C ______________
D ______________
E ______________
F ______________
G ______________
H ______________
I ______________
J ______________
K ______________
L ______________
M ______________
N ______________
O ______________
P ______________
Q ______________
R ______________

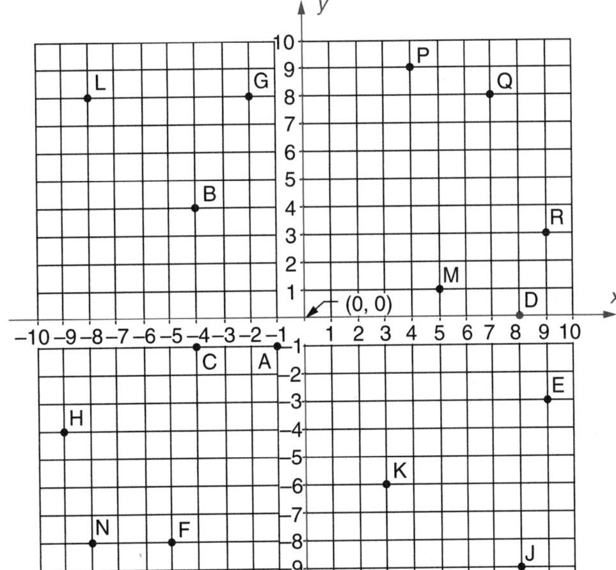

Figure 75-5

Sketching Point Locations: Three-Axis Cartesian Coordinate System

In problems 6 and 7, make sketches similar to the three-axis Cartesian coordinate system shown in Figure 75-3. For each of the given points show the approximate locations of the points (*x*, *y*, and *z* coordinates) as shown in Figure 75-3. The coordinates can be considered as either inch or millimeter units with approximate distances sketched to any scale.

6. Point A: (6, 3, 2)
   Point B: (−5, −2, 2)
   Point C: (−7, 5, 3)

7. Point D: (−5.5, 3.6, −1.8)
   Point E: (8.3, −7.6, 2.6)
   Point F: (6.4, 4.4, −3.8)

UNIT 76

Control Systems, Absolute Positioning, Incremental Positioning

Objective *After studying this unit you should be able to*

- Program position (dimension) from engineering drawings using point-to-point two-axis control systems. Both absolute and incremental positioning are applied.

Types of Systems

Some CNC machines are designed so they can be programmed for as many as six axes. On milling machines, the *x*-axis is the longest axis (most travel) perpendicular to the spindle. The *y*-axis is the shortest axis (least travel) perpendicular to the spindle. Movement parallel to the spindle is given in relation to the *z*-axis. Other axes involve rotation and tilting motions.

Programming

A program consists of many functions. The *x, y,* and *z* movements are only one function of a program. The program includes all functions required to machine a part. Using a definite format composed of all numbers and letters, a programmer writes commands such as sequencing of tools and cutting speeds and feeds. Preparatory functions are coded, such as modal commands specifying inch and metric units and absolute or incremental positioning. Miscellaneous functions such as tool changes, turning the spindle on and off, and calling for coolant are also coded.

An extensive study of CNC programming is required to write a complete manual or computer-assisted program. In this text, the purpose of presenting programming is to provide a very basic understanding of tool locations. The topic is limited to basic principles of location programming for vertical spindle milling machines controlling only *x*-axis and *y*-axis motions. The *z*-axis motions and all other functions are not considered.

Control systems are either continuous path or point-to-point. Machining centers are continuous path machines. They are capable of linear and circular interpolation. With linear interpolation the motions of two or more axes are coordinated with each other for angular milling cuts. Circular interpolation is the coordination of axes to give path in cutting an arc.

Some milling machines are point-to-point machines. They are usually restricted to drilling, boring, and nonangular milling. The continuous path system is more complex than the point-to-point system. Continuous path programming is also more complex than point-to-point programming. The part to be machined is positioned on the machine table. The movement of a CNC milling machine as it machines holes in a part is similar to conventional machinery. As previously stated, the purpose of the simple program applications in this book is to provide only a very basic understanding of programming tool locations. An in-depth presentation of programming is a study within itself and requires a textbook dealing exclusively with programming.

Tool Positioning (Coordinate) Systems

Most machine controls can operate with both incremental and absolute positioning commands. By means of a code, the program tells the control the type of positioning to be used. Most programs are written with absolute positioning. *Absolute positioning,* also called *absolute coordinates,* always directs the control where tool locations are relative to the origin (program zero point). *Incremental positioning,* also called *incremental coordinates,* always directs the control where the tool is located from the tool's immediate previous location.

Absolute Positioning (Absolute Coordinates)

Tool locations (coordinates) are given from a reference point called the origin or zero point. The origin is the point where the *x, y,* and *z* coordinates all are zero (0, 0, 0). The location of the origin is determined by the programmer. The origin is often located at the corner of a part or the center of a hole. To repeat, absolute positioning always directs the control where the tool locations are relative to the origin (program zero point). The following examples show point-to-point systems using absolute positioning with two axes on a vertical spindle milling machine.

Example 1 Figure 76-1 shows a part as it is location dimensioned on an engineering drawing before programming for CNC. All dimensions are in inches. The hole locations (x and y coordinates) are to be programmed using the absolute positioning system. The bottom left corner of the workpiece is established as the origin (0, 0).

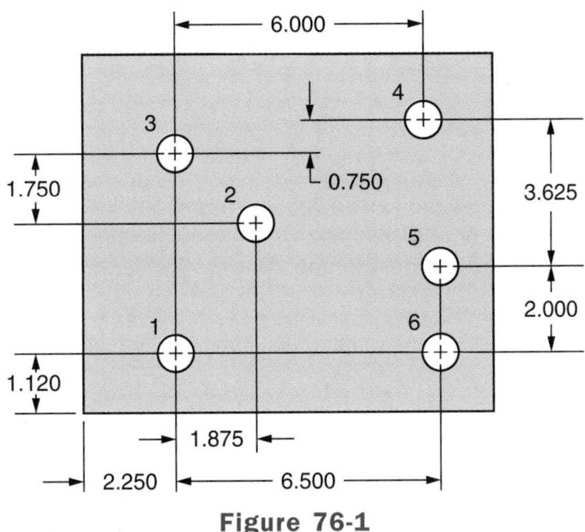

Figure 76-1

The position of the workpiece origin and the machine table are shown in Figure 76-2.

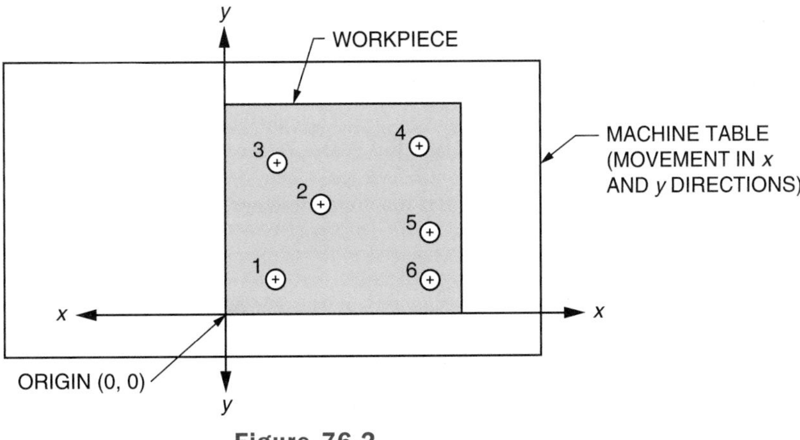

Figure 76-2

All the hole locations are programmed from the origin. Generally, the order in which the holes are machined is that which requires the least amount of machine movement. The coordinates of the hole locations from (0, 0) are listed in the following table.

Hole 1: $x = 2.250"$
$y = 1.120"$

Hole 2: $x = 2.250" + 1.875" = 4.125"$
$y = 1.120" + 2.000" + 3.625" - 0.750" - 1.750" = 4.245"$

Hole 3: $x = 2.250"$
$y = 4.245" + 1.750" = 5.995"$

Hole 4: $x = 2.250" + 6.000" = 8.250"$
$y = 5.995" + 0.750" = 6.745"$

Hole 5: $x = 2.250" + 6.500" = 8.750"$
$y = 1.120" + 2.000" = 3.120"$

Hole 6: $x = 8.750"$
$y = 1.120"$

Hole	*x*	*y*
1	2.250"	1.120"
2	4.125"	4.245"
3	2.250"	5.995"
4	8.250"	6.745"
5	8.750"	3.120"
6	8.750"	1.120"

Example 2 Figure 76-3 shows a part as it is location dimensioned before programming for CNC. All dimensions are in millimeters. The hole locations (*x* and *y* coordinates) are to be programmed using the absolute positioning system. The center of the bored hole is established as the origin (0, 0).

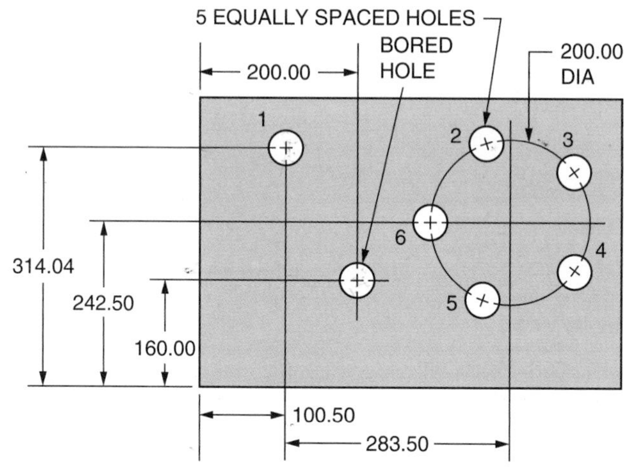

Figure 76-3

The position of the workpiece origin and the machine table are shown in Figure 76-4.

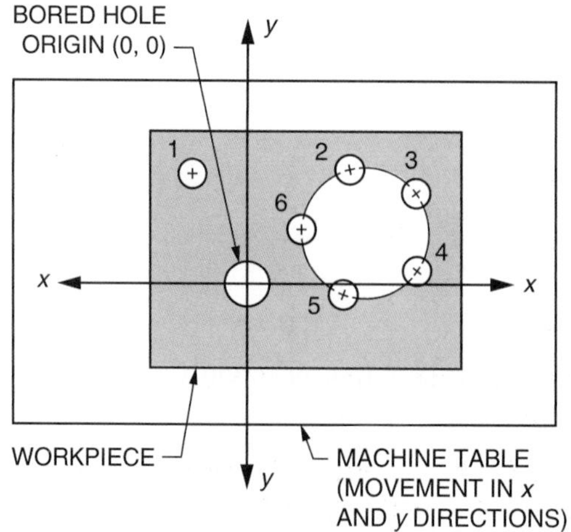

Figure 76-4

All hole locations are programmed from the origin. The coordinates of the hole locations from (0, 0) are listed in the table following Hole 6 on page 524.

Hole 1: $x = -200.00 \text{ mm} + 100.50 \text{ mm} = -99.50 \text{ mm}$
$y = 314.04 \text{ mm} - 160.00 \text{ mm} = 154.04 \text{ mm}$

For Hole 2, use Figure 76-5.

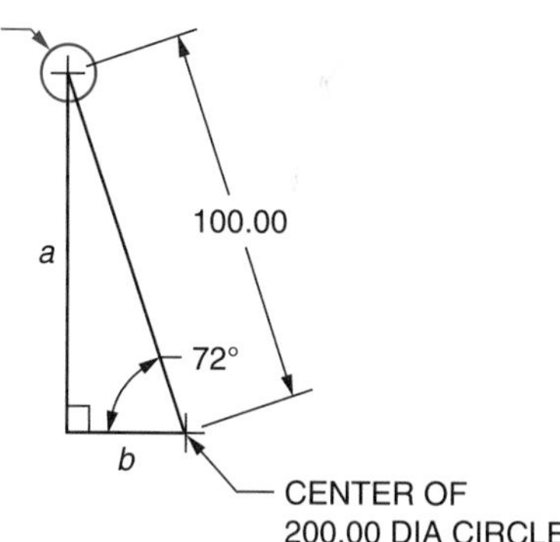

Figure 76-5

Hole 2: x = x distance to the center of the 200.00-mm diameter circle $- b$

y = y distance to the center of the 200.00-mm diameter circle $+ a$

Calculate the number of degrees between two consecutive holes on the 200.00-mm diameter circle.

$$\frac{360^\circ}{5} = 72^\circ$$

From the center of the 200.00-mm diameter circle, calculate a and b dimensions.

$$\sin 72^\circ = \frac{a}{100.00 \text{ mm}} \qquad \cos 72^\circ = \frac{b}{100.00 \text{ mm}}$$

$$0.95106 = \frac{a}{100.00 \text{ mm}} \qquad 0.30902 = \frac{b}{100.00 \text{ mm}}$$

$$a = 95.11 \text{ mm} \qquad b = 30.90 \text{ mm}$$

$x = -200.00 \text{ mm} + 100.50 \text{ mm} + 283.50 \text{ mm} - 30.90 \text{ mm}$
$= 153.10 \text{ mm}$
$y = 242.50 \text{ mm} - 160.00 \text{ mm} + 95.11 \text{ mm} = 177.61 \text{ mm}$

For Hole 3, use Figure 76-6.

Hole 3: x = x distance to the center of the 200.00-mm diameter circle $+ b$
y = y distance to the center of the 200.00-mm diameter circle $+ a$

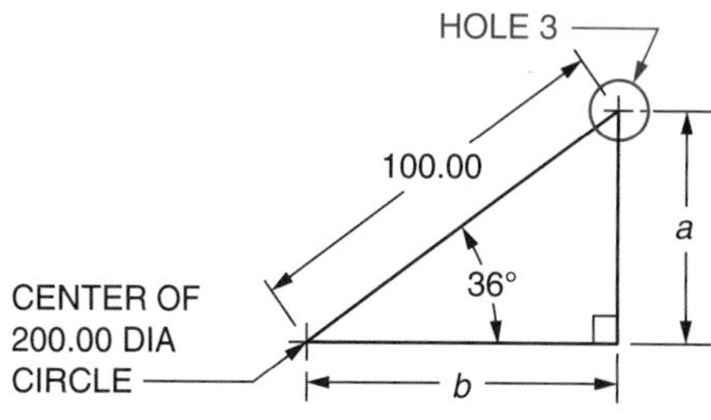

Figure 76-6

From the center of the 200.00-mm diameter circle calculate the angle formed by the horizontal centerline and Hole 3 (see Figure 76-6).

$$180^\circ - 2(72^\circ) = 36^\circ$$

Calculate a and b dimensions.

$$\sin 36^\circ = \frac{a}{100.00 \text{ mm}} \qquad \cos 36^\circ = \frac{b}{100.00 \text{ mm}}$$

$$0.58779 = \frac{a}{100.00 \text{ mm}} \qquad 0.80902 = \frac{b}{100.00 \text{ mm}}$$

$$a = 58.78 \text{ mm} \qquad b = 80.90 \text{ mm}$$

$x = -200.00 \text{ mm} + 100.50 \text{ mm} + 283.50 \text{ mm} + 80.90 \text{ mm}$
$= 264.90 \text{ mm}$

$y = 242.50 \text{ mm} - 160.00 \text{ mm} + 58.78 \text{ mm} = 141.28 \text{ mm}$

For Hole 4, use Figure 76-7.

Hole 4: $x = 264.90$ mm (the same as x of Hole 3)

y = y distance to the center of the 200.00-mm diameter circle $- a$

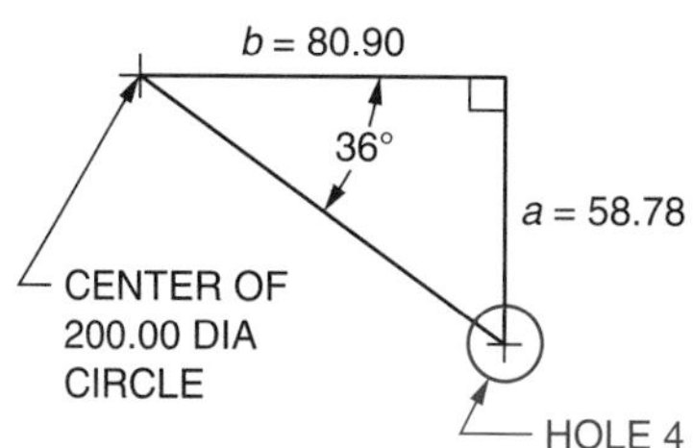

Figure 76-7

From the center of the 200.00-mm diameter circle, calculate the angle formed by the horizontal centerline and Hole 4 as in Figure 76-7.

$$3(72^\circ) - 180^\circ = 36^\circ$$

Since both Hole 4 and Hole 3 are projected 36° from the horizontal, the a and b dimensions of Hole 4 are the same as Hole 3.

x = 264.90 mm

y = 242.50 mm − 160.00 mm − 58.78 mm = 23.72 mm

For Hole 5, use Figure 76-8.

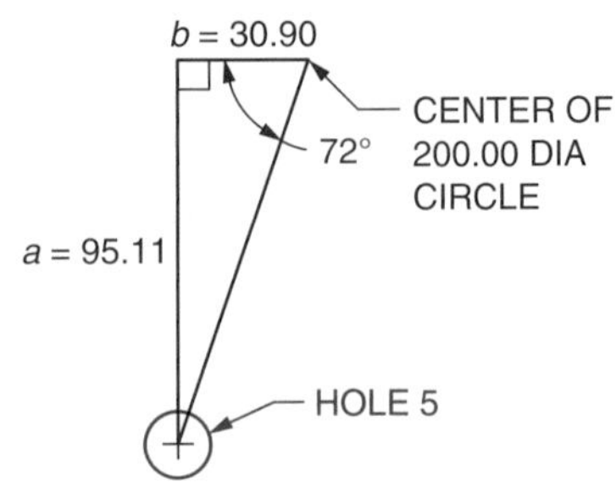

Figure 76-8

Hole 5: x = 153.10 mm (the same as x of Hole 2)

y = y distance to the center of the 200.00-mm diameter circle − a

Since both Hole 5 and Hole 2 are projected 72° from the horizontal, the a and b dimensions of Hole 5 (see Figure 76-8) are the same as Hole 2.

x = 153.10 mm

y = 242.50 mm − 160.00 mm − 95.11 mm = −12.61 mm

Hole 6: x = x distance to the center of the 200.00-mm diameter circle − 100.00 mm

x = −200.00 mm + 100.50 mm + 283.50 mm − 100.00 mm = 84.00 mm

y = 242.50 mm − 160.00 mm = 82.50 mm

This table lists the coordinates of the hole locations from (0, 0).

Hole	x	y
1	−99.50 mm	154.04 mm
2	153.10 mm	177.61 mm
3	264.90 mm	141.28 mm
4	264.90 mm	23.72 mm
5	153.10 mm	−12.61 mm
6	84.00 mm	82.50 mm

Incremental Positioning (Incremental Coordinates)

In incremental positioning, each location is given from the immediate previous location. The location of a hole is considered the origin (0, 0) of the x and y axes. From this origin, x and y distances are given to the next hole. Each new location in turn becomes the origin for the x and y distances to the next hole. The direction of travel, positive and negative, must be noted and is based upon the Cartesian coordinate system just as it was with absolute positioning. The first hole is located from the established first origin, while each subsequent hole is located from the hole directly preceding it. Each hole becomes the origin for the next hole to be machined.

Example Figure 76-9 shows a part as it is location dimensioned before programming for CNC. All dimensions are in inches. The hole locations (x and y coordinates) are to be programmed using the incremental positioning system. The bottom left edge is the established first origin (0,0). (This same part was used to illustrate absolute positioning).

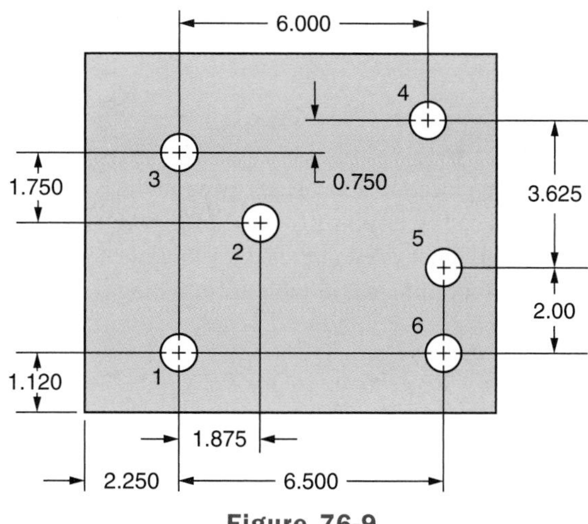

Figure 76-9

The position of the workpiece origin and the machine table are shown in Figure 76-10.

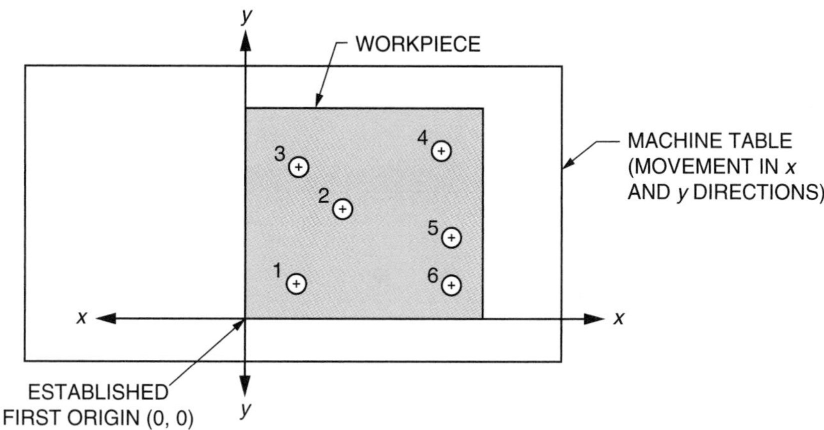

Figure 76-10

The first hole is located from the established first origin. Then each hole is the origin for the next hole to be machined. The x and y locations for Hole 1 are identical to those using absolute positioning. This is true for the first hole only. The following table lists the coordinates using incremental positioning.

Hole 1: $x = 2.250''$
$y = 1.120''$

Hole 2: $x = 1.875''$
$y = 2.000'' + 3.625'' - 0.750'' - 1.750'' = 3.125''$

Hole 3: $x = -1.875''$
$y = 1.750''$

Hole 4: $x = 6.000''$
$y = 0.750''$

Hole 5: $x = 6.500'' - 6.000'' = 0.500''$
$y = -3.625''$

Hole 6: $x = 0''$
$y = -2.000''$

Hole	x	y
1	2.250"	1.120"
2	1.875"	3.125"
3	−1.875"	1.750"
4	6.000"	0.750"
5	0.500"	−3.625"
6	0"	−2.000"

Application

Program Absolute and Incremental Positioning

Program the hole locations of the following part drawings. The location dimensions given in the tables are taken from drawings before programming for CNC. The origins (0, 0) used for programming are shown on the drawings. Use the hole location dimensions in the tables to write program hole locations. Write the program hole locations (coordinates) in table form listing the holes in sequence similar to the tables in this unit using:

a. Absolute Positioning

b. Incremental Positioning

Use this figure for #1, #2, and #3.

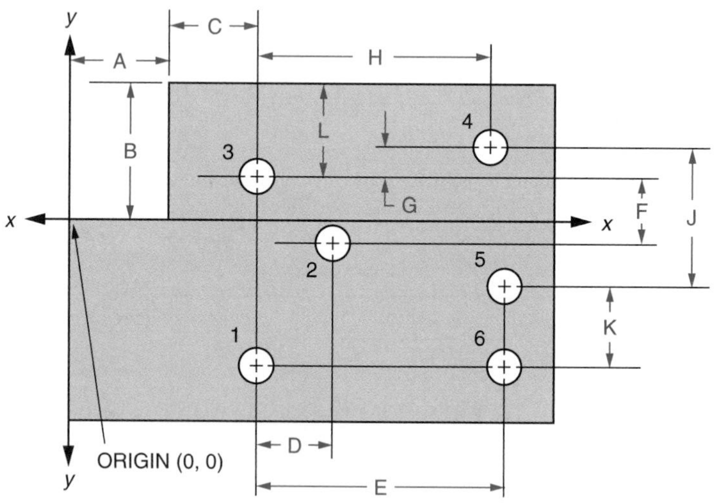

	LOCATION DIMENSIONS										
	A	B	C	D	E	F	G	H	J	K	L
1*	4.000	5.230	3.400	3.100	9.850	2.460	1.230	9.230	5.540	3.080	3.690
2*	6.000	7.846	5.077	4.615	14.769	3.692	1.846	13.846	8.308	4.615	5.538
3**	100.00	130.77	84.62	76.92	246.15	61.54	30.77	230.77	138.46	76.92	92.31

* All dimensions are in inches.

** All dimensions are in millimeters.

Use this figure for #4, #5, and #6.

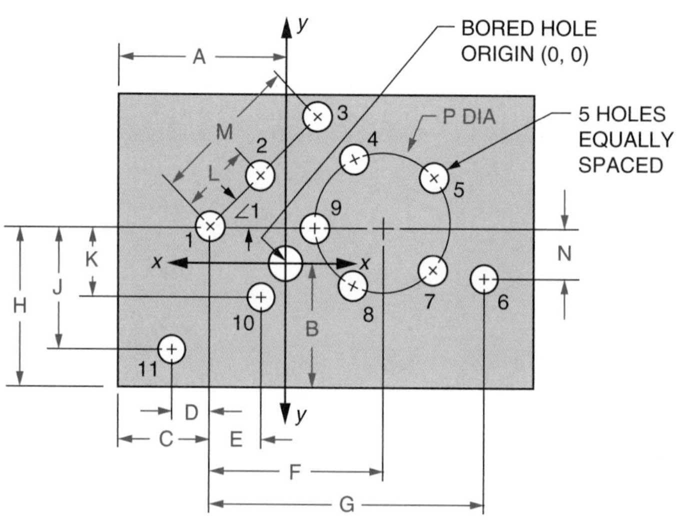

	LOCATION DIMENSIONS											
	A	**B**	**C**	**D**	**E**	**F**	**G**	**H**	**J**	**K**	**L**	**M**
4*	10.000	8.000	5.175	1.300	3.250	14.250	22.100	12.500	9.150	5.150	5.500	11.250
5*	10.000	8.000	5.250	1.412	3.562	14.400	22.250	12.750	9.375	5.270	5.600	11.300
6**	170.00	130.00	122.40	30.00	30.00	300.40	450.00	249.30	269.30	218.70	104.00	228.00

	N	**P DIA**	**∠1**
4*	4.625	10.000	42°0'
5*	4.850	10.200	43°0'
6**	95.10	196.00	41.75°

* All dimensions are in inches.

** All dimensions are in millimeters.

Use this figure for #7, #8, and #9.

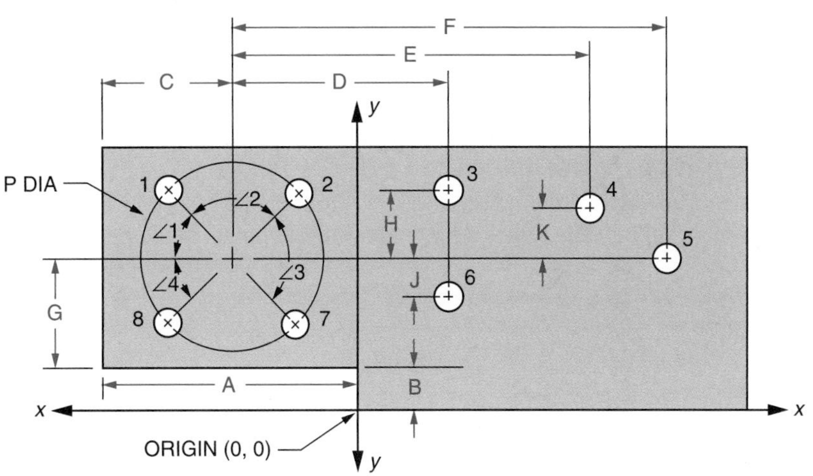

	LOCATION DIMENSIONS														
	A	**B**	**C**	**D**	**E**	**F**	**G**	**H**	**J**	**K**	**P DIA**	**∠1**	**∠2**	**∠3**	**∠4**
7*	18.000	5.000	10.185	13.700	19.215	26.750	7.500	5.750	3.170	4.250	12.200	75°45'	55°30'	95°15'	20°10'
8*	19.000	5.000	10.520	14.020	19.570	27.380	7.615	5.912	2.602	4.508	12.400	77°10'	57°15'	93°25'	15°0'
9*	400.00	80.00	196.30	255.00	378.34	521.40	142.50	106.80	59.50	79.68	224.00	72.67°	61.50°	98.33°	18.33°

* All dimensions are in inches.

** All dimensions are in millimeters.

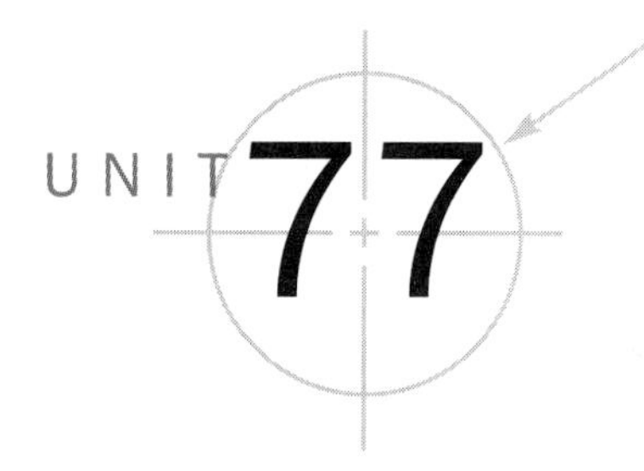

Location of Points: Polar Coordinate System

Objectives ***After studying this unit you should be able to***

- Locate points in a polar coordinate system.
- Plot points in a polar coordinate system.

While the Cartesian coordinate system is quite suitable for most CNC programs, there are times when it can become tedious and unproductive, because trigonometric calculations take too much time. Bolt hole patterns or any arrangement of holes along an arc benefit from another method for locating points: the polar coordinate system.

Location of Points: Polar Coordinate System

Each point in the *polar coordinate system* in located using the *center point* or *center of rotation*. As with the Cartesian coordinate system, two numbers are needed to locate each point. A point is located in reference to the center point by giving the point in r and θ (theta) values. The r value is always given first and represents the distance of the point from the center point. θ is an angle of the point, measured from the 0° axis (or 3 o'clock position), as indicated in Figure 77-1. The following examples locate points in the polar coordinate system.

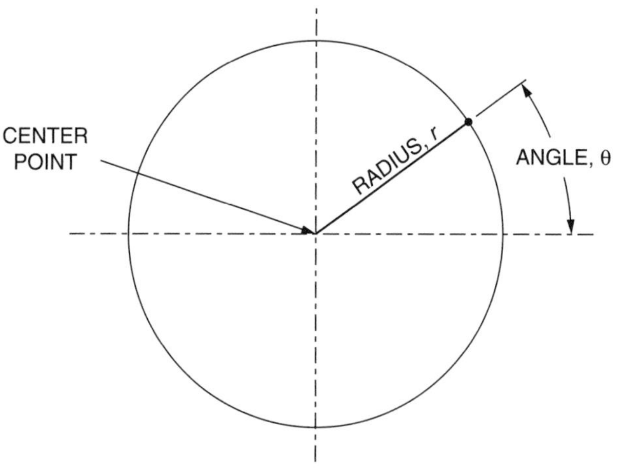

Figure 77-1

Examples

1. Locate point A, which has polar coordinates of (5, 30°)

   The r value is 5 units from the center point and rotated 30° from the 0° axis, as shown in Figure 77-2.

2. Locate point B, which has polar coordinates of (2, 135°).

   The r value is 2 units from the center point and rotated 135° from the 0° axis. Notice that while this is in Quadrant II of the Cartesian coordinate system, since r is the distance from the center point, there are no negative values.

3. Locate point C, which has polar coordinates of (3, 250°).
   The r value is 2 units from the center point and rotated 250° from the 0° axis.
4. Locate point D, which has polar coordinates of (1.5, 330°).
   The r value is 1.5 units from the center point and rotated 330° from the 0° axis.

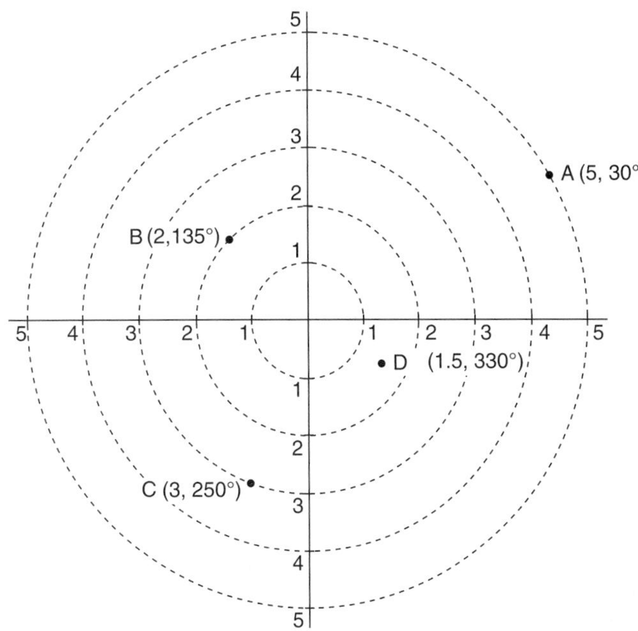

Figure 77-2

Tool locations (coordinates) with polar coordinates are given from the center point. The location of the center point is usually the last point programmed before the polar mode is activated. The following example gives the polar coordinates of the equally spaced holes shown in Figure 77-3. All dimensions are in millimeters.

There are six equally spaced holes, so each hole is $\frac{360°}{6} = 60°$ apart. Since hole 1 is at the 3 o'clock position, its angle coordinate is thus 0°. Each of the other holes will be 60° counterclockwise from the previous hole.

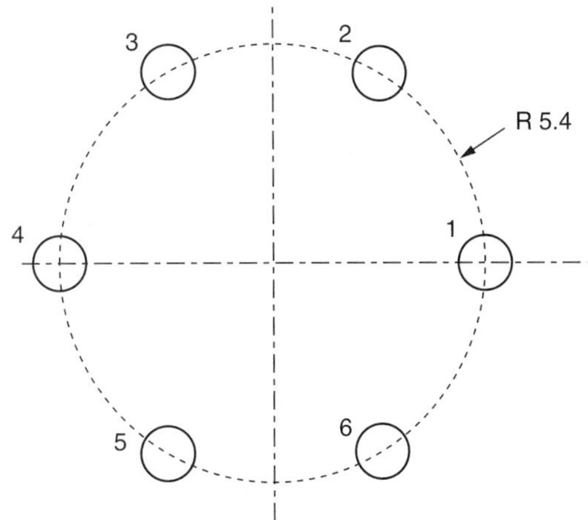

Figure 77-3

Hole	Coordinates	
	r	θ
1	5.4 mm	0°
2	5.4 mm	60°
3	5.4 mm	120°
4	5.4 mm	180°
5	5.4 mm	240°
6	5.4 mm	300°

Application

Identifying Points: Polar Coordinate System

1. Determine the polar coordinates of points A through H in Figure 77-4.

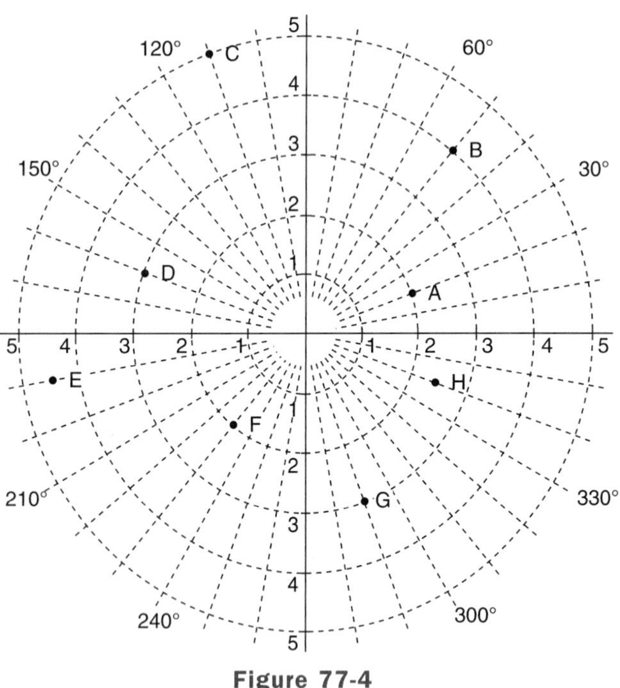

Figure 77-4

2. Determine the polar coordinates of points A through H in Figure 77-5.

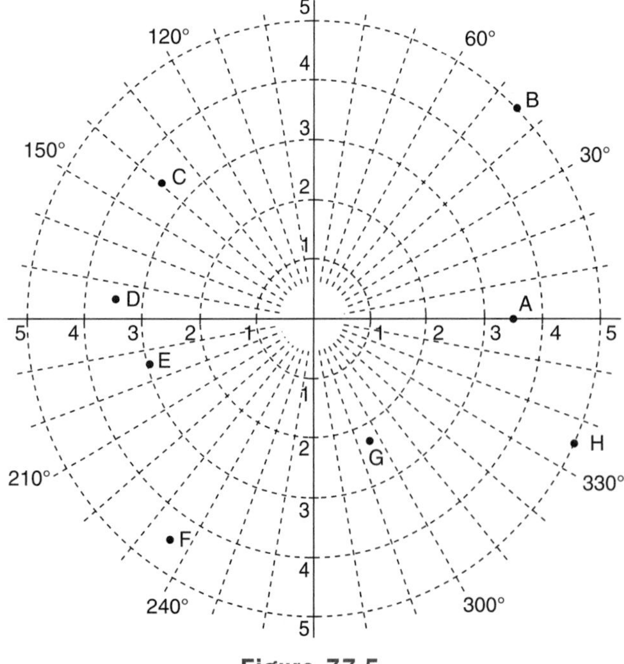

Figure 77-5

Plotting Points: Polar Coordinate System

3. Plot the following points on the polar coordinate graph paper in Figure 77-6.

A: (2, 10°)

B: (4, 70°)

C: (2.5, 120°)

D: (4.5, 215°)

E: (1.5, 255°)

F: (4.25, 295°)

G: (3.75, 345°)

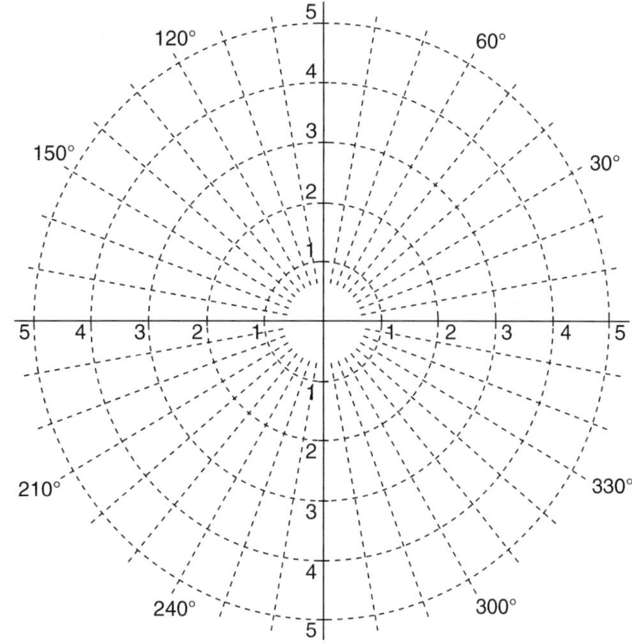

Figure 77-6

Applications: Polar Coordinate System

4. Figure 77-7 shows eight equally spaced holes on a bolt circle circumference. Give the polar coordinates of each of the holes. All dimensions are in inches.

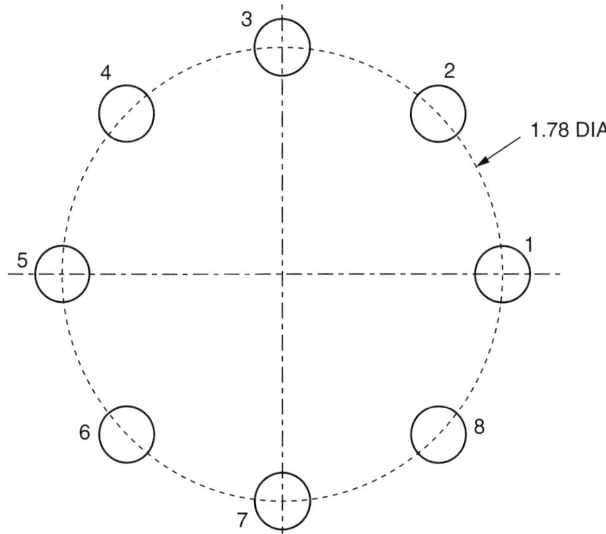

Figure 77-7

Hole	Coordinates	
	r	θ
1		
2		
3		
4		
5		
6		
7		
8		

5. Figure 77-8 shows five equally spaced holes on a bolt circle circumference. Give the polar coordinates of each of the holes. All dimensions are in millimeters.

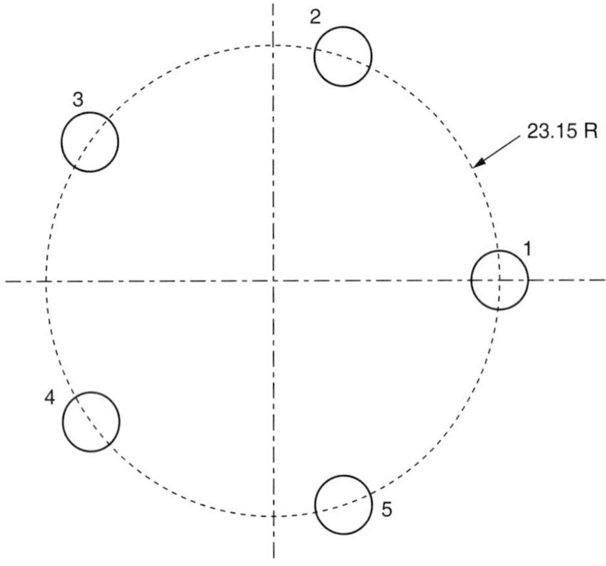

Figure 77-8

Hole	Coordinates	
	r	θ
1		
2		
3		
4		
5		

6. Figure 77-9 shows five equally spaced holes on a bolt circle circumference. Give the polar coordinates of each of the holes. All dimensions are in millimeters.

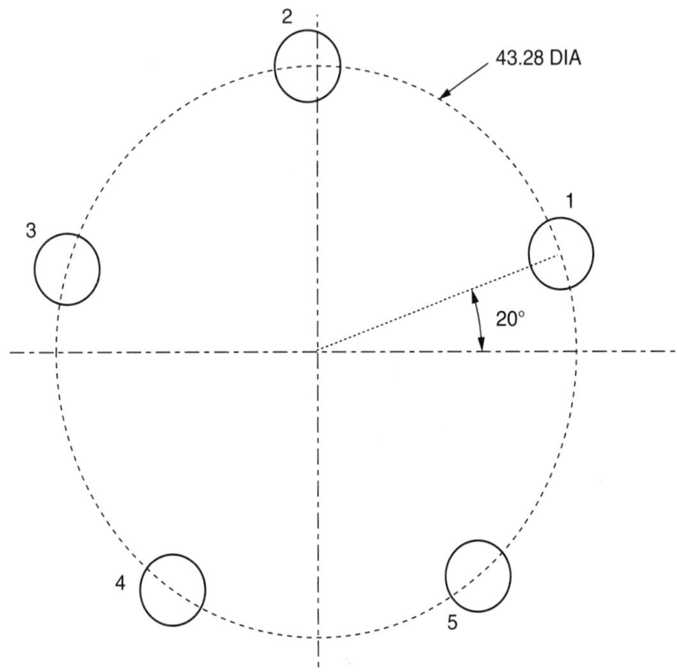

Figure 77-9

Hole	Coordinates	
	r	θ
1		
2		
3		
4		
5		

7. Figure 77-10 shows three equally spaced holes on a bolt circle circumference. Give the polar coordinates of each of the holes. All dimensions are in inches.

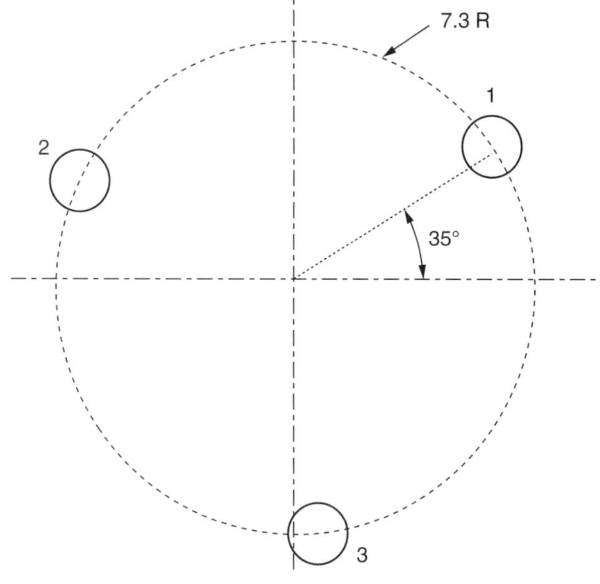

Figure 77-10

Hole	Coordinates	
	r	θ
1		
2		
3		

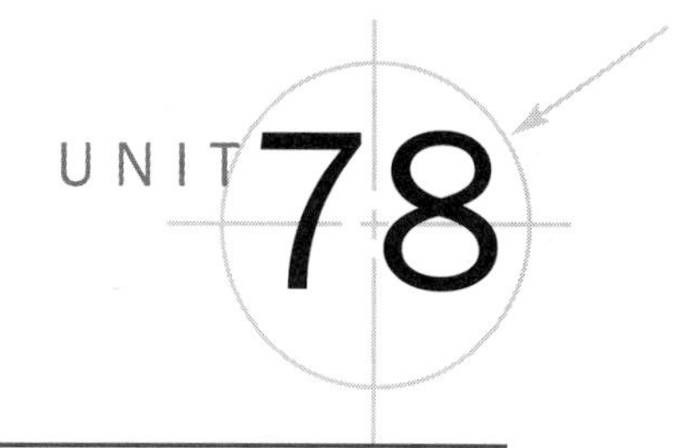

Binary Numeration System

Objectives ***After studying this unit you should be able to***

- Express binary numbers as decimal numbers.
- Express decimal numbers as binary numbers.

The mathematical system that uses only the digits 0 and 1 is called the *binary numeration system*. The two symbols, zero and one, are the base of any digital computer from personal computers to mainframe computers. The binary numeration system is fundamental to all electronic computers regardless of their size or purpose.

On early NC machines, program data was transferred to either punched paper tape or magnetic tape. Holes were punched in the paper tape in rows and columns. The tape was fed through a tape reader, which converted the tape codes to electrical signals. A hole punched in the tape (binary digit one) signaled an open circuit. The absence of a hole in the tape (binary digit zero) signaled a closed circuit. Part of a simplified binary-decimal system tape is shown in Figure 78-1 on page 537. Early NC control systems that used tape commands have been replaced by CNC controls. A few tape command NC control systems are still in use.

The computer of all CNC machines is based on the binary system as are computer aided drafting/design (CAD), computer aided manufacturing (CAM) and computer integrated manufacturing (CIM). The microscopic electronic switches in a computer's central processor assume only two states, ON (binary system one) or OFF (binary system zero). The switches are called transistors. If no charge is applied, current cannot flow and the transistor is OFF. If a positive charge is applied, the transistor is turned ON. An integrated circuit, usually a silicon chip, is made up of thousands of transistors.

The smallest unit of information in a computer, which is equivalent to a single zero or one, is call a *bit*, from *binary digit*. A group of eight bits is called a *byte*, while a *nibble* is half a byte, or four bits. The largest string of bits that a computer can handle in one operation is a *word* and the *word length* is the number of bits in a word.

Different computers use different word lengths. The original personal computers used 8 bits but most desktop computers today use 32 or 64 bits. Even though the kilo in its name indicates a kilobyte should have 1000 bytes, it has 1024. A *kilobyte* (KB) is 1024 (2^{10}) bytes. A *megabyte* (MB) is 1,048,576 or 2^{20} bytes and a *gigabyte* (GB) is 1024 megabytes or 1,073,741,824 bytes $= 2^{30}$ bytes.

Structure of the Decimal System

An understanding of the structure of the decimal system is helpful in discussing the binary system. The elements of a mathematical system are the base of the system, the particular digits used, and the locations of the digits with respect to the decimal point (place value). The decimal number system uses ten symbols called *digits* to form numbers. The digits in the decimal number system are 0, 1, 2, 3, 4, 5, 6, 7, 8, and 9.

The place values, and the place names, most often used with the decimal numbers, are shown in the table below. Notice that as we move from right to left, each place value is 10 times the value of the place to its immediate right.

Place Values of Decimal Numbers

Place name		Ten thousands	Thousands	Hundreds	Tens	Units or ones		Tenths	Hundredths	Thousandths	Ten thousandths	
Place value		10^4	10^3	10^2	10^1	10^0	.	10^{-1}	10^{-2}	10^{-3}	10^{-4}	
	. . .	10000	1000	100	10	1	.	0.1	0.01	0.001	0.0001	. . .
							Decimal point					

An analysis of the number 64,216 shows this structure.

6		4		2		1		6		**Number**
$10^4 = 10{,}000$		$10^3 = 1000$		$10^2 = 100$		$10^1 = 10$		$10^0 = 1$		**Place Value**
$6 \times 10^4 =$ $6 \times 10{,}000 =$ 60,000		$4 \times 10^3 =$ $4 \times 1000 =$ 4000		$2 \times 10^2 =$ $2 \times 100 =$ 200		$1 \times 10^1 =$ $1 \times 10 =$ 10		$6 \times 10^0 =$ $6 \times 1 =$ 6		**Value**
60,000	+	4000	+	200	+	10	+	6	=	64,216

Examples Analyze the following numbers.

1. $16 = 1(10^1) + 6(10^0) = 10 + 6$ Ans
2. $216 = 2(10^2) + 1(10^1) + 6(10^0) = 200 + 10 + 6$ Ans

3. $4216 = 4(10^3) + 2(10^2) + 1(10^1) + 6(10^0) = 4000 + 200 + 10 + 6$ Ans
4. $64{,}216 = 6(10^4) + 4(10^3) + 2(10^2) + 1(10^1) + 6(10^0)$
   $= 60{,}000 + 4000 + 200 + 10 + 6$ Ans

The same principles of structure hold true for numbers that are less than one. A number less than one can be expressed by using negative exponents. A number with a negative exponent is equal to its positive reciprocal. When the number is inverted and the negative exponent changed to a positive exponent, the result is as follows.

$$10^{-1} = \frac{1}{10^1} = 0.1$$

$$10^{-2} = \frac{1}{10^2} = \frac{1}{100} = 0.01$$

$$10^{-3} = \frac{1}{10^3} = \frac{1}{1000} = 0.001$$

$$10^{-4} = \frac{1}{10^4} = \frac{1}{10{,}000} = 0.0001$$

An analysis of the number 0.8502 shows this structure.

•8		5		0		2		Number
$10^{-1} = 0.1$		$10^{-2} = 0.01$		$10^{-3} = 0.001$		$10^{-4} = 0.0001$		**Place Value**
$8 \times 10^{-1} =$ $8 \times 0.1 =$ 0.8		$5 \times 10^{-2} =$ $5 \times 0.01 =$ 0.05		$0 \times 10^{-3} =$ $0 \times 0.001 =$ 0		$2 \times = 10^{-4}$ $2 \times .0001 =$ 0.0002		**Value**
0.8	+	0.05	+	0	+	0.0002	=	0.8502

Structure of the Binary System

The same principles of structure apply to the binary system as to the decimal system. The binary system is built upon the base 2 and uses only the digits 0 and 1. The binary system is built on the powers of the base 2; each place value is twice as large as the place value directly to its right.

Place Values of Binary Numbers

Place value	2^{10}	2^9	2^8	2^7	2^6	2^5	2^4	2^3	2^2	2^1	2^0	.	2^{-1}	2^{-2}	2^{-3}	2^{-4}
Decimal value	1024	512	256	128	64	32	16	8	4	2	1	.	0.5	0.25	0.125	0.0625

Binary point ↑

A binary number is usually written with a subscript of "2" or "two" so it is clear that it is not a decimal number. Thus, the binary number 11001101 should be written either as 11001101_2 or 11001101_{two}. Long binary numbers are often written in nibbles to make them easier to read. The number 11001101_2 would be written in nibbles as $1100\,1101_2$.

Expressing Binary Numbers as Decimal Numbers

Numbers in the decimal system are usually shown without a subscript. If a number is written without a subscript, it is understood the number is in the decimal system. In certain instances, for clarity, decimal numbers are shown with the subscript 10. The following examples show the method of expressing binary numbers as equivalent decimal numbers. Remember that 0 and 1 are the only digits in the binary system.

Examples Express each binary number as an equivalent decimal number.

1. $11_2 = 1(2^1) + 1(2^0) = 2 + 1 = 3_{10}$ Ans
2. $111_2 = 1(2^2) + 1(2^1) + 1(2^0) = 4 + 2 + 1 = 7_{10}$ Ans
3. $11101_2 = 1(2^4) + 1(2^3) + 1(2^2) + 0(2^1) + 1(2^0)$
   $= 16 + 8 + 4 + 0 + 1 = 29_{10}$ Ans
4. $101.11_2 = 1(2^2) + 0(2^1) + 1(2^0) + 1(2^{-1}) + 1(2^{-2})$
   $= 4 + 0 + 1 + 0.5 + 0.25 = 5.75_{10}$ Ans

Expressing Decimal Numbers as Binary Numbers

The following examples show the method of expressing decimal numbers as equivalent binary numbers.

Example 1 Express 25_{10} as an equivalent binary number.

Determine the largest power of 2 in 25; $2^4 = 16$. There is one 2^4. Subtract 16 from 25;

$$25 - 16 = 9.$$

Determine the largest power of 2 in 9; $2^3 = 8$. There is one 2^3. Subtract 8 from 9;

$$9 - 8 = 1.$$

Determine the largest power of 2 in 1; $2^0 = 1$. There is one 2^0. Subtract 1 from 1;

$$1 - 1 = 0.$$

There are no 2^2 and 2^1. The place positions for these values must be shown as zeros.

$$25_{10} = 1(2^4) + 1(2^3) + 0(2^2) + 0(2^1) + 1(2^0)$$
$$25_{10} = \quad 1 \qquad 1 \qquad 0 \qquad 0 \qquad 1$$
$$25_{10} = 11001_2 \quad \text{Ans}$$

Example 2 Express 11.625_{10} as an equivalent binary number.

$$2^3 = 8;\ 11.625 - 8 = 3.625$$
$$2^1 = 2;\ 3.625 - 2 = 1.625$$
$$2^0 = 1;\ 1.625 - 1 = 0.625$$
$$2^{-1} = 0.5;\ 0.625 - 0.5 = 0.125$$
$$2^{-3} = 0.125;\ 0.125 - 0.125 = 0$$

There are no 2^2 and 2^{-2}.

$$11.625_{10} = 1(2^3) + 0(2^2) + 1(2^1) + 1(2^0).+ 1(2^{-1}) + 0(2^{-2}) + 1(2^{-3})$$
$$11.625_{10} = \quad 1 \qquad 0 \qquad 1 \qquad 1\ . \qquad 1 \qquad 0 \qquad 1$$
$$11.625_{10} = 1011.101_2 \quad \text{Ans}$$

Part of a simplified binary-decimal system tape in the vertical form of an early NC control system for the decimal number 243 is shown in Figure 78-1. The decimal system is used for place location, but each digit of the vertically positioned decimal number is converted to a binary number.

	Decimal Number	Vertical Binary-Decimal Number
10^2	2	10
10^1	4	100
10^0	3	11

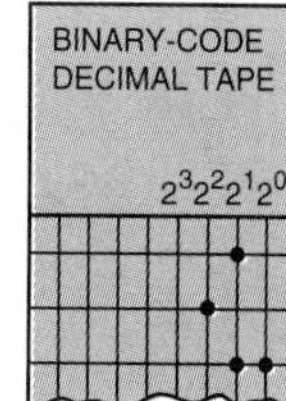

Figure 78-1

Application

Structure of the Decimal System

Analyze the following numbers.

1. 265
2. 2855
3. 90,500
4. 0.802
5. 23.023
6. 105.009
7. 4751.107
8. 3006.0204
9. 163.0643

Expressing Binary Numbers as Decimal Numbers

Express the following binary numbers as decimal numbers.

10. 10_2 ________
11. 1_2 ________
12. 100_2 ________
13. 101_2 ________
14. 1101_2 ________
15. 1111_2 ________
16. 10100_2 ________
17. 1011_2 ________
18. 11000_2 ________
19. 10101_2 ________
20. 101010_2 ________
21. 110101_2 ________
22. 111010_2 ________
23. 0.1_2 ________
24. 0.1011_2 ________
25. 11.11_2 ________
26. 11.01_2 ________
27. 10.000_2 ________
28. 1111.11_2 ________
29. 1001.0101_2 ________
30. 10011.0101_2 ________

Expressing Decimal Numbers as Binary Numbers

Express the following decimal numbers as binary numbers.

31. 14 ________
32. 100 ________
33. 87 ________
34. 23 ________
35. 43 ________
36. 4 ________
37. 105 ________
38. 98 ________
39. 1 ________
40. 6 ________
41. 51 ________
42. 270 ________

43. 0.5 ____________
44. 0.125 ____________
45. 0.375 ____________
46. 10.5 ____________
47. 81.75 ____________
48. 19.0625 ____________
49. 101.25 ____________
50. 1.125 ____________
51. 163.875 ____________

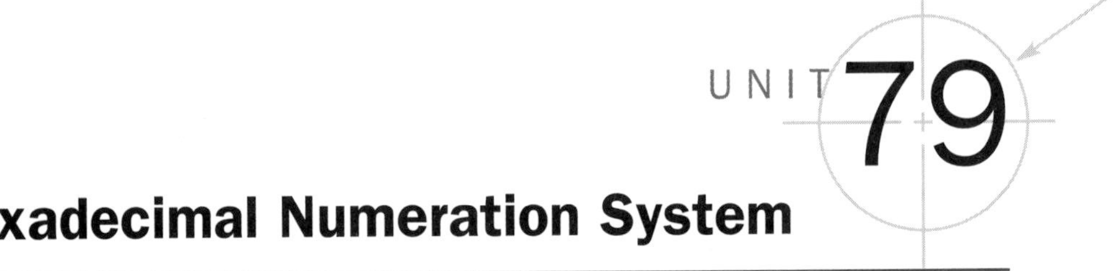

Hexadecimal Numeration System

Objectives ***After studying this unit you should be able to***

- Express hexadecimal numbers as decimal numbers.
- Express decimal numbers as hexadecimal numbers.

The big problem with the binary system is that it takes so many digits to represent a number. For example, the number 245 requires eight binary digits since $245_{10} = 1010\ 1111_2$. When working with large numbers, binary numbers become too hard to work with.

Engineers who designed computers wanted a way of expressing numbers using fewer digits and that was easy to convert to binary numbers. The *hexadecimal* (base 16) *numeration system* solved these problems.

The hexadecimal, or hex, numbers are based on a nibble. The binary numbers in a nibble range from 0000 to 1111 or from the decimal numbers 0 to 15. The digits 0 to 9 and the capital letters A to F are used for the hexadecimal numbers. The table below shows the decimal, binary, and hexadecimal numbers.

Decimal	Binary	Hexadecimal
0	0000	0
1	0001	1
2	0010	2
3	0011	3
4	0100	4
5	0101	5
6	0110	6
7	0111	7

Decimal	Binary	Hexadecimal
8	1000	8
9	1001	9
10	1010	A
11	1011	B
12	1100	C
13	1101	D
14	1110	E
15	1111	F

Converting Binary Numbers to Hexadecimal Numbers

To convert a binary number as a hexadecimal number, group the number into nibbles starting at the binary point. Add zeros as needed to fill out the groups. Assign each group the appropriate hex number from the table above.

Examples

1. Convert the binary number 111011100101100010 to hexadecimal.
   Grouping the binary number in nibbles we get

   11 1011 1001 0110 0010

   or, after zeros are added,

   0011 1011 1001 0110 0010

   Converting each nibble to hex produces

   3 B 9 6 2

   So, the binary number 111011100101100010 is equivalent to 3B962, or

   $$111011100101100010_2 = 3B962_{16}$$

2. Convert 1101101.1110101_2 to hexadecimal.
   Grouping in nibbles from the binary point, we get

   110 1101 . 1110 101

   or, after zeros are added,

   0110 1101 . 1110 1010

   Convert each nibble to hex

   6 D . E A

   The binary number 1101101.1110101 is equivalent to the hexadecimal number of 6D.EA.

Converting Hexadecimal Numbers to Binary Numbers

To convert a hexadecimal number to a binary number, reverse the above procedure

Example Convert the hexadecimal number 7A3.C2 to a binary number.
Write the nibble that corresponds to each hexadecimal symbol.

7	A	3	.	C	2
0111	1010	0011	.	1100	0010

Thus, $7A3.C2_{16} = 11110100011.1100001_2$.

Converting Hexadecimal Numbers to Decimal Numbers

As with decimal and binary numbers, each hexadecimal digit has a place value expressed as the base, 16, raised to the position number. The place value of the first digit to the left of the hex point is

$$16^0 = 1$$

and the place value of the first digit to the right of the hex point is

$$16^{-1} = \frac{1}{16}$$

To convert from hexadecimal to decimal, first replace each hex digit with its decimal equivalent. Next, write the number in expanded notation and multiply each hex digit by its place value. Add the resulting products.

Example Convert the hexadecimal number A9.E to its decimal equivalent.

Replace the hex digit A with 10 and the hex digit E with 14. Write the number in expanded form.

A		9	.	E	
10		9	.	14	
(10×16^2)	+	(9×16^1)	+	(14×16^{-1})	
2560	+	144	+	0.875	= 2704.875

So, $A9.E_{16} = 2704.875$.

Converting Decimal Numbers to Hexadecimal Numbers

The following example shows the method for expressing decimal numbers as hexadecimal numbers. To convert a decimal number to hexadecimal, keep dividing by 16 and convert each remainder to its hex.

Example Convert the decimal number 4974 to its hexadecimal equivalent.

$4974 \div 16 = 310$ with remainder 14 ↑
$310 \div 16 = 19$ with remainder 6
$19 \div 16 = 1$ with remainder 3
$1 \div 16 = 0$ with remainder 1 Read up

Changing the 14 to the hex digit E and reading up, we see that $4974 = 136E_{16}$.

Application

Converting Binary Numbers to Hexadecimal Numbers

Express the following binary numbers as hexadecimal numbers.

1. 1101_2 ____________
2. 1011_2 ____________
3. 110011_2 ____________
4. 1010011_2 ____________
5. 10100111010011_2 ____________
6. 11100110110011_2 ____________
7. 1101011.110111_2 ____________
8. 100101011.1010111_2 ____________
9. 10010101011.101010011_2 ____________
10. $1000010101011.10010010011_2$ ____________
11. $1100100101001011.1001001001_2$ ____________
12. 11101001001.1010010001_2 ____________

Converting Hexadecimal Numbers to Binary Numbers

Express the following hexadecimal numbers as binary numbers.

13. 24_{16} ____________
14. 53_{16} ____________
15. $9A_{16}$ ____________
16. $B3_{16}$ ____________
17. $C27_{16}$ ____________
18. $F71B_{16}$ ____________
19. $BF.3A_{16}$ ____________
20. $A7.B5_{16}$ ____________
21. $27C.D7_{16}$ ____________
22. $E73.6B_{16}$ ____________
23. $2B.02B_{16}$ ____________
24. $F0E.9D5_{16}$ ____________

Converting Hexadecimal Numbers to Decimal Numbers

Express the following hexadecimal numbers as decimal numbers.

25. $F7_{16}$ ____________
26. $3E0_{16}$ ____________
27. $7B.E_{16}$ ____________
28. $3F9.A8_{16}$ ____________
29. $B0B.08_{16}$ ____________
30. 573.8_{16} ____________
31. $7FF0.4_{16}$ ____________
32. $9DE2.B_{16}$ ____________

Converting Decimal Numbers to Hexadecimal Numbers

Express the following decimal numbers as hexadecimal numbers.

33. 47 ____________
34. 93 ____________
35. 143 ____________
36. 137 ____________
37. 963.25 ____________
38. 1278.75 ____________
39. 65 328 ____________
40. 57 905 ____________

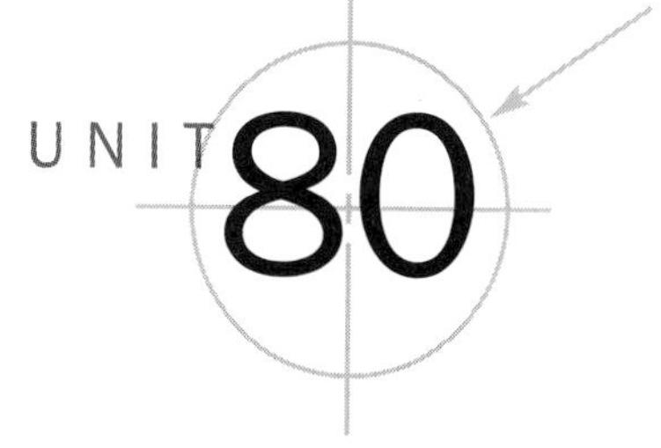

BCD (Binary Coded Decimal) Numeration Systems

Objectives ***After studying this unit you should be able to***

- Express BCD (Binary Coded Decimal) numbers as decimal numbers.
- Express decimal numbers as BCD (Binary Coded Decimal) numbers.

You are now familiar with the binary, decimal, and hexadecimal number systems. The single digit values for hex are the numbers 0–F and represent the values 0–15 in the decimal system. Each hex digit occupies a nibble when written in the binary system.

The binary equivalent of the decimal system is called Binary Coded Decimal or BCD and also occupies a nibble. In BCD, the binary patterns 1010 (decimal number 10) through 1111 (decimal number 15) do not represent valid BCD numbers, and cannot be used.

There are several types of BCD systems as shown in the table below. Each uses 4 bits, or a nibble, to represent one of the decimal digits 0 through 9. The value of each bit in columns 2, 3, and 4 is indicated by the heading of the column.

Decimal	8421	2421	5211	Excess-3
0	0000	0000	0000	0011
1	0001	0001	0001	0100
2	0010	0010	0011	0101
3	0011	0011	0101	0110
4	0100	0100	0111	0111

Decimal	8421	2421	5211	Excess-3
5	0101	1011	1000	1000
6	0110	1100	1010	1001
7	0111	1101	1100	1010
8	1000	1110	1110	1011
9	1001	1111	1111	1100

In the 5211 Code, two positions have a weight of 1. That makes it possible for two different patterns to represent the same decimal digit. But, only the pattern shown in the table above is assigned to that digit.

8421 Code/BCD Code

When one speaks of the BCD system the 8421 Code is the one that is usually meant. The values of the place values in each nibble are 2^3, 2^2, 2^1, and 2^0 (or 8, 4, 2, and 1).

Example Determine the decimal value for the BCD number 0101.

This has the decimal value of 5 because $0101 = (0 \times 8) + (1 \times 4) + (0 \times 2) + (1 \times 1) = 0 + 4 + 0 + 1 = 5$.

Conversion from Decimal Numbers to BCD

Conversion from decimal to BCD is straightforward. Each digit of the decimal number is assigned a byte and converts from 0 through 9 to 0000 through 1001.

Example Determine the BCD value for the decimal number 7529.

Since there are four digits in the decimal number 7529, there are four bytes in its BCD number. They are:

Thousands	Hundreds	Tens	Units
7	5	2	9
0111	0101	0010	1001

Thus we see that the decimal number 7529 has the BCD number 0111 0101 0010 1001.

Conversion from BCD to Decimal Numbers

To change a BCD number to a decimal number, separate the BCD number into nibbles, and write the decimal equivalent of each nibble. It may be necessary to add zeros as needed to fill out the groups.

Examples

1. Convert the BCD number 110110 to its decimal value.

   Since no specific BCD code is indicated, we can assume that this is the 8421 BCD code. Group the number into nibbles. Add zeros as needed to fill out the groups. Assign each group the appropriate decimal number from the table above.

   Grouping the binary number in nibbles we get

   11 0110

or, after zeros are added,

0011 0110

Converting each nibble to its decimal digit

3 6

So, the BCD number 110110 is equivalent to the decimal number 36.

2. Convert the BCD number 10111.001001 to its decimal value.

   Again, since no specific BCD code is indicated, we can assume that this is the 8421 BCD code.

   Group the binary number in nibbles.

   1 0111 . 0010 01

   Add zeros to complete the nibbles.

   0001 0111 . 0010 0100

   Convert each nibble to its decimal equivalent.

   1 7 . 2 4

   The BCD number 10111.001001 is equivalent to the decimal number 17.24.

2421 BCD Code

The 2421 code is a weighted code. The values of the place values in each nibbles are 2, 4, 2, and 1.

Example Determine the decimal value for the 2421 code number 1101.

This has the decimal value of 7 because $1101 = (1 \times 2) + (1 \times 4) + (0 \times 2) + (1 \times 1) = 2 + 4 + 0 + 1 = 7$.

Conversion of Decimal Numbers to 2421 Code Numbers

Example Determine the 2421 code number for the decimal number 7529.

Since there are four digits in the decimal number 7529, there are four bytes in its 2421 number. They are:

7	5	2	9
1101	1011	0010	1111

Thus we see that the decimal number 7529 has the 2421 number 1101 1011 0010 1111.

5211 BCD Code

The 5211 code is another weighted code. The values of the place values in each nibble are 5, 2, 1, and 1.

Example Determine the decimal value for the 2421 code number 1110.

This has the decimal value of 8 because $1110 = (1 \times 5) + (1 \times 2) + (1 \times 1) + (0 \times 1) = 5 + 2 + 1 + 0 = 8$.

Conversion of Decimal Numbers to 5211 Code Numbers

Example Determine the 5211 code number for the decimal number 7529.

Since there are four digits in the decimal number 7529, there are four bytes in its 5211 number. They are:

7	5	2	9
1100	1000	0011	1111

Thus we see that the decimal number 7529 has the 5211 number 1100 1000 0011 1111.

Excess-3 BCD Code

The Excess-3 code is a nonweighted code. The code gets its name from the fact that this binary codes is the corresponding 8421 code plus 0011 (or the binary value for 3).

Conversion of Decimal Numbers to Excess-3 Code Numbers

Example Determine the Excess-3 code number for the decimal number 7529.

Since there are four digits in the decimal number 7529, there are four bytes in its Excess-3 number. Referring to the table on pages 541 and 542, they are:

7	5	2	9
1010	1000	0101	1100

Thus we see that the decimal number 7529 has hte Excess-3 number 1010 1000 0101 1100.

Application

Conversion from Decimal Numbers to BCD

Express the following decimal numbers as BCD (8421) numbers.

1. 73 __________
2. 57 __________
3. 246 __________
4. 315 __________
5. 18.93 __________
6. 47.25 __________
7. 36.195 __________
8. 624.37 __________

Conversion from BCD to Decimal Numbers

Express the following BCD (8421) numbers as decimal numbers.

9. 1001 __________
10. 111 __________
11. 101 0101 __________
12. 1001 0011 __________
13. 111 0110.0100 __________
14. 1001 0000.0101 0011 __________
15. 11 0110 0101.0000 0111 __________
16. 101 1000 0011.0111 1001 __________

Conversion from Decimal Numbers to 2421 Code Numbers

Express the following decimal numbers as 2421 code numbers.

17. 72 __________
18. 129 __________
19. 362.9 __________
20. 74.685 __________

Conversion from 2421 Code Numbers to Decimal Numbers

Express the following 2421 code numbers as decimal numbers.

21. 100 1101 ____________
22. 1110 1011 ____________
23. 1100 0000.1101 ____________
24. 10 0100.1110 1100 0000 0011 ____________

Conversion from Decimal Numbers to 5211 Code Numbers

Express the following decimal numbers as 5211 code numbers.

25. 36 ____________
26. 297 ____________
27. 45.26 ____________
28. 106.31 ____________

Conversion from 5211 Code Numbers to Decimal Numbers

Express the following 5211 code numbers as decimal numbers.

29. 101 1110 ____________
30. 111 1101 ____________
31. 1110 0011.0111 ____________
32. 101 1010.1110 1100 0111 ____________

Conversion from Decimal Numbers to Excess-3 Code Numbers

Express the following decimal numbers as Excess-3 code numbers.

33. 72 ____________
34. 512 ____________
35. 93.86 ____________
36. 217.04 ____________

Conversion from Excess-3 Code Numbers to Decimal Numbers

Express the following Excess-3 code numbers as decimal numbers.

37. 101 1011 ____________
38. 111 1100 ____________
39. 1010 0110.0011 0101 ____________
40. 101 1001 1000.0110 1100 1001 ____________

UNIT 81

Achievement Review—Section Nine

Objective *You should be able to solve the exercises and problems in this Achievement Review by applying the principles and methods covered in Units 75–80.*

1. Using graph paper, draw an x- and a y-axis and plot the following coordinates.

   A = (6, −8) C = (−2, 0) E = (−7, −7)
   B = (−3, 9) D = (0, −8) F = (3, 3)

2. Refer to the points plotted on the illustrated Cartesian coordinate plane. Write the x and y coordinates of the following points, AM.

A = ______________

B = ______________

C = ______________

D = ______________

E = ______________

F = ______________

G = ______________

H = ______________

I = ______________

J = ______________

K = ______________

L = ______________

M = ______________

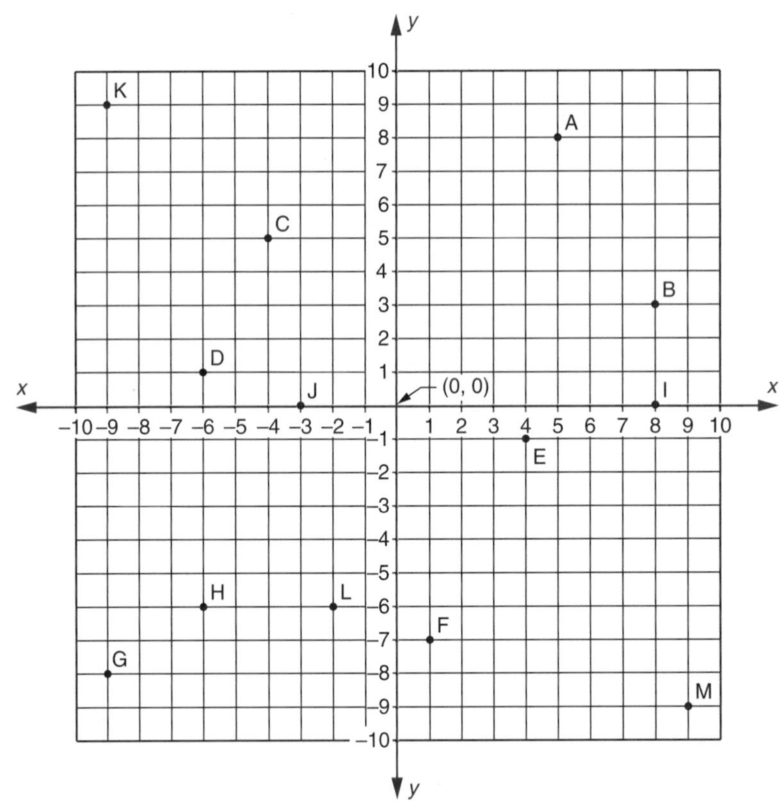

3. Write the program hole locations (coordinates) in table form. List the holes in sequence similar to the tables in Unit 76. All dimensions are in inches. Use

a. absolute dimensioning
b. incremental dimensioning

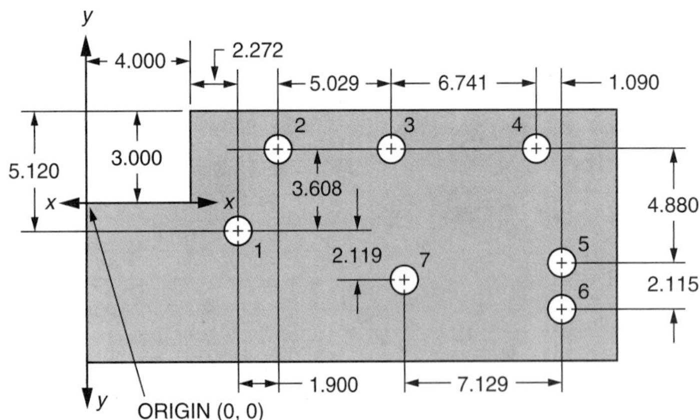

4. Write the program hole locations (coordinates) in table form. List the holes in sequence similar to the tables in Unit 76. All dimensions are in millimeters. Use

   a. absolute dimensioning

   b. incremental dimensioning

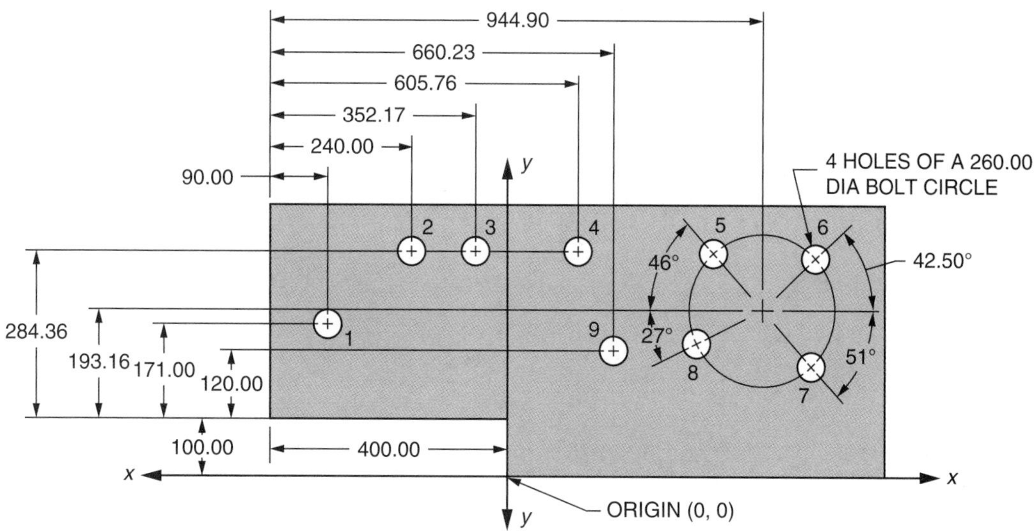

5. Figure 81-1 shows four equally spaced holes on a bolt circle circumference. Give the polar coordinates of each of the holes. All dimensions are in millimeters.

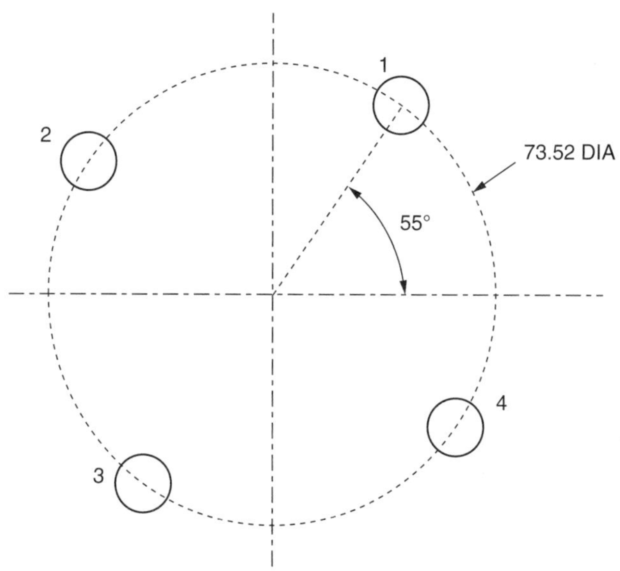

Figure 81-1

	Coordinates	
Hole	***r***	***θ***
1		
2		
3		
4		

6. Express the following binary numbers as decimal numbers.

   a. 1_2 ____________

   b. 111_2 ____________

   c. 10101_2 ____________

   d. 11.01_2 ____________

   e. 1001.1001_2 ____________

7. Express the following decimal numbers as binary numbers.

a. 7 __________
b. 32 __________
c. 157 __________
d. 0.125 __________
e. 74.25 __________

8. Express the following binary numbers as hexadecimal numbers.

a. 110001_2 __________
b. 10110101_2 __________
c. 1100101011_2 __________
d. 1011011101111_2 __________
e. 111110111.0111_2 __________
f. 110101001101.11101_2 __________

9. Express the following hexadecimal numbers as binary numbers.

a. $2B_{16}$ __________
b. $C5A_{16}$ __________
c. $27.1A_{16}$ __________
d. $E0.0D_{16}$ __________

10. Express the following hexadecimal numbers as decimal numbers.

a. $C2_{16}$ __________
b. $5B_{16}$ __________
c. $A20D_{16}$ __________
d. $17E3_{16}$ __________

11. Express the following decimal numbers as hexadecimal numbers.

a. 75 __________
b. 243 __________
c. 927 __________
d. 4231 __________

12. Express the following decimal numbers as BCD (8421) numbers.

a. 763 __________
b. 94 __________

13. Express the following BCD (8421) numbers as decimal numbers.

a. 11 1001 0010 __________
b. 101 0111 1000 110 __________

14. Express the decimal number 357 as a

a. 2421 code number __________
b. 5211 code number __________
c. Excess-3 code number __________

United States Customary and Metric Units of Measure

CUSTOMARY UNITS

CUSTOMARY UNITS OF LINEAR MEASURE
1 foot (ft) = 12 inches (in)
1 yard (yd) = 3 feet (ft)
1 yard (yd) = 36 inches (in)

CUSTOMARY UNITS OF AREA MEASURE
1 square foot (sq ft or ft^2) = 144 square inches (sq in or in^2)
1 square yard (sq yd or yd^2) = 9 square feet (sq ft or ft^2)
1 square yard (sq yd or yd^2) = 1728 square inches (sq in or in^2)

CUSTOMARY UNITS OF VOLUME MEASURE
1 cubic foot (cu ft or ft^3) = 1728 cubic inches (cu in)
1 cubic yard (cu yd or yd^3) = 27 cubic feet (cu ft)

METRIC UNITS

METRIC UNITS OF LINEAR MEASURE
1 millimeter (mm) = 0.001 meter (m)
1 centimeter (cm) = 0.01 meter (m)
1 decimeter (dm) = 0.1 meter (m)
1000 millimeters (mm) = 1 meter (m)
100 centimeters (cm) = 1 meter (m)
10 decimeters (dm) = 1 meter (m)

METRIC UNITS OF AREA MEASURE
1 square millimeter (mm^2) = 0.000 001 square meter (m^2)
1 square centimeter (cm^2) = 0.0001 square meter (m^2)
1 square decimeter (dm^2) = 0.01 square meter (m^2)
1 000 000 square millimeters (mm^2) = 1 square meter (m^2)
10 000 square centimeters (cm^2) = 1 square meter (m^2)
100 square decimeters (dm^2) = 1 square meter (m^2)

METRIC UNITS OF VOLUME MEASURE
1 cubic millimeter (mm^3) = 0.000 000 001 cubic meter (m^3)
1 cubic centimeter (cm^3) = 0.000 001 cubic meter (m^3)
1 cubic decimeter (dm^3) = 0.001 cubic meter (m^3)
1 000 000 000 cubic millimeters (mm^3) = 1 cubic meter (m^3)
1 000 000 cubic centimeters (cm^3) = 1 cubic meter (m^3)
1000 cubic decimeters (dm^3) = 1 cubic meter (m^3)

METRIC-CUSTOMARY LINEAR EQUIVALENTS (CONVERSION FACTORS)	
Metric to Customary Units	Customary to Metric Units
1 millimeter (mm) = 0.03937 inch (in)	1 inch (in) = 25.4 millimeters (mm)
1 centimeter (cm) = 0.3937 inch (in)	1 inch (in) = 2.54 millimeters (mm)
1 meter (m) ≈ 39.37 inches (in)	1 foot (ft) ≈ 0.3048 meter (m)
1 meter (m) ≈ 3.2808 feet (ft)	1 yard (yd) ≈ 0.9144 meter (m)
1 kilometer (km) ≈ 0.6214 mile (mi)	1 mile (mi) ≈ 1.609 kilometers (km)

METRIC-CUSTOMARY AREA CONVERSIONS
1 square inch (sq in or in^2) = 6.4516 cm^2
1 square foot (sq ft or ft^2) ≈ 0.0929 m^2
1 square yard (sq yd or yd^2) ≈ 0.8361 m^2

METRIC-CUSTOMARY VOLUME CONVERSIONS
1 cubic inch (cu in or in^3) ≈ 16 387 mm^3
1 cubic inch (cu in or in^3) ≈ 16.387 cm^3
1 cubic foot (cu ft or ft^3) ≈ 0.0283 m^3
1 cubic yard (cu yd or yd^3) ≈ 0.7645 m^3
1 mm^3 ≈ 0.000 061 cubic inch (cu in or in^3)
1 cm^3 ≈ 0.061 024 cubic inch (cu in or in^3)

Principles of Plane Geometry

Principles of Plane Geometry

Note: The page where the principle can be found is noted after each principle.

1. If two lines intersect, the opposite or vertical angles are equal. (p. 300)
2. If two parallel lines are intersected by a transversal, the alternate interior angles are equal. (p. 300)

   If two lines are intersected by a transversal and a pair of alternate interior angles are equal, the lines are parallel.
3. If two parallel lines are intersected by a transversal, the corresponding angles are equal. (p. 300)

   If two lines are intersected by a transversal and a pair of corresponding angles are equal, the lines are parallel. (p. 301)
4. Two angles are either equal or supplementary if their corresponding sides are parallel. (p. 301)
5. Two angles are either equal or supplementary if their corresponding sides are perpendicular. (p. 301)
6. The sum of the angles of any triangle is equal to 180°. (p. 307)
7. Two triangles are similar if their sides are respectively parallel. (p. 314)
   - Two triangles are similar if their sides are respectively perpendicular.
   - Within a triangle, if a line is drawn parallel to one side, the triangle formed is similar to the original triangle.
   - In a right triangle, if a line is drawn from the vertex of the right angle perpendicular to the opposite side, the two triangles formed and the original triangle are similar.
8. In an isosceles triangle, an altitude to the base bisects the base and the vertex angle. (p. 315)

   In an equilateral triangle, an altitude to any side bisects the side and the vertex angle.
9. In a right triangle, the square of the hypotenuse is equal to the sum of the squares of the other two sides or legs. (p. 315)
10. The sum of the interior angles of a polygon of N sides is equal to (N − 2) times 180°. (p. 317)
11. In the same circle or in equal circles, equal chords cut off equal arcs. (p. 325)
12. In the same circle or in equal circles, equal central angles cut off equal arcs. (p. 325)
13. In the same circle or in equal circles, two central angles have the same ratio as the arcs that are cut off by the angles. (p. 325)
14. A line drawn from the center of a circle perpendicular to a chord bisects the chord and the arc cut off by the chord. The perpendicular bisector of a chord passes through the center of a circle. (p. 326)

    The perpendicular bisector of a chord passes through the center of a circle.
15. A line perpendicular to a radius at its extremity is tangent to the circle. A tangent is perpendicular to a radius at its tangent point. (p. 326)

16. Two tangents drawn to a circle from a point outside the circle are equal. The angle at the outside point is bisected by a line drawn from the point to the center of the circle. (p. 327)
17. If two chords intersect inside a circle, the product of the two segments of one chord is equal to the product of the two segments of the other chord. (p. 327)
18. A central angle is equal to its intercepted arc. (p. 331)

    An angle formed by two chords that intersect inside a circle is equal to one-half the sum of its two intersecting arcs. (p. 332)
19. An angle formed by a tangent and a chord at the tangent point is equal to one-half of its intercepted arc. (p. 333)
20. An angle formed at a point outside a circle by two secants, two tangents, or a secant and a tangent is equal to one-half the difference of the intercepted arcs. (p. 334)
21. If two circles are either internally or externally tangent, a line connecting the centers of the circles passes through the point of tangency and is perpendicular to the tangent line. (p. 336)

Formulas for Areas (A) of Plane Figures

Note: The page where the formula and its application can be found is noted after each formula.

Rectangle	$A = lw$ (p. 361): l = length, w = width
Parallelogram	$A = bh$ (p. 362): b = base, h = height
Trapezoid	$A = \frac{1}{2}h(b_1 + b_2)$ (p. 363): h = height, $b_1 + b_2$ = bases
Triangle	$A = \frac{1}{2}bh$ (p. 370): b = base, h = height
	$A = \sqrt{s(s-a)(s-b)(s-c)}$ (p. 371): a, b, and c = sides where $s = \frac{1}{2}(a + b + c)$
Circle	$A = \pi r^2$ (p. 374): r = radius
	$A = \frac{\pi d^2}{4} \approx 0.7854d^2$ (p. 374): d = diameter
Sector	$A = \frac{\theta}{360°}(\pi r^2)$ (p. 375): θ = central angle, r = radius

Formulas for Volumes (V) of Solid Figures

Note: The page where the formula and its application can be found is noted after each formula.

Figure	Volume (V)
Prism	$V = A_B h$ (p. 384) A_B = area of base h = height
Right Circular Cylinder	$V = A_B h$ (p. 385) A_B = area of base h = height
Regular Pyramid	$V = \frac{1}{3}A_B h$ (p. 393) A_B = area of base h = height
Right Circular Cone	$V = \frac{1}{3}A_B h$ (p. 393) A_B = area of base h = height
Frustum of a Regular Pyramid	$V = \frac{1}{3}h\left(A_B + A_b + \sqrt{A_B A_b}\right)$ (p. 394) h = height A_B = area of larger base A_b = area of smaller base
Frustum of a Right Circular Cone	$V = \frac{1}{3}\pi h(R^2 + r^2 + Rr)$ (p. 394) h = height R = radius of larger base r = radius of smaller base
Sphere	$V = \frac{4}{3}\pi r^3$ (p. 399) r = radius of sphere

Trigonometry

TRIGONOMETRIC FUNCTIONS		
Function	Symbol	Definition of Function Using a Right Triangle
sine of Angle A	sin A	$\sin A = \frac{\text{opp side}}{\text{hyp}} = \frac{a}{c}$
cosine of Angle A	cos A	$\cos A = \frac{\text{adj side}}{\text{hyp}} = \frac{b}{c}$
tangent of Angle A	tan A	$\tan A = \frac{\text{opp side}}{\text{adj side}} = \frac{a}{b}$
cotangent of Angle A	cot A	$\cot A = \frac{\text{adj side}}{\text{opp side}} = \frac{b}{a}$
secant of Angle A	sec A	$\sec A = \frac{\text{hyp}}{\text{adj side}} = \frac{c}{b}$
cosecant of Angle A	csc A	$\csc A = \frac{\text{hyp}}{\text{opp side}} = \frac{c}{a}$

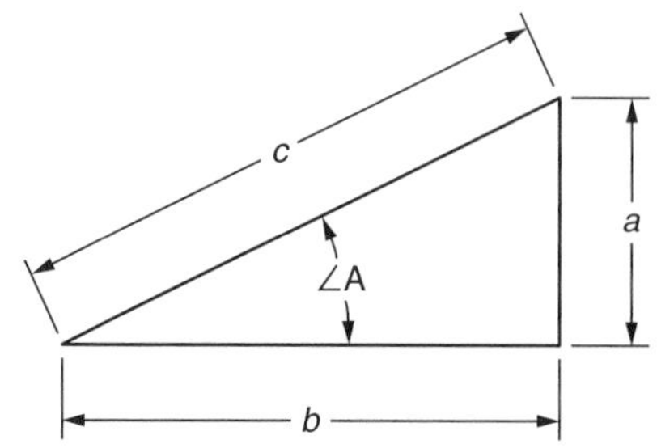

Law of Sines

The Law of Sines states that in any triangle, the sides are proportional to the sines of the opposite angles.

In reference to the triangle shown, the formula is stated:

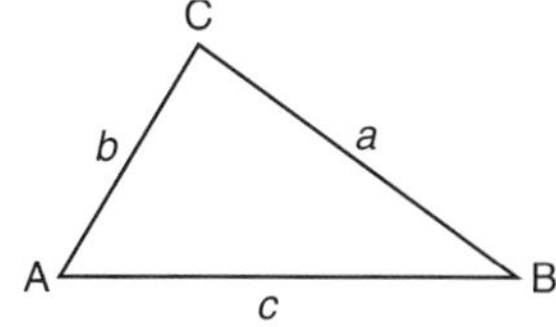

$$\frac{a}{\sin A} = \frac{b}{\sin B} = \frac{c}{\sin C}$$

Law of Cosines (Given Two Sides and the Included Angle)

In any triangle, the square of any side is equal to the sum of the squares of the other two sides minus twice the product of these two sides multiplied by the cosine of their included angle.

In reference to the triangle shown the formula is stated:

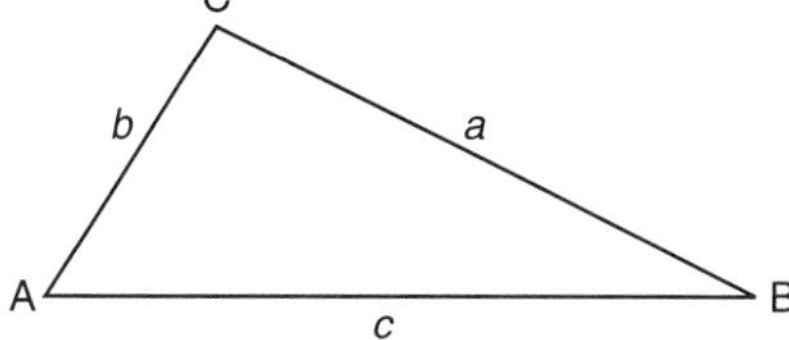

$$a^2 = b^2 + c^2 - 2bc(\cos A)$$
$$b^2 = a^2 + c^2 - 2ac(\cos B)$$
$$c^2 = a^2 + b^2 - 2ab(\cos C)$$

Law of Cosines (Given Three Sides)

In any triangle, the cosine of an angle is equal to the sum of the squares of the two adjacent sides minus the square of the opposite side, divided by twice the product of the two adjacent sides.

In reference to the triangle shown:

$$\cos A = \frac{b^2 + c^2 - a^2}{2bc}$$

$$\cos B = \frac{a^2 + c^2 - b^2}{2ac}$$

$$\cos C = \frac{a^2 + b^2 - c^2}{2ab}$$

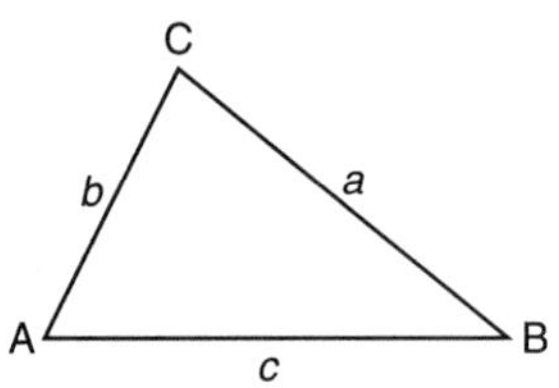

Answers to Odd-Numbered Applications

SECTION 1 Common Fractions and Decimal Fractions

UNIT 1 Introduction to Common Fractions and Mixed Numbers

1. A = $\frac{3}{32}$

   B = $\frac{7}{32}$

   C = $\frac{3}{8}$

   D = $\frac{19}{32}$

   E = $\frac{27}{32}$

   F = 1

3. a. $\frac{1}{16}$

   b. $\frac{3}{16}$

   c. $\frac{7}{16}$

   d. $\frac{5}{16}$

   e. $\frac{16}{16} = 1$

   f. $\frac{1}{32}$

   g. $\frac{1}{48}$

   h. $\frac{3}{64}$

   i. $\frac{1}{160}$

   j. $\frac{1}{256}$

5. a. $\frac{3}{4}$

   b. 3

   c. $\frac{3}{5}$

   d. 6

   e. $\frac{1}{4}$

   f. $\frac{7}{3}$

   g. 3

   h. $\frac{13}{3}$

   i. $\frac{1}{6}$

   j. $\frac{2}{15}$

7. a. $\frac{6}{8}$

   b. $\frac{21}{36}$

   c. $\frac{24}{60}$

   d. $\frac{51}{42}$

   e. $\frac{100}{45}$

   f. $\frac{84}{18}$

   g. $\frac{56}{128}$

   h. $\frac{78}{48}$

   i. $\frac{210}{160}$

9. a. $1\frac{2}{3}$

   b. $10\frac{1}{2}$

   c. $1\frac{1}{8}$

   d. $21\frac{3}{4}$

   e. 8

   f. $1\frac{3}{124}$

   g. $3\frac{31}{32}$

   h. $3\frac{12}{15} = 3\frac{4}{5}$

   i. $16\frac{2}{3}$

   j. $14\frac{11}{16}$

   k. $128\frac{1}{2}$

   l. $6\frac{17}{64}$

11.

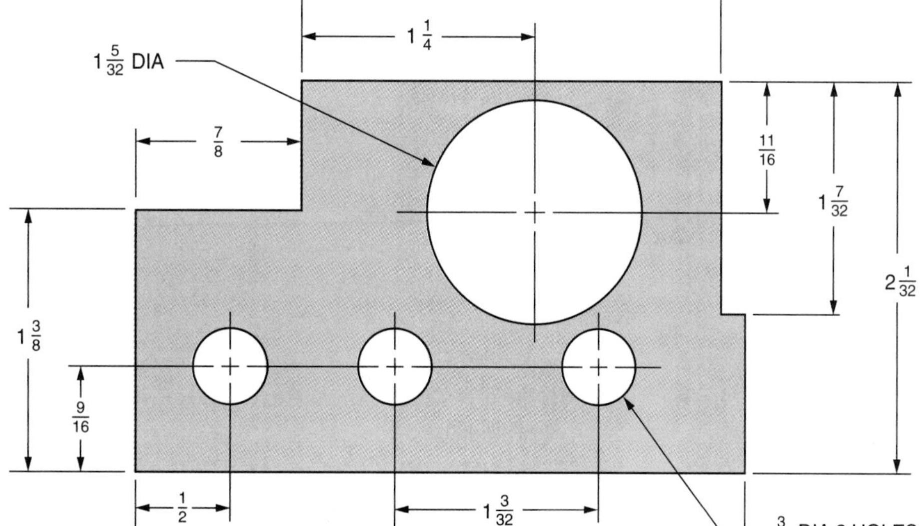

UNIT 2 Addition of Common Fractions and Mixed Numbers

1. 12

3. 48

5. $\frac{6}{12}, \frac{9}{12}, \frac{5}{12}$

7. $\frac{18}{20}, \frac{5}{20}, \frac{12}{20}, \frac{4}{20}$

9. A = $1\frac{1}{16}$"

9. B = $\frac{59}{64}$"

C = $1\frac{9}{16}$"

D = $\frac{11}{16}$"

E = $3\frac{11}{64}$"

F = $\frac{13}{16}$"

11. A = $2\frac{9}{32}$"

B = $3\frac{7}{8}$"

C = $4\frac{1}{4}$"

D = $2\frac{19}{32}$"

E = $3\frac{3}{8}$"

11. F = $1\frac{21}{64}$"

G = $4\frac{45}{64}$"

13. $5\frac{17}{60}$ h

UNIT 3 Subtraction of Common Fractions and Mixed Numbers

1. a. $\frac{11}{32}$

b. $\frac{1}{4}$

c. $\frac{13}{25}$

d. $\frac{31}{64}$

e. $\frac{23}{64}$

1. f. $\frac{29}{48}$

3. A = $\frac{7}{32}$"

B = $\frac{19}{32}$"

C = $\frac{25}{64}$"

D = $\frac{3}{16}$"

3. E = $\frac{7}{16}$"

F = $\frac{13}{32}$"

5. A = $\frac{15}{32}$"

B = $\frac{21}{32}$"

5. C = $\frac{7}{16}$"

D = $\frac{15}{32}$"

E = $\frac{9}{32}$"

F = $\frac{5}{16}$"

5. G = $1\frac{7}{32}$"

H = $\frac{33}{64}$"

I = $\frac{19}{32}$"

UNIT 4 Multiplication of Common Fractions and Mixed Numbers

1. a. $\frac{1}{9}$

b. $\frac{1}{8}$

c. $\frac{65}{512}$

1. d. $\frac{3}{10}$

e. $13\frac{1}{2}$

f. $\frac{1}{8}$

3. a. $\frac{1595}{2048}$"

b. $\frac{7}{256}$"

5. a. $10\frac{1}{2}$

5. b. $25\frac{43}{64}$

c. $11\frac{9}{16}$

d. $6\frac{13}{32}$

5. e. $\frac{201}{256}$

f. $37\frac{1}{3}$

7. $5\frac{87}{128}$ square inches

UNIT 5 Division of Common Fractions and Mixed Numbers

1. $\frac{8}{7}$

3. $\frac{8}{25}$

5. A = 6 threads

B = $8\frac{7}{16}$ threads

C = $3\frac{1}{2}$ threads

5. D = 7 threads

E = $6\frac{7}{8}$ threads

F = $6\frac{1}{2}$ threads

G = $2\frac{13}{16}$ threads

7. 5 cuts

9. 240 revolutions

11. $\frac{4}{5}$ foot

13. $4\frac{70}{93}$ lb

15. a. Lead: $\frac{21}{80}$", Pitch: $\frac{21}{160}$"

b. Lead: $\frac{5}{16}$", Pitch: $\frac{5}{32}$"

c. Lead: $\frac{5}{8}$", Pitch: $\frac{5}{16}$"

d. Lead: $\frac{19}{64}$", Pitch: $\frac{19}{128}$"

UNIT 6 Combined Operations of Common Fractions and Mixed Numbers

1. a. $\frac{7}{16}$
   b. $2\frac{1}{16}$
   c. $5\frac{33}{50}$
   d. $28\frac{1}{2}$
   e. $33\frac{49}{64}$
   f. $3\frac{1}{2}$
   g. $5\frac{3}{4}$
   h. $28\frac{3}{8}$
   i. $4\frac{8}{21}$
   j. $20\frac{77}{87}$

3. a. $B = 4\frac{7}{8}''$
   $E = 5\frac{1}{16}''$
   b. $A = \frac{9}{64}''$
   $G = 6\frac{3}{4}''$
   c. $C = 1\frac{1}{16}''$
   $D = 5\frac{19}{32}''$
   d. $B = 3\frac{29}{32}''$
   $E = 5\frac{5}{8}''$
   e. $A = \frac{19}{32}''$
   $G = 7\frac{27}{64}''$
   f. $C = \frac{63}{64}''$
   $D = 5\frac{55}{64}''$

5. $7\frac{29}{32}''$

7. $3\frac{1}{5}$ min

9. $\frac{57}{64}''$

11. $3\frac{3}{4}$ lb

UNIT 7 Computing with a Calculator: Fractions and Mixed Numbers

(All answers are given in the unit.)

UNIT 8 Introduction to Decimal Fractions

1. A = 0.3
   B = 0.5
   C = 0.8
   D = 0.92
   E = 0.04
3. A = 0.0025
   B = 0.006
   C = 0.007
   D = 0.0073
   E = 0.0004
5. 0.01
7. 10
9. 0.1
11. 100
13. 0.001
15. seven thousandths
17. thirty-five ten-thousandths
19. one and five tenths
21. sixteen and seven ten-thousandths
23. thirteen and one hundred three thousandths
25. 0.3
27. 4.00005
29. 10.2
31. 20.71
33. 0.0007
35. 0.43
37. 0.0999
39. 0.01973

UNIT 9 Rounding Decimal Fractions and Equivalent Decimal and Common Fractions

1. 0.632
3. 0.240
5. 0.04
7. 0.7201
9. 0.000
11. 0.6875
13. 0.6250
15. 0.6667
17. 0.0800
19. 0.2188
21. 0.5714
23. 0.3333
25. a. 0.125
    b. 0.6154
27. $\frac{1}{8}$
29. $\frac{3}{4}$
31. $\frac{11}{16}$
33. $\frac{3}{1000}$
35. $\frac{251}{500}$
37. $\frac{7}{16}$
39. $\frac{8717}{10{,}000}$
41. $\frac{3}{100}$
43. $\frac{237}{1000}$
45. $\frac{9}{200}$
47. $\frac{7}{8}$
49. a. $\frac{1}{4}$
    b. $\frac{9}{16}$
    c. $\frac{1}{10}$
    d. $\frac{3}{80}$
    e. $\frac{3}{8}$

UNIT 10 Addition and Subtraction of Decimal Fractions

1. a. 15.775
   b. 0.14095
   c. 1.295
   d. 5.129
   e. 381.357
   f. 4.444
   g. 94.2539
   h. 0.1101
   i. 5.7787
   j. 328.963

3. 4.1758"

5. (Other combinations may total certain thicknesses.)
   a. 0.010" + 0.004"
   b. 0.015" + 0.010" + 0.008"
   c. 0.015" + 0.006"
   d. 0.015" + 0.012" + 0.008" + 0.003"
   e. 0.008" + 0.003"
   f. 0.015" + 0.012" + 0.010" + 0.003" + 0.002"
   g. 0.015" + 0.012" + 0.002"
   h. 0.015" + 0.012" + 0.010" + 0.008" + 0.004"

7. A = 12.82 mm
   B = 27.02 mm
   C = 6.58 mm
   D = 20.00 mm
   E = 10.58 mm
   F = 7.39 mm

UNIT 11 Multiplication of Decimal Fractions

1. a. 0.0563
   b. 3.3
   c. 6
   d. 1.6718

3. Dia A = 31.763 mm
   Dia B = 19.199 mm
   Dia C = 12.847 mm
   Dia D = 22.571 mm
   Dia E = 6.741 mm

5. a. 0.15 in
   b. 0.075 in
   c. 0.1125 in
   d. 3.0 mm
   e. 7.5 mm
   f. 13.5 mm

UNIT 12 Division of Decimal Fractions

1. a. 1.597
   b. 2.56
   c. 0.0100
   d. 10,000.000
   e. 11.367
   f. 4.29
   g. 135.53
   h. 0.0062

3. A = 11.75 mm
   B = 5.91 mm
   C = 12.46 mm
   D = 10.95 mm

5. 26 complete bushings

7. 0.063 mm

9. 0.125"

11. 38.50 mm

UNIT 13 Powers

1. 39.304

3. 100,000,000

5. $2\frac{2}{3}$

7. 4.41

9. 64

11. 532.22 mm^2

13. 114.49 m^2

15. $\frac{9}{16}$ sq in

17. $14\frac{1}{16}$ sq in

19. $189\frac{1}{16}$ sq in

21. 8741.82 mm^3

23. 2744 mm^3

25. $\frac{1}{27}$ cu in

27. $3\frac{3}{8}$ cu in

29. $\frac{27}{64}$ cu in

31. 764 mm^2

33. 270 sq in

35. 0.1 cu in

37. 0.2 cu in

39. 184.2 mm^3

41. 16 cu in

43. 10 cu in

45. 11 cu in

47. 385 mm^3

49. 0 cu in

51. 329 mm^2

53. 1 cu in

55. 5265 lb

UNITS 14 Roots

1. 6
3. $\frac{2}{9}$
5. $\frac{3}{4}$
7. 12
9. 4
11. a. 6 mm
11. b. 4 in
    c. 8 in
    d. 10 mm
    e. 1 in
13. a. D = 3 in
    b. D = 6 mm
    c. D = 2 in
    d. D = 1 in
13. e. D = 10 mm
15. 19.77
17. 1.871
19. 4.42
21. 0.0857
23. 6
25. 14
27. 16
29. 1333.835
31. a. D = 7.45 mm
    b. D = 6.08 in
    c. D = 21.91 mm
    d. D = 1.07 in
33. D = 1.5 in

UNIT 15 Table of Decimal Equivalents and Combined Operations of Decimal Fractions

1. 0.78125
3. 0.34375
5. 0.078125
7. $\frac{5}{16}$
9. $\frac{13}{64}$
11. $\frac{49}{64}$
13. $\frac{1}{2}$
15. $\frac{13}{16}$
17. 14.1
19. 25.12
21. 7.24
23. 10.57
25. 16.21
27. 0.084 mm
29. a. C = 9.02 mm
    b. C = 8.74 mm
    c. C = 5.48 mm
31. H = 0.077"

UNIT 16 Computing with a Calculator: Decimals

(All answers are given in the unit.)

UNIT 17 Achievement Review—Section One

1. a. $\frac{12}{32}$
   b. $\frac{70}{100}$
   c. $\frac{16}{64}$
   d. $\frac{72}{128}$
3. a. $2\frac{1}{2}$
   b. $4\frac{1}{5}$
   c. $18\frac{3}{4}$
   d. $3\frac{19}{32}$
   e. $5\frac{9}{64}$
5. a. $\frac{8}{32}, \frac{6}{32}, \frac{9}{32}$
5. b. $\frac{28}{64}, \frac{10}{64}, \frac{9}{64}$
   c. $\frac{70}{100}, \frac{75}{100}, \frac{36}{100}, \frac{65}{100}$
7. a. $\frac{5}{16}$
   b. $\frac{2}{5}$
   c. $1\frac{245}{256}$
   d. $25\frac{23}{40}$
   e. $20\frac{5}{8}$
   f. $\frac{3}{4}$
   g. $3\frac{1}{3}$
   h. 48
7. i. $6\frac{3}{4}$
7. j. $\frac{93}{280}$
9. 51 complete pieces
11. 8 min
13. $A = 3\frac{5}{16}"$
    $B = 2\frac{15}{16}"$
    $C = 3\frac{15}{64}"$
    $D = 3\frac{15}{16}"$
    $E = 3\frac{15}{32}"$
15. a. 0.3
    b. 0.026
    c. 9.026
    d. 5.0081
17. a. 0.75
    b. 0.875
17. c. 0.667
    d. 0.08
    e. 0.65
19. a. 1.587
    b. 6.4274
    c. 12.3069
    d. 9.1053
    e. 23.4077
    f. 0.356
    g. 0.1444
    h. 0.001
    i. 0.0022
    j. 0.002
21. a. 6.76
    b. 0.125
    c. 0.000036
    d. $\frac{9}{25}$
    e. 32.768

23. a. 19.47
 b. 0.935
 c. 0.632
 d. 6.780

25. a. $\frac{15}{32}$
 b. $\frac{49}{64}$
 c. $\frac{1}{32}$
 d. $\frac{31}{32}$

27. A = 1.299 mm
 B = 0.812 mm
 C = 0.325 mm
 D = 0.162 mm
 E = 0.188 mm
 F = 0.375 mm

29. 0.12 mm

31. 12.6 in

SECTION 2 Ratio, Proportion, and Percentage

UNIT 18 Ratio and Proportion

1. $\frac{2}{7}$
3. $\frac{2}{11}$
5. $\frac{6}{23}$
7. $\frac{13}{9}$
9. $\frac{4}{3}$
11. $\frac{1}{10}$
13. a. $\frac{2}{1}$
 b. $\frac{2}{3}$
 c. $\frac{3}{2}$
 d. $\frac{3}{5}$
 e. $\frac{2}{7}$
 f. $\frac{7}{1}$
 g. $\frac{7}{3}$
 h. $\frac{5}{2}$
 i. $\frac{2}{1}$
 j. $\frac{3}{7}$
15. 0.5
17. 35
19. 12
21. 17.5
23. 2.25
25. 8.2
27. $\frac{2}{3}$
29. $\frac{5}{12}$
31. $31\frac{1}{2}$
33. 16.5
35. 12.978
37. a. 12 in
 b. $1\frac{1}{8}$ in
 c. 72.9 mm
 d. 32.4 mm
39. a. 8.031 in
 b. 1.124 in
 c. 4.016 in
 d. 2.720 in
 e. 2.808 in
 f. 7.950 in
 g. 1.125 in
 h. 1.575 in
 i. 5.300 in

39 j. 1.686 in
 k. 9.450 in
 l. 2.040 in
 m. 6.300 in
 n. 0.843 in
 o. 0.094 in
 p. 3.744 in
41. A = 0.85 in
 B = 1.02 in
 C = 1.36 in
 D = 1.53 in

UNIT 19 Direct and Inverse Proportions

1. a. 1.50 mm
 b. 2.59 mm
 c. 1.16 mm
 d. 2.33 mm
 e. 1.93 mm
3. a. 0.990 in
 b. 0.763 in
 c. 79.403 mm
 d. 12.966 mm
 e. 0.429 in
5. direct proportion; 1650 parts
7. direct proportion; 0.48 kg
9. a. 240 rpm
 b. 157.5 rpm
 c. 28 teeth
 d. 25 teeth
 e. 166.2 rpm

UNIT 20 Introduction to Percents

1. 44%
3. 25%
5. 35%
7. 4%
9. 0.8%
11. 207.6%
13. 0.02%
15. 25%
17. 15%
19. 53.125%
21. 159%
23. 1462.5%
25. 0.82
27. 0.03
29. 0.2776
31. 2.249
33. 0.0473
35. 0.0075
37. 0.0237
39. 0.3725
41. $\frac{1}{2}$
43. $\frac{5}{8}$
45. $\frac{4}{25}$
47. $1\frac{9}{10}$
49. $\frac{9}{500}$
51. $\frac{9}{1000}$

UNIT 21 Basic Calculations of Percentages, Percents, and Rates

1. 16
3. 120
5. 78.15
7. 101.4
9. 37.47
11. 392
13. 7.14
15. 0.13
17. 0.99
19. 5.38
21. 50%
23. 37%
25. 118.95%
27. 155.46%
29. 40%
31. 30.77%
33. 42.86%
35. 154.55%
37. 24.49%
39. 50%
41. 150
43. 320
45. 170
47. 184.55
49. 4.17
51. 270.57
53. 42.93
55. 0.5
57. 3.90
59. 28.87
61. 75%
63. 15.60
65. 19.05%
67. 153.99
69. 57.99
71. 3.38

UNIT 22 Percent Practical Applications

1. 56%
3. 6.8 hr
5. 1500 units
7. 12%
9. 262.5 ft
11. Copper: 725 lb
    Tin: 500 lb
    Manganese: 19 lb
    Other: 6 lb
13. 1%
15. 9/16 Rework: 2.7%
    9/16 Scrap: 3.6%
    9/17 Rework: 1.8%
    9/17 Scrap: 3.0%
    9/18 Rework: 3.1%
    9/18 Scrap: 2.0%
17. $3264
19. a. 13%
    b. 69%
    c. 110%
21. 936 castings
23. 18 lb
25. 5.6%
27. 52.63%
29. 8.7 hp
31. 3189 pieces
33. 1750 pieces
35. **Job 1** Labor Cost: 47%
    Material Cost: 22%
    Overhead Cost: 31%
    Job 2 Labor Cost: 32%
    Material Cost: 37%
    Overhead Cost: 31%
    Job 3 Labor Cost: 42%
    Material Cost: 31%
    Overhead Cost: 27%

UNIT 23 Achievement Review—Section Two

1. $\frac{15}{32}$
3. $\frac{6}{23}$
5. $\frac{7}{11}$
7. $\frac{1}{2}$
9. $\frac{3}{5}$
11. a. cost to selling price: $\frac{5}{8}$, cost to profit: $\frac{5}{3}$
    b. cost to selling price: $\frac{7}{12}$, cost to profit: $\frac{7}{5}$
    c. cost to selling price: $\frac{6}{11}$, cost to profit: $\frac{6}{5}$
    d. cost to selling price: $\frac{51}{110}$, cost to profit: $\frac{51}{59}$
13. a. 19.2
    b. 10.8
    c. $\frac{1}{3}$
    d. 14
    e. 32.5
    f. 0.778
    g. 8.282
    h. 1.535
15. a. 100%
    b. 150%
15 c. 275%
    d. 50%
17. a. 0.19
    b. 0.007
    c. 0.0075
    d. 3.103
19. a. 9
    b. 1.27
    c. 87.36
    d. 5.68
    e. 22.90
    f. 275.6
    g. 4
    h. 1.57
21. a. 33.33
    b. 16.47
    c. 223.68
    d. 59.97
    e. 3.90
    f. 41.18
    g. 0.61
23. a. 0.38 kg
    b. 0.63 kg
25. a. 23 kg
    b. 8.6 kg
    c. 11.5 kg
27. 17 lb

SECTION 3 Linear Measurement: Customary (English) and Metric

UNIT 24 Customary (English) Units of Measure

1. a. 8 ft
   b. 10.25 ft
   c. 42 in
   d. 14.4 in
   e. 45 in
   f. 4 yd
   g. 6.25 ft
   h. 24 ft
   i. 12.6 ft
   j. 9 yd
   k. 17 yd
   l. 12 in
   m. 21.5 ft
   n. 92 in
   o. 7.2 in
   p. 46.75 yd
   q. 9.25 yd
   r. 15.5 ft
   s. 62 ft
   t. 111 in
3. 6 complete lengths
5. a. 10 in
   b. 17 in
   c. 7 in
   d. 3 in
   e. $15\frac{5}{8}$ in
   f. $14\frac{3}{8}$ in
   g. $3\frac{1}{8}$ in
   h. $3\frac{9}{16}$ in
7. a. 8 ft 4 in
   b. 1 ft 8 in
   c. 23 ft 3 in
   d. 3 ft 7 in
   e. 19 ft $3\frac{5}{8}$ in
   f. 2 ft $3\frac{15}{16}$ in
   g. 64 ft $7\frac{11}{16}$ in
   h. 2 ft $7\frac{9}{16}$ in

UNIT 25 Metric Units of Linear Measure

1. a. 29 mm
   b. 157.8 mm
   c. 21.975 cm
   d. 9.783 cm
   e. 97 cm
   f. 170 mm
   g. 0.153 m
   h. 6.73 m
   i. 0.093 cm
   j. 0.8 mm
   k. 8.6 mm
   l. 104.6 cm
   m. 300.3 mm
   n. 87.684 cm
   o. 2.039 m
   p. 0.0347 m
   q. 49 mm
   r. 732.1 cm
   s. 63.77 mm
   t. 934 mm
3. 52 mm
5. a. 1.457 in
   b. 4.992 in
   c. 6.811 in
   d. 0.331 in
   e. 94.488 in
   f. 3.543 in
   g. 26.246 ft
   h. 33.464 ft
   i. 28.976 in
   j. 1.341 in
   k. 22.165 in
   l. 2.187 yd
   m. 1.476 ft
   n. 2.559 ft
7. 53.7 in
9. A = 15.75 mm
   B = 28.58 mm
   C = 327.03 mm
   D = 25.10 mm
   E = 3.30 mm
   F = 12.70 mm
   G = 25.00 mm
   H = 2.38 mm
   I = 17.46 mm
   J = 9.53 mm

UNIT 26 Degree of Precision, Greatest Possible Error, Absolute Error, and Relative Error

1. a. 0.1"
   b. 4.25"
   c. 4.35"
3. a. 0.001"
   b. 4.0775"
   c. 4.0785"
5 a. 0.001"
   b. 15.8845"
   c. 15.8855"
7. a. 0.001"
   b. 11.0025"
   c. 11.0035"
9. a. 0.01"
   b. 7.005"
   c. 7.015"
11. a. 0.1"
    b. 6.05"
    c. 6.15"
13. a. 0.01 mm
    b. 26.865 mm
    c. 26.875 mm

15. a. 0.01 mm
    b. 117.055 mm
    c. 117.065 mm

17. a. 0.01 mm
    b. 48.005 mm
    c. 48.015 mm

19. a. 0.01 mm
    b. 6.995 mm
    c. 7.005 mm

21. a. 0.001 mm
    b. 8.0005 mm
    c. 8.0015 mm

	Greatest Possible Error (inches)	ACTUAL LENGTH Smallest Possible (inches)	ACTUAL LENGTH Largest Possible (inches)
23.	0.025	5.275	5.325
25.	0.0005	0.7525	0.7535
27.	0.00005	0.93685	0.93695

	Greatest Possible Error (millimeters)	ACTUAL LENGTH Smallest Possible (millimeters)	ACTUAL LENGTH Largest Possible (millimeters)
29.	0.5	63.5	64.5
31.	0.25	98.25	98.75
33.	0.005	13.365	13.375

35. Absolute Error: 0.020 in
    Relative Error: 0.052%

37. Absolute Error: 0.200°
    Relative Error: 1.575%

39. Absolute Error: 0.140 mm
    Relative Error: 0.587%

41. Absolute Error: 0.030 mm
    Relative Error: 0.857%

43. Absolute Error: 0.010°
    Relative Error: 0.995%

45. Absolute Error: 0.026 in
    Relative Error: 0.142%

UNIT 27 Tolerance, Clearance, and Interference

1. a. $\frac{1}{32}$"
   b. $\frac{1}{8}$"
   c. 16.73"
   d. 0.911"
   e. 0.0003"
   f. 11.003"

3. a. Max. Limit = 4.643"
   Min. Limit = 4.640"
   b. Max. Limit = 5.932"
   Min. Limit = 5.927"
   c. Max. Limit = 2.004"
   Min. Limit = 2.000"
   d. Max. Limit = 4.6729"
   Min. Limit = 4.6717"
   e. Max. Limit = 1.0884"
   Min. Limit = 1.0875"
   f. Max. Limit = 28.16 mm
   Min. Limit = 28.10 mm
   g. Max. Limit = 43.98 mm
   Min. Limit = 43.94 mm
   h. Max. Limit = 118.73 mm
   Min. Limit = 118.66 mm
   i. Max. Limit = 73.398 mm
   Min. Limit = 73.386 mm
   j. Max. Limit = 45.115 mm
   Min. Limit = 45.106 mm

5. a. 0.943" ± 0.005"
   b. 1.687" ± 0.001"
   c. 2.998" ± 0.002"
   d. 0.069" ± 0.004"
   e. 4.1880" ± 0.0007"
   f. 0.9984" ± 0.0037"
   g. 1.0006" ± 0.0004"
   h. 8.4660" ± 0.0011"
   i. 44.31 mm ± 0.01 mm
   j. 10.02 mm ± 0.04 mm
   k. 64.92 mm ± 0.03 mm
   l. 38.016 mm ± 0.028 mm
   m. 124.9915 ± 0.0085 mm
   n. 43.078 mm ± 0.013 mm
   o. 98.8835 mm ± 0.0045 mm

7. All dimensions are in millimeters.

		Basic Dimension	Maximum Diameter (Max. Limit)	Minimum Diameter (Min. Limit)	Maximum Interference (Allowance)	Minimum Interference
a.	DIA A	20.73	20.75	20.71	0.09	0.01
	DIA B	20.68	20.70	20.66		
b.	DIA A	32.07	32.09	32.05	0.10	0.02
	DIA B	32.01	32.03	31.99		
c.	DIA A	12.72	12.74	12.70	0.11	0.03
	DIA B	12.65	12.67	12.63		

9. All dimensions are in millimeters.

		Basic Dimension	Maximum Diameter (Max. Limit)	Minimum Diameter (Min. Limit)	Maximum Interference (Allowance)	Minimum Interference
a.	DIA A DIA B	87.58 87.50	87.61 87.53	87.55 87.47	0.14	0.02
b.	DIA A DIA B	9.94 9.85	9.97 9.88	9.91 9.82	0.15	0.03
c.	DIA A DIA B	130.03 129.96	130.06 129.99	130.00 129.93	0.13	0.01

11. 18.20 mm

13. Max. thickness = 2.88 mm
    Min. thickness = 2.82 mm

15. Holes 5 and 6 are out of tolerance.

UNIT 28 Customary and Metric Steel Rules

1. a. $\frac{3}{32}$"
   b. $\frac{5}{16}$"
   c. $\frac{1}{2}$"
   d. $\frac{5}{8}$"
   e. $\frac{3}{4}$"
   f. $\frac{29}{32}$"
   g. $1\frac{3}{32}$"
   h. $1\frac{5}{16}$"
   i. $\frac{5}{64}$"
   j. $\frac{7}{32}$"
   k. $\frac{3}{8}$"
   l. $\frac{35}{64}$"
   m. $\frac{51}{64}$"

1. n. $\frac{31}{32}$"
   o. $1\frac{15}{64}$"
   p. $1\frac{29}{64}$"

3. a. $\frac{1}{4}$"
   b. $\frac{9}{16}$"
   c. $\frac{1}{2}$"
   d. $\frac{1}{2}$"
   e. $\frac{11}{32}$"
   f. $2\frac{7}{32}$"
   g. $\frac{7}{32}$"
   h. $\frac{3}{4}$"
   i. $\frac{23}{32}$"
   j. $\frac{1}{2}$"

3. k. $\frac{15}{32}$"
   l. $\frac{3}{8}$"
   m. $\frac{5}{32}$"
   n. $4\frac{29}{32}$"

5. a. 0.12"
   b. 0.22"
   c. 0.40"
   d. 0.62"
   e. 0.80"
   f. 1.04"
   g. 1.32"
   h. 1.42"
   i. 0.11"
   j. 0.23"
   k. 0.38"
   l. 0.57"
   m. 0.84"
   n. 1.07"
   o. 1.29"
   p. 1.45"

7. A = 0.54"
   B = 0.42"
   C = 1.38"
   D = 1.18"
   E = 0.34"
   F = 0.28"
   G = 1.00"
   H = 0.22"
   I = 0.10"

9. a. 46 mm
   b. 70 mm
   c. 20 mm
   d. 82 mm
   e. 10 mm
   f. 23 mm
   g. 25 mm
   h. 121 mm
   i. 17 mm
   j. 22 mm
   k. 36 mm
   l. 10 mm
   m. 52 mm
   n. 6 mm

UNIT 29 Customary Vernier Calipers and Height Gages

1. a. 2.641"
   b. 3.376"
   c. 2.021"
   d. 0.508"
   e. 4.788"
   f. 2.991"
   g. 1.581"
   h. 1.098"

3.

	A (inches)	B (inches)	C
a.	3.225	3.250	17
b.	2.875	2.900	2
c.	4.825	4.850	14
d.	0.600	0.625	11
e.	4.350	4.375	19
f.	0.075	0.100	9
g.	7.850	7.875	7
h.	1.625	1.650	21
i.	4.025	4.050	9
j.	0.000	0.025	22
k.	3.325	3.350	8
l.	5.975	6.000	24
m.	0.275	0.300	3
n.	0.950	0.975	15

5. a. 1.909"
   b. 4.620"
   c. 7.969"
   d. 0.439"
   e. 2.779"
   f. 6.459"
   g. 3.612"
   h. 8.391"

UNIT 30 Customary Micrometers

1. 0.589"
3. 0.736"
5. 0.808"
7. 0.738"
9. 0.157"
11. 0.949"
13. 0.441"
15. 0.153"
17. 0.424"
19. 0.038"
21. 0.983"

	Barrel Scale Setting (inches)	Thimble Scale Setting (inches)
23.	0.375–0.400	0.012
25.	0.950–0.975	0.023
27.	0.075–0.100	0.004
29.	0.025–0.050	0.013
31.	0.425–0.450	0.002

33. 0.3637"
35. 0.0982"
37. 0.3105"
39. 0.1448"
41. 0.5157"
43. 0.2749"
45. 0.3928"
47. 0.9717"
49. 0.3004"
51. 0.0009"
53. 0.8594"

	Barrel Scale Setting (inches)	Thimble Scale Setting (inches)	Vernier Scale Setting (inches)
55.	0.775–0.800	0.009–0.010	0.0006
57.	0.000–0.025	0.008–0.009	0.0003
59.	0.300–0.325	0.000–0.001	0.0001
61.	0.800–0.825	0.000–0.001	0.0008
63.	0.975–1.000	0.014–0.015	0.0004

UNIT 31 Customary and Metric Gage Blocks

One combination for each dimension is given. A number of different combinations will produce the given dimensions.

1. 0.1008", 0.113", 0.650", 3.000"
3. 0.1002", 0.122", 0.900", 2.000"
5. 0.1009", 0.125", 0.050"
7. 0.123", 0.850", 3.000", 4.000"
9. 0.250", 1.000", 2.000", 3.000", 4.000"
11. 0.1007", 0.125", 0.650", 4.000"
13. 0.1001", 0.112", 0.050"
15. 0.140", 0.950", 4.000"
17. 0.1009", 0.128", 0.750", 2.000"
19. 0.1007", 0.127", 0.550", 3.000", 4.000"
21. 0.1006", 0.134", 0.200", 2.000", 3.000", 4.000"
23. 0.103", 0.900", 1.000", 4.000"
25. 0.1008", 0.149", 0.450"
27. 1.003 mm, 1.07 mm, 2 mm, 10 mm

29. 1.09 mm, 5 mm, 60 mm, 90 mm
31. 1.007 mm, 1.7 mm, 1 mm, 40 mm
33. 1.06 mm, 1.4 mm, 4 mm, 70 mm
35. 1.06 mm, 1.8 mm, 1 mm, 10 mm
37. 1.001 mm, 1.07 mm, 4 mm
39. 1.009 mm, 1.09 mm, 7 mm, 30 mm
41. 1.005 mm, 6 mm, 60 mm
43. 1.007 mm, 1 mm
45. 1.03 mm, 2 mm, 20 mm, 80 mm, 90 mm
47. 1.004 mm, 1.8 mm, 8 mm
49. 1.005 mm, 1.05 mm, 1.5 mm, 2 mm, 50 mm

UNIT 32 Achievement Review—Section Three

1. a. 6.75 ft
   b. 75 in
   c. 28.8 ft
   d. 27 mm
   e. 800 mm
   f. 21.8 cm
3. 5 complete lengths
5.

	Greatest Possible Error	ACTUAL LENGTH	
		Smallest Possible	Largest Possible
a.	0.01"	4.27"	4.29"
b.	0.00005"	0.83665"	0.83675"
c.	0.01 mm	46.15 mm	46.17 mm
d.	0.005 mm	16.445 mm	16.455 mm

7. a. Max. Limit: 1.719"
   Min. Limit: 1.709"
   b. Max. Limit: 4.0688"
   Min. Limit: 4.0676"
   c. Max. Limit: 5.9055"
   Min. Limit: 5.9047"
   d. Max. Limit: 64.99 mm
   Min. Limit: 64.83 mm
   e. Max. Limit: 173.003 mm
   Min. Limit: 172.990 mm
9. a. 0.0040"
   b. 0.0012"
   c. 0.0020"
   d. 0.0006"
   e. 0.0028"
   f. 0.0018"
   g. 0.0021"
   h. 0.0009"
11. a. $\frac{3}{32}''$
   b. $\frac{11}{32}''$
   c. $\frac{9}{16}''$
   d. $\frac{25}{32}''$
   e. $\frac{29}{32}''$
   f. $1\frac{3}{32}''$
   g. $1\frac{7}{32}''$
   h. $1\frac{11}{32}''$
   i. $\frac{3}{64}''$
   j. $\frac{13}{64}''$
   k. $\frac{23}{64}''$
   l. $\frac{17}{32}''$
   m. $\frac{49}{64}''$
   n. $\frac{61}{64}''$
   o. $1\frac{7}{32}''$
   p. $1\frac{25}{64}''$
13. a. 6 mm
   b. 19 mm
   c. 29 mm
   d. 43 mm
   e. 49 mm
   f. 57 mm
   g. 66 mm
   h. 74 mm
   i. 4.5 mm
   j. 11 mm
   k. 21.5 mm
   l. 28.5 mm
   m. 45.5 mm
   n. 54 mm
   o. 65.5 mm
   p. 72.5 mm
15. a. (1) 0.558"
   (2) 0.089"
   (3) 0.679"
   (4) 0.638"
   b. (1) 0.3023"
   (2) 0.2855"
   (3) 0.0732"
   (4) 0.4180"
17. a. 1.03 mm, 1.5 mm, 5 mm, 60 mm
   b. 1.02 mm, 1.2 mm, 3 mm, 30 mm, 90 mm
   c. 1.002 mm, 1.09 mm, 3 mm, 80 mm
   d. 1.004 mm, 1.07 mm, 1.2 mm, 10 mm
   e. 1.006 mm, 1.06 mm, 4 mm, 60 mm
   f. 1.004 mm, 1.3 mm, 1 mm, 40 mm
   g. 1.008 mm, 1.09 mm, 1.9 mm, 6 mm, 90 mm
   h. 1.001 mm, 1.07 mm, 5 mm, 10 mm, 90 mm

SECTION 4 Fundamentals of Algebra

UNIT 33 Symbolism and Algebraic Expressions

1. $6x + y$
3. $21 - b$
5. r/s
7. xy/m^2
9. a. $2\frac{1}{2}R$
   b. $2\frac{3}{4}R$
   c. $2\frac{1}{4}R$
   d. $6\frac{1}{4}R$
11. $n - p - t$
13. a. 52
    b. 20
    c. 6
    d. 4
    e. 1
    f. 2
15. a. 151
    b. 96
    c. 14.5
    d. 8
    e. 126
    f. 108
17. a. 112.5 sq in
    b. 10.6 in
19. a. 14.7 in
    b. 35.8 sq in
21. 32.4 sq in
23. 27.2 mm^2
25. 62.7 sq in
27. a. 5.1 in
    b. 215.5 cu in

UNIT 34 Signed Numbers

1. a. (+)9
   b. (+)5
   c. (+)6
   d. (−)6
   e. (−)10
   f. (+)7
   g. (−)20
   h. (−)10
   i. (+)3
   j. (−)8
   k. (−)11
   l. (−)6
   m. (+)17.5
   n. (−)17.5
   o. (+)6.5
   p. (+)1.5
   q. $(-)5\frac{1}{4}$
   r. $(-)5\frac{3}{4}$
3. a. −25, −18, −1, 0, +2, +4, +17
   b. −21, −19, −5, −2, 0, +5, +13, +27
   c. −25, −10, −7, 0, +7, +10, +14, +25
   d. −14.9, −3.6, −2.5, 0, +0.3, +15, +17
   e. $-16, -13\frac{7}{8}, -3\frac{5}{8}, +6, +14\frac{1}{8}$
5. a. 23
   b. 30
   c. 25
   d. −23
   e. −33
   f. 7
   g. −8
   h. −1
   i. −6
   j. −22
   k. −13
   l. $-3\frac{1}{8}$
   m. $-13\frac{5}{16}$
   n. −14.47
   o. 0.43
   p. 1
   q. −39.62
   r. 31.25
   s. −28.9
   t. −14.06
7. a. −24
   b. 24
   c. −30
   d. 30
   e. −35
   f. 28
   g. 0
   h. −32.5
   i. 0.32
   j. 0.036
   k. $-1\frac{1}{8}$
   l. 0
   m. −8
   n. −8
   o. 0
   p. −7350.488
   q. 10.6
   r. −0.221
   s. 0.384
   t. −0.3
9. a. 4
   b. 8
   c. −8
   d. −64
   e. 16
   f. −32
   g. 36
   h. −125
   i. 64
   j. 2.56
   k. −0.064
   l. 0.647
   m. 2.496
   n. −0.614
   o. 0.389
   p. $-\frac{8}{27}$
   q. −0.830
   r. 0.003
   s. −1.749
   t. 0.001
11. a. 3
    b. 9
    c. 2
    d. 4
    e. −2
    f. 2
    g. −5
    h. 5
    i. 42.103
    j. 0.155
    k. 4.002
    l. 0.060
13. 14
15. 4
17. 21
19. 142
21. 9
23. 9.672
25. 0.009
27. 2
29. −0.5
31. 14
33. 4.569
35. 0.135

UNIT 35 Algebraic Operations of Addition, Subtraction, and Multiplication

1. $19y$
3. $-22xy$
5. 0
7. -10 pt
9. $15.2a^2b$
11. $1\frac{1}{4}xy$
13. $-2.91gh^3$
15. $11P$
17. $5P + 2P^2$
19. $7ab^2 - 2a^2b - a^2b^2$
21. $-1\frac{7}{8}xy$
23. $6.666M$
25. $-3T + 2T^2$
27. $-7a^2x$
29. A. $2.3x$
    B. $3.8x$
    C. $6.1x$
    D. $4.0x$
    E. $7.2x$
    F. $3.1x$
    G. $1.1x$
31. $2a - 11m$
33. $3xy^2 + 3x^2y$
35. $-2x^3 - 7x^2 + 4x + 12$
37. 0
39. $-0.4c + 3.6cd + 3.7d$
41. $2xy$
43. $-2xy$
45. $-10a^2$
47. $12mn^3$
49. $1\frac{1}{4}x^2$
51. $-13a + 7a^2$
53. $-2ax^2$
55. d^2t^2
57. $3x - 21$
59. 0
61. $x^2 + 3xy$
63. 0
65. $3a^3 - 1.3a^2 + a$
67. $-d^2 - 2dt + dt^2 + 4$
69. $8.08e + 15.76f + 10.03$
71. x^3
73. $56a^4b^3c^3$
75. 0
77. $3d^8r^4$
79. $0.21x^7y^4$
81. 0
83. $-3.36bc$
85. $-2x^8y^6$
87. $-49a^4b^4$
89. $-x^4y^2$
91. $-10x^2y^3 + 15x^5y$
93. $-8a^4b^5 + 2a^3b^4 + 4a^3b^2$
95. $-4dt - 4t^2 + 4$
97. $3x^3 + 27x + 7x^2 + 63$
99. $10a^3x^6 + 5ab^2x^4 + 2a^2bx^4 + b^3x^2$

UNIT 36 Algebraic Operations of Division, Powers, and Roots

1. $2x$
3. -1
5. 0
7. $-6H$
9. 3.7
11. $5cd$
13. $8g^2h$
15. xz^2
17. $4P^2V$
19. $\frac{1}{4}FS^2$
21. $8x^2 + 12x$
23. $-3x^5y + 2xy^3$
25. $-15a - 25a^4$
27. $-2cd + 5c^2d + 1$
29. $3a^2x + ax^2 - 2$
31. $4a - 6a^2c - 8c^2$
33. $9a^2b^2$
35. $8x^6y^3$
37. $-27c^9d^6e^{12}$
39. $49x^8y^{10}$
41. $a^9b^3c^6$
43. $-x^{12}y^{15}z^3$
45. $0.064x^9y^3$
47. $18.49M^4N^4P^2$
49. $-512a^{12}b^{18}c^3$
51. $0.36d^6e^6f^{12}$
53. $9x^4 - 30x^2y^3 + 25y^6$
55. $25t^4 - 60t^2x + 36x^2$
57. $0.16d^4t^6 - 0.16d^2t^4 + 0.04t^2$
59. $\frac{4}{9}c^4d^2 + c^3d^3 + \frac{9}{16}c^2d^4$
61. $a^{16}b^4 + 2a^8b^2x^6y^3 + x^{12}y^6$
63. m^3n^2s
65. $9x^4y^3$
67. $-3x^2y^4$
69. $0.4a^4cf^3$
71. $\frac{1}{4}xy$
73. $-4d^2t^3$
75. $2h^2$
77. $4a\sqrt[3]{c}$
79. $\frac{3}{4}ac\sqrt{b}$
81. $-2a\sqrt[5]{b^3}$
83. $9b - 15b^2 + c - d$
85. $-ab - a^2b + a$
87. $-16 - xy$
89. $1 - r$
91. $-2x + 24$
93. $6 + c^2d$
95. $6a^2 - 6b$
97. $3b$
99. $7y^6 + 15$

101. $2\frac{2}{3}d$
103. $100a - 5a^4b^6$
105. $5f^2 + 6f^2h$
107. 8×10^4
109. 9.76×10^5
111. 1.5×10^{-2}
113. 2×10^{-1}
115. 3.9×10^{-4}
117. 1.75×10^{-3}
119. 160,000
121. 5,090,000
123. 0.0000632
125. 0.000003123
127. 0.0007321
129. 0.0209
131. 1.61×10^{-6}
133. 3.20×10^{-10}
135. 1.01×10^6
137. -4.77×10^{13}
139. 4.61×10^7
141. -4.38×10^{10}
143. -2.61×10^{-7}
145. 4.30×10^{12}
147. 1.02×10^3

UNIT 37 Introduction to Equations

1. 12
3. 11
5. 4
7. 12
9. 5
11. 0.5", 1", 3"
13. 1.115 mm
15. 50 mm
17. 12°
19. 36°
21. $\frac{1"}{2}$
23. a. $\frac{3"}{4}$
    b. $\frac{1"}{2}$
    c. $1\frac{3"}{4}$
25. 3
27. 7
29. 6
31. 16
33. 84
35. 3
37. 48
39. 20

UNIT 38 Solution of Equations by the Subtraction, Addition, and Division Principles of Equality

1. 7
3. 19
5. 4
7. 43
9. –53
11. 43
13. –50
15. 18.8
17. 0
19. $-1\frac{5}{8}$
21. $-1\frac{1}{4}$
23. $-23\frac{1}{8}$
25. 18"
27. $\frac{13"}{16}$
29. 37.61 mm
31. 0.1008"
33. $7\frac{11"}{32}$
35. 4.4286"
37. 0.1653"
39. 34
41. −10
43. 135
45. 28
47. 83
49. 78
51. 9.3
53. −3.69
55. −0.005
57. −4.89
59. $\frac{1}{2}$
61. $-16\frac{5}{32}$
63. 18.052
65. $8\frac{7"}{16}$
67. $4\frac{1"}{2}$
69. 53.3 mm
71. 830 mm
73. 23
75. −3
77. 6
79. 9
81. 2.3
83. 0
85. 20
87. −1.8
89. 19.75
91. 32
93. −72
95. $-4\frac{1}{2}$
97. 0.2
99. 21.03"
101. 21.75°
103. 124.94 mm
105. 63.33 r/min
107. 14
109. 11
111. −22
113. 48.1995
115. 16.14
117. $\frac{3}{17}$
119. 0.27
121. −19
123. −17.101

UNIT 39 Solution of Equations by the Multiplication, Root, and Power Principles of Equality

1. 30
3. 63
5. 27
7. 0
9. 21.5
11. 23.4
13. −6
15. 0
17. 0.0624
19. $3\frac{3}{4}$
21. 2
23. $\frac{3}{4}$
25. 435.12 mm
27. 7.0711"
29. 0.032"
31. 163.8 mm
33. 4
35. 9
37. 4
39. 12
41. 100
43. $\frac{3}{5}$
45. $\frac{4}{7}$
47. $-\frac{1}{2}$
49. 0.2
51. 1.659
53. 0.497
55. 1.673
57. a. 6 in
    b. $\frac{5}{8}$ ft
    c. 1.2 m
    d. 8.044 m
    e. 0.221 ft
59. 36
61. 1.44
63. 0.6724
65. 4.913
67. 0
69. −32
71. −0.216
73. 0.001
75. $\frac{1}{256}$
77. $\frac{25}{64}$
79. 23.591
81. 0.480
83. a. 11.56 sq in
    b. 0.563 sq ft
    c. 0.425 m^2
    d. 4.674 mm^2
    e. 1.664 in
85. −5
87. 36
89. −0.001
91. 0.001
93. $\frac{9}{64}$
95. $\frac{4}{5}$
97. −26.016
99. 0.340
101. 0.9

UNIT 40 Solution of Equations Consisting of Combined Operations and Rearrangement of Formulas

1. 9
3. 7
5. 2
7. 1
9. 9
11. −0.67
13. 4.8
15. 7
17. 30.5
19. 3
21. 4
23. −1
25. 9
27. 0.77
29. 9.99
31. 0.893
33. 27,066.929
35. 0.5
37. 1939.655
39. 0.093
41. 133.690
43. 12.341
45. a. $a = \frac{A}{b}$
    b. $b = \frac{A}{a}$
    c. $a = \sqrt{d^2 - b^2}$
    d. $b = \sqrt{d^2 - a^2}$
47. a. $D_O = \sqrt{FW^2 + D^2}$
    b. $D = \sqrt{D_O^2 - FW^2}$
    c. $d = 2a + 2C - D_O$
    d. $a = (D_O - 2C + d) \div 2$
49. a. $D = M + 1.5155P - 3W$
    b. $P = \frac{D + 3W - M}{1.5155}$
    c. $W = \frac{M - D + 1.5155P}{3}$
51. a. $D = \frac{L - 1.57d - 2x}{1.57}$
    b. $d = \frac{L - 1.57D - 2x}{1.57}$
    c. $x = \frac{L - 1.57D - 1.57d}{2}$
53. a. $S = \frac{Ca}{C - F}$
    b. $C = \frac{Ca + SF}{S}$
55. a. $h = 5.84$ cm
    b. $h = 0.847$ cm
57. 3.45
59. 27.5
61. 20.407
63. 19.635
65. 2.053
67. –1.5
69. 3

UNIT 41 Applications of Formulas as to Cutting Speed, Revolutions per Minute, and Cutting Time

1. 57 fpm
3. 90 fpm
5. 100 fpm
7. 130 m/min
9. 106 m/min
11. 111 rpm
13. 43 rpm
15. 2037 rpm

17. 477 r/min
19. 3626 r/min
21. 4.8 min
23. 11.4 min
25. 45 m/min
27. 153 fpm
29. 183 rpm
31. 3820 r/min
33. 87 rpm
35. 1.6 min
37. 264 min
39. 0.29 inch per revolution
41. 45 h
43. 14.6 h
45. 154 rpm
47. 1493 rpm
49. 230 rpm
51. 800 rpm
53. 414 rpm
55. 436 rpm
57. 343 rpm

UNIT 42 Applications of Formulas to Spur Gears

1. 2
3. 0.6283 inch
5. 3.7143 inches
7. 0.1745 inch
9. 14 teeth
11. 7.25 inches
13. 1.2047 inches
15. 0.1818 inch
17. 0.0785 inch
19. 0.0351 inch
21. 0.3082 inch
23. 0.2112 inch
25. 0.2222 inch
27. 15
29. 7
31. 26
33. 0.1429 inch
35. 0.0964 inch
37. 0.2857 inch
39. 2.7239 inches
41. 0.0143 inch
43. 23 teeth
45. 0.0015 inch
47. 0.0038 inch
49. 0.0060 inch
51. 6.7821 inches
53. 1.9375 inches
55. a. 117 mm
    b. 20.421 mm
    c. 130 mm
    d. 6.5 mm
    e. 13 mm
    f. 10.210 mm
57. a. 25 mm
    b. 7.854 mm
    c. 30 mm
    d. 2.5 mm
    e. 5 mm
    f. 3.927 mm
59. a. 260 mm
    b. 31.417 mm
    c. 280 mm
    d. 10 mm
    e. 20 mm
    f. 15.708 mm
61. 30 teeth
63. 8.169 mm

UNIT 43 Achievement Review—Section Four

1. a. $x + y - c$
   b. $ab \div d$
   c. $2M - P^2$
3. a. −37
   b. 16
   c. −14.4
   d. −72
   e. 0.78
   f. −6
   g. 32
   h. 36
   i. −125
   j. −3
   k. $\frac{1}{16}$
   l. 1.448
   m. 40.085
   n. −0.510
5. a. 21
   b. 38
   c. 14
   d. 32.2
   e. –39.3
   f. −1.4
   g. −12
   h. 5.8
   i. 74.052
   j. −6.784
   k. −0.333
   l. 3
   m. 9
   n. $\frac{-2}{3}$
   o. 166.204
   p. 202.572
   q. −0.305
   r. 0.887
7. a. 4.243
   b. 1.56
   c. 290.948
   d. 6.089
   e. 2.709
9. a. 19.2
   b. 10.8
   c. $1\frac{1}{3}$
   d. 7
   e. 4.050
   f. −22.5
   g. −10
   h. 2.5
   i. 4
   j. 2.483
11. a. 8.19×10^{-8}
    b. -5.51×10^{1}
    c. 6.32×10^{5}
    d. 1.25×10^{-9}
13. a. 151 fpm
    b. 1273 r/min
    c. 3.57 min
    d. 3.5
    e. 0.5393 in
    f. 8.5989 inches

SECTION 5 Fundamentals of Plane Geometry

UNIT 44 Lines and Angular Measure

1. a. parallel
   b. perpendicular
   c. oblique
3. a. ||
   b. ⊥
   c. °
   d. '
   e. "
5. 67°51'
7. 117°42'
9. 93°09'
11. 6°28'
13. 77°40'
15. 212°04'16"
17. 44°26'38"
19. 103°0'32"
21. 89°54'20"
23. 19°53'50"
25. 107.75°
27. 87.27°
29. 56.80°
31. 2.32°
33. 79.98°
35. 57.1458°
37. 98.3403°
39. 2.1203°
41. 61.2017°
43. 76°52'
45. 244°08'
47. 46°42'12"
49. 540°
51. 16°09'
53. 109°21'09"
55. 44°
57. 21°59'35"
59. 97°03'59"
61. 87°57'
63. 110°51'05"
65. 84°
67. 270°15'
69. 43°30'
71. 68°30'
73. 51°25'43"
75. 161°10'25"

UNIT 45 Protractors—Simple Semicircular and Vernier

1. ∠A = 25°
   ∠B = 42°
   ∠C = 57°
   ∠D = 77°
   ∠E = 93°
   ∠F = 11°
   ∠G = 27°
   ∠H = 46°
   ∠I = 76°
   ∠J = 87°
3. The third angle measures 28°.
5. ∠1 = 29°
   ∠2 = 133°
   ∠3 = 29°
   ∠4 = 58°
   ∠5 = 39°
   ∠6 = 27°
   ∠7 = 122°
   ∠8 = 31°
5. ∠9 = 72°
   ∠10 = 103°
   ∠11 = 64°
   ∠12 = 48°
   ∠13 = 150°
   ∠14 = 19°
7. 19°45'
9. 50°15'
11. 20°15'
13. 20°30'
15. a. 47°
   b. 14°
   c. 73°
   d. 85°
   e. 22°11'
   f. 44°41'
   g. 68°17'
   h. 11°40'33"
   i. 30°59'1"

UNIT 46 Types of Angles and Angular Geometric Principles

1. a. ∠A, ∠BAF, ∠FAB
   b. ∠B, ∠ABC, ∠CBA
   c. ∠3, ∠BCD, ∠DCB
   d. ∠4, ∠CDE, ∠EDC
   e. ∠5, ∠DEF, ∠FED
   f. ∠6, ∠AFE, ∠EFA
3. a. acute
   b. right
   c. right
   d. acute
3. e. acute
   f. obtuse
   g. straight
   h. acute
   i. right
   j. reflex
   k. straight
5. a. ∠3 and ∠6, ∠4 and ∠5
   b. ∠1 and ∠6, ∠2 and ∠5, ∠3 and ∠8, ∠4 and ∠7
7. a. ∠2 = 148°, ∠3,=32°, ∠4 = 148°
   b. ∠2 = 144°41', ∠3 = 35°19', ∠4 = 144°41'
9. a. ∠1, ∠2, ∠5 ∠7, ∠9, ∠11, ∠13, ∠15 = 109°
   ∠3, ∠4, ∠6, ∠8, ∠10, ∠12, ∠14 = 71°
   b. ∠1, ∠2, ∠5, ∠7, ∠9, ∠11, ∠13, ∠15 = 93°08'
   ∠3, ∠4, ∠6, ∠8, ∠10, ∠12, ∠14 = 86°52'
11. a. ∠2 = 67°, ∠3 = 113°
   b. ∠2 = 74°12', ∠3 = 105°48'

UNIT 47 Introduction to Triangles

1. isosceles
3. scalene
5. right
7. equilateral
9. 180°
11. a. 28°
    b. 29°42'47"
13. a. 17.3"
    b. 17.3"
15. a. 81°30'
    b. 77°22'30"
17. a. 11°
    b. 46°
19. a. 48°
    b. 79°
21. a. ∠A
    b. ∠C
    c. ∠B
23. a. ∠D
    b. ∠E
    c. ∠2

UNIT 48 Geometric Principles for Triangles and Other Common Polygons

1. Pairs A, B, D, and F
3. a. 55.86 mm
   b. 93.85 mm
5. a. 52°42'
   b. 37°18'
7. a. 79°
   b. 11°
   c. 11°
9. a. 4.909 in
   b. 2.640 in
11. a. 72.5 mm
    b. 113.6 mm
13. a. 118.30 mm
    b. 118.30 mm
15. a. 15 in
    b. 5 in
17. a. 960 mm
    b. 576 mm
19. a. 3.779 in
    b. 4.281 in
21. x = 187.75 mm
    y = 191.66 mm
23. a. 65°
    b. 96°

UNIT 49 Introduction to Circles

1. a. Chord
   b. Diameter or Chord
   c. Radius
   d. Center
3. a. Sector
   b. Segment
   c. Radius
   d. Radius
   e. Chord
   f. Arc
5. a. 20.420 in
   b. 94.248 mm
   c. 116.868 mm
   d. 18.410 in
   e. 11.141 in
   f. 69.391 mm
   g. 52.044 mm
   h. 1.222 in
7. 211.12 mm
9. 167.98 in
11. a. 4.090"
    b. 3.980"
13. a. 100°
    b. 60°
15. 21.23 mm
17. a. 118°56'
    b. 114°29'
19. a. 80°
19. b. 1.44"
21. a. (1) 48°39'
       (2) 24°03'
    b. (1) 41°40'
       (2) 31°02'
23. a. 0.114 in
    b. 0.472 in
25. a. 168.75 mm
    b. 210.94 mm

UNIT 50 Arcs and Angles of Circles, Tangent Circles

1. 5.498 in
3. 243°30' or 243.506°
5. 14.921 in
7 a. (1) 76°
     (2) 31°
     (3) 134°
  b. (1) 63.76°
     (2) 43.24°
     (3) 99.76°
9. a. (1) 41°
      (2) 139°
   b. (1) 37°30'
      (2) 142°30'
11. a. (1) 94°
       (2) 16°
    b. (1) 51°30'
       (2) 35°46'
13. a. (1) 58°
       (2) 90°
13. b. (1) 56°28'
       (2) 90°
15. a. (1) 46°25'
       (2) 28°38'
    b. (1) 43°56'
       (2) 30°0'
17. a. (1) 11°
       (2) 24°
    b. (1) 1°
       (2) 16°
19. a. 41.82 mm
    b. 84.48 mm
21. a. 1.910 in
    b. 1.844 in
23. a. 22°
    b. 30°54'
25. a. 75°
    b. 82°30' or 82.50°

UNIT 51 Fundamental Geometric Constructions

(All problems are constructed.)

UNIT 52 Achievement Review—Section Five

1. a. 123°41'
   b. 62°29'
   c. 42°19'13"
   d. 13°16'
   e. 109°32'
   f. 22°11'9"
   g. 43°30'
   h. 25°50'
3. 145°38'51"
5. 64°8'31"
7. 103.6453°
9. a. 10°30'
9. b. 19°45'
   c. 39°30'
11. a. 139°
    b. 80°28'
    c. 76°56'33"
13. a. (1) 39°43'
       (2) 1 ft
    b. (1) 60°
       (2) 9.6 in
       (3) 4.8 in
    c. (1) 17°30'
       (2) 72°30'
15. 198°
17. 114.59 mm
19. a. 3.262 in
    b. 8.200 in
21. a. 9.023 in
    b. 261.16°
23. $\angle 1 = 25^\circ$
    $\angle 2 = 67^\circ$
    $\angle 3 = 48^\circ$
    $\angle 4 = 29^\circ$
    $\angle 5 = 48^\circ$
23. $\angle 6 = 86^\circ$
    $\angle 7 = 94^\circ$
    $\angle 8 = 126^\circ$
    $\angle 9 = 54^\circ$
    $\angle 10 = 88^\circ$
25. $\angle 1 = 32^\circ$
    $\angle 2 = 58^\circ$
    $\angle 3 = 17^\circ$
    $\angle 4 = 15^\circ$
    $\angle 5 = 32^\circ$
25. $\angle 6 = 20^\circ$
    $\angle 7 = 38^\circ$
    $\angle 8 = 21^\circ$
    $\angle 9 = 69^\circ$
    $\angle 10 = 112^\circ$
27. 0.4984 in.
29. OA = 70.88 cm
    OB = 36.06 cm
    OC = 27.59 cm
31. $\angle 1 = 56°01'$
33. $x = 5.7$ in

SECTION 6 Geometric Figures: Areas and Volumes

UNIT 53 Areas of Rectangles, Parallelograms, and Trapezoids

1. 1.36 sq ft
3. 5.09 sq yd
5. 1.8 sq yd
7. 38.7 sq ft
9. 5 cm^2
11. 3.825 m^2
13. 2,300,000 mm^2
15. 90.2 mm^2
16. 7500 mm^2
17. 17,200 cm^2
19. 3.3 m^2
21. 1.1 sq in
23. 201.8 ft^2
25. 51 sq in
27. 4.5 in
29. 0.5 m
31. 14.0 mm
33. 58.6 mm
35. 0.9 ft
37. 4.88 sq ft
39. $180
41. 17.8 mm
43. 104 mm^2
45. 18.7 in
47. 0.2 m
49. 16.9 in
51. 380.8 mm^2
53. 49.7 mm
55. 1.72 sq in
57. 546 mm^2
59. 104 sq in
61. 10.7 ft
63. 0.5 m
65. 602.1 sq in
67. 0.1 m^2
69. 0.5 m
71. a. 22.5 sq in
    b. 6.71 in
73. height = 692.8 mm
    lower base = 692.8 mm
    upper base = 519.6 mm

UNIT 54 Areas of Triangles

1. 178.5 mm^2
3. 0.2 m
5. 0.2 m^2
7. 28 mm
9. 83.3 sq in
11. 11.0 in
13. 11.6 sq in
15. 4.3 sq in
17. 1.2 sq ft
19. 31.3 mm^2
21. 350 mm^2
23. 3680 mm^2
25. 2744 cm^2

UNIT 55 Areas of Circles, Sectors, and Segments

1. 153.9 sq in
3. 188.7 sq in
5. 1.0 ft
7. 11.0 mm
9. 0.4 sq ft
11. 3.4 in
13. 14.1 lb
15. 1.3 kg
17. 24,414 lb/sq in
19. 104.7 cm^2
21. 81.7 sq in
23. 21.8 mm
25. 287.3°
27. 671.2 cm^2
29. 169.9°
31. 3829 mm^2
33. 91.9 sq in
35. 0.83 sq ft
37. 57.5 mm^2
39. 0.13 m^2
41. 124.9 mm^2
43. 133.5 sq in

UNIT 56 Volumes of Prisms and Cylinders

1. 2.50 cu ft
3. 4.33 cu yd
5. 7.47 cu yd
7. 562 cu in
9 2.7 cm^3
11. 78,000 mm^3
13. 75,000 cm^3
15. 730,200,000 mm^3
17. 35,400 mm^3
19. 9850 cm^3
21. 16.7 cu in
23. 8.3615 cu in
25. 1000 cu in
27. 66 lb
29. 493 cm^3
31. a. 368 cu in
    b. 104 lb
33. 0.017 m^3
35. 653.6 cu in
37. 10.67 gal
39. 210.3 cu in
41. a. 4.69 cu in
    b. 0.32 lb/cu in
43. 109 lb
45. 7.56 in
47. 1.94 in
49. 67.18 cm^2
51. 1.8 in
53. 1.99 sq in

UNIT 57 Volumes of Pyramids and Cones

1. 1013.93 cu in
3. 352.8 cu in
5. 328 cm^3
7. 0.80 L
9. 0.04 cu in
11. 13.88 in
13. 20 in
15. 20 cm
17. 22.96 in
19. 363 cu in
21. 70.7 lb
23. 4060 cm^3
25. a. 17.9 cu in
    b. 4.6 lb
27. 234.6 cm^3

UNIT 58 Volumes of Spheres and Composite Solid Figures

1. 41.63 cm^3
3. 229.82 cu in
5. 457.48 cu in
7. 2226.09 cm^3
9. 0.037 lb
11. 120.7 cm^3
13. 5.04 in
15. 103 lb
17. 144.9 lb
19. 94.7 lb
21. 4.61 lb

UNIT 59 Achievement Review—Section Six

1. 2475 sq ft
3. 104 mm^2
5. 20 sq in
7. 157.5 sq in
9. 19.5 mm
11. 3.9 in
13. 300.3 cm^2
15. 8.7 in
17. 15 in
19. 60.5 sq in
21. 2037 cm^2
23. 642.4 cm^2
25. 31.7 sq in
27. 85.9°
29. 72.5 sq in
31. 7.05 sq in
33. 19.5 sq in
35. 2160 cu in
37. 7490 mm^3
39. 70.1 cm^3
41. 1672 cm^3
43. 1.8 ft
45. 215.72 cu in
47. 1.26 gal
49. 1487.33 cm^3
51. 0.33 cu in
53. 11.57 in
55. 635 cm^3
57. 5.92 lb
59. 457.6 grams
61. 30.3 cu in
63. 0.02 cu ft
65. 11.4 cm
67. $3.54
69. 2.23 cu ft

SECTION 7 Trigonometry

UNIT 60 Introduction to Trigonometric Functions

1. r is hyp
   x is adj
   y is opp
3. a is adj
   b is hyp
   c is hyp
5. a is hyp
   b is opp
   c is adj
7. d is hyp
   m is opp
   p is adj
9. h is adj
   k is hyp
   l is opp
11. m is opp
   p is hyp
   s is adj
13. m is hyp
   r is adj
   t is adj
15. f is opp
   g is hyp
   h is adj
17. $\sin\angle 1 = \frac{y}{r}$
   $\cos\angle 1 = \frac{x}{r}$
   $\tan\angle 1 = \frac{y}{x}$
   $\cot\angle 1 = \frac{x}{y}$
   $\sec\angle 1 = \frac{r}{x}$
   $\csc\angle 1 = \frac{r}{y}$
19. $\sin\angle 1 = \frac{k}{g}$
   $\cos\angle 1 = \frac{h}{g}$
   $\tan\angle 1 = \frac{k}{h}$
   $\cot\angle 1 = \frac{h}{k}$
   $\sec\angle 1 = \frac{g}{h}$
   $\csc\angle 1 = \frac{g}{k}$
21. $\sin\angle 1 = \frac{r}{s}$
   $\cos\angle 1 = \frac{p}{s}$
   $\tan\angle 1 = \frac{r}{p}$
   $\cot\angle 1 = \frac{p}{r}$
   $\sec\angle 1 = \frac{s}{p}$
   $\csc\angle 1 = \frac{s}{r}$
23. Group 1: a, b, d
   Group 2: a, c
   Group 3: a, c, d
25. 0.60182
27. 0.95106
29. 0.99756
31. 0.45492
33. 0.99674
35. 0.76661
37. 0.01501
39. 0.49083
41. 0.18173
43. 0.11985
45. 0.95097
47. 0.34966
49. 0.76557
51. 0.06086
53. 2.88168
55. 1.78829
57. 1.32849
59. 8.31904
61. 1.06600
63. 1.03059
65. 56.82°
67. 30.68°
69. 75.84°
71. 1.45°
73. 5.50°
75. 28.67°
77. 67.30°
79. 39.52°
81. 50.95°
83. 48.16°
85. 9.70°
87. 76°39'
89. 67°7'
91. 87°43'
93. 47°3'
95. 89°43'
97. 17°47'
99. 46°8'
101. 45°0'
103. 46°10'
105. 74°22'
107. 84°12'

UNIT 61 Analysis of Trigonometric Functions

1. a. side y and side r are almost the same length
   b. side x is very small compared to side r
   c. side x is very small compared to side y
3. a. side y is very small compared to side r
   b. side x and side r are almost the same length
   c. side x is very large compared to side y
5. a. 45°
   b. 1.000 . . .
   c. 1.000 . . .
7. a. 1.000 . . .
   b. 0
   c. 0
   d. 1.000 . . .
9. tan 18°
11. cot 36°

13. csc 22°
15. cos 81°19'
17. csc 40.45°
19. sec 55°
21. cos 41°
23. csc 8°
25. sec 39°
27. cos 90°
29. sin 77.8°
31. sin 53°54'
33. cos 84.11°
35. tan 90°
37. sin 46°41'
39. cos 0°1'
41. sin 40°
43. cot 45°
45. sec 43°
47. cos 75°
49. tan 2°40'
51. cos 89.0°

UNIT 62 Basic Calculations of Angles and Sides of Right Triangles

1. 36°28'
3. 42°2'
5. 59.24°
7. 52.88°
9. a. 22°21'
   b. 67°39'
11. a. 31°30'
11. b. 58°30'
13. 5.706 in
15. 50.465 in
17. 136.16 mm
19. 75.51 mm
21. a. 2.229 in
    b. 2.049 in
23. a. 7.285 in
    b. 2.480 in
25. a. 17°30'
    b. 55.237 in
    c. 52.680 in
27. a. 82.79 mm
    b. 28.02 mm
27. c. 71.30°
29. a. 16°30'
    b. 1.579 in
    c. 5.559 in
31. a. 9°10'
    b. 1.099 in
    c. 6.812 in

UNIT 63 Simple Practical Machine Applications

1. a. 5.7358 in
   b. 2.2778 in
   c. 5.9949 in
   d. 1.6906 in
   e. 4.7767 in
   f. 6.9883 in
   g. 0.0582 in
1. h. 0.4246 in
   i. 3.3956 in
3. 8°48'
5. 13.12 mm
7. 51°25'
9. 255.48 mm
11. 28.19 mm
13. 0.897 in
15. 5.42 mm
17. 2.763 in
19. 2.577 in
21. 54.28 mm
23. 259.05 mm
24. a. 0.866
    b. 0.575
    c. 0.500
    d. 0.300
    e. 0.207

UNIT 64 Complex Practical Machine Applications

1. 3.748 in
3. 14°13'
5. 0.564 in
7. 44°21'
9. 298.85 mm
11. 4.499 in
13. 5.408 in
15. 37°26'
17. 2.290 in
19. 3.46 in
21. 58°50'
23. 49.73°
25. 4.227 in
27. 0.667 in
29. 29°40'
31. 1.433 in

UNIT 65 The Cartesian Coordinate System

1. sin 120° = 0.8660
   cos 120° = −0.5000
   tan 120° = −1.7321
   cot 120° = −0.5774
   sec 120° = −2.0000
   csc 120° = 1.1547
3. sin 260° = −0.9848
   cos 260° = −0.1736
   tan 260° = 5.6713
   cot 260° = 0.1763
   sec 260° = −5.7588
   csc 260° = −1.0154
5. sin 300° = −0.8660
   cos 300° = 0.5000
   tan 300° = −1.7321
   cot 300° = −0.5774
   sec 300° = 2.0000
   csc 300° = −1.1547

7. sin 208°50' = −0.4823
cos 208°50' = −8.760
tan 208°50' = 0.5505
cot 208°50' = 1.8165
sec 208°50' = −1.1415
csc 208°50' = −2.0735

9. sin 146°10' = 0.5568
cos 146°10' = −0.8307
tan 146°10' = −0.6703
cot 146°10' = −1.4919
sec 146°10' = −1.2039
csc 146°10' = 1.7960

11. sin 313.17° = −0.7293
cos 313.17° = 0.6842
tan 313.17° = −1.0660
cot 313.17° = −0.9381
sec 313.17° = 1.4616
csc 313.17° = −1.3711

UNIT 66 Oblique Triangles: Law of Sines and Law of Cosines

1. 5.078 in
3. 3.533 in
5. 33°54'
7. 18°16'
9. 128.73 mm
11. two solutions
13. two solutions
15. one solution
17. two solutions
19. 97.70 mm
21. 71.48°
23. 27°2'
25. 36.96 mm
27. a. 1.163 in
b. 42°19'
29. a. 364.34 mm
b. 14.52°
31. a. 94°44'
b. 49°20'
33. 71.98°
35. 9.046 in
37. 44°38'
39. 2.202 in
41. 70°36'

UNIT 67 Achievement Review—Section Seven

1. *a* is opp
b is adj
c is hyp
3. *m* is hyp
s is adj
p is opp
5. 0.3746
7. 0.7766
9. 0.4493
11. 0.1293
13. 45°50'
15. 80°10'
17. 50°36'
19. 46.76°
21. 70.10°
23. cot 41°41'
25. tan 9.53°
27. 18.827 in
29. 54.06°
31. a. 58°1'
b. 31°59'
c. 17.685 in
33. 6.60°
35. 0.594 in
37. 0.278 in
39. 58.50°
41. Ref. ∠ = 223° − 180° = 43°
sin 223° = −sin 43° = −0.6820
cos 223° = −cos 43° = −0.7314
tan 223° = tan 43° = 0.9325
cot 223° = cot 43° = 1.0724
sec 223° = −sec 43° = −1.3673
csc 223° = −csc 43° = −1.4663
43. 185.999 mm
45. a. 110°30'
b. 14.784 in
c. 8.558 in
47. 125.10°
49. 7.859 in
51. 33°27'
53. 107°49'
55. 125°44'

SECTION 8 Compound Angles

UNIT 68 Introduction to Compound Angles

1. a. 3.680 in
b. 65°57'
3. a. 7.777 in
b. 56°5'
5. a. 1.897 in
b. 65°4'

UNIT 69 Drilling and Boring Compound-Angular Holes: Computing Angles of Rotation and Tilt Using Given Lengths

1. a. 38°22'
   b. 49°39'
3. a. 43°9'
   b. 49°6'
5. a. 41°5'
   b. 49°38'
7. a.

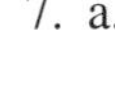

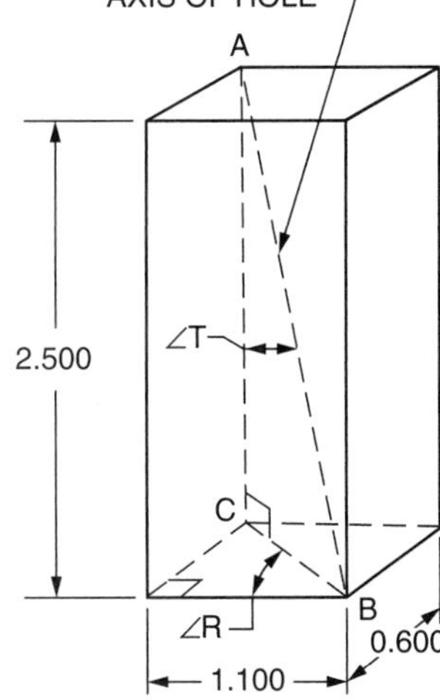

   b. 28°37'
   c. 26°37'
9. a.

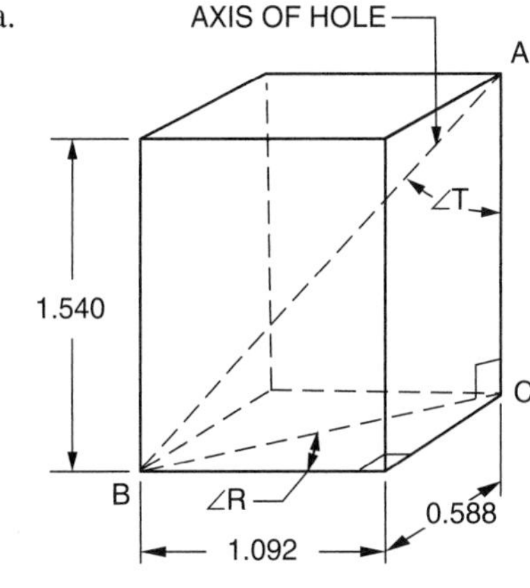

   b. 28°18'
   c. 38°51'

UNIT 70 Drilling and Boring Compound-Angular Holes: Computing Angles of Rotation and Tilt Using Given Angles

1. a. 37°52'
   b. 48°46'
3. a. 28.76°
   b. 34.43°
5. a. 32°31'
   b. 47°3'
7 a.

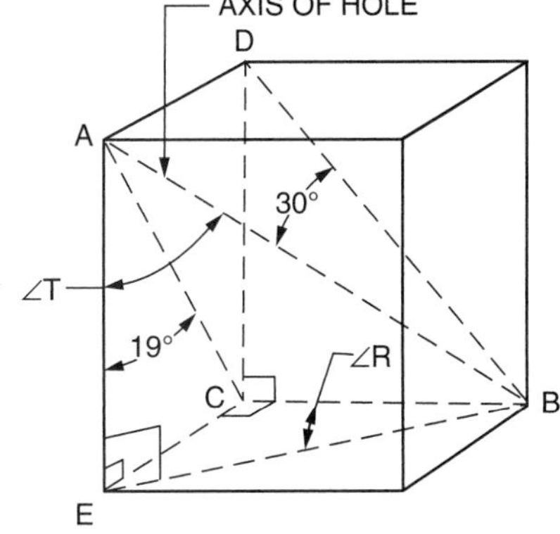

   b. 30°49'
   c. 33°54'
9. a.

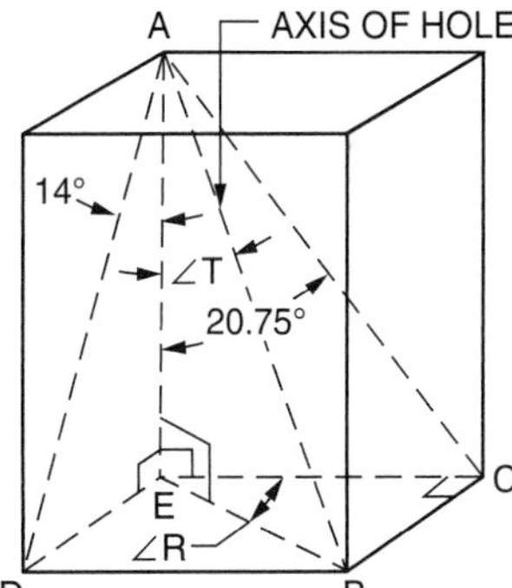

   b. 33.35°
   c. 24.40°
11. a. 20°30'
    b. 42°52'
13. a. 21.89°
    b. 27.21°
15. a. 28.76°
    b. 32.94°

UNIT 71 Machining Compound-Angular Surfaces: Computing Angles of Rotation and Tilt

1. a. 57°6'
   b. 49°0'
3. a. 58°8'
   b. 31°52'
5. a.

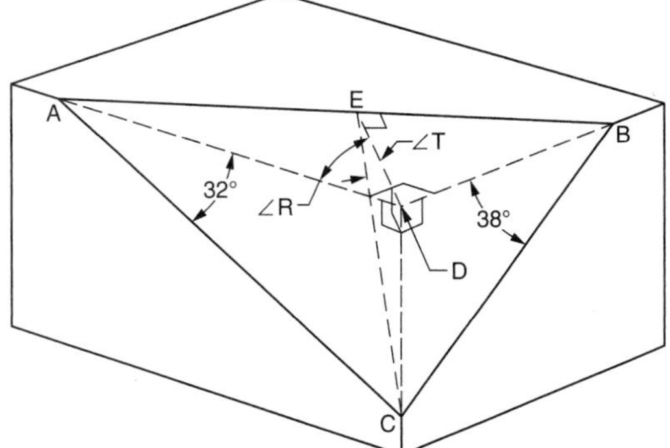

5. b. 51°21'
   c. 45°1'

7. a.

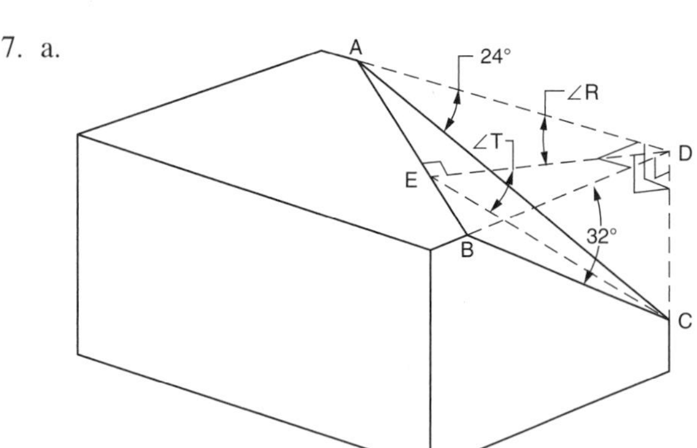

b. 54.53°
c. 37.50°

9. a. 50°01'
b. 51°34'

11. a. 46.85°
b. 43.04°

13. a. 38°48'
b. 40°56'

UNIT 72 Computing Angles Made by the Intersection of Two Angular Surfaces

1. a. 32°14'
b. 37°18'

3. a. 32°53'
b. 45°12'

5. a.

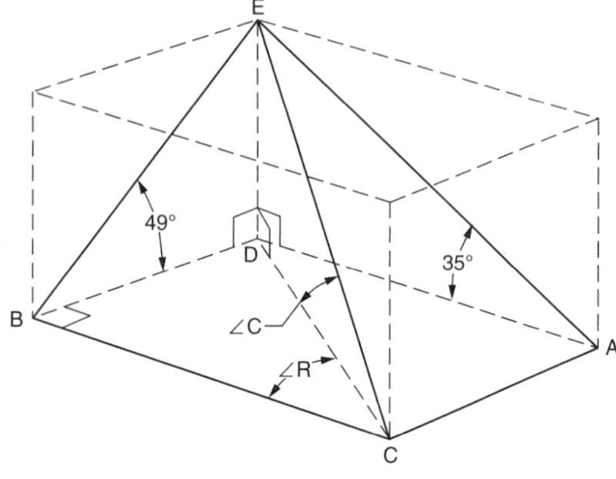

b. 31°20'
c. 30°53'

7. a.

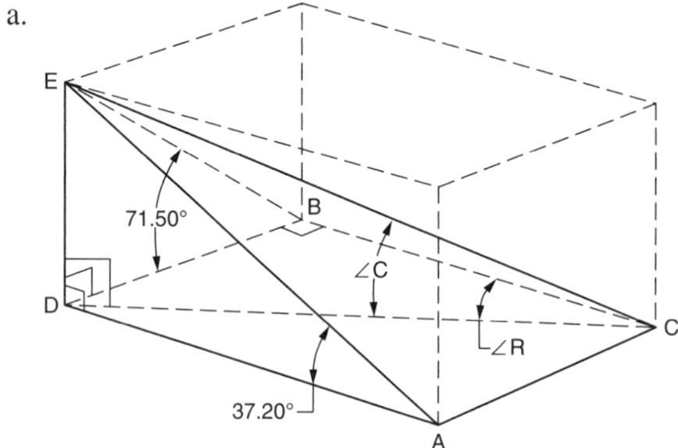

b. 14.25°
c. 36.34°

9. a. 37°7'
b. 30°5'

11. a. 46.59°
b. 30.20°

13. a. 17°36'
b. 15°40'

UNIT 73 Computing Compound Angles on Cutting and Forming Tools

1. 29°9'

3. 26°34'

5. a.

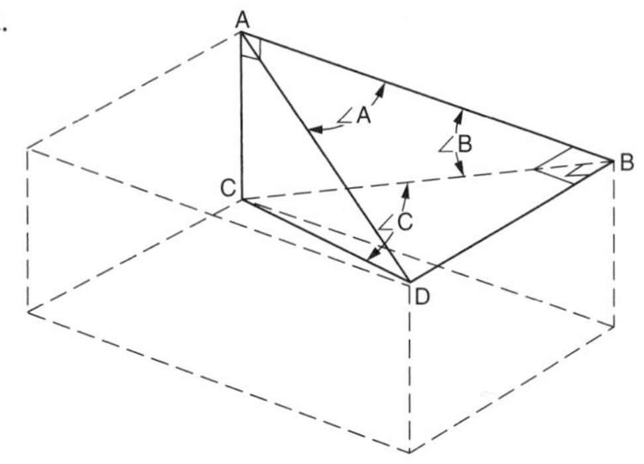

b. 23°16'

7. a.

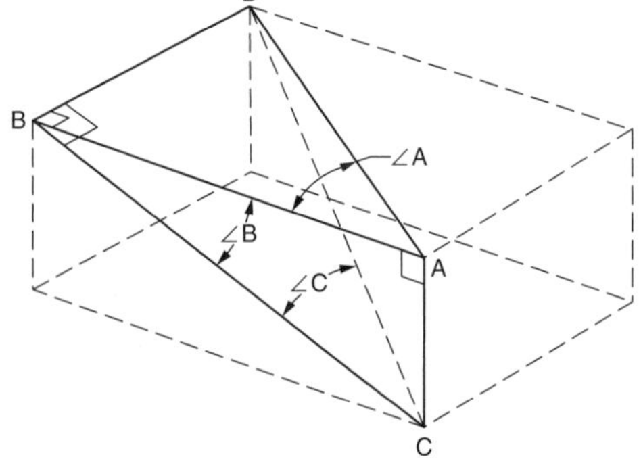

b. 29°27'

9. 32°13'

11. 25°20'

13. 30°23'

15. 33°29'

17. a.

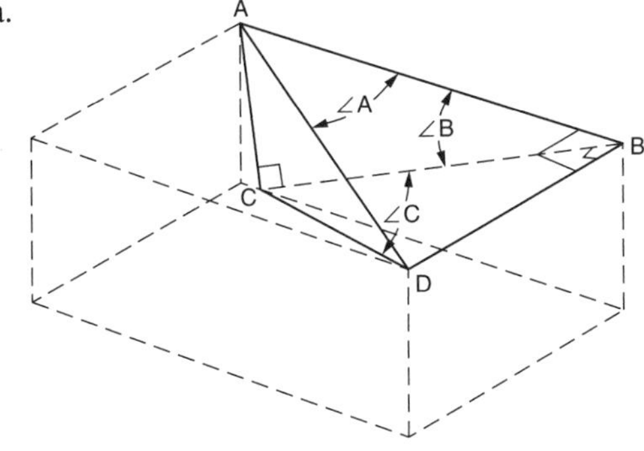

b. 30°52'

19. a.

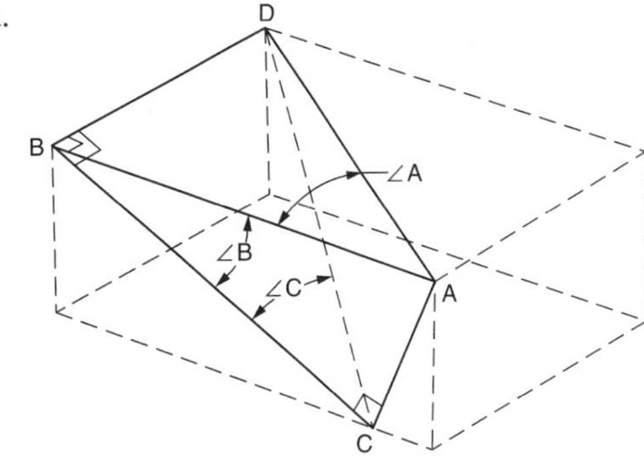

b. 32°54'

21. 25°16'

23. 34°16'

UNIT 74 Achievement Review—Section Eight

1. a. 30°4'
   b. 44°56'

3. a. 55.53°
   b. 50.00°

5. 28°21'

SECTION 9 Computer Numerical Control (CNC)

UNIT 75 Introduction to Computer Numerical Control (CNC)

1. See Instructor's Guide.

3. See Instructor's Guide. A triangle is formed.

5. A: (−1, −1)
   B: (−4, 4)
   C: (−4, 1)
   D: (8, 0)
   E: (9, −3)
   F: (−5, −8)
   G: (−2, 8)
   H: (−9, −4)
   J: (8, −9)
   K: (3, −6)
   L: (−8, 8)
   M: (5, 1)
   N: (−8, −8)
   P: (4, 9)
   Q: (7, 8)
   R: (9, 3)

7.

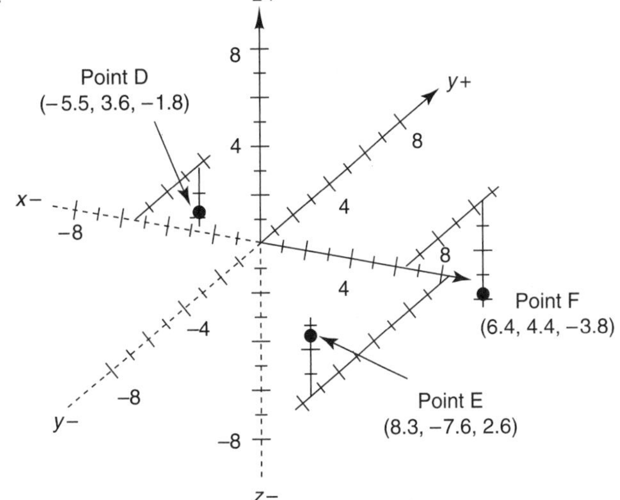

UNIT 76 Control Systems, Absolute Positioning, Incremental Positioning

1. a. Absolute Positioning

Hole	x	y
1	7.400"	−5.850"
2	10.500"	−0.920"
3	7.400"	1.540"
4	16.630"	2.770"
5	17.250"	−2.770"
6	17.250"	−5.850"

b. Incremental Positioning

Hole	x	y
1	7.400"	−5.830"
2	3.100"	4.920"
3	-3.100"	2.460"
4	9.230"	1.230"
5	0.620"	−5.540"
6	0.000"	−3.070"

3. a. Absolute Positioning

Hole	x	y
1	184.62 mm	−146.15 mm
2	261.54 mm	−23.08 mm
3	184.62 mm	38.46 mm
4	415.39 mm	69.23 mm
5	430.77 mm	−69.23 mm
6	430.77 mm	−146.15 mm

b. Incremental Positioning

Hole	x	y
1	184.62 mm	−146.15 mm
2	76.92 mm	123.07 mm
3	−76.92 mm	61.54 mm
4	230.77 mm	30.77 mm
5	15.38 mm	−138.46 mm
6	0.00 mm	−76.92 mm

5. a. Absolute Positioning

Hole	x	y
1	−4.750"	4.750"
2	−0.654"	8.569"
3	3.514"	12.457"
4	8.074"	9.600"
5	13.776"	7.748"
6	17.500"	−0.100"
7	13.776"	1.752"
8	8.074"	−0.100"
9	4.550"	4.750"
10	−1.188"	−0.520"
11	−6.162"	−4.625"

b. Incremental Positioning

Hole	x	y
1	−4.750"	4.750"
2	4.096"	3.819"
3	4.168"	3.888"
4	4.560"	−2.857"
5	5.702"	−1.852"
6	3.724"	−7.848"
7	−3.724"	1.852"
8	−5.702"	−1.852"
9	−3.524"	4.850"
10	−5.738"	−5.270"
11	−4.974	−4.105"

7. a. Absolute Positioning

Hole	x	y
1	−9.317"	18.412"
2	−3.793"	17.086"
3	5.885"	18.250"
4	11.400"	16.750"
5	18.935"	12.500"
6	5.885"	9.330"
7	−3.616"	8.075"
8	−13.541"	10.397"

b. Incremental Positioning

Hole	x	y
1	−9.317"	18.412"
2	5.524"	−1.326"
3	9.678"	1.164"
4	5.515"	−1.500"
5	7.535"	−4.250"
6	−13.050"	−3.170"
7	−9.501"	−1.255"
8	−9.925"	2.322"

9 a. Absolute Positioning

Hole	x	y
1	−237.06 mm	329.42 mm
2	−125.66 mm	302.84 mm
3	51.30 mm	329.30 mm
4	174.64 mm	302.18 mm
5	317.70 mm	222.50 mm
6	51.30 mm	163.00 mm
7	−136.30 mm	133.05 mm
8	−310.02 mm	187.28 mm

b. Incremental Positioning

Hole	x	y
1	−237.06 mm	329.42 mm
2	111.40 mm	−26.58 mm
3	176.96 mm	26.46 mm
4	123.34 mm	−27.12 mm
5	143.06 mm	−79.68 mm
6	−266.40 mm	−59.50 mm
7	−187.60 mm	−29.95 mm
8	−173.72 mm	54.23 mm

UNIT 77 Location of Points: Polar Coordinate Systems

1. A: (2, 20°)
   B: (4, 50°)
   C: (5, 110°)
   D: (3,160°)
   E: (4.5, 190°)
   F: (2, 230°)
   G: (3, 290°)
   H: (2.5, 340°)

3. 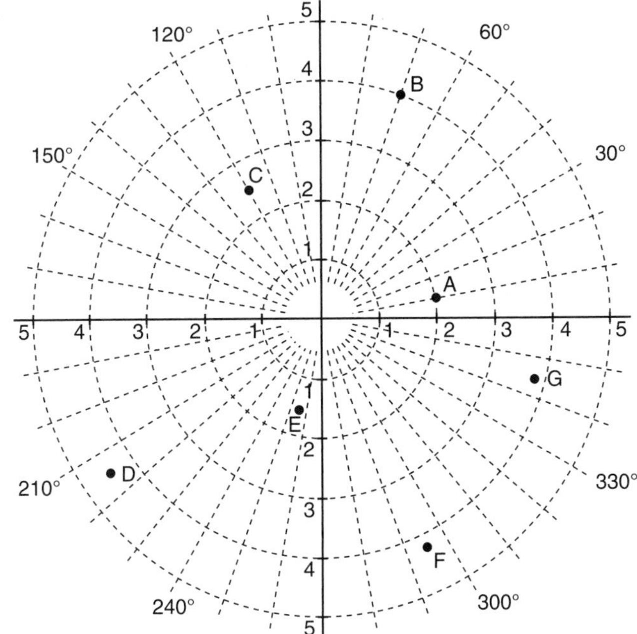

5.

	Coordinates	
Hole	r	θ
1	23.15 mm	0°
2	23.15 mm	72°
3	23.15 mm	144°
4	23.15 mm	216°
5	23.15 mm	288°

7.

	Coordinates	
Hole	r	θ
1	7.3 in	35°
2	7.3 in	155°
3	7.3 in	275°

UNIT 78 Binary Numeration System

1. $2(10^2) + 6(10^1) + 5(10^0)$
   $200 + 60 + 5 = 265$

3. $9(10^4) + 0(10^3) + 5(10^2) + 0(10^1) + 0(10^0)$
   $90{,}000 + 0 + 500 + 0 + 0 = 90{,}500$

5. $2(10^1) + 3(10^0) + 0(10^{-1}) + 2(10^{-2}) + 3(10^{-3})$
   $20 + 3 + 0 + 0.02 + 0.003 = 23.023$

7. $4(10^3) + 7(10^2) + 5(10^1) + 1(10^0) + 1(10^{-1}) + 0(10^{-2}) + 7(10^{-3})$
   $4000 + 700 + 50 + 1 + 0.1 + 0 + 0.007 = 4751.107$

9. $1(10^2) + 6(10^1) + 3(10^0) + 0(10^{-1}) + 6(10^{-2}) + 4(10^{-3}) + 3(10^{-4})$
   $100 + 60 + 3 + 0 + 0.06 + 0.004 + 0.0003 = 163.0643$

11. 1_{10}

13. 5_{10}

15. 15_{10}

17. 11_{10}

19. 21_{10}

21. 53_{10}

23. 0.5_{10}

25. 3.75_{10}

27. 2.000_{10}

29. 9.3125_{10}

31. 1110_2

33. 1010111_2

35. 101011_2

37. 1101001_2

39. 1_2

41. 110011_2

43. 0.1_2

45. 0.011_2

47. 1010001.11_2

49. 1100101.01_2

51. 10100011.111_2

UNIT 79 Hexadecimal Numeration System

1. D_{16}
3. 33_{16}
5. $29D3_{16}$
7. $6B.DC_{16}$
9. $4AB.A98_{16}$
11. $C949.924_{16}$
13. 100100_2
15. 10011010_2
17. 110000100111_2
19. 10111111.0011101_2
21. 1001111100.11010111_2
23. 101011.000000101011_2
25. 247
27. 123.875
29. 2827.03125
31. 32,752.25
33. $2F_{16}$
35. $8F_{16}$
37. $3C3.4_{16}$
39. $FF30_{16}$

UNIT 80 BCD (Binary Coded Decimal) Numeration Systems

1. 0111 0011
3. 0010 0100 0110
5. 0001 1000.1001 0011
7. 0011 0110.0001 1001 0101
9. 9
11. 53
13. 76.4
15. 265.07
17. 1101 0010
19. 0011 1100 0010.1111
21. 47
23. 30.7
25. 0101 1010
27. 0111 1000.0011 1010
29. 38
31. 82.4
33. 1010 0101
35. 1100 0110.1011 1001
37. 28
39. 73.01

UNIT 81 Achievement Review—Section Nine

1. Coordinates to be plotted
3. a. Absolute Positioning

Hole	x	y
1	6.272"	−2.120"
2	8.172"	1.488"
3	13.201"	1.488"
4	19.942"	1.488"
5	21.032"	−3.392"
6	21.032"	−5.507"
7	13.903"	−4.239"

b. Incremental Positioning

Hole	x	y
1	6.272"	−2.120"
2	1.900"	3.608"
3	5.029"	0
4	6.741"	0
5	1.090"	−4.880"
6	0	−2.115"
7	−7.129"	1.268"

5. a. 1
   b. 7
   c. 21
   d. 3.25
   e. 9.5625
7.

Hole	Coordinates	
	r	θ
1	36.76 mm	55°
2	36.76 mm	145°
3	36.76 mm	235°
4	36.76 mm	325°

9. a. 11011_2
   b. 110001011010_2
   c. 100111.0001101_2
   d. 11100000.00001101_2
11. a. $4B_{16}$
    b. $F3_{16}$
    c. $39F_{16}$
    d. 1088_{16}
13. a. 392
    b. 5786

Index

A

B

E

O

P

Q

R

S

T

U

V

W

Z